# Three Approaches to
# ELECTRON CORRELATION IN ATOMS

A CHEMISTRY–PHYSICS INTERFACE

YALE SERIES IN THE SCIENCES

General Editors, Oktay Sinanoğlu and Edward U. Condon

**International Advisory Board**

Aage Bohr, *Director, Niels Bohr Institute, Copenhagen, Denmark*
Sydney Brenner, *Medical Research Council, Laboratory of Molecular Biology, Cambridge University, England*
Vernon W. Hughes, *Professor of Physics, Yale University, New Haven, Conn., U.S.A.*
William Nierenberg, *Director, Scripps Institution of Oceanography, La Jolla, Calif., U.S.A.*
Lars Onsager, *Professor of Theoretical Chemistry, Yale University, New Haven, Conn., U.S.A.*
Charles C. Price, *Professor of Chemistry, University of Pennsylvania, Philadelphia, Pa., U.S.A.*
Frederic M. Richards, *Professor of Molecular Biophysics and Biochemistry, Yale University, New Haven, Conn., U.S.A.*
Pol Swings, *Professor of Astrophysics, University of Liège, Belgium*
Richard N. Thomas, *Fellow, Joint Institute for Laboratory Astrophysics, Boulder, Colo., U.S.A.*
Yusuf F. Vardar, *Professor of Botany, Ege University, Izmir, Turkey*

A CHEMISTRY–PHYSICS INTERFACE

Three Approaches to

# ELECTRON CORRELATION IN ATOMS

by


**Oktay Sinanoğlu**
Professor of Theoretical Chemistry, Yale University

and

**Keith A. Brueckner**
Professor of Theoretical Physics,
Institute of Pure and Applied Physical Sciences,
University of California, La Jolla


(including a number of reprints by various authors)

NEW HAVEN AND LONDON, YALE UNIVERSITY PRESS, 1970

Library of Congress catalog card number: 76-89666
Standard book number: 300-01147-4
set in IBM Selectric Press Roman type by Science Press, Inc.,
and printed in the United States of America by
The Murray Printing Company, Forge Village, Mass.
Distributed in Great Britain, Europe, Asia, and
Africa by Yale University Press Ltd., London; in
Canada by McGill University Press, Montreal; and
in Mexico by Centro Interamericano de Libros
Académicos, Mexico City.

# PREFACE

Prediction of atomic properties is central to many fields. The astrophysicist looks for accurate transition probabilities, the atomic physicist for an understanding of resonances, the theoretical chemist for ways of predicting the forces between atoms. Most aspects of atomic theory are understood in terms of orbitals, i.e. independent motion of electrons in one another's average fields. But, as is well known now, for quantitative theory this is not sufficient. The many electrons of an atom are constantly perturbed by one another, and their motions are correlated so that the wave function must reflect somehow the role of the relative electron coordinates. The term "electron correlation" was first used by Wigner (1934) for the electrons of a metal. Quantitative studies of atomic properties require the study and introduction of electron correlation into wave functions and energies.

That this realm "beyond orbitals" should have interested chemists, atomic physicists, "many-body"-ists and field theorists, and astrophysicists is therefore not surprising. Nor is it surprising that diverse excursions were taken into this realm, at different times, with different backgrounds and tools. The result has been several new approaches to the problem.

After an incubation period since about 1960, there is at present growing interest in this interdisciplinary field. Over the past several years we have heard many of our colleagues remark on the need to bridge the communication gap across the theoretical chemistry literature and the literature of theoretical physics. Since more and more students of theoretical chemistry are becoming familiar with theoretical physics techniques, and more atomic physicists are finding developments in the theoretical chemistry literature of interest, we hope our attempt here will be found helpful by both groups.

We have aimed at providing a background for the recent approaches while indicating their current status and the new problems that arise. Each chapter provides a discussion followed by papers from the original literature. In spite of considerable development, the subject still affords many avenues of investigation which may be considered at the beginning stage. Correlation in heavy atoms, in scattering, new computational techniques, more group theory seem around the corner.

We thank our colleagues for their many helpful comments, which have contributed to our remarks in the text and to the selection of papers. Our thanks go also to the publishers and authors who have permitted the reprinting of their journal articles. Grants from the National Science Foundation, the Alfred P. Sloan Foundation (O. S.), and partial support from the Institute of Pure and Applied Physical Sciences at La Jolla, which enabled us to collaborate on this project, are gratefully acknowledged.

O. Sinanoğlu and K. A. Brueckner
New Haven, Connecticut, and La Jolla, California

# Contents

# Chapter I

# *Introduction*

It is convenient and physically meaningful to separate the problem of the wave function and energy of an atom with N electrons into two parts: (1) the *self-consistent field* (SCF) orbitals of the Hartree–Fock (H. F.) method, and (2) the remaining correlation part. This of course is not the only way to separate the problem into parts, nor is it even clear at the outset that the problem should be divided at all. Various more "classical" approaches have looked at the N-electron problem without such a subdivision. Hylleraas in 1929 dealt with helium directly in terms of a power series in $r_1$, $r_2$, and $r_{12}$, not in two distinct and explicit stages. The well-known Bacher–Goudsmit method (1934) relates total energies to one another.

Should it be indeed convenient to have the two-stage problem, why take the H. F. SCF orbitals as the start, other than that they are readily available for many atoms? In other many-body problems the same type of H. F. orbitals, often, are not convenient. At times they are not useful starting points. In the uniform electron gas, the H. F. orbitals are the trivial solution for translational invariance, i.e., plane waves. In nuclear matter or for finite nuclei, the ordinary H. F. method leads to infinite integrals if the nucleon–nucleon potential has a hard core. The hard core necessitates in nuclear physics the use of specialized methods suitable for nuclei and especially nuclear matter and developed by Brueckner and collaborators (1953 and later), by Bethe and Goldstone (1957), and others. The hard core has no counterpart in atomic structure theory.

The ordinary H. F. SCF method, on the other hand, when it comes to atoms, is free of the basic limitations and difficulties it has for a uniform electron gas and for nuclei. Clearly, the N-electron atom is distinct physically and requires distinct physical approaches and computational methods.

Since about 1960, new methods have been developed for the treatment of the atomic N-electron problem. The present book is on these new methods and the understanding of correlation effects they lead to so far.

Where relativistic effects are small, and for ground states, the *correlation energy*, $E_{corr}$, is defined as

$$E_{corr} \equiv E_{nr} - E_{HF} \qquad (1)$$

where $E_{nr}$ is the exact ground-state eigenvalue of the N-electron nonrelativistic Hamiltonian, and $E_{HF}$ is the Hartree–Fock energy.

Along with most of the work in the field so far, we confine ourselves to atoms where relativistic effects (of the Breit type) are small enough corrections that they can be simply added to Eq. (1). Aside from the magnetic interactions, however, the reader should note Fröman's (1959) remark that relativistic corrections to the total energy already become comparable to the correlation energy in second-row atoms.

We start with a brief section on orbitals, Chapter II, selecting only a few of the most important features of the methods relevant to the treatment of electron correlation. For heavier atoms it is often found convenient to replace the exchange terms with approximations derived from the electron gas. Most relativistic SCF calculations have been carried out with the Slater-type approximate exchange, for example. The treatment of exchange and the more general problem of the relationship between electron density and energy have been considered in detail by Kohn and his co-workers in papers included in Chapter III. This work is of importance also in contrasting correlation in the electron gas (see Chapter VI) versus the atom.

In Chapter IV we show briefly how correlation affects the magnitudes of various atomic properties. The effects are large, so that without them a quantitative theory of atomic structures would not be possible.

Also in Chapter IV, we discuss briefly what approaches were available that went beyond the H. F. method for N-electron atoms before the development of the new methods, which started about 1960. The earlier approaches have been treated in detail in Slater's book, *Quantum Theory of Atomic Structure*, Vols. 1 and 2 (McGraw-Hill, New York, 1960) and in reviews by Löwdin [*Rev. Mod. Phys.* **32**, 328 (1960); Chapter IV-3 in this book; also *Advan. Chem. Phys.* 2, 207 (1959)].

In the early approaches, attempts are made to obtain overall N-electron trial functions via the variational principle. Though often going considerably beyond Hartree–Fock and including some features of correlation, in these approaches correlation per se is not treated as a separate physical problem. The H. F. part is often not separated out of the N-electron trial functions.

The three recent approaches, the main topics of this book, deal with correlation effects systematically starting from different physical points of view. They are:

1. The theory of pair correlations (1960, 1961) and the "many-electron theory of atoms and molecules" (MET) (1962 to date) developed by Sinanoğlu and co-workers (Chapters V, VII, and IX).

2. The atom viewed as a nonuniform electron gas based on the results and techniques of Brueckner and co-workers (Chapter VI; cf. Chapter IV).

3. Diagrammatic many-body perturbation techniques first applied to atoms by Kelly in 1963, and other field theoretic techniques (Chapter VIII).

In the first approach, Sinanoğlu's papers show that, in spite of the long range of the interelectronic Coulomb repulsion, the correlation energies of N-electron ground-state atoms are given predominantly by the sum of N(N - 1)/2 decoupled "pair correlations." In fact, for some of the pairs (but only some of them, such as the $1s^2$ or 1s2s pairs, but not $2s^2$, for example) even second order in energy suffices. The theory by Sinanoğlu provides ways for the evaluation of the pair correlations individually. Of the order of not N(N - 1)/2, but only about N, distinct "He-like" problems have to be solved with variational minimum principles analogous to those for the different states of helium atom. In the perturbation version of the theory, these were given as first-order Schrödinger equations (Chapter V-3) only. In the more general form of the theory (Chapters V and VII), the forms of the exact N-electron wave function and energy were obtained and contain a finite number of correlation terms involving 1, 2, 3, . . ., n, . . . N-electron correlations at a time. "Subvariational" principles are introduced for the study of each of these effects by themselves or in various combinations. It would be important, for example, to calculate the three- and more-electron remainders in the correlation energy by these methods to see to what accuracy the sum of the pair correlations alone is sufficient. Approach I has been extended recently to excited states revealing additional types of correlation effects and enabling the quantitative prediction of properties such as excitation energies, electron affinities, and transition probabilities.

The second approach, the electron gas, uses modern many-body perturbation theory to obtain exact solutions for the energy of the high-density weakly nonuniform electron gas as a functional of the density. This approach is highly instructive as a background for the consideration of the atomic problem, because the exact treatment possible for these limiting conditions shows clearly the nature of screening, the associated collective phenomena of plasma oscillation, and the effect of nonuniform density.

In the high-density *uniform* electron gas, the correlation energy obtained by Gell-Mann and Brueckner (Chapter VI-2) results from a summation of "ring diagrams" which involve progressively larger numbers of electrons. Each order of perturbation would yield divergent results associated with small momentum transfer and hence with the long-range part of the Coulomb interaction, the order of divergence increasing with the order of perturbation. Thus multiple electron correlations are essential.

The reconciliation of the apparent dilemma, the importance of the higher-order correlations in the uniform high-density electron gas versus the dominance found in Approach I of only the decoupled pair correlations in the atom, poses an interesting problem, particularly as both electron systems are of high density. The answer lies in the strong nonuniformity of the electronic cloud in the atom and explains why early attempts to introduce the correlation energy of the uniform electron gas into the atom via the local density failed, as also commented upon by Slater.

In their studies of the nonuniform electron gas, Brueckner and Ma (Chapter VI-3) start as in the uniform electron gas, but determine the effect of the external ($-Z/r_i$) potential of the nucleus on the correlation energy. This line of reasoning indicates how the uniform electron gas situation gradually shifts into the atomic situation as Z and the nonuniformity is increased. The nonuniformity turns out to have more pronounced effects than anticipated, because the correlation range at the average atomic density is approximately equal to the atomic radius. Thus the use of a gradient expansion to determine nonuniformity corrections is slowly convergent and one obtains only the first few terms of an asymptotic series in the density gradient. Nevertheless, from these terms a semiempirical formula is obtained allowing an estimate of nonuniformity corrections to the correlation energy of actual atoms. The formula agrees with the "experimental" correlation energy estimates to within a few percent for several rows of atoms.

Sinanoğlu in Chapter VII-2 has noted a different aspect of the nonuniformity in comparing the uniform electron gas with the atomic electrons. Since correlation is defined as what is left over after the H. F. SCF method, it is really caused not by the full Coulomb repulsions,

$$\sum_{i>j}^{N} r_{ij}^{-1}$$

but by the sum of "fluctuation potentials,"

$$\sum_{i>j}^{N} m_{ij}$$

the difference between $r_{ij}^{-1}$ and the Coulomb (and exchange) repulsion averages over H. F. spin orbitals $i(\mathbf{x}_i)$ and $j(\mathbf{x}_j)$. This "fluctuation potential" exhibits very different behavior in the uniform electron gas versus in the atom, owing purely to the different geometries of the two systems (see the graphs of $m_{ij}$ in Chapter VII-2).

Approach III, the techniques of modern diagrammatic many-body perturbation theory, and some field theoretic techniques are discussed in Chapter VIII.

Second quantization formalism provides a convenient and powerful way with which to carry out the algebra of many-particle matrix elements. Its use is not restricted to diagrammatic and field theoretic techniques, although it is most used in these techniques where each term acquires a simple physical interpretation and visualization. Second quantization formalism as used in the diagrammatic, field theoretic techniques and in the many-body Brueckner–Goldstone perturbation theory† makes use of a complete (infinite) one-electron basis set as in the infinite configuration-interaction (C. I.) series. In fact, the use of the second quantization techniques in conventional C. I. series would probably offer some advantages. For example, the reduction of arbitrary matrix elements could probably be conveniently computerized this way. An infinite basis set, nevertheless, regardless of the way it enters, retains its difficulties if it is slowly convergent, requiring extensive summations over discrete terms.

This particular difficulty does not come up (although surely many other kinds do) in many-body systems of uniform but infinite extent. In that case, the one-body basis set consists of plane waves, each basis set characterized by the magnitude and direction of the linear momentum $\mathbf{p} = \hbar\mathbf{k}$ and by an $\alpha, \beta$ intrinsic spin component. The sums over the basis set are then integrals (which can often diverge, however, as in individual orders of Rayleigh–Schrödinger perturbation theory in the uniform electron gas) over momentum transfers $\mathbf{q} = \mathbf{k}_1 - \mathbf{k}_2$.

In papers included in Chapter VIII, Kelly noted that the discrete C. I. type of summation problem could be handled without undue difficulty for ground-state atoms. At that time (1963) and earlier, most basis sets in use were the virtual H. F. orbitals generated as a by-product of a matrix H. F. calculation. Kelly observed that these virtual orbitals, except an occasional one or two, by and large lie in the continuum. Instead of them, therefore, it should be possible to use a purely continuous spectrum. The one or two discrete, below-continuum, virtual level contributions are simply added on. The infinite sums are thus replaced by integrals which Kelly evaluated numerically on the computer. In this way, the Brueckner–Goldstone diagrammatic perturbation theory was made practicable for atomic calculations.

Contributions of different diagrams have been calculated by Kelly for the Be atom (Chapter VIII-3) and later for oxygen (Chapter VIII-5) to different orders. The quantitative results are in agreement with those of Approach I as to the degree of significance of different effects, although it is difficult to draw a one-to-one correspondence between some of the terms of the two approaches. In the theory MET of Sinanoğlu, the exact N-electron wave function is in a form containing a *finite* number of terms. The terms correspond to "*correlation functions*" $\hat{U}_{ijk\cdots}(x_1, x_2, x_3, \ldots)$ involving $n \leqslant N$ electrons at a time. Calculations to successively higher accuracy are carried out by "subvariational" principles (cf. VPA, Chapter VII), so that at each step one has the upper bound either to some portion of the correlation energy or to the total energy (the two alternating). In the Brueckner–Goldstone perturbation theory all effects are in a form expanded in different orders of perturbation and infinite basis sets with no upper or lower bound property. A formal comparison of the two approaches to see what diagrams are implicit in the MET functions or energy terms has recently been attempted by J. Čížek [*Advan. Chem. Phys.* **14**, 75 (1969)].

Some of the correlation effects, such as the $n \geqslant 3$ electron remainders in the energy of MET (or terms included in the remainders), have not been explicitly calculated in any of the approaches. A more quantitative knowledge of at least the leading terms in these remainders would be of considerable importance in extending the results of the approaches above to heavier atoms and to molecules.

Throughout our comments in the book, we attempt to show what some of the difficulties and some of the unknowns are in connection with the approaches discussed, in addition to aspects that now seem quite clear. It will be interesting to see how far these theories and methods will apply and yield new information on various atomic phenomena and on molecular problems.

†This is a form of perturbation theory particularly suited to systems of infinite extent and infinitely many degrees of freedom, eliminating the difficulties in such systems of Rayleigh–Schrödinger and Wigner–Brillouin perturbation theories. It is not to be confused with other work by Brueckner and by Bethe and Goldstone referred to by these names, which is on the theory of nuclear matter concerning the nucleon–nucleon hard-core problem.

# Chapter II

# Orbitals in the Ground and Excited States of Atoms—Hartree-Fock Theory

## II-1

The H. F. wave function, especially for closed-shell systems, gives the charge distribution well. The orbitals, therefore, are very useful in themselves, but they also are necessary as a starting point for the remaining correlation effects. Not only is correlation energy by definition the departure from H. F. energy, but the H. F. orbitals also have a special physical significance as compared to other orbitals, as we shall see in later chapters.

For closed-shell systems, there is one commonly used H. F. method. The H. F. wave function $\phi_0$ is a single determinant,

$$\phi_0 = \mathcal{A}(123\cdots N) \tag{1}$$

of spin orbitals $i \equiv i(\mathbf{x}_i)$. The equations for the orbitals are solved either numerically (Chapter II-4) or by an expansion in terms of Slater-type orbitals (STO's) with exponents varied (Chapter II-3). Each way has its own advantages and has been used widely. Automatic computer programs have been developed for both (including excited states) and been made available.

For excited states there are at least nine variants† of the H. F. method, depending on which symmetry restrictions are placed on the orbitals and how the off-diagonal Lagrange multipliers (cf. Chapter II-3) are handled. In Roothaan's formulation‡ for (paper II-3) individual spin orbitals are still restricted to belong to the irreducible representations of the $O(3)^L \otimes SU(2)^S$ group [the "restricted H. F. method" (RHF)]. The orbitals of this method have been given for several rows of atoms. In heavier atoms, however, a relativistic H. F. method is required. Several relativistic H. F. results are available in the literature, a topic outside our scope here. More of the calculations, however, are of the Hartree–Fock–Slater type with approximate exchange, owing to the difficulty of the inclusion of full exchange.

The H. F. method is insufficient when it comes to properties that depend on energy differences. Chapter II-2 (Hartree, Hartree, and Swirles) contains early examples of numerical H. F. methods on several oxygen states and ions. The term splitting ratio for ground configurations calculated by the H. F. method with full exchange frequently disagrees with the experimental, as shown in Chapter II-2. The Hartree, Hartree, and Swirles paper also gave an extension of the H. F. method to include additional configurations. This method was developed by A. Jucys and co-workers as well, especially during the 1950s. It has seen further development recently, in several forms, in the work of several groups (Wahl, Roothaan and Hinze, Moser and Bagus, Clementi, Adams, and others). Hartree, Hartree, and Swirles in Chapter II-2 showed

†See, for example, O. Sinanoğlu and D. F. Tuan, *Ann. Rev. Phys. Chem.* **15**, 251-280 (1964). ‡For a more extensive treatment of Roothaan's Open-Shell Hartree-Fock method, see: C. C. J. Roothaan and P. S. Bagus, in *Methods in Computational Physics,* vol. 2, pp. 47-94 (Academic Press, N.Y., 1963).

that in atoms the main orbitals obtained by H. F. alone or by the more difficult "multiconfigurational SCF" (MCSCF) differed little. The energy improvements could be obtained nearly as well with the H. F. followed by ordinary configuration-interaction (H. F. + C. I.). In the oxygen results such C. I. with essentially one more configuration, although the dominant one, still left the term splitting ratio very much in error. The large discrepancy is due to two other types of correlation effects, which the reader will note in Chapter IX-4.

The MCSCF method becomes essential in the calculation of chemical interatomic forces. Without a several-configuration SCF the molecule would not usually dissociate into proper atomic states. By the MCSCF method on the computer, Wahl has calculated proper dissociation curves for molecules such as $Li_2$ and $Na_2$. He obtains also films of the actual (MCSCF $\phi_0$) charge distributions as the molecule dissociates.

For the treatment of electron correlation in a general non-closed-shell state, either the RHF or a kind of MCSCF involving a well-defined set of configurations is a physically meaningful and formally convenient starting point (Chapter IX-2). In the latter case, the orbitals are still restricted to the $O(3) \otimes SU(2)$ symmetry, and the starting wave function may be called GRHF, "general restricted Hartree–Fock."

II-2 Reprinted from *Philosophical Transactions of the Royal Society of London* **A238**, 229 (1939)

# SELF-CONSISTENT FIELD, INCLUDING EXCHANGE AND SUPERPOSITION OF CONFIGURATIONS, WITH SOME RESULTS FOR OXYGEN

*By* D. R. HARTREE, F.R.S., W. HARTREE AND BERTHA SWIRLES

(*Received* 13 *December* 1938)

## 1. Introduction

The calculation of approximate wave functions for the normal configurations of the ions $O^{+++}$, $O^{++}$, $O^{+}$, and neutral O, and the calculation of energy values from the wave functions, was carried out some years ago by Hartree and Black (1933). In this work, the one-electron radial wave functions were calculated by the method of the self-consistent field without exchange, but exchange terms were included in the calculation of the energy from these radial wave functions. In the energy calculations, the same radial wave functions were taken for each of the spectral terms arising from a single configuration;* consequently the ratios between the calculated intermultiplet separations were exactly those given by Slater's (1929) theory of complex spectra.† The ratios between the observed intermultiplet separations, however, depart considerably from these theoretical values (for example, we have for $O^{++}$

$$({}^1D-{}^1S)/({}^3P-{}^1D), \text{ calc. } 3:2, \text{ obs. } 1\cdot04:1),$$

although the energies of the individual terms, and particularly the intermultiplet separation between the lower terms, show quite a good agreement with the observed values.

Approximate calculations by Levy (1934) showed that when exchange terms were included, as in Fock's (1930) equations, the resulting differences between the wave functions for the different spectral terms would not be nearly large enough to account for the observed deviations of the ratios of intermultiplet separations from the ratios given by Slater's theory. This made it appear likely that these deviations are due to appreciable matrix components between the normal configuration and some excited configurations, so that the wave functions for the lowest states include appreciable multiples of the wave functions for higher configurations superposed on that for the normal configuration, a situation which we will describe by the term "superposition

* Strictly, the departure from spherical symmetry in the charge distribution of an incomplete $(2p)$ group would give rise to small differences between the radial wave functions for different spectral terms arising from a single configuration, even when only terms in the equations arising from Coulomb interactions are taken into account, and those arising from exchange interactions are omitted. But these differences are of the same order as those due to exchange and are smaller than the main effect of exchange, so there seems no significance in including them so long as exchange terms are excluded.

† See also Condon and Shortley (1935, chaps. 6, 7). For the comparison of ratios of intermultiplet separations, observed and calculated from Slater's theory, see Condon and Shortley (1935, chap. 7, § 5).

of configurations". [Condon and Shortley (1935) use the term "configuration interaction" for this situation, and its effects on spectral terms and lines are often referred to as "perturbations". But as the terms "interaction" and "perturbation" are both used in different senses from these in atomic mechanics, there seems to be a danger of confusion in their use in this sense. Hence we have preferred to coin the term "superposition of configurations" rather than use "configuration interaction".]

An estimate by Hartree and Swirles (1937) of the magnitude of the effect of superposition of configurations on the energy values, using the wave functions of Hartree and Black, seemed, however, to show that this effect would be about twice too large to account for the observed intermultiplet separations. The position seemed not altogether satisfactory, and with the development in the meantime of practical methods of evaluating solutions of Fock's equations, it seemed desirable to carry out calculations for oxygen, and to revise the calculation of energy values using the resulting wave functions.

When working with any approximation to the structure of an atom in which the wave function for the whole atom is built up from one-electron wave functions, there are two steps in the calculation of energy values at which approximation is involved, first in the way in which the one-electron wave functions are determined, and secondly in the way in which the wave function for the whole atom is taken as built up out of them. In the work of Levy and of Hartree and Swirles, the improved approximation resulting from the inclusion of effects of exchange and of the superposition of configurations was introduced in the second step only. Their inclusion in this step is the more important, since the radial wave functions given by any reasonable approximation will probably not be very greatly different from the best possible wave functions, and by the variation principle the energy value is not sensitive to small variations in the wave functions. But the intermultiplet separations are rather a refinement of the theory; as will be seen from the discussion of results, the second decimal in the ratio of intermultiplet separations involves the fifth significant figure in the calculated energy values. It seemed worth while, therefore, to find whether the use of the best obtainable wave functions would greatly improve the agreement with the experimental energy values and particularly with the ratios of observed intermultiplet separations.

The solution of Fock's equations for the self-consistent field with exchange but without superposition of configurations was carried out for $O^{---}$, $O^{--}$, $O^{-}$, neutral O, and $O^{-}$. It is possible also to include terms arising from superposition of configurations, as well as those arising from exchange, in the equations for the one-electron wave functions, though the solution of the resulting equations is much more lengthy than the solution of Fock's equations. However, to see the effect on the wave functions, and on the energy calculated from them, it was decided to attempt the solution of these equations in one case at least, and $O^{-}$ was chosen since only one of the spectral terms arising from the normal configuration appeared likely to be seriously affected by the superposition of higher configurations.

It was found that although the coefficient of the wave function for the higher configuration was quite large (about 0·2), the one-electron wave functions calculated from the generalized Fock equations, including exchange and superposition of the higher configuration, differed surprisingly little from those calculated from the solution of Fock's equations, including exchange only. Also the results of the calculation of energy values showed the discrepancy already shown by the estimates of Hartree and Swirles; the effect of the superposition of configurations, as far as we have taken it, is nearly twice as much as is required to account for the observed deviation from Slater's ratios of intermultiplet separations.

The small effect on the one-electron wave functions, and the small improvement in the energy values, resulting from the use of the generalized Fock equations for $O^+$ including superposition of configurations, seemed not enough to justify the carrying out of the extensive calculations which would be needed to obtain the solutions of the similar equations for the other states of ionization of oxygen.

## 2. Generalization of Fock's equations to include superposition of configurations

We will consider in particular the normal configuration $(1s)^2\,(2s)^2\,(2p)^q$, which we will call "configuration $A$", of an atom of atomic number $N$. The general argument, and general forms of the results, will apply to any incomplete 8-shell outside closed groups, but it is simpler and shorter to present the argument for a particular case.

Superposition of configurations can only occur if the two configurations have the same parity ($\Sigma l$ odd or even), and the terms arising from the two configurations must (in Russell-Saunders coupling) have the same $L$ and the same $S$.* The only configuration of the same parity as the normal configuration which does not involve a change of principal quantum number is $(1s)^2\,(2p)^{q+2}$, which we will call configuration $B$. This involves the excitation of two electrons; other configurations, involving the excitation of one electron with change of principal quantum number, namely $(1s)^2\,(2s)\,(2p)^q\,(3s)$ and $(1s)^2\,(2s)^2\,(2p)^{q-1}\,(3p)$, may lie not much higher in energy, but the spatial extension of their wave functions will be greater; their overlap with configuration $A$, and hence the extent to which they are superposed on it, will probably be less than that of configuration $B$. Further, their inclusion would increase not only the complexity but the number of equations to be solved, and, as a first step at any rate, it seemed best to consider configurations $A$ and $B$ only.

Let $\Psi_A$ be the normalized wave function, including exchange, for an $(L, S)$ term arising from configuration $A$, and $\Psi_B$ be that for the same $(L, S)$ term arising from configuration $B$; these will be taken as sums of determinants of one-electron wave functions of central-field type,† and they will necessarily be orthogonal.

* See Condon and Shortley (1935), chap. 15, § 1 (p. 366).

† For example, see Hartree and Swirles (1937).

By "superposition of configurations" we mean that we take the linear combination

$$\Psi = (\Psi_A + \mu\Psi_B)/(1+\mu^2)^{\frac{1}{2}} \tag{1}$$

as an approximation to the wave function of the whole atom, where $\mu$ is a parameter whose value is to be determined by the variation principle. It does not seem necessary to use the same radial wave functions in $\Psi_A$ and $\Psi_B$, but it seems doubtful if there would be any significance in not doing so; also the already large amount of work involved in evaluating a solution of the equations including superposition of configurations would be greatly increased if different $(2p)$ wave functions were taken in $\Psi_A$ and $\Psi_B$. So we have been content with the approximation in which the radial wave functions in $\Psi_B$ are taken as the same as those in $\Psi_A$.

For $\Psi$ given by (1), the energy integral

$$E = \int \Psi^* H \Psi \, d\tau \Big/ \int \Psi^* \Psi \, d\tau$$

is given by

$$(1+\mu^2)\,E = H_{AA} + \mu(H_{AB} + H_{BA}) + \mu^2 H_{BB},$$

where $H_{AB} = \int \Psi_A^* H \Psi_B \, d\tau$, etc. Writing $E_A$, $E_B$ for the energy integrals for configurations $A$, $B$ alone, respectively, and $E_{AB}$ for the real part of $H_{AB}$, this becomes

$$(1+\mu^2)\,E = E_A + 2\mu E_{AB} + \mu^2 E_B. \tag{2}$$

TABLE I. COEFFICIENTS IN ENERGY FORMULAE, FOR WAVE FUNCTION

$$\Psi = (\Psi_A + \mu\Psi_B)/(1+\mu^2)^{\frac{1}{2}},$$

where $\Psi_A$ is wave function for configuration $(1s)^2(2s)^2(2p)^q$,
$\Psi_B$ is wave function for configuration $(1s)^2(2p)^{q+2}$.

| Integral | $E_A$ | $E_{AB}$ | $E_B$ | $(1+\mu^2)\,E$ |
|---|---|---|---|---|
| $I(2s)$ | 2 | 0 | 2 | $2(1+\mu^2)$ |
| $I(2s)$ | 2 | 0 | 0 | 2 |
| $I(2p)$ | $q$ | 0 | $q+2$ | $q+(q+2)\,\mu^2$ |
| $F_0(1s, 1s)$ | 1 | 0 | 1 | $1+\mu^2$ |
| $F_0(1s, 2s)$ | 4 | 0 | 0 | 4 |
| $F_0(1s, 2p)$ | $2q$ | 0 | $2(q+2)$ | $2[q+(q+2)\,\mu^2]$ |
| $F_0(2s, 2s)$ | 1 | 0 | 0 | 1 |
| $F_0(2s, 2p)$ | $2q$ | 0 | 0 | $2q$ |
| $F_0(2p, 2p)$ | $\frac{1}{2}q(q-1)$ | 0 | $\frac{1}{2}(q+2)(q+1)$ | $\frac{1}{2}[q(q-1)+(q+2)(q+1)\,\mu^2]$ |
| $F_2(2p, 2p)$ | $\beta_A$ | 0 | $\beta_B$ | $\beta_A+\beta_B\mu^2$ |
| $G_0(1s, 2s)$ | $-2$ | 0 | 0 | $-2$ |
| $G_1(1s, 2p)$ | $-\frac{1}{3}q$ | 0 | $-\frac{1}{3}(q+2)$ | $-\frac{1}{3}[q+(q+2)\,\mu^2]$ |
| $G_1(2s, 2p)$ | $-\frac{1}{3}q$ | $\gamma$ | 0 | $-\frac{1}{3}q+2\gamma\mu$ |

Values of $\beta_A$, $\beta_B$, $\gamma$

| $q$ | 6 | 5 | 4 | | | 3 | | | 2 | | | 1 |
|---|---|---|---|---|---|---|---|---|---|---|---|---|
| Term | $^1$S | $^2$P | $^3$P | $^1$D | $^1$S | $^4$S | $^2$D | $^2$P | $^3$P | $^1$D | $^1$S | $^2$P |
| $\beta_A$ | $-\frac{6}{5}$ | $-\frac{4}{5}$ | $-\frac{3}{5}$ | $-\frac{9}{25}$ | 0 | $-\frac{3}{5}$ | $-\frac{6}{25}$ | 0 | $-\frac{1}{5}$ | $\frac{1}{25}$ | $\frac{2}{5}$ | 0 |
| $\beta_B$ | — | — | — | — | $-\frac{6}{5}$ | — | — | $\frac{4}{5}$ | $-\frac{3}{5}$ | $-\frac{9}{25}$ | 0 | 0 |
| $\gamma$ | 0 | 0 | 0 | 0 | $\frac{1}{3}\sqrt{3}$ | 0 | 0 | $\frac{1}{3}\sqrt{2}$ | $\frac{1}{3}$ | $\frac{1}{3}$ | $\frac{2}{3}$ | $\frac{1}{3}\sqrt{2}$ |

Written out in full, the expression of $(1-\mu^2)E$ consists of multiples of the integrals of the type

$$I(nl) = -\tfrac{1}{2}\int P(nl\,|\,r)\left[\frac{d^2}{dr^2}+\frac{2N}{r}-\frac{l(l-1)}{r^2}\right]P(nl\,|\,r)\,dr, \tag{3}$$

$$F_k(nl, n'l') = \iint P^2(nl\,|\,r)\,P^2(n'l'\,|\,r')\,\frac{r_a^k}{r_b^{k-1}}\,dr\,dr', \tag{4}$$

$$G_k(nl, n'l') = \iint P(nl\,|\,r)\,P(n'l'\,|\,r)\,P(nl\,|\,r')\,P(n'l'\,|\,r')\,\frac{r_a^k}{r_b^{k-1}}\,dr\,dr' \tag{5}$$

$r_a$ being the smaller and $r_b$ the larger of $r$ and $r'$, with coefficients given by Table I. In this table the coefficients in the columns $E_A$ and $E_B$ and the values of $\beta$ have been found by Slater's method,* and the values of $\gamma$ are taken from the paper by Hartree and Swirles (1937).

The value of this expression for $E$ has to be stationary for independent variations of the radial wave functions $P$ and of the parameter $\mu$. Formal differentiation with respect to $P(2s)$ and $P(2p)$ respectively, using the expressions (7)–(10) of D. R. and W. Hartree (1936$b$), gives equations for these functions, namely,

$$\left[\frac{d^2}{dr^2}-\frac{2}{r}\{N-2Y_0(1s, 1s)-Y_0(2s, 2s)-qY_0(2p, 2p)\}-\epsilon_{2s,2s}\right]P(2s)$$

$$+\frac{2Y_0(1s, 2s)}{r}P(1s)-\tfrac{1}{3}(q-6\gamma\mu)\frac{Y_1(2s, 2p)}{r}P(2p) = 0, \tag{6}$$

$$\left[\frac{d^2}{dr^2}-\frac{2}{r}\left\{N-2Y_0(1s, 1s)-\frac{2q}{q+(q+2)\mu^2}Y_0(2s, 2s)-\frac{q(q-1)+(q-2)(q-1)\mu^2}{q+(q+2)\mu^2}Y_0(2p, 2p)\right.\right.$$

$$\left.\left.-\frac{2(\beta_A-\beta_B\mu^2)}{q-(q-2)\mu^2}Y_2(2p, 2p)\right\}-\frac{2}{r^2}-\epsilon_{2p,2p}\right]P(2p)$$

$$+\frac{2}{3}\frac{Y_1(1s, 2p)}{r}P(1s)-\frac{2}{3}\frac{q-6\gamma\mu}{q-(q-2)\mu^2}\frac{Y_1(2s, 2p)}{r}P(2s) = 0. \tag{7}$$

From the condition that the value of $E$ must also be stationary for variations of $\mu$, we obtain from (2)

$$\mu = (E-E_A)/E_{AB},$$

or, eliminating $E$ between this result and (2),

$$\mu/(1-\mu^2) = -E_{AB}/(E_B-E_A). \tag{8}$$

Equations (6)–(8) form a generalization of Fock's equations to include terms (those in $\mu$) arising from the superposition of configurations $A$ and $B$.

* See Slater (1929); Condon and Shortley (1935), chaps. 6, 7.

The method of solving these equations is first to estimate a value of $\mu$, and solve (6) and (7) using this value; then to calculate the right-hand side of (8) from the solution, and hence to obtain a "final" value of $\mu$. The estimate of $\mu$ is altered by trial and error until the final value so obtained agrees with it. This requires a complete solution of the extended Fock's equations (6) and (7) for each trial value of $\mu$, so that the work involved in a single solution of the equations including superposition of configurations is as much as in several solutions of Fock's equations including exchange only. Fortunately the radial wave functions $P$ and the final value of $\mu$ are not very sensitive to the estimate of $\mu$, and the process of approximation to the correct value of $\mu$ is quite rapidly convergent. Even so, the amount of work involved is considerable.

If the superposition of other configurations, such as

$$(1s)^2\,(2s)\,(2p)^q\,(3s) \quad \text{or} \quad (1s)^2\,(2s)^2\,(2p)^{q-1}\,(3p),$$

were taken into account, both the form of the equations and the process of deriving them would be more complicated, there would also be more equations (one extra for each new radial wave function such as (3*s*) or (3*p*), and possibly more, since it might then be desirable to take different (2*s*) or (2*p*) wave functions for the different configurations), and there would be further conditions of the form (8) to be satisfied; and the numerical work might be too formidable to be practicable.

## 3. Energy formulae

In calculating energies, it is usually convenient to make use of the equations satisfied by the radial functions $P$ to eliminate some of the integrals from the energy formula, and this is the case here. The values of $E_A$ and $E_B$ are wanted in two contexts, both in the calculation of $E$ from (2), and in the calculation of the "final" value of $\mu$ from (8).

To a sufficient approximation, the variations of the (1*s*) wave function can be neglected; this is certainly an amply good enough approximation as far as intermultiplet separations are concerned, and even between different states of ionization the differences are small and, by the variation principle, the contribution to the energy from the $(1s)^2$ group is insensitive to small variations of the (1*s*) wave functions from that of $O^{-6}$. Then we can consider the energy of the (2*s*) (2*p*) shell due to its interaction with the $O^{-6}$ core and its internal interactions only; and on substituting for $d^2P/dr^2$ in $I(2s)$ and $I(2p)$ from (6), (7) we get the energies expressed as the sum of multiples of integrals and of the $\epsilon$ parameters with coefficients given in Table II.

Here the column headed $E_A(\mu = 0)$ refers to the energy expression without superposition of configurations; the other columns give the coefficients in the respective expressions when the terms arising from the superposition of configurations are included both in the equations from which the wave functions are calculated, and in the energy formula (1); they do not apply if such terms are only included in the latter formula.

TABLE II. COEFFICIENTS IN ENERGY FORMULAE AFTER SUBSTITUTION FOR $d^2P/dt^2$ IN $I(2s)$ AND $I(2p)$

| Integral | Coefficient in $E_A\ (\mu = 0)$ | $E_A$ | $E_B$ | $E_B - E_A$ | $(1+\mu^2)E$ / $E_A + 2\mu E_{AB} + \mu^2 E_B$ |
|---|---|---|---|---|---|
| $F_0(2s, 2s)$ | $-1$ | $-1$ | $0$ | $+1$ | $1$ |
| $F_0(2s, 2p)$ | $-2q$ | $-\dfrac{2q^2}{q+(q+2)\mu^2}$ | $-\dfrac{2q(q+2)}{q+(q+2)\mu^2}$ | $-\dfrac{4q}{q+(q+2)\mu^2}$ | $-2q$ |
| $F_0(2p, 2p)$ | $-\frac{1}{2}q(q-1)$ | $-\frac{1}{2}q(q-1)-\dfrac{2q(q+2)\mu^2}{q+(q+2)\mu^2}$ | $-\frac{1}{2}(q+1)(q+2)+\dfrac{2q(q+2)}{q+(q+2)\mu^2}$ | $\dfrac{3q-(q+2)\mu^2}{q+(q+2)\mu^2}$ | $-\frac{1}{2}[q(q-1)+(q+1)(q+2)\mu^2]$ |
| $F_2(2p, 2p)$ | $-\beta_A$ | $\beta_A-\dfrac{2q(\beta_A+\beta_B\mu^2)}{q+(q+2)\mu^2}$ | $\beta_B-\dfrac{2(q+2)(\beta_A+\beta_B\mu^2)}{q+(q+2)\mu^2}$ | $\beta_B-\beta_A-\dfrac{4(\beta_A+\beta_B\mu^2)}{q+(q+2)\mu^2}$ | $-[\beta_A+\beta_B\mu^2]$ |
| $G_1(2s, 2p)$ | $\frac{1}{3}q$ | $-2\gamma\mu+\dfrac{q(q-6\gamma\mu)}{3[q+(q+2)\mu^2]}$ | $\dfrac{(q+2)(q-6\gamma\mu)}{3[q+(q+2)\mu^2]}$ | $2\gamma\mu+\dfrac{2(q-6\gamma\mu)}{3[q+(q+2)\mu^2]}$ | $\frac{1}{3}q - 2\gamma\mu$ |
| $\epsilon_{2s,2s}$ | $-1$ | $-1$ | $0$ | $1$ | $-1$ |
| $\epsilon_{2p,2p}$ | $-\frac{1}{2}q$ | $-\frac{1}{2}q$ | $-\frac{1}{2}(q+2)$ | $-1$ | $-\frac{1}{2}[q+(q+2)\mu^2]$ |

The small number of integrals involved in Table II, compared with the number in Table I, shows the practical advantage of working with the modified energy formulae given by the substitution for $d^2P/dr^2$ in the integrals.

It should be noted that, from the way in which the coefficients in the formula for $E_B - E_A$ in Table II were obtained, it follows that in calculating the "final" value of $\mu$ from (8), using this formula for $E_B - E_A$, it is correct to use the *estimated* value of $\mu$ in calculating these coefficients, rather than any better approximation to the "final" value.

## 4. Procedure

The procedure used for solving both the Fock equations with exchange only, and the generalized Fock equations (6), (7) for each estimate of $\mu$, was very similar to that used for the $(2s)(2p)^3$P excited state of Be (see D. R. and W. Hartree (1936*a*)). There were two main modifications based on the experience of work subsequent to that on Be; first, the present work was done throughout with the equations for normalized wave functions, the values of $P/r^{l+1}$ at $r = 0$ being adjusted by trial so that the solutions when found were normalized, and secondly, that the work was started from estimates of the $P$'s not of $Z_k$'s.

As already explained, when the superposition of configurations is taken into account, this solution of (6), (7) has to be carried out for a number of trial values of $\mu$, until relation (8) is satisfied by the solution.

For each stage of ionization the solution of Fock's equations with exchange only was carried out first for the highest and for the lowest of the three spectral terms arising from the normal configuration. Then radial wave functions $P$ were interpolated, linearly in $\beta$, for the intermediate term, and these estimates were always found to be adequate (cf. Table I). For all but $O^-$, the work was done to three decimals in $P$ over the main part of the range, but for $O^-$ it was improved to four decimals, as the three-decimal results did not seem likely to be accurate enough to show conclusively the effect of superposition of configurations.

A rather thorough attempt was made to obtain a solution for $O^{--}$, but without success; it seems rather well established that, to the accuracy of this approximation to the structure of a many-electron atom, the $O^{--}$ ion cannot exist in the free state.

As already mentioned, the only solution of the generalized Fock equations, including the superposition of configurations, was done for $O^-$, for which the terms in $E$ and in the equations are given by putting $q = 3$ in Tables I, II and equations (6), (7). The normal configuration gives $^4$S, $^2$D and $^2$P terms, of which only the last, the highest, is subject to the superposition of the $(2p)^5$ configuration. The value of $\mu$, estimated from (8) using values of $E_A$, $E_B$ calculated from the results of the solution of Fock's equations was $-0{\cdot}18$; four approximations were made, and the result of the final approximation was $-0{\cdot}202$ (the minus sign is of no significance, it depends on the convention of sign adopted for the radial wave functions).

## 5. Results. Wave functions

The results for $O^-$ are given in Table III, and those for the other states of ionization in Table IV.

For $O^-$ all three normalized radial wave functions, namely (1s), (2s) and (2p), are tabulated for the lowest term ($^4S$, $\beta = -\frac{3}{5}$) arising from the normal configuration, the (2s) and (2p) wave functions being given to the full four decimals retained in the calculations. For the other terms ($^2D$, $\beta = -\frac{6}{25}$ and $^2P$, $\beta = 0$, the latter calculated without superposition of configurations) the differences $\Delta P$ from the wave functions for the lowest term, namely,

$$\Delta P = P_N \text{ (upper term)} - P_N \text{ (lower term)} \tag{9}$$

are tabulated in units of the fourth decimal, to show how very small is the departure, over this range, of the normalized wave function $P_N$ from a linear dependence on $\beta$. If $P_N$ were exactly linear in $\beta$, the numbers in the columns headed $\Delta\beta = \frac{9}{25}$ would be 0·6 of those in the corresponding columns headed $\Delta\beta = \frac{3}{5}$; actually the maximum departure is hardly a unit in the fourth decimal for $P(2s)$ and 2 in the fourth decimal for $P(2p)$, and this is hardly outside the range of possible effect of accumulation of rounding off errors in the integration. The values of the Lagrange parameter $\epsilon_{nl,\,nl}$ show a corresponding linear dependence on $\beta$.

For the upper state ($^2P$, $\beta = 0$) the normalized radial wave functions $P_N$ themselves, calculated without superposition of configurations (columns headed $\mu = 0$), are also tabulated, for comparison with the values calculated including the terms in the equations arising from superposition of configurations (columns headed $\mu = -0{\cdot}202$), and the differences $\Delta' P$ arising from the effect of superposition of configurations are given. These differences are small, less than 0·4 % of the maximum $|P|$ for both wave functions; they would hardly have been established with certainty by calculations to three decimals, since in the third decimal they would be hardly greater than the possible effect of accumulation of rounding off errors in the numerical integration.

The results for other states of ionization are tabulated in a more compact form, and to three decimals only, in Table IV. The wave functions $P_N$ are given in full for the lowest term only; for $O^{--}$ and O neutral, for which the normal configuration gives three terms, the difference $\Delta P$ from the lowest term (cf. (9)) are given for the upper term only, for the intermediate term they can be taken as proportional to $\beta$ in all cases. This is justified by the comparison of the four-decimal results for $O^-$, and had been established by solution of the equations for the intermediate terms for the other states of ionization, before these four-decimal results for $O^-$ were calculated.

The effect of superposition of configurations on the wave functions for $O^-$ and on the energy (cf. §6) was considered too small to justify the extensive calculations which would be required to obtain the corresponding effects for the other states of ionization; for $O^{++}$ all three terms $^3P$, $^1D$ and $^1S$ are affected, and there seemed no object in carrying out the calculation for one of these and not for the others.

## TABLE III. $O^+$ NORMALIZED RADIAL WAVE FUNCTIONS

| | | Without superposition of configurations | | | | | | ²P term, $\beta = 0$: Without superposition of configurations | | ²P term, $\beta = 0$: With superposition of configurations | | | |
|---|---|---|---|---|---|---|---|---|---|---|---|---|---|
| | | ⁴S term, $\beta = -3/5$ | | ²D term, $\Delta\beta = 9/25$ | | ²P term, $\Delta\beta = 3/5$ | | $\mu = 0$ | | $\mu = -0{\cdot}202$ | | $\Delta'\mu = -0{\cdot}202$ | |
| | | | | | | Table of $P_N/r^{l+1}$ and $\Delta P_N/r^{l+1}$ | | | | | | | |
| $r$ | $P(1s)/r$ | $P(2s)/r$ | $P(2p)/r^2$ | $\Delta P(2s)/r$ | $\Delta P(2p)/r^2$ | $\Delta P(2s)/r$ | $\Delta P(2p)/r^2$ | $P(2s)/r$ | $P(2p)/r^2$ | $P(2s)/r$ | $P(2p)/r^2$ | $\Delta' P(2s)/r$ | $\Delta' P(2p)/r^2$ |
| 0·00 | 43·17 | 10·20$_5$ | 16·34 | 3 | − 13 | 4$_5$ | − 21$_5$ | 10·25 | 16·12$_5$ | 10·31$_5$ | 16·09$_5$ | 6$_5$ | − 3 |
| 0·01 | 39·86 | 9·41$_5$ | 15·70$_5$ | 3 | − 12$_5$ | 4$_5$ | − 20$_5$ | 9·46 | 15·50 | 9·52 | 15·47 | 6 | − 3 |
| 0·02 | 36·80 | 8·68 | 15·10 | 2$_5$ | − 12$_5$ | 4 | − 20 | 8·72 | 14·90 | 8·77$_5$ | 14·87 | 5$_5$ | − 3 |
| 0·03 | 33·99 | 7·99 | 14·52 | 2$_5$ | − 12$_5$ | 4 | − 19 | 8·03 | 14·33 | 8·08 | 14·30 | 5 | − 3 |
| | | | | | | Table of $P_N$ and $\Delta P_N$ | | | | | | | |
| | $P(1s)$ | $P(2s)$ | $P(2p)$ | $\Delta P(2s)$ | $\Delta P(2p)$ | $\Delta P(2s)$ | $\Delta P(2p)$ | $P(2s)$ | $P(2p)$ | $P(2s)$ | $P(2p)$ | $\Delta' P(2s)$ | $\Delta' P(2p)$ |
| 0·00 | 0 | 0 | 0 | | | | | 0 | 0 | 0 | 0 | | |
| 0·01 | 0·398$_5$ | 0·0942 | 0·0015$_5$ | 2$_5$ | | 4 | | 0·0946 | 0·0015$_5$ | 0·0952 | 0·0015$_5$ | 6 | |
| 0·02 | 0·736 | 0·1737 | 0·0060$_5$ | 4 | − 0$_5$ | 7 | − 1 | 0·1744 | 0·0059$_5$ | 0·1754$_5$ | 0·0059$_5$ | 10$_5$ | |
| 0·03 | 1·020 | 0·2398 | 0·0130$_5$ | 6 | − 1 | 10 | − 1$_5$ | 0·2408 | 0·0129 | 0·2423 | 0·0128$_5$ | 15 | − 0$_5$ |
| 0·04 | 1·256 | 0·2939 | 0·0223$_5$ | 8 | − 2 | 13 | − 3 | 0·2952 | 0·0220$_5$ | 0·2970 | 0·0220 | 18 | − 0$_5$ |
| 0·06 | 1·609 | 0·3710 | 0·0466 | 10 | − 4 | 16 | − 7 | 0·3726 | 0·0459$_5$ | 0·3749 | 0·0459 | 23 | − 0$_5$ |
| 0·08 | 1·834 | 0·4133 | 0·0769 | 10 | − 7 | 17 | − 11 | 0·4150 | 0·0758 | 0·4176 | 0·0757 | 26 | − 1 |
| 0·10 | 1·962 | 0·4277 | 0·1117 | 11 | −10 | 18 | − 16 | 0·4295 | 0·1101 | 0·4322 | 0·1100 | 27 | − 1 |
| 0·12 | 2·015 | 0·4199 | 0·1497 | 11 | −13 | 17 | − 21 | 0·4216 | 0·1476 | 0·4243 | 0·1474 | 27 | − 2 |
| 0·14 | 2·015 | 0·3947 | 0·1900 | 9 | −16 | 15 | − 27 | 0·3962 | 0·1873 | 0·3988 | 0·1870 | 26 | − 3 |
| 0·16 | 1·974 | 0·3559 | 0·2316 | 8 | −20 | 13 | − 33 | 0·3572 | 0·2283 | 0·3596 | 0·2280 | 24 | − 3 |
| 0·18 | 1·905 | 0·3068 | 0·2739 | 6 | −24 | 10 | − 40 | 0·3078 | 0·2699 | 0·3100 | 0·2696 | 22 | − 3 |
| 0·20 | 1·818 | 0·2500 | 0·3164 | 4 | −28 | 7 | − 47 | 0·2507 | 0·3117 | 0·2526 | 0·3114 | 19 | − 3 |
| 0·25 | 1·555 | +0·0876 | 0·4204 | 0 | −39 | 0 | − 64 | +0·0876 | 0·4140 | +0·0887 | 0·4136 | 11 | − 4 |
| 0·30 | 1·281 | −0·0852 | 0·5177 | − 5 | −51 | − 8 | − 83 | −0·0860 | 0·5094 | −0·0858 | 0·5091 | + 2 | − 3 |
| 0·35 | 1·028 | −0·2533 | 0·6053 | −10 | −62 | −17 | −102 | −0·2550 | 0·5951 | −0·2555 | 0·5949 | − 5 | − 2 |
| 0·40 | 0·810 | −0·4080 | 0·6818 | −14 | −72 | −24 | −118 | −0·4104 | 0·6700 | −0·4116 | 0·6700 | −12 | 0 |
| 0·45 | 0·630 | −0·5447 | 0·7469 | −17 | −82 | −30 | −135 | −0·5477 | 0·7334 | −0·5495 | 0·7337 | −18 | + 3 |
| 0·50 | 0·485 | −0·6614 | 0·8004 | −20 | −90 | −34 | −148 | −0·6648 | 0·7856 | −0·6672 | 0·7863 | −24 | 7 |

| | | | | | | | | | | | | | |
|---|---|---|---|---|---|---|---|---|---|---|---|---|---|
| 0·6 | 0·280 | −0·8354 | 0·8755 | −22 | −100 | −37 | −165 | −0·8391 | 0·8590 | −0·8420 | 0·8603 | −29 | 13 |
| 0·7 | 0·159 | −0·9382 | 0·9137 | −21 | −103 | −36 | −170 | −0·9418 | 0·8967 | −0·9448 | 0·8985 | −30 | 18 |
| 0·8 | 0·089 | −0·9842 | 0·9224 | −17 | −98 | −28 | −162 | −0·9870 | 0·9062 | −0·9899 | 0·9084 | −29 | 22 |
| 0·9 | 0·050 | −0·9875 | 0·9087 | −13 | −85 | −21 | −142 | −0·9896 | 0·8945 | −0·9919 | 0·8968 | −23 | 23 |
| 1·0 | 0·028 | −0·9608 | 0·8787 | −7 | −67 | −11 | −113 | −0·9619 | 0·8674 | −0·9635 | 0·8696 | −16 | 22 |
| 1·1 | 0·015$_5$ | −0·9140 | 0·8375 | −1 | −45 | −1 | −77 | −0·9141 | 0·8298 | −0·9150 | 0·8317 | −9 | 19 |
| 1·2 | 0·008$_5$ | −0·8548 | 0·7891 | +4 | −22 | +7 | −38 | −0·8541 | 0·7853 | −0·8541 | 0·7867 | 0 | 14 |
| 1·4 | 0·002$_5$ | −0·7203 | 0·6822 | 12 | +26 | 20 | +40 | −0·7183 | 0·6862 | −0·7169 | 0·6866 | +14 | +4 |
| 1·6 | 0·001 | −0·5861 | 0·5743 | 17 | 68 | 28 | 110 | −0·5833 | 0·5853 | −0·5807 | 0·5846 | 26 | 7 |
| 1·8 | | −0·4654 | 0·4744 | 19 | 99 | 31 | 161 | −0·4623 | 0·4905 | −0·4590 | 0·4888 | 33 | 17 |
| 2·0 | | −0·3629 | 0·3860 | 19 | 119 | 31 | 196 | −0·3598 | 0·4056 | −0·3562 | 0·4032 | 36 | −24 |
| 2·2 | | −0·2791 | 0·3105 | 17 | 128 | 29 | 212 | −0·2762 | 0·3317 | −0·2726 | 0·3289 | 36 | −28 |
| 2·4 | | −0·2123 | 0·2474 | 15 | 130 | 26 | 216 | −0·2097 | 0·2690 | −0·2063 | 0·2659 | 34 | 31 |
| 2·6 | | −0·1600 | 0·1955 | 13 | 126 | 22 | 211 | −0·1578 | 0·2166 | −0·1547 | 0·2134 | 31 | 32 |
| 2·8 | | −0·1196 | 0·1535 | 10 | 117 | 17 | 197 | −0·1179 | 0·1732 | −0·1151 | 0·1701 | 28 | 31 |
| 3·0 | | −0·0889 | 0·1198 | 9 | 106 | 15 | 180 | −0·0874 | 0·1378 | −0·0851 | 0·1349 | 23 | −29 |
| 3·2 | | −0·0657 | 0·0931 | 7 | 94 | 12 | 160 | −0·0645 | 0·1091 | −0·0626 | 0·1065 | 19 | −26 |
| 3·4 | | −0·0484 | 0·0720 | 6 | 82 | 10 | 140 | −0·0474 | 0·0860 | −0·0458 | 0·0837 | 16 | −23 |
| 3·6 | | −0·0354 | 0·0555 | 4 | 70 | 7 | 120 | −0·0347 | 0·0675 | −0·0334 | 0·0655 | 13 | −20 |
| 3·8 | | −0·0258 | 0·0426 | 3 | 60 | 5 | 103 | −0·0253 | 0·0529 | −0·0243 | 0·0511 | 10 | −18 |
| 4·0 | | −0·0188 | 0·0326 | 2$_5$ | 51 | 4 | 87 | −0·0184 | 0·0413 | −0·0176 | 0·0398 | 8 | −15 |
| 4·5 | | −0·0084 | 0·0166 | 0$_5$ | 31 | 1 | 53 | −0·0083 | 0·0219 | −0·0078 | 0·0210 | 5 | −9 |
| 5·0 | | −0·0037 | 0·0083 | 0$_5$ | 20 | 0$_5$ | 32 | −0·0036$_5$ | 0·0115 | −0·0034 | 0·0110 | 2$_5$ | 5 |
| 5·5 | | −0·0016$_5$ | 0·0042 | | 12 | | 18 | −0·0016$_5$ | 0·0060 | −0·0015 | 0·0057 | 1$_5$ | −3 |
| 6·0 | | −0·0007$_5$ | 0·0021 | | 7 | | 10$_5$ | −0·0007$_5$ | 0·0031$_5$ | −0·0006$_5$ | 0·0029$_5$ | 1 | −2 |
| 6·5 | | −0·0003$_5$ | 0·0010$_5$ | | 4 | | 5$_5$ | −0·0003$_5$ | 0·0016 | −0·0003 | 0·0015 | 0$_5$ | −1 |
| 7·0 | | −0·0001$_5$ | 0·0005 | | 2$_5$ | | 3 | −0·0001$_5$ | 0·0008 | −0·0001$_5$ | 0·0007$_5$ | | −0$_5$ |
| 7·5 | | −0·0000$_5$ | 0·0003 | | 1 | | 1 | −0·0000$_5$ | 0·0004 | −0·0000$_5$ | 0·0004 | | |
| 8·0 | | | 0·0001$_5$ | | 0$_5$ | | 0$_5$ | | 0·0002 | | 0·0002 | | |
| 8·5 | | | 0·0000$_5$ | | | | 0$_5$ | | 0·0001 | | 0·0001 | | |
| 9·0 | | | 0·0000$_5$ | | | | | | 0·0000$_5$ | | 0·0000$_5$ | | |
| | | | | $\Delta\epsilon$ | $\Delta\epsilon$ | $\Delta\epsilon$ | $\Delta\epsilon$ | | | | | $\Delta'\epsilon$ | $\Delta'\epsilon$ |
| $\epsilon_{nl,nl}$ | 42·7 | 3·603 | 2·653$_5$ | +30 | −165 | +50$_5$ | −272 | 3·653$_5$ | 2·381$_5$ | 3·743 | 2·425 | 89$_5$ | 43$_5$ |

TABLE IV. NORMALIZED RADIAL WAVE FUNCTIONS FOR $O^{+++}$, $O^{++}$, O NEUTRAL, AND O

Table of $P_N/r^{l+1}$ and $\Delta P_N/r^{l+1}$

| | $O^{+++}$ | | | $O^{++}$ | | | | | O neutral | | | | | O | | |
|---|---|---|---|---|---|---|---|---|---|---|---|---|---|---|---|---|
| | $^2$P term, $\beta = 0$ | | | $^2$P term, $\beta = -1/5$ | | | $^1$S term, $\Delta\beta = 3/5$ | | $^2$P term, $\beta = -3/5$ | | | $\Delta\beta = 3/5$ | | $^2$P term, $\beta = -4/5$ | | |
| $r$ | $P(1s)/r$ | $P(2s)/r$ | $P(2p)/r^2$ | $P(1s)/r$ | $P(2s)/r$ | $P(2p)/r^2$ | $\Delta P(2s)/r$ | $\Delta P(2p)/r^2$ | $P(1s)/r$ | $P(2s)/r$ | $P(2p)/r^2$ | $\Delta P(2s)/r$ | $\Delta P(2p)/r^2$ | $P(1s)/r$ | $P(2s)/r$ | $P(2p)/r^2$ |
| 0·00 | 43·24 | 11·33 | 19·81 | 43·20 | 10·74 | 18·01 | 6 | −39 | 43·16 | 9·77$_5$ | 14·63 | 7 | −15 | 43·15 | 9·51 | 13·12 |
| 0·01 | 39·92 | 10·45 | 19·04 | 39·89 | 9·91 | 17·31 | 5$_5$ | −38 | 39·85 | 9·02 | 14·06 | 6$_5$ | −15 | 39·84 | 8·77$_5$ | 12·61 |
| 0·02 | 36·86 | 9·63 | 18·30 | 36·83 | 9·13$_5$ | 16·64 | 5 | −36 | 36·80 | 8·31$_5$ | 13·52 | 6 | −14 | 36·79 | 8·09 | 12·12 |
| 0·03 | 34·05 | 8·87 | 17·60 | 34·02 | 8·41 | 16·01 | 4$_5$ | −35 | 33·99 | 7·65$_5$ | 13·00 | 5$_5$ | −14 | 33·98 | 7·45 | 11·66 |

Table of $P_N$ and $\Delta P_N$

| $r$ | $P(1s)$ | $P(2s)$ | $P(2p)$ | $P(1s)$ | $P(2s)$ | $P(2p)$ | $\Delta P(2s)$ | $\Delta P(2p)$ | $P(1s)$ | $P(2s)$ | $P(2p)$ | $\Delta P(2s)$ | $\Delta P(2p)$ | $P(1s)$ | $P(2s)$ | $P(2p)$ |
|---|---|---|---|---|---|---|---|---|---|---|---|---|---|---|---|---|
| 0·00 | 0 | 0 | 0 | 0 | 0 | 0 | | | 0 | 0 | 0 | | | 0 | 0 | 0 |
| 0·01 | 0·399 | 0·104$_5$ | 0·002 | 0·399 | 0·099 | 0·001$_5$ | 0$_5$ | | 0·398$_5$ | 0·090 | 0·001$_5$ | 1 | | 0·398$_5$ | 0·088 | 0·001 |
| 0·02 | 0·737 | 0·192$_5$ | 0·007$_5$ | 0·737 | 0·182$_5$ | 0·006$_5$ | 1 | | 0·736 | 0·166 | 0·005$_5$ | 1$_5$ | | 0·736 | 0·162 | 0·005 |
| 0·03 | 1·021 | 0·266 | 0·016 | 1·020 | 0·252 | 0·014$_5$ | 1$_5$ | | 1·020 | 0·230 | 0·011$_5$ | 1$_5$ | | 1·020 | 0·223 | 0·010$_5$ |
| 0·04 | 1·258 | 0·326 | 0·027 | 1·257 | 0·309 | 0·024$_5$ | 2 | −0$_5$ | 1·256 | 0·282 | 0·020 | 2 | | 1·256 | 0·274 | 0·018 |
| 0·06 | 1·612 | 0·411 | 0·056$_5$ | 1·610 | 0·390 | 0·051$_5$ | 2$_5$ | −1 | 1·609 | 0·355 | 0·042 | 2$_5$ | −0$_5$ | 1·609 | 0·346 | 0·037$_5$ |
| 0·08 | 1·837 | 0·458 | 0·093 | 1·835 | 0·435 | 0·084$_5$ | 2$_5$ | −1$_5$ | 1·833 | 0·396 | 0·069 | 2$_5$ | −1 | 1·833 | 0·385 | 0·061$_5$ |
| 0·10 | 1·965 | 0·473 | 0·135 | 1·963 | 0·449 | 0·123 | 3 | −2$_5$ | 1·961 | 0·410 | 0·100 | 3 | −1$_5$ | 1·960 | 0·399 | 0·090 |
| 0·12 | 2·018 | 0·464 | 0·181 | 2·016 | 0·441 | 0·165 | 3 | −3$_5$ | 2·015 | 0·403 | 0·134 | 2$_5$ | −1$_5$ | 2·014 | 0·392 | 0·120 |
| 0·14 | 2·018 | 0·435 | 0·230 | 2·016 | 0·414 | 0·209 | 2$_5$ | −4$_5$ | 2·015 | 0·379 | 0·170 | 2$_5$ | −2 | 2·014 | 0·369 | 0·153 |
| 0·16 | 1·977 | 0·391 | 0·280 | 1·975 | 0·373 | 0·255 | 2$_5$ | −6 | 1·974 | 0·342 | 0·208 | 2$_5$ | −2$_5$ | 1·974 | 0·333 | 0·186 |
| 0·18 | 1·908 | 0·335 | 0·331 | 1·906 | 0·321 | 0·301 | 2$_5$ | −7 | 1·905 | 0·295 | 0·246 | 2 | −3 | 1·905 | 0·287 | 0·220 |
| 0·20 | 1·819 | 0·270 | 0·381 | 1·818 | 0·260 | 0·348 | 2 | −8$_5$ | 1·818 | 0·241 | 0·284 | 1$_5$ | −3$_5$ | 1·818 | 0·235 | 0·254 |
| 0·25 | 1·555 | +0·087 | 0·505 | 1·555 | +0·088 | 0·462 | 1$_5$ | −11$_5$ | 1·555 | +0·086 | 0·377 | +0$_5$ | −5 | 1·555 | +0·084 | 0·339 |
| 0·30 | 1·280 | −0·107 | 0·619 | 1·281 | −0·095 | 0·567 | +0$_5$ | −14 | 1·281 | −0·079 | 0·465 | −1 | −6 | 1·281 | −0·076 | 0·418 |
| 0·35 | 1·026 | −0·295 | 0·720 | 1·027 | −0·272 | 0·662 | −0$_5$ | −17$_5$ | 1·028 | −0·240 | 0·544 | −2 | −7$_5$ | 1·029 | −0·232 | 0·489 |
| 0·40 | 0·807 | −0·467 | 0·806 | 0·809 | −0·435 | 0·743 | −1 | −20 | 0·811 | −0·388 | 0·614 | −3 | −9 | 0·811 | −0·376 | 0·553 |
| 0·45 | 0·626 | −0·618 | 0·876 | 0·628 | −0·579 | 0·811 | −1$_5$ | −22 | 0·631 | −0·519 | 0·674 | −4 | −10$_5$ | 0·631 | −0·504 | 0·607 |
| 0·50 | 0·481 | −0·745 | 0·931 | 0·483 | −0·701 | 0·866 | −2 | −24 | 0·486 | −0·632 | 0·724 | −4$_5$ | −12 | 0·486 | −0·613 | 0·653 |

| | | | | | | | | | | | | | | | | |
|---|---|---|---|---|---|---|---|---|---|---|---|---|---|---|---|---|
| 0·6 | 0·276 | −0·928 | 0·999 | 0·278 | −0·879 | 0·939 | −2$_5$ | −26 | 0·281 | −0·800 | 0·797 | 5 | 13$_5$ | 0·282 | 0·779 | 0·722 |
| 0·7 | 0·155 | −1·027 | 1·019 | 0·157 | −0·981 | 0·970 | −2$_5$ | −25$_5$ | 0·160 | −0·902 | 0·839 | −5 | 14 | 0·161 | 0·880 | 0·763 |
| 0·8 | 0·086 | −1·058 | 1·002 | 0·088 | −1·021 | 0·967 | −2 | −22$_5$ | 0·090 | −0·951 | 0·855 | −5 | 14 | 0·091 | −0·929 | 0·782 |
| 0·9 | 0·047 | −1·041 | 0·959 | 0·049 | −1·016 | 0·940 | −1$_5$ | −17$_5$ | 0·050$_5$ | −0·960 | 0·852 | −4 | −13 | 0·051 | −0·940 | 0·785 |
| 1·0 | 0·026 | −0·990 | 0·899 | 0·027 | −0·978 | 0·896 | −1 | −11$_5$ | 0·028$_5$ | −0·941 | 0·835 | −2$_5$ | 12 | 0·028$_5$ | −0·924 | 0·776 |
| 1·1 | 0·014 | −0·919 | 0·828 | 0·015 | −0·921 | 0·840 | 0 | −4$_5$ | 0·016 | −0·902 | 0·808 | −1$_5$ | −10$_5$ | 0·016 | −0·890 | 0·759 |
| 1·2 | 0·007$_5$ | −0·837 | 0·752 | 0·008 | −0·851 | 0·778 | +0$_5$ | +2$_5$ | 0·009 | −0·851 | 0·774 | 0 | −7$_5$ | 0·009 | −0·843 | 0·735 |
| 1·4 | 0·002 | −0·664 | 0·600 | 0·002$_5$ | −0·697 | 0·648 | 1 | 14 | 0·003 | −0·732 | 0·696 | +2 | 3 | 0·003 | −0·734 | 0·678 |
| 1·6 | 0·000$_5$ | −0·505 | 0·463 | 0·000$_5$ | −0·549 | 0·523 | 1$_5$ | 23$_5$ | 0·001 | −0·610 | 0·613 | 3$_5$ | +2 | 0·001 | −0·621 | 0·616 |
| 1·8 | | −0·372 | 0·348 | | −0·421 | 0·413 | 1$_5$ | 29$_5$ | | −0·499 | 0·533 | 4$_5$ | 6$_5$ | | −0·517 | 0·555 |
| 2·0 | | −0·268 | 0·257 | | −0·316 | 0·320 | 1$_5$ | 32 | | −0·401 | 0·459 | 5 | 10$_5$ | | −0·424 | 0·497 |
| 2·2 | | −0·189 | 0·186 | | −0·233 | 0·244 | 1 | 32$_5$ | | −0·319 | 0·391 | 4$_5$ | 13$_5$ | | −0·346 | 0·444 |
| 2·4 | | −0·132 | 0·133 | | −0·170 | 0·185 | 1 | 30 | | −0·252 | 0·332 | 4$_5$ | 15$_5$ | | −0·281 | 0·396 |
| 2·6 | | −0·091 | 0·094 | | −0·123 | 0·138 | 1 | 27 | | −0·198 | 0·280 | 4 | 17 | | −0·227 | 0·352 |
| 2·8 | | −0·061 | 0·066 | | −0·088 | 0·102 | 0$_5$ | 24 | | −0·154 | 0·236 | 3$_5$ | 17$_5$ | | −0·183 | 0·314 |
| 3·0 | | −0·041 | 0·046 | | −0·062 | 0·075$_5$ | 0$_5$ | 20$_5$ | | −0·120 | 0·198 | 3 | 17$_5$ | | −0·147 | 0·279 |
| 3·2 | | −0·028 | 0·031$_5$ | | −0·044 | 0·055 | 0$_5$ | 17$_5$ | | −0·093 | 0·165 | 2$_5$ | 17$_5$ | | −0·118 | 0·249 |
| 3·4 | | −0·018 | 0·021$_5$ | | −0·031 | 0·040 | 0$_5$ | 14$_5$ | | −0·071 | 0·138 | 2 | 17 | | −0·095 | 0·222 |
| 3·6 | | −0·012 | 0·014$_5$ | | −0·021$_5$ | 0·029 | 0$_5$ | 11 | | −0·055 | 0·115 | 1$_5$ | 16 | | −0·077 | 0·198 |
| 3·8 | | −0·008 | 0·010 | | −0·015 | 0·021$_5$ | | 8$_5$ | | −0·042 | 0·095 | 1 | 15 | | −0·062 | 0·177 |
| 4·0 | | −0·005$_5$ | 0·007 | | −0·010 | 0·015$_5$ | | 7 | | −0·032$_5$ | 0·079 | 1 | 13$_5$ | | −0·049$_5$ | 0·158 |
| 4·5 | | −0·002 | 0·002$_5$ | | −0·004 | 0·007 | | 3 | | −0·016$_5$ | 0·049 | 0$_5$ | 11 | | −0·028$_5$ | 0·119 |
| 5·0 | | −0·000$_5$ | 0·001 | | −0·001$_5$ | 0·003 | | 2 | | −0·008$_5$ | 0·030 | 0$_5$ | 8$_5$ | | −0·016$_5$ | 0·090 |
| 5·5 | | | 0·000$_5$ | | −0·000$_5$ | 0·001 | | 1 | | −0·004$_5$ | 0·018$_5$ | 0$_5$ | 6$_5$ | | −0·009$_5$ | 0·068$_5$ |
| 6·0 | | | | | | 0·000$_5$ | | 0$_5$ | | −0·002$_5$ | 0·011 | | 5 | | −0·005$_5$ | 0·052 |
| 6·5 | | | | | | | | 0$_5$ | | −0·001 | 0·007 | | 3$_5$ | | −0·003 | 0·039$_5$ |
| 7·0 | | | | | | | | | | −0·000$_5$ | 0·004 | | 2$_5$ | | −0·001$_5$ | 0·030$_5$ |
| 8 | | | | | | | | | | | 0·001$_5$ | | 1$_5$ | | −0·000$_5$ | 0·018 |
| 9 | | | | | | | | | | | 0·000$_5$ | | 1 | | | 0·010$_5$ |
| 10 | | | | | | | | | | | | | 0$_5$ | | | 0·006$_5$ |
| 11 | | | | | | | | | | | | | | | | 0·004 |
| 12 | | | | | | | | | | | | | | | | 0·002$_5$ |
| 13 | | | | | | | | | | | | | | | | 0·001$_5$ |
| 14 | | | | | | | | | | | | | | | | 0·001 |
| 15 | | | | | | | | | | | | | | | | 0·000$_5$ |
| | | | | | | | $\Delta\epsilon$ | $\Delta\epsilon$ | | | | $\Delta\epsilon$ | $\Delta\epsilon$ | | | |
| $\epsilon_{nl,nl}$ | 46·7 | 6·49 | 5·72 | 44·6 | 4·96 | 4·09 | +0·04 | −0·48 | 41·5 | 2·48$_1$ | 1·26$_2$ | + ·06$_8$ | −·15$_1$ | 40·4 | 1·624 | 0·259 |

## 6. Results. Energies

In calculating energy values, using the expression for the energy in which the coefficients are given by Table II, we require the values of various $F$ and $G$ integrals, and the diagonal Lagrange parameters $\epsilon$ for the $(2s)$ and $(2p)$ wave functions. Values of these integrals and parameters, calculated from the wave functions discussed in § 5 are given in Table V, and the energy values calculated from them, with some other values for comparison, are given in Tables VI and VII.

Table V. Oxygen. $F$ and $G$ integrals and $\epsilon$ parameters

| | $O^{+++}$ | $O^{++}$ | | $O^{+}$ | | ($\mu = 0$) | ($\mu = -0{\cdot}202$) | O | | $O^{-}$ |
|---|---|---|---|---|---|---|---|---|---|---|
| | $^2$P | $^3$P | $^1$S | $^4$S | $^2$D | $^2$P | $^2$P | $^3$P | $^1$S | $^2$P |
| $\beta$ | 0 | $-1/5$ | $2/5$ | $-3/5$ | $-6/25$ | 0 | 0 | $-3/5$ | 0 | $-4/5$ |
| $F_0(2s, 2s)$ | 0·920 | 0·876 | 0·878 | 0·8336 | 0·8356 | 0·8369 | 0·8401 | 0·797 | 0·802 | 0·771 |
| $F_0(2s, 2p)$ | 0·954 | 0·898 | 0·884 | 0·8401 | 0·8348 | 0·8308 | 0·8340 | 0·773 | 0·766 | 0·705 |
| $F_0(2p, 2p)$ | 1·004 | 0·930 | 0·897 | 0·8530 | 0·8398 | 0·8309 | 0·8333 | 0·754 | 0·736 | 0·651 |
| $F_2(2p, 2p)$ | 0·473 | 0·434 | 0·413 | 0·3908 | 0·3828 | 0·3776 | 0·3793 | 0·336 | 0·325 | 0·278 |
| $G_1(2s, 2p)$ | 0·579 | 0·548 | 0·540 | 0·5176 | 0·5120 | 0·5090 | 0·5109 | 0·472 | 0·465 | 0·421 |
| $\epsilon_{2s,2s}$ | $6{\cdot}48_5$ | 4·96 | 5·00 | 3·603 | 3·633 | $3{\cdot}653_5$ | 3·743 | $2{\cdot}48_1$ | $2{\cdot}54_9$ | 1·625 |
| $\epsilon_{2p,2p}$ | 5·72 | 4·09 | 3·61 | $2{\cdot}653_5$ | $2{\cdot}488_5$ | $2{\cdot}381_5$ | 2·425 | $1{\cdot}26_2$ | $1{\cdot}10_9$ | 0·259 |

We will consider the results for $O^+$ first, as this is the ion for which full calculations, including the effect of superposition of configurations on the radial wave functions, were carried out, and also because for this ion the calculations were carried to a greater numerical accuracy. Four decimals are given in the values of the $F$ and $G$ integrals, but as in the calculation of $E$ some of these integrals occur with a coefficient 6, the fourth decimal has no significance in the values of $E$, which are given to three decimals, of which the last is not reliable to a unit.

Table VI. $O^+$ Energies

| | Calculated total energy of $(2s)$ $(2p)$ shell (a) | (b) | (c) | (d) | (e) | Observed $E(O^+)$ $-E(O^{++}, {}^3P)$ |
|---|---|---|---|---|---|---|
| $^4$S, $\beta = -3/5$ | −15·265 | −15·265 | −15·265 | | | −1·290 |
| $^2$D, $\beta = -6/25$ | −15·126 | −15·126 | −15·126 | | | −1·167 |
| $^2$P, $\beta = 0$ | −15·031 | −15·073 | −15·077 | | | −1·105 |
| $^2$D − $^4$S | 0·139 | 0·139 | 0·139 | 0·129 | 0·129 | 0·123 |
| $^2$P − $^2$D | 0·095 | 0·053 | 0·049 | 0·087 | 0·046 | 0·062 |
| $^2$P − $^4$S | 0·234 | 0·192 | 0·188 | 0·216 | 0·175 | 0·185 |
| ($^2$P − $^2$D) / ($^2$D − $^4$S) | 0·68 | 0·38 | 0·35 | 0·67 | 0·35 | 0·50 |

*Methods of calculation*

(a) Including exchange but not superposition of configurations, either in calculation of radial wave functions or in energy formula.

(b) Wave functions calculated with exchange, superposition of configurations included in energy formula *only*.

(c) Exchange and superposition of configurations included both in calculation of radial wave functions and in energy formula.

(d), (e) Corresponding to (a), (b) but with wave functions calculated without exchange.

Total energy values $E$ for the $(2s)$ $(2p)$ shell, calculated in three different ways, are given in Table VI. Those in column $(c)$ are to be regarded as the most accurate, as both exchange and superposition of configurations have been included in the calculation both of the radial wave functions and of the energy from those wave functions. The energy values in columns $(a)$ and $(b)$ have been calculated from the wave functions obtained by the solution of Fock's equations with exchange but without the superposition of configurations, which has also been omitted in the energy formula in obtaining the results in column $(a)$, but included, in the energy formula *only*, in obtaining the results in column $(b)$.

The inclusion of superposition of configurations in the calculation of the energy values in column $(b)$, and in the corresponding energy values for the other states of ionization which will be considered later, requires the solution of the equation

$$(E_A - E)(E_B - E) = (E_{AB})^2 \tag{10}$$

(see Hartree and Swirles 1937, equation (4)). An approximation to the root near $E_A$ is

$$E_A - (E_{AB})^2/(E_B - E_A) \tag{11}$$

(Hartree and Swirles 1937, formula (6)), but it is easy to find the root more accurately by writing (10) in the form

$$E = E_A - (E_{AB})^2/[(E_B - E_A) + (E_A - E)], \tag{12}$$

and using an iterative method, which converges very rapidly for the values of $E_B - E_A$ and $E_{AB}$ concerned. An expression for $E_B - E_A$, convenient for use in this context, is given by putting $\mu = 0$ in the column headed $E_B - E_A$ in Table II. Since $E_A - E$ is positive, (12) shows that the use of the first approximation (11) gives an overestimate of $E - E_A$; the degree of this overestimate, in the case of $O^+$ $^2$P, is ·002 which is just appreciable to the accuracy of the results in Table VI.

Comparison of columns $(b)$ and $(c)$ shows that the full integration of the extended Fock equations, including the effect of the superposition of configurations as well as the effect of exchange, gives a final energy lower by only 0·004, or 1 part in 4000, than the value obtained by neglecting the difference between the radial wave functions calculated with and without the terms expressing the effect of superposition of configurations. The smallness of this improvement in the energy values is, of course, related to the minimal property of the wave functions calculated from the integration of the extended Fock equations, and to the small differences between the wave functions calculated with and without the superposition of configurations. As pointed out in § 5, those differences are less than 1 in 200 of the maximum $|P|$, and it is not surprising that the relative energy difference is of the order of the square of this relative difference in the wave functions. As already mentioned, this improvement in the energy values was regarded as not sufficient to justify the calculation of wave functions with superposition of configurations for the other states of ionization. At the same time, the very close agreement between the energy values for the $^2$P state in columns $(b)$ and $(c)$ justifies the use of (12) as giving an adequate approximation to the contribution to the energy from the effect of the superposition of the higher configuration considered, and a similar method has been used for estimating these contributions for the other states of ionization.

In columns (*d*) and (*e*) of Table VI are given intermultiplet separations calculated using the wave functions without exchange and (*d*) without, (*e*) with, the inclusion of superposition of configurations in the energy calculations. They are quoted from Hartree and Black (1933) and Hartree and Swirles (1937) respectively.*

The last column of the table gives the observed energy values reckoned from the normal state of $O^{--}$. Accurate total energy values for the (2*s*) (2*p*) shell cannot be given on account of the lack of a well determined ionization energy for $O^{---}$.†

Comparison of the observed intermultiplet separations with those calculated shows that whereas for the original results of Hartree and Black (column (*d*)) the calculated separation between the lower two terms, $^2D-{}^4S$, is about right and the upper term ($^2P$) a good deal too high, and, as shown by column (*e*) and noted by Hartree and Swirles, the effect of the superposition of configurations was to over-correct the upper term, now the extreme term separation $^2P-{}^4S$ is about right, and the intermediate $^2D$ is relatively too high.

As far as the ratio of the intermultiplet separations is concerned, the last line of the table shows that the final result of the present calculations (column (*c*)) is very much the same as that obtained by the rougher approximation of Hartree and Swirles (column (*e*)). The simple Slater theory (1929) gives a value 2/3 for the ratio $(^2P-{}^2D)/(^2D-{}^4S)$, the observed value for $O^-$ is 0·51, whereas the best calculated value is 0·37. Thus the effect of superposition of configurations, as far as we have taken it, over-corrects the value given by the simple Slater theory by a factor of nearly 2.

The residual differences between the calculated intermultiplet separations in column (*c*) and the observed separations in column (*e*) are presumably due to the combined effect of the superposition of other configurations such as $(2s)(2p)^3(3s)$ and $(2s)^2(2p)^2(3p)$, which together give a number of terms of the same parity, *L*, and *S* as the normal configuration, and of the dependence of the wave function for the whole atom on the mutual distances $r_{ij}$ between the individual electrons. It does not seem possible to take either of these into account without a great deal of additional elaboration both in the algebraical and in the numerical work.

Table VII gives values for the total energy of the (2*s*) (2*p*) group calculated from wave functions obtained by the solution of Fock's equations including exchange only, and (*a*) without, (*b*) with, the inclusion of the effect of superposition of configurations in the energy formula. The results (*b*), where they differ from the corresponding results (*a*), were calculated from (12).

Except for $O^-$ the third decimal in the calculated results is hardly significant. For

* In these papers, the ionization energy of the $^4S$ term is given as 1·301, and the $^2D-{}^4S$ separation correspondingly as 0·134. We are indebted to Dr C. W. Ufford for pointing out the error.

† See Bacher and Goudsmit 1932, Table O IV, p. 343.

$O^{---}$ and $O^{--}$ the third decimal in the $\epsilon$ values is not established by the three decimal calculations of the radial wave functions, and for neutral O and $O^-$ some of the $F$ integrals, which are only accurate to three decimals, are multiplied by 8 or 10. A third decimal has been retained throughout Table VII, however, as it is significant in some of the differences between entries in the table, for example in the differences between columns *b* and *c*.

TABLE VII. TOTAL AND IONIZATION ENERGIES

| | | Total energies of (2s) (2p) shell | | Ionization energies | | | | | | | |
|---|---|---|---|---|---|---|---|---|---|---|---|
| | | (*a*) | (*b*) | (*a*) | diff. | (*b*) | diff. | (*c*) | diff. | Observed | diff. |
| $O^{---}$ | ²P | −11·980 | −12·036 | | | | | | | | |
| $O^{--}$ | ³P | −13·996 | −14·021 | 2·016 | 109 | 1·985 | 109 | 1·988 | 99 | 2·025 | 91 |
| | ¹D | −13·887 | −13·912 | 1·907 | 163 | 1·886 | 105 | 1·889 | 148 | 1·934 | 105 |
| | ¹S | −13·724 | −13·837 | 1·744 | | 1·781 | | 1·741 | | 1·829 | |
| $O^-$ | ⁴S | −15·265 | −15·265 | 1·269 | 139 | 1·244 | 139 | 1·258 | 129 | 1·290 | 123 |
| | ²D | −15·126 | −15·126 | 1·130 | 95 | 1·105 | 53 | 1·129 | 87 | 1·167 | 62 |
| | ²P | −15·031 | −15·073 | 1·035 | | 1·052 | | 1·042 | | 1·105 | |
| O | ³P | −15·679 | −15·679 | 0·414 | 76 | 0·414 | 76 | 0·416 | 73 | 0·500 | 72 |
| | ¹D | −15·603 | −15·603 | 0·338 | 113 | 0·338 | 86 | 0·343 | 109 | 0·428 | 81 |
| | ¹S | −15·490 | −15·517 | 0·225 | | 0·252 | | 0·234 | | 0·347 | |
| $O^-$ | ²P | −15·698 | −15·698 | 0·019 | | 0·019 | | | | 0·08 | |

(*a*) Calculated including exchange but not superposition of configurations.

(*b*) Calculated including exchange in evaluation of wave functions and energies, and superposition of configurations in evaluation of energies *only*.

(*c*) Calculated from wave functions evaluated without exchange (see Hartree and Black 1933, Table V).

Values of the ionization energies corresponding to the various values of the total energy are also given, and, for comparison, those originally calculated by Hartree and Black (column (*c*)) and the observed values.‡ It will be noted that although (presumably) the better wave functions give a lower value of the *total* energy, several of the calculated *ionization* energies in column (*b*) show rather worse agreement with the observed values than those in column (*c*). But where the normal configuration gives three terms, the uppermost is always brought into better agreement with the observed term.

The result for $O^-$ shows that, to the accuracy of these calculations, $O^-$ is not only a possible structure but is energetically stable. Further, the differences between the calculated and observed ionization energies for the other states of ionization leave little doubt that a more accurate calculation, if it could be carried out, would give a higher positive value of the ionization energy of $O^-$ in general agreement with Lozier's (1934) experimental value, though a rather higher value, perhaps between 0·12 and 0·15, might be expected.

‡ The value for $O^-$ is that quoted by Massey (1938, chap. I, § 4·21) from the work of Lozier (1934).

The intermultiplet separations for neutral oxygen in column (*b*) show a fairly good agreement with the observed values. For the other states of ionization the agreement is better than that shown by those in column (*c*) but not to any very marked extent, so that the present calculations have not provided a complete quantitative account of the deviations from Slater's values for the ratios of these separations.

Another point in this connexion, which remains unexplained, is the close agreement of the ratios of the intermultiplet separations with Slater's values in the case of sulphur. The (3*s*) (3*p*) shell of sulphur, in each state of ionization, would be expected to have very much the same properties, qualitatively and roughly quantitatively, as the (2*s*) (2*p*) shell in the corresponding state of ionization of oxygen, and it seems surprising that the effects of superposition of configuration should not be comparable with those in oxygen.

## 7. Summary

In the simplest approximation to the structure of a many electron atom, each term is regarded as arising from a single configuration of one-electron wave functions. When, to this approximation, two or more configurations give terms of the same parity, $L$, and $S$, not greatly different in energy, an appreciably better approximation may be obtained by taking, as the wave function for each term, a linear combination of the wave functions for the various configurations. The term "superposition of configurations" is used to denote the use of such linear combinations.

The effect of this superposition of configurations can be taken into account in the calculation of energy values only, or in the determination of the one-electron wave functions also. The latter requires an extension of Fock's equations, and in the present paper this extension is worked out for the superposition of the $(np)^{q+2}$ configuration on the normal $(ns)^2\,(np)^q$ configuration of atoms with an incomplete 8-shell. The solution of these equations involves a set of calculations, each of which is similar to a complete solution of Fock's equations, for a set of trial values of a parameter expressing the relative amplitudes of the wave functions for the two configurations.

For $O^+$, the normal configuration $(2s)^2\,(2p)^3$ gives $^4S$, $^2D$, and $^2P$ terms, of which only the highest ($^2P$) is subject to the superposition of the $(2p)^5$ configuration. A full solution of the extended equations has been worked out for this case. The effect of superposition of configurations on the radial wave functions is found to be small, less than 1 in 200 of their greatest values, and the consequent effect on the total energy of the (2*s*) (2*p*) shell very small, about 1 in 4000. This small improvement of the energy value did not seem to justify the solution of the extended equation for other states of ionization, and for these solutions of the ordinaryFock equations only have been evaluated, though superposition of configurations has been taken into account in calculating energy values.

## References


Bacher, R. F. and Goudsmit, S. 1932 "Atomic Energy States." McGraw Hill.
Condon, E. V. and Shortley, G. H. 1935 "Theory of Atomic Spectra." Camb. Univ. Press.
Fock, V., 1930 *Z. Phys.*, **61**, 126–48.
Hartree, D. R. and Black, M. M. 1933 *Proc. Roy. Soc.* A, **139**, 311–35.
Hartree, D. R. and W. 1936*a* *Proc. Roy. Soc.* A, **154**, 588–607.
— — 1936*b* *Proc. Roy. Soc.* A, **156**, 45–62.
Hartree, D. R. and Swirles, B. 1937 *Proc. Camb. Phil. Soc.* **33**, 240–9.
Levy, S. 1934 M.Sc. thesis, Manchester.
Lozier, W. W. 1934 *Phys. Rev.* **46**, 268.
Massey, H. S. W. 1938 "Negative Ions." Camb. Univ. Press.
Slater, J. C. 1929 *Phys. Rev.* **34**, 1293–1322.

**II-3** Reprinted from *Reviews of Modern Physics* **32**, 179 (1960)

# Self-Consistent Field Theory for Open Shells of Electronic Systems*

C. C. J. ROOTHAAN

*Laboratory for Molecular Structure and Spectra, Department of Physics, University of Chicago, Chicago 37, Illinois*

## 1. REVIEW OF THE CLOSED-SHELL THEORY

IT is well known that the Hartree-Fock equation for a closed-shell ground state of an electronic system can be derived from the variational principle.[1,2] The wave function is put forward as an antisymmetrized product of one-electron functions or orbitals $\varphi_i$, each orbital being doubly occupied:

$$\begin{aligned}\Phi &= (\varphi_1\alpha)(\varphi_1\beta)\cdots(\varphi_n\alpha)(\varphi_n\beta)\\ &= [(2n)!]^{\frac{1}{2}}(\varphi_1\alpha)^{[1}(\varphi_1\beta)^{2}\cdots(\varphi_n\alpha)^{2n-1}(\varphi_n\beta)^{2n]}\\ &= [(2n)!]^{-\frac{1}{2}}\begin{vmatrix} (\varphi_1\alpha)^1 & (\varphi_1\beta)^1 & \cdots & (\varphi_n\alpha)^1 & (\varphi_n\beta)^1 \\ (\varphi_1\alpha)^2 & (\varphi_1\beta)^2 & \cdots & (\varphi_n\alpha)^2 & (\varphi_n\beta)^2 \\ \cdots & \cdots & \cdots & \cdots & \cdots \\ \cdots & \cdots & \cdots & \cdots & \cdots \\ (\varphi_1\alpha)^{2n-1} & (\varphi_1\beta)^{2n-1} & \cdots & (\varphi_n\alpha)^{2n-1} & (\varphi_n\beta)^{2n-1} \\ (\varphi_1\alpha)^{2n} & (\varphi_1\beta)^{2n} & \cdots & (\varphi_n\alpha)^{2n} & (\varphi_n\beta)^{2n} \end{vmatrix}\end{aligned} \quad (1)$$

---

* Work assisted by a grant from the National Science Foundation and by Wright Field Air Development Center, under contract with the University of Chicago.

[1] D. R. Hartree, Repts. Progr. Phys. **11**, 113 (1948); *The Calculation of Atomic Structures* (John Wiley & Sons, Inc., New York, 1957).

[2] C. C. J. Roothaan, Revs. Modern Phys. **23**, 69 (1951). The notation and terminology introduced in that paper are adhered to in the present paper.

The orbitals are conveniently collected in a row vector, namely,

$$\boldsymbol{\varphi}=(\varphi_1\,\varphi_2\cdots\varphi_n), \tag{2}$$

and may be assumed, without loss of generality, to form an orthonormal set

$$\langle\varphi_i|\varphi_j\rangle=\int\bar{\varphi}_i\varphi_j dV=\delta_{ij}. \tag{3}$$

Different orbital sets can yield essentially the same total wave function, namely, if a set $\boldsymbol{\varphi}'$ is obtained from $\boldsymbol{\varphi}$ by a unitary transformation

$$\boldsymbol{\varphi}'=\boldsymbol{\varphi}\mathbf{U},\quad \mathbf{U}^*\mathbf{U}=\mathbf{E}, \tag{4}$$

the total wave function transforms according to

$$\Phi'=\Phi\,\mathrm{Det}^2(\mathbf{U}), \tag{5}$$

so that $\Phi'$ represents the same physical situation as $\Phi$.

In Eq. (1) the factor $[(2n)!]^{\frac{1}{2}}$ is chosen so that the orthonormality condition for the orbitals (3) implies that the total wave function $\Phi$ is normalized:

$$\langle\Phi|\Phi\rangle=\int\cdots\int\bar{\Phi}\Phi dV^1\cdots dV^{2n}=1. \tag{6}$$

The expectation value of the energy of the wave function (1) is

$$\langle\Phi|\mathcal{H}|\Phi\rangle=\int\cdots\int\bar{\Phi}\mathcal{H}\Phi dV^1\cdots dV^{2n}, \tag{7}$$

where the total Hamiltonian $\mathcal{H}$ is given by

$$\mathcal{H}=\sum_\mu H^\mu+\tfrac{1}{2}\sum_{\mu\neq\nu}(1/r^{\mu\nu}); \tag{8}$$

$H^\mu$ is the nuclear field plus kinetic energy operator for the $\mu$th electron, and $r^{\mu\nu}$ the distance between the $\mu$th and the $\nu$th electron. By using (8) in (7), the expectation value becomes a sum of integrals over the orbitals, namely,

$$\langle\Phi|\mathcal{H}|\Phi\rangle=2\sum_i H_i+\sum_{ij}(2J_{ij}-K_{ij}), \tag{9}$$

where

$$\begin{aligned} H_i&=\langle\varphi_i|H|\varphi_i\rangle,\\ J_{ij}&=\langle\varphi_i|J_j|\varphi_i\rangle=\langle\varphi_j|J_i|\varphi_j\rangle,\\ K_{ij}&=\langle\varphi_i|K_j|\varphi_i\rangle=\langle\varphi_j|K_i|\varphi_j\rangle, \end{aligned} \tag{10}$$

and

$$\begin{aligned} J_i^\mu\varphi^\mu&=\left[\int\bar{\varphi}_i^\nu\varphi_i^\nu(1/r^{\mu\nu})dV^\nu\right]\varphi^\mu,\\ K_i^\mu\varphi^\mu&=\left[\int\bar{\varphi}_i^\nu\varphi^\nu(1/r^{\mu\nu})dV^\nu\right]\varphi_i^\mu. \end{aligned} \tag{11}$$

$J_i$ and $K_i$ are commonly called the Coulomb and exchange operators, respectively, associated with the orbital $\varphi_i$; they are defined in Eqs. (11) by how they operate on an *arbitrary* one-electron function $\varphi$.

The Hartree-Fock equation is obtained by requiring that the orbitals minimize the expectation value of the energy. Those orbitals satisfy

$$F\boldsymbol{\varphi}=\boldsymbol{\varphi}\boldsymbol{\varepsilon}, \tag{12}$$

where the *Hartree-Fock Hamiltonian* $F$ is given by

$$F=H+\sum_i(2J_i-K_i), \tag{13}$$

and $\boldsymbol{\varepsilon}$ is a Hermitian matrix of Lagrangian multipliers, which are introduced by the orthonormality constraints (3). The operator $F$, which is defined in terms of the orbital set $\boldsymbol{\varphi}$, is easily shown to be invariant when $\boldsymbol{\varphi}$ is subjected to the transformation (4). Accordingly, the set $\boldsymbol{\varphi}'$ satisfies

$$F\boldsymbol{\varphi}'=\boldsymbol{\varphi}'\boldsymbol{\varepsilon}',$$

where

$$\boldsymbol{\varepsilon}'=\mathbf{U}^*\boldsymbol{\varepsilon}\mathbf{U}.$$

We may therefore single out that set of orbitals for which $\boldsymbol{\varepsilon}$ becomes diagonal, so that all the orbitals satisfy

$$F\varphi_i=\epsilon_i\varphi_i. \tag{14}$$

Equation (14) is a pseudo-eigenvalue problem, inasmuch as the operator $F$ is defined in terms of the solutions of (14). Ordinarily, Eq. (14) is called the Hartree-Fock equation, although that term is sometimes used for Eq. (12).

The total energy can be expressed in terms of the *orbital energies* $\epsilon_i$ and the one-electron integrals $H_i$, namely,

$$E=\sum_i(H_i+\epsilon_i). \tag{15}$$

The orbital energies also have a direct physical interpretation, namely, $-\epsilon_i$ is approximately equal to the ionization potential for removal of an electron occupying $\varphi_i$.

## 2. OPEN-SHELL THEORY

We consider open-shell wave functions of the following specifications:

(1) The total wave function is, in general, a sum of several antisymmetrized products, each of which contains a (doubly occupied) closed-shell core $\boldsymbol{\varphi}_C$, and a partially occupied open shell chosen from a set $\boldsymbol{\varphi}_O$, the different antisymmetrized products containing different subsets of $\boldsymbol{\varphi}_O$. The combined set of orbitals $\boldsymbol{\varphi}$ is defined by

$$\boldsymbol{\varphi}=(\boldsymbol{\varphi}_C,\boldsymbol{\varphi}_O), \tag{16}$$

and is assumed to be orthonormal, so that the two sets $\boldsymbol{\varphi}_C$ and $\boldsymbol{\varphi}_O$ are orthonormal and mutually orthogonal. In referring to the individual orbitals, we use the indices $k$, $l$ for the closed-shell orbitals, $m$, $n$ for the open-shell orbitals, and $i$, $j$ for orbitals of either set.

(2) The expectation value of the energy is given by

$$E=2\sum_k H_k+\sum_{kl}(2J_{kl}-K_{kl})$$
$$+f[2\sum_m H_m+f\sum_{mn}(2aJ_{mn}-bK_{mn})$$
$$+2\sum_{km}(2J_{km}-K_{km})], \quad (17)$$

where $a$, $b$, and $f$ are numerical constants depending on the specific case. The first two sums in Eq. (17) represent the closed-shell energy, the next two sums the open-shell energy, and the last sum the interaction energy of the closed and open shell. The number $f$ is, in general, the fractional occupation of the open shell, that is, it is equal to the number of *occupied* open-shell spin orbitals divided by the number of *available* open-shell spin orbitals; obviously $0<f<1$. The numbers $a$ and $b$ differ for different states of the same configuration. In Sec. 5 we examine which states of atoms and molecules satisfy the restrictions just stated.

Before applying the variational principle to Eq. (17) to derive the equations for the orbitals, we define the Coulomb operators $J_C$ and $J_O$ associated with the closed and open shells, respectively, the total Coulomb operator $J_T$, and similar exchange operators $K_C$, $K_O$, $K_T$ by

$$J_C=\sum_k J_k, \quad J_O=f\sum_m J_m, \quad J_T=J_C+J_O,$$
$$K_C=\sum_k K_k, \quad K_O=f\sum_m K_m, \quad K_T=K_C+K_O. \quad (18)$$

We also introduce the following (Hermitian) *Coulomb* and *exchange coupling operators* associated with the orbital $\varphi_i$:

$$L_i\varphi=\langle\varphi_i|J_O|\varphi\rangle\varphi_i+\langle\varphi_i|\varphi\rangle J_O\varphi_i,$$
$$M_i\varphi=\langle\varphi_i|K_O|\varphi\rangle\varphi_i+\langle\varphi_i|\varphi\rangle K_O\varphi_i, \quad (19)$$

and corresponding closed-shell, open-shell, and total Coulomb and exchange coupling operators by

$$L_C=\sum_k L_k, \quad L_O=f\sum_m L_m, \quad L_T=L_C+L_O,$$
$$M_C=\sum_k M_k, \quad M_O=f\sum_m M_m, \quad M_T=M_C+M_O. \quad (20)$$

All the operators (18) and (20) are invariant under the transformation

$$\varphi_C'=\varphi_C\mathbf{U}_C, \quad \varphi_O'=\varphi_O\mathbf{U}_O, \quad (21)$$

where $\mathbf{U}_C$ and $\mathbf{U}_O$ are unitary matrices of the appropriate dimensions. The closed-shell transformation of (21) does not alter the total wave function(s), except for a possible phase factor; the open-shell transformation of (21), in general, transforms degenerate total wave functions among each other.

The effect of the closed- and open-shell coupling operators, when they operate on the occupied orbitals, is of particular interest. These results are easily obtained from Eqs. (19) by specifying $\varphi=\varphi_k$ or $\varphi=\varphi_m$, and summing the index on $L$ or $M$ over the closed or open shells; this yields

$$(L_C-J_O)\varphi_k=\sum_l \varphi_l\langle\varphi_l|J_O|\varphi_k\rangle,$$
$$(M_C-K_O)\varphi_k=\sum_l \varphi_l\langle\varphi_l|K_O|\varphi_k\rangle,$$
$$L_C\varphi_m=\sum_l \varphi_l\langle\varphi_l|J_O|\varphi_m\rangle,$$
$$M_C\varphi_m=\sum_l \varphi_l\langle\varphi_l|K_O|\varphi_m\rangle,$$
$$L_O\varphi_k=f\sum_n \varphi_n\langle\varphi_n|J_O|\varphi_k\rangle, \quad (22)$$
$$M_O\varphi_k=f\sum_n \varphi_n\langle\varphi_n|K_O|\varphi_k\rangle,$$
$$(L_O-fJ_O)\varphi_m=f\sum_n \varphi_n\langle\varphi_n|J_O|\varphi_m\rangle,$$
$$(M_O-fK_O)\varphi_m=f\sum_n \varphi_n\langle\varphi_n|K_O|\varphi_m\rangle.$$

We turn now to the variational problem. We have to minimize the energy (17) with respect to the orbitals, subject to the constraints (3). We put the variation of the energy equal to zero, and add to it the variations of the constraints (3), each one multiplied with a Lagrangian multiplier. If we designate the Lagrangian multiplier for the $ij$th constraint (3) with $-2\theta_{ji}$, the resulting equation becomes, after some manipulation,

$$2\sum_k\langle\delta\varphi_k|(H+2J_C-K_C+2J_O-K_O)\varphi_k-\sum_j\varphi_j\theta_{jk}\rangle$$
$$+2\sum_k\langle\delta\bar{\varphi}_k|(H+2J_C-K_C+2J_O-K_O)\bar{\varphi}_k-\sum_j\bar{\varphi}_j\theta_{kj}\rangle$$
$$+2\sum_m\langle\delta\varphi_m|f(H+2J_C-K_C$$
$$+2aJ_O-bK_O)\varphi_m-\sum_j\varphi_j\theta_{jm}\rangle$$
$$+2\sum_m\langle\delta\bar{\varphi}_m|f(H+2J_C-K_C$$
$$+2aJ_O-bK_O)\varphi_m-\sum_j\bar{\varphi}_j\theta_{mj}\rangle=0. \quad (23)$$

Equation (23) now should hold for *any* variations $\delta\varphi_i$ with suitable values for the Lagrangian multipliers; hence,

$$(H+2J_C-K_C+2J_O-K_O)\varphi_k=\sum_j\varphi_j\theta_{jk}=\sum_j\varphi_j\bar{\theta}_{kj},$$
$$f(H+2J_C-K_C+2aJ_O-bK_O)\varphi_m \quad (24)$$
$$=\sum_j\varphi_j\theta_{jm}=\sum_j\varphi_j\bar{\theta}_{mj}.$$

We conclude that the Lagrangian multipliers form a Hermitian matrix, $\theta_{ij}=\bar{\theta}_{ji}$, and rewrite Eqs. (24) in the form

$$(H+2J_C-K_C+2J_O-K_O)\varphi_k$$
$$=\sum_l\varphi_l\theta_{lk}+\sum_n\varphi_n\theta_{nk}, \quad (25a$$

$$f(H+2J_C-K_C+2aJ_O-bK_O)\varphi_m=\sum_l \varphi_l\theta_{lm}+\sum_n \varphi_n\theta_{nm}. \quad (25b)$$

In the closed-shell case, the orbitals can always be subjected to a unitary transformation (4), which brings the matrix of Lagrangian multipliers into diagonal form; when this has been done, the orbitals all satisfy the pseudo-eigenvalue equation (14). In the open-shell case, however, we have available only the transformation (21) which transforms the open and closed shells within themselves. Such a transformation can eliminate only the off-diagonal multipliers $\theta_{lk}$ and $\theta_{nm}$, but not the multipliers $\theta_{nk}$ and $\theta_{lm}$ which couple the closed and open shells. There is one exception to this, namely, when the closed- and open-shell orbitals have no common symmetry. In that case, $\theta_{nk}$ and $\theta_{lm}$ vanish automatically, and Eqs. (25) become equivalent to two pseudo-eigenvalue problems of the type (14), one for the closed and one for the open shell; the two Hartree-Fock Hamiltonians are different, and each one is defined in terms of both the inner- and outer-shell orbitals. For this case, our SCF formalism yields the same equations which Hartree obtains by minimizing the total energy expression for *one component* of the degenerate state, and subsequently averaging the field over all directions.[1] A recent molecular example of this case is the LCAO-SCF calculations on $B_2$ by Padgett and Griffing.[3]

At first sight it seems that the general open-shell case cannot be reduced to pseudo-eigenvalue problems, due to the off-diagonal multipliers $\theta_{nk}$ and $\theta_{lm}$. Hartree,[1] and later Nesbet,[4] made further approximations in the SCF equations so that they *did* obtain pseudo-eigenvalue problems. Recently, Lefebvre[5] formulated a method by which the LCAO form of Eqs. (25) can be solved directly, without attempting to obtain pseudo-eigenvalue problems. An application of the method was given by Brion et al.[6] in a calculation on the NO molecule.

We present here an alternative to Lefebvre's method which *does* reduce to pseudo-eigenvalue problems. This is achieved by re-expressing the closed-open shell coupling terms in Eqs. (25), with the aid of the coupling operators (20), in such a way that those terms can be absorbed into the left-hand sides of Eqs. (25).

We multiply Eqs. (25a, b) by $\bar{\varphi}_m$ and $\bar{\varphi}_k$, respectively, and integrate; the results are

$$\langle\varphi_m|H+2J_C-K_C+2J_O-K_O|\varphi_k\rangle=\theta_{mk}, \quad (26a)$$

$$f\langle\varphi_k|H+2J_C-K_C+2aJ_O-bK_O|\varphi_m\rangle=\theta_{km}. \quad (26b)$$

By multiplying Eq. (26a) by $-f/(1-f)$, the complex conjugate of (26b) by $1/(1-f)$, and adding, we obtain, since $\theta_{mk}=\bar{\theta}_{km}$,

$$\theta_{mk}=-f\langle\varphi_m|2\alpha J_O-\beta K_O|\varphi_k\rangle, \quad (27)$$

where

$$\alpha=(1-a)/(1-f), \quad \beta=(1-b)/(1-f). \quad (28)$$

By using the third through the sixth Eqs. (22) and Eq. (27), one easily establishes

$$(2\alpha L_O-\beta M_O)\varphi_k=-\sum_n \varphi_n\theta_{nk},$$
$$f(2\alpha L_C-\beta M_C)\varphi_m=-\sum_l \varphi_l\theta_{lm}. \quad (29)$$

Equations (29) now allow us to re-express the coupling terms between the closed and open shells in Eqs. (25) in the desired manner; the orbitals satisfy

$$F_C\boldsymbol{\varphi}_C=\boldsymbol{\varphi}_C\boldsymbol{\eta}_C, \quad (30a)$$

$$F_O\boldsymbol{\varphi}_O=\boldsymbol{\varphi}_O\boldsymbol{\eta}_O, \quad (30b)$$

where

$$F_C=H+2J_C-K_C+2J_O-K_O+2\alpha L_O-\beta M_O,$$
$$F_O=H+2J_C-K_C+2aJ_O-bK_O+2\alpha L_C-\beta M_C, \quad (31)$$

and $\boldsymbol{\eta}_C$, $\boldsymbol{\eta}_O$ are Hermitian matrices with elements $\theta_{lk}$ and $\theta_{nm}/f$, respectively. We still have available arbitrary unitary transformations for the closed and open shells separately; these can be chosen so that the matrices $\boldsymbol{\eta}_C$ and $\boldsymbol{\eta}_O$ become diagonal. Hence, there is (at least) one set of orbitals which satisfies

$$F_C\varphi_k=\eta_k\varphi_k, \quad F_O\varphi_m=\eta_m\varphi_m. \quad (32)$$

The total energy also can be expressed, in this case, in terms of the orbital energies $\eta_i$ and the one-electron integrals $H_i$, namely,

$$E=\sum_k(H_k+\eta_k)+f\sum_m(H_m+\eta_m), \quad (33)$$

which can easily be verified by using Eqs. (10), (17), (22), (31), and (32).

The analogy with the closed-shell case would be complete if $-\eta_k$ and $-\eta_m$ were approximately equal to the ionization potentials for removal of an (the) electron from $\varphi_k$ or $\varphi_m$, respectively. This is not true in general, however; we take up this matter in a subsequent paper.

When solving the SCF equations (32), care has to be exercised in picking the correct solutions. Let us assume that we have a set of trial closed- and open-shell orbitals which are reasonably close to the SCF solution. We construct the operators $F_C$ and $F_O$ and determine their eigenfunctions. The eigenfunctions of $F_C$ are usually those belonging to the *lowest* eigenvalues, say, $p$ in number. Since $F_C$ and $F_O$ are physically little different, the eigenfunctions of $F_O$ belonging to *its* $p$ lowest eigenvalues are expected to resemble closely the closed-shell orbitals and obviously must be rejected as open-shell orbitals; instead, we must take for the latter

[3] A. A. Padgett and V. Griffing, J. Chem. Phys. **30**, 1286 (1959).
[4] R. K. Nesbet, Proc. Roy. Soc. (London) **A230**, 312 (1955).
[5] R. Lefebvre, J. chim. phys. **54**, 168 (1957).
[6] H. Brion, C. Moser, and M. Yamazaki, J. Chem. Phys. **30**, 673 (1959).

those eigenfunctions belonging to the $p+1$ eigenvalue and up.

Clearly, a formulation of the SCF problem in which the closed- and open-shell orbitals are solutions of the *same* eigenvalue equation would be desirable. This is indeed possible. From the first two and the last two Eqs. (22) we find

$$[2\alpha(L_C-J_O)-\beta(M_C-K_O)]\boldsymbol{\varphi}_C=\boldsymbol{\varphi}_C\boldsymbol{\zeta}_C, \quad (34a)$$

$$[2\alpha(L_O-fJ_O)-\beta(M_O-fK_O)]\boldsymbol{\varphi}_O=\boldsymbol{\varphi}_O\boldsymbol{\zeta}_O, \quad (34b)$$

where

$$\zeta_{kl}=\langle\varphi_k|2\alpha J_O-\beta K_O|\varphi_l\rangle, \quad (35a)$$

$$\zeta_{mn}=f\langle\varphi_m|2\alpha J_O-\beta K_O|\varphi_n\rangle. \quad (35b)$$

By adding Eqs. (34a) and (34b) to Eqs. (30a) and (30b), respectively, we obtain two equations of the same form as (30), with new operators on the left-hand sides and new matrices on the right-hand sides. The two operators are identical and equal to

$$F=H+2J_T-K_T+2\alpha(L_T-J_O)-\beta(M_T-K_O), \quad (36)$$

which follows easily from Eqs. (18), (20), and (28). Hence, the orbitals satisfy

$$F\boldsymbol{\varphi}_C=\boldsymbol{\varphi}_C\boldsymbol{\varepsilon}_C, \quad F\boldsymbol{\varphi}_O=\boldsymbol{\varphi}_O\boldsymbol{\varepsilon}_O, \quad (37)$$

where $\boldsymbol{\varepsilon}_C$ and $\boldsymbol{\varepsilon}_O$ are the new matrices $\boldsymbol{\eta}_C+\boldsymbol{\zeta}_C$ and $\boldsymbol{\eta}_O+\boldsymbol{\zeta}_O$, respectively. Now Eqs. (37) can be diagonalized by a transformation of the type (21), so that the orbitals can be chosen to satisfy

$$F\varphi_i=\epsilon_i\varphi_i. \quad (38)$$

The set $\varphi_i$ which satisfies Eq. (38) is equivalent to, but not identical with, the set $\varphi_k$, $\varphi_m$ satisfying Eqs. (32): the two sets are connected by a transformation of the type (21). Also, the eigenvalues of Eq. (38) are not identical with the eigenvalues of Eqs. (32); as a matter of fact, the two sets are not even approximately the same. If, we attempt to express the total energy in terms of $\epsilon_i$ and $H_i$, analogous to Eq. (33), we find that we need some additional terms:

$$E=\sum_k(H_k+\epsilon_k)+f\sum_m(H_m+\epsilon_m)$$
$$-f\sum_{km}(2\alpha J_{km}-\beta K_{km})-f^3\sum_{mn}(2\alpha J_{mn}-\beta K_{mn}). \quad (39)$$

## 3. SYMMETRY CONSIDERATIONS

The open-shell treatment developed in the preceding section is constructed so that the SCF equations permit *symmetry orbitals* as solutions, that is, the orbitals can be grouped in sets, each set transforming under symmetry operations according to an irreducible representation of the symmetry group. This state of affairs is assured by the expression for the total energy, Eq. (17). We assume that the orbitals do belong in sets to irreducible representations, and include in the summations in Eq. (17) only *complete* degenerate sets. This is obvious for the closed-shell contribution; for the open-shell contribution the partial occupation of a degenerate set is taken into account by summing over all the members of that set and introducing the fractional occupation number $f$. This is tantamount to writing down the total energy as the *average* expectation value for all the degenerate total wave functions of the state under consideration. Obviously, our variational procedure for determining the orbitals asks for the best orbitals to represent a *set* of degenerate total wave functions, in contrast to the customary procedure which asks for the best orbitals to represent a *single component* of a set of degenerate total wave functions.

Having assumed that the orbitals belong in sets to irreducible representations, it is easily seen that the Coulomb and exchange operators (18), the Coulomb and exchange coupling operators (20), and consequently the Hartree-Fock Hamiltonian operators (31) and (36), are totally symmetrical. As a result, the orbitals which are the solutions of the eigenvalue problems (32) and (38) are symmetry orbitals. We therefore see that our assumption about the symmetry properties of the orbitals is a consistent one, and we can say with confidence that the SCF equations have symmetry orbitals as *a* solution; that such a solution corresponds to an *absolute minimum*, not just a *stationary value*, of the total energy is a plausible assumption, although this is difficult to prove in general. We proceed from this assumption.

The requirement that the SCF orbitals are symmetry orbitals has usually been felt as necessary.[1,4] However, in the past, this was not rigorously possible with the customary method where the best orbitals were determined for just *one* component of the open-shell total wave functions; in order to satisfy this requirement, additional approximations had to be made once the SCF equations had been obtained. In the present treatment, no such additional approximations are called for.

To make the best possible use of the symmetry properties of the orbitals, we introduce the notation $\varphi_{i\lambda\alpha}$ in lieu of $\varphi_i$. In general, $\lambda$ (or $\mu$, $\nu\cdots$) refers to the irreducible representation, or *symmetry species*; $\alpha$ (or $\beta$, $\gamma\cdots$) refers to the *subspecies*, that is, it labels the individual members of the degenerate set that transforms according to the representation $\lambda$; and $i$ (or $j$, $k$, $l$, $m$, $n$) is a numbering index which labels orbitals which cannot be distinguished by symmetry any more. As before, we use $k$, $l$ for closed-shell orbitals, $m$, $n$ for open-shell orbitals, and $i$, $j$ for orbitals of either closed or open shell. The notation $i\lambda\alpha$ is analogous to the familiar *nlm* in atomic spectra.

## 4. LCAO FORM OF THE OPEN-SHELL THEORY

The solution of the Hartree-Fock equations for relatively simple problems, like small molecules, presents formidable mathematical difficulties. The reason is that they are partial differential equations in three dimensions, a very small class of which permits solution in terms of standard functions, usually by separation of variables. Actually, only for atoms, where this separation is possible and leads to ordinary differential equations for the radial functions, have they been solved. For molecules (molecular orbitals) and solids (Bloch functions), it appears more profitable to put forward an expansion of each orbital $\varphi_i$ in terms of a given set of suitable *basis functions* $\chi_p$:

$$\varphi_i = \sum_p \chi_p C_{pi} = \boldsymbol{\chi} \mathbf{c}_i; \tag{40}$$

obsiously, $\boldsymbol{\chi}$ is the row vector which collects all the basis functions, and $\mathbf{c}_i$ is the column vector with components $C_{pi}$. Since the occupied orbitals must form an orthonormal set, the vectors $\mathbf{c}_i$ must satisfy the constraints

$$\mathbf{c}_i^* \mathbf{S} \mathbf{c}_j = \delta_{ij}, \tag{41}$$

where the *overlap matrix* $\mathbf{S}$ is given by

$$S_{pq} = \langle \chi_p | \chi_q \rangle. \tag{42}$$

The expansion in terms of basis functions leads to a variational problem for the vectors $\mathbf{c}_i$, namely, the determination of the best set $\boldsymbol{\varphi}$ that can be constructed from a given basis set $\boldsymbol{\chi}$ by minimizing the total energy. This process was originally conceived for molecules to construct somewhat crude molecular orbitals from the atomic orbitals contributed by the individual atoms, and accordingly called the LCAO (linear combination of atomic orbitals) SCF method. It is, however, not necessary to restrict the process to such relatively crude applications; rather, we wish to use it as a practical method to obtain the fully optimized Hartree-Fock orbitals by making the basis set $\boldsymbol{\chi}$ sufficiently large and flexible. That this is possible without having to take an impractically large basis set has been shown in a number of recent calculations on atoms[7] and the $H_2$ molecule.[8] In fact, even for atoms this process is very desirable inasmuch as it yields the Hartree-Fock orbitals with considerably less labor than by numerical integration of the differential equations.

To take full advantage of the symmetry properties, we first reformulate Eqs. (40)–(42) by introducing *symmetry basis functions* $\chi_{p\lambda\alpha}$, analogous to the $\varphi_{i\lambda\alpha}$ of the previous section. Each occupied orbital of a given species and subspecies is a linear combination of the basis orbitals of the same species and subspecies only, and the expansion coefficients are independent of the subspecies, in order to guarantee that a degenerate set of occupied orbitals has the correct transformation properties. Hence

$$\varphi_{i\lambda\alpha} = \sum_p \chi_{p\lambda\alpha} C_{\lambda pi} = \boldsymbol{\chi}_{\lambda\alpha} \mathbf{c}_{\lambda i}; \tag{43}$$

clearly, $\boldsymbol{\chi}_{\lambda\alpha}$ is the row vector collecting all the basis functions of species $\lambda$ and subspecies $\alpha$, and $\mathbf{c}_{\lambda i}$ is the column vector with components $C_{\lambda pi}$.

For the scheme that follows, it is convenient to normalize the vectors $\mathbf{c}_{\lambda i}$ not to unity, but to *half the number of electrons* they represent, namely,

$$\mathbf{c}_{\lambda k}^* \mathbf{S}_\lambda \mathbf{c}_{\lambda l} = d_\lambda \delta_{kl}, \quad \mathbf{c}_{\lambda k}^* \mathbf{S}_\lambda \mathbf{c}_{\lambda m} = 0, \quad \mathbf{c}_{\lambda m}^* \mathbf{S}_\lambda \mathbf{c}_{\lambda n} = f d_\lambda \delta_{mn}, \tag{44}$$

where $d_\lambda$ is the dimension of the representation $\lambda$, and the matrix $\mathbf{S}_\lambda$ is defined by[9]

$$S_{\lambda pq} = d_\lambda^{-1} \sum_\alpha \langle \chi_{\lambda p\alpha} | \chi_{\lambda q\alpha} \rangle. \tag{45}$$

In addition to the matrices $\mathbf{S}_\lambda$ we need the matrices $\mathbf{H}_\lambda$ of the bare nuclear field Hamiltonian, the elements of which are, analogous to Eq. (45), given by[9]

$$H_{\lambda pq} = d_\lambda^{-1} \sum_\alpha \langle \chi_{p\lambda\alpha} | H | \chi_{q\lambda\alpha} \rangle. \tag{46}$$

Each matrix which enters the Hartree-Fock Hamiltonian has the same structure, splitting up in smaller matrices, one for each species. In some instances we need to consider such matrices as vectors, called *supervectors*; we construct the supervector corresponding to a matrix by writing down all the components, say $H_{\lambda pq}$, in dictionary order on $\lambda pq$, and denote this supervector by the symbol $\mathbf{H}$. Such a supervector can be operated on by *supermatrices*, the rows and columns of which are denoted by the components of $\mathbf{H}$. In particular, we need the *Coulomb* and *exchange supermatrices* $\mathfrak{J}$ and $\mathfrak{K}$ defined by

$$\begin{aligned} \mathcal{J}_{\lambda pq,\mu rs} &= (d_\lambda d_\mu)^{-1} \sum_{\alpha\beta} \int\int \bar{\chi}_{p\lambda\alpha}{}^1 \bar{\chi}_{r\mu\beta}{}^2 (1/r_{12}) \times \chi_{q\lambda\alpha}{}^1 \chi_{s\mu\beta}{}^2 dV^1 dV^2, \\ \mathcal{K}_{\lambda pq,\mu rs} &= (d_\lambda d_\mu)^{-1} \sum_{\alpha\beta} \int\int \bar{\chi}_{p\lambda\alpha}{}^1 \bar{\chi}_{r\mu\beta}{}^2 (1/r_{12}) \times \chi_{s\mu\beta}{}^1 \chi_{q\lambda\alpha}{}^2 dV^1 dV^2; \end{aligned} \tag{47}$$

they are symmetrical for the exchange of indices $\lambda pq \leftrightarrow \mu rs$, and Hermitian for the simultaneous exchange $p \leftrightarrow q$, $r \leftrightarrow s$.

We now apply the variational principle to the total energy in order to obtain the SCF equations for the vectors $\mathbf{c}_{\lambda i}$. This derivation closely parallels the derivation of the SCF equations in Sec. 2. We restrict ourselves here to writing down the results, and only for the scheme which is formally the simplest, namely, the

[7] C. C. J. Roothaan, L. M. Sachs, A. W. Weiss, Revs. Modern Phys. 32, 186 (1960), this issue; R. E. Watson, Tech. Rept. No. 12, Solid State and Molecular Theory Group, Massachusetts Institute of Technology (1959).

[8] W. Kolos and C. C. J. Roothaan, Revs. Modern Phys. 32, 219 (1960), this issue.

[9] Actually, all the terms under the summation signs in Eqs. (45) and (46) are equal.

scheme employing the combined Hartree-Fock Hamiltonian.

From the supermatrices $\mathfrak{J}$ and $\mathfrak{K}$ we compute two other supermatrices, namely,

$$\mathfrak{P} = 2\mathfrak{J} - \mathfrak{K}, \quad \mathfrak{Q} = 2\alpha\mathfrak{J} - \beta\mathfrak{K}. \tag{48}$$

Next, from a set of trial vectors $\mathbf{c}_{\lambda i}$ which satisfy the constraints (44), we compute the closed-shell, open-shell, and total *density matrices*

$$\mathbf{D}_{C\lambda} = \sum_k \mathbf{c}_{\lambda k}\mathbf{c}_{\lambda k}^*, \quad \mathbf{D}_{O\lambda} = \sum_m \mathbf{c}_{\lambda m}\mathbf{c}_{\lambda m}^*,$$
$$\mathbf{D}_{T\lambda} = \mathbf{D}_{C\lambda} + \mathbf{D}_{O\lambda}. \tag{49}$$

These density matrices give rise to supervectors, as explained previously; with the supermatrices (48), they permit the evaluation of two contributions to the Hartree-Fock Hamiltonian, namely,

$$\mathbf{P} = \mathfrak{P}\mathbf{D}_T, \quad \mathbf{Q} = \mathfrak{Q}\mathbf{D}_O. \tag{50}$$

Considering the supervector $\mathbf{Q}$ now as a collection of matrices $\mathbf{Q}_\lambda$, we compute the last contribution to the Hartree-Fock Hamiltonian from

$$\mathbf{R}_\lambda = d_\lambda^{-1}(\mathbf{S}_\lambda\mathbf{D}_{T\lambda}\mathbf{Q}_\lambda + \mathbf{Q}_\lambda\mathbf{D}_{T\lambda}\mathbf{S}_\lambda). \tag{51}$$

The Hartree-Fock Hamiltonian is then given by

$$\mathbf{F}_\lambda = \mathbf{H}_\lambda + \mathbf{P}_\lambda - \mathbf{Q}_\lambda + \mathbf{R}_\lambda, \tag{52}$$

and the corresponding eigenvalue problem is

$$\mathbf{F}_\lambda\mathbf{c}_{\lambda i} = \epsilon_{\lambda i}\mathbf{S}_\lambda\mathbf{c}_{\lambda i}. \tag{53}$$

Equation (53) is then solved, yielding a new set of trial vectors; the process is then repeated until the assumed and calculated vectors agree to a prescribed limit of accuracy.

Finally, the total energy may be calculated from scalar products between supervectors, namely,

$$E = (\mathbf{H}+\mathbf{F})^\dagger\mathbf{D}_T - \mathbf{Q}^\dagger(\mathbf{D}_C + f\mathbf{D}_O), \tag{54}$$

which is analogous to Eq. (39).

TABLE I. Coefficients for the configurations $\pi^N$, $\delta^N$ of linear molecules.

| $N$ | 1 | 2 | | | 3 |
|---|---|---|---|---|---|
| $f$ | $\frac{1}{4}$ | | $\frac{1}{2}$ | | $\frac{3}{4}$ |
| $\pi^N$ | $^2\Pi$ | $^3\Sigma^-$ | $^1\Delta$ | $^1\Sigma^+$ | $^2\Pi$ |
| $\delta^N$ | $^2\Delta$ | $^3\Sigma^-$ | $^1\Gamma$ | $^1\Sigma^+$ | $^2\Delta$ |
| $a$ | 0 | 1 | $\frac{1}{2}$ | 0 | 8/9 |
| $b$ | 0 | 2 | 0 | $-2$ | 8/9 |
| $\alpha$ | $\frac{4}{3}$ | 0 | 1 | 2 | 4/9 |
| $\beta$ | $\frac{4}{3}$ | $-2$ | 2 | 6 | 4/9 |

TABLE II. Coefficients for the configurations $p^N$ of atoms.

| $N$ | 1 | 2 | | | 3 | | | 4 | | | 5 |
|---|---|---|---|---|---|---|---|---|---|---|---|
| $f$ | $\frac{1}{6}$ | | $\frac{1}{3}$ | | | $\frac{1}{2}$ | | | $\frac{2}{3}$ | | $\frac{5}{6}$ |
| $p^N$ | $^2P$ | $^3P$ | $^1D$ | $^1S$ | $^4S$ | $^2D$ | $^2P$ | $^3P$ | $^1D$ | $^1S$ | $^2P$ |
| $a$ | 0 | $\frac{3}{4}$ | 9/20 | 0 | 1 | $\frac{4}{5}$ | $\frac{2}{3}$ | $\frac{15}{16}$ | 69/80 | $\frac{3}{4}$ | 24/25 |
| $b$ | 0 | $\frac{3}{2}$ | $-\frac{3}{10}$ | $-3$ | 2 | $\frac{4}{5}$ | 0 | 9/8 | 27/40 | 0 | 24/25 |
| $\alpha$ | 6/5 | $\frac{3}{8}$ | 33/40 | $\frac{3}{2}$ | 0 | $\frac{2}{5}$ | $\frac{2}{3}$ | $\frac{3}{16}$ | 33/80 | $\frac{3}{4}$ | 6/25 |
| $\beta$ | 6/5 | $-\frac{3}{4}$ | 39/20 | 6 | $-2$ | $\frac{2}{5}$ | 2 | $-\frac{3}{8}$ | 39/40 | 3 | 6/25 |

## 5. APPLICABILITY OF THE PRESENT THEORY

Clearly, the open-shell scheme developed in the previous sections is of practical value if and only if the total energy can be represented by Eq. (17), with certain values for $a$, $b$, and $f$. There are three important classes of states which satisfy this limitation:

(1) The half-closed shell: the open shell consists of singly occupied, complete degenerate sets of orbitals, and all the spins are parallel.[10] This is the ground state of this configuration; the wave function is (at least for the two components with all spin functions equal) a single determinant, and $f=\frac{1}{2}$, $a=1$, $b=2$, so that $\alpha=0$, $\beta=-2$, from Eqs. (28). Examples of atomic states in this class are C $1s^22s2p^3$, $^5S$; Cr $1s^22s^22p^63s^23p^64s3d^5$, $^7S$. A general molecular example is the lowest excited triplet of a molecule with a closed-shell ground state, the excitation being *from* a nondegenerate *to* a nondegenerate orbital.

(2) *All* the states arising from the configurations $\pi^N$, $\delta^N, \cdots, 1 \leqslant N \leqslant 3$, of a linear molecule: Table I summarizes the constants applicable to each such state.

(3) *All* the states arising from the configurations $p^N$, $1 \leqslant N \leqslant 5$, of an atom: Table II summarizes the constants applicable to each state.

It is a relatively simple matter to extend the open-shell theory just presented in such a way that other important classes of atomic states can be accommodated, as for instance, the $d^N$ configurations for the transition elements. We postpone such generalizations for the present, and include whatever new treatments may be necessary with the actual applications planned for the future.

[10] This case has often been handled by allowing the orbitals for opposite spins to be spatially different. See J. A. Pople and R. K. Nesbet, J. Chem. Phys. **22**, 571 (1954); G. W. Pratt, Jr., Phys. Rev. **102**, 1303 (1956). This method leads to a wave function which is a mixture of different multiplets.

II-4 Reprinted from *Canadian Journal of Physics* **41**, 1895 (1963)

# NUMERICAL SOLUTION OF THE HARTREE–FOCK EQUATIONS

CHARLOTTE FROESE
*Department of Mathematics, University of British Columbia, Vancouver, British Columbia*
Received February 15, 1963

ABSTRACT

Procedures for solving the Hartree–Fock equations on an automatic computer such as the IBM 7090 are described. Particular attention is given to the question of numerical accuracy and to the problem of devising automatic procedures.

## 1. INTRODUCTION

Many atomic properties such as transition probabilities, oscillator strengths, X-ray scattering factors, and electron affinities can be evaluated from radial wave functions. Approximate descriptions of molecular and solid-state phenomena can also be obtained from these functions.

The Hartree–Fock wave functions, which are solutions of a system of nonlinear integrodifferential equations, represent the best radial functions for a state function consisting of an antisymmetric product of one-particle functions (see Hartree (1957) for a derivation of these equations). Analytic approximations to these functions can be computed with methods developed by Roothaan (1951, 1960). However, the accuracy of the result depends on the particular choice of a basis set. For small atoms it is apparently a simple matter to optimize the basis set, but for large atoms, where the basis set must be more extensive, the task becomes formidable. Here numerical methods are considerably faster.

Numerical procedures for solving the Hartree–Fock equations have been described by many authors. Hartree (1927) was one of the first to give a detailed account of the numerical procedures for equations without exchange and later Hartree and Black (1933), and also Hartree and Hartree (1935), modified these slightly for equations with exchange. During the next 15 years these methods, with minor variations, were used to solve a large number of atomic structure calculations; a survey of these calculations has been compiled by Hartree (1948). When automatic computing machines became available, new techniques also became feasible and many programs using different methods were written. Procedures without exchange are described, for example, by Ridley (1955*a*); procedures with exchange, by Piper (1956), Froese (1957), and Worsley (1958), and procedures for relativistic calculations by Mayers (1957). In each of these cases, however, the computer still imposed certain restrictions on the procedures so that the program could not be fully automatic. The operator remained in control and by monitoring the calculation made many of the decisions. Also, a considerable effort was still required to obtain initial estimates as accurately as possible (Ridley 1955*b*; Hartree 1955; Douglas, Hartree, and Runciman 1955).

With computers such as the IBM 7090 the situation is entirely different. Storage no longer is a problem. In addition, the speed of the machine is such that wave functions for neon, for example, can be computed in about one and one-half minutes, for cadmium in about four minutes, and for doubly ionized cerium in about five minutes. Obviously, in such cases it is extremely important to have a process which is entirely automatic as well as easy to use. In particular, the user should not be expected to spend an undue amount of time obtaining initial estimates, nor should the user have to supply the program with a large number of parameters, as was the case in many of the earlier programs.

With such computers, it is also feasible to consider larger atoms with many more wave functions. In doing so, new problems arise. If the number of wave functions is $n$, then, as $n$ increases, the number of exchange terms increases faster than $n^2$. As a result it is extremely important to have a method for computing exchange terms as efficiently as possible. Another approach which has been investigated to some extent (Iutsis, Vizbaraite, Batarunas, and Kavetskis 1958) is to include the effect of configuration interaction in the equations. In this case the number of equations increases without an undue increase in the number of exchange terms. The problems with respect to these calculations have not yet been thoroughly investigated. Nor has an attempt been made so far to perform relativistic calculations with exchange. For such problems, an ideal procedure should meet the following requirements:

1. It should be fully automatic.
2. It should be flexible, yet require a minimum amount of information from the user.
3. The numerical procedures should be efficient as well as accurate.

With these criteria in mind, numerical procedures were investigated and then a program written for solving the Hartree–Fock equations.* For configurations $(1s)^2(2s)^2(2p)^6(nl)$ of Al III with $(nl) = (3s), (3p), (3d), (4s), (4p), (4d), (4f)$, the computing time varied from one to three minutes depending on the "stability" of the configuration. Besides some parameters which have been standardized but are left free for the sake of flexibility, the only information required by the program was the atomic number, the number of wave functions, the $nl$ quantum numbers, the number of electrons associated with each wave function, and an estimate of a screening number for each wave function. By assuming that wave functions are to a first approximation screened hydrogenic functions, all required initial estimates are computed as well as many of the usual parameters. In some other instances, special procedures have been devised so that the process is automatic as, for example, the special "tail" procedure.

The numerical procedures used in the program will be described in this paper. First the procedures for solving the hydrogen equation will be evaluated. Since the analytic solution is known in this case, methods can be compared and the accuracy tested at each stage. These procedures also form the basis for the solution of the Hartree–Fock equations. In the section pertaining to these equations, procedures for obtaining initial estimates and computing potential

*The program is written in the FORTRAN programming language for the IBM 7090. Copies of the program as well as a write-up describing the data preparation are available upon request.

and exchange terms will be described. A further section will discuss the overall iteration scheme and describe a procedure for accelerating convergence and even, in some instances, for making a divergent iteration converge.

## 2. NUMERICAL SOLUTION OF THE HYDROGEN EQUATION

### *2.1. Comparison of Numerical Methods for Outward Integration*

The solutions to the Hartree–Fock equations are, to a first approximation, similar to those of the hydrogen equation

$$P''+\left(\frac{2}{r}-\epsilon-\frac{l(l+1)}{r^2}\right)P = 0. \tag{1}$$

It is reasonable to assume that the method which is most accurate and most efficient for this equation will also be the best for the Hartree–Fock equation.

To avoid numerous changes of interval length, it is convenient to use the logarithmic variable, namely

$$\rho = \log_e r.$$

Then, with $\bar{P} = r^{-\frac{1}{2}}P$ as dependent variable,

$$\frac{d^2}{d\rho^2}\bar{P} = \{r^2\epsilon-2r+(l+\tfrac{1}{2})^2\}\bar{P}. \tag{2}$$

This is a special case of the more general second-order equation

$$y'' = f(x, y),$$

where the first-order derivative is missing.

Equation (2) was solved numerically by the four different methods listed below.

*1. Kutta Method*

$$\begin{aligned}
y'_{n+1} &= y_n'+h(k_1+2k_2+2k_3+k_4)/6,\\
y_{n+1} &= y_n+h(y_n'+h(k_1+k_2+k_3)/6,\\
k_1 &= f(x_n, y_n),\\
k_2 &= f(x_n+h/2,\ y_n+hy_n'/2),\\
k_3 &= f(x_n+h/2,\ y_n+hy_n'/2+h^2k_1/4),\\
k_4 &= f(x_n+h,\ y_n+hy_n'+h^2k_2/2).
\end{aligned}$$

*2. Nystrom Method*

$$\begin{aligned}
y'_{n+1} &= y_n'+h(k_1+4k_2+k_3)/6,\\
y_{n+1} &= y_n+h(y_n'+h(k_1+2k_2)/6),\\
k_1 &= f(x_n, y_n),\\
k_2 &= f(x_n+h/2,\ y_n+hy_n'/2+h^2k_1/8),\\
k_3 &= f(x_n+h,\ y_n+hy_n'+h^2k_2/2).
\end{aligned}$$

*3. Kutta–$\delta^2$*

$$\begin{aligned}
y_{n+1} &= 2y_n-y_{n-1}+(h^2/12)(k_1+10k_2+k_3),\\
k_1 &= f(x_{n-1}, y_{n-1}),\\
k_2 &= f(x_n, y_n),\\
k_3 &= f(x_n+h,\ 2y_n-y_{n-1}+h^2k_2).
\end{aligned}$$

*4. Numerov*

$$y_{n+1} = \left(2y_n - y_{n-1} + \frac{h^2}{12}(10y_n'' + y_{n-1}'' + g_{n+1})\right) \Big/ \left(1 - \frac{h^2}{12} f_{n+1}\right),$$

where the differential equation is assumed to have the special form

$$y'' = f(x)y + g(x).$$

Figure 1 indicates the relative accuracy of these methods. The accuracy in the region of oscillation tended to vary; however, in the region of the tail the order of increasing accuracy nearly always was Kutta, Nystrom, Numerov, Kutta-$\delta^2$. In actual fact the accuracy of the solution does not depend greatly on the method. The ratio of errors of one method compared to another is less than a factor 10, and in all instances the Numerov and Kutta-$\delta^2$ methods have similar accuracy.

Consequently accuracy alone does not indicate one method to be definitely

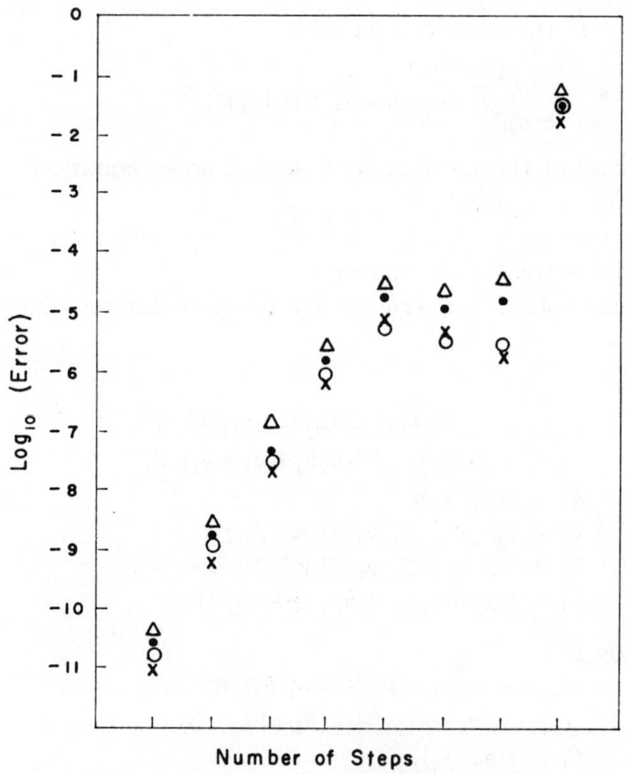


FIG. 1. A comparison of errors during an outward integration of a 6*f* wave function with $h = 0.125$.

| | Method | Number of evaluations/step |
|---|---|---|
| △ | Kutta | 4 |
| ● | Nystrom | 3 |
| × | Kutta-$\delta^2$ | 2 |
| ○ | Numerov | 1 |

superior to another. Looking at the formulae we notice that the Kutta procedure requires four evaluations of the derivative; the Nystrom, three; the Numerov, one; and the Kutta–$\delta^2$, two. On this basis the Numerov process is faster and requires only half the storage space of the first two. Thus the Numerov process appears to be the "best" for the hydrogenic wave equation and it was selected for the solution of equation (2).

This method, however, proved to be sensitive to the arithmetic procedure. To minimize the number of arithmetic operations, a function

$$a = \left(1-\frac{h^2}{12}f\right)$$

was defined so that when $g(x) = 0$

$$y_{n+1} = \{(12-10a_n)y_n - a_{n-1}y_{n-1}\}/a_{n+1}.$$

This results in a very simple algorithm but a digit of accuracy was lost in the energy adjustment because of the loss of significant digits when evaluating $(12-10a_n)$ in floating point form. The algorithm based on

$$a = \frac{h^2}{12}f$$

was found to be satisfactory.

The errors during an outward integration grow rapidly after the last zero of $d^2P/dr^2$. Thus outward integration is stable only to the point which is approximately $r = 2n^2$ and a special tail procedure must be used for solutions accurate beyond this point.

*2.2. Tail Procedure*

The Numerov process is based on the formula

$$\delta^2 y_n = h^2\left(y_n'' + \frac{\delta^2}{12} y_n''\right), \tag{3}$$

which is solved for $y_{n+1}$. However, equation (3) may also be considered as a system of equations solved by a special Gauss-elimination method.

Consider the general linear second-order equation

$$y'' = f(x)y + g(x)$$

for which $g(x) \to 0$ as $x \to \infty$ with boundary conditions

$$y(b) = y_0$$

and

$$y(\infty) = 0.$$

Let

$$x_n = b + nh,$$

$$y_n = y(x_n).$$

Application of equation (3) leads to the system of equations

$$
\text{(4)}\qquad
\begin{bmatrix}
d_1 & a_2 & & & \\
a_1 & d_2 & a_3 & & \\
 & a_2 & d_3 & & \\
 & & & \cdot & \\
 & & & & \cdot
\end{bmatrix}
\begin{bmatrix}
y_1 \\ y_2 \\ \cdot \\ \cdot \\ \cdot \\ \cdot
\end{bmatrix}
=
\begin{bmatrix}
-a_0y_0+(h^2/12)\delta^2g_1 \\ (h^2/12)\,\delta^2g_2 \\ \cdot \\ \cdot \\ \cdot \\ \cdot
\end{bmatrix},
$$

where

$$a_i = 1-\frac{h^2}{12}g_i,$$

$$d_i = -\left(2+10\frac{h^2}{12}f_i\right).$$

To eliminate elements below the diagonal, define

$$l_1 = d_1,$$

$$c_1 = -a_0y_0+\frac{h^2}{12}\delta^2g_1,$$

and

$$
\text{(5)}\qquad
\left.
\begin{aligned}
l_i &= d_i-a_ia_{i-1}/l_{i-1} \\
c_i &= \frac{h^2}{12}\delta^2g_i-c_{i-1}a_{i-1}/l_{i-1}
\end{aligned}
\right\}\; i = 2, 3, \ldots .
$$

From the asymptotic behavior of the solutions, for some large value of $N$,

$$\text{(6)}\qquad y_{N+1} \approx e^{-\sqrt{f_N}\,h}\, y_N.$$

Suppose now that the system of equations is to be computed up to and including the point $y_N$. Then the equivalent triangular form of (4) is

$$
\begin{bmatrix}
l_1 & a_2 & & & & \\
 & l_2 & a_3 & & & \\
 & & l_3 & a_4 & & \\
 & & & \cdot & & \\
 & & & & \cdot & \\
 & & & \cdot & \cdot & a_N \\
 & & & & \cdot & l_N
\end{bmatrix}
\begin{bmatrix}
y_1 \\ y_2 \\ y_3 \\ \cdot \\ \cdot \\ \cdot \\ y_N
\end{bmatrix}
=
\begin{bmatrix}
c_1 \\ c_2 \\ \cdot \\ \cdot \\ \cdot \\ \cdot \\ c_N-a_{N+1}y_{N+1}
\end{bmatrix}
$$

On substitution for $y_{N+1}$ from the asymptotic form (6), the last equation may be solved for $y_N$, namely

$$\text{(7)}\qquad y_N = c_N/(l_N+a_{N+1}e^{-\sqrt{f_N}\,h}).$$

Back-substitution into the system of equation yields

$$\text{(8)}\qquad y_i = (c_i-a_{i+1}y_{i+1})/l_i, \quad i = N-1, N-2, \ldots 2, 1.$$

The difficulty so far is the choice of $N$. In the case of the Hartree–Fock equations the range over which the solution extends depends to a large extent on the eigenvalue $\epsilon$, and is not known in advance. It can be shown, however, that for $f(x) > 0$ and $h$ sufficiently small, $l_i \to -1$ as $i$ increases. Substituting this result into equation (8) and noting that $a_i = 1+O(h^2)$, we obtain

$$y_{i+1}-y_i \approx c_i$$

or

$$hy_i' \approx c_i.$$

In other words, if $a_i$, $l_i$, and $c_i$ are generated for $i = 1, 2, 3, \ldots$, and the first value of $i$ for which $|c_i| < \tau$, say, determines the value of $N$, then a solution is generated over the range for which $y' > \tau/h$. Thus the range is determined automatically and depends on the solution. This method was used successfully for solving the Hartree–Fock equations.

The tail procedure is stable in the region where $f(x) > 0$; if this condition is not satisfied some values of $l_i$ may become very small with the result (as shown by equation (8)) that errors in previous values are amplified. The Numerov procedure, however, is stable when $f(x) < 0$, the factor in the denominator being $>1$ in this case, so that errors tend to be damped out. For the hydrogen equation $f(x) > 0$ for a short range near the origin, then becomes negative in the region of oscillation, and finally becomes positive again at a point $b$ near the mean radius. As a result these two methods complement each other.

The Numerov method was used for integration over the range $(r_0, b)$, where $r_0$ is a point near the nucleus, followed by the tail procedures for the remaining range.

### *2.3. Energy Adjustment*

In changing from the Numerov procedure to the tail procedure, equation (3) has not been applied to the point at the join, namely $y_0$. Cooley (1961) has shown that the amount by which the equation is *not* satisfied, which is called the residual, can be used to adjust the energy. His derivation was for the homogeneous equation but may be extended to the nonhomogeneous case.

The residual at $b$ in this case is

$$\begin{aligned}\text{Res} &= \delta^2 y_0 - h^2\{y_0''+(\delta^2/12)y_0''\}\\ &= a_{-1}y_{-1}+d_0y_0+a_1y_1-(h^2/12)(g_{-1}+10g_0+g_1),\end{aligned}$$

with the notation as indicated in Fig. 2.

Let $\epsilon^m$ be the $m$th estimate of the eigenvalue. Then, according to Cooley, the energy adjustment for equation (2) is

$$\epsilon^{m+1} = \epsilon^m + \bar{P}_0\text{Res}/(h^2\textstyle\sum r^2\bar{P}^2).$$

Since the normalization integral is required for other reasons as well, the adjustment process was changed to

$$\epsilon^{m+1} = \epsilon^m + \bar{P}_0\,\text{Res}\Big/\left(h\int_0^\infty P^2 dr\right).$$

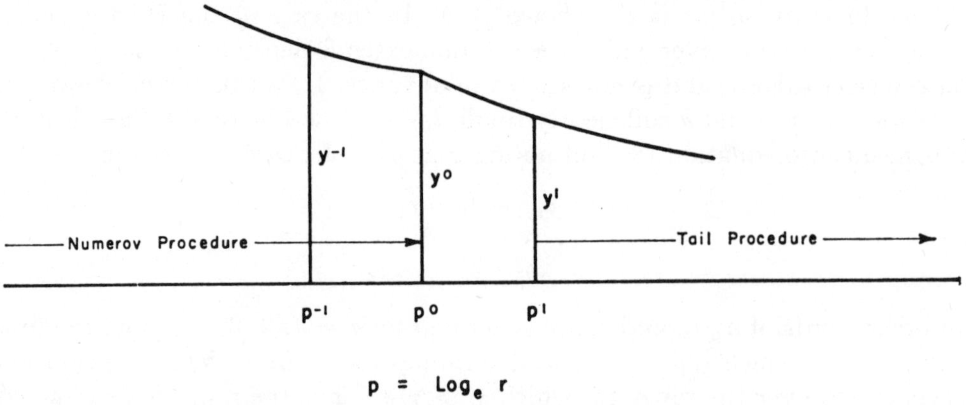


FIG. 2. Notation for the procedure at the join.

This process converged slightly faster than that suggested by Ridley (1955*a*),

$$\epsilon^{m+1} = \epsilon^{m} + (P'_{\text{in}} - P'_{\text{out}})P_0 \Big/ \int_0^\infty P^2 dr,$$

usually reducing five iterations to four and sometimes three.

Table I shows the convergence for a (1*s*) wave function starting with an initial estimate of $\epsilon^0 = 0.9$.

TABLE I

Energy adjustment for (1*s*) wave function; $h = 0.09375$, $\rho_0 = \log_e r_0 = -2.5$

| | Ridley | Cooley |
|---|---|---|
| 0 | 0.9 | 0.9 |
| 1 | 0.985305 | 0.9917152 |
| 2 | 0.999125 | 0.9999487 |
| 3 | 0.999964 | 1.000003 |
| 4 | 1.000003 | |

### 2.4. *Accuracy*

Equation (2) was solved for several wave functions and several different values of $h$. Table II summarizes these results. As can be expected, the results for the inner electrons are more accurate than those for the outer electrons. For $h = 0.0625$ the energy adjustment is surprisingly good. The values of the mean radius are probably a better indication of the number of significant digits; these range from five digits for 1*s*, 2*s* electrons to four digits for 6*s*, 6*p* electrons. Results with $h = 0.09375$ are almost as accurate for 1*s*, 2*s*, 2*p* electrons but errors for 6*s*, 6*p* have increased by almost a factor of 10.

## 3. SOLUTION OF HARTREE–FOCK EQUATIONS WITH EXCHANGE

The Hartree–Fock equations are nonlinear, integrodifferential equations in which both the potential term $Y(r)$ and the exchange term $X(r)$ depend on the solution. They are of the form (Hartree 1957)

TABLE II

Summary of results for hydrogen equation

| $(nl)$ | $\epsilon_{nl,nl}$ | Error | Max. error in $P(nl)$ | Mean radius | Error |
|---|---|---|---|---|---|
| (i) $h = 0.0625$, $\rho_0 = \log_e r_0 = -3.0$ | | | | | |
| 1s | 1.0000003 | 0.3 E-6 | 0.3 E-5 | 1.499987 | 0.1 E-4 |
| 2s | 0.2500004 | 0.4 E-6 | 0.1 E-5 | 5.999958 | 0.4 E-4 |
| 6s | 0.0277787 | 0.1 E-5 | 0.5 E-5 | 53.997898 | 0.2 E-2 |
| 2p | 0.2500000 | | 0.1 E-5 | 4.999969 | 0.3 E-4 |
| 4p | 0.0625003 | 0.3 E-6 | 0.1 E-5 | 22.999794 | 0.2 E-3 |
| 6p | 0.0277786 | 0.1 E-5 | 0.4 E-5 | 52.996747 | 0.3 E-2 |
| 3d | 0.1111111 | | 0.1 E-5 | 10.500022 | 0.2 E-4 |
| 5d | 0.0400003 | 0.3 E-6 | 0.1 E-5 | 34.499503 | 0.5 E-3 |
| 4f | 0.0625000 | | 0.1 E-5 | 17.999978 | 0.2 E-4 |

*Orthogonality Integrals*

| | | | | | | | | |
|---|---|---|---|---|---|---|---|---|
| 1s | 2s | 0.2 E-5 | 2p | 4p | 0.3 E-5 | 3d | 5d | 0.2 E-5 |
| 1s | 6s | 0.1 E-7 | 2p | 6p | 0.2 E-5 | | | |
| 2s | 6s | 0.6 E-7 | 4p | 6p | 0.6 E-6 | | | |

| $(nl)$ | $\epsilon_{nl,nl}$ | Error | Max. error in $P(nl)$ | Mean radius | Error |
|---|---|---|---|---|---|
| (ii) $h = 0.09375$, $\rho_0 = \log_e r_0 = -3.0$ | | | | | |
| 1s | 1.0000008 | 0.8 E-6 | 0.4 E-5 | 1.500012 | 0.1 E-4 |
| 2s | 0.2500010 | 0.1 E-5 | 0.2 E-5 | 6.000051 | 0.5 E-4 |
| 6s | 0.0277826 | 0.2 E-4 | 0.2 E-4 | 53.984766 | 0.2 E-1 |
| 2p | 0.2500003 | 0.3 E-6 | 0.1 E-5 | 5.000011 | 0.1 E-4 |
| 4p | 0.0625016 | 0.2 E-5 | 0.6 E-5 | 22.998835 | 0.1 E-2 |
| 6p | 0.0277818 | 0.2 E-4 | 0.2 E-4 | 52.987227 | 0.1 E-1 |
| 3d | 0.1111113 | 0.2 E-6 | 0.1 E-5 | 10.499923 | 0.8 E-4 |
| 5d | 0.0400016 | 0.2 E-5 | 0.7 E-5 | 34.498083 | 0.2 E-2 |
| 4f | 0.0625001 | 0.1 E-6 | 0.1 E-5 | 17.999700 | 0.3 E-3 |

*Orthogonality Integrals*

| | | | | | | | | |
|---|---|---|---|---|---|---|---|---|
| 1s | 2s | 0.2 E-5 | 2p | 4p | 0.8 E-6 | 3d | 5d | 0.4 E-7 |
| 1s | 6s | 0.2 E-6 | 2p | 6p | 0.1 E-6 | | | |
| 2s | 6s | | 4p | 6p | 0.4 E-6 | | | |

$$\frac{d^2}{dr^2}P(nl;r)+\left\{\frac{2}{r}(Z-Y(r))-\epsilon_{nl,nl}-\frac{l(l+1)}{r^2}\right\}P(nl;r) = \frac{2}{r}X(r)+\sum_{n'}{}' \epsilon_{nl,n'l}P(n'l;r) \tag{9}$$

with boundary conditions $P(0) = P(\infty) = 0$. The normalization and orthogonality conditions are respectively

$$\int_0^\infty P^2(nl;r)dr = 1 \quad \text{and} \quad \int_0^\infty P(nl;r)P(n'l;r)dr = 0.$$

In the present program, $\epsilon_{nl,n'l}$ was assumed to be zero so that the orthogonality conditions may not be satisfied except in the case of complete groups. Inclusion of the orthogonality condition may actually lead to a contradiction as shown by Sharma and Coulson (1962) for the 1S excited state of He.

The self-consistent field (SCF) procedure is to estimate the wave functions, compute $Y(r)$ and $X(r)$, then solve the differential equations for the wave functions, and iterate until the results are "self-consistent". With the exchange term present the differential equations are nonhomogeneous and solutions exist for every value of $\epsilon$. By selecting $\epsilon$, solving both the nonhomogeneous and homogeneous equations by the process described for the hydrogen equation, and assuming an arbitrary slope at the origin, a linear combination can be found such that the differential equation and the boundary conditions are satisfied over the entire range. At the same time the initial slope has been determined. The normalization condition must still be satisfied and $\epsilon$ can be adjusted to satisfy this condition.

Although this method has advantages in some ways, it cannot be used when the exchange term is to be neglected. In order to have a general procedure which could be applied to the Hartree equations as well, the fact that equation (9) is nonhomogeneous was not used. Instead, the differential equation transformed to a logarithmic variable was solved by an iterative method similar to that for the hydrogen equation. In other words, an initial slope was assumed at the origin, an initial estimate of energy obtained, and the energy adjusted as in the hydrogen equation until the differential equations and the boundary conditions were satisfied. The slope was then adjusted to satisfy the normalization conditions.

### *3.1. Initial Estimate of Wave Functions*

Since the self-consistent field procedure is an iterative one, initial estimates of wave functions are required and obviously the amount of computing depends on the accuracy of these initial estimates. Interpolation or extrapolation from existing results is one way of getting fairly accurate initial estimates, but since these are usually tabulated, the interpolation with respect to the atomic number $Z$ is accompanied by an interpolation with respect to the variable $r$. A much more convenient method is to assume the estimates to be screened hydrogenic functions. These can be evaluated directly at the required values of $r$ from their analytic expression. All interpolation is thus avoided. It requires a fairly reliable estimate of the screening number.

If a sufficient number of results exist for a given configuration, use can be made of the Hartree procedure (1955) of plotting $\sigma$ vs. $\bar{r}$ using the limiting values, $\sigma_0$, and the limiting slope, $d\sigma/d\bar{r}$; then finding the point on the curve for which equation $\bar{r} = \bar{r}_H/(Z-\sigma)$ holds. When only the limiting values of $\sigma_0$, and $d\sigma/d\bar{r}$ are available (these have been tabulated for a large number of configurations (Froese 1963)), assume

$$\sigma = \sigma_0+\bar{r}d\sigma/d\bar{r}.$$

Since, by definition, $\bar{r} = \bar{r}_H/(Z-\sigma)$, $\bar{r}$ can be eliminated so that

$$\sigma = \tfrac{1}{2}[Z+\sigma_0-\{(Z-\sigma_0)^2-4\bar{r}_H d\sigma/d\bar{r}\}^{\frac{1}{2}}].$$

For the result to be real, the expression under the square-root sign must be positive. If negative, the procedure was to replace it by zero. In addition, there is an upper bound to the screening, namely the number of electrons in the core. If the above formula gave a value larger then this number, the screening number was replaced by the upper bound. This tends to happen when computing screening for highly excited states, for example, $\sigma_{6p}$ of $(1s)^2(2s)^2(2p)^6(6p)$ configuration. With such initial estimates SCF procedures will converge though an acceleration technique may be necessary in the case of neutral atoms or excited states. The inner electrons are not sensitive to the choice of $\sigma$; one integration of the differential equation seems to correct for inaccuracies in $\sigma$. In the case of the $6p$ electron outside the 10-electron core, however, a screening equal to $\sigma_0$ was not adequate whereas a screening computed by the process described above using the limiting slope converged, provided wave functions were accelerated according to the technique to be discussed later.

*3.2. Computation of $y^k(nl, n'l'; r)$*

Hartree–Fock equations differ from the hydrogen equation in that potential and exchange terms are present which involve the wave functions themselves. These are defined in terms of the functions $y^k(nl, n'l'; r)$ where

$$y^k(nl, n'l'; r) = \int_0^r P(nl; r_1)P(n'l'; r_1)(r_1/r)^k dr_1 + \int_r^\infty P(nl; r_1)P(n'l'; r_1)(r/r_1)^{k+1} dr_1.$$

In other words, the function can be defined as the sum of two indefinite integrals. As a rule the function is not evaluated from these integrals, but rather as a pair of differential equations. Let us introduce the function

$$z^k(nl, n'l'; r) = \int_0^r P(nl; r_1)P(n'l'; r_1)(r_1/r)^k dr_1.$$

Then the differential equations in terms of the logarithmic variable $\rho$ are

$$dz^k/d\rho = rP(nl; r)P(n'l'; r) - kz^k \tag{10}$$

and

$$dy^k/d\rho = (k+1)y^k - (2k+1)z^k. \tag{11}$$

The boundary conditions are $z^k(0) = 0$ and $y^k(\infty) = z^k(\infty)$; therefore, to compute $y^k$, equation (10) is integrated outward for increasing values of $\rho$ followed by an inward integration of equation (11).

Both the above equations have simple integrating factors (Worsley 1958), namely $e^{k\rho}$ and $e^{(2k+1)\rho}$ respectively, so that

$$d(z^k e^{k\rho}) = rP(nl; r)P(n'l'; r)e^{k\rho}\, d\rho$$

and

$$d(y^k e^{-(k+1)}\rho) = -(2k+1)z^k e^{-(k+1)\rho}\, d\rho.$$

Integrating these equations over two intervals so that Simpson's rule may be applied directly to the integral, we are led to the following algorithm for computing $y^k$:

$$\rho_m = \rho_0 + mh,$$

$$r_m = \exp(\rho_m),$$

$$f_m = r_m P(nl; r_m) P(n'l'; r_m),$$

$$z^k_m = z^k_{m-2} e^{-2hk} + (h/3)(f_m + 4e^{-hk} f_{m-1} + e^{-2hk} f_{m-2}),$$

$$m = 2, 3, \ldots, M,$$

and

$$y^k_m = y^k_{m+2} e^{-2h(k+1)} + (2k+1)(h/3)(z^k_m + 4e^{-h(k+1)} z^k_{m+1} + e^{-2h(k+1)} z^k_{m+2}),$$

$$m = M-2, M-3, \ldots, 2, 1.$$

The $z^k$'s were computed for increasing values of $m$ with $z^k_1$ and $z^k_2$ evaluated from a series expansion; the $y^k$'s, for decreasing values of $m$ with $y^k_{m-1} = z^k_{m-1}$ and $y^k_m = z^k_m$.

For a sufficiently small $r$

$$z^k \approx 1/r^k \int_0^r A r^{l+l'+k+2} dr,$$

which is equal to

$$(1/r^k) A r^{l+l'+k+3}/(l+l'+k+3).$$

Thus

$$z^k_m \approx f_m/(l+l'+k+3), \qquad m = 1, 2.$$

This algorithm is both fast and accurate. Speed is essential in this case since when computing Hartree–Fock equations with exchange for large atoms, the number of exchange terms becomes large very rapidly. Consequently, a large amount of the time is spent in calculating $y^k$ functions. The following table is an indication of the accuracy.

TABLE III

Maximum absolute errors in $y^k(nl, n'l'; r)$

| | $h = 0.09375$ | $h = 0.078125$ |
|---|---|---|
| $y^0(1s, 1s; r)$ | 1.8 E-6 | 8.0 E-7 |
| $y^0(2p, 2p; r)$ | 3.2 E-6 | |

*3.3. Estimates of Energy Parameter*

To solve the differential equation, an initial estimate of $\epsilon$ is required. Assume the estimate of $P$ is an eigenfunction of equation (9). Then

$$\epsilon = \int_0^\infty P[P'' + \{2(Z-Y)/r - l(l+1)/r^2\}P - 2X/r]dr.$$

If the estimates are screened hydrogenic functions, then

$$P'' = \{-2(Z-\sigma)/r + (Z-\sigma)^2/n^2 + l(l+1)/r^2\}P.$$

Substituting for $P''$ leads to the estimate

(12) $$\epsilon^0 = (Z-\sigma)^2/n^2 + 2\int_0^\infty P\{(\sigma-Y)P - X\}/r\, dr.$$

This formula was used to obtain an initial estimate of $\epsilon$ when more reliable estimates were not readily available.

In the SCF iteration process, $\epsilon$ may also be adjusted for changes in the potential and exchange term. Let $i$ represent the $i$th SCF iteration. If the wave functions $P^i$ are solutions of equations with

$$\epsilon = \epsilon^i, \qquad Y = Y^i, \qquad X = X^i,$$

then the initial estimates for $\epsilon^{i+1}$ with $Y = Y^{i+1}$, $X = X^{i+1}$ are

$$\epsilon^{i+1} = \epsilon^i + 2\int_0^\infty P^i\{(Y^i - Y^{i+1})P^i + (X^i - X^{i+1})\}/r\,dr.$$

Since the energy adjustment process is fairly rapid the latter procedure was never used. Instead, the simpler estimate $\epsilon^{i+1} = \epsilon^i$ was used.

### *3.4. Estimates of Initial Slope*

The slope is related directly to the parameter

$$a_0 = P(nl; r)/r^{l+1}, \qquad r \to 0.$$

It is this parameter which is adjusted in a calculation to satisfy the normalization condition. In the case of hydrogenic wave functions

$$a_0 = N_{nl}(2Z/n)^{l+1}, \tag{13}$$

where $N_{nl}$ is the normalization constant, namely,

$$N_{nl} = \frac{\{Z(n+l)!/(n-l-1)!\}^{\frac{1}{2}}}{n(2l+1)!}. \tag{14}$$

When initial estimates of $a_0$ are not available, the program computes estimates according to equation (13) with $Z$ replaced by $Z - 0.6\sigma$. The latter value was used rather than $Z - \sigma$ because, near the origin, the wave functions as a rule are not screened as much as at the mean radius.

## 4. SCF ITERATION PROCEDURES

The SCF procedure requires a systematic way of deciding the order in which the differential equations are to be solved. Several are possible, but for large atoms a very flexible procedure is required. The method finally adopted was to associate with each wave function a number which indicates the degree of self-consistency, namely

$$\Delta P = \max_{0 \leqslant r < \infty} |P^i_{\text{estimate}} - P^i_{\text{solution}}|.$$

At the same time the wave functions are numbered in the order of decreasing energy parameters: those with the smaller energy parameters tend to be less stable in that small variations in the estimates result in large changes in the solution of the differential equations. Normally these equations have to be solved more frequently than the others. The iteration process is divided into two stages which are executed alternately. In the first, the equations are each solved once in the order of decreasing stability. At the end of this stage the estimates of the inner wave functions are normally considerably better than those for outer wave functions. In the second stage, the order is determined by $\Delta P$: the equation solved next is the one with largest $\Delta P$. Because this selection process may ignore certain wave functions which appear to be self-consistent, the number of equations selected in this manner is limited to $n(n+1)/2$, where $n$ is the number of wave functions. These two stages are repeated until the maximum $\Delta P$ is less than some prescribed tolerance, but always terminating with stage 1.

### *4.1. Accelerating Convergence*

The SCF procedure requires a sequence of estimates, $\{P^i_{\text{estimate}}\}$, from which it produces a sequence of solutions of the differential equations, $\{P^i_{\text{solution}}\}$. Normally $P^{i+1}_{\text{estimate}} = P^i_{\text{solution}}$, that is, the solution of the differential equation is taken as the next estimate, but this process may not always converge as, for example, in the case of $Na^-$. Hartree (1957) suggests that in such cases

$$P^{i+1}_{\text{estimate}} = \tfrac{1}{2}(P^i_{\text{estimate}}+P^i_{\text{solution}}).$$

This is a particular choice of a more general method where

$$P^{i+1}_{\text{estimate}} = cP^i_{\text{estimate}}+(1-c)P^i_{\text{solution}}. \tag{15}$$

A method was devised for determining a value of $c$ automatically which is near to the optimum in that it produces the most rapid convergence. Thus in some instances a value of $c \neq 0$ will accelerate convergence; in others it will cause an otherwise divergent SCF procedure to converge.

A well-known method for solving $X-f(X) = 0$ is the Newton–Raphson method which can be written in the form

$$X^{i+1} = cX^i+(1-c)f(X^i),$$

where

$$c = f'(X^i)/(f'(X^i)-1). \tag{16}$$

With a difference approximation, the Hartree–Fock equations may be represented by a system of nonlinear equations,

$$P(r_j) \equiv P_j = f_j(P_1, P_2, \ldots, P_N), \qquad j = 1, 2, \ldots, N,$$

or simply

$$P = f(P).$$

The generalized Newton method in this case would replace $c$ by a matrix (Todd 1962). But by introducing a measure of the difference between two functions and assuming the derivative to be represented approximately by the ratio of the change in $f(P)$ to the change in $P$, we estimate that

$$\begin{aligned} f'(P) &\simeq \overline{\Delta f(P)}/\overline{\Delta P} \\ &= \int P^2\Delta f(P)dr/\int P^2\Delta P dr. \end{aligned}$$

This latter value replaced $f'(X)$ in equation (16) to determine $c$.

It was found that the values of $c$ computed in this manner were reasonable provided the initial estimates were not screened hydrogenic functions, and provided the differences were not due to errors in the numerical process as is the case when wave functions are nearly self-consistent. It was also found that convergence was faster if $c$ was allowed to vary. As a result it was recomputed after every two iterations.

To what extent does this process improve the convergence? In the case of helium where normally one would iterate with $c = 0$, requiring seven iterations, the above process reduced the number of iterations to four. How close to the optimum is the value of $c$ computed by this method? To answer this question, the $2p$ wave function of $F^-$ was computed for a set of $c$ values, held constant throughout each SCF iteration: the results are given in Table IV. Starting with an initial estimate of $c = 0.4$, the above process computed $c = 0.491$, 0.499, 0.407, 0.372 and converged in 10 iterations.

TABLE IV

| $c$ | Number of iterations |
|---|---|
| 0 | Sequence diverged |
| 0.2 | Sequence diverged |
| 0.3 | >15 |
| 0.4 | 11 |
| 0.5 | 11 |
| 0.6 | 15 |
| 0.7 | >15 |

Thus this method of determining $c$ significantly improves the convergence of the usual SCF procedure.

## 5. CONCLUSIONS

By standardizing and using the screened hydrogenic functions as reference functions, it is possible to write a program which is entirely automatic in the majority of cases. In addition, if the program computes the coefficients in the expressions for the potential and exchange terms from an expression for the energy of the term under consideration, then only the deviation from the average energy of the configuration is required as input data rather than a list of coefficients for each wave function. As a result the amount of data preparation is kept to a minimum.

With procedures such as those suggested here, the statement made by Piper (1956) that a computer faster than the IBM 650 would not be worthwhile no longer applies.

## ACKNOWLEDGMENTS

This work was done while at the Harvard College Observatory and was supported by a National Science Foundation Grant. I wish to thank Professor David Layzer for making all the necessary arrangements for my use of the computing facilities and, together with Dr. Margaret N. Lewis, for the interest they have shown in this work.

## REFERENCES


COOLEY, J. W. 1961. Math. of Comp. **15**, 363.
DOUGLAS, A. S., HARTREE, D. R., and RUNCIMAN, W. A. 1955. Proc. Camb. Phil. Soc. **51**, 486.
FROESE, C. 1957. Proc. Camb. Phil. Soc. **53**, 206.
——— 1963. Can. J. Phys. **41**, 50.
HARTREE, D. R. 1927. Proc. Camb. Phil. Soc. **24**, 89 and 111.
——— 1948. Repts. Progr. Phys. **11**, 113.
——— 1955. Proc. Camb. Phil. Soc. **51**, 684.
——— 1957. The calculation of atomic structures (J. Wiley and Sons).
HARTREE, D. R. and BLACK, M. H. 1933. Proc. Roy. Soc. (London), A, **139**, 311.
HARTREE, D. R. and HARTREE, W. 1935. Proc. Roy. Soc. (London), A, **150**, 9.
IUTSIS, A. P., VIZBARAITE, IA. I., BATARUNAS, I. V., and KAVETSKIS, V. I. 1958. Akad. Nauk Litovskoi S.S.R., Trudy, Ser. B, No. 2, 3.
MAYERS, D. F. 1957. Proc. Roy. Soc. (London), A, **241**, 93.
PIPER, W. W. 1956. Trans. A.I.E.E. **75**, Part 1, 152.
RIDLEY, E. C. 1955*a*. Proc. Camb. Phil. Soc. **51**, 702.
——— 1955*b*. Proc. Camb. Phil. Soc. **51**, 693.
ROOTHAAN, C. C. J. 1951. Revs. Modern Phys. **23**, 69.
——— 1960. Revs. Modern Phys. **32**, 179.
TODD, J. (*Editor*). 1962. Survey of numerical analysis (McGraw-Hill), p. 515.
SHARMA, C. S. and COULSON, C. A. 1962. Proc. Phys. Soc. **80**, 81.
WORSLEY, B. H. 1958. Can. J. Phys. **36**, 289.

# Chapter III

# *Local Electron Density and Approximate Exchange*

## III-1

The solutions of the Hartree-Fock equations, especially for larger atoms, is complicated by the nonlocal nature of the exchange potential. To simplify this, Slater (Chapter III-2) averaged over the exchange terms and then replaced them with an approximate local potential which depends on the local electron density. The result is derived assuming that the local density, $\rho(\mathbf{r}_i)$, behaves like the density in a uniform free electron gas. The resulting equations are known as the Hartree-Fock-Slater (HFS) method and have been applied to many atoms.

The approximation of the H. F. densities in terms of results based on the electron gas, and to what extent this can be done, is of fundamental interest to the correlation problem because much is known on correlation in the uniform electron gas.

Kohn and co-workers bridged the sizable gap between the free electron gas and the atomic cloud with another significant step, by studying the density and its gradients in an electron gas in the field of an external potential $v(\mathbf{r})$. Kohn and Sham (Chapter III-3) showed that Slater's exchange is correct up to and excluding terms of order $|\nabla|^2$. They also showed that Slater's value, which is correct for the average over the filled electronic states, should be reduced by a constant factor of two thirds when calculating the density by variational methods, starting from the Kohn density functional formalism. Kohn and co-workers' density functional method, on the other hand, leads to an exchange correction to the density which in principle can include terms of order $|\nabla|^2$. Tong and Sham (Chapter III-4) compare densities and energies obtained by the Hartree-Fock, Hartree-Fock-Slater, and those with the Kohn and Sham method. They find that the Kohn and Sham method yields good radial densities. Either type of approximate exchange, however, is bound to give sizable energy errors since basically the atomic cloud is much too nonuniform.

Chapter III-3 and 4 also include calculations where densities are atomic and correlation energies per electron are from the uniform electron gas. Several authors (references in Chapter III-3) have attempted to introduce correlation energies into atoms this way, usually integrating over a uniform electron correlation energy interpolated between the low-density (Wigner, Chapter IV-2) and high-density (Gell-Mann and Brueckner, Chapter VI-2) limits. The resulting correlation energies for atoms, however, are too large by about a factor of 2 (Chapter III-4). Ma and Brueckner (Chapter IV-3) and Sinanoğlu (Chapter VII-2) have given simple physical reasons why large density gradients in atoms lead to correlations which are expected to differ markedly from those in a uniform electron gas.

Where calculations of a large number of states would be required, as in astrophysical and aerospace applications, it is useful to have much simpler estimation methods than the Hartree-Fock. Such a method ("scaled Thomas-Fermi") has been given by Stewart and Rotenberg (Chapter III-5). The method yields many transition probabilities and compares favorably in the orbitals obtained with Hartree-Fock-Slater.

III-2 Reprinted from *Physical Review* **81**, 385 (1951)

# A Simplification of the Hartree-Fock Method

J. C. SLATER
*Massachusetts Institute of Technology,* Cambridge, Massachusetts*
(Received September 28, 1950)

It is shown that the Hartree-Fock equations can be regarded as ordinary Schrödinger equations for the motion of electrons, each electron moving in a slightly different potential field, which is computed by electrostatics from all the charges of the system, positive and negative, corrected by the removal of an exchange charge, equal in magnitude to one electron, surrounding the electron whose motion is being investigated. By forming a weighted mean of the exchange charges, weighted and averaged over the various electronic wave functions at a given point of space, we set up an average potential field in which we can consider all of the electrons to move, thus leading to a great simplification of the Hartree-Fock method, and bringing it into agreement with the usual band picture of solids, in which all electrons are assumed to move in the same field. We can further replace the average exchange charge by the corresponding value which we should have in a free-electron gas whose local density is equal to the density of actual charge at the position in question; this results in a very simple expression for the average potential field, which still behaves qualitatively like that of the Hartree-Fock method. This simplified field is being applied to problems in atomic structure, with satisfactory results, and is adapted as well to problems of molecules and solids.

## I. INTRODUCTION

THE Hartree-Fock equations[1] furnish the best set of one-electron wave functions for use in a self-consistent approximation to the problem of the motion of electrons in the field of atomic nuclei. However, they are so complicated to use that they have not been employed except in relatively simple cases. It is the purpose of the present paper to examine their meaning sufficiently closely so that we can see physically how to set up a simplification, which still preserves their main features. This simplified method yields a single potential in which we can assume that the electrons move, and we shall show the properties of this field for problems not only of single atoms but of molecules and solids, showing that it leads to a simplified self-consistent method for handling atomic wave functions, easy enough to apply so that we can look forward to using it even for heavy atoms.

## II. THE HARTREE-FOCK EQUATIONS AND THEIR MEANING

It is well known that the Hartree equations are obtained by varying one-electron wave functions $u_1(x)$, $u_2(x)$, $\cdots u_n(x)$, in such a way as to make the energy $\int u_1^*(x_1)\cdots u_n^*(x_n)Hu_1(x_1)\cdots u_n(x_n)dx_1\cdots dx_n$ an extreme, where $H$ is the energy operator of a problem involving $n$ electrons in the field of certain nuclei, and where the functions $u_i$ are required to be normalized. Similarly the Hartree-Fock equations, as modified by Dirac,[2] are obtained by varying the $u_i$'s so as to make the energy

$$\frac{1}{n!}\int\begin{vmatrix} u_1^*(x_1)\cdots u_1^*(x_n) \\ \cdots\cdots\cdots\cdots \\ u_n^*(x_1)\cdots u_n^*(x_n)\end{vmatrix} H \begin{vmatrix} u_1(x_1)\cdots u_1(x_n) \\ \cdots\cdots\cdots\cdots \\ u_n(x_1)\cdots u_n(x_n)\end{vmatrix} dx_1\cdots dx_n$$

an extreme, where in this latter expression the $u$'s are assumed to be functions depending on coordinates and spin, and where the integrations over the $dx$'s are interpreted to include summing over the spins. The Hartree-Fock equations can then be written in the form

$$H_1u_i(x_1)+\left[\sum_{k=1}^{n}\int u_k^*(x_2)u_k(x_2)(e^2/4\pi\epsilon_0r_{12})dx_2\right]u_i(x_1)$$
$$-\sum_{k=1}^{n}\left[\int u_k^*(x_2)u_i(x_2)(e^2/4\pi\epsilon_0r_{12})dx_2\right]u_k(x_1)$$
$$=E_iu_i(x_1). \quad (1)$$

Here $H_1$ is the kinetic energy operator for the electron of coordinate $x_1$, plus its potential energy in the field of

* The work described in this paper was supported in part by the Signal Corps, the Air Materiel Command, and the ONR, through the Research Laboratory of Electronics of M.I.T.

[1] J. C. Slater, Phys. Rev. **35**, 210 (1930); V. Fock, Z. Physik **61**, 126 (1930); L. Brillouin, *Les Champs Self-Consistents de Hartree et de Fock*, Actualités Scientifiques et Industrielles No. 159 (Hermann et Cie., 1934); D. R. Hartree and W. Hartree, Proc. Roy. Soc. **A150**, 9 (1935); and many other references.

[2] P. A. M. Dirac, Proc. Cambridge Phil. Soc. **26**, 376 (1930).

all nuclei; $e^2/4\pi\epsilon_0 r_{12}$ is the Coulomb potential energy of interaction between electrons 1 and 2, expressed in mks units; to get the corresponding formula in Gaussian units we omit the factor $4\pi\epsilon_0$, and to get it in atomic units we replace $e^2/4\pi\epsilon_0$ by 2. The $u_i$'s as before are assumed to depend on spin as well as coordinates, and the integrations over $dx_2$ include summation over spin, so that the exchange terms, the last ones on the left side of Eq. (1), automatically vanish unless the functions $u_i$ and $u_k$ correspond to spins in the same direction.

The Hartree-Fock equations in the form given present an appearance which seems to differ from the ordinary one-electron type of Schrödinger equation, and for this reason it is ordinarily thought that they cannot be given a simple physical interpretation. This assumption arises partly from the paper of Dirac[2], in which they are interpreted in a rather involved way. The second term on the left of (1) is simple: it is clearly the Coulomb potential energy, acting on the electron at position $x_1$, of all the electronic charge, including that of the $i$th wave function whose wave equation we are writing. The last term on the left, the exchange term, however, is peculiar, in that is is multiplied by $u_k(x_1)$ rather than by $u_i(x_1)$. It must somehow correct for the fact that the electron does not act on itself, which it would be doing if this term were omitted. In the Hartree, as opposed to the Hartree-Fock, equations, this is obvious. There the last term differs from that in the Hartree-Fock equations only in that all terms in the summation are omitted except the $i$th; the exchange term in that case then merely cancels the term in $k=i$ from the Coulomb interaction found in the second term. The main point of our discussion is to show that an equally simple interpretation of this term can be given in the Hartree-Fock equations.

Let us first state in words what the interpretation proves to be; then we can more easily describe the way in which the equations lead to it. We can subdivide the total density of all electrons into two parts, $\rho_+$ from those with plus spins, $\rho_-$ from those with minus spins; the two together add to the quantity $-e\sum(k=1\cdots n)\times u_k^*(x)u_k(x)$, where $e$ is the magnitude of the electronic charge. Then we can show that the Hartree-Fock Eq. (1) for a wave function $u_i$ which happens to correspond to an electron with a plus spin is an ordinary Schrödinger equation for an electron moving in a perfectly conventional potential field. This field is calculated by electrostatics from all the nuclei, and from a distribution of electronic charge consisting of the whole of $\rho_-$, but of $\rho_+$ corrected by removing from the immediate vicinity of the electron, whose wave function we are finding, a correction or exchange charge density whose total amount is just enough to equal a single electronic charge. That is, this corrected charge distribution equals the charge of $n-1$ electrons, as it should. The exchange charge density equals just $\rho_+$ at the position of the electron in question, gradually falling off as we go away from that point. We can get a rough idea of the distance in which it has fallen to a small value by replacing it by a constant density $\rho_+$ inside a sphere of radius $r_0$, zero outside the sphere. We have $\frac{4}{3}\pi r_0^3|\rho_+|=e$, or

$$r_0=(3e/4\pi|\rho_+|)^{\frac{1}{3}}. \qquad (2)$$

The situation is then much as if the corrected charge density equaled the actual total electronic charge density outside this sphere, but was only $\rho_-$ within the sphere; there is a sort of hole, sometimes called the Fermi[3] or exchange hole, surrounding the electron in question, consisting of a deficiency of charge of the same spin as the electron in question. Actually, of course, this exchange hole does not have a sharp boundary, but the charge density of the same spin as the electron in question gradually builds up as we go away from this electron. Similar statements hold for the field acting on an electron of minus spin.

The exchange hole clearly is different for wave functions of the two spins, provided $\rho_+$ and $\rho_-$ are different; examination proves further that it is different for each different wave function $u_i$. It is this difference which leads to the complicated form of the Hartree-Fock equation; and the simplification which we shall introduce in a later section is that of using sort of an averaged exchange hole for all the electrons. The difference between the exchange charge for two wave functions $u_i$ corresponding to the same spin is not great, however. We have already seen that the radius $r_0$ which we obtain by assuming a hole of constant density depends only on $\rho_+$ (for a plus spin), and hence is the same for all $u_i$'s of that spin. Thus the exchange holes for different $u_i$'s of the same spin will only show small differences. We shall later examine these differences for the case of a free electron gas, and show that they are really not large. It is this small dependence on $u_i$ which will make it reasonable to use an averaged exchange charge in the simplified method which we shall suggest later.

To agree with the qualitative description which we have just given, we then expect the exchange charge density at point $x_2$, producing a field acting on the electron at $x_1$ whose wave function $u_i(x_1)$ we are determining by the Hartree-Fock Eq. (1), to integrate over $dx_2$ to $-e$ (a single electronic charge), and to be equal when $x_2$ approaches $x_1$ to the quantity

$$-e\sum_{\substack{k=1\\ \text{spin } k=\text{spin } i}}^{n} u_k^*(x_1)u_k(x_1). \qquad (3)$$

We shall now show that this is the case.

[3] E. Wigner and F. Seitz, Phys. Rev. **43**, 804 (1933); *ibid.* **46**, 509 (1934). The discussion of Wigner and Seitz was one of the first to show a proper understanding of the main points taken up in the present paper, which must be understood to represent a generalization and extension of previously suggested ideas, rather than an entirely new approach. See also L. Brillouin, J. de Phys. et le Radium, **5**, 413 (1934) for a discussion somewhat similar to the present one.

To show it, we rewrite (1) in the equivalent form[4]

$$H_1u_i(x_1)+\left[\sum_{k=1}^{n}\int u_k^*(x_2)u_k(x_2)(e^2/4\pi\epsilon_0r_{12})dx_2\right]u_i(x_1)$$

$$-\left[\sum_{k=1}^{n}\frac{\int u_i^*(x_1)u_k^*(x_2)u_k(x_1)u_i(x_2)(e^2/4\pi\epsilon_0r_{12})dx_2}{u_i^*(x_1)u_i(x_1)}\right]u_i(x_1)$$

$$=E_iu_i(x_1). \quad (4)$$

The exchange term now appears as the product of a function of $x_1$, times the function $u_i(x_1)$; thus it has the standard form of a potential energy term in a one-electron Schrödinger equation. This exchange potential energy is the potential energy, at the position of the first electron, of the exchange charge density,

$$-e\sum_{k=1}^{n}\frac{u_i^*(x_1)u_k^*(x_2)u_k(x_1)u_i(x_2)}{u_i^*(x_1)u_i(x_1)}, \quad (5)$$

located at the position $x_2$ of the second electron. We note as we expect that the exchange charge density depends on the position of the first electron, as well as the second, and also on the quantum state $i$ in which this first electron is located. We note, furthermore, that the total charge is that of a single electron. To show this, we integrate the exchange charge density (5) over $dx_2$, and find at once, on account of the orthogonality of the $u_i$'s (which follows from the Hartree-Fock equations) that the integral over all space is $-e$. Furthermore, as $x_2$ approaches $x_1$, we see at once that the exchange charge density approaches the correct value (3), where the restriction that the spins of $i$ and $k$ must be equal arises from (1), where an exchange term $u_k^*(x_2)u_i(x_2)$ is automatically zero unless this condition is satisfied. Thus we have shown that the exchange charge density (5) satisfies all the conditions necessary to justify our qualitative discussion of its behavior. In a later section, where we work out detailed values for the free-electron case, we can examine its properties more in detail.

The great difference between the Hartree and the Hartree-Fock methods is the fact that in the Hartree-Fock method the exchange hole or correction charge appropriate for an electron at $x_1$ moves around to follow that electron; in the Hartree method it does not, the correction charge depending only on the index $i$ of the wave function $u_i$. If our problem is a single atom, this is not very important, but in a crystal, for instance a metal, the difference is profound. Thus consider a periodic lattice, in which the one-electron functions $u_i$ are modulated plane waves, corresponding to $1/N$ of an electronic charge on each of the $N$ atoms of the crystal. In the Hartree scheme, the potential acting on the electron in the wave function $u_i$ is that of all electrons, minus this charge corresponding to $1/N$ of an electron on each atom. This correction charge is so spread out that its effect on the potential field is completely negligible, and each electron acts as if it were in the field of all electrons, thus finding itself in the field of a neutral atom when near any of the nuclei of the metal. On the other hand, with the Hartree-Fock equations, the exchange charge is located near the position $x_1$ of the electron in question, moving around with it, so that when this electron is on a given atom, the exchange charge is removed largely from that atom, leaving it in the form of a positive ion, which, as our physical intuition tells us, is the correct situation.

### III. AVERAGED EXCHANGE CHARGE

We have seen that the exchange charges for different wave functions $u_i$ corresponding to the same spin are not very different from each other, since in every case they reduce to the same value when $x_2=x_1$, and integrate to the same value over all space. Furthermore, in a system containing equal or approximately equal numbers of electrons with both spins, $\rho_+$ and $\rho_-$ will be at least approximately the same, so that exchange charges for different $u_i$'s even of opposite spins will be nearly the same. It then seems clear that we shall make no very great error if we use a weighted mean of the exchange charge density, weighting over $i$, for each value of $x_1$. The result of this will be that we shall have a single potential field to use for the Schrödinger equation for each $u_i$, simplifying greatly the application of the Hartree-Fock method. Let us first set up this average exchange charge and the consequent averaged exchange potential, then give some discussion of their properties and uses.

The probability that an electron at $x_1$ should be in the state $i$ is evidently $u_i^*(x_1)u_i(x_1)/[\sum_j u_j^*(x_1)u_j(x_1)]$. We can then use this quantity as a weighting factor to weight the exchange charge density (5). When we do this, we find as the average exchange charge density the quantity[5]

$$-e\frac{\sum_{j=1}^{n}\sum_{k=1}^{n}u_j^*(x_1)u_k^*(x_2)u_k(x_1)u_j(x_2)}{\sum_{j=1}^{n}u_j^*(x_1)u_j(x_1)}. \quad (6)$$

Using this average exchange charge density, we come to the following Schrödinger equations for the $u_i$'s, as

[4] J. C. Slater and H. M. Krutter, Phys. Rev. 47, 559 (1935); particularly p. 564, where this same method is used in discussing the Thomas-Fermi method.

[5] J. C. Slater, Rev. Mod. Phys. 6, 209 (1934), particularly p. 267, where this same expression is used for similar purposes, but without pointing out that it is the weighted mean of the exchange charge density found in the Hartree-Fock equations.

substitutes for the Hartree-Fock equations:

$$H_1u_i(x_1)+\left\{\sum_{k=1}^{n}\int u_k^*(x_2)u_k(x_2)(e^2/4\pi\epsilon_0r_{12})dx_2 - \frac{\sum_{j=1}^{n}\sum_{k=1}^{n}\int u_j^*(x_1)u_k^*(x_2)u_k(x_1)u_j(x_2)(e^2/4\pi\epsilon_0r_{12})dx_2}{\sum_{j=1}^{n}u_j^*(x_1)u_j(x_1)}\right\}u_i(x_1)=E_iu_i(x_1). \quad (7)$$

The wave functions $u_i$, and energy values $E_i$, as determined from these equations, will not be quite so accurate as those determined from the Hartree-Fock equations; but they will at least be much better than those found from the Hartree equations, particularly for the case of the crystal, and they have the great advantage that they are all solutions of the same potential problem. This automatically brings one good feature, which the solutions possess in common with solutions of the Hartree-Fock equations, but which solutions of the Hartree equations do not have: the functions $u_i$ are all orthogonal to each other.

There is one aspect of Eqs. (7) which is very important. In the last few years there has been a great development of the energy-band theory of semiconductors. This is all based on the hypothesis that we can build up a model of a solid in which each electron moves independently in a potential field which is made up from the nuclei, and all other electrons except the one in question. The electric field derived from this potential is sometimes called the motive field acting on the electron. Each wave function corresponds to a definite energy level, and the Fermi statistics are applied to the distribution of the electrons in these levels. The soundest way to set up this potential acting on each electron is by the Hartree-Fock method, but we see by our present discussion that this implies a different potential energy or motive for each electron, or each $u_i$. If we wish to have a single motive field appropriate for all electrons, the best thing we can do is to use the weighted mean suggested in the present section. Thus Eqs. (7) may well be taken to be the basis of the ordinary form of the energy-band theory of solids.

In many problems, we are interested in cases of degeneracy, not merely in evaluating the wave function of a single nondegenerate stationary state. Thus we may be solving a problem of multiplet structure in an atom or molecule, or discussing ferromagnetism in a solid. In such a case we start with a number of degenerate or approximately degenerate energy levels, corresponding to different orientations of orbit or spin, or in some cases (as in the hybridization of atomic orbitals) corresponding to different total or azimuthal quantum numbers, and then carry out perturbations. If we take the Hartree-Fock scheme literally, we shall use different potentials for finding the $u_i$'s of each of these various unperturbed functions. It is highly desirable in such cases, in the interests of simplicity, to modify the method so as to use the same potential function for the calculation of each wave function. This may involve even more averaging than is contemplated in setting up Eqs. (6) and (7). As one illustration, Hartree's use of a spherical potential for discussing atomic structure is an example of this procedure; this involves averaging over all orientations of the various orbital angular momenta of electrons which are not in closed shells. Whether we are using the Hartree scheme or the present simplification of the Hartree-Fock scheme, such averaging over orientations seems certainly desirable. Again, in studying ferromagnetism, the potentials to use, according to the scheme of the present paper, will depend on the net magnetization, or on the number of electrons of each spin. It is much simpler to handle such a problem, however, by using a single potential function, and that will usually be chosen to be that representing the unmagnetized state, with equal numbers of plus and minus spins.

In all these cases which we have just been discussing, we use one-electron wave functions which are slightly less accurate than those found by the Hartree-Fock scheme. When we apply perturbation methods, we must remember this, computing the matrix components of the exact energy operator with respect to these somewhat incorrect wave functions, remembering the wave equation (for instance (7)) which they actually satisfy. Nondiagonal matrix components of energy between these somewhat inaccurate wave functions will be somewhat larger than those between exact Hartree-Fock functions. Nevertheless they will still not be very large, for the wave functions are still quite accurate; the slight decrease in exactness is much more than made up by the simplicity of the method. The energy values computed by averaging the exact energy operator over the wave function will be very nearly the same as for Hartree-Fock functions, on account of the theorem that the mean value of energy over an incorrect wave function has errors only of the second order of small quantities.

## IV. THE EXCHANGE CHARGE FOR THE FREE-ELECTRON CASE

The calculations of exchange charge and exchange potential which we have been describing in general language can be carried out exactly for the case of a free-electron gas, as is well known. In this section we shall give the results, as an illustration of the general case. Then we shall point out in the next section that by using a free-electron approximation we can get an exchange potential much simpler than that of Eq. (7), which still is accurate enough for many purposes.

Let us have a free-electron gas with $n$ electrons in the volume $V$, half of them of each spin; the volume is assumed to be filled with a uniform distribution of positive charge, just enough to make it electrically neutral. The electrons are assumed to obey the Fermi statistics. Then by elementary methods we find that they occupy energy levels with uniform density in momentum space, out to a level whose energy is $P^2/2m=(h^2/2m)(3n/8\pi V)^{2/3}$, corresponding to a maximum momentum $P=h(3n/8\pi V)^{1/3}$. The de Broglie wavelength

$$d=h/P=(8\pi V/3n)^{1/3} \tag{8}$$

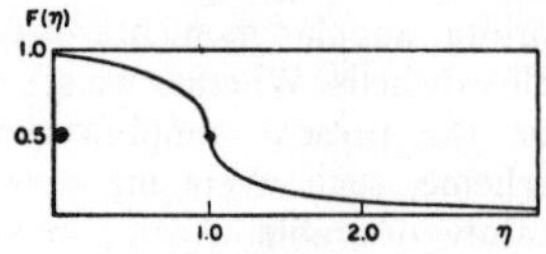


FIG. 1. $F(\eta)$ as function of $\eta$ (from Eq. (11)).

associated with this maximum momentum is clearly related to the radius $r_0$ of the exchange hole, which we introduced in Eq. (2). When we notice that $|\rho_+|$ which appeared there equals $ne/2V$, we see that

$$d=(4\pi/3)^{1/3}r_0. \tag{9}$$

We can now state some of the principal results of the application of this model to the exchange charge density and exchange energy. The exchange potential energy can be conveniently stated in terms of the ratio $\eta=p/P$ of the magnitude of the momentum of the electron to the maximum momentum corresponding to the top of the Fermi distribution. It is[2]

$$\text{exchange potential energy}=(e^2/4\pi\epsilon_0)(4P/h)F(\eta)$$
$$=(6/\pi)^{2/3}(e^2/4\pi\epsilon_0 r_0)F(\eta), \tag{10}$$

where

$$F(\eta)=\frac{1}{2}+\frac{1-\eta^2}{4\eta}\ln[(1+\eta)/(1-\eta)]. \tag{11}$$

The function $F(\eta)$ is shown in Fig. 1. It goes from unity when $\eta=0$, for an electron of zero energy, to $\frac{1}{2}$ when $\eta=1$, at the top of the Fermi distribution. We see that this exchange potential energy is of the form which we should expect. If we had a sphere of radius $r_0$, filled

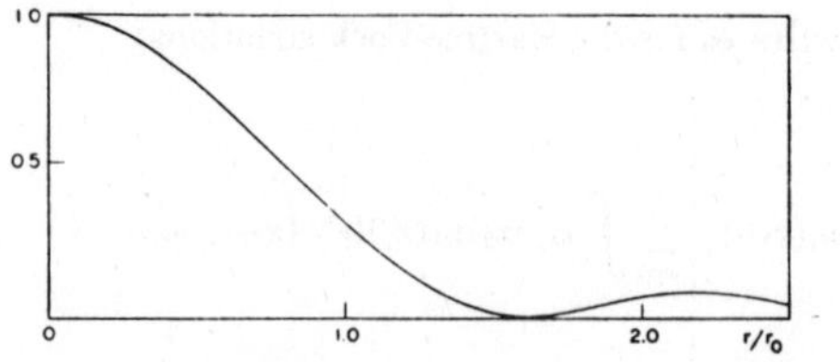


FIG. 2. Exchange charge density (divided by $\rho/2$) plotted as a function of $r/r_0$, from Eq. (12), where $r_0$ is given by Eq. (9).

with uniform charge density $|\rho_+|=ne/2V$, the potential energy of an electronic charge at the center of the sphere would be $\frac{3}{2}(e^2/4\pi\epsilon_0 r_0)$, while the value from Eq. (10) is $1.54(e^2/4\pi\epsilon_0 r_0)$ at the bottom of the Fermi band, half this value at the top. Thus this simple model of an exchange hole of constant charge density gives a qualitatively correct value for the exchange potential, and rather accurate quantitative value; and the extreme difference between top and bottom of the band corresponds only to a factor of 2 in the exchange potential.

If now we average over-all wave functions, we find that the properly weighted average of $F(\eta)$ is $\frac{3}{4}$. Thus the exchange potential energy of the averaged exchange charge[6] is $(\frac{3}{4})(6/\pi)^{2/3}(e^2/4\pi\epsilon_0 r_0)$. This can also be found from the averaged exchange charge density. This charge density is[3]

$$\frac{\rho}{2}\left[\frac{3\sin(r/d)-(r/d)\cos(r/d)}{(r/d)^3}\right]^2, \tag{12}$$

where $\rho$ is the total charge density of electrons, $d$ is given by Eqs. (8) and (9), and $r$ is the distance from point $x_1$, where the electron whose wave function we are computing is located, to $x_2$, where we are finding the exchange charge density. This function (12) is shown in Fig. 2, plotted as a function of $r/r_0$, and we see that it does in fact represent a density which equals $\rho/2$ when $r=0$, and falls to small values at approximately $r=r_0$. The potential energy of an electron at the center of this averaged exchange charge distribution is just the value $(\frac{3}{4})(6/\pi)^{2/3}(e^2/4\pi\epsilon_0 r_0)$ previously given.

## V. USE OF THE FREE-ELECTRON APPROXIMATION FOR THE EXCHANGE POTENTIAL

From the argument of Sec. III, it is clear that the exchange charge density (6), and the corresponding potential appearing in (7), must depend on the density of electronic charge, but not greatly on anything else. Thus in no case will we expect it to be very different from what we should have in a free-electron gas of the same charge density. We may then make a further approximation and simplification, beyond that of Sec. III; we may approximate the averaged exchange potential by what we should have in a free-electron gas

[6] F. Bloch, Z. Physik 57, 545 (1929) gave the first derivation of this value.

of the same density, as given in Sec. IV.[7] Thus, combining (10) and (2), we have

exchange potential energy

$$= -\tfrac{3}{4}(6/\pi)^{\frac{1}{3}}(e^2/4\pi\epsilon_0 r_0)$$
$$= -3(e^2/4\pi\epsilon_0)(3n/8\pi V)^{\frac{1}{3}}, \quad (13)$$

where we are now to interpret $n/V$ as the local density of electrons, a function of position. If we recall that this is $\sum(k)u_k^*(x)u_k(x)$, we finally have as our simplified Schrödinger equation for the one-electron functions $u_i$, to replace (7),

$$H_1 u_i(x_1)+\Big[\sum(k)\int u_k^*(x_2)u_k(x_2)(e^2/4\pi\epsilon_0 r_{12})dx_2$$
$$-3(e^2/4\pi\epsilon_0)\Big\{\frac{3}{8\pi}\sum(k)u_k^*(x_1)u_k(x_1)\Big\}^{\frac{1}{3}}\Big]u_i(x_1)$$
$$=E_i u_i(x_1). \quad (14)$$

This equation is in practice a very simple one to apply. We solve it for each of the wave functions $u_i$, then find the total charge density arising from all these wave functions, and can at once calculate the potential energy, including the exchange term, to go into (14), so as to check the self-consistency of the solution. Here, as before, we change to Gaussian units by omitting $4\pi\epsilon_0$, and atomic units by changing $e^2/4\pi\epsilon_0$ to 2.

One result of this formulation of the self-consistent problem is of immediate interest. In a periodic potential problem such as a crystal, it is obvious that the total charge density will have the same periodicity as the potential. Thus the corrected potential of Eq. (14) will also be periodic, and hence the functions $u_i$ will be modulated according to Block's theorem. In other words, such modulated functions are the only type which can follow from a proper application of our simplification of the Hartree-Fock method to a periodic potential problem.

Our general method is applicable to any problem of atoms, molecules, or solids. It is easy to give it a very explicit formulation for the case of atoms, which can then be used for the self-consistent treatment of atomic structure. Let the electrostatic potential of the nucleus, and of all electrons, at distance $r$ from the nucleus of a spherical atom, be $Z_p(r)e/4\pi\epsilon_0 r$. Then the charge density is given by Poisson's equation as $\rho=-\epsilon_0\nabla^2(Z_p e/4\pi\epsilon_0 r)$. When we express the Laplacian in spherical coordinates, this gives at once $\rho=-(e/4\pi)(1/r)d^2Z_p/dr^2$. This is the quantity which is expressed as $-e\sum(k)u_k^*(x)u_k(x)$. Thus the exchange potential energy becomes $-3(e^2/4\pi\epsilon_0)[(\frac{3}{32}\pi^2)(1/r)d^2Z_p/dr^2]^{\frac{1}{3}}$, and, finally, the total potential energy, for use in the Schrödinger equation for $u_i(x_1)$, is

$$-\frac{e^2}{4\pi\epsilon_0 r}\left[Z_p+3\left(\frac{3}{32\pi^2}\right)^{\frac{1}{3}}\left(r^2\frac{d^2Z_p}{dr^2}\right)^{\frac{1}{3}}\right]. \quad (15)$$

To carry out a self-consistent solution for an atom, using this simplified method, we then find a $Z_p$ such that the wave functions $u_i$, determined from a single Schrödinger equation using the potential energy (15), determined from $Z_p$, add to give a charge density which would lead by Poisson's equation to a potential energy $-e^2Z_p/4\pi\epsilon_0 r$.

In order to check the applicability of the method Mr. George W. Pratt is carrying out a self-consistent solution of the ion $Cu^+$ by this method. This ion was chosen, as being the heaviest one for which solutions by both the Hartree and the Hartree-Fock methods are available for comparison. The solution has gone far enough to show that the wave functions and energy parameters $E_i$ determined from it are not far from those found by the Hartree and the Hartree-Fock methods. The discrepancies come principally from large values of $r$, where the charge density is small, and our free-electron approximation for exchange is not very good. Over most of the range of $r$, however, the approximation seems very satisfactory. Detailed results will be reported later. The great advantages of this method for numerical calculation are clear from this example which has been worked out. Actual calculation is simpler than for the original Hartree scheme, since only one potential function need be computed, and can be used for all wave functions. The wave functions have the advantage of being orthogonal; and they possess a considerable part of the accuracy which the solutions of the Hartree-Fock equations possess, in contrast to the Hartree equations. It is to be hoped that they will make enough simplification so that it will be possible to carry out calculations for more complicated atoms than have yet been attempted by the Hartree-Fock method. At the same time the method should prove valuable in setting up solutions for molecules and solids.

[7] This method of treating the exchange potential as if the electrons formed part of a free-electron gas is similar to what is done in the Thomas-Fermi method with exchange (see Dirac (reference 2), Slater and Krutter (reference 4), and L. Brillouin, *L'Atome de Thomas-Fermi et la Méthode du Champ "Self-Consistent,"* Actualités Scientifiques et Industrielles No. 160 (Hermann et Cie., 1934)).

III-3 Reprinted from *Physical Review* **140**, A1133 (1965)

# Self-Consistent Equations Including Exchange and Correlation Effects*

W. Kohn and L. J. Sham
*University of California, San Diego, La Jolla, California*
(Received 21 June 1965)

From a theory of Hohenberg and Kohn, approximation methods for treating an inhomogeneous system of interacting electrons are developed. These methods are exact for systems of slowly varying or high density. For the ground state, they lead to self-consistent equations analogous to the Hartree and Hartree-Fock equations, respectively. In these equations the exchange and correlation portions of the chemical potential of a uniform electron gas appear as additional effective potentials. (The exchange portion of our effective potential differs from that due to Slater by a factor of $\frac{2}{3}$.) Electronic systems at finite temperatures and in magnetic fields are also treated by similar methods. An appendix deals with a further correction for systems with short-wavelength density oscillations.

## I. INTRODUCTION

In recent years a great deal of attention has been given to the problem of a homogeneous gas of interacting electrons and its properties have been established with a considerable degree of confidence over a wide range of densities. Of course, such a homogeneous gas represents only a mathematical model, since in all real systems (atoms, molecules, solids, etc.) the electronic density is nonuniform.

It is then a matter of interest to see how properties of the homogeneous gas can be utilized in theoretical studies of inhomogeneous systems. The well-known methods of Thomas-Fermi[1] and the Slater[2] exchange hole are in this spirit. In the present paper we use the formalism of Hohenberg and Kohn[3] to carry this approach further and we obtain a set of self-consistent equations which include, in an approximate way, exchange and correlation effects. They require only a knowledge of the true chemical potential, $\mu_h(n)$, of a homogeneous interacting electron gas as a function of the density $n$.

We derive two alternative sets of equations [Eqs. (2.8) and (2.22)] which are analogous, respectively, to the conventional Hartree and Hartree-Fock equations, and, although they also include correlation effects, they are no more difficult to solve.

The local effective potentials in these equations are unique in a sense which is described in Sec. II. In particular, we find that the Slater exchange-hole potential, besides its omission of correlation effects, is too large by a factor of $\frac{3}{2}$.

Apart from work on the correlation energy of the homogeneous electron gas, most theoretical many-body studies have been concerned with elementary excitations and as a result there has been little recent progress in the theory of cohesive energies, elastic constants, etc., of real (i.e., inhomogeneous) metals and alloys. The methods proposed here offer the hope of new progress in this latter area.

In Secs. III and IV, we describe the necessary modifications to deal with the finite-temperature properties and with the spin paramagnetism of an inhomogeneous electron gas.

Of course, the simple methods which are here proposed in general involve errors. These are of two general origins[4]: a too rapid variation of density and, for finite systems, boundary effects. Refinements aimed at reducing the first type of error are briefly discussed in Appendix II.

## II. THE GROUND STATE

### A. Local Effective Potential

It has been shown[3] that the ground-state energy of an interacting inhomogeneous electron gas in a static potential $v(\mathbf{r})$ can be written in the form

$$E=\int v(\mathbf{r})n(\mathbf{r})\,d\mathbf{r}+\frac{1}{2}\int\int\frac{n(\mathbf{r})n(\mathbf{r}')}{|\mathbf{r}-\mathbf{r}'|}\,d\mathbf{r}\,d\mathbf{r}'+G[n]\,, \tag{2.1}$$

where $n(\mathbf{r})$ is the density and $G[n]$ is a universal functional of the density. This expression, furthermore, is a minimum for the correct density function $n(\mathbf{r})$. In this section we propose first an approximation for $G[n]$, which leads to a scheme analogous to Hartree's method but contains the major part of the effects of exchange and correlation.

We first write

$$G[n]\equiv T_s[n]+E_{xc}[n]\,, \tag{2.2}$$

where $T_s[n]$ is the kinetic energy of a system of noninteracting electrons with density[5] $n(\mathbf{r})$ and $E_{xc}[n]$ is, by our definition, the exchange and correlation energy of an interacting system with density $n(\mathbf{r})$. For an arbitrary $n(\mathbf{r})$, of course, one can give no simple exact expression for $E_{xc}[n]$. However, if $n(\mathbf{r})$ is sufficiently slowly varying, one can show[3] that

$$E_{xc}[n]=\int n(\mathbf{r})\epsilon_{xc}(n(\mathbf{r}))\,d\mathbf{r}\,, \tag{2.3}$$

* Supported in part by the U. S. Office of Naval Research.
[1] L. H. Thomas, Proc. Cambridge Phil. Soc. **23**, 542 (1927); E. Fermi, Z. Physik **48**, 73 (1928).
[2] J. C. Slater, Phys. Rev. **81**, 385 (1951).
[3] P. Hohenberg and W. Kohn, Phys. Rev. **136**, B864 (1964); referred to hereafter as HK.
[4] W. Kohn and L. J. Sham, Phys. Rev. **137**, A1697 (1965).
[5] For such a system it follows from HK that the kinetic energy is in fact a unique functional of the density.

where $\epsilon_{xc}(n)$ is the exchange and correlation energy per electron of a uniform electron gas of density $n$. Our sole approximation consists of assuming that (2.3) constitutes an adequate representation of exchange and correlation effects in the systems under consideration. We shall regard $\epsilon_{xc}$ as known from theories of the homogeneous electron gas.[6]

From the stationary property of Eq. (2.1) we now obtain, subject to the condition

$$\int \delta n(\mathbf{r})\, d\mathbf{r}=0, \tag{2.4}$$

the equation

$$\int \delta n(\mathbf{r})\left\{\varphi(\mathbf{r})+\frac{\delta T_s[n]}{\delta n(\mathbf{r})}+\mu_{xc}(n(\mathbf{r}))\right\} d\mathbf{r}=0; \tag{2.5}$$

here

$$\varphi(\mathbf{r})=v(\mathbf{r})+\int \frac{n(\mathbf{r}')}{|\mathbf{r}-\mathbf{r}'|}d\mathbf{r}', \tag{2.6}$$

and

$$\mu_{xc}(n)=d(n\epsilon_{xc}(n))/dn \tag{2.7}$$

is the exchange and correlation contribution to the chemical potential of a uniform gas of density $n$.

Equations (2.4) and (2.5) are precisely the same as one obtains from the theory of Ref. 3 when applied to a system of noninteracting electrons, moving in the given potential $\varphi(\mathbf{r})+\mu_{xc}(n(\mathbf{r}))$. Therefore, for given $\varphi$ and $\mu$, one obtains the $n(\mathbf{r})$ which satisfies these equations simply by solving the one-particle Schrödinger equation

$$\{-\tfrac{1}{2}\nabla^2+[\varphi(\mathbf{r})+\mu_{xc}(n(\mathbf{r}))]\}\psi_i(\mathbf{r})=\epsilon_i\psi_i(\mathbf{r}), \tag{2.8}$$

and setting

$$n(\mathbf{r})=\sum_{i=1}^{N}|\psi_i(\mathbf{r})|^2, \tag{2.9}$$

where $N$ is the number of electrons.

It is physically very satisfactory that $\mu_{xc}$ appears in Eq. (2.8) as an additional effective potential so that gradients of $\mu_{xc}$ lead to forces on the electron fluid in a manner familiar from thermodynamics.

Equations (2.6)-(2.9) have to be solved self-consistently: One begins with an assumed $n(\mathbf{r})$, constructs $\varphi(\mathbf{r})$ from (2.6) and $\mu_{xc}$ from (2.7), and finds a new $n(\mathbf{r})$ from (2.8) and (2.9). The energy is given by

$$E=\sum_1^N \epsilon_i-\frac{1}{2}\int\int\frac{n(\mathbf{r})n(\mathbf{r}')}{|\mathbf{r}-\mathbf{r}'|}d\mathbf{r}\,d\mathbf{r}' +\int n(\mathbf{r})[\epsilon_{xc}(n(\mathbf{r}))-\mu_{xc}(n(\mathbf{r}))]\,d\mathbf{r}. \tag{2.10}$$

The results of our procedure are exact in two limiting cases:

(*a*) *Slowly varying density.* This regime is characterized by the condition $r_s/r_0 \ll 1$, where $r_s$ is the Wigner-Seitz radius and $r_0$ is a typical length over which there is an appreciable change in density. In this case, as shown in HK, we can expand the true exchange and correlation energy as follows:

$$E_{xc}[n]=\int \epsilon_{xc}(n)n\,d\mathbf{r} +\int \epsilon_{xc}^{(2)}(n)|\nabla n|^2\,d\mathbf{r}+\cdots, \tag{2.11}$$

where $\epsilon_{xc}^{(2)}$ is the exchange and correlation portion of the second term in the energy expansion in powers of the gradient operator. In this regime we may similarly expand $T_s[n]$ in the form

$$T_s[n]=\int \tfrac{3}{10}(3\pi^2 n)^{2/3}n\,d\mathbf{r} +\int t^{(2)}(n)|\nabla n|^2\,d\mathbf{r}+\cdots. \tag{2.12}$$

From HK, expecially Sec. III 2, we have the following expression for the energy:

$$E_v[n]=\int v(\mathbf{r})n(\mathbf{r})\,d\mathbf{r}+\frac{1}{2}\int\frac{n(\mathbf{r})n(\mathbf{r}')}{|\mathbf{r}-\mathbf{r}'|}d\mathbf{r}\,d\mathbf{r}' +\int g_0(n)\,d\mathbf{r}+\int g_2^{(2)}(n)|\nabla n|^2\,d\mathbf{r}+\cdots, \tag{2.13}$$

where

$$g_0(n)=\{\tfrac{3}{10}(3\pi^2 n)^{2/3}+\epsilon_{xc}(n)\}n, \tag{2.14}$$

and

$$g_2^{(2)}(n)=\{\epsilon_{xc}^{(2)}(n)+t^{(2)}(n)\}n. \tag{2.15}$$

Since in our approximation (2.3), the $|\nabla|^2$ term of Eq. (2.11) is neglected, it is clear that for a gas of slowly varying density our expression (2.10) for the energy has errors of the order $|\nabla|^2$, or equivalently, of the order $r_0^{-2}$.

Surprisingly, our procedure determines the density with greater accuracy, the errors being of order $|\nabla|^4$. This is shown in Appendix I.

At this point a comparison of our procedure and that of Slater[2] may be appropriate. For one thing, Slater's original work does not include correlation effects.[7] But even the exchange correction is different from ours. To obtain Slater's exchange correction, one may begin by writing the Hartree-Fock exchange operator in the form of an equivalent potential acting on the $k$th wave function

$$v_{xk}(\mathbf{r})=-\sum_{k'=1}^{N}\int\frac{\psi_k^*(\mathbf{r})\psi_{k'}^*(\mathbf{r}')\psi_{k'}(\mathbf{r})\psi_k(\mathbf{r}')}{|\mathbf{r}-\mathbf{r}'|}d\mathbf{r}' \Big/ \psi_k^*(\mathbf{r})\psi_k(\mathbf{r}), \tag{2.16}$$

[6] For a review see D. Pines, *Elementary Excitations in Solids* (W. A. Benjamin, Inc., New York, 1963).

[7] Subsequent to the original paper by Slater, there have been several attempts to add correlation corrections: S. Olszewski, Phys. Rev. 121, 42 (1961); J. E. Robinson, F. Bassani, B. S. Knox, and J. R. Schrieffer, Phys. Rev. Letters 9, 215 (1962); W. A. Harrison, Phys. Rev. 136, A1107 (1964); S. Lundqvist and C. W. Ufford, Phys. Rev. 139, A1 (1965).

where the symbols $\mathbf{r}$ and $\mathbf{r}'$ are understood to include electron spin coordinates and integration is understood to include summation over spin coordinates. One next assumes that the wave functions can be approximated by plane waves which results in

$$v_{xk}(\mathbf{r}) = -\frac{k_F(\mathbf{r})}{\pi}\left[1+\frac{k_F^2(\mathbf{r})-k^2}{2kk_F(\mathbf{r})}\ln\left|\frac{k+k_F(\mathbf{r})}{k-k_F(\mathbf{r})}\right|\right], \quad (2.17)$$

where $k_F(\mathbf{r}) \equiv \{3\pi^2 n(\mathbf{r})\}^{1/3}$. Finally, one averages $v_{xk}$ over the occupied state $k$, which results in

$$v_x(\mathbf{r}) = -(3/2\pi)\{3\pi^2 n(\mathbf{r})\}^{1/3}. \quad (2.18)$$

In our procedure (neglecting correlation) we obtain, in place of Slater's $v_x$

$$\mu_x(\mathbf{r}) = -(1/\pi)\{3\pi^2 n(\mathbf{r})\}^{1/3}, \quad (2.19)$$

smaller by a factor of $\frac{2}{3}$. From the discussion in Appendix I, it follows that while $\mu_x$ gives the exchange correction of the density correct to order $|\nabla|^2$, inclusive, $v_x$ [as indeed any other function of $n(\mathbf{r})$] leads to errors of order $|\nabla|^2$. The same comment applies to any extension of Slater's exchange to include correlation in the self-consistent potential.

We may note that our result is equivalent to taking, not the average of (2.17), but rather its value at $k = k_F(\mathbf{r})$; i.e., the effective exchange potential for a state at the top of the Fermi distributions. This is physically understandable since density adjustments come about by redistribution of the electrons near the Fermi level.

(*b*) ***High density.*** This regime is characterized by the condition $r_s/a_0 \ll 1$, where $a_0$ is the Bohr radius. In this case, the entire exchange and correlation energy is smaller than the kinetic energy by a factor of order $(r_s/a_0)$ and hence our inaccuracy in representing these portions becomes negligible.

The reader will have noticed that while in Eq. (2.3) we approximate the exchange and correlation energy by the expression valid for a slowly varying density, we made no approximation for the kinetic-energy functional $T_s[n]$ of Eq. (2.2). This procedure is responsible for the exactness of the high-density limit, even when the density is rapidly varying, such as in the vicinity of an atomic nucleus.

We now make a few further remarks about our approximation. If in Eq. (2.2), we had approximated $T_s[n]$ by its form appropriate to a system of slowly varying density,

$$T_s[n] \to \int \tfrac{3}{10}(3\pi^2 n)^{2/3} n\, d\mathbf{r}, \quad (2.20)$$

we would have been led to the generalization of the Thomas-Fermi method suggested by Lewis.[8] This method shares with the Thomas-Fermi method two shortcomings: (1) It leads to an infinite density near an atomic nucleus, and (2) it does not lead to quantum density oscillations,[4] such as the density fluctuations due to atomic shell structures. By not making the replacement (2.20), we avoid both of these shortcomings.

Let us now qualitatively discuss the appropriateness of our procedure for various classes of electronic systems.

In atoms and molecules one can distinguish three regions: (1) A region near the atomic nucleus, where the electronic density is high and therefore, in view of case (*b*) above, we expect our procedure to be satisfactory. (2) The main "body" of the charge distribution where the electronic density $n(\mathbf{r})$ is relatively slowly varying, so that our approximation (2.3) for $\epsilon_{xc}$ is expected to be satisfactory as discussed in case (a) above. (3) The "surface" of atoms and the overlap regions in molecules. Here our approximation (2.3) has no validity and therefore we expect this region to be the main source of error. We do not expect an accurate description of chemical binding. In large atoms, of course, this "surface" region becomes of less importance. (The surface is more satisfactorily handled in the nonlocal method described under *B* below.)

For metals, alloys, and small-gap insulators we have, of course, no surface problem and we expect our approximation (2.3) to give a good representation of exchange and correlation effects. In large-gap insulators, however, the actual correlation energy will be considerably reduced compared to that of a homogeneous electron gas of the same density.

### B. Nonlocal Effective Potential

Instead of the Hartree-type procedure discussed in Sec. IIA it is also possible to obtain a scheme which includes exchange effects exactly. We write in place of Eq. (2.3)

$$E_{xc}[n] = E_x[n] + \int n(\mathbf{r})\epsilon_c(n(\mathbf{r}))\, d\mathbf{r} \quad (2.21)$$

where $E_x[n]$ is the exchange energy of a Hartree-Fock system of density $n(\mathbf{r})$ and $\epsilon_c(n)$ is the correlation energy per particle of a homogeneous electron gas. Applying this ansatz in conjunction with Eq. (2.2) and the stationary property of (2.1) leads to the following system of equations:

$$\{-\tfrac{1}{2}\nabla^2 + \varphi(\mathbf{r}) + \mu_c(\mathbf{r})\}\psi_i(\mathbf{r}) - \int \frac{n_1(\mathbf{r},\mathbf{r}')}{|\mathbf{r}-\mathbf{r}'|}\psi_i(\mathbf{r}')\, d\mathbf{r}' = \epsilon_i\psi_i(\mathbf{r}), \quad (2.22)$$

where

$$\mu_c = d(n\epsilon_c)/dn, \quad (2.23)$$

$$n_1(\mathbf{r},\mathbf{r}') = \sum_{j=1}^{N} \psi_j(\mathbf{r})\psi_j^*(\mathbf{r}'), \quad (2.24)$$

and $\varphi(\mathbf{r})$, $n(\mathbf{r})$ are defined as before, Eqs. (2.6) and (2.9).

[8] H. W. Lewis, Phys. Rev. 111, 1554 (1958).

The energy is now

$$E=\sum_1^N \epsilon_i-\frac{1}{2}\int\int\frac{n(\mathbf{r})n(\mathbf{r}')}{|\mathbf{r}-\mathbf{r}'|}d\mathbf{r}\,d\mathbf{r}'+\frac{1}{2}\int\int\frac{n_1(\mathbf{r},\mathbf{r}')n_1(\mathbf{r}',\mathbf{r})}{|\mathbf{r}-\mathbf{r}'|}d\mathbf{r}\,d\mathbf{r}'+\int n(\mathbf{r})\{\epsilon_c(n(\mathbf{r}))-\mu_c(n(\mathbf{r}))\}\,d\mathbf{r}. \quad (2.25)$$

This procedure may be regarded as a Hartree-Fock method corrected for correlation effects. It is no more complicated than the uncorrected Hartree-Fock method but, because of the nonlocal operator appearing in Eq. (2.22), very much more complicated than the method described in Sec. IIA. Since at least exchange effects are now treated exactly we must expect, in general, more accurate results than from the method of Sec. IIA. In particular, near the surface of an atom the effective potential now is correctly $(-1/r)$ whereas in Sec. IIA it approaches zero much faster. Even here, however, correlation effects are not correctly described near the surface.

## III. FREE ENERGY; SPECIFIC HEAT

We can generalize the consideration of the ground state to finite temperature ensembles by using the finite temperature generalization of Eq. (2.1) given by Mermin.[9] He has shown that the grand canonical potential can be written in the form

$$\Omega=\int v(\mathbf{r})n(\mathbf{r})\,d\mathbf{r}+\frac{1}{2}\int\frac{n(\mathbf{r})n(\mathbf{r}')}{|\mathbf{r}-\mathbf{r}'|}d\mathbf{r}\,d\mathbf{r}'+G[n]-\mu\int n(\mathbf{r})\,d\mathbf{r}, \quad (3.1)$$

where $G[n]$ is a unique functional of the density at a given temperature $\tau$ and $\mu$ is the chemical potential. For the correct $n$ this quantity is a minimum.

In analogy with (2.2) we now write

$$G[n]=G_s[n]+F_{xc}[n]; \quad (3.2)$$

here

$$G_s[n]\equiv T_s[n]-\tau S_s[n], \quad (3.3)$$

where $T_s[n]$ and $S_s[n]$ are, respectively, the kinetic energy and entropy of noninteracting electrons with density $n(\mathbf{r})$ at a temperature $\tau$; and $F_{xc}[n]$ is, by definition, the exchange and correlation contribution to the free energy. For the latter quantity, we make the approximation

$$F_{xc}[n]=\int n(\mathbf{r})f_{xc}(n(\mathbf{r}))\,d\mathbf{r}, \quad (3.4)$$

where $f_{xc}(n)$ is the exchange and correlation contribution to the free energy per electron of a uniform electron gas of density $n$; i.e.,

$$f_{xc}(n)\equiv f(n)-f_0(n), \quad (3.5)$$

where $f$ and $f_0$ are the free energies per electron of an interacting and noninteracting gas, respectively.

$$0=\varphi(\mathbf{r})+(\delta G_s[n]/\delta n(\mathbf{r}))+\mu_{xc}(n(\mathbf{r}))-\mu, \quad (3.6)$$

where $\varphi(\mathbf{r})$ is given, as before, by Eq. (2.6) and

$$\mu_{xc}(n)\equiv d(nf_{xc}(n))/dn. \quad (3.7)$$

Equation (3.6) is identical to the corresponding equation for a system of noninteracting electrons in the effective potential $\varphi+\mu_{xc}$. Its solution is therefore determined by the following system of equations:

$$\{-\tfrac{1}{2}\nabla^2+\varphi(\mathbf{r})+\mu_{xc}(n(\mathbf{r}))\}\psi_i=\epsilon_i\psi_i, \quad (3.8)$$

and

$$n(\mathbf{r})=\sum_{i=1}^N|\psi_i(\mathbf{r})|^2/\{e^{(\epsilon_i-\mu)/k\tau}+1\}. \quad (3.9)$$

$\mu$ is determined as usual by the total number of particles from Eq. (3.9). This value also represents our approximation for the chemical potential of the interacting system.

Of special interest for metals and alloys is the low-temperature heat capacity. This may be obtained by making an expansion, in powers of $\tau$, of the above system of equations. An equivalent, but more convenient, method is as follows: From thermodynamics and Eq. (3.1) we have

$$S[n]\equiv-\frac{\partial}{\partial\tau}(\Omega+\mu N)_V=-\int\left\{\varphi(\mathbf{r})+\frac{\delta G}{\delta n(\mathbf{r})}\right\}\times\left(\frac{\partial n(\mathbf{r})}{\partial\tau}\right)_V d\mathbf{r}-\left(\frac{\partial G[n]}{\partial\tau}\right)_{n(\mathbf{r}),V} \quad (3.10)$$

The integral vanishes because of the stationary property of $\Omega$, so that

$$S[n]=-(\partial G[n]/\partial\tau)_{n(\mathbf{r}),V}. \quad (3.11)$$

The same argument, applied to a system of noninteracting electrons of density $n(\mathbf{r})$, gives

$$S_s[n]=-(\partial G_s[n]/\partial\tau)_{n(\mathbf{r}),V}. \quad (3.12)$$

Combining Eqs. (3.11), (3.12), (3.2), and (3.4), we obtain

$$S[n]=S_s[n]+\int n(\mathbf{r})(\partial f_{xc}(n)/\partial\tau)_{n(\mathbf{r}),V}d\mathbf{r}. \quad (3.13)$$

For small $\tau$ it is well known that $S_s$ is given by

$$S_s[n]=V\tfrac{1}{3}\pi^2k^2\tau g_s(\mu), \quad (3.14)$$

where $g_s$ is the single-particle density of states in the effective potential $\varphi+\mu_{xc}$ at zero temperature; further,

$$(\partial f_{xc}(n)/\partial\tau)_{n(\mathbf{r}),V}=\tfrac{1}{3}\pi^2k^2\tau[g(\mu_h(n))-g_0(\mu_0(n))], \quad (3.15)$$

[9] N. D. Mermin, Phys. Rev. 137, A1441 (1965).

where $\mu_h(n)$ and $\mu_0(n)$ are, respectively, the chemical potentials of an interacting and a noninteracting homogeneous gas of density $n$, and $g$ and $g_0$ are the respective densities of states.[10]

It follows immediately that the low-temperature heat capacity is given by

$$C_v=\gamma\tau\,, \tag{3.16}$$

where

$$\gamma=\tfrac{1}{3}\pi^2k^2\left[Ng_s(\mu)+\int n(\mathbf{r})\{g(\mu_h(n))-g_0(\mu_0(n))\}\,d\mathbf{r}\right]. \tag{3.17}$$

We shall not present a treatment, analogous to Sec. IIB, in which exchange effects are included exactly. The development is straightforward but leads to a well-known divergence in the low-temperature specific heat.

## IV. SPIN SUSCEPTIBILITY

To obtain a theory of the spin susceptibility of an electron gas, we first extend the theory of HK to include the effects of spin interaction with an external magnetic field. The result is that if we take the field in the $z$ direction and write the magnetic-moment density as

$$m(\mathbf{r})=-(1/2c)\langle 0|\psi_\uparrow^*(\mathbf{r})\psi_\uparrow(\mathbf{r})-\psi_\downarrow^*(\mathbf{r})\psi_\downarrow(\mathbf{r})|0\rangle\,, \tag{4.1}$$

the ground-state energy can be written in the form

$$E_{v,H}=\int\{v(\mathbf{r})n(\mathbf{r})-H(\mathbf{r})m(\mathbf{r})\}\,d\mathbf{r}+\frac{1}{2}\int\frac{n(\mathbf{r})n(\mathbf{r}')}{|\mathbf{r}-\mathbf{r}'|}\,d\mathbf{r}\,d\mathbf{r}'+G[n(\mathbf{r}),m(\mathbf{r})]\,, \tag{4.2}$$

where $G$ is a universal functional of $n$ and $m$, and the correct $m(\mathbf{r})$, $n(\mathbf{r})$ make (4.2) a minimum.

For small $m$ we expand $G$ in the form

$$G=G[n]+\frac{1}{2}\int G(\mathbf{r},\mathbf{r}';[n])m(\mathbf{r})m(\mathbf{r}')\,d\mathbf{r}\,d\mathbf{r}'+\cdots\,; \tag{4.3}$$

the linear term vanishes for a paramagnetic system in which $m\equiv 0$ when $H\equiv 0$. From the stationary property of (4.2) we find, for small $H$, that $n$ is unchanged to first order and that

$$-H(\mathbf{r})+\int G(\mathbf{r},\mathbf{r}';[n])m(\mathbf{r}')d\mathbf{r}'=0\,, \tag{4.4}$$

where $n$ is the zero-field density. We now formally invert this equation, which gives

$$m(\mathbf{r})=\int G^{-1}(\mathbf{r},\mathbf{r}';[n])H(\mathbf{r}')d\mathbf{r}'\,. \tag{4.5}$$

For a uniform field this gives for the susceptibility

$$\chi[n]=\frac{1}{V}\frac{\partial}{\partial H}\int m(\mathbf{r})\,d\mathbf{r}=\int G^{-1}(\mathbf{r},\mathbf{r}';[n])\,d\mathbf{r}\,d\mathbf{r}'\,. \tag{4.6}$$

[10] J. M. Luttinger, Phys. Rev. 119, 1153 (1960).

So far everything is formal and exact. We now write, in the spirit of the previous sections,

$$G^{-1}(\mathbf{r},\mathbf{r}';[n])\equiv G_s^{-1}(\mathbf{r},\mathbf{r}';[n])+G_{xc}^{-1}(\mathbf{r},\mathbf{r}';[n])\,. \tag{4.7}$$

The second term we approximate as for a slowly varying gas, which gives

$$\chi[n]=\chi_s[n]+\frac{1}{V}\int[\chi(n(\mathbf{r}))-\chi_0(n(\mathbf{r}))]\,d\mathbf{r}\,, \tag{4.8}$$

where

$$\chi_s[n]=(1/2c)^2(N/V)\times g_s(\mu)\,, \tag{4.9}$$

and $\chi(n)$, $\chi_0(n)$ are, respectively, the susceptibilities for uniform systems with and without interactions.

## APPENDIX I: GRADIENT EXPANSION OF THE DENSITY

In this Appendix we show that for a system of slowly varying density our procedure gives the density correct to order $|\nabla|^2$ inclusive. When dealing with such a system we may proceed in two entirely equivalent ways: (1) We can solve the self-consistent equations, Eqs. (2.8) and (2.9), for $n(\mathbf{r})$, and (2) we can go back to the underlying variational principle (2.5), make a gradient expansion and determine $n(\mathbf{r})$ directly. We shall here follow the second route to estimate the errors in $n(\mathbf{r})$.

From (2.5) and the expansion (2.12) of $T_s[n]$, we obtain

$$\mu=\varphi(\mathbf{r})+\mu_h(n)-t^{(2)\prime}(n)|\nabla n|^2-2t^{(2)}(n)\nabla^2 n+O(\nabla^4)\,, \tag{A1.1}$$

where $\mu$ is the chemical potential [cf. HK, Eq. (68)]. Note however that because of our approximation of keeping only the first term in (2.11), some other contributions of order $|\nabla|^2$ are missing in (A1.1).

To solve (A1.1), let us write the external charge density as

$$n_{\text{ext}}(\mathbf{r})\equiv f_0(\mathbf{r}/r_0)\,, \tag{A1.2}$$

where $r_0\to\infty$ (slow spatial variation), and try the ansatz

$$n(\mathbf{r})=n_0(\mathbf{r})+n_1(\mathbf{r})\,, \tag{A1.3}$$

where

$$n_0(\mathbf{r})=f_0(\mathbf{r}/r_0) \tag{A1.4}$$

exactly neutralizes the external charge and $n_1$ is assumed to approach zero as $r_0\to\infty$. Neglecting, for the moment, the terms of order $|\nabla|^2$ in (A1.1) and substituting (A1.3) into (A1.1), we obtain

$$\mu=\int\frac{n_1(\mathbf{r}')}{|\mathbf{r}-\mathbf{r}'|}\,d\mathbf{r}'+\mu_h(n_0)+n_1(\mathbf{r})\mu_h'(n_0)+O(n_1^2)\,. \tag{A1.5}$$

Now define

$$\mathbf{R}\equiv\mathbf{r}/r_0\,, \tag{A1.6}$$

and write

$$n_1(\mathbf{r})\equiv f_1(\mathbf{R})\,. \tag{A1.7}$$

With this notation, (A1.5) becomes

$$\mu = r_0^2 \int \frac{f_1(\mathbf{R}')}{|\mathbf{R}-\mathbf{R}'|} d\mathbf{R}' + \mu_h(f_0(\mathbf{R})) + f_1(\mathbf{R})\mu_h'(f_0(\mathbf{R})) + O(f_1^2). \quad (A1.8)$$

We may now write

$$f_1(\mathbf{R}) = (1/r_0^2) f_1^{(2)}(\mathbf{R}) + (1/r_0^4) f_1^{(4)}(\mathbf{R}) + \cdots, \quad (A1.9)$$

and

$$\mu = \mu^{(0)} + (1/r_0^2)\mu^{(2)} + \cdots. \quad (A1.10)$$

The first term of Eq. (A1.9) is correctly determined by Eq. (A1.8) and not affected either by the inclusion of terms of order $\nabla^2$ in (A1.5) or by the terms of order $f_1^2$ in (A1.8). Hence, in spite of the errors of order $\nabla^2$ in (A1.1), the density given by our procedure is correct to order $1/r_0^2$ or $|\nabla|^2$, inclusive. Equation (A1.8) shows that this curious result stems from the infinite range of the Coulomb interaction.

## APPENDIX II: EFFECT OF RAPID DENSITY OSCILLATION ON EXCHANGE AND CORRELATION

In Eq. (2.3), we approximated $E_{xc}[n]$ by the first term in the gradient expansion. In actual physical systems, there are quantum density oscillations[4] whose effects on exchange and correlation are not included in the approximation (2.3). Now we put forward a correction to (2.3) to include such effects.

In HK, the gradient expression for the energy functional is partially summed such that it is also correct for a system of almost constant density[1] even when the density fluctuations are of short wavelength[11]:

$$G[n] = \int g_0(n(\mathbf{r}))\, d\mathbf{r} - \frac{1}{2}\int K(\mathbf{r}-\mathbf{r}'; n(\bar{\mathbf{r}})) \times \{n(\mathbf{r}) - n(\mathbf{r}')\}^2\, d\mathbf{r}\, d\mathbf{r}', \quad (A2.1)$$

where $K(\mathbf{r}-\mathbf{r}'; n)$ is determined by the polarizability of a homogeneous electron gas at density $n$, and $\bar{\mathbf{r}} = \frac{1}{2}(\mathbf{r}+\mathbf{r}')$. To the same approximation,

$$E_{xc}[n] = \int n(\mathbf{r})\epsilon_{xc}(n(\mathbf{r}))\, d\mathbf{r} - \frac{1}{2}\int K_{xc}(\mathbf{r}-\mathbf{r}'; n(\bar{\mathbf{r}})) \times \{n(\mathbf{r}) - n(\mathbf{r}')\}^2\, d\mathbf{r}\, d\mathbf{r}' \quad (A2.2)$$

where $K_{xc}$ is the difference between $K$ of the interacting homogeneous gas and that of the noninteracting gas at the same density. We believe that for an infinite system, such as a metal or an alloy, the second term on the right-hand side of (A2.2) accounts adequately for the effect of rapid density change on exchange and correlation.

This $E_{xc}[n]$ again leads to a set of Hartree-type equations like Eq. (2.8), with an addition to the effective potential given by

$$-\frac{1}{2}\int \{\partial K_{xc}(\mathbf{r}'; n(\mathbf{r}))/\partial n(\mathbf{r})\} \times \{n(\mathbf{r}+\tfrac{1}{2}\mathbf{r}') - n(\mathbf{r}-\tfrac{1}{2}\mathbf{r}')\}^2\, d\mathbf{r}' - 2\int K_{xc}(\mathbf{r}-\mathbf{r}'; n(\bar{\mathbf{r}}))\{n(\mathbf{r}) - n(\mathbf{r}')\}\, d\mathbf{r}'. \quad (A2.3)$$

Note that in the random-phase approximation $K_{xc}$ vanishes. Hence, in a calculation which includes the effective potential (A2.3), we need reliable estimates of $K_{xc}$, calculated beyond the random-phase approximation, which are not available at present.

The addition of (A2.3) to the effective potential obviously makes the solution of the self-consistent equations much more difficult. However, assuming that the modification of $n(\mathbf{r})$ produced by this term is small, one may calculate $n(\mathbf{r})$ and $E$ first without including it, and then, because of the stationary property, Eq. (2.5), one can obtain the correction to the energy by evaluating the second term in (A2.2) with the unmodified density.

*Note added in proof.* We should like to point out that it is possible, formally, to replace the many-electron problem by an *exactly* equivalent set of self-consistent one-electron equations. This is accomplished quite simply by using the expression (2.2) [without the approximation (2.3)] in the energy variational principle. This leads to a set of equations, analogous to Eqs. (2.4)–(2.9), but with $\mu_{xc}(n)$ replaced by an effective one-particle potential $v_{xc}$, defined formally as

$$v_{xc}(\mathbf{r}) \equiv \delta E_{xc}[n]/\delta n(\mathbf{r}).$$

Of course, an explicit form of $v_{xc}$ can be obtained only if the functional $E_{xc}[n]$, which includes all many-body effects, is known. This effective potential will reproduce the exact density and the exact total energy is then given by

$$E = \sum_1^N \epsilon_i - \frac{1}{2}\int\int \frac{n(\mathbf{r})n(\mathbf{r}')}{|\mathbf{r}-\mathbf{r}'|}\, d\mathbf{r}\, d\mathbf{r}' + E_{xc}[n] - \int v_{xc}(\mathbf{r})\, n(\mathbf{r})\, d\mathbf{r}.$$

Of course, if we make the approximation (2.3) for $E_{xc}$ the above exact formulation reverts to the approximate theory of Sec. II.

[11] The second term of HK, Eq. (83) is in error; it should be

$$-\frac{1}{2}\int K(\mathbf{r}'; n(\mathbf{r}))\{n(\mathbf{r}+\tfrac{1}{2}\mathbf{r}') - n(\mathbf{r}-\tfrac{1}{2}\mathbf{r}')\}^2\, d\mathbf{r}'.$$

The kernel $K$ has the same meaning as in HK.

III-4 Reprinted from *Physical Review* 144, 1 (1966)

# Application of a Self-Consistent Scheme Including Exchange and Correlation Effects to Atoms*

B. Y. Tong† and L. J. Sham
*University of California, San Diego, La Jolla, California*
(Received 3 November 1965)

Self-consistent schemes including approximations to exchange and correlation proposed by Kohn and Sham are applied to computing atomic energies and densities. These quantities, with and without the correlation correction, are obtained and compared with the results of calculations using the Slater exchange hole or the Hartree-Fock method and with experimental values. The present method, without correlation, gives slightly better results for energies and substantially better results for densities than Slater's method. This was anticipated in the general theory. The correlation corrections of the present scheme are not very good, presumably because the electronic density in atoms has too rapid a spatial variation.

## I. METHOD

A SELF-consistent-field method has been proposed by Kohn and Sham[1] which takes into account approximately exchange and correlation effects. We present here the calculation of some atomic properties using this scheme and compare the results with experimental measurements in the hope of throwing some light on the applicability of this self-consistent method to atoms. If, in this scheme, we take out terms corresponding to the correlation effect, we have an approximation to the Hartree-Fock method. Slater's exchange hole[2] presents another approximation. The difference between Slater's method and ours has been discussed in Ref. 1. Comparison of the numerical results of both methods with the Hartree-Fock calculation is presented here.

Our computation consists in solving the set of self-consistent equations (in atomic units):

$$\{-\tfrac{1}{2}\nabla^2+\varphi(\mathbf{r})+v_{xc}(\mathbf{r})\}\psi_i(\mathbf{r})=\epsilon_i\psi_i(\mathbf{r}), \qquad (1)$$

where the electrostatic potential for a nucleus with charge $Z$ is given by

$$\varphi(\mathbf{r})=-\frac{Z}{r}+\int\frac{n(\mathbf{r}')}{|\mathbf{r}-\mathbf{r}'|}d\mathbf{r}', \qquad (2)$$

the density for an atom with $N$ electrons is

$$n(\mathbf{r})=\sum_{i=1}^{N}|\psi_i(\mathbf{r})|^2, \qquad (3)$$

and $v_{xc}(\mathbf{r})$ represents the effects of exchange and correlation,[1]

$$v_{xc}(\mathbf{r})=\mu_x(n(\mathbf{r}))+\mu_c(n(\mathbf{r})). \qquad (4)$$

$\mu_x(n)$ and $\mu_c(n)$ are, respectively, the exchange and correlation contributions to the chemical potential of a system of interacting electrons with uniform density $n$. It is well known that

$$\mu_x(n)=-(3n/\pi)^{1/3}. \qquad (5)$$

For $\mu_c(n)$, we interpolate between the low-density formula of Wigner and the high-density one of Gell-Mann and Brueckner.[3] The results as a function of $r_s$ are shown in Fig. 1.

The total energy is given by

$$E=\sum_{i=1}^{N}\epsilon_i-\frac{1}{2}\int\int\frac{n(\mathbf{r})n(\mathbf{r}')}{|\mathbf{r}-\mathbf{r}'|}d\mathbf{r}d\mathbf{r}' +\int n(\mathbf{r})\{\epsilon_{xc}(n(\mathbf{r}))-v_{xc}(n(\mathbf{r}))\}d\mathbf{r}, \qquad (6)$$

* Supported by the U. S. Office of Naval Research.
† Permanent address: Department of Mathematics, University of Hong Kong, Hong Kong.
[1] W. Kohn and L. J. Sham, Phys. Rev. 140, A1133 (1965).
[2] J. C. Slater, Phys. Rev. 81, 385 (1951).
[3] D. Pines, *Elementary Excitations in Solids* (W. A. Benjamin, Inc., New York, 1963).

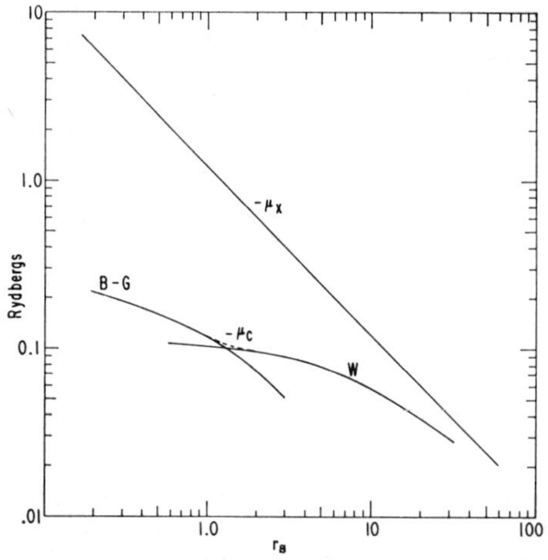


FIG. 1. Exchange and correlation contributions to the chemical potential of the homogeneous electron gas.

where $\epsilon_{xc}(n)$ is the exchange and correlation energy per electron of the system of interacting electrons with uniform density $n$:

$$\epsilon_{xc}(n)=\epsilon_x(n)+\epsilon_c(n)\,, \quad (7)$$

and

$$\epsilon_x(n)=-\tfrac{3}{4}(3n/\pi)^{1/3}, \quad (8)$$

$\epsilon_c(n)$ being obtained by an interpolation scheme as $\mu_c(n)$.

Results of computation by the scheme described above will be designated by KS-XC. If we want to drop the correlation effects, then in place of Eq. (4) we use

$$v_x(\mathbf{r})=\mu_x(n(\mathbf{r})) \quad (9)$$

and in Eq. (6), $\epsilon_{xc}$ and $v_{xc}$ have to be replaced by $\epsilon_x$ and $v_x$. This scheme will be denoted by KS-X. In the Slater approximation,

$$v_x(\mathbf{r})=2\epsilon_x(n(\mathbf{r}))=\tfrac{3}{2}\mu_x(n(\mathbf{r}))\,. \quad (10)$$

This scheme will be denoted by S.

TABLE I. Total energies (in rydbergs).

| Atoms | Expt | KS-XC | KS-X | S | HF |
|---|---|---|---|---|---|
| He | − 5.807 | − 5.651 | − 5.447 | − 5.404 | − 5.723356[b] |
| Ne | − 257.880 | − 256.349 | − 254.986 | − 254.764 | − 257.09396[b] |
| Ar | −1055.21 | −1051.709 | −1049.027 | −1048.702 | −1053.63410[b] |
| Kr | | −5499.947 | −5493.686 | −5492.976 | −5504.1086[b] |
| Li | − 14.956 | − 14.656 | − 14.349 | − 14.303 | − 14.865452[b] |
| Na | − 324.521 | − 322.768 | − 321.258 | − 321.041 | − 323.71714[b] |
| K | −1199.97 | −1196.215 | −1193.387 | −1193.061 | −1198.32894[b] |
| Rb | | −5872.321 | −5865.889 | −5865.185 | |
| $Rb^+$ | | −5872.002 | −5865.618 | −5864.930 | −5876.440[c] |
| O | − 150.069[a] | − 148.861 | − 147.851 | − 147.696 | − 149.5388[a] |

[a] References 6 and 7. Averaged over spin multiplet.
[b] Reference 6.
[c] R. E. Watson and A. J. Freeman, Phys. Rev. **124**, 1117 (1961).

TABLE II. Exchange energies (in rydbergs).

| Atoms | $E_x=\int\epsilon_x n d\mathbf{r}$ | $\Delta E_x=E_{HF}-E_{KS\text{-}X}$ | $\Delta E_x/(E_x+\Delta E_x)$ (%) |
|---|---|---|---|
| He | − 1.707 | −0.275 | 13.88 |
| Ne | −21.873 | −2.115 | 8.82 |
| Ar | −55.551 | −4.607 | 7.66 |
| Be | − 4.557 | −0.700 | 13.32 |
| O | −14.388 | −1.768 | 10.94 |
| Li | − 2.961 | −0.516 | 14.84 |
| Na | −25.362 | −2.464 | 8.86 |
| K | −60.222 | −4.942 | 7.58 |
| $Ne^+$ | −20.856 | −2.198 | 9.53 |
| $Ar^+$ | −54.849 | −4.607 | 7.75 |
| $Na^+$ | −25.182 | −2.419 | 8.76 |
| $K^+$ | −60.063 | −4.918 | 7.57 |

Herman and Skillman[4] have done extensive calculations using Slater's exchange and published the programs which they have used. We just modify their non-relativistic program for the appropriate $v_{xc}(\mathbf{r})$ or $v_x(\mathbf{r})$. It should be noted that we have *not* used the tail correction which Herman and Skillman proposed. It appears from independent calculations by Liberman et al.,[5] that in the case of KS-X, tail correction actually worsens agreement with the exact Hartree-Fock result for both total energy and density distribution.

## II. RESULTS AND DISCUSSIONS

In Table I we present the total energies of several atoms and ions. The column marked Expt actually represents results of experimental values minus the relativistic correction, both taken from Clementi's

TABLE III. Correlation energies (in rydbergs).

| Atoms | $-(E_{KS\text{-}XC}-E_{KS\text{-}X})$ | $-E_c$ (Clementi)[a] |
|---|---|---|
| He | 0.204 | 0.0842 |
| Ne | 1.363 | 0.786 |
| Ar | 2.682 | 1.582 |
| Kr | 6.261 | |
| Li | 0.307 | 0.0906 |
| Na | 1.510 | 0.806 |
| K | 2.828 | 1.64 |
| Rb | 6.432 | |
| O | 1.010 | 0.530[b] |
| $Ne^+$ | 1.257 | 0.656 |
| $Ar^+$ | 2.584 | |
| $Na^+$ | 1.446 | 0.792 |
| $K^+$ | 2.768 | |

[a] Reference 7. [b] Averaged over spin multiplet.

[4] F. Herman and S. Skillman, *Atomic Structure Calculations* (Prentice-Hall, Inc., Englewood Cliffs, New Jersey, 1963).
[5] R. D. Cowan, A. Larson, D. Liberman, J. B. Mann, and J. Waber, following paper, Phys. Rev. **144**, 5 (1966).

TABLE IV. Ionization potential (in rydbergs) of one electron from a neutral atom.

| Atoms | Experiment[a] | KS-XC | KS-X | Analytic Hartree–Fock[b] |
|---|---|---|---|---|
| He | 1.807 | 1.938 | 1.834 | |
| Ne | 1.5848 | 1.657 | 1.551 | 1.4605 |
| Ar | 1.1582 | 1.183 | 1.085 | 1.0855 |
| Kr | 1.0289 | 1.06 | 0.973 | 0.9754 |
| Li | 0.3962 | 0.396 | 0.331 | 0.3926 |
| Na | 0.3777 | 0.387 | 0.325 | 0.3636 |
| K | 0.3190 | 0.330 | 0.271 | 0.2947 |
| Rb | 0.3070 | 0.32 | 0.272 | |

[a] *American Institute of Physics Handbook*, edited by D. E. Gray (McGraw-Hill Book Company, Inc., New York, 1963), 2nd ed., pp. 7–14.
[b] Reference 6.

work.[6,7] The column HF denotes Hartree-Fock calculations with sources as indicated. Notice that the energies of KS-X are very close to, though slightly better than, those of those of S in comparison with HF. With some correlation effect taken into account, the energies of KS-XC are in fair agreement with experiment but are not as good as HF. Most of the error is due to our imperfect approximation of the exchange part present in KS-XC.

Let us denote by ES the electrostatic approximation in which $v_{xc}(\mathbf{r})$ is put equal to zero. This is not identical with the Hartree approximation except in the limit of large $N$, since $\varphi(\mathbf{r})$ in Eq. (2) includes the interaction of an orbital with itself. Let us define $E_{HF}-E_{ES}$ as the exchange energy. Then $E_{KS\text{-}X}-E_{ES}$ is our approximation to the exchange energy. This is approximately given by

$$E_{KS\text{-}X}-E_{ES}\approx\int n(\mathbf{r})\epsilon_x(n(\mathbf{r}))d\mathbf{r}, \tag{11}$$

where $n(\mathbf{r})$ is $n_{KS\text{-}X}(\mathbf{r})$. Table II shows that KS-X scheme approximates the exchange energy to within about 10%, better for larger atoms.

The correlation energy in KS-XC may roughly be taken as $E_{KS\text{-}XC}-E_{KS\text{-}X}$. Comparison with experiment is shown in Table III. Again, by experimental values,

TABLE V. Ionization of oxygen atoms (in rydbergs).

| Ionization process | Expt | KS-XC | SSM[a] | KS-X | HF[b] |
|---|---|---|---|---|---|
| $O \to O^+$ | 1.1646 | 1.215 | 1.009 | 1.115 | 1.0712 |
| $O^+ \to O^{2+}$ | 2.4401 | 2.469 | 2.581 | 2.361 | 2.3697 |
| $O^{2+} \to O^{3+}$ | 3.9522 | 3.955 | 4.037 | 3.843 | 3.9085 |
| $O^{3+} \to O^{4+}$ | 5.6878 | 5.644 | 5.691 | 5.530 | 5.6740 |
| $O^{4+} \to O^{5+}$ | 8.3721 | 8.344 | 8.372 | 8.233 | 8.1594 |
| $O^{5+} \to O^{6+}$ | 10.1519 | 10.004 | 10.155 | 9.898 | 10.1338 |
| $O^{6+} \to O^{7+}$ | 54.3419 | 54.647 | } 118.392 | 54.456 | 54.2222 |
| $O^{7+} \to O^{8+}$ | 64.0458 | 62.585 | } | 62.415 | 64.0000 |
| $E_{TOT}(O) =$ | −150.1564 | −148.861 | −150.237 | −147.851 | −149.5388 |

[a] Reference 8. [b] Reference 7.

[6] E. Clementi, C. C. J. Roothaan, and M. Yoshimine, Phys. Rev. **127**, 1618 (1962); E. Clementi, J. Chem. Phys. **38**, 996 (1963); **38**, 1001 (1963); **41**, 295 (1964); **41**, 303 (1964); IBM J. Res. Develop. **9**, 2 (1965).
[7] E. Clementi, J. Chem. Phys. **38**, 2248 (1963); **39**, 175 (1963); **42**, 2783 (1965); IBM J. Res. Develop. **9**, 2 (1965).

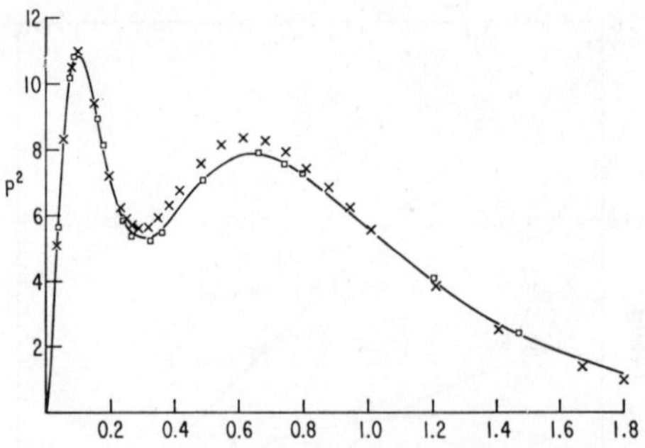


FIG. 2. Radial density of Ne atom. Solid line, KS-XC; □, HF from Ref. 10; ×, S.

we mean the real ones minus the estimates of relativistic corrections. Our approximation is about a factor of 2 too large for small atoms, but better for larger atoms.

The ionization energy of an atom can simply be obtained as the difference between total energies before and after ionization. Thus, the ionization energy on removing one electron from a neutral atom is $E(N)$-$E(N$-1). Results for several atoms are presented in Table IV. Here no attempt is made to remove estimates of relativistic effects from the experimental measurements since such effects in total energies change little on ionizing an atom.[6,8] Our approximation for the ionization energy appears to be in good shape. In Table V, we present the calculated ionization energies as we strip the oxygen atom of its electrons one by one. The experimental values[9] quoted are averaged over various spin multiplets. The column SSM represents semi-empirical results including relativistic effects.[8] That our approximation in Eqs. (4) or (9) is worse for smaller number of electrons is borne out in Table V.

TABLE VI. Expectation values $\langle r^2\rangle$ of $r^2$.

| Atoms | Experiment[a] | KS-XC | KS-X | S | HF |
|---|---|---|---|---|---|
| He | 2.4 | 2.597 | | | 2.386[b] |
| Ne | 8.4 – 9.8 | 9.893 | 10.038 | | 9.3682[c] |
| Ar | 14.2 –24.9 | 26.460 | 26.810 | 23.307 | 26.03[d] |
| Kr | 34.8 –36.9 | 39.847 | 40.331 | 35.570 | 39.530[e] |
| $Li^+$ | 0.88– 5.5 | 0.938 | | 0.831 | 0.8924[b] |
| $Na^+$ | 6.3 –13.1 | 6.615 | 6.674 | 5.914 | 6.4130[c] |
| $K^+$ | 17.2 –23.4 | 19.658 | 19.825 | 17.806 | 19.5358[c] |
| $Rb^+$ | 27.2 –39.5 | 31.853 | 32.101 | 29.117 | |
| $Ar^+$ | | 20.272 | 20.459 | 18.278 | 20.1742[c] |
| $Kr^+$ | | 31.995 | 32.266 | 29.139 | |

[a] Reference 15. Diamagnetic susceptibility $\chi=c\langle r^2\rangle$, where $c=7.927\times10^{-7}$ when $\chi$ is measured in ($cm^3\ mol^{-1}$).
[b] C. L. Pekeris, Phys. Rev. **115**, 1216 (1959); **126**, 143 (1962); **126**, 1470 (1962).
[c] P. S. Bagus, Phys. Rev. **139**, A619 (1965).
[d] Reference 13.
[e] *Added in proof.* J. B. Mann (private communication).

[8] C. W. Scherr, J. N. Silverman, and F. A. Matsen, Phys. Rev. **127**, 830 (1962).
[9] J. C. Slater, *Quantum Theory of Atomic Structure* (McGraw-Hill Book Company, Inc., New York, 1960), Vol. 1, Table 15–15.

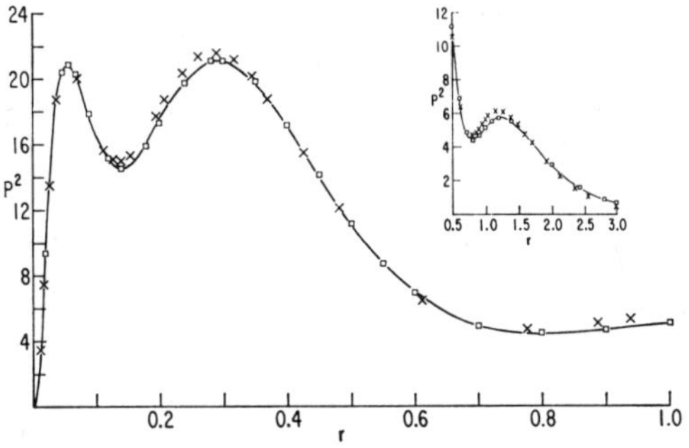

FIG. 3. Radial density of Ar atom. Solid line, KS-XC; □, HF from Ref. 12; ×, S.

In Figs. 2–4 are plotted the radial densities of Ne, Ar, and Kr; i.e.,

$$P^2(r) = 4\pi r^2 n(r). \qquad (12)$$

The densities calculated from the two schemes KS-XC and KS-X are so close as to be indistinguishable in the figures. In other words, our correlation term does not affect the density much. Our Ne curve agrees well with the HF calculation by Worsley[10] which is also presented in Fig. 2. Both depart somewhat from an earlier HF calculation by Brown.[11] Our density for Ar agrees well with HF.[12] Since our approximation should get better for larger atoms, our good density results for Ne and Ar encourage us to expect that our Kr density should be better still. However, our density for Kr does not

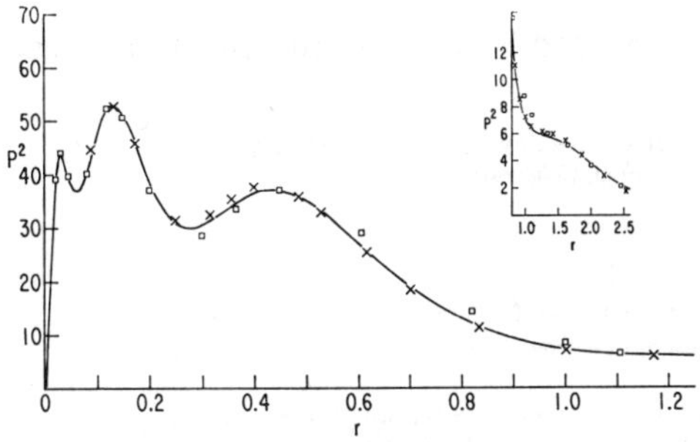

FIG. 4. Radial density of Kr atom. Solid line, KS-XC; □, HF from Ref. 13; ×, S.

[10] B. H. Worsley, Can. J. Phys. **36**, 28 (1958).
[11] F. W. Brown, Phys. Rev. **44**, 214 (1933).
[12] D. R. Hartree and W. Hartree, Proc. Roy. Soc. (London) **A166**, 450 (1938).

have the same good agreement with the HF calculation by Worsley.[13] The small discrepancies show that either our expectation was not met or Worsley's results are not sufficiently self-consistent because of the large number of orbitals involved in the calculation.[14] Density distributions by the Slater scheme are peaked more near the nucleus as it should be since the Slater potential is more attractive. Also the tails fall more rapidly. Herman and Skillman's modification[4] nearly preserves the Slater curve with a minute flattening of the tail.

A good measure of the shape of the density distribution is $\langle r^2 \rangle$. This quantity is also proportional to the experimentally measurable diamagnetic susceptibility. Values in Table VI (see Ref. 15) confirm our conclusions in the preceding paragraph that (i) there is little difference between the densities from KS-XC or KS-X, (ii) there is good agreement between KS-X and HF, and (iii) $\langle r^2 \rangle$ from S is smaller than that calculated by KS-X. Unfortunately, the uncertainties in experimental values of $\langle r^2 \rangle$ are too great to permit any definite conclusion in comparing various calculations with experiment.

In conclusion, we can say that our computation is simple compared with Hartree-Fock and configuration-interaction schemes and that our results are quite good, especially for large atoms. Our exchange approximation is satisfactory. Comparison of numerical results of KS-X and S confirms the inference about these two schemes drawn in Ref. 1, namely, that the total energies are close but that our density should be a distinct improvement over that obtained by the Slater scheme. Our approximation to the correlation energy is poor. We are trying to understand precisely why this error is fractionally so much larger than that of the exchange energy.

**ACKNOWLEDGMENTS**

We are indebted to Professor W. Kohn for constant guidance. We thank Dr. D. A. Liberman and Dr. R. Cowan for sending their results on Kr for the KS-X scheme for checking with our own results.

[13] B. H. Worsley, Proc. Roy. Soc. (London) **A247**, 390 (1958).
[14] *Note added in proof.* Dr. J. B. Mann has since performed the Hartree-Fock calculation for the density of Kr whose agreement with our result is much improved as compared to Worsley's and is as good as the agreement between HF and KS-X for Ne. We are grateful to Dr. J. B. Mann for sending us his results.
[15] Landolt-Börnstein, *Zahlenwerte und Funktionen aus Physik, Chemie, Astronomie, Geophysik, und Technik* (Springer-Verlag, Berlin, 1950), 6 Auflage, Band 1.

III-5 Reprinted from *Physical Review* **140**, A1508 (1965)

# Wave Functions and Transition Probabilities in Scaled Thomas-Fermi Ion Potentials*

JOHN C. STEWART†
*General Atomic Division, General Dynamics Corporation, John Jay Hopkins Laboratory for Pure and Applied Science, San Diego, California*

AND

MANUEL ROTENBERG‡
*University of California at San Diego, La Jolla, California*

(Received 5 April 1965; revised manuscript received 26 July 1965)

A semiempirical method for computing atomic and ionic wave functions is described. The method uses a one-electron potential obtained from a scaled Thomas-Fermi ionic charge distribution. The scale factors are chosen to make the one-electron binding energies agree with experimental ionization potentials. A number of oscillator strengths have been evaluated and are found to agree well with results of alternative methods of calculation. The method has a wider range of applicability than that of Bates and Damgaard, and calculations are much shorter than those using the self-consistent-field methods.

## I. INTRODUCTION

WE consider in this paper an approximate and rapid method for obtaining radial wave functions in many-electron atoms and ions. Our main goal in this work is the calculation of electric-dipole transition probabilities ("oscillator strengths"), although other applications (collision cross sections, spin-orbit parameters, Slater integrals, etc.) suggest themselves. In applications such as opacity calculations, which require many hundreds of wave functions for transition probabilities, speed of computation becomes an important consideration. Our particular concern, therefore, is to evaluate wave functions via some method which circumvents the lengthy "self-consistent-field"

* This work was supported primarily by General Atomic and partly by the Advanced Research Projects Agency under Contract No. DA-31-124-ARO-D-257 with the University of California.
† Visiting Fellow, 1965–66, Joint Institute for Laboratory Astrophysics, Boulder, Colorado.
‡ Consultant to General Atomic.

(SCF) methods, since SCF rapidly gets lengthier as the number of occupied electron orbitals increases, and it becomes prohibitively lengthy if one contemplates evaluating many oscillator strengths in many stages of ionization of many elements. Current methods of computing oscillator strengths are discussed briefly in Sec. II.

Our method may be regarded as a hybrid of the semiempirical method of Bates and Damgaard[1] and the nonempirical method of Thomas[2] and Fermi.[3] The basic approximation we make is that any one-electron wave function can be represented adequately by a solution of the one-electron Schrödinger equation in which the potential is that produced by a Thomas-Fermi ion with the correct nuclear charge, the correct number of electrons, and some value (*not* in general the Thomas-Fermi value) of the mean radius of the charge distribution. The Thomas-Fermi ion, in other words, is uniformly and isotropically dilated or contracted to force some parameter of the radial wave function to have a prescribed value. In the present work, this parameter is the energy parameter in the one-electron wave equation, which we take equal to the experimental ionization potential of the electron in question; other choices are, of course, possible. With this choice the oscillator strength of any transition can be evaluated from a knowledge of the initial and final energy levels referred to the energy of the parent ion, as in the Bates-Damgaard method; unlike their method, however, our method does not neglect the part of the dipole matrix element arising from small radial distances and so is not restricted to initial and final states in which the active electron lies mostly outside the passive electrons. The price paid for this wider range of applicability is a brief machine calculation instead of a hand calculation; the machine calculation, however, is at least an order of magnitude faster than any of the SCF methods known to us. Our method is described in Sec. III.

Our particular choice of potential is based on a number of empirical considerations, besides the obvious one of ease of handling: good numerical fit with potentials obtained from the SCF methods, the resulting close agreement of radial wave functions, and, ultimately, reasonable agreement with the dipole matrix elements obtained by other methods. The comparison with other theoretical and experimental results is exhibited and discussed in Sec. IV. Conclusions and future work are summarized briefly in Sec. V.

## II. CURRENT METHODS

Methods for obtaining wave functions of many-electron atoms and ions may be divided into two broad classes: first-principle methods and semiempirical methods. The former class, of which the Hartree,[4] Hartree-Fock,[5] Hartree-Fock-Slater,[6] and "charge-expansion"[7] methods are examples, attempts to find approximate solutions to the many-electron Schrödinger equation which satisfy a criterion of variational optimization within a restricted class of trial functions chosen in advance. The quantity which is optimized is always the total energy of the atom or ion. Although the total energy is usually given quite accurately by the variational methods, it is not ordinarily the quantity which one wishes to evaluate. One usually wants the ionization potential of an outer electron, a transition probability (square of the off-diagonal matrix element of an operator such as the electric-dipole moment), etc. These quantities are usually much more sensitive to the behavior of the wave function in the outer parts of the atom or ion than is the total energy, so that, as has been pointed out many times, approximate wave functions which give excellent total energies need not give even fair transition probabilities.

In contrast to first-principle methods, semiempirical methods require that the approximate wave functions satisfy some constraints taken from experiment, such as one-electron energy levels or ionization potentials. Among the semiempirical methods, the "Coulomb approximation" of Bates and Damgaard[1] (which has been extended to photoelectric transitions by Burgess and Seaton[8]) leads to a relatively simple computational scheme and has been widely applied with considerable success,[9] often yielding oscillator strengths which agree with experiment better than the self-consistent-field values do. The Bates-Damgaard (BD) method utilizes the experimental binding energy of the active electron to construct an approximate radial wave function which is accurate at large radial distances: a Coulomb wave function with the right behavior at infinity and the right energy. The length form of the electric-dipole matrix element, which emphasizes large radial distances, is then evaluated using two such wave functions. The resulting transition probability depends on the initial and final energies and angular momenta of the active electron, but not on the details of the atomic potential at small radii.

The basic postulate of the BD method is that nearly all of the dipole-length matrix element is contributed by the wave function at radii where the active electron is outside the "core" (the charge distribution of the rest of the electrons and the nucleus); the contribution

[1] D. R. Bates and A. Damgaard, Phil. Trans. Roy. Soc. London **A242**, 101 (1949).
[2] L. H. Thomas, Proc. Cambridge Phil. Soc. **23**, 542 (1927).
[3] E. Fermi, Atti Accad. Nazl. Lincei, Rend., Classe sci. fis., mat. e nat. **6**, 602 (1927); Z. Physik **48**, 73 (1928).
[4] D. R. Hartree, Proc. Cambridge Phil. Soc. **24**, 89 (1928).
[5] See, for example, D. R. Hartree, *The Calculation of Atomic Structures* (John Wiley & Sons, Inc., New York, 1957).
[6] J. C. Slater, Phys. Rev. **81**, 385 (1951); J. C. Slater, *Quantum Theory of Atomic Structure* (McGraw-Hill Book Company, Inc., New York, 1960), Vols. I and II; see also Ref. 31.
[7] D. Layzer, Ann. Phys. (N. Y.) **8**, 271 (1959); C. M. Varsavsky, Astrophys. J., Suppl. Ser. 6, No. 53, 75 (1961); A. Naqvi, J. Quant. Spectry. Radiative Transfer 4, 597 (1964).
[8] A. Burgess and M. J. Seaton, Monthly Notices Roy. Astron. Soc. **120**, 121 (1960).
[9] H. R. Griem, *Plasma Spectroscopy* (McGraw-Hill Book Company, Inc., New York, 1964), Chap. 3.

from smaller radii is either neglected (by using a cutoff radius) or severely approximated (by introducing a modifying factor into the approximate wave function, which prevents the radial integral from diverging at the origin). The postulate is violated, and the BD method suspect, in two cases: (a) when the active electron is in an inner orbital with very small amplitude outside the core or (b) when the outer contribution to the matrix element is very small because of cancellation so that the value of the inner contribution is important. The method described in the next section is designed to remain accurate even when the postulate of the BD method is violated.

## III. THE SCALED THOMAS-FERMI METHOD

The derivation of the Thomas-Fermi (TF) statistical model for atoms and ions has been treated extensively by Gombás[10] and will not be repeated here. Several authors have evaluated one-electron wave functions and energy levels in the TF potential (or approximations to it); the most comprehensive set of results is due to Latter,[11] who obtained one-electron energies for neutral atoms distributed through the periodic table. The potential experienced by the electron whose wave function is sought (the "valence" electron) is produced by the nucleus and the rest of the electrons (the "core") and reduces at large radii to the Coulomb potential of the core charge. Thus, an approximate potential for all radii is obtained if we replace the core by a TF ion whose net charge is the core charge and whose nucleus has the actual nuclear charge. (Latter used, instead, a neutral TF atom and a somewhat unphysical charge shell to get the asymptotic Coulomb potential.) The one-electron radial wave equation for an electron of angular momentum $l$ and principal quantum number $n$ is then

$$d^2P_{nl}/dr^2+[2Z\psi(r/\mu)/r+E_{nl}-l(l+1)/r^2]P_{nl}(r)=0, \quad (1)$$

where $r$ is in Bohr radii, E is in Rydbergs, and

$$\psi(x)=\varphi(x)+qx/x_0 \quad (x\leq x_0)$$
$$=q \quad (x\geq x_0);$$

$\varphi$ is the TF function, which satisfies

$$d^2\varphi/dx^2=\varphi^{3/2}x^{-1/2};$$
$$\varphi(0)=1;$$
$$\varphi(x_0)=0;$$
$$q=-x_0 d\varphi/dx|_{x_0};$$
$$x=r/\mu; \quad (2)$$
$$\mu=0.8853Z^{-1/3};$$
$$q=z/Z;$$
$$z=\text{core charge};$$
$$Z=\text{atomic number}.$$

[10] P. Gombás, *Die Statistische Theorie des Atoms* (Springer-Verlag, Vienna, 1949).
[11] R. Latter, Phys. Rev. **99**, 510 (1955).

The radial wave function $P(r)$ satisfies the usual boundary and normalization conditions:

$$P_{nl}(0)=0;$$
$$P_{nl}(\infty)=0 \text{ and } \int_0^\infty P_{nl}^2dr=1, \text{ if } E<0; \quad (3)$$
$$P(r) \underset{r\to\infty}{\longrightarrow} \cos(kr+\delta(r)), \quad \text{if } E=k^2>0.$$

For bound states, $n$ is defined as $l+1+N$, where $N$ is the number of nodes excluding those at zero and infinity.

If the bound-state energies $E_{nl}$ are determined as eigenvalues of Eq. (1), they do not of course coincide with experimental values. We can render the model semiempirical if we replace $\mu$ by $\alpha_{nl}\mu$, where the scale factor $\alpha_{nl}$ plays the role of an eigenvalue to be determined by setting $E_{nl}$ equal to the experimental one-electron binding energy. The introduction of $\alpha_{nl}$ corresponds to a uniform dilatation or contraction of the TF core in an effort to compensate for all the omissions and approximations of the TF model (exchange, correlation, relativity, etc.), and its justification is primarily the quality of the resulting wave functions and oscillator strengths. We note in passing that radial scale factors in the TF model have been employed by Fock[12] in connection with a virial theorem and by March[13] as a variational parameter in an approximate correction for exchange.

With $\alpha_{nl}$ determined as above, the parallel between our method (which we will call the "scaled TF" or STF method) and the BD method is very close—at radii outside the core radius ($r>\alpha\mu x_0$), the STF wave function is, apart from normalization, identical with the BD wave function.

We have found that $\alpha_{nl}$ is a very weak function of the quantum number $n$ (see Table II). It is therefore a simple matter to extrapolate $\alpha_{nl}$ to a value to use in the Schrödinger equation for unknown levels or continuum wave functions. Hence the method is also readily applicable to the calculation of photoionization cross sections, including those for x rays.

With the radial wave functions obtained as above, the dipole absorption oscillator strength for a transition from $nl$ to $n'l'$ is given by the standard relations

$$gf=\tfrac{1}{3}(E'-E)\mathcal{S}(\mathfrak{M})\mathcal{S}(\mathfrak{L})\sigma^2,$$

$$\sigma^2=\frac{1}{4l_>^2-1}\left|\int_0^\infty P_{nl}P_{n'l'}rdr\right|^2,$$

where $l_>=\max(l,l')$, and the angular factors $\mathcal{S}(\mathfrak{M})$ and $\mathcal{S}(\mathfrak{L})$ can be obtained from tables computed by Goldberg[14] and reproduced by Allen,[15] or from formulas of

[12] V. Fock, Physik Z. Sowjetunion **1**, 747 (1932).
[13] N. H. March, Advan. Phys. **6**, 1 (1957).
[14] L. Goldberg, Astrophys. J. **82**, 1 (1935); **84**, 11 (1936).
[15] C. W. Allen, *Astrophysical Quantities* (The Athlone Press, University of London, London, 1963), 2nd edition.

TABLE I. Oscillator strengths for some atoms and ions.[a]

| $Z$ | Atom or ion | Transition | $gf_{STF}$ | $gf_{BD}$ | $gf_{Expt}$ | Other | $\alpha^b$ | $\alpha'$ |
|---|---|---|---|---|---|---|---|---|
| 3 | Li I | $2s$-$2p$ | 1.46 | 1.49 | ... | 1.49[c] 1.59[d] | 1.30 | 1.12 |
| 10 | Ne I | $(2p^5)2p$-$3s$ | 0.193 | ... | ... | 0.163[f] 0.188[e] | 0.995 | 1.027 |
| | | $2p$-$4s$ | 0.031 | ... | ... | 0.026[f] 0.029[e] | 0.995 | 1.010 |
| | | $2p$-$5s$ | 0.011 | ... | ... | 0.009[f] 0.008[e] | 0.995 | 1.007 |
| | | $2p$-$6s$ | 0.005 | ... | ... | 0.004[f] 0.003[e] | 0.995 | 1.006 |
| | | $2p$-$3d$ | 0.052 | ... | ... | 0.037[f] 0.036[e] | 0.995 | 1.199 |
| | | $2p$-$4d$ | 0.026 | ... | ... | 0.020[f] 0.025[e] | 0.995 | 1.146 |
| | | $2p$-$5d$ | 0.015 | ... | ... | 0.011[f] 0.018[e] | 0.995 | 1.154 |
| | | $2p$-$6d$ | 0.009 | ... | ... | 0.007[f] 0.012[e] | 0.995 | 1.159 |
| 11 | Na I | $3s$-$3p$ | 1.95 | 1.87 | ... | 1.94[c] | 0.95 | 0.97 |
| 19 | K I | $4s$-$4p$ | 2.10 | 1.98 | 1.98[d] 2.1[g] | 2.08[c] | $1.13_5$ | $1.13_5$ |
| 20 | Ca I | $3d4s(^3D_3)$-$3d4p(^3F_4)$ | 2.70 | 2.45 | 1.1[d] 2.0[g] | ... | 1.19 | 1.24 |
| 21 | Sc I | $3d^24s(^4F_{9/2})$-$3d^24p(^4G_{11/2})$ | 3.58 | 3.20 | 2.6[d] 2.3[g] | ... | 1.17 | 1.21 |
| 22 | Ti I | $3d^34s(^5F_5)$-$3d^34p(^5G_6)$ | 3.94 | 3.46 | 3.8[d] 3.7[g] | ... | 1.14 | 1.18 |
| 23 | V I | $3d^44s(^6D_{9/2})$-$3d^44p(^6F_{11/2})$ | 3.80 | 3.33 | 4.0[d] 3.0[g] | ... | 1.11 | 1.13 |
| 24 | Cr I | $3d^54s(^7S_3)$-$3d^54p(^7P_4)$ | 2.73 | 2.35 | 0.6[d] 0.5[g] | ... | 1.08 | 1.11 |
| 25 | Mn I | $3d^64s(^6D_{9/2})$-$3d^64p(^6F_{11/2})$ | 3.76 | 3.24 | 5.3[d] 9.3[g] | ... | 1.05 | 1.07 |
| 26 | Fe I | $3d^74s(^5F_5)$-$3d^74p(^5G_6)$ | 4.13 | 3.56 | 4.1[d] 4.4[g] | ... | 1.02 | 1.03 |
| 27 | Co I | $3d^84s(^4F_{9/2})$-$3d^84p(^4G_{11/2})$ | 3.78 | 3.29 | 3.2[d] 4.6[g] | ... | $0.98_5$ | 1.00 |
| 28 | Ni I | $3d^94s(^3D_3)$-$3d^94p(^3F_4)$ | 2.68 | 2.31 | 0.8[d] 1.0[g] | ... | 0.96 | 0.99 |
| 29 | Cu I | $3d^{10}4s(^2S_{1/2})$-$3d^{10}4p(^2P_{3/2})$ | 1.24 | 1.08 | 0.8[d] 0.64[g] | ... | 0.93 | 0.95 |
| 55 | Cs I | $6s$-$6p_{1/2}$ | 0.753 | 0.697 | 2.38[h] | 2.26[c] 0.788[i] | 1.14 | 1.135 |
| | | $6s$-$6p_{3/2}$ | 1.56 | 1.50 | | 1.63[i] | 1.14 | 1.119 |
| | | $6s$-$7p_{1/2}$ | 0.012 | 0.013 | 0.032[h] | 0.054[c] 0.0057[i] | 1.14 | 1.127 |
| | | $6s$-$7p_{3/2}$ | 0.040 | 0.034 | | 0.0348[i] | 1.14 | 1.111 |
| | | $6s$-$8p_{1/2}$ | 0.00233 | 0.00259 | 0.00616[h] | 0.013[c] 0.00063[i] | 1.14 | 1.125 |
| | | $6s$-$8p_{3/2}$ | 0.00933 | 0.00982 | | 0.00698[i] | 1.14 | 1.109 |
| 80 | Hg I | $6p(^3P)$-$7s(^3S)$ | 1.41 | 0.77 | 0.7[d] 1.46[j] | | 1.00 | 1.04 |
| | | $6p(^3P_1)$-$6d(^3D_1)$ } $6p(^3P_1)$-$6d(^3D_2)$ | 1.46 | 1.15 | 0.9[j] | | 1.00 | 1.03 |

[a] When only one strength is given for a doublet, it is the combined value.
[b] $\alpha$ and $\alpha'$ are the scale factors for the lower and upper states, respectively.
[c] Anderson and Zilitis, Ref. 18.
[d] Allen, Ref. 15.
[e] Kelly, Ref. 19.
[f] Cooper, Ref. 20.
[g] Corliss and Bozman, Ref. 21.
[h] Kvater and Meister, Ref. 22.
[i] Stone, Ref. 23.
[j] Penkin, Ref. 24.

Rohrlich[16] or Kelly.[17] The above expression refers to $gf$ for a spectral line; $gf$ values for multiplets or transition arrays are sums over the $gf$'s of the constituent lines.

For photoionization transitions, the upper-state wave function belongs to the continuum; the cross section per atom in the lower state, in units of $\pi\times$(Bohr radius)$^2$, is

$$A=\frac{4}{3}\frac{1}{137}(E'-E)\sigma^2\frac{1}{kg}\sum_{L'S'}\mathcal{S}(\mathfrak{M}),$$

where $k^2=E'=$kinetic energy of photoelectron, in Rydbergs; $g$ is the statistical weight of the lower state; and $\sigma^2$ is as before but with $P_{n'l'}$ replaced by a continuum wave function normalized to asymptotic amplitude unity as in Eq. (3). The above expressions (and all the results of the present paper) neglect configuration mixing, departures from $LS$ coupling, and rearrangement of the core electrons.

## IV. RESULTS AND DISCUSSION

In order to make a preliminary assessment of the STF method, we have used it to compute a number of oscillator strengths in various atoms and ions. Our choice of test cases is intended to be representative rather than comprehensive; we have included examples of heavy and light, simple and complex atomic systems. The examples are chosen, in part, for availability of comparison data. In Table I (which cites Refs. 18–24) we have tabulated some values, for neutral atoms, of the quantity $gf$, where $g$ is the statistical weight of the lower state and $f$ is the absorption oscillator strength. Some of the transitions treated are single lines, others are multiplets, and still others are transition arrays, as indicated by the presence or absence of the quantum numbers $L$, $S$, and $J$. In all cases, the $gf$ values we give are sums over quantum numbers which do not appear explicitly. For each transition, we give in the STF column, our value of $gf$; in the next column, the BD value; in the "Expt" column, experimental values; in the "Other" column, values calculated by other

[16] F. Rohrlich, Astrophys. J. **129**, 441 (1959); **129**, 449 (1959).
[17] P. S. Kelly, Astrophys. J. **140**, 1247 (1964).

[18] E. M. Anderson and V. A. Zilitis, Opt. i Spectroskopiya **16**, 382 (1964); **16**, 177 (1964) [English transl.: Opt. Spectry. (USSR) **16**, 211 (1964); **16**, 99 (1964)].
[19] P. S. Kelly, J. Quant. Spectry. Radiative Transfer **4**, 117 (1964).
[20] J. W. Cooper, Phys. Rev. **128**, 681 (1962).
[21] C. H. Corliss and W. R. Bozman, Natl. Bur. Std. (U.S.), Monograph 53 (1962).
[22] G. S. Kvater and T. G. Meister, Vestn. Leningr. Univ., Ser. Fiz. i Khim. **9**, 137 (1952), as referred to in Ref. 18.
[23] P. M. Stone, Phys. Rev. **127**, 1151 (1962).
[24] N. D. Penkin, J. Quant. Spectry. Radiative Transfer **4**, 41 (1964).

semiempirical methods or by the Hartree-Fock-Slater method; and in the last two columns, the values of the scale factors $\alpha$ and $\alpha'$, corresponding respectively to the lower and upper states of the transition. Each scale factor was determined, as described in Sec. III, by setting the energy parameter in the one-electron radial wave equation equal to the experimental ionization potential[25] of the electron which makes the transition (referred to the appropriate parent term); for transition arrays, the ionization potential was taken as a weighted mean over $L$ and $S$. The same energy parameters were used in the BD calculations shown here.

We consider first the results for the alkali metals (Li, Na, K, Cs) in Table I. The strengths of the resonance lines ($ns \rightarrow np$) of these elements agree closely with the results of other methods; this is hardly surprising since the large oscillator strengths of these lines are rather insensitive to the details of the wave functions. On the other hand, strengths of the higher lines in the principal series ($ns \rightarrow n'p$) are extremely sensitive to the wave functions because of strong cancellation in the radial integrals. In particular, we note that in Cs I the computed oscillator strengths of $6s_{1/2}$-$7p_{3/2}$ and $6s_{1/2}$-$7p_{1/2}$ are in the ratio 3.3/1 even though the $7p_{3/2}$ and $7p_{1/2}$ binding energies differ by only 2%. If the radial integrals for these transitions were identical, the strengths would have the ratio 2/1; the departure from this value is observed experimentally.[26] The close agreement of our $gf$ values with those of BD indicates that the contributions from the wave functions inside the core are insignificant in CsI. In the region outside the core, the form of the wave function depends almost entirely on the one-electron binding energy. The radial integral, as we have seen, is very sensitive to changes of the order of 1% in the one-electron binding energy. This accuracy is not often attained by SCF methods.

In Hg, we see that our values differ appreciably from those of BD. This is undoubtedly because the states involved have binding energies large enough to nearly invalidate BD's own criterion that the effective principal quantum number should not be less than about $l+\frac{1}{2}$.

The data for the iron group ($Z=20$ through 29) illustrate the application of the STF method to complex atoms. Throughout this group three low-lying configurations of even parity ($3d^{x+1}$, $3d^x4s$, and $3d^{x-1}4s^2$) are present. Strong configuration mixing is to be expected,[27] and may produce large departures of the line strengths from those computed from any pure-configuration model. Therefore, it is somewhat surprising that the over-all agreement with experiment is as good as it is.

Neon is included as an example which cannot be

[25] Charlotte E. Moore, *Atomic Energy Levels*, Natl. Bur. Std. (U. S.), Circ. 467, Vol. I (1949), Vol. II (1952), Vol. III (1958).
[26] E. U. Condon and G. H. Shortley, *The Theory of Atomic Spectra* (Cambridge University Press, New York, 1959), p. 376.
[27] Reference 26, p. 330.

TABLE II. Selected oscillator strengths and scale factors in the sodium isoelectronic sequence.[a]

| $Z$ | Ion | Transition | $f_{STF}$ | $f_{BD}$ | $\alpha$ | $\alpha'$ |
|---|---|---|---|---|---|---|
| 11 | Na I | $3s$-$3p$ | 0.977 | 0.934 | 0.943 | 0.963 |
| | | $3s$-$4p$ | 0.151(−1) | 0.143(−1) | | 0.958 |
| | | $3s$-$5p$ | 0.236(−2) | 0.201(−2) | | 0.957 |
| 12 | Mg II | $3s$-$3p$ | 0.940 | 0.891 | 1.008 | 1.004 |
| | | $3s$-$4p$ | 0.230(−3) | 0.124(−3) | | 0.997 |
| | | $3s$-$5p$ | 0.100(−2) | 0.129(−2) | | 0.995 |
| 13 | Al III | $3s$-$3p$ | 0.866 | 0.826 | 1.043 | 1.027 |
| | | $3s$-$4p$ | 0.113(−1) | 0.105(−1) | | 1.019 |
| | | $3s$-$5p$ | 0.681(−2) | 0.746(−2) | | 1.017 |
| 14 | Si IV | $3s$-$3p$ | 0.796 | 0.758 | 1.067 | 1.042 |
| | | $3s$-$4p$ | 0.302(−1) | 0.286(−1) | | 1.033 |
| 17 | Cl VII | $3s$-$3p$ | 0.627 | 0.607 | 1.139 | 1.093 |
| | | $3s$-$4p$ | 0.102 | 0.091 | | 1.088 |
| 20 | Ca X | $3s$-$3p$ | 0.525 | 0.507 | 1.140 | 1.087 |
| | | $3s$-$4p$ | 0.145 | 0.143 | | 1.078 |
| 23 | V XIII | $3s$-$3p$ | 0.450 | 0.439 | 1.217 | 1.135 |
| | | $3s$-$4p$ | 0.195 | 0.182 | | 1.126 |
| 26 | Fe XVI | $3s$-$3p$ | 0.399 | 0.394 | 1.259 | 1.154 |
| | | $3s$-$4p$ | 0.226 | 0.210 | | 1.144 |
| 29 | Cu XIX | $3s$-$3p$ | 0.363 | 0.358 | 1.305 | 1.173 |

[a] Note: $\alpha$ and $\alpha'$ are the scale factors for the lower and upper states, respectively.

treated by the BD method because the effective principal quantum number for the $2p$ orbital is too small; here we obtain generally good agreement with Cooper's[20] simplified self-consistent-field results and with Kelly's[19] Hartree-Fock-Slater results.

In Table II we exhibit the behavior of a few STF oscillator strengths and scale factors along the sodium-like isoelectronic sequence, again based on experimental energy levels. Table II serves to illustrate the smooth behavior of $\alpha$ along such a sequence or along a Rydberg series, and hence the possibility of extrapolating $\alpha$'s to unknown levels. For comparison, the BD oscillator strengths are also shown.

Table III (citing Ref. 28) shows STF photoionization cross sections from the ground state of atomic oxygen. In these calculations the continuum radial wave functions were evaluated using values of $\alpha$ determined from bound states of the corresponding angular momenta. For comparison, cross sections computed from Hartree-Fock wave functions using the length and velocity forms of the dipole matrix elements[28] are quoted.

We have compared the $2p$ orbital of the ground state of oxygen obtained in the STF potential with orbitals obtained by the Hartree,[29] Hartree-Fock,[30] and Hartree-Fock-Slater[31] methods. When the $\alpha$ in the STF method was adjusted to give the same energy parameter as the HF method (1.26 Ry) the HF and STF wave functions

[28] A. Dalgarno, R. J. W. Henry, and A. L. Stewart, Planetary Space Sci. **12**, 235 (1964).
[29] D. R. Hartree and M. M. Black, Proc. Roy. Soc. (London) **A139**, 311 (1933).
[30] D. R. Hartree, W. Hartree, and B. Swirles, Phil. Trans. Roy. Soc. London **A238**, 229 (1939).
[31] F. Herman and S. Skillman, *Atomic Structure Calculations* (Prentice-Hall, Inc., Englewood Cliffs, New Jersey, 1963).

were found to agree within 5% in the tail ($r>4$ Bohr radii); at smaller radii, the agreement was considerably better. The maxima of both wave functions occurred at 0.85 Bohr radii.

In Table IV is a comparison of wave functions obtained by the various methods at one point in the tail; the ratios of the wave functions are typical for $4<r<8$. It is important to note that the energy parameter in each case is the one obtained by the respective method. This table is shown to emphasize how sensitive the asymptotic amplitudes of the wave functions are with respect to relatively small changes in the energy. By way of contrast, all these functions are within 5% of one another at 1 Bohr radius.

Since the TF model is a rather simplified picture of the atom, one may well ask to what extent its omissions and approximations can be compensated by introduction of the single parameter $\alpha$. Our use of the STF, rather than the unmodified TF, potential rests simply on the numerical observation that (in the cases we have examined) the STF gives a reasonably good fit to SCF potentials, and is usually significantly better than the TF. This observation is supported by comparisons of the resulting wave functions and transition probabilities with SCF results.

We consider first the "screening function" $\psi$ which appears in the radial wave equation (1). In the numerical SCF methods this function is replaced by a function $U(r)$ which is constructed from previously estimated wave functions. Specifically, for a configuration with $N_i$ electrons in the $i$th orbital $(n_il_i)$, the screening function $U_i$ in the $i$th radial wave equation is[6]

$$U_i(r)=1+\frac{1}{Z}\Bigg\{-\sum_j N_jY_0(jjr)+AY_0(iir) + B\frac{3r}{4}\left[\frac{6}{\pi^2r^2}\sum_j N_jP_j^2(r)\right]^{1/3} + C\Bigg[Y_0(iir)+\frac{N_i-1}{4l_i+1}\sum_{k\geq 2} c^k(ii)Y_k(iir) + \sum_{j\neq i}\frac{N_jP_j(r)}{P_i(r)}\left(\frac{\sum_k c^k(ij)Y_k(ijr)}{2[(2l_i+1)(2l_j+1)]^{1/2}} - (r/2)\delta_{l_il_j}\lambda_{ij}\right)\Bigg]\Bigg\},$$

where $(A,B,C)=(1,0,0)$ in the H (Hartree) method
$=(0,1,0)$ in the HFS (Hartree-Fock-Slater) method
$=(0,0,1)$ in the HF (Hartree-Fock, configuration average) method;

$$Y_k(ijr)=\int_0^r P_i(r')P_j(r')(r'/r)^k dr' + \int_r^\infty P_i(r')P_j(r')(r/r')^{k+1}dr';$$

$$c^k(ij)=c^k(l_i0;l_j0)=\tfrac{1}{2}[(2l_i+1)(2l_j+1)]^{1/2}C_{l_il_jk}; \qquad \text{(Ref. 32)}$$

$\lambda_{ij}$ = off-diagonal Lagrange multiplier

and, in each case, $-E_i$ is the computed ionization potential in the Koopmans ("unrelaxed-core") approximation. The quantity $(Z-\sum_j N_j+1)/Z$ is, of course, the (core charge/nuclear charge) ratio $q$.

The potentials corresponding to all of the above screening functions [and to the STF screening function $U_{\mathrm{STF}}(r)=\psi(r/\alpha\mu)$] reduce to the nuclear potential as $r\to 0$, i.e., $U(0)=1$. As $r\to\infty$, $U_{\mathrm{H}}$ and $U_{\mathrm{STF}}$ approach $q$, and $U_{\mathrm{HF}}$ does the same when the $i$th orbital is the outermost so that $P_j/P_i\to 0$. The HFS method, which uses an averaged exchange potential obtained from the *local* electron density (as in the Thomas-Fermi-Dirac model), fails to eliminate completely the self-interaction potential. Accordingly, $U_{\mathrm{HFS}}\neq 1$ for one-electron atoms, and in general $U_{\mathrm{HFS}}\to q-1/Z$ as $r\to\infty$. In recent extensive applications[31,19] the procedure adopted was to replace $U_{\mathrm{HFS}}$ by $q$ whenever $U_{\mathrm{HFS}}<q$; in the rest of this paper the designation HFS implies this replacement.

In the analytic Hartree-Fock (AHF) method,[17,33,34] the orbitals are not obtained as numerical solutions of wave equations with the screening functions $U_{\mathrm{HF}}$, but are restricted to prescribed analytic forms. The wave equation which such an orbital does satisfy can be obtained readily by differentiating the analytic orbital twice. The resulting screening function $U_{\mathrm{AHF}}$, may or

TABLE III. Photoelectric cross sections at various wavelengths from the ground state of atomic oxygen, in units of $10^{-18}$ cm$^2$.[a]

| λ(Å) | $\mathrm{HF}_L$ | $\mathrm{HF}_V$ | STF |
|---|---|---|---|
| 550 | 12.4 | 10.0 | 14.3 |
| 420 | 11.4 | 9.0 | 11.2 |
| 290 | 7.8 | 6.0 | 5.7 |
| 175 | 3.2 | 2.5 | 1.9 |

[a] Note: $\mathrm{HF}_L$ and $\mathrm{HF}_V$ refer to the use of length and velocity matrix elements, respectively, taken between Hartree-Fock wave functions as reported in Ref. 28.

TABLE IV. Comparison of $2p$ orbitals in the ground state of oxygen at $r=6$ obtained by various methods.

| Method | $-E_{2p}$(Ry) | $P_{2p}(r=6)$ |
|---|---|---|
| Hartree-Fock[c] | 1.26 | 0.011 |
| Scaled Thomas-Fermi | 1.165[a] | 0.013 |
| Hartree-Fock-Slater[d] | 1.04 | 0.016 |
| Hartree[b] | 0.94 | 0.025 |

[a] This energy parameter is the experimental ionization potential averaged over $L$ and $S$.
[b] See Ref. 29.
[c] See Ref. 30.
[d] See Ref. 31.

[32] See Ref. 26, p. 182.
[33] C. C. J. Roothaan and P. S. Bagus, *Methods in Computational Physics* (Academic Press Inc., New York, 1963), Vol. II.
[34] C. C. J. Roothaan and P. S. Kelly, Phys. Rev. **131**, 1177 (1963).

may not agree closely with the $U_{HF}$ constructed by integration from the same set of orbitals, depending on the flexibility of the trial functions employed and on the size of the wave function at the point where the comparison is made; the agreement tends to be worst at very small and very large radii where the wave function is small and has little weight in the total energy. Unfortunately these regions may contribute appreciably to transition probabilities.[17] The "charge-expansion" method[7] may be regarded as a rather highly restricted AHF scheme using scaled hydrogenic orbitals, and corresponds to $U_{AHF}=Z_i/Z$, where the effective charge $Z_i$ is a variationally determined constant for each orbital. This scheme naturally leads, in general, to rather poor wave functions except for inner orbitals or highly stripped ions.

It is illuminating to compare the various screening functions for a specific example. Figure 1 exhibits this comparison for the $2p$ orbital in the $1s^2\,2s^2\,2p$ configuration of O IV; we have used the AHF orbitals of Roothaan and Kelly[34] to construct $U_H$, $U_{HFS}$, and $U_{HF}$ by integration, and $U_{AHF}$ by differentiation. We also show $U_{STF}$ with $\hat{\alpha}=1.67$ (which yields the experimental $E_{2p}$) and with $\alpha=1$ (which yields the unmodified TF potential). Figure 2 shows the corresponding quantities for the $3d$ orbital in the $1s^2\,2s^2\,3d$ configuration, where Kelly's AHF orbitals[17] were used and $\alpha=1.96$. The reason for using the same wave functions throughout is to exhibit directly the HF exchange contribution $(U_{HF}-U_H)$ and its HFS approximation $(U_{HFS}-U_H)$.

Examining Figs. 1 and 2, we see that the STF and HFS potentials appear to be of comparable quality as approximations to the HF potential. Both decrease somewhat too steeply in the outer part of the ion. The Hartree potential, and especially the TF potential, are consistently smaller than the HF potential; a radial scaling of the Hartree potential would bring it quite close to the HF. The AHF potential is indistinguishable (in the graphs) from the HF, except at large radii for the $2p$ orbital and small radii for the $3d$ orbital.

The $2p$ and $3d$ orbitals are nodeless; for orbitals with nodes, the HF and AHF potentials in general have poles at the nodes of the orbital, and so cannot be fitted in detail by any smooth potential such as the STF or HFS. Nevertheless the resulting wave functions can be quite similar, as illustrated by the $4p$ orbitals for neutral krypton[31,35] shown in Fig. 3.

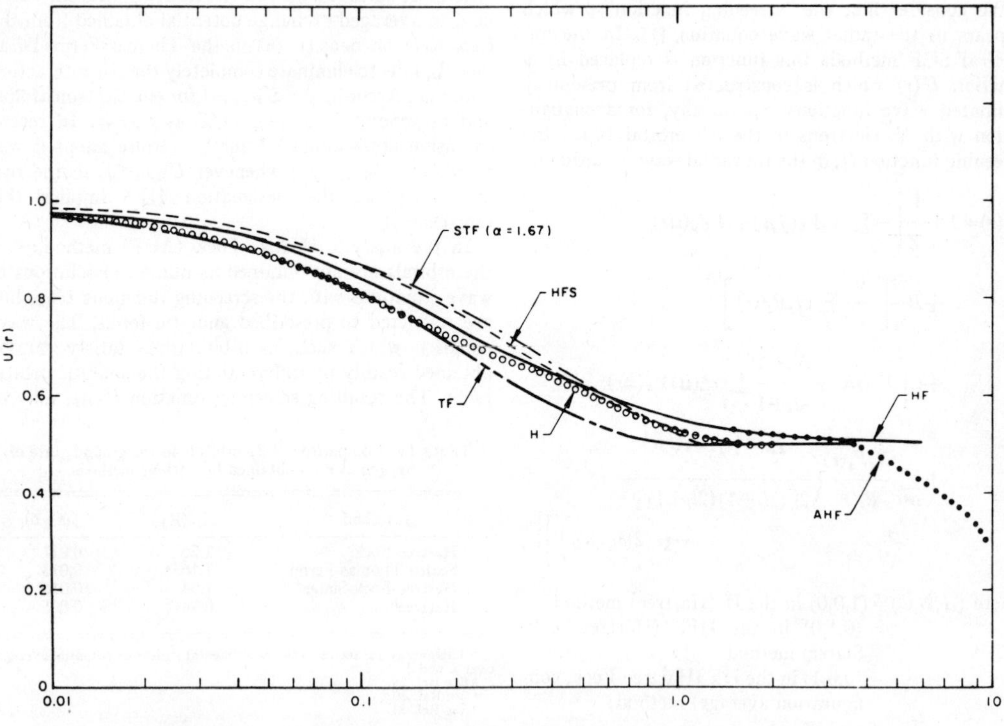


FIG. 1. Screening functions for the $2p$ orbital in the $1s^2\,2s^2\,2p$ configuration of O IV. Where the analytic Hartree-Fock (AHF) function (dotted curve) is not shown, it is indistinguishable from the Hartree-Fock (HF) function.

[35] B. H. Worsley, Proc. Roy. Soc. (London) A247, 390 (1958).

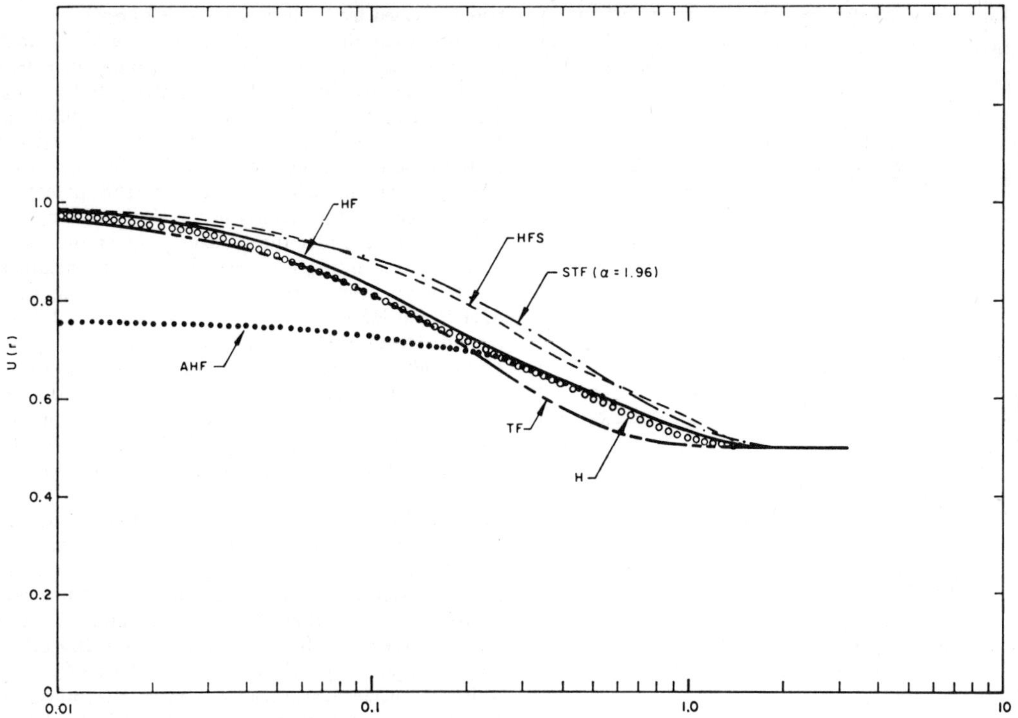


FIG. 2. Screening functions for the $3d$ orbital in the $1s^2\,2s^2\,3d$ configuration of O IV. Where the AHF function (dotted curve) is not shown, it is indistinguishable from the HF function.

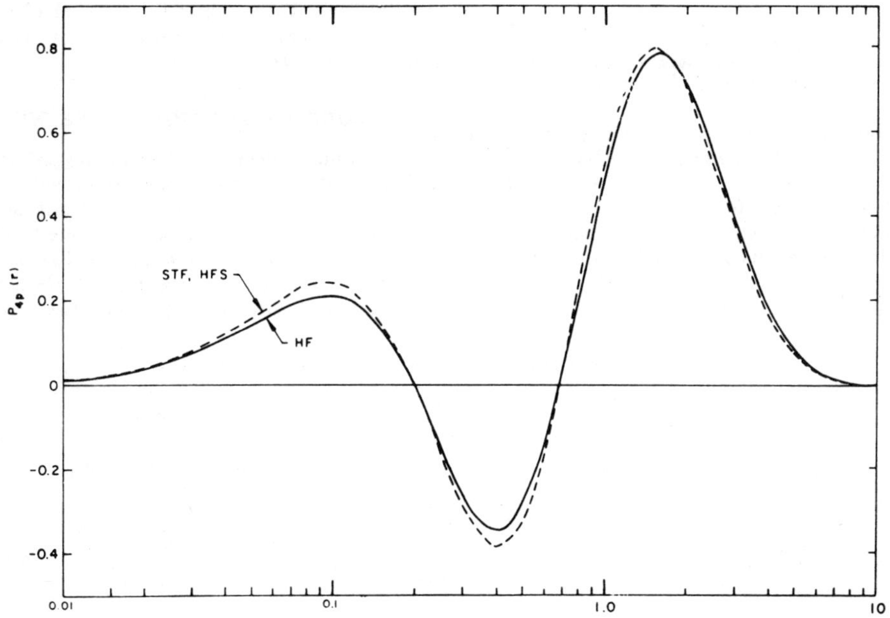


FIG. 3. $4p$ orbital in the ground state of neutral krypton ($Z=36$): Hartree-Fock (HF) by Worsley (Ref. 35), Hartree-Fock-Slater (HFS) by Herman and Skillman (Ref. 31), scaled Thomas-Fermi (STF) with $\alpha=1.02$. The difference between HFS and STF is too small to show on the graph.

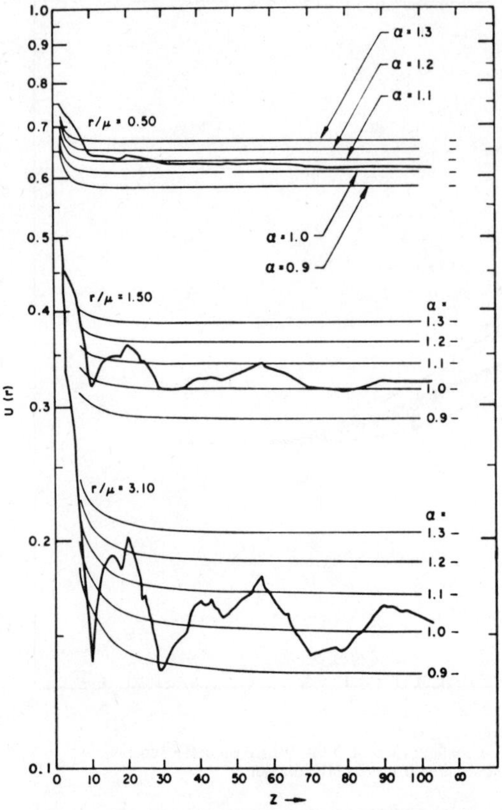


FIG. 4. Hartree-Fock-Slater (HFS) (Ref. 31) and scaled Thomas-Fermi (STF) screening functions for neutral atoms at $r/\mu = 0.50$, 1.50, 3.10. The heavy curves give the HFS values, and the light curves (labeled with values of $\alpha$) the STF values.

We consider next the comparison of $U_{\mathrm{STF}}$ and $U_{\mathrm{HFS}}$ for neutral atoms. Although no HF or AHF results are available, to our knowledge, beyond $Z=47$, the HFS potentials and orbitals for the ground states of all neutral atoms have been computed by Herman and Skillman.[31] Their results for $U_{\mathrm{HFS}}$ at three selected values of $r/\mu$ are plotted against $Z$ in Fig. 4, together with $U_{\mathrm{STF}}$ for several values of $\alpha$. (The neutral TF potential used by Latter corresponds to $\alpha=1$, $Z=\infty$.) The peaks and dips in $U_{\mathrm{HFS}}$ as a function of $Z$, which reflect the influence of shell structure, show the extent to which $\alpha$ must fluctuate if $U_{\mathrm{STF}}$ is forced to fit $U_{\mathrm{HFS}}$. The quality of fit obtained in a few cases is shown in Fig. 5. The STF potentials, with $\alpha$'s which correspond to the experimental ionization potential of the outermost electron in each atom, fit the HFS potentials closely except in the vicinity of the HFS cutoff radius.

Finally we consider the transitions $2p$-$3s$ and $2p$-$3d$ in oxygen and nitrogen in several stages of ionization, which have been investigated by Kelly using both AHF [17] and HFS [19] orbitals. Table V lists his values of the radial factor $\sigma^2$ for the transition arrays $1s^2\,2s^2\,2p^N$—$1s^2\,2s^2\,2p^{N-1}\,3s$ and $1s^2\,2s^2\,2p^N$—$1s^2\,2s^2\,2p^{N-1}\,3d$, with $N=4$ to 1 in oxygen and $N=3$ to 1 in nitrogen. Each array (except when $N=1$) consists of several multiplets; we have averaged Kelly's AHF results over the multiplets in each array, weighted with the relative multiplet strengths $\mathcal{S}(\mathfrak{M})$. The HFS method yields one value of $\sigma^2$ per array. Our STF results were calculated using as one-electron binding energies the appropriately weighted means of the experimental values; the corresponding values of $\alpha$ appear in the table. Some BD values, calculated from the same energies, are also shown; these result from extrapolating the BD tables into the region $l+\frac{1}{2}<n^*<l+1$ [where $n^*=z/\sqrt{(-E)}$], which is considered marginally valid by BD; where no BD values appear, it is because $n^*<l+\frac{1}{2}$ for the $2p$ orbital. Unfortunately, no measured strengths of these ultraviolet transitions are known to us. Using the AHF values as a standard of comparison, we find for every case in Table V that the STF value of $\sigma^2$ is closer (though not always much closer) to the AHF than either HFS or BD is.

## V. CONCLUSIONS AND FUTURE WORK

The results presented in Sec. IV suggest that the STF method is accurate enough to provide a useful complement to the BD method. Roughly speaking, we may divide the transitions of interest into two classes: those (Class I) to which the BD method is thought to be applicable (because $n^*\geq l+1$ for the initial and final states of the active electron) and those

TABLE V. Radial integrals for nitrogen and oxygen.

| | | | | $\sigma^2(2p,3s)$ | | | | $\sigma^2(2p,3d)$ | | | |
|---|---|---|---|---|---|---|---|---|---|---|---|
| Ion | $\alpha(2p)$ | $\alpha(3s)$ | $\alpha(3d)$ | STF | AHF[a] | HFS[b] | BD | STF | AHF[a] | HFS[b] | BD |
| N I | 1.27 | $1.40_5$ | 1.53 | 0.197 | 0.166 | 0.243 | ··· | 0.0121 | 0.0071 | 0.0138 | ··· |
| N II | 1.45 | 1.64 | 1.75 | 0.101 | 0.112 | 0.142 | ··· | 0.036 | 0.030 | 0.043 | ··· |
| N III | 1.65 | 1.84 | $1.92_5$ | 0.056 | 0.069 | 0.090 | 0.038 | 0.044 | 0.048 | 0.056 | 0.040 |
| O I | 1.16 | 1.24 | $1.28_5$ | 0.137 | 0.109 | 0.151 | ··· | 0.0050 | 0.0028 | 0.0055 | ··· |
| O II | 1.32 | 1.46 | 1.50 | 0.080 | 0.082 | 0.098 | ··· | 0.021 | 0.016 | 0.024 | ··· |
| O III | 1.48 | 1.65 | 1.70 | 0.048 | 0.056 | 0.066 | 0.025 | 0.030 | 0.028 | 0.035 | 0.024 |
| O IV | 1.67 | 1.88 | 1.96 | 0.031 | 0.037 | 0.046 | 0.022 | 0.033 | 0.036 | 0.040 | 0.032 |

[a] See Ref. 17.
[b] See Ref. 19.

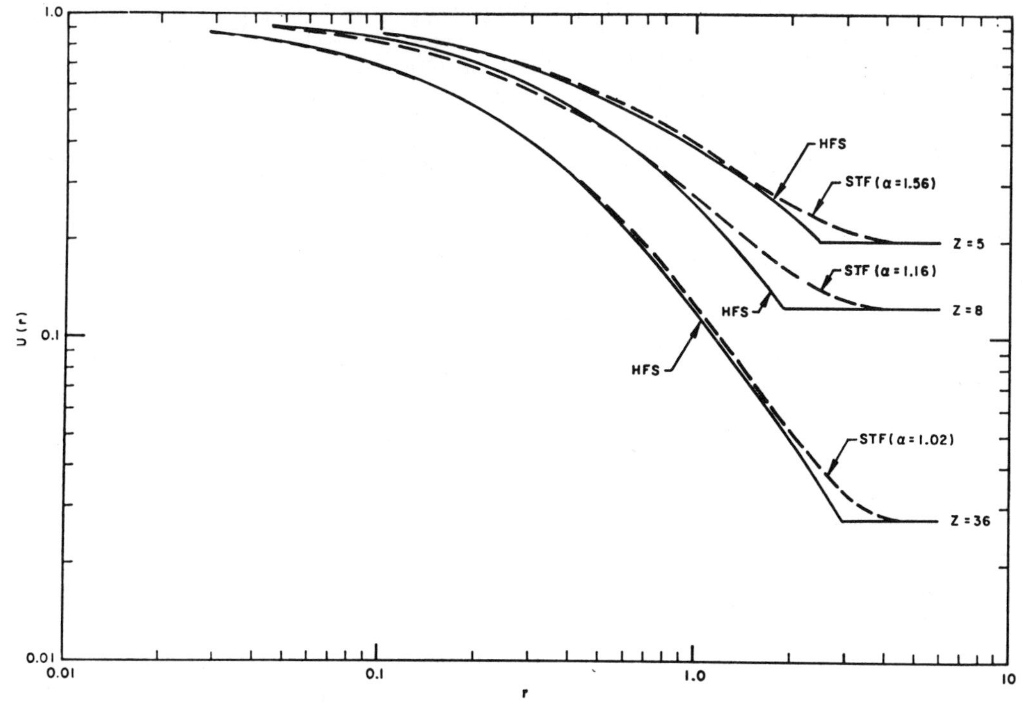


FIG. 5. Hartree-Fock-Slater (HFS) (Ref. 31) and scaled Thomas-Fermi (STF) screening functions for neutral atoms of boron ($Z=5$), oxygen ($Z=8$), and krypton ($Z=36$).

(Class II) to which it is inapplicable or marginally applicable. In Class I the STF and BD methods agree rather closely, as expected; in Class II the STF method appears to be, at least, competitive in accuracy with the HFS method as an approximation to the AHF method.

On the basis of the extensive experience of Herman and Skillman[31] and Clementi,[36] representative computing times (IBM-7090) per configuration may be estimated: about one minute for the HFS method and one hour for the AHF method at moderate atomic number, with a trend toward longer times for higher atomic numbers. In contrast, the STF method takes about five seconds per orbital (ten seconds for a transition probability). In view of these numbers and the fact that applications in astrophysics and plasma physics exist[37] which require very large numbers of oscillator strengths, we cannot regard the speed advantage of the STF method as insignificant for large-scale computation. Its most useful application is probably to Class-II transitions, which ordinarily lie in the experimentally difficult ultraviolet or soft x-ray region and are inaccessible to the BD method. For Class-I transitions the STF method does not seem to offer any advantage over the BD method except perhaps near the class boundary $n^*=l+1$; a simple program we have used which interpolates in the BD tables takes 1/15 second for a transition probability.

The need for experimental energies in the STF method is somewhat alleviated by the observation we have already mentioned: The scale factor $\alpha$ varies smoothly along Rydberg series and isoelectronic sequences and is always of the order of unity, permitting accurate extrapolation. The quantum defect is also smooth but has a far greater range of variation.

We also note from Tables I, II, and V that the $\alpha$'s involved in a transition are usually within a few percent of each other. This suggests that, as a simplifying approximation, the two $\alpha$'s be set equal. We would then be able to produce a universal line-strength table similar to BD but extended in scope. Instead of their parameters $n^*$ and $n^{*\prime}$ for a given $l$ and $l'$, we would use $n$, $n'$, $n^*$, and $n^{*\prime}$ for a given $l$ and $l'$. The extra parameters $n$ and $n'$ are discrete and take on only a few values.

## ACKNOWLEDGMENTS

The authors are indebted to Mrs. E. Metzner, Mrs. M. Bryant, and Dr. B. W. Roos for their invaluable aid in the computations.

[36] E. Clementi, J. Chem. Phys. 41, 303 (1964).
[37] J. Quant. Spectry. Radiative Transfer 4, No. 5 (1964); 5, No. 1 (1965).

## APPENDIX

### A. Scaled Thomas-Fermi Potential for the Positive Ion

The scaled Thomas-Fermi (STF) potential of a positive ion as seen by an electron is given (in Rydbergs) by

$$\begin{aligned} -V(r) &= (2Z/r)[\varphi(x)+qx/x_0] \quad (x\leq x_0), \\ &= (2Z/r)q \quad (x\geq x_0), \end{aligned} \tag{A1}$$

where $r$ is the distance in Bohr radii and $x$ is the distance in scaled TF units, which is related to the actual distance in Bohr radii by

$$r=\alpha\mu x, \tag{A2}$$

with $\mu=\frac{1}{4}(9\pi^2/2Z)^{1/3}=0.88534Z^{-1/3}$. The fractional degree of ionization is $q=(Z-N)/Z$, where $Z$ is the nuclear charge and $N$ is the number of screening electrons in the ion. The factor $\alpha$ in (A2) is the scaling factor for the charge distribution, and the ionic radius in scaled TF units is $x_0$.

The TF function $\varphi(x)$ in Eq. (A1) satisfies the differential equation

$$\varphi''(x)=x^{-1/2}\varphi(x)^{3/2} \tag{A3}$$

subject to the boundary conditions

$$\varphi(0)=1 \tag{A4}$$

and

$$\varphi(x_0)=0, \quad \varphi'(x_0)=-q/x_0. \tag{A5}$$

The problem of solving for $\varphi(x)$ is not trivial because $x_0$ as a function of $q$ is not initially available.

We have made a high-accuracy polynomial fit of $x_0(q)$ which allows the boundary conditions (A5) to be met as initial conditions so that when (A3) is integrated from $x_0$ back to 0, $\varphi(0)=1$ is automatically satisfied to better than 1 part in $10^5$. To eliminate the singularity in (A3) for numerical solution, the substitution $x=y^2$ is made.

TABLE VI. Coefficients for determining the ionic radius $x_0$ (in TF units) as a function of the fractional degree of ionization $q$. [See Eqs. (A6)–(A11)].[a]

| $n=$ | $a_n$ | $b_n$ | $c_n$ |
|---|---|---|---|
| 0 | 9.959960(−9) | 2.402529(−1) | 2.039856(−2) |
| 1 | 2.959563 | −4.751733 | −2.060645 |
| 2 | 1.853800(−1) | −3.441959 | −9.042510(−1) |
| 3 | 7.709556 | −6.138401 | −3.031483(−1) |
| 4 | −5.57687(1) | −7.127922 | −3.978742(−2) |
| 5 | 4.724630(2) | −6.352223 | 1.407012(−2) |
| 6 | −3.063916(3) | −4.261793 | 8.716138(−3) |
| 7 | 1.503024(4) | −2.135743 | 2.182259(−3) |
| 8 | −5.522799(4) | −7.861645(−1) | 2.367851(−5) |
| 9 | 1.516092(5) | −2.061712(−1) | 7.862225(−7) |
| 10 | −3.086518(5) | −3.644020(−2) | 0 |
| 11 | 4.587142(5) | −3.888973(−3) | 0 |
| 12 | −4.830562(5) | −1.892519(−4) | 0 |
| 13 | 3.412177(5) | 0 | 0 |
| 14 | −1.449479(5) | 0 | 0 |
| 15 | 2.798640(4) | 0 | 0 |

[a] The integers in parentheses are the powers of 10 by which the coefficients must be multiplied.

In order to maintain high accuracy throughout the physically interesting range $q=(1,0.01)$, it was found necessary to split the range into three regions: $1.0\geq q>0.5$, $0.5\geq q>0.1$, $0.1\geq q>0.01$. These regions will be identified as I, II, and III, respectively.

In region I, we have

$$x_0^{\mathrm{I}}=\sum_{n=0}^{15} a_n\xi_{\mathrm{I}}^n, \tag{A6}$$

where

$$\xi_{\mathrm{I}}=(1-q)^{2/3}; \quad 1.0\geq q>0.5. \tag{A7}$$

In region II,

$$x_0^{\mathrm{II}}=\sum_{n=0}^{12} b_n\xi_{\mathrm{II}}^n, \tag{A8}$$

where

$$\xi_{\mathrm{II}}=\ln q; \quad 0.5\geq q>0.1. \tag{A9}$$

Finally, in region III

$$x_0^{\mathrm{III}}=\exp\left(\sum_{n=0}^{9} c_n\xi_{\mathrm{III}}^n\right), \tag{A10}$$

where

$$\xi_{\mathrm{III}}=\ln q; \quad 0.1\geq q>0.01. \tag{A11}$$

The coefficients $a_n$, $b_n$, $c_n$ are given in Table VI.

### B. Numerical Solution of the Schrödinger Equation

Probably the most efficient way to solve numerically a second-order ordinary differential equation with first derivatives absent is to use Numerov's method.[38] This scheme approximates the solution to the Schrödinger equation:

$$p''(r)=K(r)p(r), \tag{B1}$$

where

$$K(r)=l(l+1)/r^2+V(r)-E, \tag{B2}$$

through the recursion relation

$$p_{n+1}=\frac{2+(5h^2/6)K(r_n)}{1-(h^2/12)K(r_{n+1})}p_n - \frac{1-(h^2/12)K(r_{n-1})}{1-(h^2/12)K(r_{n+1})}p_{n-1}, \tag{B3}$$

where $h$ is the step size.

There is much to be gained, however, in changing the independent variable to

$$y=\sqrt{r} \tag{B4}$$

since, in the kind of fields we are dealing with, the nodes of the wave function become nearly equally spaced in the $y$ scale so that the step size need not be constantly changed to maintain numerical accuracy.

[38] R. W. Hamming, *Numerical Methods for Scientists and Engineers* (McGraw-Hill Book Company, Inc., New York, 1962).

The disadvantage to such a change of independent variable is that instead of Eq. (B1) there now appears an equation with a first derivative,

$$p''(y)-y^{-1}p'(y)=4y^2K(y^2)p(y), \qquad \text{(B5)}$$

which is not amenable to solution by Numerov's method. A change of dependent variable eliminates the first derivative:

$$p(y)=y^{1/2}q(y), \qquad \text{(B6)}$$

so that we now have

$$q''(y)=\tilde{K}(y)q(y), \qquad \text{(B7)}$$

where

$$\tilde{K}(y)=[16l(l+1)+3]/4y^2+4y^2[V(y^2)-E], \qquad \text{(B8)}$$

which is of the required form. Equation (B7) is now solved using the recursion scheme (B3), using $\tilde{K}(y)$ instead of $K(r)$, and the final solution is obtained through the transformation (B6).

### C. Normalization of the Continuous Wave Functions

The continuous wave functions are integrated numerically from the origin starting with an arbitrary slope, with the consequence that the amplitude does not reach $\pm 1$ asymptotically, as required. At energies of several Rydbergs, this is not a problem since the wave function quickly reaches its asymptotic amplitude and could be renormalized by inspection. At a fraction of a Rydberg, however, the asymptotic amplitude is not attained except at distances which are prohibitively distant for numerical integration.

We have found, using recently published tables of Coulomb wave functions[39] as a check, that it is possible to apply the WKB approximation to normalize our continuum wave functions with sufficient accuracy.

Let $r_0=\alpha\mu x_0$ be the ionic radius. At distances $r>r_0$, continuous wave functions must be a linear combination of $F(r)$ and $G(r)$, the regular and irregular Coulomb wave functions, respectively:

$$P(r)=\alpha F(r)+\beta G(r). \qquad \text{(C1)}$$

Since we require that $P(r)$ reach unit amplitude at infinity, and since $F(r)$ and $G(r)$ are already normalized to behave asymptotically like sine and cosine, $P(r)$ will be properly normalized by dividing it by

$$(\alpha^2+\beta^2)^{1/2}.$$

To find $\alpha^2+\beta^2$, a point $R>r_0$ is found, by interpolating between points of the numerical solution if necessary, where

$$P(R)=\alpha F(R)+\beta G(R)=0. \qquad \text{(C2)}$$

The derivative (with respect to $r$) is found here also:

$$D(R)\equiv\alpha F'(R)+\beta G'(R). \qquad \text{(C3)}$$

We make use of the Wronskian, $FG'-GF'=\sqrt{E}$, to find that

$$\alpha^2+\beta^2=D^2(F^2+G^2)/E. \qquad \text{(C4)}$$

We have found that the WKB approximation

$$F^2(R)+G^2(R)\approx\left[\frac{E}{E+(2z/R)-[l(l+1)/R^2]}\right]^{1/2} \qquad \text{(C5)}$$

holds extremely well, provided

$$R>\max\{r_0,\ (l+1)^2/z,\ 5/z\}. \qquad \text{(C6)}$$

These are modest requirements. Equation (C5) is then accurate to better than 1% even in the limit $E\rightarrow 0$.

[39] A. R. Curtis, *Royal Society Mathematical Tables* (University Press, Cambridge, England, 1964), Vol. 11.

# Chapter IV

# *Role of Electron Correlation in Atomic Properties and Some Early Approaches*

## IV-1

How important is electron correlation on properties such as transition probabilities, excitation energies, and electron affinities? A few magnitudes are shown in Table 1. Clearly correlation is a major effect on such properties, especially on small energy differences such as the electron affinity. The energy required to separate a molecule into neutral atoms is another property strongly dependent on correlation. The H. F. molecular orbital method makes the fluorine molecule $F_2$ come out highly unstable, for example. It was the role in the cohesive energies of metals that prompted Wigner (Chapter IV-2) to single out and study "electron correlation" in the first place. Other chemical effects, such as those on the heats of isomerization of hydrocarbon molecules, were stressed by K. S. Pitzer [*Advan. Chem. Phys.* **2**, 59ff. (1959)], who studied the correlations between nonoverlapping charge distributions as long-range van der Waals–London dispersion forces.

Wigner, in his classic paper, looked at some of the physical aspects of correlation and some

**Table 1.** Importance of Electron Correlation in Atomic Properties

| | Hartree–Fock | H.F. + correlation | Experiment |
|---|---|---|---|
| Electron affinity of oxygen, eV | $-0.54^a$ | $1.24^b$ | $1.465^c$ |
| SD/DP term splitting ratio of $1s^2 2p^2 (Z = 8)$ | $1.47^a$ | $3.66^b$ | $3.15^d$ |
| Multiplet oscillator strength: $1s^2 2s^2 2p^2 (^3P)$ to $1s^2 2s2p^3 (^3D)(Z = 7)$ | $0.236^e$ | $0.100^{f,b}$ | 0.101 $(\pm 0.006)^g$ |
| Binding energy of $F_2$ molecule, eV | $-1.37^h$ | – | $1.68^i$ |

[a] C. C. J. Roothaan and P. S. Kelly, *Phys. Rev.* **131**, 1177 (1963).
[b] O. Sinanoğlu, in *Atomic Physics*, Plenum Press, New York, 1969.
[c] L. Branscomb et al., *Phys. Rev.* **111**, 504 (1958).
[d] Using C. Moore, *Atomic Energy Levels*, Vol. I, National Bureau of Standards, Washington, D.C., 1949.
[e] Calculated from wave function of [a] using dipole length operator.
[f] For dipole length operator; Ref. [b].
[g] L. Heroux, *Phys. Rev.* **153**, 156 (1967).
[h] A. C. Wahl, *J. Chem. Phys.* **41**, 2600 (1964). Note opposite sign to experimental.
[i] R. Iczkowski and J. Margrave, *J. Chem. Phys.* **30**, 403 (1959); zero-point correction from A. G. Gaydon, *Dissociation Energies*, Chapman & Hall, London, 1953.

of the magnitudes in atoms (see Sec. 3 of Chapter IV-2). He noted mainly the correlations between electrons of opposite spin in doubly occupied orbitals and pointed out that this quantity seemed to be about $\frac{1}{2}$ eV per electron in systems as diverse as the He atom and the Na metal. He added, however (last paragraph of Sec. 3, p. 1005), that the electron densities were quite different and implied some sort of compensation between two effects. Löwdin [*Advan. Chem. Phys.* **2**, 207ff. (1959)] has emphasized the near Z independence of the roughly -1.1-eV correlation error in helium-like ions. This near independence from density and Z turns out to be true only for some of the correlation effects in atoms, not for others (cf. Chapter 5).

In Chapter IV-2 Wigner also introduced a "correlation factor" method [cf. his Eqs. (2) and (4)] applicable to N-electron systems, in addition to obtaining the low-density limit of the correlation energy of a uniform electron gas.

An extensive review of the methods available for treating electron correlation in atoms and molecules, in the period 1934–1960, is given by Löwdin [*Advan. Chem. Phys.* **2**, 207ff. (1959)] [see also *Rev. Mod. Phys.* **32**, 328 (1960); Chapter IV-3]. In these approaches the emphasis is on computational ways for obtaining the full N-electron wave function $\psi(\mathbf{x}_1, \mathbf{x}_2, \ldots, \mathbf{x}_N)$ by the variation method, $E \leqslant \langle\psi \mid H \mid \psi\rangle / \langle\psi \mid \psi\rangle$ (for ground states).

In the configuration-interaction (C. I.) method, the full wave function $\psi$ is expanded in terms of an ordered set of Slater determinants constructed from a one-electron basis set. As discussed by Löwdin in Chapter IV-3, the method converges slowly. It becomes rapidly very difficult for systems with more electrons. Even with a truncated, finite basis set, the number of determinants rapidly reaches into the thousands for a system like the Ne atom or the $CH_4$ molecule, with the result that only a small fraction of the correlation is usually obtained. This is true even if one starts with the H. F. determinant and diagonalizes the large Hamiltonian matrix to second order only, thereby omitting also the single excitations, which then vanish (to this order) by the theorem of Brillouin [*Actualites Sci. Ind.*, No. 159 (1934)] and Møller and Plesset [*Phys. Rev.* **46**, 618 (1934)]. Research in this area has concentrated on improving convergence with better basis sets. In recent years there has been considerable progress along these lines with more quantitative results made possible, as in the carbon calculations of Weiss (Chapter IX-3).

An early way to circumvent the convergence problem of C. I. was found by Hylleraas (1929), who used the $|\mathbf{r}_1 - \mathbf{r}_2| = r_{12}$ interelectronic distance directly in a trial function of He atom. Variations on this method, the latest ones, for example, by C. Schwartz [*Atomic Physics* (Proceedings of the 1968 New York International Conference on Atomic Physics), Plenum Press, New York 1969], still yield the most accurate calculations on two-electron atoms. There has been considerable interest, therefore, in extending the method to N-electron ($N > 2$) systems.

Löwdin in Chapter IV-3 discusses the introduction of a "correlation factor," or function of the interelectronic coordinates, $g(r_{12}, r_{13}, \ldots, r_{(N-1)N})$, into the trial function. The use of a "correlation factor" in general has proved very difficult. It was not clear what form this factor should take beyond helium. Also, the Hamiltonian matrix elements invariably turned out very cumbersome. This was still true if one took a simpler correlation factor, of the Wigner–Jastrow type [Jastrow, *Phys. Rev.* **98**, 1479 (1955)], such as the one used by Szász [*Z. Naturforsch.* **15a**, 909 (1960)]. Szász took $g(r_{12}, r_{13}, \ldots, r_{(N-1)N}) =$

$\left(1 + \sum_{i>j}^{N} \sum_{m,n,\ell} c_{ij}^{m,n,\ell} (r_i - r_j)^{2m} (r_i + r_j)^n r_{ij}^{\ell}\right)$ and made a calculation for the Be atom using

the factor $(1 + c_1 r_{12} + c_2 r_{34})$ but obtained only about half of the actual correlation energy. His calculations involve the total expectation value of the Hamiltonian and concentrate on the evaluation of the resulting matrix elements and the integrals which contain several $r_{ij}$ coordinates [Szász, *J. Chem. Phys.* **35**, 1072 (1961)].

An assumption implicit in such attempts to use a correlation factor is that all electrons correlate in a larger atom as in the He $1s^2$ case. This and the related approach of minimizing the total energy with an N-electron trial function have perhaps at their base the consideration of the long range of the Coulomb repulsions. In reality, electrons in different orbitals have quite different types of correlation behavior. Even with just s orbitals, the $2s^2$ electrons correlate very differently from the $1s^2$ ones. Whereas the $1s^2$ correlation is very nearly independent of Z, the correlation energy of the $1s^2 2s^2$ beryllium-like ions increases linearly with Z, as noted by Linderberg and Shull, Layzer, R. E. Watson, and others. The distinct behavior of outer electrons, caused by strongly mixing nearby configurations of the same principal quantum number, seems to have been recognized early by A. Jucys and his school. It appears that in extending an early work by Fock, Veselov, and Petrashen [*Soviet Phys. JETP* **10**, 723 (1940)] on the problem of two electrons outside closed shells,† the Vilnius school developed and used the MCSCF method, discussed in Chapter II.

Although it is on molecules, we include in this chapter an early paper by Hurley, Lennard-Jones, and Pople (Chapter IV-4). The approach, often known as the "geminal" method, has been discussed extensively in the chemical literature on several molecules and on the Be atom as a test case. In the method one assumes that only correlations within doubly occupied orbitals are important. For each such double occupation, a "geminal" function $\Lambda_K(x_i, x_j)$ is written and the product is antisymmetrized. Thus one has

$$\Psi_G \equiv \mathcal{Q}\{\Lambda_K(x_1, x_2)\Lambda_L(x_3, x_4)\Lambda_M(x_5, x_6)\cdots\}$$

or for Be ($1s^2\ 2s^2$; $^1S$ state),

$$\Psi_G \equiv \mathcal{Q}\{\Lambda_{1s^2}(x_1, x_2)\Lambda_{2s^2}(x_3, x_4)\}$$

Further, it is assumed that the geminals satisfy a "strong-orthogonality" condition,

$$\int \Lambda^*_K(x_i, x_j)\Lambda_L(x_i, x_k)dx_i = 0$$

For a molecule such as $CH_4$, $\Psi_G$ consists of a product of one carbon $1s^2$ and four (C–H) bond geminals. The geminals, therefore, are conceptually like the electron-pair bonds of Lewis, Langmuir, and Pauling and appeal to chemical intuition.

For a calculation, the *total energy of the system* is minimized with the geminals expanded in C. I. series. It has been shown by Löwdin [*J. Chem. Phys.* **35**, 78 (1961)] and others that the strong orthogonality is too restrictive a condition. It requires that a one-electron basis set be subdivided into mutually exclusive subsets, each subset assigned to one geminal. It is particularly evident in atoms that each part of a system such as Be should make use of the same complete one-electron basis set in a C. I. series, and so the above assumption cannot be generally satisfactory.

The geminal approach has had a stimulating influence in the field, and further work on it is described in the book by R. G. Parr (*Quantum Theory of Molecular Electronic Structure*,

†The FVP work showed how correlation may be introduced into the specific, but interesting, case of two electrons outside closed shells, as in the $(ns)^2$ shells of atoms like Be, Mg, and Ca. Here the two outer electrons move in the H. F. SCF potential of the ions $Be^{2+}$, $Mg^{2+}$, etc. The outer electrons' wave function includes new orbitals made self-consistent not only with other orbitals, but also with the correlation part of their wave function.

Benjamin, New York, 1963). The most serious defect of the method, however, which probably cannot be remedied without altering the theory altogether, is the lack of interorbital correlations. As discussed in Chapter VIII, it has been shown by Sinanoğlu and co-workers that correlations in doubly occupied orbitals alone amount to less than 50 percent of the total correlation energy of atoms like neon and a saturated molecule like $CH_4$ (tetrahedrally localized geminals) or a molecule with delocalized electrons like benzene. The remaining correlations occur naturally and by derivation in the Approach I discussed in the present book, but they are not a part of the "geminal" approaches.

Considerable progress has resulted from calculations using the various approaches mentioned above in terms of computational techniques, experience in the selection of effective basis sets, and development of computer programs. Some of these techniques are essential to any other approach which may depart from the earlier ones in basic physical aspects, but still require the same computational, group theoretic, and other tools at some stage. For more details on the approaches and techniques up to about 1960 the reader is referred to Slater's book, *Quantum Theory of Atomic Structure*, Vols. I and II (McGraw-Hill, New York, 1960), which also contains extensive bibliographies.

IV-2 Reprinted from *Physical Review* **46**, 1002 (1934)

# On the Interaction of Electrons in Metals

E. WIGNER, *Princeton University*

(Received October 15, 1934)

The energy of interaction between free electrons in an electron gas is considered. The interaction energy of electrons with parallel spin is known to be that of the space charges plus the exchange integrals, and these terms modify the shape of the wave functions but slightly. The interaction of the electrons with antiparallel spin, contains, in addition to the interaction of uniformly distributed space charges, another term. This term is due to the fact that the electrons repell each other and try to keep as far apart as possible. The total energy of the system will be decreased through the corresponding modification of the wave function. In the present paper it is attempted to calculate this "correlation energy" by an approximation method which is, essentially, a development of the energy by means of the Rayleigh-Schrödinger perturbation theory in a power series of $e^2$.

## 1.

THE attempt has been made in previous work[1] to give a more general expression for the wave function of free electrons in metals than that provided by Hartree's method of the self-consistent field[2,3] or Fock's equations. The form of the wave function assumed in Fock's equations for a system of $2n$ electrons, occupying $n$ doubly-degenerate states is

$$\frac{1}{n!}\begin{vmatrix} \psi_1(x_1) & \cdots & \psi_1(x_n) \\ \cdot & & \cdot \\ \cdot & & \cdot \\ \cdot & & \cdot \\ \psi_n(x_1) & \cdots & \psi_n(x_n) \end{vmatrix} \cdot \begin{vmatrix} \psi_1(y_1) & \cdots & \psi_1(y_n) \\ \cdot & & \cdot \\ \cdot & & \cdot \\ \cdot & & \cdot \\ \psi_n(y_1) & \cdots & \psi_n(y_n) \end{vmatrix}, \quad (1)$$

where $x$ stands for three Cartesian coordinates of electrons with upward spin, and $y$ for those of electrons with downward spin. The $\psi_\nu$ are the solutions of a Schrödinger equation in which the potential of the charge distribution of the other electrons enters as well as the potential arising from the ions.

In a metal the charge distribution of all electrons is practically unaltered by removing one so that the second quantity may be replaced by the former and the potential for a given electron at the point $u$ is given by adding to the Coulomb field of the ions the fields of all electrons with parallel and with antiparallel spin. The former distribution may be obtained by inserting $u$ for $x_n$ in (1) and integrating over all coordinates except $x_1$ and $u$, while the latter is obtained by a similar operation with the exception that the integration should be carried out over all coordinates except $y_1$ and $u$.

Actually, it had been shown in[1,4] that the wave functions $\psi_\nu$ of the free electrons in a Na-lattice are very nearly plane waves $e^{2\pi i\nu\cdot x/L}$ where $L$ is the cube edge of the crystal and $\nu$ stands for a set of three integers, $\nu\cdot x$ denotes the scalar product of $\nu$ and $x$. Hence the charge distribution of the electrons with opposite spin is practically uniform, that of the electrons with parallel spin uniform with a "hole" around $u$.[5]

In no wave function of the type (1) is there a statistical correlation between the positions of electrons with antiparallel spin. The purpose of the aforementioned generalization of (1) is to allow for such correlations. This will lead to an improvement of the wave function and, therefore, to a lowering of the energy value. This energy gain will be called "correlation energy."

## 2.

The new form of the wave function, assumed in[1] was

$$\frac{1}{n!}\begin{vmatrix} \psi_1(y_1\cdots y_n;x_1) & \cdots & \psi_1(y_1\cdots y_n;x_n) \\ \cdot & & \cdot \\ \cdot & & \cdot \\ \cdot & & \cdot \\ \psi_n(y_1\cdots y_n;x_1) & \cdots & \psi_n(y_1\cdots y_n;x_n) \end{vmatrix} \cdot \begin{vmatrix} \psi_1(y_1) & \cdots & \psi_1(y_n) \\ \cdot & & \cdot \\ \cdot & & \cdot \\ \cdot & & \cdot \\ \psi_n(y_1) & \cdots & \psi_n(y_n) \end{vmatrix}, \quad (2)$$

[1] E. Wigner and F. Seitz, Phys. Rev. **46**, 509 (1934).
[2] D. R. Hartree, Proc. Camb. Phil. Soc. **24**, 89 (1928).
[3] J. C. Slater, Phys. Rev. **35**, 210, 1930; V. Fock, Zeits. f. Physik **61**, 126 (1930).
[4] J. C. Slater, Phys. Rev. **45**, 794 (1934); A. Sommerfeld and H. Bethe, Geiger-Scheel's Handbuch der Physik, Vol. **24**, 2nd part, 2nd edition, p. 406.
[5] E. Wigner and F. Seitz, Phys. Rev. **43**, 804 (1933).

which contains the functions $\psi_\nu(y_1\cdots y_n; x)$ which are different functions of $x$ for different configurations $y_1, \cdots, y_n$ of the electrons with opposite spin, instead of the $\psi_\nu(x)$. It is proposed to find the best wave function, i.e., that with the lowest total energy of the form[6] (2). The $y_1, \cdots, y_n$ in $\psi_\nu(y_1\cdots y_n; x)$ are to be viewed merely as parameters. (Cf. Eq. (23) ref. 1.) The total energy of the wave function (2) was previously calculated and relative to the solution of Fock's equations yielded the following energy

$$E_1 = \frac{1}{n!}\int dy|y|^2\left\{\sum_\nu \epsilon_\nu - \frac{h^2}{2m}\sum_{\nu\nu'}\sum_\kappa \left|\int \psi_{\nu'}(y; x_1)^* \frac{\partial\psi_\nu(y; x_1)}{\partial y_\kappa}dx_1\right|^2\right\}, \tag{3}$$

where $|y|$ denotes the second determinant of (1), the summation $\kappa$ runs over all $3n$ coordinates $y$ and that over $\nu$ and $\nu'$ over all occupied states, i.e., over all indices occurring in the wave functions in (1) or (2); $\psi_\nu(y; x)$ stands for $\psi_\nu(y_1\cdots y_n; x)$ and $dy$ for $dy_1\cdots dy_n$. The quantities $\epsilon_\nu$ are the integrals

$$\epsilon_\nu(y_1\cdots y_n) = \int \psi_\nu(y; x_1)^*\{V - (h^2/2m)(\Delta_{x_1} + \Delta_{y_1} + \cdots + \Delta_{y_n})\}\psi_\nu(y; x_1)dx_1$$
$$+ \int \psi_\nu(x_1)^*(h^2/2m)\Delta_{x_1}\psi_\nu(x_1)dx_1, \tag{3a}$$

where again $V(y_1\cdots y_n; x)$ is the difference between the potentials at the point $x$ of a charge distribution corresponding to $\psi_1(y), \psi_2(y), \cdots, \psi_n(y)$, on the one hand, and point charges at $y_1, \cdots, y_n$, on the other.

It is necessary now to assume for $\psi_\nu(y_1, \cdots, y_n; x)$ the form

$$\psi_\nu(y; x) = \psi_\nu(x)\{1 + f_\nu(y_1 - x) + f_\nu(y_2 - x) + \cdots + f_\nu(y_n - x)\} \tag{4}$$

and that if $\nu$ and $\nu'$ are both occupied states

$$\int \psi_{\nu'}(y; x_1)^*(\partial\psi_\nu(y; x_1)/\partial y_\kappa)dx_1 = 0 \tag{4a}$$

so that the second term in (3) vanishes. Both (4) and (4a) will turn out to be correct in the approximation to be used. By means of (4) it is possible to transform (3a) so as to get rid of all derivatives with respect to the $y$, after which one may minimize (3) by minimizing $\epsilon_\nu(y_1, \cdots, y_n)$ for every combination of the $y$. This would lead to a differential equation for the $\psi_\nu(y_1\cdots y_n; x)$ in which the $y$ would be merely parameters. (The solutions $\psi_\nu(x)$ of Fock's equations which also enter into this equation are supposed to be known.) The result of the transformation is especially simple if one uses for $\psi_\nu(x) = \epsilon^{2\pi i\nu\cdot x/L}$, namely,

$$\epsilon_\nu(y_1\cdots y_n) = \int \psi_\nu(y; x)^*\{V - (h^2/m)(\Delta_x - (2\pi i\nu/L)\,\text{grad}_x)\}\psi_\nu(y; x)dx. \tag{3b}$$

In addition to the energy contribution (3) which is negative and was calculated in[1] there is a further one which is generally positive. This arises from the fact that the probabilities of the relative distances of electrons with upward spin are changed by the transition from the $\psi_\nu(x)$ to the $\psi_\nu(y_1, \cdots, y_n; x)$. Since the latter will be large for $x$'s, which lie in regions comparatively free from $y$'s, the distribution of the $x$'s will not be uniform throughout space and they will be nearer together than they were under the previous assumption.

[6] The form (2) of the wave function is certainly not the correct one. It does not belong even to one single multiplicity but is a linear combination of functions of different "multiplicities" (belonging to different representations of the symmetric group). If, however, the functions $\psi_\nu(y_1\cdots y_n; x)$ are not too different from the functions $\psi_\nu(x)$, they all belong to very low multiplicities. This is the only case, anyway, in which the present approximation is good and it can be expected that the real wave function is in this case near to (2).

In order to evaluate this energy change, one may first calculate the probability of two electrons with parallel spin being at the points $x_1$ and $x_2$, respectively, if the complete wave function is given by (1). Under the approximate assumption that $\psi_\nu(x) = \psi_0(x)e^{2\pi i\nu\cdot x/L}$ one obtains for this in the way given in reference 5,

$$|\psi_0(x_1)\psi_0(x_2)|^2 g(|x_1 - x_2|), \tag{5}$$

where

$$g(r) = 1 - 9\left(\frac{\sin(r/d) - (r/d)\cos(r/d)}{(r/d)^3}\right)^2 \tag{5a}$$

is the probability of the distance $|x_1 - x_2| = r$ for free electrons with parallel spin,

$$d = (v_0/3\pi^2)^{\frac{1}{3}} = (4/9\pi)^{\frac{1}{3}} r_s = 0.521 r_s \tag{6}$$

is $1/(2\pi)$ times the wave-length of the fastest electron, $v_0$ is the atomic volume and $r_s$ the radius of the sphere with this volume. If we make the assumption $\psi_\mu(y, x) = \psi_\nu(y, x)e^{2\pi i(\mu-\nu)\cdot x/L}$, an expression of the form (5) is valid for the wave function (2) as well as for (1) and the change of mutual potential energy of the electrons with upward spin arising from the transition from (1) to (2) is

$$E_2 = \frac{1}{n!}\int dy|y|^2 \sum_\nu \epsilon_\nu'(y_1\cdots y_n), \tag{7}$$

where

$$\epsilon_\nu'(y_1\cdots y_n) = \int dx_1 dx_2 (ne^2/2r)\{|\psi_\nu(y; x_1)\psi_\nu(y; x_2)|^2 - |\psi_\nu(x_1)\psi_\nu(x_2)|^2\} g(r). \tag{7a}$$

The $\frac{1}{2}$ enters once again because the interaction of a pair of electrons should be counted once only. In order to be able to evaluate the integral (7a), $g(r)$ has been replaced by

$$1 - e^{-1.6r/d}(1 + 1.6r/d + 1.2(r/d)^2), \tag{8}$$

which, as is shown in Fig. 1, runs rather near to $g(r)$.

### 3.

The task of Section 4 will be to calculate the wave functions $\psi_\nu(y_1\cdots y_n; x)$ which minimize the sum of expressions (3b) and (7a) and to calculate $E_1 + E_2$ corresponding to these wave functions. Before doing this, however, an estimate of the order of magnitude of the effect to be expected should be given. This can be taken from calculations of atomic spectra by the method of Fock's equation or Hartree's field, and their comparison with experimental results. The best result in this connection seems to be that on the normal state of He, where Fock's equation is identical with Hartree's. The discrepancy here is[7] $0.077Ry$ (Rydberg units) for both electrons, or 12 Cal. per electron. This must be the amount of correlation energy in He.

The situation is somewhat more complicated in the calculation of the terms of $O^{++}$, $O^+$, and O by Hartree and Black,[8] since Hartree's method has been used instead of Fock's, and also because in the latter cases more than two electrons play important rôles. For $O^{++}$ the differences between observed values (in brackets) and theory are

$^3P(4.050)0.074$; $^1D(3.868)0.090$; $^1S(3.658)0.176$.

If we denote the radial wave function for the electrons by $P(r)$, these terms correspond to the linear combinations $P(r)P(r')$ multiplied by

$$xy' - yx'; \quad (x+iy)(x'+iy'); \quad xx'+yy'+zz'.$$

In the first case the correlation energy is very small, since the electrons are probably far away

[7] Cf. D. R. Hartree and A. L. Ingman, Mem. of the Manchester Lit. and Phil. Soc. **77**, **69**, 87 (1933).

[8] D. R. Hartree and M. M. Black, Proc. Roy. Soc. **A139**, 311 (1933).

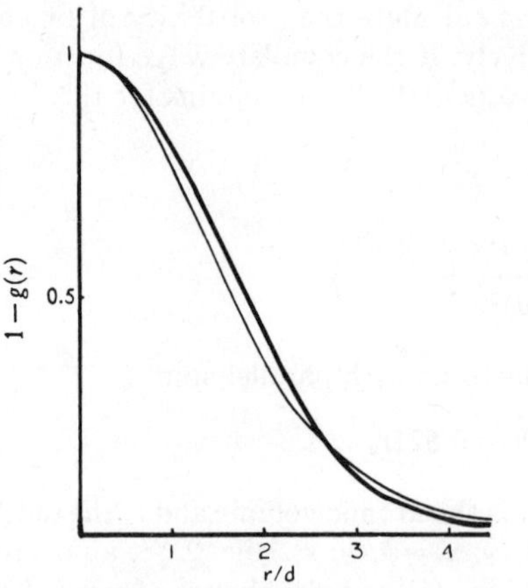


FIG. 1.

from each other anyway, because their spins are parallel. In such cases in the theory of metals we do not take into account any correlation energy at all. The discrepancy in this case may, to a considerable extent, be due to the use of Hartree's instead of Fock's method, and certainly would be further diminished by the use of the latter. On the other hand, in the case of singlet's, Hartree's equations correspond much more closely to Fock's equation and the increased discrepancy of about $0\cdot 050 Ry$ is probably due to the neglect of the correlation energy. It is not quite clear, however, why it is so much greater in the $^1S$ than in the $^1D$ term.

A comparison of experimental and theoretical values in $O^+$ and O points in a similar direction, the correlation energy is smaller though in these cases by a factor of the order 2. It is evident that it must diminish eventually if one goes over to more and more loosely bound electrons, since because of the lower electron densities the total interaction energy diminishes and the correlation energy is only the non-appearance of part of this.

If one goes over to a metal like Na, on first sight the effect could be expected (because of the low electron density) to be much smaller than in He, about as great as in O. This will not be quite so, however, because the fluctuations in the potential of the electrons with downward spin will be greatly increased by their great number. The effect to be calculated in the next section will be about equal, therefore, to the effect per electron in He.

## 4.

Although the actual wave functions $\psi_\nu(x)$ in the Na lattice are actually different from plane waves $\epsilon^{2\pi i\nu\cdot x/L}$ in the Hartree-Fock approximation, we shall use plane waves for $\psi_\nu(x)$ in the subsequent calculation. Since only integrals over the unperturbed functions occur in the perturbation calculation, this will not introduce a great error, because the $\psi_\nu(x)$ are extremely near to plane waves in much the greatest part of the volume.

The whole following calculation will be performed in the approximation which corresponds to the second approximation in the Rayleigh-Schrödinger perturbation theory. The wave functions will be developed into a Fourier series

$$\psi_\nu(y;x) = L^{-\frac{3}{2}}\left(e^{2\pi i\nu\cdot x/L} + \sum_\mu \alpha_{\nu\mu} e^{2\pi i\mu\cdot x/L}\right), \quad (9)$$

$\epsilon^{2\pi i\nu\cdot x/L}$ being taken as unperturbed function and $V(y_1, \cdots, y_n; x)$ the perturbation. The $\alpha_{\nu\mu}$, which are functions of the $y$, are supposed to be small, so that third order terms of $\alpha$ and $V$ will be neglected for the energy, and second order terms for the wave function.

For the actual calculation it can be seen, first of all, that one can replace a set of $\psi_\nu(y;x)$ by any orthogonal linear combination of them without affecting the final result. The orthogonality condition between $\psi_\nu(y;x)$ and $\psi_{\nu'}(y;x)$ gives

$$\alpha_{\nu\nu'} + \alpha_{\nu'\nu}^* + \sum_\mu \alpha_{\nu\mu}\alpha_{\nu'\mu}^* = 0. \quad (10)$$

By Schmidt's method one can build a set of orthogonal $\psi_\nu(y;x)$ such that $\alpha_{\nu\nu'} = 0$ for $\nu < \nu'$

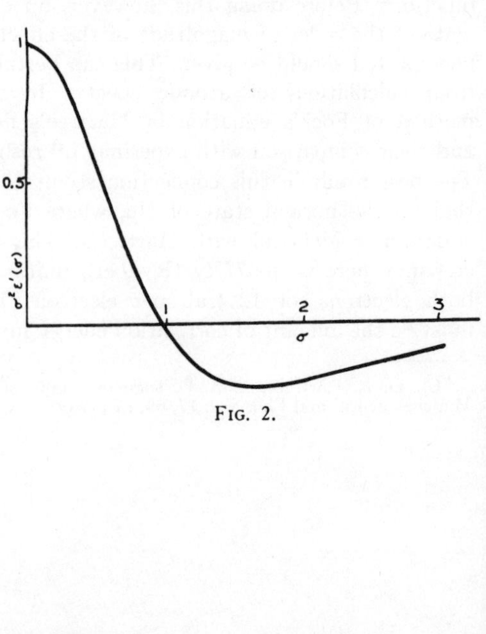


FIG. 2.

if $\nu'$ is occupied. It then follows under omission of the last term in (10) that $\alpha_{\nu\nu'}=0$ if $\nu$ and $\nu'$ are both occupied states, whence the summation over $\mu$ in (9) must be extended only over the unoccupied states. It is then seen at once that (4a) is satisfied in our approximation.

In order to calculate $\epsilon_\nu(y_1\cdots y_n)$ one must first calculate the matrix elements of $V(y_1, \cdots, y_n; x)$. The matrix elements $V_{\nu\nu}(y_1, \cdots, y_n; x)=V_{00}$ will not be zero for all values of the $y$, but they evidently will not change the wave functions and merely give a contribution to the energy. The average value over all configurations of this contribution is zero, however, since the mean value of the potential of different configurations of charges is equal to the potential of the mean charges. Therefore, we shall set $V_{\nu\nu}(y_1, \cdots, y_n; x)=0$. For $\mu\neq\nu$ we have

$$\begin{aligned} V_{\nu\mu}(y_1\cdots y_n; x) &= L^{-3}\int e^{2\pi i(\mu-\nu)\cdot x/L}V(y; x)dx \\ &= -(1/4\pi^2(\mu-\nu)^2L)\int e^{2\pi i(\mu-\nu)\cdot x/L}\Delta V(y; x)dx \\ &= (e^2/\pi(\mu-\nu)^2L)\sum_{\kappa=1}^{n} e^{2\pi i(\mu-\nu)\cdot y_\kappa/L} \end{aligned} \tag{11}$$

because of Poisson's equation.

For $\epsilon_\nu$, (3b) yields

$$\epsilon_\nu(y_1\cdots y_n)=\sum_\mu(\alpha_{\nu\mu}V_{\nu\mu}+\alpha_{\nu\mu}^*V_{\nu\mu}^*)+\sum_{\mu\mu'}\alpha_{\nu\mu}^*\alpha_{\nu\mu'}V_{\mu\mu'}+\sum_\mu\frac{4\pi^2h^2}{mL^2}(\mu^2-\mu\cdot\nu)\,|\alpha_{\nu\mu}|^2, \tag{12}$$

In order to calculate the $\epsilon_\nu'(y_1\cdots y_n)$ by means of (7a), the charge distribution for $\psi_\nu(y; x)$ will first be found:

$$|\psi_\nu(y; x)|^2=L^{-3}+L^{-3}\sum_\mu(\alpha_{\nu\mu}e^{2\pi i(\mu-\nu)\cdot x/L}+\alpha_{\nu\mu}^*e^{2\pi i(\nu-\mu)\cdot x/L})+L^{-3}\sum_{\mu\mu'}\alpha_{\nu\mu}^*\alpha_{\nu\mu'}e^{2\pi i(\mu-\mu')\cdot x/L}. \tag{13}$$

Now $g(r)$ is 1 minus the function of the "hole," which decreases rapidly with increasing $r$. The 1, inserted into (7a), simply gives the energy difference of a uniform charge distribution and that corresponding to $\psi_\nu(y; x)$, namely,

$$4\pi ne^2L^{-3}\left\{\sum_\mu\frac{|\alpha_{\nu\mu}|^2}{4\pi^2(\mu-\nu)^2/L^2}+R\sum_{\substack{2\nu-\mu\\ \text{unocc.}}}\frac{\alpha_{\nu\mu}\alpha_{\nu 2\nu-\mu}}{4\pi^2(\mu-\nu)^2/L^2}\right\}, \tag{14a}$$

in which the terms higher than the second order in $\alpha$ are omitted, and $R$ means that the real part of the following expression is to be taken. The second sum must be taken only over those $\mu$'s for which $2\nu-\mu$ is unoccupied, while $\mu$ itself is an unoccupied state in *all* summations. For the calculation of the other part of (7a), arising from the function of the hole, one sets $x_1+r$ for $x_2$ in (7a), introduces (13), and performs the integration over $x_1$. The result

$$2L^{-3}R\{\sum_\mu|\alpha_{\nu\mu}|^2+\sum_{\substack{2\nu-\mu\\ \text{unocc.}}}\alpha_{\nu\mu}\alpha_{\nu 2\nu-\mu}\}e^{2\pi i(\nu-\mu)\cdot r/L}$$

must be multiplied with $ne^2/2r$ and the function of the hole $g(r)-1$, which must be taken from (8), and integrated over $r$. Setting $\sigma=2\pi(\mu-\nu)d/L$, this yields

$$-\frac{4\pi d^2ne^2}{2.56L^3}R\Big(\sum_\mu|\alpha_{\nu\mu}|^2+\sum_{\substack{2\nu-\mu\\ \text{unocc.}}}\alpha_{\nu\mu}\alpha_{\nu 2\nu-\mu}\Big)\cdot\left(\frac{1}{1+\sigma^2/2.56}+\frac{1.063}{(1+\sigma^2/2.56)^2}+\frac{3.75}{(1+\sigma^2/2.56)^3}\right). \tag{14b}$$

Added to (14a) this gives with help of the relation $4\pi r_s^3/3=v_0$

$$\epsilon_\nu'(y_1\cdots y_n)=\frac{2^{\frac{1}{3}}}{3^{\frac{1}{3}}\pi^{\frac{2}{3}}}\frac{e^2}{r_s}R\left(\sum_\mu|\alpha_{\nu\mu}|^2+\sum_{\substack{2\nu-\mu\\ \text{unocc.}}}\alpha_{\nu\mu}\alpha_{\nu 2\nu-\mu}\right)\epsilon'(\sigma),$$

$$\epsilon'(\sigma)=\frac{1}{\sigma^2(1+\sigma^2/2.56)}-\frac{0.415}{(1+\sigma^2/2.56)^2}-\frac{1.465}{(1+\sigma^2/2.56)^3}. \tag{14}$$

$\sigma^2\epsilon'(\sigma)$ is given graphically in Fig. 2.

This formula for the increase of potential energy between electrons with upward spin is not exact and may be viewed as containing two parts: First (14a), the increase of the potential of the space charges, due to the less even charge distribution for $\psi_\nu(y;x)$ than for $\psi_\nu(x)$. This increase is lowered by the second part (14b), caused by the greater efficiency of the Fermi hole in a non-uniform charge distribution. The first neglection was made in setting $\psi_\mu(y;x)=\epsilon^{2\pi i(\mu-\nu)\cdot x/L}\psi_\nu(y;x)$ when calculating $\epsilon_\nu'$. This tends to increase the $\epsilon_\nu'$ especially for those $\nu$, for which it is large anyway, because it overemphasizes the unevenness in the charge distribution. The second neglection was to keep no terms higher than the second order terms in $\alpha$. This certainly decreases $\epsilon_\nu'$, because part of the uneveness in the charge density is due to the higher terms, especially to those of the fourth order. Finally, the normalization constant, which is smaller than 1, enters in the second power in (14) and only in the first in (12). Its omission again increases (14). In the whole these errors will about compensate.

The final quantity to be minimized is, after omission of the higher order terms,

$$\epsilon_\nu+\epsilon_\nu'=R\sum_\mu\{2\alpha_{\nu\mu}V_{\nu\mu}+(t_{\nu\mu}+t_{\nu\mu}')|\alpha_{\nu\mu}|^2+t_{\nu\mu}'\alpha_{\nu\mu}\alpha_{\nu 2\nu-\mu}\}, \tag{15}$$

where

$$t_{\nu\mu}=\frac{4\pi^2h^2}{mL^2}(\mu^2-\mu\cdot\nu),\qquad t_{\nu\mu}'=\frac{2^{\frac{1}{3}}}{3^{\frac{1}{3}}\pi^{\frac{2}{3}}}\frac{e^2}{r_s}\epsilon'\left(\frac{2\pi(\mu-\nu)d}{L}\right). \tag{15a}$$

$V_{\nu\mu}$ is given in (11), $\epsilon'(\sigma)$ in (14), $\nu$ is an occupied state, the last term in (15) should be taken only if $2\nu-\mu$ is an unoccupied state. By setting the derivative of (15) with respect to $\alpha_{\nu\mu}^*$ equal to zero, one obtains

$$V_{\nu\mu}^*+(t_{\nu\mu}+t_{\nu\mu}')\alpha_{\nu\mu}=0 \tag{16a}$$

if $2\nu-\mu$ is occupied. If it is unoccupied

$$V_{\nu\mu}^*+(t_{\nu\mu}+t_{\nu\mu}')\alpha_{\nu\mu}+t_{\nu\mu}'\alpha_{\nu 2\nu-\mu}=0,\qquad V_{\nu\mu}^*+(t_{\nu 2\nu-\mu}+t_{\nu\mu}')\alpha_{\nu 2\nu-\mu}^*+t_{\nu\mu}'\alpha_{\nu\mu}=0. \tag{16b}$$

The last equation is obtained by differentiating (15) with respect to $\alpha_{\nu 2\nu-\mu}$ and considering that $V_{\nu 2\nu-\mu}=V_{\nu\mu}^*$ and $t_{\nu 2\nu-\mu}'=t_{\nu\mu}'$. Solving (16), one finds if $2\nu-\mu$ is occupied

$$\alpha_{\nu\mu}=-V_{\nu\mu}^*/(t_{\nu\mu}+t_{\nu\mu}') \tag{17a}$$

and

$$\alpha_{\nu\mu}=-V_{\nu\mu}^*t_{\nu 2\nu-\mu}/[(t_{\nu\mu}+t_{\nu\mu}')t_{\nu 2\nu-\mu}+t_{\nu\mu}t_{\nu\mu}'] \tag{17b}$$

if $2\nu-\mu$ is unoccupied, while $\mu$ is, of course, always unoccupied. These formulas show that the $\psi_\nu(y_1,\cdots,y_n;x)$ do have the form (4) with

$$f_\nu(y-x)=-\frac{e^2}{\pi L}\left(\sum_{\substack{2\nu-\mu\\ \text{occ.}}}\frac{e^{2\pi i(\nu-\mu)\cdot(y-x)/L}}{(\mu-\nu)^2(t_{\nu\mu}+t_{\nu\mu}')}+\sum_{\substack{2\nu-\mu\\ \text{unocc.}}}\frac{e^{2\pi i(\nu-\mu)\cdot(y-x)/L}}{(\mu-\nu)^2(t_{\nu\mu}+t_{\nu\mu}'+t_{\nu\mu}t_{\nu\mu}'/t_{\nu 2\nu-\mu})}\right). \tag{18}$$

Inserting (17) into (15) one obtains for the total energy

$$\epsilon_\nu+\epsilon_\nu'=-\sum_{\substack{2\nu-\mu\\ \text{occ.}}}\frac{|V_{\nu\mu}|^2}{t_{\nu\mu}+t_{\nu\mu}'}-\sum_{\substack{2\nu-\mu\\ \text{unocc.}}}\frac{|V_{\nu\mu}|^2t_{\nu 2\nu-\mu}}{(t_{\nu\mu}+t_{\nu\mu}')t_{\nu 2\nu-\mu}+t_{\nu\mu}t_{\nu\mu}'}. \tag{19}$$

Instead of the second term one could write half of the sum of the terms for $\mu$ and $2\nu-\mu$, which makes it somewhat more symmetric.

## 5.

It would be rather difficult to perform the summation over $\mu$ in (19) for an arbitrary set of $y_1, \cdots, y_n$. Fortunately only the mean value of (19) with the weight $|y|^2/n!$ is needed and this can be computed quite easily. Since the $t$ do not depend on the $y$, one finds

$$\frac{1}{n!}\int |V_{\nu\mu}|^2|y|^2dy=\frac{e^4}{\pi^2(\mu-\nu)^4L^2}\sum_{\kappa,\lambda=1}^{n}\int e^{2\pi i(\mu-\nu)\cdot(y_\kappa-y_\lambda)/L}|y|e^{2\pi i(\nu_1y_1+\cdots+\nu_ny_n)/L}dy$$

$$=\frac{e^4}{\pi^2(\mu-\nu)^4L^2}\left(\sum_{\kappa=1}^{n}1-\sum_{\kappa\neq\lambda}\delta(\nu-\mu+\nu_\kappa-\nu_\lambda)\right). \qquad (20)$$

The first term comes from $\kappa=\lambda$, the second from $\kappa\neq\lambda$. In the second, the summation over $\lambda$ can be carried out, and it yields 1 if $\nu-\mu+\nu_\kappa$ is an occupied state, zero otherwise. The whole bracket is, therefore, equal to the number of occupied states $\nu'$, for which $\nu+\nu'-\mu$ is not occupied. Fig. 3 shows a cross section of the $\nu'$-space through the origin and the point $\nu-\mu$. The radius of the circles is $(3n/4\pi)^{1/3}=L/2\pi d$. The sphere through the weak circle contains the points $\nu-\mu+\nu'$ and the hatched part is unoccupied. As a consequence, it is

$$\frac{1}{n!}\int |V_{\nu\mu}|^2|y|^2dy=\frac{e^4n}{\pi^2(\mu-\nu)L^2}\eta(2\pi(\mu-\nu)d/L),\qquad \eta(\sigma)=\begin{cases}1 & \text{for } |\sigma|>2\\ \dfrac{3|\sigma|}{4}-\dfrac{|\sigma|^3}{16} & \text{for } |\sigma|<2\end{cases} \qquad (21)$$

and hence

$$F_\nu=\frac{1}{n!}\int(\epsilon_\nu+\epsilon_\nu')|y|^2dy=-\frac{2Ry}{3\pi^3}\int_{\substack{|\sigma+\rho|>1\\|\sigma-\rho|<1}}\frac{\eta(\sigma)d\sigma}{\sigma^4[\sigma^2+\sigma\cdot\rho+c\epsilon'(\sigma)]}-\frac{2Ry}{3\pi^3}\int_{\substack{|\sigma+\rho|>1\\|\sigma-\rho|>1}}\frac{\eta(\sigma)d\sigma}{\sigma^2[\sigma^4-(\sigma\cdot\rho)^2+2c\sigma^2\epsilon'(\sigma)]}. \qquad (22)$$

Here $\rho=2\pi\nu d/L$, $\sigma=2\pi(\mu-\nu)d/L$ and the summation over $\mu$ has been replaced by an integration over $\sigma$. The total energy is expressed in Rydberg units. The constant $c$ is

$$c=(2^{\frac{1}{3}}/3^{\frac{1}{3}}\pi^{\frac{2}{3}})(e^2md^2/h^2r_s)=0.1106r_s/a_0 \qquad (22a)$$

when expressed as function of the radius of the "$s$-sphere" in Bohr units $a_0$.

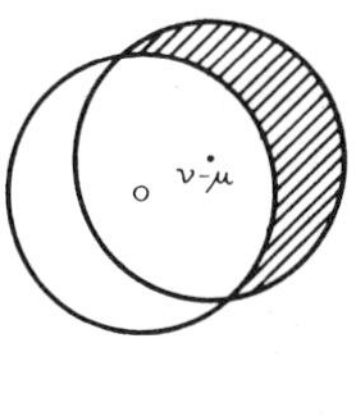


FIG. 3. Cross section of $\nu'$ space.

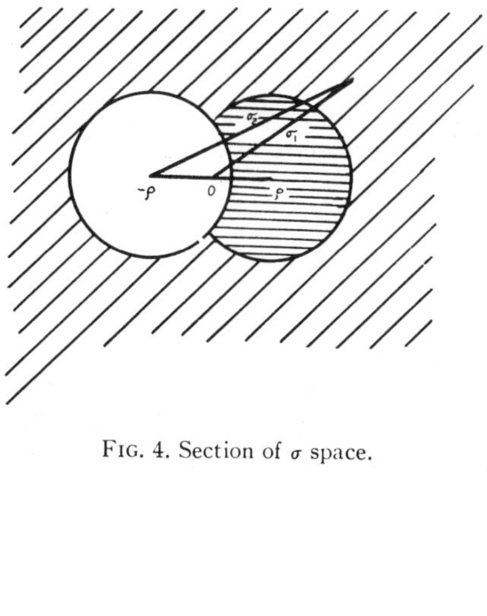


FIG. 4. Section of $\sigma$ space.

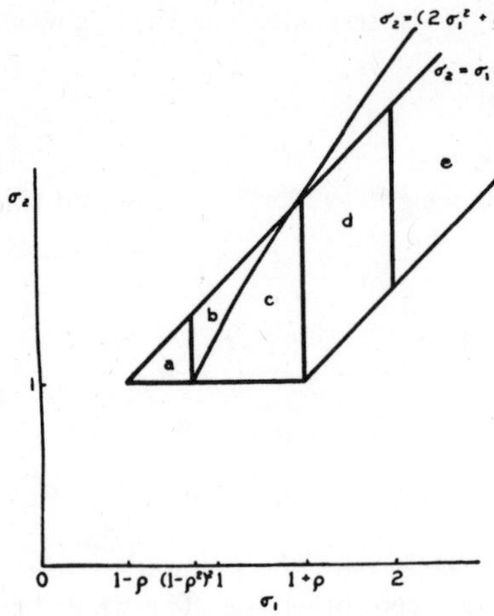


FIG. 5. *Note:* $(1-\rho^2)^2$ should read $(1-\rho^2)^{\frac{1}{2}}$.

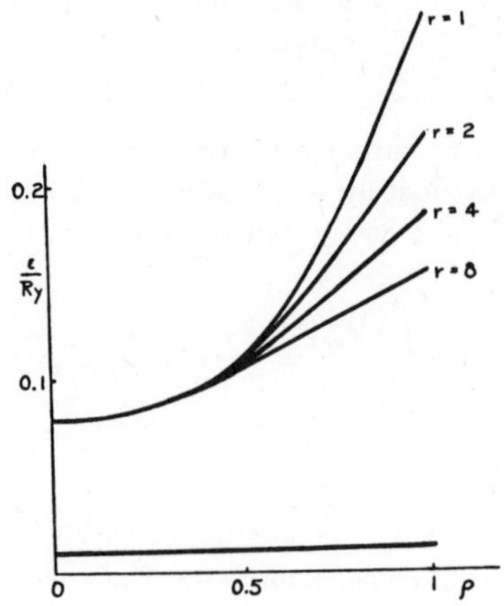


FIG. 6.

In order to perform the integration of (22), one may first introduce elliptic coordinates $\sigma_1=|\sigma|$, $\sigma_2=|\sigma+\rho|=|2\pi\mu d/L|$ in the $\sigma$-space, using as centers the origin and the point $-\rho$ (cf. Fig. 4). The first integral is to be extended over the horizontally, the second over the obliquely hatched region. The first will be a sum of two integrals (a) and (b) in Fig. 5, in both of which $\eta(\sigma)=3\sigma_1/4-\sigma_1^3/16$. The second integral is a sum of three integrals (c), (d) and (e), and in the first two of them the same expression is valid for $\eta(\sigma)$ while $\eta(\sigma)=1$ in the last. For $d\sigma$ one writes $2\pi\sigma_1 d\sigma_1\sigma_2 d\sigma_2/\rho$, $\sigma_1^2$ for $\sigma^2$, and $\sigma\cdot\rho=\frac{1}{2}(\sigma_2^2-\sigma_1^2-\rho^2)$. The integration over $\sigma_2$ can be carried out simply and gives the five integrals

$$-F_\rho=\frac{4Ry}{3\pi^2\rho}\int_{1-\rho}^{(1-\rho^2)^{\frac{1}{2}}}\ln\frac{2c\epsilon'(\sigma_1)+2\sigma_1^2+2\sigma_1\rho}{2c\epsilon'(\sigma_1)+\sigma_1^2+1-\rho^2}\left(\frac{3\sigma_1}{4}-\frac{\sigma_1^3}{16}\right)\frac{d\sigma_1}{\sigma_1^3}$$

$$+\frac{4Ry}{3\pi^2\rho}\int_{(1-\rho^2)^{\frac{1}{2}}}^{1+\rho}\ln\frac{2c\epsilon'(\sigma_1)+2\sigma_1^2+2\sigma_1\rho}{2c\epsilon'(\sigma_1)+3\sigma_1^2+\rho^2-1}\left(\frac{3\sigma_1}{4}-\frac{\sigma_1^3}{16}\right)\frac{d\sigma_1}{\sigma_1^3}$$

$$+\frac{4Ry}{3\pi^2\rho}\int_{(1-\rho^2)^{\frac{1}{2}}}^{1+\rho}\ln\frac{2\sigma_1 v+\sigma_1^2+\rho^2-1}{2\sigma_1 v-\sigma_1^2-\rho^2+1}\left(\frac{3\sigma_1}{4}-\frac{\sigma_1^3}{16}\right)\frac{d\sigma_1}{\sigma_1^2 v}$$

$$+\frac{4Ry}{3\pi^2\rho}\int_{1+\rho}^{2}\ln\frac{v+\rho}{v-\rho}\left(\frac{3\sigma_1}{4}-\frac{\sigma_1^3}{16}\right)\frac{d\sigma_1}{\sigma_1^2 v}+\frac{4Ry}{3\pi^2\rho}\int_2^\infty\ln\frac{v+\rho}{v-\rho}\frac{d\sigma_1}{\sigma_1^2 v},$$

where $v$ stands for $(2c\epsilon'(\sigma_1)+\sigma_1^2)^{\frac{1}{2}}$. A calculation of this quantity for $r_s=4$ shows that it is practically equal $\sigma_1$ if $\sigma_1>2$ and the last integral can be evaluated accordingly. It yields

$$\frac{4Ry}{3\pi^2\rho}\left\{\left(\frac{1}{8}-\frac{1}{2\rho^2}\right)\ln\frac{2+\rho}{2-\rho}+\frac{1}{2\rho}\right\}\approx\frac{Ry}{9\pi^2}\left(1+\frac{\rho^2}{20}\right).$$

The other integrals were evaluated numerically and the results are plotted against $\rho$. Fig. 6 shows the plots for $r_s=1, 2, 4, 8$. From this, the mean value of the correlation energy which is the mean value of $-\frac{1}{2}F_\rho$ with the weight $\rho^2$, was calculated, and plotted against $r_s$. The $\frac{1}{2}$ enters, because the whole energy correction is present only for half of the electrons, that is, those with upward spin. In Fig. 7 the upper curve represents the values calculated in this way. The energy is given in multiples of $e^2/r_s$.

### 6.

One must remember, of course, that the preceding calculation is only an approximate one. Even if one confines oneself to wave functions of the form (2), the upper curve of Fig. 7 gives the correlation energy in first approximation only. The neglections are due to three causes: first to the use of an unnormalized wave function, second to the neglection of the terms with higher than the second power of $\alpha$ in $\epsilon_\nu$, when going over from (12) to (14a) and third to the non-complete orthogonality of the wave functions employed. Our approximation is good if the $\alpha$'s are small. One gets an idea about the accuracy of the approximation by calculating $\sum_\mu \int |\alpha_{\nu\mu}|^2 |y|^2 dy/n!$ (which is the reciprocal square of the normalization constant minus one), though an idea only, since $\sum_\mu |\alpha_{\nu\mu}|^2$ should really be small for all configurations of the $y$, not only its mean value. A calculation of the former quantity shows[9] that it stays well below one, except for large $r_s$ and for $\rho$ which are very near to 1. It is in these cases that our approximation must be expected to break down.

The real value of the correlation energy will be smaller in these cases than the calculated one. The correction with the normalization constant could easily be taken into account, as has been shown at another place.[10] It always decreases the correlation energy, not very much, however, as the magnitude of the normalization constant or the formulas in[10] show. The second neglection is probably more dangerous and also more laborious to correct. It has been done for one point ($r_s = 4$) only and for this one very roughly. The second neglection also increased the calculated value of the correlation energy, since it amounts to taking $1-2f$ instead of $(1-f)^2$ for the probability of the electron being at a certain point, and the minima of $1-2f$ are much deeper than those of $(1-2f)^2$ (the maxima are lower but less important).

If the electrons had no kinetic energy, they would settle in configurations which correspond to the absolute minima of the potential energy. These are closed-packed lattice configurations, with energies very near to that of the body-centered lattice. Here, every electron is very nearly surrounded with a spherical hole of radius $r_s$ and the potential energy is smaller than in the random configuration by the amount $0.75 = e^2/r_s$. This would be the sum of the correlation energy and that due to the Fermi hole. Since the latter one is,[11, 1] $0.458e^2/r_s$, the maximum amount of the correlation energy is $0.292e^2/r_s$. This value will be attained only if the kinetic energy can be neglected, i.e., for $r_s = \infty$, and represents the asymptote to the real correlation energy curve, which is attempted to be drawn into Fig. 7. It appears to run much higher than one would have thought without calculation. I believe it to be in error everywhere by less than 20 percent.

The dotted line at $0.142e^2/r_s$ corresponds to a correlation energy as great as assumed in the first calculation[5] giving $0.6e^2/r_s$ together with the energy of the Fermi hole.

The calculated constants of the Na lattice with the correlation energy of Fig. 7 and the other quantities as in reference 1 are as follows: lattice constant 4.62A as compared with the observed value of 4.23A. The binding energy associated with this is 26.1 Calories, to be compared with the observed value of 26.9. The calculated value of the binding energy for the observed lattice

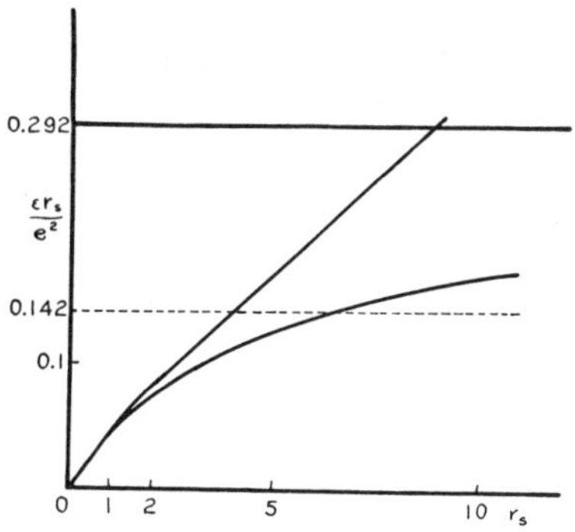


FIG. 7.

[9] The greatest part of the numerical work has been done by Dr. M. Vermes of Budapest. A table of the calculated values is given here:

| $\rho$ | $r=1$ | 4 | 8 |
|---|---|---|---|
| 0 | 0·006 | 0·10 | 0·45 |
| 0·4 | 0·009 | 0·14 | 0·54 |
| 1 | 0·04 | 0·30 | 0·94 |

[10] To appear shortly in the Bull. of the Hung. Acad.

[11] F. Bloch, Zeits. f. Physik 57, 545 (1929).

constant is 22.3 Cal. As far as the lattice constant goes, one must remember, however, that both the correlation energy and that due to the Fermi hole are calculated for a flat wave function and the wave functions are flat only for $r_s=4$ and its neighborhood.

The magnitude of the correlation energy is important for questions of paramagnetism and ferromagnetism as well as for questions of lattice energy. It modifies Bloch's original theory on the ferromagnetism of free electrons[11] in such a way that it yields ferromagnetism in fewer cases than in its original form.[12] I hope to return to this question at another time.

I wish to express my gratitude to Dr. F. Seitz for his kind help in connection with the preparation of this manuscript.

[12] A paper of S. Schubin and S. Wonsowsky, Proc. Roy. Soc. **A145**, 159 (1934) which appeared recently, points in the same direction.

 Reprinted from *Reviews of Modern Physics* **32**, 328 (1960)

# Expansion Theorems for the Total Wave Function and Extended Hartree-Fock Schemes*

PER-OLOV LÖWDIN

*Quantum Chemistry Group, Uppsala University, Uppsala, Sweden*

## 1. EXPANSION THEOREMS

### (a) Superposition of Configurations

IN order to solve the Schrödinger equation $\mathcal{H}\Psi=E\Psi$ for a system of $N$ electrons, it is often convenient to use expansion theorems for the total wave function $\Psi=\Psi(\mathbf{x}_1,\mathbf{x}_2,\cdots\mathbf{x}_N)$ where $\mathbf{x}_i=(\mathbf{r}_i,\zeta_i)$ is the combined space-spin coordinate for electron $i$. For this purpose, one may introduce a certain orthonormal and complete set of discrete one-electron functions or spin-orbitals $\psi_k(\mathbf{x})$ as a basis and assume the existence of an expansion theorem

$$\psi(\mathbf{x})=\sum_k c_k\psi_k(\mathbf{x}) \qquad (1)$$

for every normalizable function $\psi(\mathbf{x})$ of a single electronic coordinate. For a function of several electronic coordinates, one can then repeat the use of (1) for one coordinate at a time, which leads to an expansion of the form

$$\Psi(\mathbf{x}_1,\mathbf{x}_2,\mathbf{x}_3,\cdots)=\sum_{klm\cdots}c_{klm}\psi_k(\mathbf{x}_1)\psi_l(\mathbf{x}_2)\psi_m(\mathbf{x}_3)\cdots. \qquad (2)$$

Every normalizable function of $N$ particle coordinates can hence be expanded in Hartree-products built up from a complete basic set. This expansion can be further simplified for an electronic system, which obeys the Pauli exclusion principle mathematically equivalent to the antisymmetry requirement $P\Psi=(-1)^p\Psi$, where $P$ is a permutation of the coordinates and $p$ its parity. Introducing the antisymmetry projection operator

$$\mathcal{O}_{AS}=(N!)^{-1}\sum_P(-1)^pP, \qquad (3)$$

fulfilling the relations

$$P\mathcal{O}_{AS}=(-1)^p\mathcal{O}_{AS},\quad (\mathcal{O}_{AS})^2=\mathcal{O}_{AS}, \qquad (4)$$

we have $\mathcal{O}_{AS}\Psi=\Psi$, and, by applying operator (3) to both members of Eq. (2), we thus obtain

$$\begin{aligned}\Psi&=\mathcal{O}_{AS}\Psi\\&=(N!)^{-1}\sum_{klm\cdots}c_{klm\cdots}\det\{\psi_k,\psi_l,\psi_m,\cdots\}\\&=\sum_{k<l<m\cdots}c_{klm\cdots}\det\{\psi_k,\psi_l,\psi_m,\cdots\}.\end{aligned} \qquad (5)$$

The relation shows that every normalizable antisymmetric wave function $\Psi$ may be expanded in Slater determinants built up from a complete basic set, and this theorem forms the basis for the method of "superposition of configurations." A selection $K$ of $N$ indices $k<l<m<\cdots$ is called an "ordered configuration" and, introducing the notations

$$\Psi_K=(N!)^{-\frac{1}{2}}\det\{\psi_k,\psi_l,\psi_m,\cdots\},\quad C_K=(N!)^{+\frac{1}{2}}c_{klm\cdots}, \qquad (6)$$

one can write expansion (5) in the condensed form

$$\Psi(\mathbf{x}_1,\mathbf{x}_2,\cdots\mathbf{x}_N)=\sum_K C_K\Psi_K(\mathbf{x}_1,\mathbf{x}_2,\cdots\mathbf{x}_N). \qquad (7)$$

In the $N$-electron configuration space, the Slater determinants $\Psi_K$ form an antisymmetric basis fulfilling the orthonormality relation $\int\Psi_K^*\Psi_L(dx)=\delta_{KL}$. With respect to this basis, the Hamiltonian $\mathcal{H}_{op}$ is represented by the matrix $\mathcal{H}_{KL}=\int\Psi_K^*\mathcal{H}_{op}\Psi_L(dx)$, where the elements are easily simplified.[1]

In order to evaluate the coefficients $C_K$ corresponding to the solutions of the Schrödinger equation $\mathcal{H}\Psi=E\Psi$, one conventionally applies the variation principle $\delta\langle\mathcal{H}\rangle_{Av}=0$, which leads to the equation system

$$\sum_L(\mathcal{H}_{KL}-E\delta_{KL})C_L=0, \qquad (8)$$

where the eigenvalues $E$ are determined by the secular equation

$$\det\{\mathcal{H}_{KL}-E\delta_{KL}\}=0. \qquad (9)$$

In principle, one can obtain any accuracy desired for an eigenfunction $\Psi$ by means of an expansion of type (7). This fact was well known in the early days of wave mechanics but seems to have been almost forgotten during quite some time until the revival a decade ago by the work of Slater, Boys, Jucys, and others.

#### *Convergency Problem*

Even if the expansion (7) formally exists as soon as the basis $\{\psi_k\}$ is complete, the convergency may be exceedingly slow, and an important problem is hence to find the particular basis which leads to a most rapid convergency. This problem was first treated by Slater[2] who constructed the basis by solving a set of integro-differential equations similar to the Hartree-Fock equations.

The convergency problem can also be approached in a different way by studying the first-order density matrix.[3] If the wave function $\Psi$ is normalized so that

* Work supported in part by the King Gustaf VI Adolf's 70-Years Fund for Swedish Culture, Knut and Alice Wallenberg's Foundation, the Swedish Natural Science Research Council, and in part by the Aeronautical Research Laboratory, Wright Air Development Center of the Air Research and Development Command, U. S. Air Force through its European Office under a contract with Uppsala University.

[1] J. C. Slater, Phys. Rev. **34**, 1293 (1929); **38**, 1109 (1931); E. U. Condon, ibid. **36**, 1121 (1930).

[2] J. C. Slater, Quarterly Progress Report of Solid-State and Molecular Theory Group at Massachusetts Institute of Technology, 6 (January 15, 1953) (unpublished); Tech. Rept. No. 3, 39 (February 15, 1953) (unpublished); Phys. Rev. **91**, 528 (1953).

[3] P. O. Löwdin, Phys. Rev. **97**, 1474 (1955); Advances in Phys. **5**, 1 (1956); *Advances in Chemical Physics*, I. Prigogine, Editor (Interscience Publishers, Inc., New York, 1959), Vol. 2, p. 207.

$\int |\Psi|^2(dx)=1$, the first-order density matrix $\gamma(\mathbf{x}_1'|\mathbf{x}_1)$ is defined by

$$\gamma(\mathbf{x}_1'|\mathbf{x}_1)=N\int \Psi^*(\mathbf{x}_1'\mathbf{x}_2\mathbf{x}_3\cdots\mathbf{x}_N)\times\Psi(\mathbf{x}_1\mathbf{x}_2\mathbf{x}_3\cdots\mathbf{x}_N)dx_2dx_3\cdots dx_N. \quad (10)$$

By means of the basic set $\psi_k$ and repeated use of (1), one can expand $\gamma(\mathbf{x}_1'|\mathbf{x}_1)$ in the form

$$\gamma(\mathbf{x}_1'|\mathbf{x}_1)=\sum_{kl}\psi_k^*(\mathbf{x}_1')\psi_l(\mathbf{x}_1)\gamma_{lk}, \quad (11)$$

where the coefficients $\gamma_{lk}$ form an Hermitean matrix $\boldsymbol{\gamma}$. By substituting (7) into (10) and carrying out the integrations, one finds relations between the quantities $\gamma_{lk}$ and the coefficients $C_K$. For the diagonal elements, one obtains the simple formula

$$\gamma_{kk}=\sum_K^{(k)}|C_K|^2, \quad (12)$$

where the summation goes over all configurations $K$ containing the specific index $k$. The diagonal element $\gamma_{kk}$ may be interpreted as the "occupation number" of the spin-orbital $\psi_k$, and we note that it depends only on $\Psi$ and $\psi_k$. Since the normalization gives $\sum_K|C_K|^2=1$, one has further the inequality $0\leq\gamma_{kk}\leq1$ which may be considered a consequence of the antisymmetry requirement and hence also of the Pauli principle.

Because of (12), there is a connection between the convergency property of expansion (7) and the diagonal elements of the matrix $\boldsymbol{\gamma}$, and we now consider the optimum case. Let $\mathbf{U}$ be the unitary matrix which brings $\boldsymbol{\gamma}$ to diagonal form:

$$\mathbf{U}^\dagger\boldsymbol{\gamma}\mathbf{U}=\mathbf{n}, \quad (13)$$

where $\mathbf{n}$ is a diagonal matrix with the eigenvalues $n_1\geq n_2\geq n_3\geq\cdots$. Introducing a new basic set $\{\chi_k\}$ by the matrix formula $\boldsymbol{\chi}=\boldsymbol{\psi}\mathbf{U}$ or $\chi_k=\sum_\alpha\psi_\alpha U_{\alpha k}$, we obtain $\boldsymbol{\gamma}=\mathbf{U}\mathbf{n}\mathbf{U}^\dagger$, $\boldsymbol{\psi}=\boldsymbol{\chi}\mathbf{U}^\dagger$, and

$$\gamma(\mathbf{x}_1'|\mathbf{x}_1)=\sum_k\chi_k(\mathbf{x}_1')\chi_k(\mathbf{x}_1)n_k. \quad (14)$$

The basic set $\chi_k$ has hence the occupation numbers $n_k$, and, since they are the eigenvalues of $\boldsymbol{\gamma}$, they have extremum properties. The first function $\chi_1$ has the highest occupation number possible, the second function $\chi_2$ has the same property within the class orthogonal to $\chi_1$, etc. The functions $\chi_k$ are called the "natural spin-orbitals" associated with the system and state under consideration.

Substituting the relation $\boldsymbol{\psi}=\boldsymbol{\chi}\mathbf{U}^\dagger$ or $\psi_k=\sum_\alpha\chi_\alpha U_{\alpha k}^\dagger$ into the Slater determinants of expansion (7) and using a theorem for evaluating the determinant of a matrix being a product of two rectangular matrices, we obtain an expansion of $\Psi$ into configurations of the natural spin orbitals:

$$\Psi=(N!)^{-\frac{1}{2}}\sum_K A_K\det\{\chi_k,\chi_l,\chi_m,\cdots\}, \quad (15)$$

which are called the "natural expansion." This series has a certain optimum convergency property.

Let us consider the matrix $\boldsymbol{\gamma}$ and all possible submatrices $\boldsymbol{\gamma}^{(r)}$ of order $r$ along the diagonal which may be obtained by truncating the basis to finite order $r$. According to a well-known separation theorem,[4] the $r$ highest eigenvalues of $\boldsymbol{\gamma}$ are in order of magnitude larger than the corresponding eigenvalues of $\boldsymbol{\gamma}^{(r)}$. The sum of the $r$ highest eigenvalues of $\boldsymbol{\gamma}$ is hence always larger than the sum of the $r$ eigenvalues of $\boldsymbol{\gamma}^{(r)}$, i.e., larger than the sum of its $r$ diagonal elements, which gives the general inequality

$$\sum_{k=1}^{r} n_k\geq\sum_{k=(1)}^{(r)}\gamma_{kk}, \quad (16)$$

where in the sum $k=(1)\cdots(r)$, one can take any $r$ indices. By using (12), we obtain finally

$$\sum_{k=1}^{r}\sum_K^{(k)}|A_K|^2\geq\sum_{k=(1)}^{(r)}\sum_K^{(k)}|C_K|^2, \quad (16')$$

for $r=1, 2, 3, \cdots$, which relation expresses the optimum convergencey property of the natural expansion.

### *Case of Two-Electron Systems*

In the special case of $N=2$, the properties of the natural expansion have been investigated in greater detail.[5] The spin function may be taken out as a separate factor and, instead of (7), one obtains for the singlet state

$${}^1\Psi(\mathbf{x}_1,\mathbf{x}_2)=2^{-\frac{1}{2}}(\alpha_1\beta_2-\beta_1\alpha_2)\sum_{k,l=1}^{\infty}C_{kl}\psi_k(\mathbf{r}_1)\psi_l(\mathbf{r}_2), \quad (17)$$

where $C_{lk}=C_{kl}$. The problem of diagonalizing the first-order density matrix $\gamma(\mathbf{x}_1'|\mathbf{x}_1)$ is essentially a problem of diagonalizing the matrix $\mathbf{C}$, and the natural expansion takes the simple form

$${}^1\Psi(\mathbf{x}_1,\mathbf{x}_2)=2^{-\frac{1}{2}}(\alpha_1\beta_2-\beta_1\alpha_2)\sum_{k=1}^{\infty}c_k\chi_k(\mathbf{r}_1)\chi_k(\mathbf{r}_2), \quad (18)$$

where $c_k=\pm n_k^{\frac{1}{2}}$. Hence the double sum in (17) has been reduced to a single sum in (18). A similar theorem holds also for the triplet state.

For $N=2$, the natural expansion (18) has in addition to (16) another important optimum convergency property. Let ${}^1\Psi_r(\mathbf{x}_1,\mathbf{x}_2)$ be any singlet wave function which may be built up from a basic orbital set of finite order $r$ and consider the total quadratic deviation between the "exact solution" and such a trial function

$$\int|\Psi_{\text{exact}}-\Psi_r|^2dx_1dx_2. \quad (19)$$

[4] E. A. Hylleraas and B. Undheim, Z. Physik **65**, 759 (1930); J. K. L. MacDonald, Phys. Rev. **43**, 830 (1933).

[5] P. O. Löwdin and H. Shull, Phys. Rev. **101**, 1730 (1956).

It may be shown that this quantity has a minimum for

$$^{1}\Psi_r(\mathbf{x}_1,\mathbf{x}_2)=2^{-\frac{1}{2}}(\alpha_1\beta_2-\beta_1\alpha_2) \times\sum_{k=1}^{r} c_k\chi_k(\mathbf{r}_1)\chi_k(\mathbf{r}_2)/\{\sum_{k=1}^{r} c_k^2\}^{\frac{1}{2}}, \quad (20)$$

which is actually the function obtained by truncating the natural expansion (18) after $r$ terms and renormalizing the result. For two-electron systems, there are particularly two types of approximate wave functions which have been discussed,[6] namely, the closed-shell or $u^2$ form and the split-shell or $(u,v)$ form having $r=1$ and $r=2$, respectively. From the general theorem follows now that $(\chi_1)^2$ is the $u^2$ function which has the smallest quadratic deviation (19) from the exact eigenfunction. The first natural orbital $\chi_1$ is hence closely related to the corresponding Hartree-Fock function and has actually only a slightly higher energy.[7] The space function

$$(n_1^{\frac{1}{2}}\chi_1^2-n_2^{\frac{1}{2}}\chi_2^2)/(n_1+n_2)^{\frac{1}{2}} \quad (21)$$

is further the optimum function for $r=2$ and is easily transformed to $(u,v)$ form[8] by putting $u=n_1^{\frac{1}{2}}\chi_1+n_2^{\frac{1}{2}}\chi_2$, $v=n_1^{\frac{1}{2}}\chi_1-n_2^{\frac{1}{2}}\chi_2$.

The natural orbitals have so far been investigated in detail for the ground state of helium,[9] for the long-range interaction between two hydrogen atoms,[10] and for the hydrogen molecule,[11] and further work is in progress.

### (b) Introduction of Correlation Factcr

The first practical application of the method of superposition of configurations was carried out by Hylleraas[12] in connection with the helium problem, and he found that if the basic orbital set contained the spherical harmonics $Y_{lm}$ in the conventional way, the expansion (7) converged comparatively slowly.[13] Hylleraas[14] suggested therefore that one should instead try to inclued the interelectronic distance $r_{12}$ explicitly in the wave function, and, for a two-electron system, $r_{12}$ could actually be chosen as one of the basic coordinates. For the helium problem this led to a fruitful development which so far has culminated in the highly accurate calculations by Kinoshita[15] and by Pekeris,[16] the latter using perimetric coordinates. For the hydrogen molecule, the corresponding development started with James and Coolidge[17] and has now reached the accurate level reported by Kolos[18] at this conference.

Unfortunately this $r_{12}$ coordinate approach cannot easily be generalized to a many-electron system. Another way of introducing $r_{12}$ in the wave functión was also suggested by Hylleraas[14] who considered space functions of the form

$$g(r_{12})u(r_1)u(r_2), \quad (22)$$

where $g(r_{12})$ is a "correlation factor." By using the simple functions $g=\exp\{\alpha r_{12}\}$ and $g=1+\alpha r_{12}$, Hylleraas showed that one could obtain surprisingly good results. Both for the helium atom[19] and the hydrogen molecule,[20] this idea has been extensively developed in the literature. In the more recent works, also quite a few nonlinear parameters have been varied, and excellent results have been obtained by the largest electronic computers available.

Here we would like to draw attention to the fact that the method with correlation factor can easily be combined with the method using superposition of configurations in such a way that very good results can be obtained without too extensive numerical calculations. If the correlation factor $g=g(r_{12})$ is nodeless and such that the function $\Psi/g$ is still normalizable, one can apply the expansion theorems (17) and (18) to this function. For the symmetric space function associated with a singlet state for a two-electron system, this gives

$$\Psi(\mathbf{r}_1,\mathbf{r}_2)=g(r_{12})\sum_{k,l=1}^{\infty} C_{kl}\psi_k(\mathbf{r}_1)\psi_l(\mathbf{r}_2) =g(r_{12})\sum_{k=1}^{\infty} c_k\chi_k'(\mathbf{r}_1)\chi_k'(\mathbf{r}_2), \quad (23)$$

where, for fixed $g$, the coefficients $C_{kl}$ may be determined

[6] See, e.g., the contributions by R. S. Mulliken, M. Kotani, G. R. Taylor, and R. G. Parr in Proceedings of the Shelter Island Conference on Quantum Mechanical Methods in Valence Theory, Physics Branch of the Office of Naval Research (unpublished) (1951).

[7] According to reference 9, the $(\chi_1)^2$ energy for the ground state of helium is $-2.861530$ a.u.$_{\text{He}}$ to be compared with the SCF energy $-2.861673$ a.u.$_{\text{He}}$.

[8] C. A. Coulson and I. Fischer, Phil. Mag. 40, 386 (1949).

[9] H. Shull and P. O. Löwdin, J. Chem. Phys. 30, 617 (1959).

[10] J. O. Hirschfelder and P. O. Löwdin, Mol. Phys. 2, 229 (1959).

[11] H. Shull, Tech. Notes No. 17 (September 15, 1958) and No. 26 (June 5, 1959) from the Uppsala Quantum Chemistry Group (to be published).

[12] E. A. Hylleraas, Z. Physik 48, 469 (1928).

[13] Recent investigations (see reference 9) indicate that the energy limit for wave functions including $f$ orbitals $(l=3)$ would be $-2.90332$ au.$_{\text{He}}$ compared to the exact value $-2.90372$ au.$_{\text{He}}$. Including angular terms up to $l=5$, D. H. Tycho, L. H. Thomas, and K. M. King, Phys. Rev. 109, 369 (1958), obtained $-2.90343$ a.u.$_{\text{He}}$, but they did not get any appreciable improvement by including additional terms up to $l=14$. A closer investigation of this convergency problem would certainly be of value.

[14] E. A. Hylleraas, Z. Physik 54, 347 (1929).

[15] T. Kinoshita, Phys. Rev. 105, 1490 (1957); Phys. Rev. 115, 366 (1959).

[16] C. L. Perkeris, Phys. Rev. 112, 1649 (1958); Phys. Rev. 115, 1216 (1959).

[17] H. M. James and A. S. Coolidge, J. Chem. Phys. 1, 825 (1933).

[18] W. Kolos and C. C. J. Roothaan, Revs. Modern Phys. 32, 205 (1960), this issue.

[19] D. R. Hartree and A. L. Ingman, Mem. Proc. Manchester Lit. & Phil. Soc. 77, 69 (1933); T. D. H. Baber and H. R. Hassé, Proc. Cambridge Phil. Soc. 33, 253 (1937); P. Pluvinage, Ann. Phys. 5, 145 (1950); C. C. J. Roothaan, Colloq. intern. centre natl. recherche sci. (Paris) 82, 49 (1958); L. C. Green, S. Matsushima, C. Stephens, E. K. Kolchin, M. M. Kohler, Y. Wang, B. B. Baldwin, and R. J. Wisner, Phys. Rev. 112, 1187 (1958); C. C. J. Roothaan and A. W. Weiss, Revs. Modern Phys. 32, 194 (1960), this issue.

[20] A. A. Frost, J. Braunstein, and W. Schwemer, J. Am. Chem. Soc. 70, 3292 (1948); W. Kolos and C. C. J. Roothaan, reference 18.

by the variation principle. By using the simple factor $g=1+\alpha r_{12}$, this method has so far been applied to the helium atom.[21]

This approach can now be generalized to a many-electron system. Let us introduce a "correlation factor" $g=g(r_{12},r_{13},r_{23},\cdots)$ which is a symmetric function of all the $N(N-1)/2$ interelectronic distances $r_{ij}$ or, still more general, an even and symmetric function of the corresponding vectors $\mathbf{r}_{ij}=\mathbf{r}_i-\mathbf{r}_j$. Let us further assume that $g$ is nodeless and such that the quotient $\Psi(\mathbf{x}_1,\mathbf{x}_2,\cdots\mathbf{x}_1)/g$ is still normalizable. By applying the expansion theorem (7) to the function $\Psi/g$, we obtain the representation

$$\Psi(\mathbf{x}_1,\mathbf{x}_2,\cdots\mathbf{x}_1)=g(r_{12},r_{13},r_{23}\cdots)\times\sum_K C_K\Psi_K(\mathbf{x}_1,\mathbf{x}_2,\mathbf{x}_3\cdots\mathbf{x}_N), \quad (24)$$

where, for fixed $g$, the coefficients $C_K$ may be determined by the variation principle. The antisymmetric wave function $\Psi$ is here described in a configuration space spanned by the basic vectors $g\Psi_K$, and the main practical difficulty is connected with the evaluation of such parts of the matrix elements $\mathcal{H}_{KL}$ where the integrand contains three or more interelectronic distances $r_{ij}$. Convenient test problems are here provided by the lithium and beryllium atoms.

In this connection, I would also like to draw attention to the important work on the lighter atoms carried out by Jucys and his group in Vilna, Lithuania, by means of the method of superposition of configurations or modifications of this scheme by introducing self-consistency requirements or convenient correlation factors. Since 1947, the group has published more than 30 papers on this subject in various journals in U.S.S.R., but unfortunately only a few of them have been translated into English.

### (c) Symmetry Properties and Projections

In studying the solution of the Schrödinger equation $\mathcal{H}\Psi=E\Psi$, we now consider the case when there exists an auxiliary operator $\Lambda$ symmetric in all coordinates which commutes with the Hamiltonian $\mathcal{H}$ so that $\Lambda\mathcal{H}=\mathcal{H}\Lambda$. For the sake of simplicity, the eigenvalue problem

$$\Lambda\Theta_l=\lambda_l\Theta_l \quad (25)$$

is assumed to have only a finite number of discrete solutions for $l=1, 2, 3, \cdots n$, where later we can let $n\rightarrow\infty$ provided our expressions remain convergent. Let $\Theta$ be an arbitrary trial function which is expandable in the form

$$\Theta=\sum_{l=1}^{n} a_l\Theta_l, \quad (26)$$

and let us now try to construct an operator $\mathcal{O}_k$ which, applied to $\Theta$, selects only the term corresponding to $l=k$:

$$\mathcal{O}_k\Theta=a_k\Theta_k. \quad (27)$$

Since one can visualize the expansion (26) by considering a Hilbert space spanned by the basic vectors $\Theta_1$, $\Theta_2, \cdots\Theta_n$, in which it is required to resolve an arbitrary vector $\Theta$ into components along the axes, one can say that the meaning of $\mathcal{O}_k$ is to take the "parallel projection" of $\Theta$ on the axis $\Theta_k$. Repeated use of $\mathcal{O}_k$ does not change the result, which leads to the relation $\mathcal{O}_k^2=\mathcal{O}_k$ characteristic for the projection operators.[22] Since further, $\mathcal{O}_k\mathcal{O}_l=0$ for $k\neq l$, one has

$$\mathcal{O}_k\mathcal{O}_l=\mathcal{O}_k\delta_{kl}. \quad (28)$$

It is immediately clear that the product operator[3]

$$\mathcal{O}_k=\prod_{l\neq k}\frac{\Lambda-\lambda_l}{\lambda_k-\lambda_l}, \quad (29)$$

has the selective property (27) desired. Since

$$(\Lambda-\lambda_l)\Theta_l\equiv 0,$$

all terms in (26) having $l\neq k$ are annihilated, whereas the term for $l=k$ survives the operation in an unchanged form. In (29) each eigenvalue $\lambda_l$ appears only once, even if it is multiple, and, if the eigenvalue $\lambda_k$ itself is degenerate, the operator $\mathcal{O}_k$ does not project on a specific axis but on the entire subspace associated with this eigenvalue.

In the following we assume that $\Lambda$ is either self-adjoint or normal, i.e., commutes with its Hermitean adjoint operator $\Lambda^\dagger$ so that $\Lambda\Lambda^\dagger=\Lambda^\dagger\Lambda$. This condition is fulfilled both by the operators corresponding to the physical observables and by the unitary operators associated with the basic symmetry operations. A normal operator $\Lambda$ may be written as the sum of a Hermitean part and an anti-Hermitean part, which commute with each other and hence have simultaneous eigenfunctions $\Theta_l$. These functions are then also simultaneous eigenfunctions to $\Lambda$ and $\Lambda^\dagger$ associated with the eigenvalues $\lambda_l$ and $\lambda_l^*$, respectively. Since eigenfunctions $\Theta_k$ and $\Theta_l$ associated with different eigenvalues are now automatically orthogonal, the operators $\mathcal{O}_k$ describe orthogonal projections. The selection effect of the operator $\mathcal{O}_k^\dagger$ is identical with that of $\mathcal{O}_k$, i.e., the operator is essentially self-adjoint. As a general rule self-adjoint projection operators select orthogonal projections, and reverse.

Since $\Lambda$ commutes with $\mathcal{H}$, the same holds for the projection operators $\mathcal{O}_k$ formed by (29) and, instead of (28), we obtain now the more useful relations

$$\mathcal{O}_k^\dagger\mathcal{H}\mathcal{O}_l=\mathcal{H}\mathcal{O}_k\delta_{kl}, \quad \mathcal{O}_k^\dagger\mathcal{O}_l=\mathcal{O}_k\delta_{kl}, \quad (30)$$

which form the basis for the practical applications of the theory. Let $\Phi_{\mathrm{I}}$ and $\Phi_{\mathrm{II}}$ be two arbitrary trial functions which are identical or not. By means of the quantum mechanical "turnover rule" and (30), we

[21] P. O. Löwdin and L. Redei, Phys. Rev. **114**, 752 (1959).

[22] J. von Neumann, *Math. Grundlagen der Quantenmechanik* (Dover Publications, New York, 1943), p. 41.

obtain for their projections

$$\int (\mathcal{O}_k\Phi_{\rm I})^* \mathcal{H}(\mathcal{O}_l\Phi_{\rm II})(dx) = \int \Phi_{\rm I}^* \mathcal{O}_k^\dagger \mathcal{H}\mathcal{O}_l\Phi_{\rm II}(dx)$$

$$= \delta_{kl}\int \Phi_{\rm I}^* \mathcal{H}\mathcal{O}_k\Phi_{\rm II}(dx), \quad (31)$$

$$\int (\mathcal{O}_k\Phi_{\rm I})^* (\mathcal{O}_l\Phi_{\rm II})(dx) = \int \Phi_{\rm I}^* \mathcal{O}_k^\dagger \mathcal{O}_l\Phi_{\rm II}(dx)$$

$$= \delta_{kl}\int \Phi_{\rm I}^* \mathcal{O}_k\Phi_{\rm II}(dx), \quad (32)$$

showing that, for $k \neq l$ the projections $(\mathcal{O}_k\Phi_{\rm I})$ and $(\mathcal{O}_l\Phi_{\rm II})$ are not only *orthogonal* but also *noninteracting* with respect to $\mathcal{H}$.

Let us now consider an antisymmetric wave function $\Psi = \Psi(\mathbf{x}_1,\mathbf{x}_2,\cdots\mathbf{x}_N)$ which is already an eigenfunction to $\Lambda$ so that $\Lambda\Psi = \lambda\Psi$ and $\mathcal{O}\Psi = \Psi$, where for a moment we have dropped index $k$. Letting the operator $\mathcal{O}$ work on both members of (7), we obtain

$$\Psi = \mathcal{O}\Psi = \sum_K C_K(\mathcal{O}\Psi_K), \quad (33)$$

showing that a simultaneous eigenfunction to $\mathcal{H}$ and $\Lambda$ may actually be expanded into the projections $\mathcal{O}\Psi_K$ of the Slater determinants $\Psi_K$. The functions $\mathcal{O}\Psi_K$ usually are not linearly independent and expansion (33) often may be considerably condensed by eliminating the redundancies. Since $\Lambda$ was assumed to be fully symmetric in the coordinates $\mathbf{x}_1$, $\mathbf{x}_2$, $\mathbf{x}_3$, $\cdots\mathbf{x}_N$ and hence ommutes with all the permutations $P$, the projection operator $\mathcal{O}$ defined by (29) commutes with the antisymmetrization operator $\mathcal{O}_{\rm AS}$ defined by (3). The antisymmetric character of $\Psi$ is hence undisturbed by the projection $\mathcal{O}$.

The existence of an auxiliary normal operator $\Lambda$, which commutes with $\mathcal{H}$ and $\mathcal{O}_{\rm AS}$, could be used to split the secular equation (9), but we note that a much more powerful approach could be obtained by splitting the basis $\{\Psi_K\}$ by the projection operators $\mathcal{O}_k$ into the orthogonal and noninteracting bases

$$\{\mathcal{O}_1\Psi_K\},\quad \{\mathcal{O}_2\Psi_K\},\quad \cdots\{\mathcal{O}_n\Psi_K\}. \quad (34)$$

This leads automatically to a splitting of the secular equation (9), since we obtain a smaller secular equation for each one of the eigenvalues $\lambda_1$, $\lambda_2$, $\cdots\lambda_n$, but for different values of $\lambda_k$, one could now actually choose different basic spin orbitals $\{\psi_j\}$ or different adjustable parameters describing this set. This new degree of freedom in the projection technique is particularly of importance in calculations using truncated sets.

Projection operators of the simple product type (29) have so far been successfully used for creating wave functions of pure total spin[23] and pure angular momenta[24] in general. They have also been used for constructing Bloch functions in the case of translational symmetry.[25] It is clear that a much more general class of projection operators or fundamental idempotents exist in the group algebra,[26] and that they could be used in exactly the same way, provided that they commute with the Hamiltonian and hence fulfill the relations (30). The antisymmetry projection operator (3) is a typical example. Many otherwise useful projection operators of this category are unfortunately ruled out since they do not commute with the Hamiltonian and $\mathcal{O}_{\rm AS}$, and particularly the treatment of degeneracies requires special consideration.[27]

One can improve the speed of convergency of expansion (33) by introducing a correlation factor $g = g(r_{12},r_{13},r_{23},\cdots)$ which is totally symmetric. This leads to the form

$$\Psi(\mathbf{x}_1,\mathbf{x}_2,\cdots\mathbf{x}_N) = g(r_{12},r_{13},r_{23},\cdots)\sum_K C_K(\mathcal{O}\Psi_K), \quad (35)$$

which is a combination of (24) and (33).

### 2. EXTENSION OF THE HARTREE-FOCK SCHEME

If the expansion (7) is truncated to a single Slater determinant $D$ and the $N$ spin orbitals involved are optimized to give the best total energy possible, one obtains the conventional Hartree-Fock scheme. Various extensions of this scheme are possible by starting out from the expansions (24), (33), and (35), and truncating them to a single dominant term. Hence we obtain, in order,

$$\Psi \approx D, \quad \Psi \approx gD,$$
$$\Psi \approx \mathcal{O}D, \quad \Psi \approx g\mathcal{O}D. \quad (36)$$

Common features of all these approximations are that the number of spin orbitals involved equals the number of electrons, that, except for an irrelevant normalization factor, the total wave function is invariant under linear transformations of these basic functions, and that one can without loss of generality choose the basis $\psi_1$, $\psi_2$, $\cdots\psi_N$ orthonormal. The fundamental quantity in all four approximations is apparently the Fock-Dirac density matrix,

$$\rho(\mathbf{x}_1,\mathbf{x}_2) = \sum_{k=1}^{N}\psi_k^*(\mathbf{x}_1)\psi_k(\mathbf{x}_2), \quad (37)$$

which is also invariant under transformations of the basis.

Let us start by considering the case $\Psi \approx \mathcal{O}D$, where the total wave function is approximated by a projection of a single Slater determinant. This wave function is

[23] P. O. Löwdin, Phys. Rev. **97**, 1509 (1955); Colloq. intern. centre natl. recherche sci. (Paris) **82**, 23 (1958).

[24] See a series of research reports from the Uppsala Quantum Chemistry Group 1956–1959 (unpublished).

[25] P. O. Löwdin, Advances in Phys. **5**, 56 (1956).

[26] See, e.g., E. P. Wigner, *Gruppentheorie und ihre Anwendung auf die Quantenmechanik der Atomspektren* (Vieweg und Sohn, Braunschweig, 1931); D. E. Littlewood, *Theory of Group Characters and Matrix Representations of Groups* (Clarendon Press, Oxford, 1950).

[27] Cf. P. O. Löwdin, reference 3, p. 290.

not only antisymmetric under permutations but is also an eigenfunction to the auxiliary operator $\Lambda$ commuting with $\mathcal{H}$ or fulfills certain additional symmetry requirements connected with $\mathcal{O}$. The function $\mathcal{O}D$ could be expressed as a sum of determinants—a "codetor" according to Boys's terminology—but the condensed projection form has several advantages, e.g., in showing the invariance under linear transformations of the basis $\{\psi_k\}$ and in connection with the use of the "turn-over rule" in (31) and (32). For the normalization integral, we obtain, according to (32),

$$c=\int|\Psi|^2(dx)=\int D^*\mathcal{O}D(dx), \tag{38}$$

and it turns out that for very important categories of projections (e.g., those connected with angular momenta), this quantity is a *constant* characteristic for the projection $\mathcal{O}$ under certain conditions on the spin orbitals $\{\psi_k\}$. By using (31), we obtain for the total energy

$$\langle\mathcal{H}\rangle=\int D^*(c^{-1}\mathcal{H}_{\text{op}}\mathcal{O})D(dx), \tag{39}$$

which is exactly the same expression as in the conventional Hartree-Fock theory except that $\mathcal{H}_{\text{op}}$ has been replaced by the composite Hamiltonian

$$\mathcal{H}_{\text{op}}'=c^{-1}\mathcal{H}_{\text{op}}\mathcal{O}. \tag{40}$$

It has been shown[28] that even in the case of complicated Hamiltonians containing many-particle interactions, the one-particle scheme based on a single determinant $D$ leads to Hartree-Fock equations of the conventional form

$$H_{\text{eff}}(1)\psi_k(\mathbf{x}_1)=\sum_{l=1}^{N}\psi_l(\mathbf{x}_1)\lambda_{lk}, \tag{41}$$

where, according to Eq. (40) in reference 28, the effective Hamiltonian $\mathcal{H}_{\text{eff}}$ depends only on the total Hamiltonian $\mathcal{H}_{\text{op}}$ and the density matrix $\rho$ defined in (37). Since $H_{\text{eff}}$ is invariant under unitary transformations of $\{\psi_k\}$, the right-hand member can further be diagonalized to the form $\epsilon_k\psi_k$. This implies that the variation problem $\delta\langle\mathcal{H}\rangle_{\text{av}}=0$ connected with (39) leads to extended Hartree-Fock equations of the type (41), where $H_{\text{eff}}'$ now depends on the operator $c^{-1}\mathcal{O}$ in addition to $\mathcal{H}_{\text{op}}$ and $\rho$. Hence the symmetry properties of the state under consideration influences now the effective Hamiltonian. Again the right-hand member can be diagonalized to the form $\epsilon_k\psi_k$, provided the normalization integral (38) remains invariant also under this unitary transformation.

So far, we have assumed that the normalization integral (38) is a constant $c$ characteristic for the projection $\mathcal{O}$. If this is not the case, one must, instead of (39), consider the more general expression

$$\langle\mathcal{H}\rangle_{\text{av}}=\int D^*(\mathcal{H}\mathcal{O})D(dx)\bigg/\int D^*\mathcal{O}D(dx), \tag{42}$$

and the variation principle $\delta\langle\mathcal{H}\rangle_{\text{av}}=0$ leads again to one-dimensional eigenvalue problems[28] which are generalizations of the type (41). This approach is mathematically more complicated, but it has the advantage that there are no extra constraints on the basic spin orbitals $\{\psi_k\}$ which could prevent an optimum lowering of the energy.

The split-shell method for treating two-electron systems, frequently discussed in the current literature,[6] may be considered as a special case of the general type $\Psi\approx\mathcal{O}D$. Two spin states are here possible—a singlet and a triplet—and, according to (29), the singlet projection operator takes the form ${}^1\mathcal{O}=1-S^2/1\cdot 2=\frac{1}{2}(1-P_{12}{}^{\zeta})$. Starting out from a single Slater determinant $D=\det\{u\alpha,v\beta\}$, one obtains

$${}^1\mathcal{O}D=\tfrac{1}{2}(\alpha_1\beta_2-\beta_1\alpha_2)(u_1v_2+u_2v_1),$$

which is the $(u,v)$ form used in the split-shell theory. By using (38), one gets for the normalization integral $c=1+|\int u^*v(dr)|^2$, which is constant $c=1$, if $u$ and $v$ are assumed to be orthogonal. In general, however, it is essential that $u$ and $v$ may be overlapping and, in such a case, the energy variation must be based on (42).

The results obtained so far on helium[6,9], lithium[29] and beryllium[30] seem very promising, and further applications on the lighter atoms are now in progress. For molecules and crystals, the exact solution of the extended Hartree-Fock equations is a rather cumbersome numerical problem which can be replaced by approximate schemes of a simpler type, like the alternant-molecular orbital method.[31]

In conclusion, we also consider the extended Hartree-Fock schemes based on the approximations $\Psi\approx gD$ and $\Psi\approx g(\mathcal{O}D)$, where $g$ is a correlation factor

$$g=g(r_{12},r_{13},r_{23},\cdots).$$

The results on the helium atom[21] and the hydrogen molecule[20] obtained by using wave functions of the form $g(u)^2$ and $g(u,v)$ and simple correlation factors $g$ seem very promising, but, in order to extend this approach practically to many-electron systems, one has to master the energy integrals containing more than two interelectronic distances $r_{ij}$. In this connection, it is inter-

[28] P. O. Löwdin, Phys. Rev. **97**, 1490 (1955).

[29] G. H. Brigman and F. A. Matsen, J. Chem. Phys. **27**, 829 (1957); R. P. Hurst, J. D. Gray, G. H. Brigman, and F. A. Matsen, Mol. Phys. **1**, 189 (1958); E. A. Burke and J. F. Mulligan, J. Chem. Phys. **28**, 995 (1958).

[30] G. H. Brigman, R. P. Hurst, J. D. Gray, and F. A. Matsen, J. Chem. Phys. **29**, 251 (1958).

[31] P. O. Löwdin, Nikko Symposium on Molecular Physics, 13 (Maruzen, Tokyo, 1954); Phys. Rev. **97**, 1509 (1955); T. Itoh and H. Yoshizumi, J. Phys. Soc. Japan **10**, 201 (1955); H. Yoshizumi and T. Itoh, J. Chem. Phys. **23**, 412 (1955); Busseiron Kenkyu **82**, 13 (1955) (in Japanese).

TABLE I. Different expansion methods and extended Hartree-Fock schemes with helium errors in kcal/mole.

| Expansion | Principal term | Helium error[a] |
|---|---|---|
| $\Psi=\Sigma_K C_K\Psi_K$ | $\Psi\approx D$ | 26.3 |
| $\Psi=\Sigma_K C_K(\mathcal{O}\Psi_K)$ | $\Psi\approx\mathcal{O}D$ | 16.0 |
| $\Psi=g\,\Sigma_K C_K\Psi_K$ | $\Psi\approx gD$ | 2.3 |
| $\Psi=g\,\Sigma_K C_K(\mathcal{O}\Psi_K)$ | $\Psi\approx g\mathcal{O}D$ | 1.2 |

[a] The errors refer to a case where $g=1+\alpha r_{12}$ and $\mathcal{O}$ stands for a spin projection only.

esting to note that Krisement[32] has pointed out that the form $\Psi\approx gD$ is closely connected both with Wigner's[33] classical theory for the electrons in an alkali metal and Bohm and Pines'[34] plasma model.

### 3. DISCUSSION

In this paper, we have tried to show how the expansion (7) of configurations could be gradually changed by the introduction of a correlation factor

$$g=g(r_{12},r_{13},r_{23},\cdots)$$

or a projection operator $\mathcal{O}$, or both. It is anticipated that this transformation would lead to an improvement in the rate of convergency, and, in Sec. 2, we have discussed the various Hartree-Fock schemes (36) which are obtained by truncating the expansions to a single dominant term. One gets an idea of the strength of the various methods from Table I, where we have listed the helium errors[9,29] in kcal/mole for the optimized truncated wave functions.

Let us now discuss the physical reasons for the obvious improvement which comes from the introduction of $\mathcal{O}$ and $g$. The conventional Hartree-Fock scheme having $\Psi\approx D$ is originally based on the independent-particle model, in which one neglects the fact that the electrons actually repel each other, so that each electron is surrounded by a "Coulomb hole" with respect to all other electrons. Since the electrons apparently try to avoid each other, there is a certain correlation between their movements, and the omission of this effect causes a so-called "correlation error." This picture is partly changed by the introduction of the antisymmetry requirement through the use of (3), since each electron is now surrounded by a "Fermi hole" with respect to all other electrons having the same spin.[35] The correlation error in the Hartree-Fock scheme can hence be essentially attributed to electrons having antiparallel spins, and 1.1 ev is a rather typical figure for a single electron pair. According to the virial theorem, this correlation error consists of an error of $-1.1$ ev in the kinetic energy, which comes out so low since the electrons are permitted to make simpler movements when they do not have to avoid each other, and an error of $+2.2$ ev in the Coulomb potential.

The pairing of electrons with antiparallel spins in the same orbital is a characteristic feature of the conventional Hartree-Fock scheme, which is mathematically convenient since it enables us to construct pure spin functions easily, but which is physically more doubtful since it introduces large correlation errors. If the pure spin functions are instead created by spin-projection operators, one can permit "different orbitals for different spins" and, since the electrons having antiparallel spins now get a possibility to avoid each other in space, the correlation error starts going down. A still better result can be obtained by also introducing other projection operators, e.g., in the helium case, the orbital angular-momentum operator creating $S$ states.

If the correlation factor $g=g(r_{12}r_{13},r_{23},\cdots)$ is chosen so that it has its smallest value when any $r_{ij}=0$, the form $\Psi\approx gD$ corresponds to the introduction of a "Coulomb hole" for both parallel and antiparallel spins. The correlation error goes then down essentially, and, by combining the use of $g$ and $\mathcal{O}$, one can apparently in the helium case approach a good "chemical" accuracy of about 1.2 kcal/mole even with the simple function $g=1+\alpha r_{12}$.

The main advantage of the Hartree-Fock schemes here described comes from the fact that they have a strong physical and chemical visuality which leads to a simple model of the system under consideration. On the other hand, the solution of the corresponding Hartree-Fock equations is usually a numerically extremely cumbersome procedure involving nonlinear integro-differential equations, and, in this connection, one should observe that it is usually considerably simpler to solve linear secular equations of the type (9). In the latter case, there are further no limitations on the accuracy obtainable, and the final results can always be simplified by going over to the natural expansions.

[32] O. Krisement, Phil. Mag. **2**, 245 (1957).
[33] E. Wigner, Phys. Rev. **46**, 1002 (1934); Trans. Faraday Soc. **34**, 678 (1938).
[34] See, e.g., D. Pines, Solid State Phys. **1**, 368 (1957).
[35] E. Wigner and F. Seitz, Phys. Rev. **43**, 804 (1933); J. C. Slater, ibid. **81**, 385 (1951).

IV-4 Reprinted from *Proceedings of the Royal Society* (London) A**220**, 446 (1953)

# The molecular orbital theory of chemical valency XVI. A theory of paired-electrons in polyatomic molecules

By A. C. Hurley,† Sir John Lennard-Jones, F.R.S.‡ and J. A. Pople

*Department of Theoretical Chemistry, University of Cambridge*

(*Received* 14 *July* 1953)

A further step is made in the development of the orbital theory of molecular structure, the object being to allow for the tendency of two electrons in the same space orbital to keep apart because of the repulsive field between them. The description of a molecule is generalized in such a way as to bracket together electrons in pairs, the simple theory of molecular orbitals being used as a starting point for a wider treatment. If two orbitals are used for each electron pair (paired electron orbitals), it is shown that the theory contains as special cases both previous results of the present series of papers and also the valence-bond wave function which is shown to be an alternative particular approximation to the general function.

## 1. Introduction

It has become evident in the development of the theory of molecular structure, given in earlier parts of this series (Lennard-Jones 1949; Lennard-Jones & Pople 1950) and elsewhere (Lennard-Jones 1952), that one of the main difficulties still to be overcome is the calculation of the effect of the electrostatic forces on the spatial arrangement of the electrons relative to each other. When electrons are assigned in pairs to orbitals, the spins being opposite, as is done in describing the ground states of most molecules, it appears that the exclusion principle plays the primary role of keeping apart those electrons which have the same kind of spin. Electrons of the same spin may thus be said to be correlated in their spatial arrangement by the property of exclusion. Electrostatic forces play only a secondary role in causing electrons of the same spin to avoid each other. Satisfactory calculations of the energy of interaction of electrons of the same spin can therefore be made by assigning the electrons to molecular orbitals because these automatically take into account the effect of exclusion. Similar calculations of the interaction of electrons of opposite spin are not so satisfactory, and the reason for this is that the exclusion principle does not affect their spatial correlation. The position of one electron relative to another of opposite spin is influenced only by the electrostatic force between them. It thus becomes necessary to consider the effect of these forces on the wave functions of the two electrons, considered as a pair. This means that the wave function of a pair cannot be regarded simply as a product of two functions, each of three co-ordinates, but must be expressed as a six-dimensional function. This function must be such as to represent the tendency of two electrons to avoid being at the same place at the same time. Round each electron there is an 'atmosphere' of electron density due to the other, and on this distribution the repulsion between the electrons has an important influence.

† Now at Commonwealth Scientific and Industrial Research Organization, Division of Industrial Chemistry, Melbourne, Australia.

‡ Now at the University College, Keele, Staffordshire.

The purpose of this paper is to develop the idea that the description of a molecule requires generalization in such a way as to bracket together electrons in pairs and yet to use the molecular orbital theory as a starting point for a logical treatment. The wave function of the molecule as a whole must be made anti-symmetrical for the interchange of any two electrons. For electrons within a pair this property of anti-symmetry is represented by a spin factor. The spatial part of the wave function for a pair is, however, symmetrical as regards spatial co-ordinates. It is this part of the wave function which calls for a more general formulation than has hitherto been made. It will be shown that a simple modification opens up the possibility of great improvements in the theory. Existing methods of dealing with molecular structure, whether by molecular orbitals or by the pairing of electrons in valence bonds, are found to be particular cases of this more general treatment.

## 2. The theory of paired-electron functions

In the molecular orbital theory, as given in earlier parts of this series, it was assumed that the total electronic wave function of a molecule in its ground state may be expressed as a determinant of one-electron orbital functions. Thus when there are $N$ spatial orbitals, each doubly occupied with electrons of opposite spin, the normalized total wave function $\Phi$ may be written

$$\Phi = \{(2N)!\}^{-\frac{1}{2}} \sum_P (-1)^P P\{\psi_1(1)\,\psi_1(2)\,\alpha(1)\,\beta(2)\,\psi_2(3)\,\psi_2(4)\,\alpha(3)\,\beta(4)\ldots \times \psi_N(2N-1)\,\psi_N(2N)\,\alpha(2N-1)\,\beta(2N)\}, \quad (2\cdot1)$$

where the summation is over all the permutations $P$ of the numbers 1, 2, ..., $2N$, $(-1)^P$ being $\pm 1$ according to the parity of the permutation. It may be assumed without loss of generality that the orbitals $\psi_1 \ldots \psi_N$ form an orthonormal set (Lennard-Jones 1949*a*),

$$\int \bar{\psi}_m \psi_n \,\mathrm{d}x = \delta_{mn} \quad (m, n = 1, \ldots, N). \quad (2\cdot2)$$

In order to introduce some form of correlation between the motions of two electrons in the same orbital, we shall replace the products $\psi_n(i)\,\psi_n(j)$ occurring in (2·1) by a more general function $\psi_{nn}(i,j)$. The total wave function now becomes

$$\Phi = \{(2N)!\}^{-\frac{1}{2}} \sum_P (-1)^P P\{\psi_{11}(1,2)\,\alpha(1)\,\beta(2)\,\psi_{22}(3,4)\,\alpha(3)\,\beta(4)\ldots \times \psi_{NN}(2N-1,2N)\,\alpha(2N-1)\,\beta(2N)\}. \quad (2\cdot3)$$

To make the treatment parallel to that of the determinantal function (2·1), we impose the following conditions on $\psi_{nn}(i,j)$:

$$\psi_{nn}(i,j) = \psi_{nn}(j,i), \quad (2\cdot4)$$

$$\iint \bar{\psi}_{nn}(i,j)\,\psi_{nn}(i,j)\,\mathrm{d}x_i\,\mathrm{d}x_j = 1, \quad (2\cdot5)$$

$$\int \bar{\psi}_{mm}(i,k)\,\psi_{nn}(i,j)\,\mathrm{d}x_i = 0 \quad (m \neq n), \quad (2\cdot6)$$

(2·5) and (2·6) being the generalization of the orthonormality condition (2·2). The condition (2·6) imposes restrictions on the functions that can be used, but (2·3) is still a powerful generalization of (2·1) which can allow for correlation between paired electrons.

The Hamiltonian $\mathscr{H}$ for the system is given by

$$\mathscr{H} = -\frac{1}{2}\sum_{i=1}^{2N} \nabla_i^2 - \sum_{i=1}^{2N}\sum_{\alpha} Z_\alpha/r_{i\alpha} + \sum_{i<j}^{2N}(1/r_{ij}). \tag{2·7}$$

Since the wave function (2·3) is normalized, the total electronic energy of the system is

$$\begin{aligned} E &= \int \bar{\Phi}\mathscr{H}\Phi \, d\tau \\ &= \{(2N)!\}^{-1}\int \sum_P (-1)^P P\{\bar{\psi}_{11}(1,2)\bar{\alpha}(1)\bar{\beta}(2)\ldots\} \\ &\qquad \times \mathscr{H}\sum_{P'}(-1)^{P'}P'\{\psi_{11}(1,2)\alpha(1)\beta(2)\ldots\}\,d\tau. \end{aligned} \tag{2·8}$$

As $\mathscr{H}$ is totally symmetric in all its variables, this expression can be reduced by permutation theory to give

$$E = \int \{\bar{\psi}_{11}(1,2)\bar{\alpha}(1)\bar{\beta}(2)\ldots\}\mathscr{H}\sum_P(-1)^P P\{\psi_{11}(1,2)\alpha(1)\beta(2)\ldots\}\,d\tau. \tag{2·9}$$

Most terms in this series vanish. All permutations which alter the spin of any electron lead to zero contributions by integration over spin space. Also, since the Hamiltonian is a sum of terms not involving more than two electrons at a time, all permutations of more than two electrons will vanish by virtue of the extended orthogonality condition (2·6). Thus the only terms that do not vanish arise from the identical permutation and all single permutations of two electrons of the same spin. If we write

$$\left.\begin{aligned} (n\|H\|n) &= \iint \bar{\psi}_{nn}(1,2)\{-\tfrac{1}{2}\nabla_1^2 - \sum_\alpha Z_\alpha/r_{1\alpha}\}\psi_{nn}(1,2)\,dx_1\,dx_2, \\ (n\|G\|n) &= \iint \bar{\psi}_{nn}(1,2)\,(1/r_{12})\,\psi_{nn}(1,2)\,dx_1\,dx_2, \\ (mn\|G\|mn) &= \iiiint \bar{\psi}_{mm}(1,2)\,\bar{\psi}_{nn}(3,4)\,(1/r_{13})\,\psi_{mm}(1,2)\,\psi_{nn}(3,4)\,dx_1\,dx_2\,dx_3\,dx_4, \\ (mn\|G^*\|nm) &= \iiiint \bar{\psi}_{mm}(1,2)\,\bar{\psi}_{nn}(3,4)\,(1/r_{13})\,\psi_{mm}(3,2)\,\psi_{nn}(1,4)\,dx_1\,dx_2\,dx_3\,dx_4, \end{aligned}\right\} \tag{2·10}$$

the expression for the electronic energy can be written

$$E = \sum_n [2(n\|H\|n) + (n\|G\|n)] + \sum_{m\neq n}[2(mn\|G\|mn) - (mn\|G^*\|nm)]. \tag{2·11}$$

Each term in (2·11) has a simple physical interpretation. $(n\|H\|n)$ represents the kinetic energy of an electron plus its potential energy due to the nuclei. $(n\|G\|n)$

is the interaction between the two electrons in $\psi_{nn}$, and $(mn \| G \| mn)$ represents the averaged or 'Coulomb' electrostatic energy between two electrons associated with different paired electron functions. The term $(mn \| G^* \| nm)$ arises from the single permutation in (2·9) and represents a generalized exchange energy. This contribution is a consequence of the correlation between electrons of the same spin imposed by the anti-symmetry principle.

It we use the simple product $\psi_n(i)\,\psi_n(j)$ for $\psi_{nn}(i,j)$, we find that

$$\left.\begin{aligned} (n \| H \| n) &= H_{nn},\\ (n \| G \| n) &= (nn \mid G \mid nn),\\ (mn \| G \| mn) &= (mn \mid G \mid mn),\\ (mn \| G^* \| nm) &= (mn \mid G \mid nm),\end{aligned}\right\} \tag{2·12}$$

where
$$\left.\begin{aligned} H_{nn} &= \int \bar{\psi}_n\{-\tfrac{1}{2}\nabla^2 - \sum_\alpha Z_\alpha/r_\alpha\}\,\psi_n\,\mathrm{d}x,\\ (mn \mid G \mid pq) &= \iint \bar{\psi}_m(1)\,\bar{\psi}_n(2)\,(1/r_{12})\,\psi_p(1)\,\psi_q(2)\,\mathrm{d}x_1\,\mathrm{d}x_2.\end{aligned}\right\} \tag{2·13}$$

The expression (2·11) for the energy then reduces to

$$E = \sum_n [2H_{nn} + (nn \mid G \mid nn)] + \sum_{m \neq n} [2(mn \mid G \mid mn) - (mn \mid G \mid nm)], \tag{2·14}$$

which is the usual expression based on the single-determinant theory (Lennard-Jones 1949*a*; Lennard-Jones & Pople 1950). The energy expression (2·11) is clearly a direct generalization of this earlier result, the various terms having a close physical correspondence.

## 3. Paired electron orbitals

In order to proceed further with the theory, it is convenient to assume that the paired electron functions $\psi_{nn}(i,j)$ can be expressed as a symmetric bilinear form of one-electron orbitals

$$\psi_{nn}(i,j) = \sum_{rs} a_{rs}\,\psi_n^r(i)\,\psi_n^s(j), \tag{3·1}$$

where
$$\int \bar{\psi}_m^r(i)\,\psi_n^s(i)\,\mathrm{d}x_i = 0 \quad (m \neq n), \tag{3·2}$$

so that the condition (2·6) may be satisfied. By a linear transformation (3·1) can be reduced to the diagonal form

$$\psi_{nn}(i,j) = \sum_r \pm\,\psi_n^r(i)\,\psi_n^r(j). \tag{3·3}$$

This transformation will not destroy the orthogonality relation (3·2). The single-determinant molecular orbital theory corresponds to using a single product instead of a series (3·3).

In dealing with the problem of two electrons in the field of two nuclei (Lennard-Jones & Pople, part IX, 1951), it was found that a simple function which introduced some degree of correlation between the electrons was

$$\psi = f_s(1)\, f_s(2) - f_a(1)\, f_a(2), \tag{3·4}$$

where $f_s$ and $f_a$ were certain orthogonal functions. The first term gives the simple molecular orbital representation and does not alone lead to correlation between the electrons. The addition of the second term gives correlation in the direction of the molecular axis. This form is not the most general one, but, in the case of the hydrogen molecule, the second term was found to be important in correlating the positions of the two electrons and so in reducing the calculated value of their mutual potential energy. This reduction proved to be considerable, a particular form for $f_s$ and $f_a$ leading to a lowering of the calculated value of the electron-electron potential energy by as much as 26 kcal/mole. This is a large correction and improves the calculated value of the dissociation energy considerably.

We propose therefore to use this form of function for all $\psi_{nn}$. It would be possible to take further terms in an expansion (3·3), but the expression with two terms will indicate the type of result to be expected. We accordingly write

$$\psi_{nn}(i,j) = \psi_n(i)\, \psi_n(j) - \psi'_n(i)\, \psi'_n(j), \tag{3·5}$$

where

$$\int \bar{\psi}_m \psi_n \, \mathrm{d}x = \int \bar{\psi}_m \psi'_n \, \mathrm{d}x = \int \bar{\psi}'_m \psi'_n \, \mathrm{d}x = 0 \quad (m \neq n). \tag{3·6}$$

The orbitals $\psi_n$ and $\psi'_n$ are incompletely defined to the extent of a linear transformation with one degree of freedom. We may use this degree of freedom to require that

$$\int \bar{\psi}_n \psi'_n \, \mathrm{d}x = 0. \tag{3·7}$$

The normalization condition (2·6) requires

$$S_n^2 + S_n'^2 = 1, \tag{3·8}$$

where

$$S_n = \int \bar{\psi}_n \psi_n \, \mathrm{d}x, \quad S'_n = \int \bar{\psi}'_n \psi'_n \, \mathrm{d}x. \tag{3·9}$$

The energy-type expressions (2·10) now take the form

$$\left.\begin{aligned}
(n \| H \| n) &= S_n H_{nn} + S'_n H'_{nn}, \\
(n \| G \| n) &= (nn \mid G \mid nn) - 2(nn \mid G \mid n'n') + (n'n' \mid G \mid n'n'), \\
(mn \| G \| mn) &= S_m S_n (mn \mid G \mid mn) + S_m S'_n (mn' \mid G \mid mn') \\
&\quad + S'_m S_n (m'n \mid G \mid m'n) + S'_m S'_n (m'n' \mid G \mid m'n'), \\
(mn \| G^* \| nm) &= S_m S_n (mn \mid G \mid nm) + S_m S'_n (mn' \mid G \mid n'm) \\
&\quad + S'_m S_n (m'n \mid G \mid nm') + S'_m S'_n (m'n' \mid G \mid n'm').
\end{aligned}\right\} \tag{3·10}$$

The total electronic energy can then be written

$$E = \sum_n \{2S_n H_{nn} + 2S'_n H'_{nn} + (nn \mid G \mid nn) - 2(nn \mid G \mid n'n') + (n'n' \mid G \mid n'n')\}$$
$$+ \sum_{m \neq n} \{S_m S_n[2(mn \mid G \mid mn) - (mn \mid G \mid nm)] + 2S_m S'_n[2(mn' \mid G \mid mn')$$
$$- (mn' \mid G \mid n'm)] + S'_m S'_n[2(m'n' \mid G \mid m'n') - (m'n' \mid G \mid n'm')]\}, \tag{3·11}$$

where the notation of equation (2·13) has been extended in an obvious manner by writing

$$\left.\begin{aligned} H'_{nn} &= \int \bar{\psi}'_n\{-\tfrac{1}{2}\nabla^2 - \sum_\alpha Z_\alpha/r_\alpha\}\psi'_n\,\mathrm{d}x, \\ (mn' \mid G \mid pq') &= \iint \bar{\psi}_m(1)\,\bar{\psi}'_n(2)\,(1/r_{12})\,\psi_p(1)\,\psi'_q(2)\,\mathrm{d}x_1\,\mathrm{d}x_2, \end{aligned}\right\} \tag{3·12}$$

and other similar expressions.

Equation (3·11) reduces to known expressions in two special cases. If we put $\psi'_n \equiv 0$ for all $n$, we get the expression for $E$ in the single determinant molecular orbital theory (equation (2·14)). Also in the case of a two electron homonuclear bond, (3·11) agrees with the expression given by Lennard-Jones & Pople (1951) for such molecules.

Up to this point neither the two-electron functions $\psi_{nn}$ nor the paired electron orbitals $\psi_n$, $\psi'_n$ have been specified in detail. Following the general methods used in earlier papers it is possible to derive a set of coupled integro-differential equations for $\psi_n$ and $\psi'_n$ by a variational method, minimizing the total electronic energy $E$ as given by equation (3·11) subject to the constraints (3·6), (3·7) and (3·8).† At a lower level of accuracy, the solutions of these equations can be approximated by suitable linear combinations of given analytical functions such as atomic orbitals of the Slater type.

When dealing with polyatomic molecules of any size, it is important to know the extent to which the paired electron functions $\psi_{nn}(i,j)$ are localized. In previous papers of this series considerable emphasis has been laid on the two equivalent versions of the single-determinant molecular orbital theory based on localized or delocalized orbitals. The transformation properties making the two descriptions equivalent no longer apply to the generalized wave function (2·3), but there are grounds for believing that the optimum forms for $\psi_{nn}$ are of a localized nature. It was shown in part IV (Lennard-Jones & Pople 1950) that the interaction between electrons in the same orbital was largest if the localized or equivalent orbital representation was used. It therefore appears likely that the generalization to (2·3) or (3·5) will be most effective in reducing the total electronic energy if we start from the localized representation.

If the paired electron functions $\psi_{nn}$ are supposed localized (so that they correspond to distinct bonds, lone pairs and inner shells), equation (2·3) provides a straightforward method of constructing a wave function which corresponds to the general picture of the chemical structure of a saturated molecule while lacking some of the disadvantages of cruder functions. Different approaches such as the single deter-

† This has, in fact, been done but the equations are somewhat involved and for this reason have not been included in this paper.

minant molecular orbital theory and the valence bond method correspond to different approximations to the two electron functions $\psi_{nn}$. In the next section we shall show how the theory of paired electron orbitals can be developed in a slightly different manner, formally similar to the type of wave function used in valence bond theory.

## 4. Paired-electron orbitals and valence-bond theory

In the previous section we developed the theory of paired electron functions by using an approximate form

$$\psi_{nn}(i,j) = \psi_n(i)\,\psi_n(j) - \psi_n'(i)\,\psi_n'(j), \tag{4·1}$$

where $\psi_n$ and $\psi_n'$ are orthogonal orbitals (not necessarily normalized). By a transformation of the type

$$\left.\begin{aligned} \chi_n &= c_n(\psi_n + \psi_n'), \\ \chi_n' &= c_n(\psi_n - \psi_n'), \end{aligned}\right\} \tag{4·2}$$

where $c_n$ is a constant, $\psi_{nn}(i,j)$ can be written in the form

$$\psi_{nn}(i,j) = \tfrac{1}{2}c_n^{-2}\{\chi_n(i)\,\chi_n'(j) + \chi_n'(i)\,\chi_n(j)\}. \tag{4·3}$$

We shall choose $c_n$ so that $\chi_n$ and $\chi_n'$ are normalized. They will not generally be orthogonal, however, and if we write

$$\Delta_n = \int \bar{\chi}_n \chi_n' \,\mathrm{d}x, \tag{4·4}$$

(4·3) becomes $$\psi_{nn}(i,j) = \{2(1+\Delta_n^2)\}^{-\frac{1}{2}}\{\chi_n(i)\,\chi_n'(j) + \chi_n'(i)\,\chi_n(j)\}. \tag{4·5}$$

This expression is formally similar to that used by Heitler & London in their original treatment of the hydrogen molecule, although the functions $\chi_n$ and $\chi_n'$ need no longer be atomic orbitals. The use of a function such as (4·5) for the two-electron problem was discussed in part IX (Lennard-Jones & Pople 1951). Here we have a method of extending the use of such a function to polyatomic molecules.

The total electronic energy may be expressed in terms of matrix elements with respect to the orbitals $\chi_n$ and $\chi_n'$. The various terms in (2·11) become

$$\left.\begin{aligned}
(n\,\|\,H\,\|\,n) &= \frac{1}{2(1+\Delta_n^2)}\{(n\,|\,h\,|\,n) + 2\Delta_n(n\,|\,h\,|\,n') + (n'\,|\,h\,|\,n')\}, \\
(n\,\|\,G\,\|\,n) &= \frac{1}{1+\Delta_n^2}\{(nn'\,|\,g\,|\,nn') + (nn'\,|\,g\,|\,n'n)\}, \\
(mn\,\|\,G\,\|\,mn) &= \frac{1}{4(1+\Delta_m^2)(1+\Delta_n^2)}\{(mn\,|\,g\,|\,mn) + (m'n\,|\,g\,|\,m'n) + (mn'\,|\,g\,|\,mn') \\
&\quad + (m'n'\,|\,g\,|\,m'n') + 2\Delta_m[(mn\,|\,g\,|\,m'n) + (mn'\,|\,g\,|\,m'n')] \\
&\quad + 2\Delta_n[(mn\,|\,g\,|\,mn') + (m'n\,|\,g\,|\,m'n')] + 4\Delta_m\Delta_n(mn\,|\,g\,|\,m'n')\}, \\
(mn\,\|\,G^*\,\|\,nm) &= \frac{1}{4(1+\Delta_m^2)(1+\Delta_n^2)}\{(mn\,|\,g\,|\,nm) + (m'n\,|\,g\,|\,nm') + (mn'\,|\,g\,|\,n'm) \\
&\quad + (m'n'\,|\,g\,|\,n'm') + 2\Delta_m[(mn\,|\,g\,|\,nm') + (mn'\,|\,g\,|\,n'm')] \\
&\quad + 2\Delta_n[(mn\,|\,g\,|\,n'm) + (m'n\,|\,g\,|\,n'm')] + 4\Delta_m\Delta_n(mn\,|\,g\,|\,n'm')\},
\end{aligned}\right\} \tag{4·6}$$

where we have written

$$\left.\begin{aligned}(n \mid h \mid n') &= \int \bar{\chi}_n\{-\tfrac{1}{2}\nabla^2 - \sum_\alpha Z_\alpha/r_\alpha\}\chi'_n\,\mathrm{d}x,\\ (mn' \mid g \mid pq') &= \iint \bar{\chi}_m(1)\bar{\chi}'_n(2)(1/r_{12})\chi_p(1)\chi'_q(2)\,\mathrm{d}x_1\,\mathrm{d}x_2,\end{aligned}\right\} \tag{4·7}$$

and other similar expressions. The symbols $h$ and $g$ have been used instead of $H$ and $G$ to distinguish these matrix elements from those with respect to $\psi$-functions used in § 3. The energy expression for the single-determinant molecular orbital function can be recovered if we put $\chi_n \equiv \chi'_n$ for all $n$.

If we impose the further condition that $\chi_n$ and $\chi'_n$ are orthogonal ($\Delta_n = 0$), the energy expression derived from (4·6) simplifies considerably. It should be emphasized, however, that this is a further restriction of the wave function which may make it inaccurate. Further the single-determinant function ceases to be a particular case of the general theory. Putting $\Delta_n = 0$ in (4·6) leads to

$$\begin{aligned}E = {} & \sum_n [(n \mid h \mid n) + (n' \mid h \mid n') + (nn' \mid g \mid nn') + (nn' \mid g \mid n'n)]\\ & + \sum_{m \neq n} [(mn \mid g \mid mn) + (mn' \mid g \mid mn') - \tfrac{1}{2}(mn \mid g \mid nm) - \tfrac{1}{2}(mn' \mid g \mid n'm)].\end{aligned} \tag{4·8}$$

The two-electron integrals $(nn' \mid g \mid n'n)$, $(mn \mid g \mid nm)$ and $(mn' \mid g \mid n'm)$ are of the exchange type. If the other terms in (4·8) are referred to as the Coulomb energy, we can write

$$\begin{aligned}E = {} & \text{Coulomb energy} + (\text{exchange integral between bonded orbitals})\\ & - \tfrac{1}{2}(\text{exchange integrals between non-bonded orbitals}),\end{aligned} \tag{4·9}$$

which is the familiar form used in the valence bond theory.

The treatment given above forms a suitable general background for a critical examination of the valence-bond method of describing molecular wave functions of saturated molecules. If the valence-bond type expression (4·5) for the paired electron function is used, the total wave function (2·3) can be expanded, apart from the normalizing factor, as a sum of $2^N$ determinants obtained from

$$\begin{aligned}\Phi_0 = \det\{&\chi_1(1)\,\alpha(1)\,\chi'_1(2)\,\beta(2)\,\chi_2(3)\,\alpha(3)\,\chi'_2(4)\,\beta(4)\ldots\\ & \times \chi_N(2N-1)\,\alpha(2N-1)\,\chi'_N(2N)\,\beta(2N)\},\end{aligned} \tag{4·10}$$

by all possible interchanges of the pairs of symbols $(\chi_n, \chi'_n)$. If these determinants are written $\Phi_\nu$ after rearranging the order of columns so that the orbitals always occur in the same order as in $\Phi_0$, then the total wave function can be written

$$\Phi = K \sum_\nu \delta_1(\nu)\,\delta_2(\nu)\ldots\delta_N(\nu)\,\Phi_\nu, \tag{4·11}$$

where

$$\delta_n(\nu) = \left\{\begin{array}{ll} +1 & \text{if } \chi_n \text{ has } \alpha \text{ spin and } \chi'_n\ \beta \text{ spin in } \Phi_\nu,\\ -1 & \text{if } \chi_n \text{ has } \beta \text{ spin and } \chi'_n\ \alpha \text{ spin in } \Phi_\nu,\end{array}\right\} \tag{4·12}$$

and $K$ is a normalizing factor. If the $\chi$ functions are atomic orbitals, equation (4·11) is just the usual expression for the bond eigenfunction of the principal valence state of a saturated molecule (Eyring, Walter & Kimball 1944). It is clear, therefore, that the valence-bond wave function is a particular version of the theory of paired electron orbitals in which atomic orbitals are supposed to be adequate approximations to the $\chi$-functions of (4·5).

In deriving an approximate energy expression from the valence-bond wave function it has been customary to neglect the non-orthogonality between the component atomic orbitals, leading eventually to an expression such as (4·8) or (4·9). This is a very crude approximation, however, for the corresponding overlap integrals may often be large. This applies particularly to atomic orbitals whose electrons are paired, for, according to qualitative versions of the theory, these are chosen so as to make the overlap *as large as possible*. Further, after the approximation has been made, the exchange terms left in (4·8) or (4·9) are two electron integrals which are definitely positive. In order to make semi-empirical versions of the valence-bond theory fit experimental data, however, it is necessary to give these integrals negative values. This makes it clear that, whatever the merits of the valence-bond wave function itself, the approximations made in deriving (4·9) are so crude that the expression is quantitatively of little value. The calculations would be more satisfactory if based on the more accurate expression obtained from (4·6).

The principal points concerning the various theories can be illustrated if we write down the most general form for the paired electron function $\psi_{nn}$ that can be constructed from two given atomic orbitals $\phi_n$ and $\phi'_n$. This is

$$\psi_{nn}(1,2) = A\phi_n(1)\,\phi_n(2) + B[\phi_n(1)\,\phi'_n(2) + \phi'_n(1)\,\phi_n(2)] + C\phi'_n(1)\,\phi'_n(2). \quad (4\cdot 13)$$

If $\psi_{nn}$ is normalized, this function contains two degrees of freedom. The single-determinant molecular orbital theory supposes that $\psi_{nn}$ can be expressed as a simple product so that $B^2 = AC$ if the molecular orbitals are to be expressed as linear combinations of atomic orbitals. The valence-bond function is more restricted involving the two conditions $A = C = 0$, but it has the advantage of giving the correct solution for large internuclear separations. The general form (4·13) is usually described as a valence-bond function including ionic structures. However, by a suitable linear transformation, (4·13) can always be put in the form (3·5) or (4·5). It is clear, therefore, that the concept of resonance between covalent and ionic structures only has significance in terms of atomic orbital approximations. When the theory is generalized by replacing the atomic orbitals by optimum functions, the covalent and ionic terms are no longer uniquely separable and the concept of resonance with ionic structures is replaced by delocalization of the orbitals.

## 5. Conclusion

The principal result of this paper has been to show how a function constructed from paired electron functions or paired electron orbitals can be used to give a satisfactory interpretation of the electronic structure of a saturated molecule. By breaking the problem down to wave functions for single bonds, the method forms a suitable basis for using our considerable knowledge of an isolated two-electron bond in a theory of many-electron polyatomic molecules. As is clear from the previous section, theories based on atomic orbitals such as the valence-bond theory are particular forms of a general treatment. As a basis for future work on saturated molecules, the authors consider that the wave function based on paired electron orbitals is more general than anything proposed hitherto. At the same time it is a logical development of previous work, being a synthesis of the simple molecular orbital theory of polyatomic molecules with the more precise treatment of a two electron bond (part IX).

One of the authors (A. C. H.) is indebted to the Commonwealth Scientific and Industrial Research Organization (Australia) for a grant.

## References

Eyring, H., Walter J. & Kimball, G. E. 1944 *Quantum chemistry*, New York: Wiley & Sons.
Lennard-Jones, J. E. 1949*a* *Proc. Roy. Soc.* A, **198**, 1 (part I).
Lennard-Jones, J. E. 1949*b* *Proc. Roy. Soc.* A, **198**, 14 (part II).
Lennard-Jones, J. E. 1952 *J. Chem. Phys.* **20**, 1024.
Lennard-Jones, J. E. & Pople, J. A. 1950 *Proc. Roy. Soc.* A, **202**, 166 (part IV).
Lennard-Jones, J. E. & Pople, J. A. 1951 *Proc. Roy. Soc.* A, **210**, 190 (part IX).

# Chapter V

# *Pair-Correlations Theory of Ground-State Atoms and the "Many-Electron Theory of Atoms and Molecules" (Approach I, Part 1)*

## V-1

The first approach, by Sinanoğlu and co-workers, has developed in three stages. In the first part (1960–1961), the first-order Schrödinger equation is solved formally for ground states and the correlation problem reduced to this order into $N(N - 1)/2$ "He-like" variation-perturbation problems. In the second part (1961–1964), a nonperturbation theory is developed to study the remaining correlation effects involving $1, 2, 3, \ldots, n, \ldots N$ electrons at a time. For this, the N-electron Hilbert space is decomposed into the subspaces of $1, 2, 3, \ldots, n, \ldots$ N electrons. In the third part, the theory is generalized to arbitrary nonclosed shell and excited states, where it shows the existence of new correlation effects in such states. The effects have been calculated for over a hundred states of the $1s^2 2s^m 2p^n$ type of the atoms boron through sodium and their ions. Using the results, various atomic properties, such as excitation energies, term splitting ratios, electron affinities, and transition probabilities, are obtained. The predicted values agree with recent experimental values to high accuracy, including those cases where traditional methods fail.

The various effects in the theory may either be calculated nonempirically, or some of them (the "dynamical," "He-like" correlations) may be obtained from some data then used in predicting other data. In the present chapter and in Chapter VII, emphasis is on the fundamental aspects of the theory and the derivation of the methods involved. Recent applications of atomic physics interest are mentioned in Chapter IX. Theory has also been applied in various ways to molecules and to intermolecular forces. The latter applications are covered in a recent book by H. Margenau and N. R. Kestner, *Theory of Intermolecular Forces* (Pergamon Press, Oxford, 1968). Molecular applications are outside the scope of the present book.

Before going into the more detailed aspects, we shall list some of the physical conclusions of a quite general nature the theory has led to, which, however, should be studied further in heavier atoms and in molecules:

1. A ground-state atomic correlation energy is given predominantly by the sum of $N(N - 1)/2$ decoupled pair-correlation energies, $\epsilon_{ij}$ (possibly to within a few percent of $E_{corr}$).

2. The major part of the correlation wave function of a ground state in $\psi = \phi_0 + \chi$, where $\langle\phi_0 | \chi\rangle = 0$, includes the possible "unlinked" products $\Sigma' \hat{u}_{ij}\hat{u}_{k\ell}\hat{u}_{mn} \ldots$ of the decoupled pair-correlation functions as well as each of them singly. The correlation induced one-electron cor-

rections to the H.F. SCF orbitals, as well as the true ("linked") $n \geqslant 3$ correlation "clusters," $\hat{U}_{ijk\cdots}$, are far less important. (Nevertheless, they may have to be included, of course, for some other property, such as in hyperfine structure.)

3. A given $\hat{u}_{ij}$ or $\epsilon_{ij}$ is affected by the rest of the electrons of an atom on two counts: (a) the total H.F. SCF potential, and (b) the "exclusion effect," $\langle \hat{u}_{ij} / k \rangle = 0$ (all k being spin orbitals of the occupied H.F. "sea").

4. Some $\epsilon_{ij}$ are quite independent of environmental effects (a) and (b), and of Z and N. They, then, are nearly the same for a whole lot of atoms and ions (and molecules) in which they occur. Such transferable values can be used in a "pair-aufbau" of atomic correlation energies. They seem to be the kind of $\epsilon_{ij}$ caused by the steep, short-range parts of the $m_{ij}$. Their $\hat{u}_{ij}$ are of the slowly convergent type in C.I. and may be called "dynamical" correlations.

5. Other $\epsilon_{ij}$ may be largely due to near-degeneracy-type strong C.I. mixing within the same shell (within the L shell, the M shell, etc.). Such $\epsilon_{ij}$ not only strongly increase with Z [Linderberg and Shull (1960)], but decrease linearly with N (Chapter VII-5) and depend strongly on state (see the question of Wigner in Chapter III-2, p. 1005, end of first paragraph, on the $^3P$, $^1D$, and $^1S$ states of $O^{2+}$ and the reasons in Chapters VII-2 and IX-4). These types "of nondynamical" $\epsilon_{ij}$ (which may be included in the MCSCF of Jucys and the "complexes" of Layzer) nevertheless contain a small amount of "dynamical" ("helium-like"; $r_{ij}$-dependent) correlation as well (Chapter VII-2).

6. All pair correlations need to be included in estimating the correlation energy, not just those in doubly occupied orbitals. The sum of interorbital correlations (such as 2s, $2p_x$, $2p_y$, $2p_z$, etc.; Chapter VII-5, 7, and 8) contribute to the total correlation energy as much as and often even more than doubly occupied orbital pair correlations. This is true even with "localized orbitals."

7. The above are mostly for ground states. In general non-closed-shell excited states (Chapter IX-2), there are *two* equally important (Chapter IX-4) "nondynamical"-type (i.e., strongly Z, N, and state specific) correlation effects: "internal correlation" (C.I. within the same shell, usually the subject of MCSCF), and a new one, "semi-internal correlation" (Chapter IX). These two effects complicate the correlation behavior in excited states. However, once they are calculated and taken out, as shown in Chapter IX, the remaining "all-external" correlations behave just like those in closed shells: made up of transferable, "dynamical," but now occupation-probability-weighted pair correlations.

8. Pair correlations in a closed-shell state consist only of "all-external" pair correlations (values given in Chapter IX). In a single determinantal non-closed-shell state, each of the $N(N-1)/2$ pair correlations is a "total pair correlation." It is made up of "internal," "semi-internal," and "all-external" contributions.

9. The following steps are involved in predicting the energy of an excited state: (a) restricted H. F. wave function and $E_{RHF}$ are obtained, (b) "internal" and "semi-internal" correlation effects are calculated by a finite C.I. procedure (automatic program available) (Chapter IX-4), and (c) "all-external" correlation energy is obtained combining group theoretic coefficients with values of "all-external" pair-correlation energies (listed in Chapter IX). The total E is the sum of (a), (b), and (c). Allowed transition probabilities are given to high accuracy by the wave function of steps (a) and (b) only.

The quantitative use of the above picture is subject to the remaining smaller effects to remain small. On the other hand, in some atomic properties, such as hyperfine structure, one may also need to go beyond the dominant effects. Both Approach I (MET) and another recent approach, i.e., one which makes use of diagrammatic and field theoretic techniques (Approach III), allow one to systematically refine the results by going to higher effects in different ways.

# 1. Reduction of the N-Electron Problem to N(N − 1)/2 "Helium-like" Problems in Ground-State Atoms—Variation-Perturbation Theory

As discussed in Chapter IV, the difficulty of methods like C.I. and "correlation factor" rapidly increases with the number of electrons, N. Accurate solutions are obtainable, on the other hand, for two-electron systems such as He and $H_2$.

It has been shown by Sinanoğlu (1961) that the first-order Schrödinger equation for nondegenerate states,

$$(H_0 - E_0)X_1 = (E_1 - H_1)\,\phi_0 \qquad \text{where } \langle \phi_0 \mid X_1 \rangle = 0 \tag{1}$$

treated by operator techniques separates into the first-order "helium-like" equations of N(N − 1) / 2 electron pairs (Chapter V-3). The equation for each spin-orbital pair $i(\mathbf{x}_i)$ and $j(\mathbf{x}_j)$ is like those of the ground and excited states of the helium atom, except that now the electrons move in the total H. F. potential of N electrons while the two "correlate" with what is left over of their mutual repulsion after their Coulomb and exchange potentials have been taken out (the "fluctuation potential"). The results of Chapter V-3 may be summarized in more compact notation as follows:

1. Given $\phi_0 = \mathcal{A}(1, 2, 3, \ldots, N)$ as the nondegenerate H. F. ground state and $H_0\phi_0 = E_0\phi_0$.

$$H_0 - E_0 = \sum_{i>1}^{N} e_i \tag{2}$$

$$e_i \equiv h_i^0 + V(x_i) - \epsilon_i \tag{3}$$

$$E_{HF} = \langle \phi_0 | H | \phi_0 \rangle = \langle \phi_0 | H_0 + H_1 | \phi_0 \rangle \equiv E_0 + E_1 \tag{4}$$

Equation (1) for the N-electron system is equivalent to the set of N(N − 1) /2 pair equations,

$$(e_i + e_j)\,\hat{u}_{ij}^{(1)} = -Qg_{ij}B(ij) \tag{5}$$

The $\hat{u}_{ij}^{(1)}$ is a pair-correlation function to first order for the antisymmetrized $[B \equiv 1/\sqrt{2}\,(1 - P_{12})]$ spin-orbital pair B(ij), with the properties

$$\hat{u}_{ij}^{(1)}(x_i, x_j) = -\hat{u}_{ij}^{(1)}(x_j, x_i) \tag{6}$$

$$Q\hat{u}_{ij}^{(1)} = \hat{u}_{ij}^{(1)} \tag{7}$$

$$\langle \hat{u}_{ij}^{(1)} | k \rangle \equiv \int \hat{u}_{ij}^{(1)}(x_1, x_2)^* k(x_1)\, dx_1 = 0 \tag{8}$$

for (k = 1, 2, 3, . . ., i, . . ., j, . . ., N) the H. F. spin orbitals.

The Q simply projects out the one-electron Fock–Dirac subspaces, $\rho_i$:

$$\rho_i = \sum_{k>1}^{N} |k\rangle \langle k| \tag{9}$$

yielding the rigorous "orbital-orthogonality" property of a $\hat{u}_{ij}^{(1)}$.

2. To first order, it does not matter if any one of the several equivalent procedures given in the Sinanoğlu (1961, 1962) papers are followed: (a) Eq. (5) is for "reducible pairs" such as $B(p_x^\alpha p_y^\beta)$, $B(1s_\alpha 2s_\beta)$, etc. One may equally use the analogous equations for "irreducible pairs," symmetry combinations such as $(1s2s, {}^1S)$, $(1s2s, {}^3S)$, $(2p^2; {}^3P)$, etc. The actual helium-atom equations would be like the "irreducible pairs" (Chapter V-3); however, for the pairs of the N-electron system, the invariant Q projection operator in Eq. (7) makes the "reducible" pairs, Eq. (5), equally valid. (b) One may solve the pair equations in (Chapter V-3) without the full Q in them and then "orbital-orthogonalize" the $\hat{u}_{ij}^{(1)}$'s to the H. F. spin orbitals by the Schmidt process. Alternatively, one may have the full "orbital orthogonality" (the full Q) in Eq. (5) from the outset. The results are equivalent to this order.

3. One may solve the inhomogeneous linear, decoupled equations, Eq. (5), for the $\hat{u}_{ij}$'s directly, *or* apply the variational principle to each. The use of the variation method is simpler here, in a sense, than the actual He atom. A pair like (1s2s) or $(2p^2)$ is a highly excited one in an actual He atom and involves the usual difficulties of the variation method when applied to excited states. Each of the pairs of Chapter V-3 or Eq. (5), on the other hand, may still be obtained by a *minimum principle*. It is proved [MET II (1962), and O. Sinanoğlu, *Phys. Rev.* **122**, 491 (1961), in conjunction with Chapter V-3] that each pair $\epsilon_{ij}^{(2)}$ is not only stationary but a minimum,

$$\epsilon_{ij}^{(2)} \leqslant 2\langle B(ij), g_{ij}\tilde{u}_{ij}^{(1)}\rangle + \langle \tilde{u}_{ij}^{(1)}, (e_i + e_j)\tilde{u}_{ij}^{(1)}\rangle \tag{10}$$

provided the trial functions $\tilde{u}_{ij}^{(1)} \longrightarrow \hat{u}_{ij}^{(1)}$ *are subject during variations to* $\langle \tilde{u}_{ij}^{(1)} | k\rangle = 0$ for $1 \leqslant k \leqslant N$, as they should be. When each minimum is achieved, one has a second-order pair-correlation energy $\epsilon_{ij}^{(2)}$, and

$$E_{corr} \approx E^{(2)} + E^{(3)} \tag{11}$$

$$E^{(2)} = \sum_{i>j}^{N} \epsilon_{ij}^{(2)} \tag{12}$$

where

$$\epsilon_{ij}^{(2)} \equiv \langle B(ij), g_{ij}\hat{u}_{ij}^{(1)}\rangle \tag{13}$$

4. Aside from the total H. F. potential $V(x_i)$ effect, the $\hat{u}_{ij}^{(1)}$ differs from a helium pair in its "orbital orthogonality" to the remaining $k \neq i, j$ H. F. "sea" orbitals as well. This is the *exclusion effect*. Using the detailed form of Q, in $\hat{u}_{ij}^{(1)} = Q\bar{u}_{ij}^{(1)}$, they may be written out explicitly as three- and four-electron "Fermi correlations" tacked on to free two-electron-ion-like correlations,

$$\bar{e}_{ij}^{(2)} \equiv \langle B(ij), g_{ij}\, \bar{u}_{ij}^{(1)}\rangle$$

Thus

$$E^{(2)} = \sum_{i>j}^{N} \langle B(ij), g_{ij}\bar{u}_{ij}^{(1)}\rangle + \text{exclusion effects} \tag{14}$$

(see Chapter V-2 and 3).

The equations above allow the application to N-electron atoms of the methods that were used by Hylleraas, Pekeris, C. Schwarz, and others on helium. The pair equations here are only slightly more difficult than those for the helium atom, in that N-electron H. F. exchange potentials appear in $e_i + e_j$ (there would be simpler exchange potential terms for, e.g., the 1s2s: ${}^3S$ helium state). For atoms, the integrals that come up when $\hat{u}_{ij}^{(1)}$ contains the $r_{ij}$ coordinate become lengthier due to the exchange terms, but are straightforward. For larger atoms and especially for molecules, the H. F. potential might well be replaced by the Slater–Kohn–Sham (see Chapter III) approximate exchange. This would simplify the calculations considerably while not greatly affecting the accuracy of the result.

It has been noted in the MET papers [Sinanoğlu (1962, 1964)] that rather than attack each of the pair equations of an atom by the same two-electron method, one might select a different method for each type of pair depending on its nature. For example, in the Be-atom case [MET calculations: Tuan and Sinanoğlu, *J. Chem. Phys.* **41**, 2677 (1964); Levine, Geller, and Taylor, *J. Chem. Phys.* **43**, 1727 (1965)], the $r_{ij}$ coordinate trial functions are used for the $1s^2$ pair, a $2s^2 \longleftrightarrow 2p^2$ mixing for the $2s^2$ pair (see below, however) and the more approximate core-polarization methods (see, e.g., Chapter V-2) for the 1s2s pairs.

It may be thought that the number of pairs to be calculated would increase still too rapidly, i.e., as $N(N-1)/2$, for larger N. Actually, there are only about N distinct pairs to be calculated due to symmetry. The rest of the $N(N-1)/2$ pairs differ in spin or spatial orientation only. In the neon atom, for example ($N = 10$), there are 11 distinct pairs rather than 45. The "irreducible pairs" are ($1s^2\ {}^1S$), ($1s2s\ {}^1S$), ($1s2s\ {}^3S$), ($2s^2\ {}^1S$), ($1s2p\ {}^1P$), ($1s2p\ {}^3P$), ($2s2p\ {}^1P$), ($2s2p\ {}^3P$), ($2p^2\ {}^3P$), ($2p^2\ {}^1D$), and ($2p^2\ {}^1S$). The "reducible pairs" in the s, $p_x$, $p_y$, $p_z$ unitary transform of the $1 \leqslant k \leqslant N$ Hartree-Fock set would be ($1s^2$), ($1s_\alpha 2_\beta$), ($1s_\alpha 2s_\alpha$), . . ., ($2p_x^\alpha 2p_x^\beta$), ($2p_x^\alpha 2p_y^\alpha$), and ($2p_x^\alpha 2p_y^\beta$).

An approximate form of the decoupled pair correlations, each to be treated as a separate object, is first obtained in Chapter V-2 by taking out "mean excitation energies" from the denominators of the usual infinite basis set expanded version of perturbation theory. The mean-excitation-energy approximation is very likely not valid quantitatively, but it does allow an interesting analogy between the pair correlations and the van der Waals–London dispersion forces between a set of atoms. The $\epsilon_{ij}^{(2)}$ becomes roughly proportional to the *mean-square fluctuation*,

$$\langle B(ij)\,|g_{12}^2 - \langle g_{12}\rangle^2\,|B(ij)\rangle$$

of the Coulomb repulsion $g_{12} = r_{12}^{-1}$ between spin orbitals i and j. Different shells or spin orbitals of a single atom then become analogous to a collection of atoms interacting with pairwise additive mean-square fluctuation London dispersion interactions. Historically, it is this analogy that led in Chapter V-2 to the theory of decoupled pair correlations in atoms by the appropriate application and analysis of second-order perturbation theory.

## 2. "Pair-Aufbau" Principle

Equation (14) suggests a "pair-aufbau" principle for total correlation energies, at least for ground states. Nominally for each atom or ion we have a set of pair correlations $\epsilon_{ij}$, such as $\epsilon_{1s^2}$, $\epsilon_{1s_\alpha 2s_\beta}$, $\epsilon_{2p_x{}^2}$, etc., plus nonadditive "environmental" effects, the main ones being the "exclusion effects." We know that a pair like $\epsilon_{1s^2}$ is nearly independent of Z and should therefore be very insensitive to the change in the total H.F. $V(x_i)$ in going from, say, $Be^{2+}$ to the $1s^2$ cores of $Be^+$ and Be. It is suggested, therefore (Chapter V-2 and the MET papers), that the $\tilde{\epsilon}_{ij}$'s be treated as semiempirical building blocks determined from a few atoms or ions by direct nonempirical calculation or from several known total correlation energies and then used over and over again in larger atoms or ions. To this semiempirical sum, $\sum_{i>j}^{N} \tilde{\epsilon}_{ij}$, estimates or calculations of exclusion effects are to be added.

As mentioned, many pairs turn out to be of the He $1s^2$ type, "dynamical," slowly convergent in C.I., with small "exclusion effects," but a few such as the $\epsilon_{2s^2}$ of Be, B, C, N, . . . , are "nondynamical" with major "exclusion effects." These nontransferable pairs, however, are easily calculable by finite C.I. The "pair-aufbau" principle, in combination with calculations of nondynamical effects, has yielded the dynamical (and other) pair-correlation values of atoms Be, B, C, N, O, F, Ne, and Na and their ions applicable to ground (Chapter VII-5) and excited states (Chapter IX-4). They result from the use of the generalization (MET) of this theory where perturbation results alone may be inadequate in a number of ways.

## 3. Some Methods of Calculation

The decoupled first-order pair equations of the Sinanoğlu theory, Eq. (5) or the equivalent ones in Chapter V-3, have been applied nonempirically by several authors. V. McKoy (to be published) was recently able to solve the equations directly using numerical techniques on the computer. Musher and Shulman (to be published) used different approximations and also obtained numerical solutions. These techniques are probably the simplest and the most powerful, capable of yielding pair functions and $\epsilon_{ij}$'s for many atoms. They apply equally to the MET generalizations of the pair equations. It would be interesting to compare the nonempirical $\epsilon_{ij}$'s thus obtained with the recent semiempirical values ("dynamical" + "nondynamical" total $\epsilon_{ij}$'s) of Chapter IX-4.

Tuan and Sinanoğlu (MET IV, 1964) and Byron and Joachain (Chapter V-5) applied the variational forms, Eq. (10), to the Be atom. As mentioned above, the former authors use the $r_{ij}$ coordinate, the pair C.I., and the core-polarization method for different pairs. The latter authors (Chapter V-5) find the use of $r_>$ and $r_<$ coordinates, instead of the $r_{12}$, convenient in the trial functions.

Both in MET IV and in Chapter V-5, pairs are calculated also to beyond the first order. In MET IV, as discussed below, the first-order equations are replaced by variational-pair equations, "exact pairs" valid to all orders in pairs but in closed form. These pairs are also upper bounds to the actual, full decoupled pairs $\epsilon_{ij} \leqslant \tilde{\epsilon}_{ij}$ of MET. Chapter V-5 extends the $\hat{u}_{ij}^{(1)}$ results by adding $\hat{u}_{ij}^{(2)}$. Thus $\hat{u}_{ij}\,(\text{MET}) \cong \hat{u}_{ij}^{(1)} + \hat{u}_{ij}^{(2)} + \ldots$, but the sum of the first few orders do not yield upper bounds to the full $\epsilon_{ij}$. Also the convergence of the perturbation series for a pair is slow where a particular pair results mainly from a strong near-degeneracy C.I., as in the $\epsilon_{2s^2}$ case. The $1s^2$ and $2s^2$ pair calculations of the two sets of authors agree to within a few percent. The most accurate pair correlation calculations on the Be atom to date, however, are probably still those of Levine, Geller, and Taylor [*J. Chem. Phys.* **43**, 1727 (1965)], who used the "exact-pair" methods of MET and extended the MET IV calculation.

The equations above apply to nondegenerate ground states. Formal solutions to Eq. (1) for a general excited state, where $\phi_0$ is the Roothaan open-shell H. F. wave function (RHF, Chapter II) with several Slater determinants, have been obtained by Silverstone and Sinanoğlu (Chapter IX-2). Decoupled pair equations result only after additional nondynamical effects have been taken out. The Q-projection operator and the "exclusion effects" get generalized. Extensions of Approach I to nonclosed shells will be discussed further in Chapter IX.

## 4. Questions on Higher-Order Effects

The construction of the correlation energy and wave function of an N-electron atom from decoupled pair solutions or $\epsilon_{ij}$ values simplifies the N-body problem. *But how decoupled are the pairs, and how valid is this approach*? The results above, and Eqs. (1) through (14), leave a number of questions unanswered:

1. How important are the $n \geqslant 3$ electron correlations that would enter the wave function in higher orders?

2. The third-order energy $E_3$ is obtained also from just the $\hat{u}_{ij}^{(1)}$'s and the $X_1$. However, this and higher-order energies contain $n \geqslant 3$ electron correlation energies (the first ones involving several different $\hat{u}_{ij}^{(1)}$'s in the same matrix element). These are corrections to the simple sum of the pair-correlations picture. If they are large, pair correlations as decoupled, often transferable entities will not be adequate.

3. The first-order wave function did not contain one-electron terms $\hat{f}_1$, but had only the $\hat{u}_{ij}^{(1)}$'s. This is because we started with $\phi_0$ as the H. F. Then $\hat{f}_i^{(1)} = 0$, but how important are such "effects of correlation" on orbitals in higher orders where they do not vanish?

4. The $\epsilon(2s^2)$ in Be, B, C, . . . , atoms is an example of a "nondynamical" correlation arising from a large mixing with $(2p^2)$. Clearly the perturbation expansion of such a C.I. matrix solution does not converge rapidly, so how can one, in general, calculate the pair correlations to all orders without having to sum perturbation series?

5. For the pair correlations $\tilde{\epsilon}_{ij}$, Eq. (13), to be usable semiempirically and to be carried over from ion to ion, they must be approximately independent of the total number of electrons N in each ion. Aside from the exclusion effects, for each N, the H. F. orbitals are somewhat different even when they are nominally of the same ($n\ell m$). This effect, too, makes an effective pair correlation dependent on N. For example, the $\epsilon_{1s^2}$ in Be, $Be^+$, and $Be^{2+}$ depend on somewhat different H. F. 1s orbitals B(12) in Eqs. (5) and (10). In spite of such effects of the atomic "medium," can one and the same set of $\epsilon_{ij}$'s still be used to construct the $E_{corr}$ of various ions or atoms?

The theory MET by Sinanoğlu (Approach I, Part 2 in Chapter VII) is primarily concerned with these questions. To investigate them, nonperturbative methods (cf. Chapter VII) are used and successive approximation methods more general than perturbation theory are developed. Once the "nondynamical" correlations are calculated fully, the theory indicates that the decoupled pair correlations should be dominant for most atomic properties in spite of additional, smaller effects. We shall reserve further discussion of the above questions to Chapter VII. We continue below with some of the aspects particularly relevant to the present chapter.

## 5. Full Correlation Equation of the Helium Atom

Consider the Schrödinger equation of the helium atom itself. For the ground state,

$$E \leqslant \frac{\langle \psi \mid H \mid \psi \rangle}{\langle \psi \mid \psi \rangle} \qquad (15)$$

or its Euler equation, $H\psi(x_1, x_2) = E\psi(x_1, x_2)$, may be rewritten for the correlation part alone, after a renormalization of $\psi$ such that

$$\psi = \phi_0 + \chi \qquad \langle \phi_0 \mid \phi_0 \rangle = 1 \qquad \langle \phi_0 \mid \chi \rangle = 0$$

With $\phi_0 = B(12) = \mathcal{A}(1s^\alpha 1s^\beta)$ the H. F. wave function, $\langle \chi \mid \chi \rangle$, is the *correlation probability*. (For atoms, $\langle \chi \mid \chi \rangle / \langle \phi_0 \mid \phi_0 \rangle$ is typically of the order of a percent or less.) With $\chi \equiv u_{12}^0(x_1, x_2)$ in the case of helium, Eq. (15) is equivalent to [see Sinanoğlu, *Advan. Chem.*

*Phys.* **6**, 368 (1964)]

$$E \leqslant E_{HF} + \frac{2\langle B(12)| m_{12} | u_{12}^0\rangle + \langle u_{12}^0 | e_1 + e_2 + m_{12} | u_{12}^0\rangle}{1 + \langle u_{12}^0 | u_{12}^0\rangle} \tag{16}$$

where

$$m_{12} \equiv \frac{1}{r_{12}} - V(x_1) - V(x_2) + \bar{J}_{12} \qquad \bar{J}_{12} \equiv \langle B(12) | r_{12}^{-1} | B(12)\rangle$$

Comparing this full variational principle for the helium exact pair correlation with the perturbation result, Eq. (10), we note two points:

1. In Eq. (16), $\langle u_{12}^0 | B(12)\rangle = 0$, but we do not have the orbital orthogonality ($\langle u_{12}^0 | 1\rangle \neq 0$) to the H.F. orbitals.

2. Equation (10) results from minimizing not the full $\tilde{\epsilon}_{12}$ as in Eq. (16), but a portion of the numerator in Eq. (16).

The helium exact $\psi$ may also be written (MET III, Chapter VII-3)

$$\psi(x_1, x_2) = \phi_0 + \chi = B(12) + u_{12}^0 = B(12) + B(\hat{f}_1 2) + B(1\hat{f}_2) + \hat{U}_{12} \tag{17}$$

where, e.g.,

$$\hat{f}_1(x_1) = \sqrt{2!} \int u_{12}^0(x_1, x_2) *2(x_2)\, dx_2 \tag{18}$$

and

$$\langle \hat{f}_i, j\rangle = 0$$
$$\langle \hat{u}_{12}, j\rangle = 0$$

The $\hat{f}_i(x_i)$ are *one-electron effects of correlation*. Equations (17) and (18) show that they are very simply calculated from a given two-electron function such as the Hylleraas one for He. The $\hat{f}_i$ resulting from the partial integration in Eq. (18) is exact to all orders. If expanded in perturbation series,

$$\hat{f}_i(x_1) = \hat{f}_i^{(1)} + \hat{f}_i^{(2)} + \cdots \tag{19}$$

the first-order term $\hat{f}_i^{(1)}$ would vanish *provided* $\phi_0$ were the H.F. wave function (Brillouin theorem).

The full $\hat{f}_i(x_i)$ is calculated from Eq. (18) in MET III (Chapter VII-3) for He. It contributes only 0.001 eV to the He correlation energy. (There are other situations, however, especially in some molecular problems, where $\hat{f}_i$ need not be small.) Thus for He, at least,

$$\hat{f}_i(x_i) \cong 0 \tag{20}$$

and, therefore,

$$u_{12}^0(x_1, x_2) \cong \hat{u}_{12}(x_1, x_2) \tag{21}$$

The $\hat{f}_i$ question, although not significant for He, is crucial to the decoupling of pairs in an N-electron system. Note also that if $\phi_0$ is not the H.F. wave function, $\hat{f}_k^{(1)} \neq 0$, as in the alternative approach discussed below.

Sinanoğlu [*Advan. Chem. Phys.* **6**, 315–412 (1964), and MET II, 1962] has obtained three different variational principles for the correlation energies of "He-like" pairs in N-electron systems. The analogous helium-atom case demonstrates the three:

Taking $u_{12}^0 = u_{12}$, we have the "exact pair" equation, Eq. (16):

$$E \cong E_{HF} + \epsilon_{12}' \leqslant E_{HF} + \frac{2\langle B(12)| g_{12} | \hat{u}_{12}\rangle + \langle \hat{u}_{12} | e_1 + e_2 + m_{12} | \hat{u}_{12}\rangle}{1 + \langle \hat{u}_{12} | \hat{u}_{12}\rangle} \tag{16'}$$

or

$$\epsilon'_{12} \leqslant \tilde{\epsilon}'_{12}$$

To remind ourselves that $\hat{u}_{12}$ is to be varied subject to $\langle \hat{u}_{12} | k \rangle = 0$, we may insert the operators Q in front of $\hat{u}_{12}$'s in Eq. (16′).

2. If a portion of the numerator is varied, we get the second-order minimum principle, Eq. (10).

3. Sinanoğlu (1964, 1962) has noted that there is another stationary part of Eq. (16), *the entire numerator*. At the time, no particular attention was paid to this third equation, because one might just as well either use the full, upper-bound exact pair, Eq. (16′), or the second-order one, Eq. (10), which is the simplest one to calculate. But recently, Sinanoğlu and Goscinski applying Löwdin's "bracketing theorem" found that the entire numerator gives yet another minimum principle, *this time a lower bound* to the pair correlation $\epsilon'_{12}$ [for the method of proof used, see *Intern. J. Quantum Chem.* **2**, 397 (1968)]. Thus we have

$$\epsilon^{L}_{12} \leqslant 2\langle B(12) | g_{12} | \hat{u}^{L}_{12} \rangle + \langle \hat{u}^{L}_{12} | e_1 + e_2 + m_{12} | \hat{u}^{L}_{12} \rangle \tag{22}$$

And the actual pair correlation $\epsilon'_{12}$, the solution of Eq. (16′), is then *bounded from above and below*,

$$\boxed{\epsilon^{L}_{12} \leqslant \epsilon'_{12} \leqslant \tilde{\epsilon}'_{12}} \tag{23}$$

We propose to call Eq. (22) *the lower-bound pair equation* as compared to Eq. (16′), the "exact pair" equation, to avoid adding to the semantic problem caused by the unspecified use of the term "pair" in a dozen different contexts and approximations.† It remains to be seen how close a lower bound $\epsilon^{L}_{12}$ is to $\epsilon'_{12}$.

## 6. Correlation Equations of "Helium-like" Pairs for N-Electron Atoms

We have seen that for some pairs the decoupling of the N-electron problem to first order [Eq. (5)] is not sufficient. Clearly higher orders are needed. However, in higher orders all electrons are coupled together by the $\hat{f}_i$, by the higher $n \geqslant 3$ electron terms in energy, and by the N-electron normalization constant, $1 + \langle X | X \rangle$. Application of the "generalized variation-perturbation approach" developed in MET (Chapter VIII) and the analysis of the coupling effects with it leads for ground states to full pair correlations $\epsilon'_{ij}$ analogous to the helium atom, Eq. (16) [Sinanoğlu, *Proc. U.S. Natl. Acad. Sci.* **47**, 1217 (1961), and subsequent papers]. Thus it is found that

$$E = E_{(\text{MET; decoupled pairs approx.})} \cong \sum_{i>j}^{N} \epsilon'_{ij} + \text{remainders}$$

where the $\epsilon'_{ij}$ are coupled to the N-electron atom only trivially, i.e., through all the occupied

†We thank Professor Bethe for noting this maze in the recent literature and making illuminating remarks as to the very different nature of the approximations that have been made in atomic theory versus those that are involved in the theory of nuclear matter, where, aside from the different techniques used, all the nucleons are coupled together in their correlations (cf. also still other contexts, such as Chapter IV-4). Similar remarks are made also in a recent critical discussion of the literature on pair correlation in atomic structure theory by E. U. Condon [*Rev. Mod. Phys.* **40**, 872 (1968)].

H. F. orbitals, and as in helium:

$$\epsilon_{ij}' \leqslant \tilde{\epsilon}_{ij}' \equiv \frac{2\langle B(ij)|g_{ij}|\hat{u}_{ij}\rangle + \langle \hat{u}_{ij}|e_i + e_j + m_{ij}|\hat{u}_{ij}\rangle}{1 + \langle \hat{u}_{ij}|\hat{u}_{ij}\rangle} \tag{24}$$

*for any i and j*, provided $\langle \hat{u}_{ij}|k\rangle = 0$ for $(k = 0, 1, 2, \ldots, i, \ldots, j, \ldots N)$ during variations $\hat{u}_{ij} \Longrightarrow (\hat{u}_{ij} + \delta\hat{u}_{ij})$ in Eq. (24).

Again, as in the helium atom, Eqs. (10), (16), and (22), three different approximations to each pair are obtained if different portions of Eq. (24) are varied which turn out stationary. *For each decoupled pair correlation* $\tilde{\epsilon}_{ij}'$ we have

$$\boxed{\epsilon_{ij}^{L} \leqslant \epsilon_{ij}' \leqslant \tilde{\epsilon}_{ij}'} \qquad [\text{any } i, j \text{ within } \phi_0 = \mathcal{A}(1\,2\,3 \cdots i \cdots j \cdots N)] \tag{25}$$

The new lower bound in Eq. (25) if it can be calculated and if it were close enough to $\epsilon_{ij}'$ would aid greatly in obtaining more definitive values of nonempirical pair correlations than have become available so far. In perturbation theory as such, one does not know how far above or below the actual $\epsilon_{ij}'$ a calculated value is. With the "exact-pair" $\tilde{\epsilon}_{ij}'$ variational calculation, one attempts to attain lower values by the minimum principle. However, the actual $\epsilon_{ij}'$ is not known (except semiempirically within possibly a few hundredths of 1 eV if relativistic-effect estimates allow this accuracy; cf. Chapter IX-4). So, in minimizing $\tilde{\epsilon}_{ij}'$ one again does not know if a low-enough value has been reached rather than a false plateau. For the actual variational calculations [Eq. (16′)], any of the methods mentioned above for $\epsilon_{ij}^{(2)}$ can be used. The numerical solutions to Eq. (5), for example, may be taken as a first iteration to the solution of the Euler equation of Eq. (16′) or as a trial function in Eq. (16′) to calculate an upper bound $\tilde{\epsilon}_{ij}'$.

# 7. Hydrogenic Orbitals—An Alternative Starting Point

We mentioned in question 5 (p. 117) that with the H. F. starting point, the pair correlations depend implicitly on the SCF orbitals which involve long-range Coulomb repulsion effects from all the electrons. These effects of the N-electron medium become explicit in a variant of the approach above which was proposed by Sinanoğlu [*Phys. Rev.* **122**, 493 (1961)] as an alternative which might have some advantages. In this variant the "bare-nucleus Hamiltonian" $H_0^0$ with $\phi_0^0$ made up of hydrogenic orbitals is taken as the starting point. The full Coulomb repulsions ($\sum_{i>j}^{N} 1/r_{ij}$) then become the perturbation.

Instead of the subdivision of $H = H_0 + H_1$ in Eqs. (1) ff., we now have

$$H = H_0^0 + H_1^0 \tag{26}$$

$$H_0^0 = \sum_{i>j}^{N} \left( -\frac{1}{2}\nabla_i^2 - \frac{Z}{r_i} \right) \tag{27}$$

$$H_1^0 = \sum_{i>j}^{-N} \frac{1}{r_{ij}} \tag{28}$$

$$H_0^0 \, \phi_0^0 = E_0^0 \, \phi_0^0 \qquad \phi_0^0 = \mathcal{A}\,(1^0 2^0 3^0 \cdots N^0) \tag{29}$$

The two-electron equations resulting from

$$(H_0^0 - E_0^0)X_1^0 = (\langle \phi_0^0 | H_1^0 | \phi_0^0 \rangle - H_1^0)\phi_0^0$$

now correspond to those of *actual* independent two-electron systems. The symmetry combinations, the "irreducible pair" equations, now become convenient because they are the ones that would normally result from, say, Hylleraas-type calculations on actual two-electron systems. One has, for example, for $1s^2$ and 1s2s pairs [Eqs. (38), (39), and (45b) in *Phys. Rev.* **122**, 493 (1961)],

$$(e_1^0 + e_2^0)\, u_{12} = (J_{12} - g_{12})B(1s_\alpha^0 1s_\beta^0) \tag{30}$$

$$(e_1^0 + e_3^0)\, u_{13} = (J_{13} - K_{13} - g_{13})B(1s_\alpha^0 2s_\alpha^0) \tag{31}$$

where "reducible" and "irreducible" pairs are the same; but for the $(1s_\alpha 2s_\beta)$ "reducible" pairs we have the irreducible forms

$$(e_2^0 + e_3^0)\; {}^3u_{23} = (J_{23} - K_{23} - g_{23})\; {}^3\phi_{23}^0 \tag{32}$$

$$(e_2^0 + e_3^0)\; {}^1u_{23} = (J_{23} + K_{23} - g_{23})\; {}^1\phi_{23}^0 \tag{33}$$

The notation is $1s_\alpha^0 \equiv 1^0$, $1s_\beta^0 \equiv 2^0$, $2s_\alpha^0 \equiv 3^0$, etc., $g_{ij} \equiv r_{ij}^{-1}$, and

$$\left.\begin{matrix} {}^3\phi_{23}^0 \\ \\ {}^1\phi_{23}^0 \end{matrix}\right\} = \frac{1}{\sqrt{2}}[B(2^0 3^0) \pm B(1^0 4^0)] \tag{34}$$

The Sinanoğlu (1961) paper notes for this alternative hydrogenic starting point: "The [main] advantage of this approach is that the two-electron solutions $u_{ij}$ can be obtained independently, and when they are obtained as a function of the atomic number, Z, as Hylleraas has done for $H^-$, He, $Li^+$, etc., they can be used to build up larger atoms. In going from Li to Be, *only one more* pair function, the one for the $(2s^2)\,^1S$ state, is needed. [For this state the $(2s)^2$, $(2p)^2$ degeneracy would be removed first.]"

The overall wave function and energy now become

$$\psi(x_1, x_2, \ldots, x_N) = \phi_0^0 + X_1^0 + X_2^0 + \cdots \tag{35}$$

$$X_1^0 = \frac{1}{\sqrt{2}} \sum_{i>j}^{N} \mathcal{A}\left[(1^0\, 2^0\, 3^0 \cdots N^0)\, \frac{u_{ij}}{i^0 j^0}\right] \tag{36}$$

where "irreducible" pairs appear in combinations like

$$u_{23}(x_2, x_3) = \frac{1}{\sqrt{2}}\, [{}^3u_{23}(x_2, x_3) + {}^1u_{23}(x_2, x_3)] \tag{37}$$

$$E = E_0^0 + E_1^0 + E_2^0 + E_3^0 + \cdots \tag{38}$$

$$E_2^0 = \sum_{i>j}^{N} \langle B(i^0 j^0) | g_{ij} | u_{ij} \rangle$$
$$+ \text{"cross-polarization terms"}$$
$$+ \text{"exclusion effects"} \tag{39a}$$

The last two nonpairwise effects also involve just the $u_{ij}$'s; e.g.,

$$\langle \mathscr{A}(123) | g_{13} | u_{12}(\mathbf{x}_1, \mathbf{x}_2)\, 3(\mathbf{x}_3) \rangle \tag{39b}$$

is a "cross-polarization" term, with the integration over $\mathbf{x}_2$ giving a correction to the starting-orbitals type of term, as in Eq. (18). The "cross-polarization" terms vanish formally because $\hat{f}_i^{(1)} = 0$ with the H. F. $\phi_0$ (Chapter V-3), but only to become implicit in the SCF orbitals.

Comparison of the $\phi_0$ versus $\phi_0^0$ starting points leads one to a way of obtaining H.F. SCF orbitals directly from wave functions $\psi$, bypassing H.F. but containing correlation as in the full Hylleraas helium wave functions.

Consider again Eq. (17), but now with hydrogenic $\phi_0^0$:

$$\psi(\mathbf{x}_1, \mathbf{x}_2) = \phi_0^0 + X^0 = B(1^0\, 2^0) + B(\hat{f}_1^0\, 2^0) + B(1^0 \hat{f}_2^0) + \hat{u}_{12}^{\text{hydr}} \tag{17$'$}$$

Again

$$\hat{f}_1^0\,(\mathbf{x}_1) = \sqrt{2!} \int X^0(\mathbf{x}_1, \mathbf{x}_2)^*\, 2^0(\mathbf{x}_2) d\mathbf{x}_2 \tag{18$'$}$$

or, in the new perturbation expansion,

$$\hat{f}_1^0(\mathbf{x}_1) = \hat{f}_1^{0(1)} + \hat{f}_1^{0(2)} + \cdots \tag{19$'$}$$

but now

$$\boxed{\hat{f}_1^0\,(\mathbf{x}_1) \not\cong \hat{f}_1^{0(1)} \neq 0} \tag{20$'$}$$

The $\hat{f}$'s of the $\phi_0$ (SCF) and $\phi_0^0$ (hydrogenic) are simply related by

$$\hat{f}_1^0(\mathbf{x}_i) \equiv \hat{f}_i^{\text{HF}}\,(\mathbf{x}_i) + \hat{f}_i^{\text{corr}}(\mathbf{x}_i) \tag{40}$$

where the hydrogenic correction $\hat{f}^0$, Eq. (18), involves a part which tries to turn the original hydrogenic $i^0(\mathbf{x}_i)$ into a SCF $i(\mathbf{x}_i)$, the $\hat{f}_i^{\text{HF}}$, and a "correlation effect on the SCF orbital," $\hat{f}_i^{\text{corr}}$. In the $\phi_0$ (SCF) case we have only

$$\hat{f}_i(\mathbf{x}_i) = \hat{f}_i^{\text{corr}}\,(\mathbf{x}_i) \tag{41}$$

of Eq. (18).

With these equations, it is possible to start, say, with the Hylleraas $\psi(\mathbf{x}_1, \mathbf{x}_2)$ and some initial, say, hydrogenic orbitals and by simply partial integrations on the $\psi(\mathbf{x}_1, \mathbf{x}_2)$ end up in a few iterations with the H.F. orbitals (if $\hat{f}_i^{\text{corr}} = 0$ as in helium, so that $\hat{f}_i^0 \cong \hat{f}_i^{\text{HF}}$) using

$$i_{\text{SCF}} \cong \frac{i^0_{\text{initial}} + \hat{f}_i^0}{\sqrt{1 + \langle \hat{f}_i^0 | \hat{f}_i^0 \rangle}} \tag{42}$$

The method [Sinanoğlu, *Rev. Mod. Phys.* **35**, 517 (1963)] has been applied by Kestner [*J. Chem. Phys.* **39**, 1361 (1963)].

This procedure would allow one (e.g., in molecules by arbitrary basis set C.I.) to calculate only the decoupled functions directly with any starting orbital set. It would then yield the MOSCF by partial integrations without a separate H.F. calculation.

The "cross-polarization" terms in Eq. (34) involve just the first-order part of Eq. (18$'$):

$$\hat{f}_i^{0(1)} = \sqrt{2!} \int u_{12}\,(\mathbf{x}_1, \mathbf{x}_2)^*\, 2^0\,(\mathbf{x}_2) d\mathbf{x}_2 \tag{43}$$

and represent the first corrections from hydrogenic $i^0(\mathbf{x}_i)$ toward Hartree-Fock $i(\mathbf{x}_i)$. The $\hat{f}_i^{0(2)HF}$ and higher-order terms as well as the renormalization term needed in Eq. (40) are not included in them.

The H.F. corrections are included all at once, with the $\phi_0$ (SCF) starting point. It has also been found (MET IV and Chapter VII-5 and IX-4) that the "dynamical" pair correlations are transferable from ion to ion in spite of the "Hartree-Fock medium" effect. For these reasons the $\phi_0$ (H.F.) start has been preferred in further extensions of the pair-correlations theory by Sinanoğlu.

The hydrogenic alternative above has been applied by Z. Horák [in *Modern Quantum Chemistry* (Istanbul Lectures), Vol. 2, Interactions, O. Sinanoğlu, ed., Academic Press, New York, 1965 (also his later papers with D. Layzer et al.)], expanding each of the terms above in a basis set, and by Chisholm and Dalgarno, using the closed-form equations (Chapter V-4). Chisholm and Dalgarno obtained the energy through second order of the ($1s^2 2s$) Li-like sequence as a function of Z. They noted that $f_i^{0(1)}$ satisfies the first-order H.F. equations and used the Dalgarno-Cohen (1961, 1963) first-order H.F. orbitals. The H.F. result they obtain this way differs from true H.F. by 0.5 eV for Z = 3. The error here is sizable, about a third of the Li correlation energy, indicating the role of $\hat{f}_i^{0(2)} + \cdots$ or possibly the renormalization in Eq. (40). This $\hat{f}_i^{HF}$ error should diminish, however, for larger Z, while $E_{corr}$ changes more slowly. The hydrogenic alternative with its explicit Z dependence provides, therefore, a promising method of calculation (cf. p. 121 above).

In concluding this chapter, we remark on another N-body problem, the N-nucleon problem of light to medium nuclei. As remarked earlier, the theory by Sinanoğlu outlined above of the reduction to N(N − 1)/2 decoupled two-body problems, each one solved by the various approximations given, cannot be applied as such to the problem of relating the properties of finite nuclei to an experimental nucleon-nucleon two-body potential (such as the "Yale potential" of G. Breit and co-workers or the Hamada–Johnson potential obtained from nucleon–nucleon scattering data). This nucleon–nucleon potential replaces in the Hamiltonian of the N-body system the $g_{12} = r_{12}^{-1}$ of the atomic problem. But aside from additional aspects such as isotopic spin and large spin-orbit forces, the two potentials are far too different. The long-range aspects of the $g_{12}$ of atoms were already commented upon and will be discussed further in Chapter VI. The real difficulty with nuclei, however, is the short-range repulsion of their $g_{ij}$'s. This makes the $\phi_0$, H.F. orbitals of the above theory a poor starting approximation leading even to infinities. This means that as noted in some of the MET papers, $\hat{f}_i$, as in Eqs. (43) and (18), must also become very large or infinite. New orbitals allowing for the necessary strong short-range correlations must be used. The treatment of this problem leads to all the pair correlations remaining after the definition of such new orbitals to be coupled implicitly together. The cancellation of the infinities is accomplished for the case of infinite nuclear matter by Brueckner's theory (1953, and subsequent papers; see e.g., in *The Many-Body Problem*, C. Dewitt, ed., Wiley, New York, 1959). Sinanoğlu has remarked in the MET papers on the double self-consistency procedures which involve nucleon orbital motions in the potentials averaged over correlated motions, which make the problem more difficult for the case of finite nuclei. In an a priori theory of finite nuclei, i.e., one that does not assume a phenomenological central potential to start with, decoupling of pairs as in the atomic theory does not seem possible. The use of some of the closed-form methods and "subvariational" principles above may nevertheless prove possible in an N-nucleon theory of the ligher nuclei.

The difficult problem of obtaining a nuclear well starting from an a priori nucleon–nucleon two-body potential is the problem treated in the theory of Brueckner. The problem of the motions of nucleons in an assumed semiempirical nuclear well is a considerably simpler one. Bhaduri, Peirls, and Tomusiak [*Proc. Phys. Soc.* **90**, 945 (1967)] have recently suggested for the latter problem, for light nuclei, the analogue of Sinanoğlu's hydrogenic starting-point pair-decoupling theory [*Phys. Rev.* **122**, 493 (1961)]. For the unperturbed $\phi_0^0$ they use harmonic oscillator well orbitals. In this case, the first-order inhomogeneous equations [analogous to Eq. (30] of the resulting effective two-body problems are simpler to solve than in the atomic case.

V-2 Reprinted from *Journal of Chemical Physics* **33**, 1212 (1960)

# Inter- and Intra-Atomic Correlation Energies and Theory of Core-Polarization

Oktay Sinanoğlu*

*Department of Chemistry and Lawrence Radiation Laboratory, University of California, Berkeley 4, California*

(Received March 18, 1960)

Atomic and molecular energies depend strongly on the correlation in the motions of electrons. Their complexity necessitates the treatment of a chemical system in terms of small groups of electrons and their interactions, but this must be done in a way consistent with the exclusion principle. To this end, a nondegenerate many-electron system is treated here by a generalized second-order perturbation method based on the classification of all the Slater determinants formed from a complete one-electron basis set. The correlation energy of the system is broken down into the energies of pairs of electrons including exchange. Also some nonpairwise additive terms arise which represent the effect of the other electrons on the energy of a correlating pair because of the Pauli exclusion principle. All energy components are written in approximate but closed forms involving only the initially occupied H.F. orbitals. Then each term acquires a simple physical interpretation and becomes adoptable for semiempirical usage. The treatment is applied in detail to two particular problems: (a) The correlation energy between an outer electron in any excited state and the core electrons, e.g., in the Li atom, is represented by a potential acting on the outer one. This potential can be regarded as the mean square fluctuation of the Hartree-Fock potential of the core, and applies even when the outer electron penetrates into the core. The magnitudes of some of the correlation effects are calculated for Li. (b) Starting from a complete one-electron basis set of SCF MO's, the energy of a molecule is separated into those of groups of electrons and of *intra*-molecular dispersion forces acting between the groups. The assumptions that are usually made in discussing dispersion forces at such short distances are then removed and generally applicable formulas are given. Some three or more electron-correlation effects and limitations in the use of "many electron group functions" for overlapping systems are also discussed.

## I. INTRODUCTION

THE total energy of a many-electron system is well approximated by independent particle theories, in particular by the Hartree-Fock (H.F.) method.[1,2] But the remaining error, which results from the tendency of the electrons to avoid one another, is usually of the order of chemically interesting quantities. This error i.e., the difference between the completely self-consistent field (SCF) H.F. solution to the energy and the exact solution of the many-electron nonrelativistic Hamiltonian, will be taken here as the precise definition[2] of "correlation energy" unless otherwise indicated. Considerable effort has been devoted to obtaining the correlation energies of atoms and molecules by variational techniques, especially by the method of configuration-interaction.[2] In spite of slow convergence, computers now allow some progress by these techniques for simpler systems, e.g., molecules from the first row of the periodic table. But, generally, it is necessary to have schemes that lend themselves to simple physical interpretation and generalization. It is also important in view of the complexity of most molecules and atoms that such schemes should be usable in a semiempirical but still well-defined way.

There exists a special class of problems where it has been possible to treat the correlation effects very simply. These involve the correlation energies between nonoverlapping charge distributions; or more precisely, in these cases different electron groups are assumed to be localized in totally isolated regions of space, each group with its own complete set of eigenfunctions. The best example is the familiar London dispersion attraction[3,4] between two atoms in their ground states.[5] This energy, which has the very useful property of being pairwise-additive for a system of atoms, also has a very simple physical interpretation. It arises from the mean square fluctuations of the instantaneous atomic dipoles.

A different case concerning correlation energy between separable groups arises in the theory of atomic spectra. In alkalilike configurations optical transitions occur by the excitation of a "series," i.e., valence, electron, outside an inner "core" giving a hydrogenlike spectrum. The H.F. SCF method for such an atom, takes the average effect of the outer orbital on the core into account, but the correlation between the instantaneous position of the series electron and the core electrons is left out. This correlation effect has come to be known under the name of "core polarization."[1] The simplification in this case arises when the series electron is in a highly excited "nonpenetrating" orbit, i.e., when the core and the valence electron can be separated. Then it is possible to represent the correlation energy between the core electrons and the series electron by an additional effective potential which is determined only by the core and acts on the outer electron in any one of its higher orbits. A classical argument indicates that at large separations this potential is given by $-\frac{1}{2}\alpha r^{-4}$ where $\alpha$ is the polarizability of the ion core and $r$ is the distance of the series electron from the nucleus. Several quantum mechanical derivations have

* University of California Lawrence Radiation Laboratory Postdoctoral Fellow 1959–60. Present address: Sterling Chemistry Laboratory, Yale University, New Haven, Connecticut.

[1] For reviews see D. R. Hartree, *Calculation of Atomic Structures* (John Wiley and Sons, Inc., New York, 1957).

[2] P. O. Löwdin, Advances in Chem. Phys. **2**, 207 (1959).

[3] F. London, Z. Physik. Chem. **B11**, 222 (1930).

[4] H. Margenau, Revs. Modern Phys. **11**, 1 (1939).

[5] In this case, the definition given for correlation energy is somewhat altered; it still applies, however, if the electrons of one system are considered to move in the self-consistent field of the other.

been given for this asymptotic potential. The "adiabatic approximation"[6] method distinguishes a tightly bound core electron from an outer one in any orbital with high quantum numbers, e.g., in the excited states of He, and replaces the exact two-electron wave functions by

$$\psi_k(\mathbf{r}_1, \mathbf{r}_2) = u_c(\mathbf{r}_1; \mathbf{r}_2) u_v^k(\mathbf{r}_2), \tag{1}$$

where $\mathbf{r}_1$ and $\mathbf{r}_2$ refer to the coordinates of the inner and outer electrons, respectively; $u_v^k(\mathbf{r}_2)$'s are the various states of the outer electron; $u_c(\mathbf{r}_1; \mathbf{r}_2)$ describes the core electron and depends only parametrically on $\mathbf{r}_2$. This approximation is very similar to the Born-Oppenheimer separation of nuclear and electronic motions in molecules. Since the outer electron is much less tightly bound than the inner one, it can be considered, in view of the virial theorem, to move slowly compared to the latter. Then the energy of the core can be determined for various fixed values of $\mathbf{r}_2$, thus depending on $\mathbf{r}_2$ parametrically; and it, in turn, acts as a potential energy for the motion of the outer electron. Bethe[7] has treated the highly excited states of He by this approach, considering the stationary charge of $r_2$ as a perturbation on the free "core" function, $u_c^0(r_1)$. This discussion should emphasize that the name "core polarization" refers to the polarization of the core by the outer electron at its instantaneous position. This *is* the entire correlation effect and should not be confused with the polarization of the core only by the smeared-out over-all charge distribution of the outer *orbital*. The latter which we shall call the "orbital average polarization" is a much smaller effect and in fact vanishes if completely SCF H.F. orbitals are used (see Sec. II).

Mayer and Mayer[8] and Van Vleck and Whitelaw[9] have treated the core-polarization energy by conventional second-order perturbation theory with hydrogen-like orbitals for the outer electron and by assuming the valence electron and the core to be essentially independent systems. This treatment shows that the analogy between core polarization and the interatomic London forces mentioned previously, is perfect. We may even be tempted to refer to the attraction arising between the outer electron and the core in the former case, as a Van der Waals force.

The particular advantage of both London forces and the use of a correlation potential, is that they both relate the interactions in the composite system to the properties of the component parts. This advantage has induced attempts at using both theories under more general conditions.

Pitzer[10] has reviewed the use of London forces not only at the usual large separations, but even within the same molecule, e.g., to estimate the correlation energies between the lonepair electrons in the halogen series, $F_2$ to $I_2$, and to account for the isomerization energies of the hydrocarbons. Similarly, Douglass[11] has assumed the feasibility of using a core-polarization potential even inside the core and has determined such a potential semiempirically by comparing the observed and H.F. energy values for the first few series levels in alkaline ions. Various other empirical attempts at obtaining such a potential for cases where the outer electron penetrates well into the core are summarized by Hartree.[1,12] If it were possible to write a correlation potential which could act over the whole range of an outer orbital coordinate another interesting application would be justified: In molecules where the inner cores may be assumed to be quite unchanged by the binding, the correlation energy between the cores and the valence electrons could be obtained simply by taking the expectation value of the core polarization potentials (obtained perhaps partly from atomic spectra) over the outer molecular orbitals. Callaway[13] did this for Li, Na, and K atoms[14] in their ground states and for their metals, but he used Bethe's[7] method and neglected the penetration effects. This type of application of the core polarization idea is of course also implied in the "Π-electron" approximation[15] used for organic molecules.

Both the use of London forces *within* a single molecule and the representation of inner-outer electron, or sigma-pi correlation effects by effective potentials when there is penetration, are open to question. Some of the difficulties involved in the case of intra-molecular London forces have already been mentioned by Coulson.[16] On the other hand, the need for the clarification of the form or the feasibility of a correlation potential for penetrating orbitals has been emphasized by Hartree.[1,12] In both cases, multipole expansions are often used at short distances and, sometimes, only the dipole terms are retained. But this is a defect whose elimination is relatively straightforward (see Secs. II and VI). All the other difficulties have to do with the exclusion principle and the validity of separating electrons into distinguishable (either by strong localization around different centers or by large differences in energy) groups.

The preceding discussion indicates that important simplifications have been possible in dealing with correlation effects whenever a system of electrons could

[6] I am indebted to Professor W. T. Simpson for a stimulating discussion on this approach.

[7] (a) H. A. Bethe, *Handbuch der Physik* (Edwards Brothers, Inc., Ann Arbor, Michigan, 1943), Vol. 24/1, p. 339; see also Bethe and E. E. Salpeter, *Encyclopedia of Physics* (Springer-Verlag, Berlin, Germany, 1957), Vol. 35, p. 223 ff; (b) G. Ludwig, Helv. Physica Acta **7**, 273 (1934).

[8] J. E. and M. G. Mayer, Phys. Rev. **43**, 605 (1933).

[9] J. H. Van Vleck and N. G. Whitelaw, Phys. Rev. **44**, 551 (1933).

[10] K. S. Pitzer, Advances in Chem. Phys. **2**, 59 (1959).

[11] A. S. Douglas, Proc. Cambridge Phil. Soc. **52**, 687 (1956).

[12] D. R. Hartree, Repts. Progr. Phys. Kyoto **11**, 113 (1946).

[13] J. Callaway, Phys. Rev. **106**, 868 (1957).

[14] The adiabatic method has also been applied to alkali by G. Veselov and I. B. Bersuker, Vestnik Leningrad Univ. Ser. Fiz. i Khim. No. 16, 55 (1957).

[15] R. Pariser and R. G. Parr, J. Chem. Phys. **21**, 466 (1953).

[16] C. A. Coulson, *Symposium on Hydrogen Bonding, Ljubljana, 1957* (Pergamon Press, New York, 1959), p. 349.

be divided up into groups in a clear-cut way, but that as soon as this condition is relaxed difficulties are encountered. Now if similar simplified treatments of correlation effects are to be possible in the general case of a system of many electrons all spread out in nearly the same region of space, the following question, which is fundamental to all quantum chemistry, must be answered:

*Given an atom or molecule with some $N$ electrons, can we consider this system in a nonarbitrary way as made up of certain groups of electrons in spite of the fact that all these electrons are indistinguishable from one another?* The presence of a periodic table of the elements and chemistry suggest that the answer should be yes, not only in single particle theories, but also after the inclusion of correlation energies. Clearly if this is to be done in a nonarbitrary way: (1) All effects of the Pauli exclusion principle must be included, (2) If we start with a method based on the expansion of a many electron function in terms of single particle functions, then all electrons must use the *same* complete basis set of one-electron orbitals. Previous separations of an electronic system into simpler systems do not satisfy the second condition and therefore neither the first. In these treatments, either a one-electron basis set is divided into two mutually exclusive subsets as in connection with the sigma-pi problem,[17] or else, antisymmetrized products of many-electron group functions satisfying generalized orthogonality conditions are used.[18] These conditions, which we shall refer to again in Sec. VI, are too restrictive and they imply the subdivision of the basis set, too.

The general problem of formulating a scheme such that the correlation energy of a system with a large number of electrons can be obtained in terms of its simpler components, can now be broken down into two phases: (a) The separation of the total correlation energy into that of small groups of electrons. (b) The evaluation of the interactions within or between these small groups. The extensions of London-force or core-polarization type treatments to systems with no strongly localized groups now become special cases of (b).

With the objectives (a) and (b) in view, we shall now present a generalized second-order perturbation theory of a many-electron system whose zero-order wave function is a single Slater determinant of Hartree-Fock orbitals. The perturbation method will be based on the use of all "ordered configurations,"[2] i.e., all unique Slater determinants that can be formed from a complete one electron basis set. By a systematic classification of all the virtual transitions represented by the ordered configurations, the correlation energy of the system will be broken down into the energies of pairs of electrons including exchange. Some non-pairwise additive terms will also arise. These represent the tendency of a correlating pair of electrons to avoid the other electrons of the system, because of the Pauli exclusion principle. These extra terms will be defined here as the "exclusion" terms and in future work they may turn out to be the most important three or more particle effects. The use of Hartree-Fock energy as a starting point is convenient, but is not a requirement. The approach presented here also provides a link with the recent "many-body" methods developed in different fields of physics.[19] Definition of "mean excitation energies" for each pair of electrons will allow the estimation of both the pair energies and the exclusion effects, using only the initially occupied spin-orbitals. This procedure, although semiquantitative, is useful for giving many simple physical interpretations, e.g., as in London forces. For instance, in Sec. II, a core-polarization potential which is applicable even near the nucleus will be derived and shown to be the mean-square fluctuation of the Hartree-Fock potential acting on an electron. A large portion of this article will be devoted to the core polarization problem, not only because it is of interest in itself, but also because a system with a "series" electron is more general than a system of closed shells, and the excited states, as well as the ground state of the former, can also be treated. Detailed derivations will be given, for convenience, with specific reference to the lithium atom, although generalization to any other appropriate atom or molecule is straightforward. Section III discusses the "exclusion" effect of an outer orbital, e.g., in Li, on the correlation energy of the core ($Li^+$) itself. In Sec. IV, numerical magnitudes of some of these effects will be calculated for Li and the penetration parts of the core-polarization potential will be examined. In Sec. V, molecules are discussed, in general, and both the use of intramolecular London-type energies and of "core polarization" justified eliminating the usual approximations mentioned previously by starting from a complete one-electron basis set of molecular orbitals, SCF MO's. In the last section, some higher-order correlation effects are mentioned; a "dispersion" energy formula including such effects, but for use only with asymptotic intermolecular forces, is derived, and the use of many-electron group functions[18] is discussed.

## II. SEPARATION OF VARIOUS CORRELATION EFFECTS AND THEORY OF CORE POLARIZATION

The basic theorem[2] of the method of "superposition of configurations" is that if $\{u_k(\mathbf{x})\}$ form a complete orthonormal basis set for the space of a single electron ($\mathbf{x}$ including both spatial and spin coordinates), than any antisymmetric $N$ electron function can be ex-

[17] P. G. Lykos and R. G. Parr, J. Chem. Phys. **24**, 1166 (1956).
[18] R. McWeeny, Proc. Roy. Soc. (London) **A253**, 242 (1959).
[19] For a brief introduction, see D. ter Haar, *Introduction to the Physics of Many-Body Systems* (Interscience Publishers, Inc., New York, 1958).

panded as

$$\psi(\mathbf{x}_1, \mathbf{x}_2, \cdots \mathbf{x}_N) = \sum_K^{\infty} C_K \psi_K(\mathbf{x}_1, \mathbf{x}_2, \cdots \mathbf{x}_N), \quad (2)$$

where $\psi_K$ represents the normalized Slater determinants

$$\psi_K(\mathbf{x}_1, \mathbf{x}_2, \cdots \mathbf{x}_N) = (N!)^{-\frac{1}{2}} \det\{u_{k_1}, u_{k_2}, \cdots u_{k_N}\}, \quad (3)$$

and $K$ runs over every unique selection of the one-electron indices $k_1 < k_2 < \cdots < k_N$, i.e., all "ordered configurations."[2] The $\psi_K$ now form a complete orthonormal basis set for the space of $N$ electrons. In this representation, the energy eigenvalues are the solutions of the secular equation

$$|\mathbf{H} - E\mathbf{I}| = 0, \quad (4)$$

with $\mathbf{H} = \{\langle \psi_K | H | \psi_L \rangle\} \equiv \{\langle K, HL \rangle\}$ and $\mathbf{I}$ the unit matrix. If any one of the nondegenerate $\psi_K$ in Eq. (2), say, $\psi_a$, is chosen as the first approximation to $\psi$, then $E$ can be written[2] rigorously as the solution of

$$E = H_{aa} + \mathbf{H}_{ab}(E\mathbf{I}_{bb} - \mathbf{H}_{bb})^{-1}\mathbf{H}_{ba}, \quad (5)$$

where the matrix $\mathbf{H}$ has been partitioned into four submatrices. A great variety of perturbation methods can be derived[2,20] from Eq. (5) by making various approximations in the exact remainder after taking $H_{aa}$ as the "unperturbed" energy. Thus, if the off-diagonal elements of $\mathbf{H}_{bb}$ are neglected and $E$ replaced by $H_{aa}$, a Schrödinger-type generalized perturbation equation is obtained. This equation, first given by Epstein,[21] is

$$E_\mu \cong H_{MM} - \sum_{K \neq M}^{\infty} [H_{MK} H_{KM} / (H_{KK} - H_{MM})]$$

$$= H_{MM} + E_M^{(2)}. \quad (6)$$

Here, among all the diagonal elements of $\mathbf{H}$, $H_{MM}$ is assumed to be the closest to the exact eigenvalue $E_\mu$. In the summation, $K$ covers the entire orthonormal many-electron basis set, as in Eq. (2), except $\psi_M$. Because of the orthonormality of $\{u_k\}$ the only $H_{MK}$ that contribute to Eq. (6) are those in which $K$ differs from $M$ [Eq. (3)] by only one or two spin-orbitals. These will be referred to as single and double virtual excitations from $M$. Further, if $\{u_k\}$ are chosen as Hartree-Fock functions, then the contribution of single virtual excitations vanishes. This use of Eq. (6) was proposed by Brillouin[22a] and Møller and Plesset[22b] for nondegenerate $\psi_M$.[22c] Extension to degenerate states has been discussed by Nesbet.[23] In this article we shall deal only with systems whose zero-order wave function can be taken as a single Slater determinant and use a more general basis set, $\{u_k\}$.

[20] See also; W. B. Riesenfeld and K. M. Watson, Phys. Rev. **104**, 492 (1956).

[21] P. S. Epstein, Phys. Rev. **28**, 695 (1926); see also; L. C. Pauling and E. B. Wilson, *Introduction to Quantum Mechanics*, (McGraw-Hill Book Company, Inc., New York, 1935), p. 191.

[22] (a) L. Brillouin, Actualites sci. et ind., **No. 159 (1934)**; **No. 71 (1933)**; (b) C. Møller and M. S. Plesset, Phys. Rev. **46**, 618 (1934); (c) see, also, footnote reference 2, p. 283.

[23] R. K. Nesbet, Proc. Roy. Soc. (London) **A230**, 312 (1955).

Consider first an $N$ electron system of closed shells. For these, the first $N$ spin-orbitals will be the occupied H.F. SCF orbitals of the system. The rest of the basis set $\{u_k\}$ $(k > N)$ may be assumed to be completed, e.g., as described by Löwdin,[22c] by taking independent functions and using the Schmidt orthogonalization process. As we shall see, in the following, the specification of $\{u_k\}$ for $k > N$ will be unnecessary. Now, consider the more general case of a system of closed shells plus an outer electron, e.g., an alkali atom. The treatment of this case will include the treatment of closed shells only, and, in addition, will allow a theory of core polarization for the outer electron in its various states. We shall return to a discussion of all closed shells in the section on molecules (Sec. V). For the more general case, if a polarization potential independent of the state of the outer electron is to be obtained, the initially occupied core orbitals must be chosen so as to be independent of the outer orbitals. The following modified basis set $\{u_k\}$ is the most suitable one for the core-polarization problem: Let the system contain $N$ core electrons and the $(N+1)st$ outer electron, and let us suppose that we are interested in, say, $n$ actual states of the outer electron including the lowest one corresponding to the ground state of the atom. Then the first $N$ spin-orbitals $u_1, u_2, \cdots, u_N$ are to be obtained from a complete H.F. SCF solution of only the closed shell part of the system after stripping the $(N+1)$st outer electron. The next $n$ orbitals $u_{(N+1)}, u_{(N+2)}, \cdots, u_{N+n}$ will be taken as the solutions of the H.F. equation for an electron in the field of the already determined *fixed* core orbitals. The rest of the basis set may be completed again, e.g., by a Schmidt orthogonalization process. The complete basis set of spin orbitals thus obtained will now be denoted by $\{u_k(\mathbf{x})\} = \{k\} = 1, 2, 3, \cdots$, all odd integers designating the spin-orbitals with spin $\alpha$, and all even integers those with spin $\beta$. For concreteness and convenience, we now continue the treatment on the simplest case, the Li atom, although extension to larger systems is straightforward. For any state $[1/(3!)^{\frac{1}{2}}]$ det $(1s_\alpha 1s_\beta u_{k\alpha})$ or det$'$ $(12k)$ [det$'$ denoting the normalized determinant] of Li, the first few spin orbitals by our choice satisfy the equations

$$h_1^0 1s(\mathbf{r}_1) + \left(\int |1s(\mathbf{r}_2)|^2 r_{12}^{-1} d\tau_2\right) 1s(\mathbf{r}_1) = \epsilon_{1s} 1s(\mathbf{r}_1), \quad (7)$$

where

$$h_1^0 \equiv -\tfrac{1}{2}\nabla_1^2 - (Z/r_1) \quad (8)$$

and

$$h_{\text{core}}^{\text{eff}} u_k \equiv h_1^0 u_k(\mathbf{r}_1) + 2\left(\int |1s(\mathbf{r}_2)|^2 r_{12}^{-1} d\tau_2\right) u_k(\mathbf{r}_1)$$

$$-\left(\int 1s^*(\mathbf{r}_2) u_k(\mathbf{r}_2) r_{12}^{-1} d\tau_2\right) 1s(\mathbf{r}_1) = \epsilon_k u_k(\mathbf{r}_1)$$

for $u_{k\alpha} \equiv k \geq 3$. The second-order correlation energy is given by Eq. (6) in conjunction with Eq. (2) and the one-electron basis set just defined. To separate this energy into parts, all the ordered configurations (with any orthonormal $\{u_k\}$) that would give matrix elements in Eq. (6) can be classified according to which one or two spin-orbitals of the initial state det′(12*k*) have been virtually excited to give the particular configuration. Consider first the ground state det′(123), i.e., $(1/\sqrt{6})$ det $(1s_\alpha 1s_\beta 2s_\alpha)$. From this state, keeping in mind that the indices $k_1$, $k_2$, $k_3$ which make up $K$ in Eq. (2) must satisfy the condition $k_3>k_2>k_1(k_1=1$ to $\infty)$, the following types of virtual excitations are possible:

Single excitations:

$$\begin{aligned} 1,2,3&\rightarrow 1,2,l &\quad l&>3\\ &\rightarrow 1,k,3 & k&>3\\ &\rightarrow m,2,3 & m&>3 \end{aligned}\qquad (9)$$

Double excitations:

$$\begin{aligned} 1,2,3&\rightarrow 1,k,l &\quad l&>k>3\\ &\rightarrow m,2,l & l&>m>3\\ &\rightarrow m,k,3 & k&>m>3. \end{aligned}\qquad (10)$$

All higher excitations $l>k>m>3$ give $H_{OK}=0$. We have labeled the indices differently depending upon which initial orbitals remain unchanged. The first thing to notice is that with our choice of the orbitals all $H_{OK}$ for the single virtual excitations of the *outer* electron[24] vanish by the *generalized* [22c] *Brillouin theorem*,[25] i.e.,

$$H_{OK}=\langle\det'(123), H\det'(12l)\rangle=\langle l, h_{\text{core}}^{\text{eff}}3\rangle$$
$$=\epsilon_3\langle l,3\rangle=0;\qquad (l\neq 3).\qquad (11)$$

On the other hand, single excitations from the core and double excitations lead to

$$\langle\det'(123), H\det'(1k3)\rangle=\langle\det'(23), g_{12}\det'(k3)\rangle$$
$$\langle\det'(123), H\det'(m23)\rangle=\langle\det'(13), g_{12}\det'(m3)\rangle\qquad (12)$$

where Eq. (7) has been used; and,

$$\langle\det'(123), H\det'(1kl)\rangle=\langle\det'(23), g_{12}\det'(kl)\rangle\qquad (13)$$

and, similarly, for the rest of Eqs. (10). Here $g_{12}=r_{12}^{-1}$ and det′ denotes $(1/\sqrt{6})$ det on the left and $(1/\sqrt{2})$ det on the right of Eqs. (12) and (13). Notice that Eqs. (12) do not yield zero, because the core orbitals 1 and 2 have been chosen as the H.F. SCF solutions of $Li^+$ rather than of Li. This point will be discussed in detail after stating Eq. (17c).

Equations (9) through (13) allow the separation of the second-order energy in Eq. (6) into correlation energies of pairs of electrons. Of course it is a general result that, whenever, orthonormal functions are used and overlap, and exchange effects are neglected, second-order energies come out "pairwise additive"[4] as in intermolecular "dispersion" forces. Three-body and higher correlations appear in higher orders of perturbation. However, we shall see shortly that, here, the Pauli exclusion principle has already introduced some many-body correlations into the second-order. The first-order energy in Eq. (6) can be written as the energy of the ion core in the field of the bare nucleus and the energy of the outer electron in the SCF field of the ion core. For the ground state

$$H_{00}=\langle\det'(123), (\sum_{i=1}^{3}h_i^0+\sum_{i>j}^{3}g_{ij})\det'(123)\rangle$$
$$=\epsilon_{\text{core}}^0(12)+\langle 3, h_{\text{core}}^{\text{eff}}3\rangle,\qquad (14)$$

where

$$\epsilon_{\text{core}}^0(12)=2\langle 1s, h_1^0 1s\rangle+\langle 1s1s, g_{12}1s1s\rangle.$$

With the systematic classification of the virtual transitions given by Eqs. (9) and (10), $E_0^{(2)}$ in Eq. (6) can be broken down into the following terms, depending upon which initial orbitals are involved:

$$E_0^{(2)}(123)=E_0^{(2)}(12)+E_0^{(2)}(23)+E_0^{(2)}(13).\qquad (15)$$

Again numbers in parentheses label the orbitals, and not the electrons; and,

$$-E_0^{(2)}(12)=\sum_{k>m>3}\frac{\langle\det'(12), g_{12}\det'(mk)\rangle^2}{\Delta(1,2,3\rightarrow m,k,3)}\qquad (16a)$$

$$-E_0^{(2)}(23)=\sum_{k>3}\frac{\langle\det'(23), g_{12}\det'(k3)\rangle^2}{\Delta(1,2,3\rightarrow 1,k,3)}+\sum_{l>k>3}\frac{\langle\det'(23), g_{12}\det'(kl)\rangle^2}{\Delta(1,2,3\rightarrow 1,k,l)}\qquad (17a)$$

$$-E_0^{(2)}(13)=\sum_{m>3}\frac{\langle\det'(13), g_{12}\det'(m3)\rangle^2}{\Delta(1,2,3\rightarrow m,2,3)}+\sum_{l>m>3}\frac{\langle\det'(13), g_{12}\det'(ml)\rangle^2}{\Delta(1,2,3\rightarrow m,2,l)}.\qquad (18a)$$

The Δ's in the denominators represent the energies of the respective virtual transitions and are given from Eq. (6) by

$$\Delta(1,2,3\rightarrow m,k,l)\equiv\langle\det'(mkl), H\det'(mkl)\rangle-\langle\det'(123), H\det'(123)\rangle.\qquad (19)$$

[24] Because of complete antisymmetry, we cannot refer to a definite electron. What we mean by "outer" electron is *any* one of the electrons occupying the particular *orbital* 3.

[25] Notice that with this generalization,[22] only the initially occupied orbital 3 must be an eigenfunction of $h_{\text{core}}^{\text{eff}}$ to make Eq. (11) vanish. There is no such condition for $l$.

In particular, when $m$ is odd and $k$ is even,

$$\Delta(1,2,3\rightarrow m,k,3)=\Delta h^0(1\rightarrow m)+\Delta h^0(2\rightarrow k)+(J_{mk}-J_{12})+(J_{k3}-J_{23})+(J-K)_{m3}-(J-K)_{13}, \quad (16b)$$

$$\Delta(1,2,3)\rightarrow 1,k,3)=\Delta h^0(2\rightarrow k)+(J_{k1}-J_{21})+(J_{k3}-J_{23}), \quad (17b)$$

$$\Delta(1,2,3\rightarrow 1,k,l)=\Delta h^0(2\rightarrow k)+\Delta h^0(3\rightarrow l)+(J_{k1}-J_{21})+(J_{kl}-J_{23})+(J-K)_{l1}-(J-K)_{31}, \quad (17c)$$

and, similarly, for Eq. (18a). $J$'s and $K$'s are the usual coulomb and exchange integrals,

$$J_{rs}=\langle rs, g_{12}rs\rangle; \qquad K_{rs}=\langle rs, g_{12}sr\rangle$$

and $\Delta h^0(r\rightarrow s)=\langle s, h^0 s\rangle-\langle r, h^0 r\rangle$. The numerical evaluation of $E^{(2)}$ for any one case is a lengthy procedure because large contributions to such sums are to be expected[2] from the continuum (or from what would take the place of the continuum) part of the basis set $\{k\}$. Our objective here, instead, is to investigate, semiquantitatively, various physical aspects of Eqs. (16a)–(18a) and their variation with respect to the different *actual* excited states of the "series" electron. (See, however, Sec. IV.)

The first part, Eq. (16a), corresponds to the correlation energy of the ion-core ($1s_\alpha 1s_\beta$) as it exists in the neutral atom. The effect of the "outer" electron by means of the exclusion principle, on this core energy will be taken up later in Sec. III. Equations (17a) and (18a) give the correlation energy between an electron occupying 3, i.e., $2s_\alpha$ and those in 1 (i.e., $1s_\alpha$) and 2 (i.e., $1s_\beta$), respectively. The first terms on the right hand sides of Eqs. (17a) and (18a) represent the inner electrons in orbitals 1 and 2 making virtual transitions in the average field of the orbital 3. These are the "orbital average polarization" terms as we referred to them in the Introduction. Notice that they arose only because 1 and 2 were chosen as the SCF orbitals of $Li^+$ without introducing the H.F. field of $2s$ into Eq. (7). [See also Eq. (12).] Actually "orbital average polarization" is a very small effect and, therefore, in Hartree-Fock calculations on alkalilike configurations the same core orbitals are often used[1,26] for the free ion and the atom as we have done for $Li^+$ and Li. On the other hand, the real correlation energy that remains after a completely SCF calculation has been made, is given by the second terms of Eqs. (17a) and (18a). These terms are due to one core electron at a time making virtual transitions *simultaneously* with the outer electron. Such double excitation effects are sometimes referred to as "dispersion" energy in analogy to London forces.[18] Here we have referred to the totality of Eqs. (17a) and (18a), including "dispersion," by "core-polarization" energy (see Introduction). Unfortunately, the possibility of confusion exists, because sometimes the "orbital average polarization" is simply referred to as "polarization."[18,23] Such "orbital average polarization" effects, which also arise in going from various "restricted" types of H.F. calculation to "unrestricted" types, have been discussed[23] for Li. As it has been shown,[27] however, these are much smaller effects than the correlation energy (in fact, about $10^{-3}$ of it), i.e., than the "core-polarization" energy given by Eqs. (17a) and (18a). Although a small effect, here, we have included the "orbital average polarization" energy in the total "core-polarization" for completeness and convenience.

The sums in Eqs. (16a) to (18a) can be put into approximate but closed forms which are physically interesting by taking out the energy denominators as "mean excitation energies" in each case. This type of approximation has been made in many different contexts since Unsöld[28] and is used, for instance, in London forces.[3,10] In the first parts of Eqs. (17a) and (18a), $\Delta$'s consist of excitations of a single core electron at a time, and we write

$$\langle\Delta(1,2,3\rightarrow 1,k,3)\rangle_{Av}{}^{k}\approx\langle\Delta(1,2,3\rightarrow m,2,3)\rangle_{Av}{}^{m}=\bar{\delta}_c, \quad (20)$$

where $\langle\ \rangle_{Av}$ denote averages. The second, i.e., "dispersion" parts, of Eqs. (17a) and (18a) involve a similar single core excitation, but, in addition, virtual transitions of the outer electron in the field of the already excited core. [See Eq. (17c).] Thus to define a "mean excitation energy" for the outer electron, an averaging over both $k$ and $l$ in Eq. (17c) is necessary, i.e.,

$$\langle\Delta(1,2,3\rightarrow 1,k,l)\rangle_{Av}{}^{k,l}\approx\langle\Delta h^0(2\rightarrow k)+(J_{k1}-J_{21})\rangle_{Av}{}^{k}+\langle\Delta h^0(3\rightarrow l)+(J-K)_{l1}-(J-K)_{31}+(J_{kl}-J_{23})\rangle_{Av}{}^{k,l}$$

and

$$\langle\Delta(1,2,3\rightarrow 1,k,l)\rangle_{Av}{}^{k,l}\approx\langle\Delta(1,2,3\rightarrow m,2,l)\rangle_{Av}{}^{m,l}\equiv\bar{\delta}_C+\bar{\delta}_V{}^*(3) \quad (21)$$

where $\bar{\delta}_V{}^*(3)$ refers to the "mean excitation energy" of the valence, i.e., the "series" electron initially occupying orbital 3. For our purposes only very rough ideas of the magnitudes of $\delta$'s are necessary. The best use of $\delta$'s would be as semiempirical parameters. If the part of the basis set $\{u_k\}$ corresponding to the continuum[2] had not made a large contribution to the second order energy sums, $\delta$'s would have been of the order of ionization potentials, $I$. But actually more than half the second order energy, e.g., in He, comes from the continuum. Thus in general $\delta$'s will be several times the ionization potentials. The use of ionization potentials is

[26] See for example: D. R. Hartree and W. Hartree, Proc. Cambridge Phil. Soc. **34**, 550 (1938).

[27] R. E. Watson and R. K. Nesbet, Mass. Inst. Technol. Quart. Progr. Rept., October 15, 1959, p. 35.

[28] A. Unsöld, Z. Physik. **43**, 563 (1927).

permissible, however, only where lower limits to energy terms are desired. Thus we write $\bar{\delta}_C > I_{(\mathrm{Li}^+)}$ and $\bar{\delta}_V^* \gtrsim I_{(\mathrm{Li})}$ and defer further discussion of $\bar{\delta}$'s until Sec. IV.

In Eq. (21), $\bar{\delta}_V^*$ represents a small fraction of $\bar{\delta}_C$; moreover, the single excitation terms in Eqs. (17a) and (17b) are small as was mentioned earlier. Thus neglecting the absence of $\bar{\delta}_V^*$ in Eq. (19) [compare Eq. (20)] all the energy denominators may be equated and Eqs. (17a) and (18a) combined into one "series" electron-core correlation energy. For the ground state of Li

$$-E_{CV}^{(2)}(3) = -E_0^{(2)}(23) - E_0^{(2)}(13)$$
$$\approx [\bar{\delta}_C + \bar{\delta}_V(3)]^{-1}[\sum_{l>k\geq 3} \langle \det'(23), g_{12} \det'(kl)\rangle^2 + \sum_{l>m\geq 3} \langle \det'(13), g_{12} \det'(ml)\rangle^2]. \quad (22)$$

For convenience, so far we have considered only the ground state [det (123)] of the Li atom and in Eqs. (9) and (10) analyzed the various types of virtual transitions that are possible from this state. Since, however, one of our objectives is to examine the validity of deriving an effective potential for the outer electron, before giving closed expressions for the sums in Eq. (22) it is necessary to consider the excited states of Li, [det'(12$n$)] where now the "series" electron occupies an orbital ($n$) different from $2s_\alpha$.

With a different state of Li, the second-order sum in Eq. (6) again involves the same complete set of "ordered configurations," except that whereas before, det'(123) had been excluded and singled out as the zero-order wave function, now the sum includes this determinant but instead excludes det'(12$n$), the new zero-order wave function. $E^{(2)}$ in Eq. (6), or in operator form

$$-E_N^{(2)} = \langle N, \{H \sum_{M\neq N} M\rangle (H_{NN} - H_{MM})^{-1} \langle M, H\} N\rangle$$
$$= \langle N, Q_N^{(2)} N\rangle, \quad (23)$$

where $N$ denotes det'(12$n$), can be put also in the form of an expectation value over the "series" electron orbital $n$. Denoting

$$N = \det'(12n) = \mathcal{A}\{n u_c(12)\}$$

where

$$u_c(12) = (1/\sqrt{2}!) \det(12),$$

and $\mathcal{A}$ is the operator antisymmetrizing $n$ with the two core electrons, we obtain

$$-E_N^{(2)} = \langle n\{u_c(12), \mathcal{A}H \sum_{M\neq N} M\rangle (H_{NN} - H_{MM})^{-1} \times \langle M, H\mathcal{A}u_c(12)\} n\rangle$$
$$= \langle n, t_{cn}^{(2)} n\rangle. \quad (24)$$

The curly brackets in the second term have been placed to indicate that integrations over the coordinates in $n$ are to be performed last. As it stands the formal core "series" electron interaction operator $t_{cn}^{(2)}$ is far from being a potential energy for $n$. First, the summation over $M$ includes all double virtual excitations of the core orbitals, (12); thus the part corresponding to Eq. (16a), the correlation energy of the core itself, has not yet been separated. As we have mentioned earlier, this part actually depends on $n$, due to the "exclusion" principle and it will be taken up later. In addition if an arbitrary one-electron basis set were to be used, $t_{cn}^{(2)}$ would depend on which $N$ had been excluded from the sum in Eqs. (23) or (24). On the other hand, with an SCF H.F. basis or the basis $\{k\}$ we have chosen above, this type of dependence is eliminated, because the "series" electron orbitals satisfy Eq. (11). Thus in $E_0^{(2)}$ single virtual excitations from det'(123), i.e., the first of the Eqs. (9), would include

$$1, 2, 3 \rightarrow 1, 2, n,$$

whereas in $E_N^{(2)}$, the same type of excitations from det'(12$n$) would exclude 1, 2, $n$ but include

$$1, 2, n \rightarrow 1, 2, 3.$$

But, both of these virtual transitions have zero matrix elements by Eq. (11) and, hence, are without effect on $t_{cn}^{(2)}$. The rest of Eqs. (9) and (10) can be generalized to any state of Li, det'(12$n$), keeping in mind the ordering of configurations in Eq. (2) by $k>l>m$ ($m=1$ to $\infty$). Some transitions that were single core transitions for one state become "core"-$n$ double transitions for another state and vice versa, as, for example, in

$$1, 2, 3 \rightarrow 1, 5, 3$$
$$1, 2, n \rightarrow 1, 3, 5 \qquad (n \neq 3, 5);$$

otherwise the *totality* of the virtual transitions involving either a single core orbital or a single core orbital and the outer one, $n$, are unchanged. Thus, after making the same approximations in the energy denominators, the core-valence electron correlation energy in Eq. (22) can be written for any "series" state of Li as

$$-E_{CV}^{(2)}(n) = -E_N^{(2)}(2n) - E_N^{(2)}(1n)$$
$$\approx [\bar{\delta}_C + \bar{\delta}_V(n)]^{-1}[\sum_{l>k\geq 3} \langle \det'(2n), g_{12} \det'(kl)\rangle^2 + \sum_{l>m\geq 3} \langle \det'(1n), g_{12} \det'(ml)\rangle^2], \quad (25)$$

where $n>2$. The independence of the indices in the above sums, from $n$, is the biggest advantage of the way the basis set $\{k\}$ was chosen previously. [See Eqs. (12) and (7)]. If the orbitals 1, 2, 3 had been taken as the completely SCF solutions of Li, the first terms in Eqs. (17a) and (18a) would have vanished and both the core orbitals 1, 2 and the summing indices in Eq. (25) would have been made to depend on the particular outer orbital. We can now anticipate putting the sums appearing in Eq. (24) into closed forms and hope to obtain the desired potential (independent of $n$) were it not for the appearance of $\bar{\delta}_V(n)$ in the denominator. $\bar{\delta}_V(n)$ as in Eqs. (21) and (22) represents the "mean *virtual* excitation" energy of the outer orbital $n$. For $n>3$, in some terms such as

$$1, 2, n \to 1, 3, 5 \quad (n>3, 5),$$

of the equations corresponding to Eqs. (17a) and (18a), $n\to 5$ would actually be a "deexcitation" and make a negative contribution to $\bar{\delta}_V(n)$. However, there are only a few such terms for low lying $n$, and the weight of any one term to the sum, e.g., in Eq. (17a), is small. As mentioned earlier, $\bar{\delta}_V(n)$ represents a small fraction of $\bar{\delta}_C$, so that it can either be neglected in comparison with $\bar{\delta}_C$ or replaced by a reasonable average,[7] $\delta_S$, for the few states det′(12$n$) of interest. Since $\bar{\delta}_V(n)$ is small, the errors made by replacing it by $\delta_S$ will be even smaller. Assuming for the moment that the sums could be put in the suitable form, the requirement $(\bar{\delta}_V(n)/\bar{\delta}_C)<1$ is necessary if any "core-polarization" potential derived from Eq. (25) is to be independent of $n$. Notice that this requirement is similar to the necessity of having the outer electron move slowly compared to the core electrons in the adiabatic approximation, except that here it does not enter in a fundamental way.

The summations in Eq. (25) can now be carried out. Since the core $(1s_\alpha 1s_\beta)$ is a closed shell, it is necessary to discuss only those $n$ that have either all $\alpha$ or all $\beta$ spin. Take all $n$ to be odd, i.e., with spin $\alpha$, and in Eq. (25) consider first the pair involving opposite spins, i.e., det′(2$n$). Then

$$\sum_{l>k\geq 3} \langle \det'(2n), g_{12} \det'(kl) \rangle^2 = \langle \det'(2n), g_{12}^2 \det'(2n) \rangle$$
$$- \sum_{l>2, k=2} \langle \det'(2n), g_{12} \det'(2l) \rangle^2$$
$$- \sum_{l>1, k=1} \langle \det'(2n), g_{12} \det'(1l) \rangle^2 \quad (26)$$

This follows from a matrix multiplication relation of the type

$$\sum_{L=1}^{\infty} \langle M, AL \rangle \langle L, BN \rangle = \langle M, (AB)N \rangle, \quad (27)$$

where $A$ and $B$ are operators acting on the space for which the set of orthonormal vectors $\{L\}=1, 2, 3, \cdots \infty$ form a complete basis. In Eq. (26) the set of all normalized two-electron determinants corresponding to the ordered configurations $l>k\geq 1$, as in Eq. (2), form a complete orthonormal basis for the space of two-electron coordinates (including spin) on which $g_{12}$ acts. The spins of 2 and $n$ being opposed, the determinantal matrix elements on the right-hand side of Eq. (26) reduce to the direct integrals only; e.g., for the ground state of Li,

$$\langle \det'(23), g_{12}^2 \det'(23) \rangle = \langle 1s2s, g_{12}^2 1s2s \rangle. \quad (28)$$

Similarly,

$$\sum_{l>2} \langle \det'(2n), g_{12} \det(2l) \rangle^2 = \sum_{l>2, (l=\text{odd})} \langle 2n, g_{12} 2l \rangle^2, \quad (29)$$

where to conserve spin, $l$ must have spin $\alpha$, i.e., be odd. The last sum can also be evaluated by carrying out the integrations in each matrix element over one set of the electron coordinates and, thus, obtaining a function $(W_{2,2})$ of the coordinates of the other electron; i.e.,

$$\langle 2n, g_{12} 2l \rangle^2 = \langle n, W_{2,2} l \rangle^2, \quad (30)$$

where

$$W_{2,2} \equiv \int |1s(1)|^2 r_{12}^{-1} d\tau_1.$$

Then, making use of Eq. (27) we get

$$\sum_{l>2, (l=\text{odd})} \langle 2n, g_{12} 2l \rangle^2 = \sum_{l>2, (l=\text{odd})} \langle n, W_{2,2} l \rangle^2$$
$$= \langle n, (W_{2,2})^2 n \rangle - \langle n, W_{2,2} 1 \rangle^2, \quad (31)$$

since $l=1, 3, 5, \cdots$ odd $\cdots \infty$ form a complete *one-electron* basis set and $W_{2,2}$ of Eq. (30) is a function of the coordinates of a single electron. In the same way, in Eq. (26)

$$\sum_{l>1} \langle \det'(2n), g_{12} \det'(1l) \rangle^2 = \sum_{l>1, (l=\text{even})} \langle 2n, g_{12} l1 \rangle^2$$
$$= \sum_{l>1, (l=\text{even})} \langle 2, W_{n,1} l \rangle^2 = \langle 2, (W_{n,1})^2 2 \rangle, \quad (32)$$

since now to conserve spin, $l$ must be even (spin $\beta$) and $l=2, 4, 6, \cdots$ even $\cdots \infty$ form a complete basis set for functions of the spatial coordinates of one electron.

Similarly for the parallel spin pair (1$n$) in Eq. (25), the use of Eq. (27) leads to

$$\sum_{l>m>3} \langle \det'(1n), g_{12} \det'(ml) \rangle^2$$
$$= \langle \det'(1n), g_{12}^2 \det'(1n) \rangle$$
$$- \sum_{l>2} \langle \det'(1n), g_{12} \det'(2l) \rangle^2$$
$$- \sum_{l>1} \langle \det'(1n), g_{12} \det'(1l) \rangle^2. \quad (33)$$

Now both of the spin orbitals 1 and $n$ have spin $\alpha$, so that

$$\langle \det'(1n), g_{12}^2 \det'(1n) \rangle = \langle 1n, g_{12}^2 1n \rangle - \langle 1n, g_{12}^2 n1 \rangle. \quad (34)$$

Also, since 2 has spin $\beta$,

$$\sum_{l>2} \langle \det'(1n), g_{12} \det'(2l) \rangle^2 = 0. \quad (35)$$

The last term of Eq. (33)

$$\sum_{l>1} \langle \det'(1n), g_{12} \det'(1l) \rangle^2$$

$$= \sum_{l>1,(l=\text{odd})} [\langle 1n, g_{12}1l \rangle - \langle 1n, g_{12}1l \rangle]^2$$

$$= \sum_{l>1,(\text{odd})} \langle n, W_{1,1}l \rangle^2 + \sum_{l>1,(\text{odd})} \langle 1, W_{n,1}l \rangle^2$$

$$-2 \sum_{l>1,(\text{odd})} \langle n, W_{1,1}l \rangle \langle l, (W_{n,1})1 \rangle \quad (36)$$

defining the one electron functions as in Eq. (30). Then, using Eq. (27) in each term of the last expression,

$$\sum_{l>1} \langle \det'(1n), g_{12} \det'(1l) \rangle^2$$

$$= \langle n, (W_{1,1})^2 n \rangle + \langle 1, (W_{n,1})^2 1 \rangle$$

$$-2 \langle n, (W_{1,1})(W_{n,1}) 1 \rangle, \quad (37)$$

and, from Eq. (33),

$$\sum_{l>m\geq 3} \langle \det'(1n), g_{12} \det'(ml) \rangle^2$$

$$= \langle 1n, g_{12}^2 1n \rangle - \langle 1n, g_{12}^2 n1 \rangle$$

$$- \langle n, (W_{1,1})^2 n \rangle - \langle 1, (W_{n,1})^2 1 \rangle$$

$$+2 \langle n, (W_{1,1})(W_{n,1}) 1 \rangle. \quad (38)$$

Before going into the correlation energy of the core itself and deriving expressions for it similar to those given above, let us examine the meaning of the various terms in the results for $E_{CV}^{(2)}(n)$, Eq. (25). The major part of the core-valence electron correlation energy is due to the pair with opposing spins, i.e., $E_N^{(2)}(2n)$, since the other $\alpha\alpha$ pair electrons are already kept apart by the Pauli exclusion principle. By Eqs. (26), and (29) through (32), the $\alpha\beta$-pair energy for some state det'$(12n)$ of Li is given by

$$-E_N^{(2)}(2n) \approx (\bar{\delta}_C + \delta_S)^{-1} [\langle 2n, g_{12}^2 2n \rangle - \langle n, (W_{2,2})^2 n \rangle$$

$$+ \langle n, (W_{2,2}) 1 \rangle^2 - \langle 2, (W_{n,1})^2 2 \rangle]. \quad (39)$$

Or, for the ground state of Li with $n=3$, replacing $1=(1s)_\alpha$, $2=(1s)_\beta$ and $3=(2s)_\alpha$,

$$-E_0^{(2)}[(1s)_\beta (2s)_\alpha] \approx (\bar{\delta}_C + \delta_S)^{-1}$$

$$\times [\langle (2s), F_{(1s)}(2s) \rangle - R_{(1s_\beta)(2s_\alpha)}], \quad (40)$$

where we have defined

$$\langle (2s), F_{(1s)}(2s) \rangle \equiv \langle (2s)(1s), g_{12}^2 (2s)(1s) \rangle$$

$$- \langle (2s), (W_{(1s)(1s)})^2 (2s) \rangle \quad (41a)$$

and

$$R_{(1s_\alpha)(1s_\beta)} \equiv \langle (1s), (W_{(2s),(1s)})^2 (1s) \rangle - \langle 1s2s, g_{12} 1s1s \rangle^2. \quad (41b)$$

Clearly $F_{(1s)}$ represents a true potential acting on the outer electron, and depends solely on the core orbital $(1s)$; it is given by

$$F_{(1s)}(\mathbf{r}_2) = \left\{ \int |1s(\mathbf{r}_1)|^2 [1/(|\mathbf{r}_2 - \mathbf{r}_1|^2)] d\tau_1 \right\}$$

$$- \left\{ \int |1s(\mathbf{r}_1)|^2 [1/(|\mathbf{r}_2 - \mathbf{r}_1|)] d\tau_1 \right\}^2, \quad (42)$$

where $\mathbf{r}_2$ and $\mathbf{r}_1$ are the position coordinates in $(2s)$ and $(1s)$, respectively, and the potential acting at $\mathbf{r}_2$ is obtained by integration over all $\mathbf{r}_1$. $F_{(1s)}(\mathbf{r}_2)$ in Eq. (42) can be identically written as

$$F_{(1s)}(\mathbf{r}_2)$$

$$= \langle (1s), [(1/r_{12}^2) - \langle (1s), (1/r_{12})(1s) \rangle_1^2](1s) \rangle_1$$

$$= \langle (1s), [(1/r_{12}) - \langle (1s), (1/r_{12})(1s) \rangle_1]^2 (1s) \rangle_1$$

$$= \langle (g_{12} - \langle g_{12} \rangle_{\text{Av},1s})^2 \rangle_{\text{Av},1s}, \quad (43)$$

where $\langle \quad \rangle_1$ means that all integrations in Eq. (43) are over the same coordinates $\mathbf{r}_1$ as in Eq. (42). In the last term we have denoted the quantum mechanical averages of the "source point" $\mathbf{r}_1$ over $(1s)$ with the symbol, $\langle \quad \rangle_{\text{Av}}$. It will be noticed that $(g_{12} - \langle g_{12} \rangle_{\text{Av},1s}$ simply represents the "instantaneous" (in the virtual sense) deviation of the electrostatic potential, produced at the point $\mathbf{r}_2$ by the electronic charge at $\mathbf{r}_1$, from the orbital average [i.e., the expectation value over $1s(\mathbf{r}_1)$] potential of the electron $(\mathbf{r}_1)$ produced at $r_2$. Thus as we have done in the theory of Van der Waals' interactions between molecules and solid surfaces,[29] $F_{(1s)}(\mathbf{r}_2)/\bar{\delta}_C$ may be called the fluctuation potential since $F_{(1s)}(\mathbf{r}_2)$ is simply the mean square fluctuation of the coulombic potential of the orbital $(1s)$ at the point $\mathbf{r}_2$.

In Eq. (40), there still exists a remainder term $R_{(1s)_\beta(2s)_\alpha}$ which cannot be put in the form of an expectation value of a potential. A close examination of the Eqs. (26) through (32) leading to Eq. (40) shows that $R_{(1s)_\beta(2s_\alpha)}$ arises because the closed inner shell $(1s_\alpha 1s_\beta)$ of Li prevents the outer electron occupying $2s_\alpha$ from making virtual transitions to the already occupied inner levels. This is one of the exclusion effects that were mentioned in the introduction and represents the nonpairwise additive effect of $(1s_\alpha)$ on the pair $(1s_\beta 2s_\alpha)$. $R_{(1s_\beta)(n)}$ has been neglected in the previous treatments of core-polarization for nonpenetrating orbitals.[7–9,13,14] However, especially for the ground state, its magnitude requires examination (See Sec. IV).

Likewise, the correlation energy of the $\alpha\alpha$ pair $(1s_\alpha 2s_\alpha)$ of the ground state of Li in Eq. (25) is derived from Eqs. (38), (40), and (41) by

$$-E_0^{(2)}(1s_\alpha 2s_\alpha) \cong (\bar{\delta}_C + \delta_S)^{-1} [\langle (2s), F_{(1s)}(2s) \rangle$$

$$- \langle (2s), F_{(1s)}^{\text{ex}}(2s) \rangle - R_{1s_\alpha 2s_\alpha}{}^{\text{ex}}] \quad (44)$$

[29] O. Sinanoğlu, Ph.D. thesis, Part I, University of California, August, 1959; O. Sinanoğlu and K. S. Pitzer, J. Chem. Phys. 32, 1279 (1960).

where

$$\langle (2s), F_{(1s)}{}^{ex}(2s)\rangle \equiv \langle 2s1s, g_{12}{}^2 1s2s\rangle - \langle 2s, (W_{1s,1s})(W_{1s,2s})1s\rangle \quad (45a)$$

and

$$R_{1s_\alpha 2s_\alpha}{}^{ex} \equiv \langle 1s, (W_{2s,1s})^2 1s\rangle - \langle 2s, (W_{1s,1s})(W_{1s,2s})1s\rangle. \quad (45b)$$

Comparison of the Eqs. (38) and (44) with (39) and (40) shows that $E_0{}^{(2)}(1s_\alpha 2s_\alpha)$, in addition to the fluctuation potential of $E_0{}^{(2)}(1s_\beta 2s_\alpha)$ also contains $F_{(1s)}{}^{ex}/(\bar{\delta}_C+\delta_S)$ or what may be called the "exchange fluctuation potential." Note that the latter is a "potential" only in exactly the same sense[2] as the exchange part of the Hartree-Fock field [see Eq. (8)] is. The "exclusion" term in Eq. (45b) is similar to that in Eq. (41b). Combining Eqs. (6), (7), (14), (15), (40), and (44), the total energy of Li in any state, $N \equiv \det'(12n)$, is given as

$$E_N(12n) \cong H_{NN}+E_N{}^{(2)} \cong \epsilon_{core}{}^0(12)+\langle n, h_{core}{}^{eff} n\rangle + E_N{}^{(2)}(12)+\langle n, U_f n\rangle + Q_n = E_N(\text{core})+\langle n, (h_{core}{}^{eff}+U_f)n\rangle + Q_n, \quad (46)$$

where

$$U_f = (2F_{(1s)} - F_{(1s)}{}^{ex})/(\bar{\delta}_C+\delta_S)$$

and

$$Q_n = (R_{1s_\beta n}+R_{1s_\alpha n}{}^{ex})/(\bar{\delta}_C+\delta_S).$$

$E_N$ (core) contains the H.F. energy [Eq. (14)] of the *free* ion $Li^+$, and, in addition, the correlation energy $E_N{}^{(2)}(12)$ [given for the ground state by Eq. (16a)] which refers not to free $Li^+$ but to the core as it exists in state $N$ of Li. The dependence of $E_N{}^{(2)}(12)$ on the outer orbital $n$ through the "exclusion" principle is discussed in Sec. III. $Q_n$ constitutes the rest of the "exclusion" effect. It depends upon the exchange charge density of $n$ with $(1s)$, so that it will be small for excited states of Li, i.e., "nonpenetrating" $n$. $U_f$ is the desired correlation (fluctuation) potential. With a larger atom, similar results can easily be written down. In general, there will be a contribution from each electron of the core to $U_f$ and $Q_n$. After estimating the "exclusion" effects, $U_f$ can be determined semi-empirically (see e.g., Douglas[11]) for instance, by leaving $(\bar{\delta}_C+\delta_S)$ as a parameter. Aside from the "exclusion" effects, Eq. (46) has the variational form for the outer electron with an effective "core-Hamiltonian." It is important to realize however that in this form $n$ cannot be varied to improve the energy even when $Q_n \sim 0$ and $E_N(\text{core}) \sim$ constant, because the result was derived for a specific choice of orbitals for $n$, namely, those satisfying the H.F. condition, Eq. (11). With any other choice, the single virtual transitions of $n$ as in Eq. (9) would lead to nonvanishing matrix elements and make a new contribution to Eq. (46). This point brings out the connection between the present treatment and the recent nuclear manybody theory.[19,30] In fact, Eq. (46) corresponds to a starting approximation of that theory with the neglect of higher-order correlations (Sec. VI) as can be seen, e.g., in the work of Bethe[31] and Rodberg.[32] Improved choices for $n$ can be made and perhaps restrictions, as in Eq. (11), removed by going to higher orders of perturbation, but such generalizations will be deferred to a future date.

## III. "EXCLUSION" EFFECT OF AN OUTER ELECTRON ON THE CORE ENERGY

The dependence of the core correlation energy $E_N{}^{(2)}(12)$ of Eq. (46) on the outer orbital $n$ in Li can be examined by a careful classification of all the "ordered configurations" entering Eq. (6) and generalization of Eq. (16a) to any $n$. In Eq. (10) some of the configurations that correspond to double virtual transitions from the core $(1s_\alpha 1s_\beta)$ when $n$ was 3 (i.e., $2s_\alpha$) become triple transitions from another initial state $\det'(12n)$ with $n>3$, and vice versa. Including all such configurations in $E^{(2)}$, one obtains the expected result that

$$-E_N{}^{(2)}(12) = \sum_{k>m\geq 3,(k,m\neq n)} \frac{\langle \det'(12), g_{12}\det'(mk)\rangle^2}{\Delta(1,2,n\to m,k,n)}; \quad (47)$$

i.e., all the double core transitions $1, 2 \to m, k$ are missing when $m$ or $k$ is the already occupied orbital $n$. This may be compared with the second-order energy of the *free* ion core $Li^+$ using the same one-electron basis set $\{k\}$ that was defined previously for Li:

$$-E_{Li^+}{}^{(2)}(12) = \sum_{k>m>2} \frac{\langle \det'(12), g_{12}\det'(mk)\rangle^2}{\Delta(1,2\to m,k)}. \quad (48)$$

The energy denominator in Eq. (47) differs from that in Eq. (48) by the presence of $n$ [See Eq. (19)]. Nevertheless a semiquantitative estimate of the variation of $E_N{}^{(2)}(12)$ with $n$ and its difference from the energy of $Li^+$ can be obtained by replacing both Δ's by one average, $\bar{\Delta}_{core}[\bar{\Delta}_{core} > (I_{Li^+}+I_{Li^{2+}})]$. Then comparing Eqs. (47) and (48),

$$(E^{(2)})_{Li\,core}(n) \equiv E_N{}^{(2)}(12) \sim E_{Li^+}{}^{(2)}(12) + \bar{\Delta}_{core}{}^{-1}\sum_{k\geq 3}^{\infty}\langle \det'(12), g_{12}\det'(nk)\rangle^2, \quad (49)$$

or using Eq. (27) as in Eq. (32),

$$E_N{}^{(2)}(12) \sim E_{Li^+}{}^{(2)} + \bar{\Delta}_{core}{}^{-1}[\langle (1s), (W_{(1s),n'})^2(1s)\rangle - \langle (1s)(n'), g_{12}(1s)(1s)\rangle^2], \quad (50a)$$

[30] K. A. Brueckner, C. A. Levinson, and H. M. Mahmoud, Phys. Rev. **95**, 217 (1954); for later references see H. Yoshizumi, Advances in Chem. Phys. **2**, 323 (1959).
[31] H. A. Bethe, Phys. Rev. **103**, 1353 (1956).
[32] L. S. Rodberg, Ann. Phys. **2**, 199 (1957).

with $n'\alpha=n$, and from Eq. (46),

$$E_N(\text{core})\cong E_{\text{Li}^+}+\bar{\Delta}_{\text{core}}^{-1}[\langle(1s),(W_{1s,n'})^2(1s)\rangle - \langle(1s)n',g_{12}(1s)(1s)\rangle^2]$$
$$\equiv E_{(\text{Li}^+)}+(R_{\text{core}-n}/\bar{\Delta}_{\text{core}}). \quad (50b)$$

Thus, the *total* energy of the core in the Li atom is given by the total energy of the *free* $\text{Li}^+$ ion including its correlation energy plus the last term which is the desired "exclusion" effect of $n'$ on the core. With any larger system, the evaluation of all such "exclusion" effects is similarly possible from a classification of ordered configurations and the use of Eq. (27) for summations.

## IV. MAGNITUDES FOR THE GROUND STATE OF LITHIUM

In Sec. II, Eq. (46) we have obtained an expression for the total energy of the Li atom in any one of its "series" states. That derivation shows, that aside from the "exclusion" effects, $U_f$ is the desired "core-polarization" potential including exchange and it may be regarded as the mean-square fluctuation of the Hartree-Fock potential of the core per unit of "mean excitation energy." Notice that $U_f$ is a complete potential and is *not* dependent on a multipole expansion of $g_{12}$. In the previous treatments[7–9,13,14] of "core polarization" (a) the "exclusion" effects, (b) the exchange part of $U_f$, i.e., $F_{1s}^{\text{ex}}[\bar{\delta}_C+\delta_S]$, has been neglected (see, however, Ludwig[7b]) and (c) after making a multipole expansion of $g_{12}$, as

$$g_{12}=1/r_{12}=1/r_>+(r_</r_>^2)(\cos\omega)+(r_<^2/2r_>^3)\times(3\cos^2-1)+\cdots \quad (51)$$

mainly the dipole term (with estimates of quadrupole terms) has been considered and the first part $r_>^{-1}$ dropped. In Eq. (51) $r_>$ denotes the greater of the two distances $r_1$ and $r_2$, and $\omega$ is the angle between the radius vectors of the two electrons. For highly excited states, i.e., with larger $n$, assumptions *a* to *c* approach validity. For instance, as the portion of the outer orbital that is inside the core becomes negligible, we get $r_>=r_2$ only, so that the $r_>^{-1}$ part of $g_{12}$ no longer contributes to $U_f$ [See Eqs. (41) and (45)]. For "penetrating" orbitals, however, such is not the case. To get an idea of the magnitudes of the previously neglected penetration and "exclusion" terms, we shall consider here the ground state of Li for which the effects should be largest. We take for $2s$ the orthogonalized Slater orbital,

$$2s^0=1.0148(\delta_2^5/3\pi)^{\frac{1}{2}}r\exp(-\delta_2 r)-0.1742(\delta_1^3/\pi)^{\frac{1}{2}}\times\exp(-\delta_1 r) \quad (52a)$$

with $\delta_1=2.65$ and $\delta_2=0.65$, which sufficiently approximates the Hartree-Fock $(2s)$ orbital of Fock and Petrashen,[33] and for $1s$,

$$1s=(\delta_1^3/\pi)^{\frac{1}{2}}\exp(-\delta_1 r) \quad (52b)$$

with $\delta_1=2.65$. The "exclusion" terms in Eqs. (41b), (45b), and (50b) involve $g_{12}$ in the $W$ integrals. These are like the usual atomic integrals[34]; upon substitution of Eq. (51) for $g_{12}$, only the $r_>^{-1}$ term contributes due to the spherical symmetry of $1s$ and $2s$. The first parts of $F_{(1s)}$ and $F_{(1s)}^{\text{ex}}$ in $U_f$ on the other hand, contain $g_{12}^2$. $U_f$ can be obtained completely, without any expansions, from Eqs. (43) and (46). However, in this article we shall consider only the penetration terms.

To compare the penetration effects with the previously considered dipole terms on an equivalent basis, we must also expand $g_{12}^2$ and not consider higher multipoles. $g_{12}^2$ can be conveniently expanded in terms of the Gegenbauer[35] polynomials, $C_n^{(1)}(\cos\omega)$.

$$g_{12}^2=1/r_{12}^2=(1/r_>^2)\sum_{n=0}^{\infty}(r_</r_>)^n C_n^{(1)}(\cos\omega). \quad (53)$$

These polynominals are analogous to the Legendre polynominals and have similar addition theorems. The first term of Eq. (53) is $r_>^{-2}$, neglected previously. It is responsible for most of the penetration effects of $2s$, as can be seen e.g., from the fact that only $r_>^{-1}$ contributes to the "exclusion" terms and will be calculated here.

With the orbitals of Eq. (52), the desired integrations can be performed analytically and yield

$$\langle 2s^0,(W_{1s,1s})^2 2s^0\rangle=\left\langle 2s^0,\left(\int|1s(r_1)|^2 r_>^{-1}d\tau_1\right)^2 2s^0\right\rangle$$
$$=0.16144(\text{a.u.})^2$$
$$\langle 1s,(W_{2s^0,1s})^2 1s\rangle=0.018537(\text{a.u.})^2$$
$$\langle 1s2s^0,g_{12}1s1s\rangle^2=\langle 1s2s^0,r_>^{-1}1s1s\rangle^2=0.013855(\text{a.u.})^2$$
$$\langle 2s^0,(W_{1s,1s})(W_{2s^0,1s})1s\rangle=0.029806(\text{a.u.})^2$$
$$\langle 2s^0 1s,r_>^{-2}2s^0 1s\rangle=0.17385(\text{a.u.})^2$$
$$\langle 2s^0 1s,r_>^{-2}1s2s^0\rangle=0.053256(\text{a.u.})^2 \quad (54a)$$

where $1(\text{a.u.})=27.202$ ev and, e.g.,

$$(1/4\pi)\int|1s(r_1)|^2 r_>^{-2}d\tau_1=(1/r_2^2)\int_0^{r_2}|1s(r_1)|^2 r_1^2 dr_1+\int_{r_2}^{\infty}|1s(r_1)|^2 dr_1. \quad (55)$$

As it was mentioned in Sec. II, ionization potentials provide lower limits to the "mean excitation energies,"

[33] V. Fock and M. J. Petrashen, Physik. Z. Sowjetunion, 8, 547 (1935).

[34] E. U. Condon and G. H. Shortley, *The Theory of Atomic Spectra* (Cambridge University Press, New York, 1957), p. 174.

[35] For a quite detailed account of these polynominals, see I. Prigogine *et al.*, *Molecular Theory of Solutions* (North-Holland Publishing Company, Amsterdam, 1957), p. 265.

and therefore, we write

$$\bar{\delta}_C+\delta_S\equiv\mathcal{K}_{CS}[I_{(\mathrm{Li})}+I_{(\mathrm{Li}^+)}]\cong 3\mathcal{K}_{CS}(\mathrm{a.u.}) \quad (54\mathrm{b})$$

and

$$\bar{\Delta}_{\mathrm{core}}\equiv\mathcal{K}_{\mathrm{core}}[I_{(\mathrm{Li}^+)}+I_{(\mathrm{Li}^{2+})}]=7.24\mathcal{K}_{\mathrm{core}}(\mathrm{a.u.}),$$

where $\mathcal{K}_{\mathrm{core}}>\mathcal{K}_{CS}>1$. Estimates show that in the ($Li^+$) ion, where two electrons are in the same orbital, $\mathcal{K}_{\mathrm{core}}$ is approximately four due to the large contribution of the continuum in this case. On the other hand, when two electrons are in different orbitals $\mathcal{K}$ may be about two. We will take $\mathcal{K}_{CS}\sim 2$. For results more quantitative than we are aiming at here, it is also possible to obtain $\mathcal{K}$ values for each of the pair sums in Eqs. (16a) to (18a), e.g., by a procedure devised by Kessler.[36]

We may also remark that, once the total correlation energy of a large system has been separated into those of pairs of electrons, the $E^{(2)}$ of each pair can be obtained in a number of ways. Here we have emphasized the semiempirical approach. One may also formulate "variation-perturbation" methods for each pair as has been possible for He[7a].

Substituting the foregoing results, Eqs. (54a, b), into Eqs. (41b), (45b), and (50b), we obtain

$$(\bar{\delta}_C+\delta_S)^{-1}R_{1s_\beta 2s_\alpha}=0.04245\mathcal{K}_{cs}^{-1}\ \mathrm{ev}, \quad (56\mathrm{a})$$

$$(\bar{\delta}_C+\delta_S)^{-1}R_{1s_\alpha 2s_\alpha}=-0.1022\mathcal{K}_{cs}^{-1}\ \mathrm{ev}, \quad (56\mathrm{b})$$

and

$$(\bar{\Delta}_{\mathrm{core}})^{-1}R_{\mathrm{core}-2s_\alpha}=0.0176\mathcal{K}_{\mathrm{core}}^{-1}\ \mathrm{ev}. \quad (56\mathrm{c})$$

Using only the first terms of $g_{12}$ and $g_{12}^2$ from Eqs. (51) and (53) in Eqs. (40) and (44) and denoting the corresponding parts of $F_{(1s)}$ and $F_{(1s)}^{\mathrm{ex}}$ by

$$\langle 2s^0, F_{r>}2s^0\rangle\equiv\langle 2s^0 1s, r_>^{-2}2s^0 1s\rangle-\langle 2s^0, (W_{1s,1s})^2 2s^0\rangle$$

$$\langle 2s^0, F_{r>}^{\mathrm{ex}}2s^0\rangle\equiv\langle 2s^0 1s, r_>^{-2}1s2s^0\rangle$$
$$-\langle 2s^0, (W_{1s,1s})(W_{1s,2s})1s\rangle \quad (57)$$

we get

$$(\bar{\delta}_C+\delta_S)^{-1}\langle 2s^0, F_{r>}2s^0\rangle=0.1125\mathcal{K}_{cs}^{-1}\ \mathrm{ev}. \quad (58\mathrm{a})$$

$$(\bar{\delta}_C+\delta_S)^{-1}\langle 2s^0, F_{r>}^{\mathrm{ex}}2s^0\rangle=0.2126\mathcal{K}_{cs}^{-1}\ \mathrm{ev}. \quad (58\mathrm{b})$$

Essentially the same values are obtained by the use of $\delta_1=2.70$ instead of 2.65 in $2s^0$ and $1s$ so that the results are not very sensitive to our specific choice of the orbital parameters in Eqs. (52).

Most of the "penetration" effects of $2s$ are included in Eqs. (56) to (58). Callaway[13] has obtained a "core-polarization" potential in Li using only the dipole part of $g_{12}$, equivalent to taking the second term of Eq. (53) in $F_{(1s)}$ and neglecting $F_{(1s)}^{\mathrm{ex}}$. He finds a contribution of 0.1 ev to $\langle 2s, U_f 2s\rangle$. Actually the results of Ludwig[7b] suggest that the exchange term in the dipole part may be negligible. By comparison, several interesting conclusions follow from the Eqs. (56)–(58). First, the "exclusion" effect of the outer orbital on the core correlation energy is only 0.0176 $\mathcal{K}_{\mathrm{core}}^{-1}$ ev with $\mathcal{K}_{\mathrm{core}}^{-1}<1$, hence negligible even for $2s$. In Eq. (50) we can then take $E_N(\mathrm{core})\approx E_{(\mathrm{Li}^+)}$. Secondly, the total contribution to the correlation energy of the $1s_\beta 2s_\alpha$ pair from the $r_>^{-1}$ terms is $(\bar{\delta}_C+\delta_S)^{-1}(\langle 2s^0, F_{r>}2s^0\rangle-R_{1s_\beta 2s_\alpha})$ or $(0.1125-0.0423=0.0702)\mathcal{K}_{CS}^{-1}$ ev. With $\mathcal{K}_{CS}^{-1}\sim\frac{1}{2}$, this is still appreciable compared to 0.1 ev, the dipole contribution from both of the core electrons.[13] Also the "orbital average polarization" effect of the orbital $2s$ appears only in the $r_>^{-1}$ part of $g_{12}$ due to the spherical symmetry of $1s$ and $2s$. This effect, however. as was mentioned earlier, is not strictly a correlation effect since it results in converting the $Li^+$ H.F. SCF orbitals to the completely H.F. SCF orbitals in Li. The term, $F_{r>}$, which we have calculated previously, on the other hand, corresponds to the fluctuation of $r_>^{-1}$ i.e., $\langle r_>^{-2}-\langle r_>^{-1}\rangle_1^2\rangle_1$, and, hence, to the inclusion of the "dispersion" effect. It is much larger than the "orbital average polarization" energy (see Sec. II). Thirdly, combining all the $r_>$ or "penetration" terms for the $1s_\alpha 2s_\alpha$ pair we find that

$$(\bar{\delta}_C+\delta_S)^{-1}(\langle 2s^0, F_{r>}2s^0\rangle-\langle 2s^0, F_{r>}^{\mathrm{ex}}2s^0\rangle-R_{1s_\alpha 2s_\alpha}^{\mathrm{ex}})$$
$$=0.00208\mathcal{K}_{CS}^{-1}\ \mathrm{ev},$$

an entirely negligible value. Thus the "Fermi hole" is very effective in keeping the electrons of the $\alpha\alpha$ pair apart and not necessitating a "Coulomb hole." Hence to obtain the over-all "core-polarization" potential we need to add the $r_>^{-1}$ terms only for the $1s_\beta 2s_\alpha$ pair. Then neglecting exchange in dipole and higher-order terms, $U_f$ in Li may be taken as

$$U_f\sim(\bar{\delta}_C+\delta_S)^{-1}[F_{r>}+2(F_{(1s)}-F_{r>})]. \quad (59)$$

## V. MOLECULES

The treatment that was given in previous sections and demonstrated in detail for the case of the Li atom can be applied to any $N$-electron system whose zero-order wave function is a single Slater determinant of H.F. orbitals. Thus, the second-order energy of most molecules can be separated into "pair correlations" and nonpairwise additive "exclusion" effects by taking the H.F. SCF molecular orbitals (MO) as the one-electron basis set $\{k\}$. These orbitals are obtainable by Roothaan's procedure.[37] Each energy component can be obtained in closed form by taking out the denominators as "mean excitation energies" for *each* electron pair. Although rather crude, this procedure has the advantage that the various energy components can then be estimated using only the *same* H.F. orbitals as in the initial single determinant. Contrary to the use of an average energy denominator for the over-all second order energy[36] here each "mean excitation energy" has

[36] P. Kessler, Compt. rend., **242**, 350 (1955).

[37] C. C. J. Roothaan, Revs. Modern Phys. **23**, 69 (1951).

a more physical basis and can be left as a semiempirical parameter especially for those electron groups that are relatively unchanged in going from one atom or molecule to another. [See also the discussion following Eq. (54b)].

More directly, two specific applications are suggested by this approach as was mentioned in the Introduction. Both of them may be demonstrated with reference to the $Li_2$ molecule, for convenience. In this molecule, the first four MO's of the one electron basis $\{k\}$ are $(\sigma_g 1s)^2(\sigma_u 1s)^2$. When the atomic orbitals (AO) that make up such inner shells do not overlap appreciably as in $Li_2$, we can perform a unitary transformation on the $(\sigma_g 1s)^2(\sigma_u 1s)^2$ part of the basis *only* and convert the MO determinant $\det'[(\sigma_g 1s)^2(\sigma_u 1s)^2]$ into the ion core description $\det'[(1s_a)^2(1s_b)^2]$, i.e., $K_aK_b$, where $a$ and $b$ refer to the two nuclei, assuming that admixture of other AO's is negligible. Then taking $(1s_a)^2(1s_b)^2$ equivalently, as the first four spin-orbitals of $\{k\}$ with the rest of the MO's unchanged, a classification of all "ordered configurations" as in Eqs. (9) and (10) into various types of virtual transitions leads to a separation of the correlation energy as in Eqs. (16)–(18). We get the *total* energy of the molecule separated as in Eq. (46) into the energy of two *free* $Li^+$'s (including their individual $E^{(2)}$'s), the energy of the two H.F. MO valence electrons $(\sigma_g 2s)^2$, each in the field of both cores including the core "fluctuation potentials" [the effective "core-hamiltonian" from one $Li^+$ is $(h_{\text{core}}^{\text{eff}}+U_f)_a$] the energy of the two valence electrons (as in $H_2$), $(E^{(2)})_{(\sigma_g 2s)^2}$, and finally the correlation energy between the two cores $K_a$ and $K_b$. There are also the "exclusion" terms associated with each of these components.

The first application concerns the "core-polarization" energy between a valence electron and the ion-cores. When these cores can be assumed quite unchanged, the expectation value of the potential, $U_f$ determined from Eq. (59) in conjunction with the "series" levels of the atom, may be calculated over the ground or an excited state valence MO in the molecule. Thus the calculation of the contribution of "core polarization" to the already small binding energy of a diatomic alkali molecule is possible. Callaway[13] has made such calculations on alkali metals and found appreciable values even with just the dipole part (see Sec. IV) of $U_f$. In this type of application the change in the "exclusion" energy $(\bar{\delta}_C+\delta_S)R_{1s_\beta n_\alpha}$ [Eq. (56a)] may also need to be estimated.

The second application, although a small effect in the case of $Li_2$, concerns the correlation energy between the two cores, $K_a$ and $K_b$ themselves. This energy which can be written in the "fluctuation" form similar to that in Eq. (41) and including the exchange part, is just the "dispersion" (plus "orbital average polarization") energy which on making a multipole expansion [or more conveniently, Gegenbauer expansion[35] as in Eq. (53) but now for two centers] for $g_{12}{}^2$ and taking the second term would simply lead to London's formula.[3] Aside from not requiring such an expansion, the approach presented here now includes the exchange terms as well as the "exclusion" effects similar to those in Eq. (50), but with the appropriate orbitals. It is particularly important to recognize that the discussion given here does *not* require the two cores to occupy completely isolated spaces, each with its own distinct basis set as has been considered necessary in previous discussions of London forces.[38] We have started from a complete set of MO's and have made assumptions only about the first four MO's, $(\sigma_g 1s)^2(\sigma_u 1s)^2$. The assumption that only the AO's, $(1s_a)$, $(1s_b)$, should not overlap appreciably is necessary so that we get into the $K_aK_b$ description. In general, such an assumption is much more plausible (and may even be improved on by considering some overlap) than the requirement of essentially complete localization of the core electrons around different centers, each group with its own distinct set of eigenfunctions.

Finally we observe that the second-order method which has been presented in detail in Sec. II, not only provides an approximate but very convenient way of estimating the energy of a many electron system, but also allows one to discuss many of the correlation effects in simple physical terms.

## VI. HIGHER-ORDER CORRELATION EFFECTS

For simplicity the treatment presented in this article has been so far confined to the second order. Here, correlations among more than two particles at a time are introduced only by the "exclusion" terms. For three and more body "Coulomb" correlations it is necessary to go to higher orders of perturbation.[39] In the system of a single electron outside closed shells, some third-order correlation effects influencing the "core-polarization" energy (e.g., in Na) can be introduced into $U_f$ as an additional potential by methods similar to those in Sec. II, or by letting the mean-excitation energy, $\bar{\delta}_C$, absorb the higher order effects semiempirically. (A similar situation usually occurs in various Van der Waals forces[10,29].)

An interesting case where higher-order correlations deserve further examination is a nondegenerate system of two electrons outside large closed shells as in Ca. Here, aside from the "exclusion" effects, there would be a $U_f$ from the core acting on each of the $4s$ electrons, and a "fluctuation potential" $(r_{12}{}^{-2}-\langle r_{12}{}^{-1}\rangle^2{}_{\text{Av},4s})/\bar{\Delta}$ [as in Eq. (41)] acting between the two $(4s)$ electrons [similar to the correlation of $(1s)^2$ in He]. However, the presence of a large polarizable core inside introduces additional effective interactions between the two outer electrons in higher orders. This can be seen by a crude

[38] See, e.g., H. C. Longuet-Higgins, Proc. Roy. Soc. (London) **A235**, 537 (1956).

[39] Actually, at least for the light atoms, the empirical work of Arai and Onishi suggests that the pairwise additivity of correlation effects may turn out to be a quite good description. See T. Arai and T. Onishi, J. Chem. Phys. **26**, 70 (1957).

but suggestive classical argument: Consider the core as a charge sphere with polarizability $\alpha$ and assume the two outer electrons to be momentarily at rest [see Eq. (1)] at distances $r_1$ and $r_2$ from the nucleus with an angle $\theta_{12}$ between them. Then, each electron induces a dipole moment of $\alpha/r_i^2$ at the core with which it interacts to yield an energy $-\alpha/2r_i^4$. This is the limiting form of $U_f$ in Eq. (46) as $r_i$ approaches infinity. But, in addition, the dipole induced by the electron at $r_1$ acts on the electron at $r_2$ giving

$$[t_{12}^{(3)} \sim -\alpha \cos\theta_{12}/r_1^2 r_2^2].$$

The introduction of such effective interactions between two electrons to account for the influence of a "medium" (the core in this case) is possible quite generally by essentially an extension of the methods given for only the second order in Sec. II. It is also interesting to note the similarity of the additional interaction, $t_{12}^{(3)}$, to the new third-order Van der Waals interaction that is introduced by a solid surface between two inert molecules adsorbed on it[29] (the new force is in addition to the London force between the two molecules alone).

In general, the correlation energy ***between*** two groups of electrons will be influenced by the ***internal*** correlation of each group, as in the example just mentioned, where the ($4s^2$) shell and the core in Ca could be considered as the two groups. The resulting higher order correlation energy can also be treated very simply for the special class of problems pointed out in the introduction. These were the cases where the two groups could be assumed to be totally separated and localized individually, each group with its own set of eigenfunctions. To see how "many electron group functions" can be used in such cases, consider, for instance, two many-electron atoms $A$ and $B$ far apart, and their mutual Van der Waals attraction. Here we can use ordinary second-order Schrödinger perturbation theory. Let $\psi_A^k$ or $\psi_B^l$ be the complete set of exact many electron eigenfunctions for each of the unperturbed "independent" systems $A$ and $B$, respectively. Then the unperturbed Hamiltonian $H_0$ equals $H_A+H_B$, the composite basis set $\psi_{kl}=\psi_A^k\psi_B^l$ and the perturbation is the total interaction between $A$ and $B$ given by

$$V_{AB}=G_{AB}+U_{AB}, \qquad (60)$$

where

$$U_{AB}\equiv\sum_{i,j}[1/(|\mathbf{r}_i^A-\mathbf{r}_j^B|)]=\sum_{i,j}(1/r_{ij}^{AB}),$$

and

$$G_{AB}\equiv\sum_{I,J}[Z_IZ_J/(|\mathbf{R}_I^A-\mathbf{R}_J^B|)].$$

$G_{AB}$ refers to the interaction between the nuclei in $A$ and the nuclei in $B$. $U_{AB}$ is the electrostatic potential between pairs of electrons, one in $A$ (at $\mathbf{r}_i^A$) and the other in $B$ (at $\mathbf{r}_j^B$). $\psi_A^k$ and $\psi_B^l$, being the set of exact eigenfunctions of the isolated systems $A$ and $B$, include the coordinates of all the electrons and even the nuclei localized at $A$ or $B$, respectively.[29] Then the intergroup "dispersion" energy is given by

$$E_{\text{disp}}^{(2)}(AB)=-\sum_{k\neq0,l\neq0}\frac{\langle\psi_A^0\psi_B^0, V_{AB}\psi_A^k\psi_B^l\rangle^2}{\delta_A^{0k}+\delta_B^{0l}}$$

$$\equiv-\sum_{k\neq0,l\neq0}\frac{\langle 00, V_{AB}kl\rangle^2}{\delta_A^{0k}+\delta_B^{0l}}, \qquad (61)$$

where $\delta_A^{0k}=E_A^k-E_A^0$ and $H_A\psi_A^k=E_A^k\psi_A^k$. Replacing the denominators by mean excitation energies and using Eq. (27) we get

$$E_{\text{disp}}^{(2)}(AB)\approx(\bar\delta_A+\bar\delta_B)^{-1}(\langle 00, V_{AB}^2 00\rangle-\langle 0, W_A^2 0\rangle - \langle 0, W_B^2 0\rangle+\langle 00, V_{AB}00\rangle^2) \qquad (62)$$

where

$$\langle 0, W_A^2 0\rangle=\langle\psi_B^0, W_A^2(\mathbf{R}_B)\psi_B^0\rangle$$

and

$$W_A(\mathbf{R}_B)=\int|\psi_A^0(\mathbf{R}_A)|^2U_{AB}d\tau_A, \qquad (63)$$

with $\mathbf{R}_A$ denoting all $\mathbf{r}_i^A$ and $\mathbf{R}_I^A$. If $A$ and $B$ are neutral systems, the last three terms of Eq. (62) may be negligible since they depend on the static charge distributions of $A$ and $B$. Then Eq. (62) takes on the form of the "fluctuation potential" of one system acting on the other; i.e.,

$$E_{\text{disp}}^{(2)}(AB)\approx-\langle 00, V_{AB}^2 00\rangle/(\bar\delta_A+\bar\delta_B). \qquad (64)$$

This description of the Van der Waals forces, that they are the result of the mean-square fluctuation of the electrostatic potential between two systems, is the generalization of the usual fluctuating dipoles picture (Sec. I). Note also that the form of Eq. (64) is the same as the potential in the core polarization problem. Similar considerations, of course, apply to the scattering of electrons[40] by atoms as well.

Now, consider only the $U_{AB}$ part of $V_{AB}$ [Eq. (60)] and in Eq. (64) write $U_{AB}^2$ in detailed form as

$$U_{AB}^2=\left(\sum_{i,j}\frac{1}{r_{ij}^{AB}}\right)^2=\sum_{i,j}\left(\frac{1}{r_{ij}^{AB}}\right)^2+\sum_{i,j\neq r,s}\left(\frac{1}{r_{ij}^{AB}r_{rs}^{AB}}\right). \qquad (65)$$

Here $i$ and $r$ designate any two electrons in $A$ and $j$ and $s$ any two electrons in $B$. The $(r_{ij}^{AB})^2$ terms in Eq. (65) involve the coordinates of only one electron at a time from each group. Their contribution to Eq. (64) can be written in terms of the first-order density matrices of $A$ and $B$. On the other hand, two electrons $i$ and $r$ from $A$ enter along with one or two electrons ($j$ and $s$) from $B$ into the $(r_{ij}^{AB}r_{rs}^{AB})$ terms and their contribution is in terms of the second-order density matrices of $A$ and $B$.[29,41] Now $\psi_A^0$ and $\psi_B^0$ were exact many-electron eigenfunctions and included the *internal* correlations of each group, respectively. Thus the preceding examination of Eq. (65) along with Eqs. (60)

[40] See A. Temkin, Phys. Rev. **107**, 1004 (1957).
[41] J. Bardeen, Phys. Rev. **58**, 727 (1940).

and (64) brings out the desired result, i.e., the effect of the internal correlations of $A$ and $B$ on the "*inter*-group" correlation energy has been taken into account in $E_{\mathrm{disp}}^{(2)}(AB)$.

If, instead of the special case considered, we now have any two groups that cannot be assumed to be separately localized, the treatment of higher-order correlations by the use of many-electron group functions is no longer straight-forward. The difficulty is again mainly due to the exclusion principle (see Introduction). To circumvent this difficulty, previous treatments have been restricted to the use of very special many electron group functions, [17,18] i.e., those satisfying "generalized orthogonality conditions." However, the subdivision of a one-electron basis set into mutually exclusive subsets, apparently implied by these conditions, is too restrictive. The degree of restriction becomes particularly apparent if we consider the Be atom as a two-shell system ($1s^2$ and $2s^2$) with correlated group functions for each shell, instead of the "sigma-pi" problem. In the former case where there is no nominal symmetry difference between the two groups, it is more evident that both groups would have to use the same spin-orbital set. On the other hand, it seems that a treatment based on a single complete basis set can be given not only for two, but also for many-body correlation effects, by going to higher orders of perturbation and classifying all possible virtual excitations as in Sec. II. Some of these excitations will now involve more than two electrons at a time.

## ACKNOWLEDGMENTS

The author wishes to thank Professor W. T. Simpson and Professor K. S. Pitzer for various helpful discussions. This research was carried out under the auspices of the U.S. Atomic Energy Commission.

V-3 Reprinted from *Proceedings of the Royal Society* (London) **A260**, 379 (1961)

# Theory of electron correlation in atoms and molecules

By Oktay Sinanoğlu†

*Department of Chemistry and Lawrence Radiation Laboratory, University of California, Berkeley, California*

(*Communicated by J. W. Linnett, F.R.S.—Received* 18 *August* 1960)

Corrections to the Hartree–Fock (H.F.) wave function, $\Phi_0$, and energy of a many-electron system are given. By the use of operator techniques in perturbation theory, the first-order w.f., $X_1'$, is obtained in terms of pair functions. These satisfy equations just like those of an actual two-electron system, except that now each electron moves in the H.F. field of the entire $N$-electron 'medium' added to the field of nuclei. Every pair function must be orthogonalized to each of the two H.F. orbitals associated with it to get the complete $X_1'$. This $X_1'$ determines both $E_2$ and $E_3$. The second-order energy, $E_2$, comes out as the sum of pair interactions and three- and four-body correlations due to the exclusion principle. The latter may be incorporated into the pair energies if each pair function is orthogonalized also to the remaining H.F. orbitals of $\Phi_0$. The approach allows the valence and inner shells, etc., to be discussed separately and extends some of the concepts of quantum chemistry based on orbital approximations so as to include correlation.

## Introduction

The best solution of an $N$-electron system in the form of a determinant of $N$ spin-orbitals is given by the Hartree–Fock method. This is a very good starting point, but not good enough to yield the small energy differences of chemistry and spectroscopy. The energy difference between the Hartree–Fock (H.F.) and the exact non-relativistic solution is called the 'correlation energy' and amounts to about 23 kcal/mole per doubly filled orbital.

The next best thing to the orbital approximation is to try to build up a many electron system from groups of electrons each containing some correlation. A natural building block to choose for this would be an electron pair, since the only satisfactory energy calculations so far have been on He and $H_2$.

As shown recently (Sinanoğlu 1960), perturbation theory allows one to talk of groups of electrons in a non-arbitrary way, i.e. by taking the indistinguishability of the electrons fully into account. We started from the occupied H.F. orbitals and classified the complete set of Slater determinants as to the virtual transitions they represent. In this way, the second-order energy, $E_2$, was separated into the energies of antisymmetrized pairs and three- and four-body Fermi correlations ('exclusion effects'). The use of ordinary Schrödinger-type perturbation theory leads to infinite sums, e.g. in $E_2$, which are very difficult to evaluate. The largest contributions to such sums come from the continuum part of the orbital basis set used (Löwdin 1959). Hence we suggested semi-empirical means of getting the pair energies. The exclusion effects were separated and obtained in closed form but approximately by replacing the energy denominators with 'mean excitation energies'.

† Present address: Sterling Chemistry Laboratory, Yale University, New Haven, Connecticut.

In this article we develop a rigorous theory to obtain the corrections to the H.F. wave function (w.f.) and energy of a many-electron system in terms of those of electron pairs and added exclusion effects. Formal perturbation theory with operator techniques is used and the infinite sums of the usual form of the Schrödinger method are avoided. We take the single determinant H.F. solution of an atom or molecule as its non-degenerate zero-order w.f., $\Phi_0$, and then give its first-order w.f., $X_1'$, rigorously in terms of pair functions. Each pair function satisfies an independent equation which can be solved separately (e.g. by variational principles) just like the ground or an excited state of any two-electron system. The only difference is that the electrons of the pair in the $N$-electron 'medium', move in a Hartree–Fock potential rather than in the field of bare nuclei. The techniques used for He and $H_2$ (e.g. w.f. with $r_{12}$ in it) can now be applied to such 'immersed' pairs. The 'immersed pair' functions as such do not determine $X_1'$ completely. It will be shown below that the exact $X_1'$ is obtained after orthogonalizing each 'immersed pair' function to its pair of the H.F. orbitals. $X_1'$ yields $E_2$ as the sum of pair energies and exclusion effects. The latter can be formally incorporated into the pair energies by orthogonalizing the new pair functions to the remaining H.F. orbitals of $\Phi_0$.

The same $X_1'$ yields the third-order energy, $E_3$, as well. Then having started from an H.F. solution, this should give the correlation energy to sufficient accuracy. The approach presented here also provides a basis for many 'chemical' concepts (Sinanoğlu 1960). It allows the core and valence electrons, sigma and pi electrons, etc., to be treated in a separate, yet non-arbitrary fashion.

We have applied the same formalism before to an $N$-electron system starting with the bare nuclei Hamiltonian instead of the Hartree–Fock solution and taking the entire interelectronic repulsion $\left(H_1 = \sum_{i>j=1}^{N} r_{ij}^{-1}\right)$ as the perturbation (Sinanoğlu 1961*a*). The pair functions in this case are totally independent of the 'medium' and correspond to the first order w.f.'s of the *actual* states of a real two-electron system. For example, Li atom is built up from the $(1s)^2\,{}^1S$, $(1s2s)\,{}^3S$ and ${}^1S$ states of $Li^+$. With $H_1 = \sum_{i>j} r_{ij}^{-1}$ the entire $X_1$ is given by the independent pair functions. The effects of the 'medium' show up in $E_2$ only as 'cross-polarization' terms (Sinanoğlu 1961*a*) which are simply added on. This advantage is lost upon starting with an H.F. $\Phi_0$ where the effects of medium become implicit in each pair. On the other hand, the H.F. $\Phi_0$ treatment has the advantage of greater accuracy.

There have been other attempts at obtaining solutions for many-electron systems in terms of pairs (Hurley, Lennard-Jones & Pople 1953) or larger groups (Lykos & Parr 1956; McWeeny 1959). These use arbitrary orthogonalization conditions which divide a complete set of one-electron orbitals into mutually exclusive subsets. Thus they violate the exclusion principle and are valid only for widely separated groups (Sinanoğlu 1960*a*) (e.g. the far ends of a long molecule).

Pair correlations have been used also in the theory of nuclear matter (Brueckner 1959) and other infinite Fermi gases. But the present form of these theories for finite systems (Eden 1958) require formidable self-consistency procedures many times more involved than those in the Hartree–Fock method.

### The perturbation method starting from a Hartree–Fock solution

Perturbation method will be used in the following form. Let $H = H_0 + H_1$ with $H_0$ the unperturbed Hamiltonian, $H_1$ some perturbation and

$$\left.\begin{aligned}(H_0+H_1)\Psi &= E\Psi,\\ H_0\Phi_0 = E_0\Phi_0;\ \langle\Phi_0,\Phi_0\rangle &= 1.\end{aligned}\right\} \tag{1}$$

$\Phi_0$ is the zero-order w.f., $E$, $\Psi$ the exact energy and eigenfunction. If we write

$$\Psi = \Phi_0 + \mathrm{X} \tag{2}$$

such that $\langle\Phi_0,\Psi\rangle = 1;\quad \langle\Phi_0,\mathrm{X}\rangle = 0,$

we get exactly

$$E = \langle\Phi_0, H\Psi\rangle = E_0 + \langle\Phi_0, H_1\Psi\rangle. \tag{3}$$

Upon expanding X and $E$ in terms of a parameter $\lambda$ and then setting $\lambda = 1$, the perturbation series

$$\left.\begin{aligned}\mathrm{X} &= \mathrm{X}_1 + \mathrm{X}_2 + \dots,\\ E &= E_0 + E_1 + E_2 + E_3 + \dots\end{aligned}\right\} \tag{4}$$

is obtained. The first order w.f., $\mathrm{X}_1$ is the solution of

$$(H_0 - E_0)\mathrm{X}_1 = (E_1 - H_1)\Phi_0, \tag{5}$$

where $E_1 = \langle\Phi_0, H_1\Phi_0\rangle,$

and it determines the energy to third order (see, for example, Bethe & Salpeter 1957):

$$E_2 = \langle\Phi_0,\quad (H_1 - E_1)\mathrm{X}_1\rangle, \tag{6}$$

$$E_3 = \langle\mathrm{X}_1,\quad (H_1 - E_1)\mathrm{X}_1\rangle. \tag{7}$$

This is general; each order of w.f., $\mathrm{X}_n$ (except $\Phi_0$), determines the energy to *two new* orders, i.e. to $E_{2n+1}$. Moreover, *if calculations are carried out to odd orders of energy, e.g. both $E_2$ and $E_3$ are obtained from* $\mathrm{X}_1$, *the resulting energy will be an upper limit to the exact E* (Morse & Feshbach 1953). The same is not true to even orders only.

In cases of degeneracy, the degenerate zero-order w.f.'s, $\Phi_0^k$, corresponding to $E_0$, must be so chosen that *each* $(E_1^k - H_1)\Phi_0^k$, where $E_1^k = \langle\Phi_0^k, H_1\Phi_0^k\rangle$, will be orthogonal to *all* $\Phi_0^l$

$$\langle\Phi_0^l,\quad (E_1^k - H_1)\Phi_0^k\rangle = 0 \tag{8}$$

(Sinanoğlu 1961*a*). Then there will be a different equation, equation (5), for each $\Phi_0^k$, and equation (8) will ensure that solutions to these equations exist.

To solve equation (5) for $\mathrm{X}_1$, one is not restricted to the Schrödinger method of expanding $\mathrm{X}_1$ in terms of the eigenfunctions of $H_0$. Other direct means of solving non-homogeneous differential equations may be used. Approximations to $\mathrm{X}_1$ can also be found by minimizing the expression (see, for example, Bethe & Salpeter 1957)

$$\begin{aligned}E_2 < E_2(\mathrm{X}_1^{\mathrm{trial}}) = 2\langle\Phi_0,\quad & (H_1 - E_1)\mathrm{X}_1^{\mathrm{trial}}\rangle\\ &+ \langle\mathrm{X}_1^{\mathrm{trial}},\quad (H_0 - E_0)\mathrm{X}_1^{\mathrm{trial}}\rangle\end{aligned} \tag{9}$$

with respect to trial functions, $X_1^{\text{trial}}$. Equation (9) applies directly if $\Phi_0$ is a ground state or the lowest state of a given symmetry. For a general excited state a modified minimization procedure has been developed (Sinanoğlu 1961 *b*).

The approach to be followed here is to analyze $X_1$ into components exactly using the inverse operator $L_0^{-1} = (H_0 - E_0)^{-1}$. The components turn out to be pair functions which satisfy two-electron equations similar to equation (5) and to which the variational principle, equation (9), or its modification (Sinanoğlu 1961 *b*) can be applied individually. We start with the zero-order w.f., the Hartree–Fock solution.

Consider an $N$-electron atom or molecule with a Hamiltonian

$$\left.\begin{aligned} H &= \sum_{i=1}^{N} h_i^0 + \sum_{i>j=1}^{N} g_{ij}, \\ h_i^0 &\equiv -\tfrac{1}{2}\nabla_i^2 - \sum_{\alpha} \frac{Z_\alpha}{R_{\alpha i}}; \quad g_{ij} \equiv \frac{1}{r_{ij}}. \end{aligned}\right\} \tag{10}$$

$Z_\alpha/R_{\alpha i}$ is the potential of electron $i$ in the field of nucleus $\alpha$, and atomic units (1 a.u. = 27·2 eV) are used. $H$ may be split into: $H = H_0 + H_1$ such that

$$H_0 = \sum_{i=1}^{N} (h_i^0 + V_i),$$

and

$$H_1 = \sum_{i>j=1}^{N} g_{ij} - \sum_{i=1}^{N} V_i. \tag{11}$$

$V_i$ is an average potential that acts on $i$ and may be non-local. In the non-degenerate case the best choice for $V_i$, i.e. one that leads to the smallest $H_1$, is the Hartree–Fock potential

$$V_i(\mathbf{x}_i) = \int \frac{\rho(\mathbf{x}_j, \mathbf{x}_j)}{r_{ij}} \mathrm{d}\tau_j - \int \frac{\rho(\mathbf{x}_j, \mathbf{x}_i)}{r_{ij}} P_{ij} \mathrm{d}\tau_j. \tag{12}$$

$P_{ij}$ exchanges $\mathbf{x}_i$ and $\mathbf{x}_j$ and $\rho(\mathbf{x}_j, \mathbf{x}_j')$ is the Fock–Dirac density (see, for example, Löwdin 1959)

$$\rho(\mathbf{x}_i, \mathbf{x}_j) = \sum_{n'=1}^{N'} n'^*(\mathbf{x}_i)\, n'(\mathbf{x}_j). \tag{13}$$

$\mathbf{x}_i$ denotes both the spatial ($\mathbf{r}_i$) and the spin co-ordinates ($\xi_i$); $n'(\mathbf{x}_i)$ are the Hartree–Fock spin-orbitals† (in molecules m.o. s.c.f.) with the orbital energies, $\epsilon_1'$, satisfying

$$h_i^{\text{eff.}}\, n'(\mathbf{x}_i) = \epsilon_i' n'(\mathbf{x}_i), \tag{14}$$

where $h_i^{\text{eff.}} \equiv h_i^0 + V_i$. Formally the same $V_i$ acts on all the $n'$ and $i$.

The zero-order w.f. is taken here as the normalized determinant

$$\Phi_0 = (1/\sqrt{N!}) \det\{1'(\mathbf{x}_1)\, 2'(\mathbf{x}_2) \ldots N'(\mathbf{x}_N)\} \equiv \mathscr{A}\{1'2'3' \ldots N'\}; \tag{15}$$

$\mathscr{A}$ is the anti-symmetrizer

$$\mathscr{A} = \frac{1}{\sqrt{N!}} \sum_P (-1)^P P. \tag{16}$$

Among the spin-orbitals $1'$ to $N'$, *all odd integers refer to H.F. orbitals multiplied by spin* $\alpha$ and *all even ones to those with spin* $\beta$; for example, in atoms $1' \equiv 1s_\alpha$, $2' \equiv 1s_\beta$, .... In equation (15), before $\mathscr{A}$ is applied, electron $\mathbf{x}_i$ occupies the spin-orbital $i'$.

† Primes will be used to refer to the properties of H.F. orbitals as distinct from the solutions of $h_1^0$.

From equation (11) we have

$$(H_0 - E_0) = \sum_{i=1}^{N} (h_i^{\text{eff.}} - \epsilon_i'), \tag{17}$$

$$E_1 = \left\langle \Phi_0, \left( \sum_{i>j} g_{ij} - \sum_{i=1}^{N} V_i \right) \Phi_0 \right\rangle = - \sum_{i>j=1}^{N} (J_{ij} - K_{ij}''), \tag{18}$$

where $\quad J_{ij} = \langle i'(\mathbf{x}_i) j'(\mathbf{x}_j), \quad g_{ij} i'(\mathbf{x}_i) j'(\mathbf{x}_j) \rangle \equiv \langle i'j', \quad g_{ij} i'j' \rangle$

and $\quad K_{ij}'' = \langle i'(\mathbf{x}_i) j'(\mathbf{x}_j), \quad g_{ij} j'(\mathbf{x}_i) i'(\mathbf{x}_j) \rangle \equiv \langle i'j', \quad g_{ij} j'i' \rangle.$

$K_{ij}'' \neq 0$ only for $i'$ and $j'$ of like spin.

The correction, $X_1'$, to the H.F. $\Phi_0$ of equation (15) must be solved from equation (5)

$$L_0 X_1' = (E_1 - H_1) \Phi_0 = \left\{ \sum_{i>j=1}^{N} (-J_{ij} + K_{ij}'' - g_{ij}) + \sum_{i=1}^{N} V_i \right\} \mathscr{A}(1'2'\ldots N') \tag{19}$$

with $L_0 = H_0 - E_0$ of equation (17). Since $L_0$ is linear, $X_1'$ and equation (19) can be split into two parts

$$X_1' = U' - W, \tag{20}$$

such that

$$L_0 U' = \left\{ \sum_{i>j=1}^{N} (J_{ij} - K_{ij}'' - g_{ij}) \right\} \Phi_0 \tag{21}$$

and

$$L_0 W = \left\{ \sum_{i>j=1}^{N} 2(J_{ij} - K_{ij}'') - \sum_{i=1}^{N} V_i \right\} \Phi_0. \tag{22}$$

Subtracting equation (22) from (21), equation (19) is obtained. Each of these equations have solutions for non-degenerate $\Phi_0$, since they separately satisfy the existence condition, equation (8).

Equation (21) involves only the two-electron repulsions, $g_{ij}$; equation (22) only the one-electron potentials, $V_i$. This is a very convenient result. Equation (21) has exactly the same form as the equation that gives the first order w.f., $X_1$, of an $N$-electron system when one takes the entire interelectronic repulsions, $\sum_{i>j=1}^{N} g_{ij}$ as the perturbation, $H_1$, and uses bare nuclei orbitals (hydrogen-like orbitals in atoms) (Sinanoğlu 1961*a*) instead of the H.F. $\Phi_0$. The main difference between equation (21) and an equation giving $X_1$, with $H_1 = \sum_{i>j} g_{ij}$ lies in $L_0$ which in the H.F. case contains the $h_i^{\text{eff.}} = h_i^0 + V_i$ instead of just the $h_i^0$ (equation (14)). Thus the methods developed previously (Sinanoğlu 1961*a*) for obtaining $X_1$ in terms of pairs apply in exactly the same way to $U'$. Moreover, as we shall show first for a two-electron system and then generally, once $U'$ is determined $W$ is obtained from it by simple orthogonalization procedures.

We give first the method of separating $U'$ into pairs. This is based on the method that was used in the separation of $X_1$ with $H_1 = \sum_{i>j} g_{ij}$ (Sinanoğlu 1961*a*).

The particular solution of equation (21) is obtained by inverting $L_0$

$$U' = L_0^{-1} \left\{ \sum_{i>j=1}^{N} (J_{ij} - K_{ij}'' - g_{ij}) \right\} \mathscr{A}(1'2'\ldots N'); \tag{23}$$

$L_0^{-1}$ is the Green function operator and by equation (17)

$$L_0^{-1} = (H_0 - E_0)^{-1} = \frac{1}{e_1' + e_2' + \ldots e_N'} \tag{24}$$

where we have defined $\quad e_i' \equiv h_i^{\text{eff.}} - \epsilon_i' + (\text{spin part})_i \quad$ (25)

and included the corresponding one-electron spin part. In terms of the complete set of the anti-symmetrized eigenfunctions of $H_0$, $L_0^{-1}$ is given by

$$L_0^{-1} = \sum_{n > \ldots l > k \geq 1}^{\infty} \frac{\mathscr{A}(kl\ldots n)\rangle\langle\mathscr{A}(kl\ldots n)}{(\epsilon_k' + \epsilon_l' + \ldots \epsilon_n') - (\epsilon_1' + \epsilon_2' + \ldots \epsilon_N')}. \tag{26}$$

$L_0^{-1}$ is a self-adjoint, linear operator and commutes with $\mathscr{A}$.

From $e_i' i(\mathbf{x}_i) = 0$ and equation (25) it follows that if, for example, $f_{12}(\mathbf{x}_1, \mathbf{x}_2)$ is any function of $\mathbf{x}_1$ and $\mathbf{x}_2$ only, then

$$(e_1' + e_2' + \ldots e_N')^{-1}\{f_{12}(\mathbf{x}_1, \mathbf{x}_2)\, 3(\mathbf{x}_3)\, 4(\mathbf{x}_4) \ldots N(\mathbf{x}_N)\}$$
$$= \{3(\mathbf{x}_3)\, 4(\mathbf{x}_4) \ldots N(\mathbf{x}_N)\}\, (e_1' + e_2')^{-1} f_{12}(\mathbf{x}_1, \mathbf{x}_2). \tag{27}$$

Commuting $\mathscr{A}$ with $\sum_{i>j} m_{ij}\, (m_{ij} \equiv J_{ij} - K_{ij}'' - g_{ij})$ and $L_0^{-1}$, this equation would allow $U'$ in equation (23) to be written as a sum of terms each involving a two-electron $q_{ij} \equiv (e_i' + e_j')^{-1} m_{ij}\{i'(\mathbf{x}_i) j'(\mathbf{x}_j)\}$ and multiplied by the remaining spin-orbitals $n'(\mathbf{x}_n) \neq i, j$. However, when each of the two-electron inverse operators, $(e_i' + e_j')^{-1}$, is expanded as in equation (26), the expressions for $q_{ij}$ become infinite. This is because $q_{ij}$ does not correspond to a first-order two-electron w.f. obtainable as a solution of $(e_i' + e_j')\, q_{ij} = m_{ij}\{i'(\mathbf{x}_i) j'(\mathbf{x}_j)\}$ similar to equation (5). The $\{i'(\mathbf{x}_i) j'(\mathbf{x}_j)\}$ will be degenerate with other two-electron states, e.g. with $\{i'(\mathbf{x}_j) j'(\mathbf{x}_i)\}$, so that the $q_{ij}$ equations do not satisfy the existence conditions, equation (8). In the overall $U'$, however, such pair degeneracies leading to infinities are removed when $\mathscr{A}$ is finally applied.

*To obtain $U'$ in terms of finite pair functions each one satisfying a two-electron non-homogeneous equation of the type, equation* (5), *the pair degeneracies must be removed before applying $L_0^{-1}$.* This is accomplished by taking those linear combinations of $(i' j')$ pair products that belong to the irreducible representations of the symmetry group of a two-electron system. It is possible to find these combinations in a way leaving the overall $\mathscr{A}\{\sum_{i>j} m_{ij}\, (1'2' \ldots N')\}$ invariant (Sinanoğlu 1961*a*) in the equation

$$U' = L_0^{-1} \mathscr{A} \left\{ \sum_{i>j=1}^{N} m_{ij}(1'2' \ldots N') \right\}. \tag{28}$$

For instance, if $i'$ and $j'$ correspond to the same spatial orbital and differ only in spin, it is sufficient to make use of

$$(1/\sqrt{2})\, \mathscr{A}\mathscr{B}_{ij} = \mathscr{A}, \tag{29}$$

where $\mathscr{B}_{ij} \equiv (1/\sqrt{2!})\,(1 - P_{ij}) = (1/\sqrt{2!})\,\det$, antisymmetrizes only $i'$ and $j'$ by interchanging $\mathbf{x}_i$ and $\mathbf{x}_j$. Then from equations (27), (28) and (29) we get, for example, for the $(1'2')$ pair

$$L_0^{-1} \mathscr{A}\{(J_{12} - g_{12})\,(1'2' \ldots N')\} = L_0^{-1}(\mathscr{A}/\sqrt{2})\,\{(J_{12} - g_{12})\, \mathscr{B}_{12}(1'2')\, 3'(\mathbf{x}_3) \ldots N'(\mathbf{x}_N)\}$$
$$= (\mathscr{A}\sqrt{2})\,\{3'(\mathbf{x}_3)\, 4'(\mathbf{x}_4) \ldots N'(\mathbf{x}_N)\, u_{12}'(\mathbf{x}_1, \mathbf{x}_2)\}, \tag{30}$$

where now $u'_{12}(\mathbf{x}_1, \mathbf{x}_2)$ is the particular solution† of

$$(e'_1 + e'_2)\, u'_{12}(\mathbf{x}_1, \mathbf{x}_2) = (J_{12} - g_{12})\, \mathscr{B}_{12}\{1'(\mathbf{x}_1)\, 2'(\mathbf{x}_2)\}, \tag{31}$$

such that $\langle u'_{12}, \mathscr{B}_{12}(1'2')\rangle = 0$. Before the last step of equation (29), $L_0^{-1}$ and $\mathscr{A}$ have been commuted.

The total $U'$ is given by

$$U' = \frac{1}{\sqrt{2}} \mathscr{A} \left\{ \sum_{i>j=1}^{N} (1'2' \ldots [i-1]' [j+1]' \ldots N')\, u'_{ij} \right\}, \tag{32}$$

where $1' = 1'(\mathbf{x}_1)$, $n' = n'(\mathbf{x}_n)$ and $u'_{ij} = u'_{ij}(\mathbf{x}_i, \mathbf{x}_j)$. If for an $(i'j')$ pair, $\mathscr{B}_{ij}(i'j')$ already belongs to an irreducible representation (or to one of its rows) of the two-electron symmetry group (a 'closed shell' pair) the corresponding $u'_{ij}$ will be the particular solution of

$$(e'_i + e'_j)\, u'_{ij} = m_{ij} \mathscr{B}_{ij}(i'j') \tag{33}$$

such that $\langle u'_{ij}, \mathscr{B}_{ij}(i'j')\rangle = 0$; $m_{ij} = J_{ij} - K''_{ij} - g_{ij}$. If on the other hand $\mathscr{B}_{ij}$ is not sufficient to remove the degeneracy of an $(i'j')$ pair, then, in the equation

$$U' = L_0^{-1} \frac{\mathscr{A}}{\sqrt{2}} \left\{ \sum_{i>j=1}^{N} m_{ij} \mathscr{B}_{ij}(1'2' \ldots N') \right\} \tag{34}$$

obtained from equations (28) and (29), $m_{ij} \mathscr{B}_{ij}(i'j')$ must be written as a linear combination of terms that belong to the mentioned irreducible representations and hence satisfy equation (8) individually.

Suppose there are $r$ such terms, $\phi^s_{ij}$, for a given $(i'j')$,

$$\frac{\mathscr{A}}{\sqrt{2}} \{m_{ij} \mathscr{B}_{ij}(1'2' \ldots i'j' \ldots N')\} = \frac{\mathscr{A}}{r} \left\{ (1' \ldots [i-1]' [j+1]' \ldots N') \sum_{s=1}^{r} m^s_{ij} \phi^s_{ij}(\mathbf{x}_i, \mathbf{x}_j) \right\}, \tag{35}$$

then

$$u'_{ij} = \frac{\sqrt{2}}{r} \sum_{s=1}^{r} u^s_{ij}, \tag{36a}$$

where $u^s_{ij}$ are the solutions of

$$(e'_i + e'_j)\, u^s_{ij} = m^s_{ij} \phi^s_{ij} \tag{36b}$$

and $m^s_{ij} = \langle \phi^s_{ij}, g_{ij} \phi^s_{ij} \rangle - g_{ij}$. The $u^s_{ij}$ must be made orthogonal to $\phi^s_{ij}$ before inserting in equations (36*a*) and (32). This procedure for obtaining the proper $u_{ij}$ has been illustrated in detail before for the case of Li atom with $H_1 = \sum_{i>j} g_{ij}$ (Sinanoğlu 1961*a*) where the $(1/\sqrt{2!}) \det (1s_\beta 2s_\alpha)$ pair leads to the $u_{ij}$ of the $(1s2s)^1S$ and $^3S$ states of $Li^+$.

Now each $u'_{ij}$ occurring in equation (32) can be obtained by solving several two-electron equations, equations (33) or (36*b*), e.g. by variational principles (Sinanoğlu 1961*b*) such as equation (9). The number of independent pair functions in $U'$ in general will be less than the total number of pairs. In the next sections, the complete $X'_i$ will be obtained by relating $W$ (equation (20)) to $U'$.

† The particular solution of equation (31) with the proper behaviour at $\mathbf{x}_1$, $\mathbf{x}_2 = 0$ and $\infty$, must be made orthogonal to $\mathscr{B}(1'2')$ before inserting it in equation (30).

*(a) Two-electron system*

Consider a non-degenerate two-electron system such as the ground state of He or $H_2$. Then $1'$ and $2'$ are the H.F. s.c.f. spin-orbitals, e.g. $1s_\alpha$, $1s_\beta$ or $(\sigma_g 1s)_\alpha$, $(\sigma_g 1s)_\beta$. Now $U' = u'_{12}$ satisfies equation (31) and $X'_1 = u'_{12} - w_{12}$. Equation (22) becomes

$$(e'_1 + e'_2)\, w_{12}(\mathbf{x}_1, \mathbf{x}_2) = [(J_{12} - V_1) + (J_{12} - V_2)]\, \mathscr{B}_{12}(1'2'). \tag{37}$$

We have written the perturbation appearing on the right-hand side of this equation as the sum of two terms: $(J_{12} - V_1)$ and $(J_{12} - V_2)$. Each one of these is a one-electron function. Changing the notation of equation (22) slightly and letting the subscripts denote the orbitals producing the potential, $V_1 \equiv V_1(\mathbf{x}_2)$ and from equation (12)

$$V_1(\mathbf{x}_2) = \langle 1'(\mathbf{x}_1), \quad g_{12}\, 1'(\mathbf{x}_1)\rangle_{\mathbf{x}_1} \equiv \int \mathrm{d}\mathbf{x}_1\, g_{12}(\mathbf{x}_1, \mathbf{x}_2)\, |1'(\mathbf{x}_1)|^2. \tag{38}$$

Inverting $(e'_1 + e'_2)$, we have

$$w_{12} = (e'_1 + e'_2)^{-1}\, \mathscr{B}_{12}[(J_{12} - V_1(\mathbf{x}_2)) + (J_{12} - V_2(\mathbf{x}_1))]\, (1'(\mathbf{x}_1)\, 2'(\mathbf{x}_2))$$

or using equation (27)

$$w_{12} = \mathscr{B}_{12}\{1'(\mathbf{x}_1)\, t_2(\mathbf{x}_2)\} + \mathscr{B}_{12}\{t_1(\mathbf{x}_1)\, 2'(\mathbf{x}_2)\} \tag{39a}$$

where, for example,

$$t_2(\mathbf{x}_2) = \frac{1}{e'_2}\, (J_{12} - V_1(\mathbf{x}_2))\, 2'(\mathbf{x}_2). \tag{39b}$$

In equation (39) $t_1$ and $t_2$ are related to $u'_{12}(\mathbf{x}_1, \mathbf{x}_2)$ simply by:

$$\left.\begin{aligned} t_1(\mathbf{x}_1) &= \sqrt{2!}\, \langle 2'(\mathbf{x}_2), \quad u'_{12}(\mathbf{x}_1, \mathbf{x}_2)\rangle_{\mathbf{x}_2}, \\ t_2(\mathbf{x}_2) &= \sqrt{2!}\, \langle 1'(\mathbf{x}_1), \quad u'_{12}(\mathbf{x}_1, \mathbf{x}_2)\rangle_{\mathbf{x}_1}, \end{aligned}\right\} \tag{40}$$

where, as in equation (38), $\langle\,\rangle_{\mathbf{x}_i}$ denotes an integration over only $\mathbf{x}_i$. To prove this we substitute the particular solution of equation (31) into equation (40)

$$\langle 2', u'_{12}\rangle_{\mathbf{x}_2} = \langle 2'(\mathbf{x}_2), \quad (e'_1 + e'_2)^{-1}\, (J_{12} - g_{12})\, \mathscr{B}_{12}(1'2')\rangle_{\mathbf{x}_2}.$$

The operations $(e'_1)^{-1}$ and $\langle\,\rangle_{\mathbf{x}_2}$ commute and $(e'_1 + e'_2)^{-1}$ is self-adjoint with respect to $\langle\,\rangle$ as well as $\langle\,\rangle_{\mathbf{x}_2}$. Thus

$$\begin{aligned} \langle 2', u'_{12}\rangle_{\mathbf{x}_2} &= \langle (e'_1 + e'_2)^{-1}\, 2'(\mathbf{x}_2), \quad (J_{12} - g_{12})\, \mathscr{B}_{12}(1'2')\rangle_{\mathbf{x}_2} \\ &= (e'_1)^{-1} \langle 2'(\mathbf{x}_2), \quad (J_{12} - g_{12})\, \mathscr{B}_{12}(1'2')\rangle_{\mathbf{x}_2}. \end{aligned} \tag{41}$$

Equation (27) has been used in the last step. Since $1'$ and $2'$ are orthonormal and $\langle 2'(\mathbf{x}_2), g_{12}\, 2'(\mathbf{x}_2)\rangle = V_2(\mathbf{x}_1)$, equation (41) leads to the desired result, equation (40)

$$\langle 2', u'_{12}\rangle_{\mathbf{x}_2} = \frac{1}{\sqrt{2!}\, e'_1}\, (J_{12} - V_2(\mathbf{x}_1))\, 1'(\mathbf{x}_1) = \frac{1}{\sqrt{2!}}\, t_1(\mathbf{x}_1).$$

Note also that

$$\langle t_1(\mathbf{x}_1), \quad 1'(\mathbf{x}_1)\rangle = \langle t_2(\mathbf{x}_2), \quad 2'(\mathbf{x}_2)\rangle = 0. \tag{42}$$

Equations (19), (20), (39a) and (40) show that the first-order correction, $X'_1$, to the H.F. solution is given by $u'_{12}$ orthogonalized (by the Schmidt process) to the H.F. spin-orbitals $1'$ and $2'$; i.e.

$$\begin{aligned} X'_1(\mathbf{x}_1, \mathbf{x}_2) = u'_{12} &- \det\{1'(\mathbf{x}_1)\, \langle 1'(\mathbf{x}_1), \quad u'_{12}(\mathbf{x}_1, \mathbf{x}_2)\rangle_{\mathbf{x}_1}\} \\ &- \det\{2'(\mathbf{x}_2)\, \langle 2'(\mathbf{x}_2), \quad u'_{12}(\mathbf{x}_1, \mathbf{x}_2)\rangle_{\mathbf{x}_2}\}, \end{aligned} \tag{43}$$

so that $$\langle X_1', 1'\rangle_{x_1} = \langle X_1', 2'\rangle_{x_2} = 0. \tag{44}$$

Substituting this result in equation (6) with $\langle X_1', \mathscr{B}_{12}(1'2')\rangle = 0$ and using equation (11) we get
$$E_2 = \langle (1\sqrt{2!})\det(1'2'),\quad g_{12}X_1'\rangle. \tag{45}$$

The $\sum_i V_i$ part of $H_1$ has vanished because of the orthogonalization conditions, equation (44). If $G_{12}\mathscr{B}_{12}(1'2') = g_{12}X_1'(\mathbf{x}_1,\mathbf{x}_2)$ is written in equation (45) and the $\mathbf{x}_1$ integration carried out first, $E_2$ becomes the expectation value, $\langle 2'(\mathbf{x}_2), V_p(\mathbf{x}_2)\,2'(\mathbf{x}_2)\rangle$, of the correlation or fluctuation (Sinanoğlu 1960) potential, $V_p(\mathbf{x}_2) = \langle 1', G_{12}\,1'\rangle_{x_1}$ acting on the electron $\mathbf{x}_2$.

In the language of the usual Schrödinger perturbation theory, i.e. when equation (32) is solved using equation (26), $u_{12}'$ contains all the double and the single virtual excitations from $\Phi_0 = \mathscr{B}_{12}(1'2')$. To get the complete solution $X_1'$ to equation (5), the contribution of the single excitations are subtracted out from $u_{12}'$ by orthogonalization to $1'$ and $2'$ (equation (43)). Thus when $\Phi_0$ is a Hartree–Fock solution, $X_1'$ consists only of the 'dispersion' (double excitations) part, the 'orbital average polarization' (single excitations) part does not appear because it is already implicit in the s.c.f. nature of $\Phi_0$ (see, for example, Sinanoğlu 1960).

### (*b*) *N-electron system*

In the case of an $N$-electron system, as in the previous case, the complete first-order correction, $X_1'$ to the H.F. $\Phi_0$ is obtained by orthogonalizing each $u_{ij}'$ in equation (32) to its H.F. orbitals $i'$ and $j'$.

To show this, we first split the sum of the H.F. potentials, $\sum_{i=1}^{N} V_i(\mathbf{x}_i)$ appearing as the perturbation in equation (22) into the Coulomb and exchange potentials of individual orbitals $i'$ acting on electrons $\mathbf{x}_j$. Substituting equation (13) in (12) gives
$$\sum_{k=1}^{N} V_k(\mathbf{x}_k) = \sum_{i>j=1}^{N} [(S_i(\mathbf{x}_j)+S_j(\mathbf{x}_i)) - (R_i''(\mathbf{x}_j)+R_j''(\mathbf{x}_i))]. \tag{46}$$

$S_i(\mathbf{x}_j)$ is the Coulomb and $R_i(\mathbf{x}_j)$ the exchange potential of the orbital $i'$. In $\Phi_0 = \mathscr{A}(1'2'\ldots N')$, $i'$ is occupied by the electron $\mathbf{x}_i$ before $\mathscr{A}$ is applied.
$$S_i(\mathbf{x}_j) = \langle i'(\mathbf{x}_i),\quad g_{ij}\,i'(\mathbf{x}_i)\rangle_{x_i}, \tag{47a}$$
$$\begin{aligned} R_i''(\mathbf{x}_j)j'(\mathbf{x}_j) &\equiv \langle i'(\mathbf{x}_i),\quad g_{ij}\,i'(\mathbf{x}_i)\,P_{ij}\rangle_{x_i} j'(\mathbf{x}_j) \\ &= \langle i'(\mathbf{x}_i),\quad g_{ij}\,j'(\mathbf{x}_i)\rangle_{x_i} i'(\mathbf{x}_j). \end{aligned} \tag{47b}$$

We write the perturbation in equation (22) as the sum of pair terms using equation (46), after commuting it with $\mathscr{A}$
$$L_0 W = \mathscr{A}\left\{\sum_{i>j=1}^{N} k_{ij}\right\}(1'2'3'\ldots N'), \tag{48a}$$
where $$k_{ij} = [J_{ij} - K_{ij}'' - S_i(\mathbf{x}_j) + R_i''(\mathbf{x}_j)] + [J_{ij} - K_{ij}'' - S_j(\mathbf{x}_i) + R_j''(\mathbf{x}_i)]. \tag{48b}$$

The form of this equation is exactly the same as equation (28) which gave $U'$; only $k_{ij}$ has replaced the $m_{ij}$. Thus $W$ is obtained in terms of the corresponding pair functions, $w_{ij}$, in the same way that $U'$ was related to $u_{ij}'$, except now the use of

$\mathscr{A} = (1/\sqrt{2})\mathscr{A}\mathscr{B}_{ij}$ suffices to do this. For reasons that will be apparent shortly, it is not necessary to resolve some of the $\mathscr{B}_{ij}(i'j')$ further in terms of $\phi^s_{ij}$ (cf. equation (35)).

We get
$$W = L_0^{-1}\frac{\mathscr{A}}{\sqrt{2}}\left\{\sum_{i>j=1}^{N} k_{ij}\mathscr{B}_{ij}(1'2'\ldots i'j'\ldots N')\right\}. \tag{49}$$

Then, commuting $\mathscr{A}$ and $L_0^{-1}$ and using equation (27) for each $k_{ij}$ term,

$$W = \frac{\mathscr{A}}{\sqrt{2}}\left\{\sum_{i>j=1}^{N} (1'2'\ldots[i-1]'[j+1]'\ldots N')\, w_{ij}\right\}, \tag{50}$$

where $w_{ij} = (e_1'+e_2')^{-1} k_{ij}\mathscr{B}_{ij}(i'j')$ (cf. equations (34)).

The $w_{ij}$'s are related to the 'immersed pair' functions $u'_{ij}$ in the same way as we showed in the two-electron system. $\mathscr{B}_{ij}$ (i.e. $(1/\sqrt{2!})$ det) commutes with $k_{ij}$ and $(e_1'+e_2')^{-1}$, so that using equation (27) and the definition of $k_{ij}$, equation (48$b$),

$$w_{ij} = \mathscr{B}_{ij}\{i'(\mathbf{x}_i)\,t_j(\mathbf{x}_j)\} + \mathscr{B}_{ij}\{t_i(\mathbf{x}_i)\,j'(\mathbf{x}_j)\}, \tag{51$a$}$$

$$t_i(\mathbf{x}_i) = (e_1')^{-1}[J_{ij} - K''_{ij} - S_j(\mathbf{x}_i) + R''_j(\mathbf{x}_i)]\, i'(\mathbf{x}_i) \tag{51$b$}$$

$i$ and $j$ are interchanged to get $t_j(\mathbf{x}_j)$. These equations are the generalization of equations (39) to any pair. Again we have

$$\left.\begin{aligned} t_i(\mathbf{x}_i) &= \sqrt{2!}\,\langle j'(\mathbf{x}_j), \quad u'_{ij}(\mathbf{x}_i,\mathbf{x}_j)\rangle_{\mathbf{x}_j},\\ t_j(\mathbf{x}_j) &= \sqrt{2!}\,\langle i'(\mathbf{x}_i), \quad u'_{ij}(\mathbf{x}_i,\mathbf{x}_j)\rangle_{\mathbf{x}_i}.\end{aligned}\right\} \tag{52}$$

Equation (52) is valid for any $w_{ij}$ whether its $u'_{ij}$ is obtained just from $\mathscr{B}_{ij}(i'j')$ (equation (33)), or from pair symmetry states, $\phi^s_{ij}$ by equations (36). The proof of the first case with $\mathscr{B}_{ij}(i'j')$ follows in exactly the same way as in the two-electron system by substituting $u'_{ij} = (e_i'+e_j')^{-1} m_{ij}\mathscr{B}_{ij}(i'j')$ in equation (52) and noting that

$$\langle i'(\mathbf{x}_i), \quad (S_j(\mathbf{x}_i) - R''_j(\mathbf{x}_i))\, i'(\mathbf{x}_i)\rangle = J_{ij} - K''_{ij}.$$

The other case is similarly verified by substituting

$$u'_{ij} = \frac{\sqrt{2}}{r}\sum_{s=1}^{r}(e_i'+e_j')^{-1} m^s_{ij}\,\phi^s_{ij}$$

and writing out $\phi^s_{ij}$ in terms of the appropriate $\mathscr{B}_{ij}(i'j')$'s at the last step.

$$\begin{aligned}\langle i'(\mathbf{x}_i), \quad u'_{ij}(\mathbf{x}_i,\mathbf{x}_j)\rangle_{\mathbf{x}_i} &= \frac{\sqrt{2}}{r}\sum_{s=1}^{r}\left\langle i'(\mathbf{x}_i), \quad \frac{1}{e_i'+e_j'} m^s_{ij}\,\phi^s_{ij}\right\rangle_{\mathbf{x}_i}\\ &= \frac{\sqrt{2}}{r}\sum_{s=1}^{r}\frac{1}{e_j'}\langle i'(\mathbf{x}_i), \quad m^s_{ij}\,\phi^s_{ij}\rangle_{\mathbf{x}_i} = \frac{1}{\sqrt{2}}\frac{1}{e_j'}(J_{ij} - K''_{ij} - S_i + R''_i)\, j'(\mathbf{x}_j)\\ &= \frac{1}{\sqrt{2}} t_j(\mathbf{x}_j).\end{aligned}$$

Finally, the overall first order w.f., $\mathrm{X}_1'$ is obtained by combining the $U'$ given by equation (32) and the $W$ of equation (50). The result is

$$\mathrm{X}_1' = U' - W = \frac{1}{\sqrt{2}}\mathscr{A}\left\{\sum_{i>j=1}^{N}(1'2'\ldots[i-1]'[j+1]'\ldots N')\,\bar{u}_{ij}\right\} \tag{53}$$

expressed in terms of $\bar{u}_{ij}$, the 'immersed pair' functions orthogonalized to the H.F. orbitals $i'$ and $j'$

$$\bar{u}_{ij}(\mathbf{x}_i, \mathbf{x}_j) = u'_{ij} - \det\{i'(\mathbf{x}_i)\langle i'(\mathbf{x}_i), \quad u'_{ij}(\mathbf{x}_i, \mathbf{x}_j)\rangle_{\mathbf{x}_i}\}$$
$$- \det\{j'(\mathbf{x}_j)\langle j'(\mathbf{x}_j), \quad u'_{ij}(\mathbf{x}_i, \mathbf{x}_j)\rangle_{\mathbf{x}_j}\} \tag{54}$$

$\bar{u}_{ij}$ corresponds to only the 'dispersion' part of $u'_{ij}$. Note that since the particular solutions, $u'_{ij}$, were obtained such that

$$\langle u'_{ij}, \quad \mathscr{B}_{ij}(i'j')\rangle = 0, \tag{55}$$

we have $\langle \Phi_0, X'_i\rangle = 0$. Also

$$\langle \bar{u}_{ij}, i'\rangle_{\mathbf{x}_i} = \langle \bar{u}_{ij}, j'\rangle_{\mathbf{x}_j} = 0. \tag{56}$$

The second-order energy of the system is given from equations (6), (11), (53) and (55) by

$$E_2 = \sqrt{(N!/2)}\langle \mathscr{A}(1'2'\ldots N'), \quad (\sum_{i>j} g_{ij})\,[3'(\mathbf{x}_3)\,4'(\mathbf{x}_4)\ldots N'(\mathbf{x}_N)\,\bar{u}_{12}(\mathbf{x}_1, \mathbf{x}_2)$$
$$+ 1'(\mathbf{x}_1)\,2'(\mathbf{x}_2)\,5'(\mathbf{x}_5)\ldots N'(\mathbf{x}_N)\,\bar{u}_{34}(\mathbf{x}_3, \mathbf{x}_4) + \ldots]\rangle, \tag{57}$$

where we have used the self-adjoint property of $\mathscr{A}$ and $\mathscr{A}^2 = (\sqrt{N!})\,\mathscr{A}$. The $\sum_i V_i$ terms of $H_i$ vanished because of equation (56). The $E_2$ just given is simply the sum of 'dispersion' energies of pairs of electrons $(i'j')$ immersed in the $N$ electron medium, plus three- and four-body Fermi correlation terms ('exclusion effects'). To see this examine, for example, the first part of equation (57) with $\bar{u}_{12}$. Out of all the permutations in $\mathscr{A} = (1/\sqrt{N!})\sum_P(-1)^P P$ only those $P$ that interchange the electrons $\mathbf{x}_i, \mathbf{x}_j, \mathbf{x}_1, \mathbf{x}_2$, in $i'(\mathbf{x}_i)j'(\mathbf{x}_j)\,\bar{u}_{12}(\mathbf{x}_1, \mathbf{x}_2)$ contribute to the part of this term involving $g_{ij}$. We have

$$\frac{\sqrt{N!}}{\sqrt{2}}\left\langle \mathscr{A}(1'2'\ldots N'), \quad \left(\sum_{i>j=1}^{N} g_{ij}\right) 3'(\mathbf{x}_3)\,4'(\mathbf{x}_4)\ldots\bar{u}_{12}(\mathbf{x}_1, \mathbf{x}_2)\right\rangle$$
$$= \langle \mathscr{B}_{12}(1'2'), \quad g_{12}\bar{u}_{12}\rangle - 2\sum_{m>2}^{N}\langle m'(\mathbf{x}_2)\,\mathscr{B}_{1m}\{1'(\mathbf{x}_1)\,2'(\mathbf{x}_m)\},$$
$$\times\, g_{1m}\,m'(\mathbf{x}_m)\,\bar{u}_{12}(\mathbf{x}_1, \mathbf{x}_2)\rangle + \sum_{l>k>2}^{N}\langle \mathscr{B}_{12}(k'l'), \quad \bar{u}_{12}\rangle$$
$$\times\,\langle \mathscr{B}_{12}(1'2'), \quad g_{12}\mathscr{B}_{12}(k'l')\rangle. \tag{58}$$

The first term on the right is the dispersion energy of the pair $(1'2)$; it depends on the other electrons of the system only through the H. F. s.c.f. potential. The second term gives the 'exclusion effect' of the occupied orbital $m'$ on the correlation of $(1'2')$. In the usual perturbation theory it arises because the virtual excitations of $(1'2')$ to $m'(2' < m' < N')$ are forbidden by the Pauli principle. The last term of equation (58) similarly represents four-body 'exclusion effects' (effect of $k'l'$ on $(1'2')$). There are no other many-body effects. Terms such as $\langle 1'(\mathbf{x}_1)\,2'(\mathbf{x}_2)\,3'(\mathbf{x}_3), \quad g_{13}\,3'(\mathbf{x}_3)\,\bar{u}_{12}(\mathbf{x}_1, \mathbf{x}_2)\rangle$ which we have called 'cross-polarization' (Sinanoğlu 1961*a*) vanish due to equation (56) in this case, where $\Phi_0$ was chosen as the Hartree–Fock solution.

From equations (57) and (58) we thus have

$$E_2 = \sum_{n>m>1}^{N} \langle \mathscr{B}_{mn}(m'n'), \quad g_{mn}\,\bar{u}_{mn}\rangle + (\text{exclusion effects}). \tag{59}$$

This is a rigorous result that had been obtained approximately before (Sinanoğlu 1960).

Equation (59) can be put into a very compact form by orthogonalizing $\bar{u}_{mn}$ to all the other *occupied* H.F. orbitals of the system. Equation (58) is equivalent to

$$\frac{\sqrt{N!}}{\sqrt{2}}\langle \mathscr{A}(1'2'\ldots N'), \quad (\sum_{i>j} g_{ij})\, 3'(\mathbf{x}_3)\, 4'(\mathbf{x}_4)\ldots N'(\mathbf{x}_N)\,\bar{u}_{12}(\mathbf{x}_1, \mathbf{x}_2)\rangle$$
$$= \langle \mathscr{B}_{12}(1'2'), \quad g_{12}\hat{u}_{12}\rangle, \tag{60}$$

where

$$\hat{u}_{12} = \bar{u}_{12}(\mathbf{x}_1, \mathbf{x}_2) - \sum_{m'>2}^{N} \det\{m'(\mathbf{x}_1)\langle m'(\mathbf{x}_1), \quad \bar{u}_{12}(\mathbf{x}_1, \mathbf{x}_2)\rangle_{\mathbf{x}_1}\}$$
$$+ \sum_{l>k>2}^{N} \langle \mathscr{B}_{12}(kl), \bar{u}_{12}\rangle \mathscr{B}_{12}(kl), \tag{61}$$

so that $\langle \hat{u}_{12}(\mathbf{x}_1,\mathbf{x}_2), m'(\mathbf{x}_1)\rangle_{\mathbf{x}_1} = 0$ for $2' < m' < N'$. Thus equation (57) or (59) becomes

$$E_2 = \sum_{n>m\geqslant 1}^{N} \langle \mathscr{B}_{mn}(m'n'), \quad g_{mn}\,\hat{u}_{mn}\rangle \tag{62}$$

and $\langle \hat{u}_{mn}(\mathbf{x}_m,\mathbf{x}_n), k'(\mathbf{x}_m)\rangle_{\mathbf{x}_m} = 0$ for $N \geqslant k' \neq m', n'$.

The $X_1'$, equation (53), that gave this $E_2$, also gives the $E_3$, by equation (7), which can be analyzed in a similar way.

## Summary of the non-empirical procedure

The steps involved in a purely non-empirical calculation of the corrections to the Hartree–Fock energy and w.f. of an $N$-electron system by the method given above are the following.

(1) Determine all the unique $u'_{ij}$ equations as in equations (33), (35) and (36) (see also Sinanoğlu 1961*a*). $u'_{ij}$'s belong to the irreducible representations of the symmetry group of a two-electron system. The number of $u'_{ij}$'s to be solved for, will usually be less than the total number of pairs, $N(N-1)/2$, due to the multiplicity of some of the pair states. For example, in Be, the only $u'_{ij}$'s needed are for $(1s^2)\,{}^1S$, $(1s2s)\,{}^1S$ and ${}^3S$, and $(2s^2)\,{}^1S$.

(2) Obtain each $u'_{ij}$ from equation (33) or from equations (36), either by a special variational principle such as equation (9) or by other approximate ways of solving these non-homogeneous two-electron partial differential equations. Orthogonalize $u'_{ij}$ to $\mathscr{B}(i'j')$ or $\phi^s_{ij}$. Many of the pairs will correspond to the excited states of a two-electron system. These can be obtained by the modified minimum principle (Sinanoğlu 1960*c*).

(3) Orthogonalize each $u'_{ij}$ to its *own* $i'$ and $j'$, so that the new pair function, $\bar{u}_{ij}$, satisfies $\langle \bar{u}_{ij}, i'(\mathbf{x}_i)\rangle_{\mathbf{x}_i} = \langle \bar{u}_{ij}, j'(\mathbf{x}_j)\rangle_{\mathbf{x}_j} = 0$.

(4) The complete first-order w.f., $X_1'$ is given by equation (53). The use of this $X_1'$ leads to $E_2$ as the sum of pair dispersion interactions and three- and four-body exclusion effects (equations (57) and (59)). Alternatively, $\bar{u}_{ij}$ may be orthogonalized by equation (61) to all the other H.F. orbitals different from $i'$ and $j'$ and occupied in $\Phi_0$. This gives the same $E_2$, but expressed as a sum of *effective pair energies*, equation (62).

(5) $X_1'$ also determines $E_3$ by equation (7). *The total energy calculated will be an upper limit to the exact E only if $E_3$ is added to $E_0 + E_1 + E_2$.*

## Discussion

We have seen in the previous sections that in an $N$-electron system each $(i'j')$ pair of orbitals is associated with one $u'_{ij}$. This means that even when electron correlation can no longer be neglected there will be some validity to talking about the orbitals $i'$ and $j'$ which merely serve as labels and imply $u'_{ij}$ as well. Thus many of the cruder and qualitative concepts of quantum chemistry which were based on the orbital approximation and neglected correlation can be justified on this basis.

The theory developed above does not have to be used in a purely non-empirical way. Its usefulness even for large molecules depends on the possibility of its semi-empirical application. The 'immersed pairs' of $X_1'$ are almost independent of the other electrons of the system. The $u'_{ij}$ equations are just like those that determine the first order w.f. of an actual two-electron system taking $H_1 = g_{12}$, except that the electrons of $u'_{ij}$ move in the H.F. s.c.f. of all $N$ electrons rather than only in the field of nuclei. Often some of the $i', j'$ orbitals, e.g. inner shells, will be quite unchanged in going from one atom or molecule to another. Then provided the change in $V_1$ is also small, the corresponding $u'_{ij}$, the $\bar{u}_{ij}$ and the pair energy $\langle \mathscr{B}_{ij}(i'j'), g_{ij}\bar{u}_{ij} \rangle$ as well as the orbitals $i', j'$ can be transferred from one system to the other.

When a pair of electrons occupy the same orbital, the solution of their $u'_{ij}$, for example, by equation (9), will be similar to the He or $H_2$ problem and should be greatly aided by the use of $r_{12}$ in the trial function, $(u'_{ij})_{\text{trial}}$.

If the electrons of a pair occupy different orbitals their correlation error will be much smaller and may be obtained by other approximate methods. When the different orbitals are equivalent to two orbitals around different centres the resulting pair interaction may be described as a generalized London force within the same molecule. On the other hand one of the orbitals may belong to an inner and the other to an outer shell. Then we may describe the outer electron as moving in a core-polarization potential. Both of these weaker correlation effects can be obtained by the adiabatic approximation or by the use of mean excitation energies (Sinanoğlu 1960).

We considered above a system with a single H.F. determinant $\Phi_0$. Many molecular problems, e.g. involving $\pi$-electrons however will have degeneracies. These cases can also be treated by the formalism of this article. The $L_0^{-1} = (\sum_i e'_i)^{-1}$ (equations (24) to (27)) acting on $\Phi_0$ would separate $X_1'$ into a closed shell part and the degenerate outer shell part. The degeneracy of the latter may then be resolved so as to satisfy equation (8). The rest of the method then applies as before.

It is a pleasure to thank Professor K. S. Pitzer for encouraging this research. I also wish to express my gratitude to the Turkish Educational Society (Türk Eğitim Derneği) of Ankara, Turkey, which indirectly made this work possible.

This research was carried out under the auspices of the U.S. Atomic Energy Commission.

## References

Bethe, H. A. & Salpeter, E. E. 1957 *Encyclopedia of Physics*, vol. 35. Berlin: Springer-Verlag.

Brueckner, K. A. 1959 *The many body problem; Grenoble Université, École d'été de physique theorique, Les Houches.* London: Methuen. New York: Wiley.

Eden, R. J. 1958 *Nuclear reactions*, vol. **1**. Amsterdam: North Holland Publishing Company.

Hurley, A. C., Lennard-Jones, J. E. & Pople, J. A. 1953 *Proc. Roy. Soc.* A, **220**, 446.

Löwdin, P. O. 1959 *Advances in chemical physics*, vol. **2**, pp. 207–322. New York: Interscience Publishers.

Lykos, P. G. & Parr, R. B. 1956 *J. Chem. Phys.* **24**, 1166.

McWeeny, R. 1959 *Proc. Roy. Soc.* A, **253**, 242.

Morse, P. M. & Feshbach, H. 1953 *Methods of theoretical Physics*, vol. **2**, pp. 1119–20. New York: McGraw Hill.

Sinanoğlu, O. 1960 *J. Chem. Phys.* **33**, 1212.

Sinanoğlu, O. 1961*a* *Phys. Rev.* (To be published); also: *Univ. Calif. Lawrence Radn Lab. Rep.* UCRL-9320.

Sinanoğlu, O. 1961*b* *Phys. Rev.* (To be published); also: *Univ. Calif. Lawrence Radn Lab. Rep.* UCRL-9316.

V-4 Reprinted from *Proceedings of the Royal Society* (London) **A290**, 264 (1966)

# An expansion method for calculating atomic properties
# VII. The correlation energies of the lithium sequence

By C. D. H. Chisholm and A. Dalgarno

*Department of Applied Mathematics, The Queen's University of Belfast*

(*Communicated by D. R. Bates, F.R.S.—Received* 23 *September* 1965)

**If the energy of an atomic system is expanded in inverse powers of the nuclear charge $Z$, the leading term of the correlation energy after degeneracies have been removed is a constant which can be expressed as a weighted sum of electron pair energies and certain non-additive terms. The pair energies may be obtained from direct two-electron variational calculations and the non-additive terms may be evaluated exactly. Calculations are carried out for the lithium $^2S$ sequence with the result that the non-relativistic eigenvalues are given by**

$$E(Z) = -1{\cdot}125Z^2+1{\cdot}02280521Z-0{\cdot}40814899+O(Z^{-1}).$$

## 1. Introduction

In recent years we have explored the usefulness of expansions in inverse powers of the nuclear charge for predicting atomic properties. For many properties the convergence is rapid and results comparable in accuracy with those of the Hartree–Fock approximation have been obtained from very simple calculations. The expansion method may be developed further to yield results superior in accuracy to those of the Hartree–Fock approximation. As an example, we consider the correlation energy of the lithium $^2S$ sequence, but the procedures are generally applicable to all atomic systems and all atomic properties.

## 2. Theory

We choose a set of units in which the scale of distance is $Z$ atomic units and the scale of energy is $Z^2$ atomic units, $Z$ being the nuclear charge of an $N$-electron atomic system. If $\mathbf{r}_i$ is the position vector of the $i$th electron, the Hamiltonian is given by

$$H = H_0+V/Z, \tag{1}$$

where

$$H_0 = -\frac{1}{2}\sum_{i=1}^{N}\nabla_i^2-\sum_{i=1}^{N}\frac{1}{r_i} \tag{2}$$

consists of one-electron terms only and

$$V = \sum_{i<j}^{N}\frac{1}{|\mathbf{r}_i-\mathbf{r}_j|} = \sum_{i<j}^{N}\frac{1}{r_{ij}} \tag{3}$$

consists of two-electron terms only. The Schrödinger equation

$$H\Psi = E\Psi \tag{4}$$

may be solved by expanding $\Psi$ and $E$ in inverse powers of $Z$ according to

$$\Psi = \sum_{n=0}^{\infty}\psi_n/Z^n, \quad E = Z^2\sum_{n=0}^{\infty}E_n/Z^n, \tag{5}$$

where

$$(H_0-E_0)\psi_0 = 0 \tag{6}$$

is the zero order unperturbed equation. Then

$$(H_0 - E_0)\psi_1 + (V - E_1)\psi_0 = 0, \tag{7}$$

$$(H_0 - E_0)\psi_2 + (V - E_1)\psi_1 - E_2\psi_0 = 0 \tag{8}$$

and

$$(H_0 - E_0)\psi_3 + (V - E_1)\psi_2 - E_2\psi_1 - E_3\psi_0 = 0, \tag{9}$$

so that

$$E_1 = \langle\psi_0|\, V\,|\psi_0\rangle, \quad E_2 = \langle\psi_0|\, V - E_1\,|\psi_1\rangle, \quad E_3 = \langle\psi_1|\, V - E_1\,|\psi_1\rangle. \tag{10}$$

The unperturbed eigenfunction $\psi_0$ is an antisymmetrical product or combination of products of hydrogenic orbitals $\phi(n_i l_i m_{l_i} m_{s_i})$ so constructed that $\psi_0(\gamma SL|N)$ is an $N$-electron eigenfunction corresponding to spin and orbital angular momenta $S$ and $L$, $\gamma$ representing the $(n_i l_i)$ values in the configuration and also any other quantum numbers that may be necessary to distinguish between states with the same values of $L$ and $S$, and the unperturbed eigenvalue $E_0$ is given by

$$E_0(N) = -\sum_{i=1}^{N} \frac{1}{2n_i^2}. \tag{11}$$

It is clear that a similar expansion of the Hartree–Fock approximation would yield identical values of $E_0(N)$ and $E_1(N)$ unless degeneracy occurs between different configurations. In such cases, the degenerate zero-order eigenfunctions must be combined linearly so that the Hamiltonian is diagonalized to first order and we suppose that

$$\langle\gamma SL|\, V\,|\gamma' SL\rangle = \delta_{\gamma\gamma'} E_1(\gamma SL|N), \tag{12}$$

where $\gamma$ now also specifies the mixture of configurations (Layzer 1959).

To determine the correlation energy, we must calculate $E_2$ and the utility of the expansion method lies in the observation that since $V$ is a sum of two-electron terms it must be possible to express the solution $\psi_1(\gamma SL|N)$ of (7) in terms of solutions of two-electron equations. We construct $\psi_0(\gamma SL|N)$ according to

$$\psi_0(\gamma SL|N) = \sum_{\gamma' S' L'} \sum_{\gamma'' S'' L''} \langle\gamma' S' L', \gamma'' S'' L''|\}\gamma SL\rangle \{\psi_0(\gamma' S' L'|N-2), \psi_0(\gamma'' S'' L''|2)\}, \tag{13}$$

where $\{\psi_0(\gamma' S' L'|N-2),\ \psi_0(\gamma'' S'' L''|2)\}$ is a vector-coupled product of normalized antisymmetric $N-2$ electron eigenfunctions $\psi_0(\gamma' S' L' M_{S'} M_{L'}|N-2)$ and normalized antisymmetric two-electron eigenfunctions $\psi_0(\gamma'' S'' L'' M_{S''} M_{L''}|2)$ and $\langle\gamma' S' L', \gamma'' S'' L''|\}\gamma SL\rangle$ are two-particle fractional parentage coefficients which ensure the antisymmetry of $\psi_0(\gamma SL|N)$. Then the substitution into (7) of

$$\psi_1(\gamma SL|N) = \mathscr{A} \sum_{\gamma' S' L'} \sum_{\gamma'' S'' L''} \langle\gamma' S' L', \gamma'' S'' L''|\}\gamma SL\rangle \{\psi_0(\gamma' S' L'|N-2), \psi_1(\gamma'' S'' L''|2)\}, \tag{14}$$

where $\{\psi_0(\gamma' S' L'|N-2),\ \psi_1(\gamma'' S'' L''|2)\}$ is a vector-coupled product of

$$\psi_0(\gamma' S' L' M_{S'} M_{L'}|N-2)$$

and first-order two-electron eigenfunctions $\psi_1(\gamma'' S'' L'' M_{S''} M_{L''}|2)$ and $\mathscr{A}$ is an antisymmetrizing operator which must be introduced to ensure that $\psi_1(\gamma SL|N)$ is totally antisymmetric, yields the two-electron helium-like equations

$$\left(-\tfrac{1}{2}\nabla_1^2 - \tfrac{1}{2}\nabla_2^2 - \frac{1}{r_1} - \frac{1}{r_2} - E_0(2)\right)\psi_1(\gamma'' S'' L''|2) + \left(\frac{1}{r_{12}} - E_1(2)\right)\psi_0(\gamma'' S'' L''|2) = 0. \tag{15}$$

As was pointed out by Bacher & Goudsmit (1934), the second-order energy $E_2(\gamma SL|N)$ consists of a weighted sum of the pair energies

$$\tfrac{1}{2}N(N-1)\sum_{\gamma'S'L'}\sum_{\gamma''S''L''}|\langle\gamma'S'L',\gamma''S''L''|\}\gamma SL\rangle|^2E_2(\gamma''S''L''|2) \tag{16}$$

that are obtained by coupling the electrons in pairs and certain non-additive contributions. Our procedure demonstrates that apart from an outside factor the coefficients of the pair energies are simply the squares of the two-particle fractional parentage coefficients (cf. Trees 1954). The non-additive contributions can be described as dispersion, cross-polarization and exclusion effects arising from double and single virtual transitions (cf. Sinanoglu 1960). They can be evaluated straightforwardly using the two-electron solutions of (15) as Sinanoglu suggests or by explicitly summing over the virtual transitions as Layzer, Horák, Lewis & Thompson (1964) have done. We have used a third method which allows the non-additive terms to be evaluated exactly. To compare the three methods we examine the calculation of $E_2$ for the lithium $(1s^22s\,^2S)$ sequence.

## 3. The lithium $(1s^22s\,^2S)$ sequence

Equation (14) for $\psi_1(1s^22s\,^2S)$ is

$$\psi_1(1s^22s\,^2S) = \mathscr{A}\{\tfrac{1}{\sqrt{3}}\{\psi_0(2s\,^2S),\psi_1(1s^2\,^1S)\}-\tfrac{1}{\sqrt{6}}\{\psi_0(1s\,^2S),\psi_1(1s2s\,^1S)\} + \tfrac{1}{\sqrt{2}}\{\psi_0(1s\,^2S),\psi_1(1s2s\,^3S)\}\}, \tag{17}$$

the correlated pair functions being solutions of the equations

$$\left(-\tfrac{1}{2}\nabla_1^2-\tfrac{1}{2}\nabla_2^2-\frac{1}{r_1}-\frac{1}{r_2}-E_0(1s^2\,^1S)\right)\psi_1(1s^2\,^1S)+\left(\frac{1}{r_{12}}-E_1(1s^2\,^1S)\right)\psi_0(1s^2\,^1S) = 0, \tag{18}$$

$$\left(-\tfrac{1}{2}\nabla_1^2-\tfrac{1}{2}\nabla_2^2-\frac{1}{r_1}-\frac{1}{r_2}-E_0(1s2s\,^{1,3}S)\right)\psi_1(1s2s\,^{1,3}S) + \left(\frac{1}{r_{12}}-E_1(1s2s\,^{1,3}S)\right)\psi_0(1s2s\,^{1,3}S) = 0. \tag{19}$$

Then if $u$, $s$ and $t$ represent respectively the space components of $\psi_1(1s^2\,^1S)$, $\psi_1(1s2s\,^1S)$ and $\psi_1(1s2s\,^3S)$ and $nl$ represents the space component of the hydrogenic orbital such that

$$\left(-\tfrac{1}{2}\nabla_1^2-\tfrac{1}{2}\nabla_2^2-\frac{1}{r_1}-\frac{1}{r_2}-E_0(1s^2\,^1S)\right)u(12) + \left(\frac{1}{r_{12}}-E_1(1s^2\,^1S)\right)1s(1)\,1s(2)=0, \tag{20}$$

$$\left(-\tfrac{1}{2}\nabla_1^2-\tfrac{1}{2}\nabla_2^2-\frac{1}{r_1}-\frac{1}{r_2}-E_0(1s2s\,^1S)\right)s(1,2) + \left(\frac{1}{r_{12}}-E_1(1s2s\,^1S)\right)(1s(1)\,2s(2)+2s(1)\,1s(2)) = 0, \tag{21}$$

$$\left(-\tfrac{1}{2}\nabla_1^2-\tfrac{1}{2}\nabla_2^2-\frac{1}{r_1}-\frac{1}{r_2}-E_0(1s2s\,^3S)\right)t(1,2) + \left(\frac{1}{r_{12}}-E_1(1s2s\,^3S)\right)(1s(1)\,2s(2)-2s(1)\,1s(2)) = 0, \tag{22}$$

the second-order energy can be written

$$E_2(1s^22s\,^2S) = E_2(1s^2\,^1S) + \tfrac{1}{2}E_2(1s2s\,^1S) + \tfrac{3}{2}E_2(1s2s\,^3S) + 4M_1 - 2M_2 - M_3 - M_4 + \tfrac{1}{2}M_5 + \tfrac{3}{2}M_6, \quad (23)$$

where

$$\left.\begin{aligned} M_1 &= \langle u(31)\,2s(2)|\,1/r_{12}\,|1s(3)\,1s(1)\,2s(2)\rangle, \\ M_2 &= \langle u(31)\,2s(2)|\,1/r_{12}\,|1s(3)\,2s(1)\,1s(2)\rangle, \\ M_3 &= \langle u(31)\,2s(2)|\,1/r_{12}\,|2s(3)\,1s(1)\,1s(2)\rangle, \\ M_4 &= \langle s(31)\,1s(2)|\,1/r_{12}\,|1s(3)\,1s(1)\,2s(2)\rangle, \\ M_5 &= \langle s(31)\,1s(2)|\,1/r_{12}\,|1s(3)\,2s(1)\,1s(2)\rangle, \\ M_6 &= \langle t(31)\,1s(2)|\,1/r_{12}|\,1s(3)\,2s(1)\,1s(2)\rangle. \end{aligned}\right\} \quad (24)$$

The approximate evaluation of the $M$ integrals can be accomplished by using variationally determined representations of the pair functions $u$, $s$ and $t$, but the procedure is laborious and much of the advantage gained from separating off the pair energies is lost. The possible accuracy is high, however. We have evaluated $M_1$ using the representation of $u$ given by Somerville & Stewart (1962) and obtain a value of $-0{\cdot}00505663$ compared to the correct value of $-0{\cdot}00505770$.

Layzer *et al.* (1964) have evaluated the combination of $M$ integrals in (23) by deriving it in the form of infinite summations. The two formulas can be related by using the sum rule procedure of Dalgarno & Lewis (1955) or by noting, for example, that the solution of (20) can be written formally as the double summation

$$u(12) = \mathbf{SS}_{m+n=1} \frac{\langle 1s(1)\,1s(2)\,|\,1/r_{12}\,|ml(1)\,nl(2)\rangle}{-1 + (1/2m^2) + (1/2n^2)}\, ml(1)\,nl(2), \quad (25)$$

so that

$$M_1 = \mathbf{S}_{n+1} \frac{\langle 1s(1)\,1s(2)\,|1/r_{12}|\,1s(1)\,ns(2)\rangle\langle 2s(1)\,ns(2)\,|1/r_{12}|\,2s(1)\,1s(2)\rangle}{-\frac{1}{2} + (1/2n^2)}, \quad (26)$$

$$M_2 = \mathbf{S}_{n+1} \frac{\langle 1s(1)\,1s(2)\,|\,1/r_{12}\,|1s(1)\,ns(2)\,\rangle\langle 2s(1)\,ns(2)|\,1/r_{12}\,|1s(1)\,2s(2)\rangle}{-\frac{1}{2} + (1/2n^2)} \quad (27)$$

and

$$M_3 = \mathbf{S}_{n} \frac{\langle 1s(1)\,1s(2)\,|\,1/r_{12}\,|2s(1)\,ns(2)\rangle\langle 2s(1)\,ns(2)|\,1/r_{12}\,|1s(1)\,1s(2)\rangle}{-\frac{7}{8} + (1/2n^2)}. \quad (28)$$

Layzer *et al.* (1964) computed these single summations by direct term-by-term addition.

Because they are summations of single electron transitions, they can be expressed in terms of the solutions of single electron differential equations similar to those for Hartree–Fock orbitals. Thus defining $g(r)$, $\sigma(r)$, and $\tau(r)$ as the well-behaved solutions of respectively

$$(-\tfrac{1}{2}\nabla^2 - (1/r) + \tfrac{1}{2})\,g(r) + \left\langle 1s(r')\left|\frac{1}{|\mathbf{r}-\mathbf{r}'|}\right|1s(r')\right\rangle 1s(r) - E_1(1s^2\,^1S)\,1s(r) = 0, \quad (29)$$

$$(-\tfrac{1}{2}\nabla^2 - (1/r) + \tfrac{1}{8})\,\sigma(r) + \left\langle 1s(r')\left|\frac{1}{|\mathbf{r}-\mathbf{r}'|}\right|1s(r')\right\rangle 2s(r) + \left\langle 1s(r')\left|\frac{1}{|\mathbf{r}-\mathbf{r}'|}\right|2s(r')\right\rangle 1s(r) - E_1(1s\,2s\,^1S)\,2s(r) = 0 \quad (30)$$

and

$$(-\tfrac{1}{2}\nabla^2-(1/r)+\tfrac{1}{8})\,\tau(r)+\left\langle 1s(r')\left|\frac{1}{|\mathbf{r}-\mathbf{r}'|}\right|1s(r')\right\rangle 2s(r)$$

$$-\left\langle 1s(r')\left|\frac{1}{|\mathbf{r}-\mathbf{r}'|}\right|2s(r')\right\rangle 1s(r)-E_1(1s2s\,{}^3S)\,2s(r)=0, \quad (31)$$

it follows from (20), (21) and (22) that

$$g(r_1)=\langle u(12)|1s(2)\rangle, \quad (32)$$

$$\sigma(r_1)=\langle s(12)|1s(2)\rangle, \quad (33)$$

$$\tau(r_1)=\langle t(12)|1s(2)\rangle, \quad (34)$$

and accordingly

$$M_1=\left\langle g(r)\,2s(r')\left|\frac{1}{|\mathbf{r}-\mathbf{r}'|}\right|1s(r)\,2s(r')\right\rangle, \quad (35)$$

$$M_2=\left\langle g(r)\,2s(r')\left|\frac{1}{|\mathbf{r}-\mathbf{r}'|}\right|2s(r)\,1s(r')\right\rangle, \quad (36)$$

$$M_4=\left\langle \sigma(r)\,1s(r)'\left|\frac{1}{|\mathbf{r}-\mathbf{r}'|}\right|1s(r)\,2s(r')\right\rangle, \quad (37)$$

$$M_5=\left\langle \sigma(r)\,1s(r')\left|\frac{1}{|\mathbf{r}-\mathbf{r}'|}\right|2s(r)\,1s(r')\right\rangle, \quad (38)$$

$$M_6=\left\langle \tau(r)\,1s(r')\left|\frac{1}{|\mathbf{r}-\mathbf{r}'|}\right|2s(r)\,1s(r')\right\rangle. \quad (39)$$

The solution $g(r)$ of (29) is simply the first order Hartree–Fock orbital for the helium ground state and it may be written (Cohen & Dalgarno 1961)

$$g(r)=\tfrac{3}{8}\phi(2r)+(\tfrac{3}{16},(1/r)+\tfrac{5}{8}r)-(3/16r)+\tfrac{1}{4})\exp(-2r)-\tfrac{1}{16}(23-12\ln 2)\exp(-r), \quad (40)$$

where

$$\phi(\alpha r)=\int_0^r\frac{\exp(-\alpha s)-1}{s}\,\mathrm{d}s. \quad (41)$$

The solutions of (30) and (31) are linear combinations of auxiliary functions listed by Cohen (1963) in his tabulation of first order Hartree–Fock orbitals for $K$ and $L$ shell electrons and appropriately normalized they may be written

$$\sigma(r)=\frac{1}{2\sqrt{2}}(2-r)\exp(-\tfrac{1}{2}r)\left\{\frac{106}{729}\phi(2r)+\left(\frac{53}{729}\frac{1}{r}+\frac{1087}{729}\frac{1}{2-r}+\frac{338}{729}\right)\right.$$

$$\left.+\left(\frac{-53}{729}\frac{1}{r}+\frac{2189}{729}\frac{1}{2-r}-\frac{25}{27}\right)\exp(-2r)-\frac{4751}{2187}+\frac{106}{729}\ln 3\right\}, \quad (42)$$

$$\tau(r)=\frac{1}{2\sqrt{2}}(2-r)\exp(-\tfrac{1}{2}r)\left\{\frac{362}{729}\phi(2r)+\left(\frac{181}{729}\frac{1}{r}+\frac{191}{729}\frac{1}{2-r}+\frac{274}{729}r\right)\right.$$

$$\left.+\left(-\frac{181}{729}\frac{1}{r}-\frac{1523}{729}\frac{1}{2-r}+\frac{7}{27}\right)\exp(-2r)-\frac{2063}{4374}-\frac{592}{729}\ln\frac{3}{2}\right\}. \quad (43)$$

Then by elementary integrations

$$\left.\begin{aligned}
M_1 &= \frac{8}{27}\ln\frac{5}{6}+\frac{1\,189\,813}{243\times 10^5} &&= -0{\cdot}005057704,\\
M_2 &= \frac{-2}{243}\ln\frac{5}{6}-\frac{105\,853}{13\,668\,750} &&= -0{\cdot}006243572,\\
M_4 &= \frac{-1696}{531\,441}\ln\frac{5}{9}+\frac{328\,064}{3\,985\,075} &&= \phantom{-}0{\cdot}010106621,\\
M_5 &= \frac{689}{118\,098}\ln\frac{5}{9}-\frac{291\,406}{4\,428\,675} &&= -0{\cdot}069229044\\
\text{and}\qquad M_6 &= \frac{2353}{118\,098}\ln\frac{5}{9}-\frac{27\,995\,926}{553\,584\,375} &&= -0{\cdot}062283241.
\end{aligned}\right\} \tag{44}$$

The evaluation of $M_3$ is less straightforward. Defining

$$f(r_1) = \langle u(12)|2s(2)\rangle, \tag{45}$$

we have from (20) that

$$(-\tfrac{1}{2}\nabla^2-(1/r)+\tfrac{7}{8})f(r)+\left\langle 1s(r')\left|\frac{1}{|\mathbf{r}-\mathbf{r}'|}\right|2s(r')\right\rangle 1s(r) = 0. \tag{46}$$

Expanding $f(r)$ in terms of the complete set of associated Laguerre polynomials according to

$$f(r) = r\sum_{k=1}^{\infty} a_k \exp(-\alpha r)\,L_k^1(2\alpha r), \tag{47}$$

we obtain with $\alpha = \frac{\sqrt{7}}{2}$,

$$a_k = \frac{16\sqrt{(2)}\,\alpha^2(k-1)!}{9(1-k\alpha)\,(k!)^3}\left\{\frac{1}{(2\alpha)^4}K_3+\frac{2}{3}\,\frac{1}{(2\alpha)^3}\,K_2\right\}, \tag{48}$$

where

$$K_m = \int_0^\infty x^m \exp(-\gamma x)\,L_k^1(x)\,\mathrm{d}x \tag{49}$$

and

$$\gamma = (\sqrt{(7)}+5)/2\sqrt{7}. \tag{50}$$

The $K$ integrals are readily evaluated using the generating function

$$\sum_{q=1}^{\infty}\frac{L_q^1(\rho)}{q!}s^q = \frac{-s\exp(-\rho s/(1-s))}{(1-s)^2}. \tag{51}$$

Then

$$M_3 = \left\langle f(r)\,1s(r')\left|\frac{1}{|\mathbf{r}-\mathbf{r}'|}\right|1s(r)\,2s(r)'\right\rangle \tag{52}$$

reduces to

$$M_3 = \frac{1}{162\alpha^6\gamma^8}\sum_{k=1}^{\infty}\frac{k}{1-k\alpha}J_k^2, \tag{53}$$

where

$$\begin{aligned}
J_k = (k+1)(k+2)\left(\frac{\gamma-1}{\gamma}\right)^{k-1} &- 2(k-1)(k+1)\left(\frac{\gamma-1}{\gamma}\right)^{k-2} + (k-2)(k-1)\left(\frac{\gamma-1}{\gamma}\right)^{k-3}\\
&+\frac{4\alpha\gamma}{3}\left\{(k+1)\left(\frac{\gamma-1}{\gamma}\right)^{k-1}-(k-1)\left(\frac{\gamma-1}{\gamma}\right)^{k-2}\right\},
\end{aligned} \tag{54}$$

negative powers being excluded. The series (53) is rapidly convergent and

$$M_3 = -0{\cdot}023758968. \tag{55}$$

Finally using

$$\left.\begin{aligned} E_2(1s^{2\,1}S) &= -0{\cdot}157666405,\\ E_2(1s2s\,^1S) &= -0{\cdot}11447618,\\ E_2(1s2s\,^3S) &= -0{\cdot}047409192 \end{aligned}\right\} \tag{56}$$

(Knight & Scherr 1963) we obtain

$$E_2(1s^22s\,^2S) = -0{\cdot}40814899, \tag{57}$$

to which the pair energies contribute $-0{\cdot}28601828$ and the non-additive terms $-0{\cdot}12213071$.

Layzer *et al.* (1964) have previously computed a value of $-0{\cdot}124930$ for the contribution of the non-additive term but their work contains a numerical error. The theoretical value of $E_2$ is in harmony with the semi-empirical value of $-0{\cdot}409$ obtained by analyses of data on the isoelectronic sequence (Layzer & Bahcall 1962; Scherr, Silverman & Matsen 1962).

## 4. Discussion

To second order the non-relativistic eigenvalue of the $(1s^22s\,^2S)$ state is given by

$$E(Z) = -1{\cdot}125Z^2 + 1{\cdot}02280521Z - 0{\cdot}40824899 \tag{58}$$

and the Hartree–Fock eigenvalue is given by

$$E^{\text{H.-F.}}(Z) = -1{\cdot}125Z^2 + 1{\cdot}02280521Z - 0{\cdot}35454903. \tag{59}$$

Thus the correlation energy tends to a constant value of $-0{\cdot}05369996$ a.u. or $1{\cdot}46$ eV as the nuclear charge increases. The correlation energy of the helium $1s^{2\,1}S$ sequence tends to a constant value of $-1{\cdot}27$ eV (Dalgarno 1960) so that the addition of a $2s$ electron increases the correlation energy by $0{\cdot}19$ eV.

Most of the correlation energy arises from the pair energies. Thus if we write

$$E_2 = E_2(1s^{2\,1}S) + \tfrac{1}{2}E_2(1s2s\,^1S) + \tfrac{3}{2}E_2(1s2s\,^3S) + \zeta, \tag{60}$$

$$E_2^{\text{H.-F.}} = E_2^{\text{H.-F.}}(1s^{2\,1}S) + \tfrac{1}{2}E_2^{\text{H.-F.}}(1s2s\,^1S) + \tfrac{3}{2}E_2^{\text{H.-F.}}(1s2s\,^3S) + \zeta^{\text{H.-F.}}, \tag{61}$$

so that the leading term of the correlation energy is

$$E_2(1s^{2\,1}S) - E_2^{\text{H.-F.}}(1s^{2\,1}S) + \tfrac{1}{2}E_2(1s2s\,^1S) - \tfrac{1}{2}E_2^{\text{H.-F.}}(1s2s\,^1S) + \tfrac{3}{2}E_2(1s2s\,^3S) - \tfrac{3}{2}E_2^{\text{H.-F.}}(1s2s\,^3S) + (\zeta - \zeta^{\text{H.-F.}}), \tag{62}$$

$\zeta - \zeta^{\text{H.-F.}}$ is $+0{\cdot}001455$ a.u. $= 0{\cdot}040$ eV while the contribution from the pair energies is $-0{\cdot}055155$ a.u. $= -1{\cdot}50$ eV.

For $Z = 3$, (59) yields an eigenvalue of $-7{\cdot}4111$ a.u., whereas the actual Hartree–Fock eigenvalue is $-7{\cdot}4327$ a.u. The value of (58) is $-7{\cdot}4647$ a.u., whereas the experimental value is $-7{\cdot}4781$ (cf. Cooper & Martin 1963). Thus (58) is more accurate than the Hartree–Fock approximation even for neutral lithium.

Since the next term in the series expansion of $E(Z)$ is $O(Z^{-1})$, the accuracy of (58) increases with increasing $Z$. For $Z = 8$, (58) yields a value of $-64{\cdot}2257$ a.u. compared to the experimental value of $-64{\cdot}2292$ a.u., so that the third-order term $E_3$ must be about $-0{\cdot}0282/Z$.

We are greatly indebted to Dr F. R. Innes for a number of valuable discussions.
The research reported in this paper has been sponsored by the Air Force Cambridge Research Laboratories OAR under Contract AF 61(052)-780 with the European Office of Aerospace Research, United States Air Force.

## References

Bacher, R. F. & Goudsmit, S. 1934 *Phys. Rev.* **46**, 948.
Cohen, M. 1963 *Proc. Phys. Soc.* **82**, 778.
Cohen, M. & Dalgarno, A. 1961 *Proc. Phys. Soc.* **77**, 165.
Cooper, J. W. & Martin, J. B. 1963 *Phys. Rev.* **131**, 1183.
Dalgarno, A. 1960 *Proc. Phys. Soc.* **75**, 439.
Dalgarno, A. & Lewis, J. T. 1955 *Proc. Roy. Soc.* A, **233**, 70.
Knight, R. E. & Scherr, C. W. 1963 *Rev. Mod. Phys.* **35**, 431.
Layzer, D. 1959 *Ann. Phys.* **8**, 271.
Layzer, D. & Bahcall, J. B. 1962 *Ann. Phys.* **17**, 177.
Layzer, D., Horák, Z., Lewis, N. M. & Thompson, D.P. 1964 *Ann. Phys.* **29**, 101.
Scherr, C. W., Silverman, J. N. & Matsen, F. A. 1962 *Phys. Rev.* **127**, 830.
Sinanoglu, O. 1960 *J. Chem. Phys.* **33**, 1212.
Sinanoglu, O. 1961 *Phys. Rev.* **122**, 493.
Somerville, W. B. & Stewart, A. L. 1962 *Proc. Phys. Soc.* **80**, 97.
Trees, R. E. 1954 *J. Res. Nat. Bur. Stand.* **53**, 35.

PRINTED IN GREAT BRITAIN AT THE UNIVERSITY PRINTING HOUSE, CAMBRIDGE
(BROOKE CRUTCHLEY, UNIVERSITY PRINTER)

V-5 Reprinted from *Physical Review* 157, 7 (1967)

# Correlation Effects in Atoms. III. Four-Electron Systems*

F. W. Byron, Jr.,†‡ and C. J. Joachain‡
*Department of Physics, University of California, Berkeley, California*
(Received 21 November 1966)

The binding energy of the ground state of four-electron systems is investigated in the framework of Hartree-Fock perturbation theory. First, we show that the original problem can be decoupled into a series of helium-like equations describing pair correlation between electrons. Then, the variational-perturbation method is applied to each of these equations through fifth order in energy. We have obtained for the correlation energies the values $E_{corr}$ (Be) $=-0.0925$ atomic units (a.u.) and $E_{corr}(B^+)=-0.1096$ a.u. The experimental numbers are $E_{corr}{}^{exp}$ (Be) $=-0.0944$ a.u. and $E_{corr}{}^{exp}$ $(B^+)=-0.1116$ a.u.

## I. INTRODUCTION

THE quantum mechanics of one- and two-electron systems can be said to be understood in complete detail,[1] with the possible exception of very small relativistic effects. However, attempts to extend the methods which have been so successful in dealing with these simplest of systems have proven to be difficult. In this paper, we wish to investigate, starting from first principles, the binding energies and wave functions of four-electron systems. Recently, a great deal of work has been done on three- and four-electron atoms,[2–7] and there are several approaches which have provided some useful insight into these problems. We shall comment on some of these approaches in what follows.

Certainly the most comprehensive attack on the many-electron problem is the Hartree-Fock method, which replaces the true many-body problem by a series of one-body problems in which each particle moves in a self-consistently determined potential generated by all the other particles. For many systems, and in particular for the four-electron atoms and ions, the Hartree-Fock equations have been solved and give accurate approximations to atomic wave functions and binding energies. In our case, since the Hartree-Fock energies and wave functions are known, the problem comes down to finding the relevant corrections to the energies and wave functions. We define

$$E_{corr}=E_{tot}-E_{HF},$$
$$\Psi_{corr}=\Psi_{tot}-\Psi_{HF},$$

where $E_{tot}$ and $\Psi_{tot}$ are, respectively, the total *nonrelativistic* energy and wave function, whereas $E_{HF}$ and $\Psi_{HF}$ are the Hartree-Fock energy and wave function of the system considered.

Since $E_{tot}$ as defined above is not directly accessible to experiment, we have estimated it as follows: We take the extremely accurate calculations of Pekeris[8] to provide us with the nonrelativistic energy of the two electron core ($1s$-$1s$) corresponding to the four-electron system in question, and then use the experimental ionization potentials to obtain the binding energies of the outer two electrons. We estimated the relativistic corrections to the experimental binding energies in the manner suggested by Watson,[2] i.e., by using a simple Dirac-type relation for the energy. For Be and $B^+$, in which we are interested in this paper, these relativistic corrections are less than 1% of the correlation energy, and therefore the precision with which they are estimated is not of great importance for our purposes. Thus, we believe that the values

$$E_{corr}{}^{exp}(\text{Be})=-0.0944 \text{ a.u.},$$
$$E_{corr}{}^{exp}(\text{B}^+)=-0.1116 \text{ a.u.},$$

are very accurate estimates of the correlation energies for these two systems.[9] It seems unlikely that the correct nonrelativistic value could differ by more than 0.5% from these numbers. It is these "experimental" results that we wish to understand in a reasonably straightforward manner.

## II. DERIVATION OF THE BASIC EQUATIONS

Since the Hartree-Fock wave functions give an excellent approximation to atomic energies, it seems reasonable to apply the Rayleigh-Schrödinger perturbation theory to our four-electron system, starting from the Hartree-Fock Hamiltonian as zeroth-order approximation. We begin by establishing some notation. We define the operators $T(\mathbf{r})$ and $V_i(\mathbf{r})$ by

$$T(\mathbf{r})f(\mathbf{r})\equiv-\tfrac{1}{2}\nabla^2 f(\mathbf{r})-(Z/r)f(\mathbf{r}), \tag{1a}$$

* Research supported in part by the U. S. Air Force Office of Scientific Research under Grant No. AFOSR-130-66.

† Alfred P. Sloan Foundation Fellow, on leave for the academic year 1966–67 from the Dept. of Physics, University of Massachusetts, Amherst, Massachusetts.

‡ Present address: Physique Théorique et Mathématique, Faculté des Sciences, Université Libre de Bruxelles, Brussels, Belgium.

[1] H. A. Bethe and E. E. Salpeter, *Quantum Mechanics of One- and Two-Electron Atoms* (Academic Press Inc., New York, 1957).

[2] R. E. Watson, Phys. Rev. **119**, 170 (1960); Ann. Phys. (N.Y.) **13**, 250 (1961).

[3] A. W. Weiss, Phys. Rev. **122**, 1826 (1961).

[4] R. Mc Weeny and B. Sutcliffe, Proc. Roy. Soc. (London) **A273**, 103 (1963).

[5] L. Szasz, Phys. Rev. **126**, 169 (1962); J. Chem. Phys. **35**, 1072 (1961); J. Math. Phys. **3**, 1147 (1962); Phys. Letters **3**, 263 (1963).

[6] H. P. Kelly, Phys. Rev. **131**, 684 (1963); **136**, B896 (1964).

[7] O. Sinanoglu, J. Chem. Phys. **36**, 706, 3198 (1962); D. F. Tuan and O. Sinanoglu, *ibid.* **41**, 2677 (1964).

[8] C. L. Pekeris, Phys. Rev. **115**, 1216 (1959).

[9] Thoughout this paper we will mean by a.u. atomic units, in which the reduced mass of the electron is set equal to one.

$$V_i(\mathbf{r})f(\mathbf{r})\equiv\int\phi_i^*(\mathbf{r}')\frac{1}{|\mathbf{r}-\mathbf{r}'|}\phi_i(\mathbf{r}')d\mathbf{r}'f(\mathbf{r})$$
$$-\int\phi_i^*(\mathbf{r}')\frac{1}{|\mathbf{r}-\mathbf{r}'|}f(\mathbf{r}')d\mathbf{r}'\phi_i(\mathbf{r}),\quad(1b)$$

where $Z$ is the nuclear charge of the system considered, and $\phi_i$ is a Hartree-Fock single-particle orbital. The subscript $i$ refers to a complete collection of one-particle quantum numbers, including spin. For our specific four-electron problem, we further define

$$\mathcal{V}(\mathbf{r})=V_{1s\uparrow}(\mathbf{r})+V_{1s\downarrow}(\mathbf{r})+V_{2s\uparrow}(\mathbf{r})+V_{2s\downarrow}(\mathbf{r}).\quad(2)$$

In a more general case $\mathcal{V}(\mathbf{r})$ would be a sum of operators $V_i$ with $i$ ranging over all filled Hartree-Fock orbitals. The spinor wave functions $\phi_{1s\uparrow}$, $\phi_{1s\downarrow}$, $\phi_{2s\uparrow}$, and $\phi_{2s\downarrow}$, whose existence is implied in our above definitions, are the solutions to four coupled equations, namely,

$$[T(\mathbf{r})+\mathcal{V}(\mathbf{r})]\phi_i(\mathbf{r})=\epsilon_i\phi_i(\mathbf{r}),(i=1s\uparrow,1s\downarrow,2s\uparrow,2s\downarrow).\quad(3)$$

Because we are dealing with systems having all angular momentum shells filled, the space parts of $\phi_{1s\uparrow}$ and $\phi_{1s\downarrow}$ are the same, i.e.,

$$\phi_{1s\uparrow}=\phi_{1s}\alpha$$

and

$$\phi_{1s\downarrow}=\phi_{1s}\beta,$$

where $\phi_{1s}$ is a function of space variables only, and $\alpha$ and $\beta$ are spinors representing spin up and spin down, respectively. Similarly,

$$\phi_{2s\uparrow}=\phi_{2s}\alpha$$

and

$$\phi_{2s\uparrow}=\phi_{2s}\beta.$$

A simple computation using Eq. (3) gives then the following two coupled integro-differential equations for $\phi_{1s}$ and $\phi_{2s}$:

$$[T(\mathbf{r})+V_{1s}^d(\mathbf{r})+2V_{2s}^d(\mathbf{r})-V_{2s}^e(\mathbf{r})]\phi_{1s}(\mathbf{r})$$
$$=\epsilon_{1s}\phi_{1s}(\mathbf{r}),\quad(4a)$$

$$[T(\mathbf{r})+V_{2s}^d(\mathbf{r})+2V_{1s}^d(\mathbf{r})-V_{1s}^e(\mathbf{r})]\phi_{2s}(\mathbf{r})$$
$$=\epsilon_{2s}\phi_{2s}(\mathbf{r}).\quad(4b)$$

Here we have defined the spin-independent operators

$$V_{1s}^d(\mathbf{r})=\int\phi_{1s}^*(\mathbf{r}')\frac{1}{|\mathbf{r}-\mathbf{r}'|}\phi_{1s}(\mathbf{r}')d\mathbf{r}',\quad(5a)$$

$$V_{1s}^e(\mathbf{r})f(\mathbf{r})=\int\phi_{1s}^*(\mathbf{r}')\frac{1}{|\mathbf{r}-\mathbf{r}'|}f(\mathbf{r}')d\mathbf{r}'\phi_{1s}(\mathbf{r}),\quad(5b)$$

with similar definitions for $V_{2s}^d(\mathbf{r})$ and $V_{2s}^e(\mathbf{r})$. The quantities $\epsilon_{1s}$ and $\epsilon_{2s}$ introduced in Eqs. (4a) and (4b) are such that $\epsilon_{1s}=\epsilon_{1s\uparrow}=\epsilon_{1s\downarrow}$ and $\epsilon_{2s}=\epsilon_{2s\uparrow}=\epsilon_{2s\downarrow}$. The object $\phi_{1s}$ is the eigenfunction corresponding to the lowest possible eigenvalue of Eq. (4a), whereas $\phi_{2s}$ is the eigenfunction corresponding to the lowest eigenvalue of Eq. (4b) which also satisfies the requirement $(\phi_{1s}|\phi_{2s})=0$. For such equations, Roothaan *et al.*[10] have given very accurate solutions in a convenient analytic form. We will use their functions, both for Be and $B^+$, throughout this work. Table I lists the relevant expectation values, where we have used the following definitions:

TABLE I. The values of important Hartree-Fock matrix elements (in a.u.).

| | Be | $B^+$ |
|---|---|---|
| $(1s\|T\|1s)$ | −7.94206 | −12.44114 |
| $(2s\|T\|2s)$ | −1.58939 | − 2.73667 |
| $V_{1s1s}$ | +2.27293 | + 2.89618 |
| $V_{1s2s}$ | +0.48107 | + 0.70260 |
| $V_{2s2s}$ | +0.34346 | + 0.50380 |
| $V_{ex}$ | +0.02536 | + 0.04614 |
| $\epsilon_{1s}$ | −4.73235 | − 8.18590 |
| $\epsilon_{2s}$ | −0.30915 | − 0.87381 |

$$V_{1s1s}=\int|\phi_{1s}(\mathbf{r})|^2|\phi_{1s}(\mathbf{r}')|^2\frac{1}{|\mathbf{r}-\mathbf{r}'|}d\mathbf{r}d\mathbf{r}',\quad(6a)$$

$$V_{1s2s}=\int|\phi_{1s}(\mathbf{r})|^2|\phi_{2s}(\mathbf{r}')|^2\frac{1}{|\mathbf{r}-\mathbf{r}'|}d\mathbf{r}d\mathbf{r}',\quad(6b)$$

$$V_{2s2s}=\int|\phi_{2s}(\mathbf{r})|^2|\phi_{2s}(\mathbf{r}')|^2\frac{1}{|\mathbf{r}-\mathbf{r}'|}d\mathbf{r}d\mathbf{r}',\quad(6c)$$

$$V_{ex}=\int\phi_{1s}^*(\mathbf{r})\phi_{1s}^*(\mathbf{r}')\frac{1}{|\mathbf{r}-\mathbf{r}'|}\phi_{2s}(\mathbf{r})\phi_{2s}(\mathbf{r}')d\mathbf{r}d\mathbf{r}'.\quad(6d)$$

Using these definitions, we see from Eqs. (4a) and (4b) that

$$\epsilon_{1s}=\langle 1s|T|1s\rangle+V_{1s1s}+2V_{1s2s}-V_{ex},\quad(7a)$$

$$\epsilon_{2s}=\langle 2s|T|2s\rangle+V_{2s2s}+2V_{1s2s}-V_{ex}.\quad(7b)$$

With this notation established, we may write the total Hamiltonian for the system as

$$H=H_0+H_1,$$

where

$$H_0=H_{HF}(\mathbf{r}_1)+H_{HF}(\mathbf{r}_2)+H_{HF}(\mathbf{r}_3)+H_{HF}(\mathbf{r}_4),\quad(8a)$$

and

$$H_1=\sum_{i>j}\frac{1}{r_{ij}}-\mathcal{V}(\mathbf{r}_1)-\mathcal{V}(\mathbf{r}_2)-\mathcal{V}(\mathbf{r}_3)-\mathcal{V}(\mathbf{r}_4).\quad(8b)$$

The operator $H_{HF}(\mathbf{r})$ is the single-particle Hartree-Fock Hamiltonian

$$H_{HF}(\mathbf{r})=T(\mathbf{r})+\mathcal{V}(\mathbf{r}).$$

[10] C. C. J. Roothaan, L. M. Sachs, and A. W. Weiss, Rev. Mod. Phys. 32, 186 (1960).

The generalization to more than four electrons is obvious. The equation

$$H_0\psi_0=E_0\psi_0 \tag{9}$$

has as its lowest eigenvalue (in the Hilbert space of totally antisymmetric functions)

$$E_0=2\epsilon_{1s}+2\epsilon_{2s}, \tag{10}$$

with the corresponding normalized spinor eigenfunction

$$\psi_0(\mathbf{r}_1,\mathbf{r}_2,\mathbf{r}_3,\mathbf{r}_4)=[1/\sqrt{(24)}]\times\mathcal{A}[\phi_{1s\uparrow}(\mathbf{r}_1)\phi_{1s\downarrow}(\mathbf{r}_2)\phi_{2s\uparrow}(\mathbf{r}_3)\phi_{2s\downarrow}(\mathbf{r}_4)], \tag{11}$$

where $\mathcal{A}$ denotes the operation of antisymmetrization in all coordinates. The quantities $\psi_0$ and $E_0$ being known, we look for the solution to

$$H\Psi=E\Psi$$

in the form

$$\Psi=\sum_{n=0}^{\infty}\psi_n, \tag{12a}$$

$$E=\sum_{n=0}^{\infty}E_n. \tag{12b}$$

This leads to the equations

$$(H_0-E_0)\psi_1+(H_1-E_1)\psi_0=0, \tag{13a}$$

$$(H_0-E_0)\psi_n+H_1\psi_{n-1}-\sum_{m=1}^{n-1}E_m\psi_{n-m}-E_n\psi_0=0 \quad (n\geqslant 2), \tag{13b}$$

and, therefore, we have[11]

$$E_1=\langle\psi_0|H_1|\psi_0\rangle, \tag{14a}$$

$$E_2=\langle\psi_0|H_1-E_1|\psi_1\rangle=-\langle\psi_1|H_0-E_0|\psi_1\rangle, \tag{14b}$$

$$E_3=\langle\psi_1|H_1-E_1|\psi_1\rangle-2E_2\langle\psi_0|\psi_1\rangle, \tag{14c}$$

$$E_4=-\langle\psi_2|H_0-E_0|\psi_2\rangle-E_2\langle\psi_1|\psi_1\rangle-2E_3\langle\psi_0|\psi_1\rangle, \tag{14d}$$

$$E_5=\langle\psi_2|H_1-E_1|\psi_2\rangle-2E_2\langle\psi_1|\psi_2\rangle-E_3\langle\psi_1|\psi_1\rangle-2E_3\langle\psi_0|\psi_2\rangle-2E_4\langle\psi_0|\psi_1\rangle. \tag{14e}$$

The expression (14a) for $E_1$ is readily evaluated using Roothaan's functions for $\phi_{1s}$ and $\phi_{2s}$ in Eq. (11). We have

$$E_1=-V_{1s1s}-4V_{1s2s}-V_{2s2s}+2V_{\text{ex}}. \tag{15}$$

The sum $E_{\text{HF}}=E_0+E_1$ is what is usually referred to as the Hartree-Fock energy. For Be, $E_{\text{HF}}(\text{Be})=-14.5730$ a.u., while for B$^+$, $E_{\text{HF}}(\text{B}^+)=-24.2376$ a.u.

We now proceed to the calculation of $E_2$, given by Eq. (14b) and therefore we have to solve Eq. (13a) for $\psi_1$. In BJI,[11] Eq. (13a) was solved for He in the case where $\psi_0$ is the ground-state solution to the Hartree-Fock equation. It was also noted there that this equation could be derived from a variational principle[1] based on the expression

$$F_1[\psi_1{}^t]=\langle\psi_1{}^t|H_0-E_0|\psi_1{}^t\rangle+2\langle\psi_1{}^t|H_1-E_1|\psi_0\rangle, \tag{16}$$

where $\psi_1{}^t$ is a (real) trial function. Varying $F_1[\psi_1{}^t]$ with respect to $\psi_1{}^t$, one obtains Eq. (13a) for the function $\psi_1$, which makes $F_1[\psi_1{}^t]$ stationary. If we are dealing with the ground state of a system, which is the case in this paper, it is easy to show using the completeness of the eigenfunctions of $H_0$ that we actually have a minimum principle, i.e., $F_1[\psi_1{}^t]$ takes on its smallest possible value for $\psi_1{}^t=\psi_1$, where $\psi_1$ satisfies Eq. (13a). This smallest value is just the expression (14b) for $E_2$. Because of the form of Eq. (13a) it is clear that we can always pick $\psi_1$ orthogonal to $\psi_0$. Throughout this paper we will constrain $\psi_1$ to satisfy this orthogonality condition. With this convention, $E_2$ is given simply by

$$E_2=\langle\psi_0|H_1|\psi_1\rangle. \tag{17}$$

One would like to build a trial wave function $\psi_1{}^t$ with a number of variable parameters in it, compute $F_1[\psi_1{}^t]$, and then vary the parameters to get a minimum. In BJI, this was done for helium, and good results were obtained for $E_2$. It might be thought that helium is a poor precedent to cite in this paper, since the possibility of many-electron overlap integrals, of the type already encountered in the study of lithium by James and Coolidge,[12] would seem to make this a prohibitively difficult program to carry out for Be. However, we will show that in fact we can reduce the problem of solving Eq. (13a) to a decoupled collection of two-electron problems, so that the techniques of BJI can be employed.

Since electron-pair correlations are expected to dominate the problem, it seems reasonable to take our trial function to be of the following form:

$$\begin{aligned}\psi_1{}^t(\mathbf{r}_1,\mathbf{r}_2,\mathbf{r}_3,\mathbf{r}_4)=\mathcal{S}[&\chi_{1s\uparrow1s\downarrow}(\mathbf{r}_1,\mathbf{r}_2)\phi_{2s\uparrow}(\mathbf{r}_3)\phi_{2s\downarrow}(\mathbf{r}_4)\\&+\chi_{1s\uparrow2s\uparrow}(\mathbf{r}_1,\mathbf{r}_3)\phi_{1s\downarrow}(\mathbf{r}_2)\phi_{2s\downarrow}(\mathbf{r}_4)\\&+\chi_{1s\uparrow2s\downarrow}(\mathbf{r}_1,\mathbf{r}_4)\phi_{1s\downarrow}(\mathbf{r}_2)\phi_{2s\uparrow}(\mathbf{r}_3)\\&+\chi_{1s\downarrow2s\uparrow}(\mathbf{r}_2,\mathbf{r}_3)\phi_{1s\uparrow}(\mathbf{r}_1)\phi_{2s\downarrow}(\mathbf{r}_4)\\&+\chi_{1s\downarrow2s\downarrow}(\mathbf{r}_2,\mathbf{r}_4)\phi_{1s\uparrow}(\mathbf{r}_1)\phi_{2s\uparrow}(\mathbf{r}_3)\\&+\chi_{2s\uparrow2s\downarrow}(\mathbf{r}_3,\mathbf{r}_4)\phi_{1s\uparrow}(\mathbf{r}_1)\phi_{1s\downarrow}(\mathbf{r}_2)]/\sqrt{(12)},\end{aligned} \tag{18}$$

where the $\chi_{ij}$ are totally antisymmetric in the space and spin coordinates. The operator $\mathcal{S}$ denotes the linear combination of all 12 symmetric permutations on $\mathbf{r}_1$, $\mathbf{r}_2$, $\mathbf{r}_3$, and $\mathbf{r}_4$. The function $\psi_1{}^t(\mathbf{r}_1,\mathbf{r}_2,\mathbf{r}_3,\mathbf{r}_4)$ is then totally antisymmetric. In fact, what we are doing here is adding to the Hartree-Fock solution $\psi_0$ a function $\psi_1{}^t$ containing all possible terms in which electrons with quantum numbers $i$ and $j$ correlate through some function $\chi_{ij}$, while all the other electrons stay in their unperturbed Hartree-Fock orbitals.

---

[11] F. W. Byron, Jr., and C. J. Joachain, Phys. Rev. **146**, 1 (1966), to be referred hereafter as BJI. In this paper the following misprints should be corrected: in Eq. (8d), the first term should be $-\langle\chi_2|H_0-E_0|\chi_2\rangle$ and in Eq. (9b) the last term should read $-2E_2\langle\chi_2{}'|\varphi_0\rangle$.

[12] H. M. James and A. S. Coolidge, Phys. Rev. **49**, 688 (1936).

Now, let us consider the functions $\chi_{ij}$. Because of the operator $\mathcal{S}$, it is clear that $\chi_{ij}$ cannot contain any components of $\phi_k(k\neq i,j)$. Indeed, if it did, such a component would vanish upon applying $\mathcal{S}$, since $\chi_{ij}$ is antisymmetric. Thus, without loss of generality, we may take[13]

$$\langle\chi_{ij}(\mathbf{r},\mathbf{r}')|\phi_k(\mathbf{r})\rangle=\langle\chi_{ij}(\mathbf{r},\mathbf{r}')|\phi_k(\mathbf{r}')\rangle=0 \quad (19)$$

for $k\neq i$, $j$.

Finally, we anticipate here that $\chi_{ij}$ contains no components of $\phi_i$ or $\phi_j$. This will be shown at the end of this section. Thus, we conclude that the $\chi_{ij}$ contain no components of any of the filled Hartree-Fock orbitals. This is one way of looking at the success of the Hartree-Fock model of the atom. Precisely those intermediate states that one would expect to contribute predominantly to $\psi_1$ (and hence to $E_2$ and $E_3$) do not occur either because of the Pauli principle [see Eq. (19)] or because of the particular form of the Hartree-Fock potentials ($\phi_i$, $\phi_j$ orthogonal to $\chi_{ij}$), as we shall see below.

Knowing that $\chi_{ij}$ is orthogonal to all the filled single-particle Hartree-Fock orbitals, we can easily derive the basic equations for the $\chi_{ij}$. Returning to our expression for $F_1[\psi_1{}^t]$ [see Eq. (16)], we find for the first term[14]

$$\langle\psi_1{}^t|H_0-E_0|\psi_1{}^t\rangle=\sum_{i,j}{}'\langle\chi_{ij}(\mathbf{r},\mathbf{r}')|H_{\mathrm{HF}}(\mathbf{r})+H_{\mathrm{HF}}(\mathbf{r}')-\epsilon_i-\epsilon_j|\chi_{ij}(\mathbf{r},\mathbf{r}')\rangle, \quad (20)$$

where we have used the fact that $H_{\mathrm{HF}}$ contains only *one-body* operators. For example, if one tried to obtain such a result for the expectation value of the total Hamiltonian $H$ between two functions having the same form as our $\psi_1{}^t$, the presence of terms like $r_{12}{}^{-1}$ would prevent this decoupling from occurring (i.e., there are more than just pair-correlation effects in atoms).

The second term in $F_1[\psi_1{}^t]$ [see Eq. (16)] is $2\langle\psi_0|H_1-E_1|\psi_1{}^t\rangle$. Although this expression does contain terms of the type $r_{12}{}^{-1}$, $\psi_1{}^t$ occurs only on one side of the inner product. A simple calculation using Eq. (15) and the orthogonality properties of the $\chi_{ij}$ yields the result

$$\langle\psi_0|H_1-E_1|\psi_1{}^t\rangle=\sum_{i,j}{}'\langle\tfrac{1}{2}\sqrt{2}[\phi_i(\mathbf{r})\phi_j(\mathbf{r}')-\phi_i(\mathbf{r}')\phi_j(\mathbf{r})]|(1/|\mathbf{r}-\mathbf{r}'|)-V_i(\mathbf{r})-V_j(\mathbf{r})-V_i(\mathbf{r}')-V_j(\mathbf{r}')-\epsilon_{ij}{}^{(1)}|\chi_{ij}(\mathbf{r},\mathbf{r}')\rangle, \quad (21)$$

where we have defined

$$\epsilon_{1s\uparrow1s\downarrow}{}^{(1)}=-V_{1s1s}, \quad (22a)$$

$$\epsilon_{1s\uparrow2s\uparrow}{}^{(1)}=\epsilon_{1s\downarrow2s\downarrow}{}^{(1)}=-V_{1s2s}+V_{\mathrm{ex}}, \quad (22b)$$

$$\epsilon_{1s\uparrow2s\downarrow}{}^{(1)}=\epsilon_{1s\downarrow2s\uparrow}{}^{(1)}=-V_{1s2s}, \quad (22c)$$

$$\epsilon_{2s\uparrow2s\downarrow}{}^{(1)}=-V_{2s2s}. \quad (22d)$$

Note that

$$E_1=\sum_{i,j}{}'\epsilon_{ij}{}^{(1)}. \quad (22e)$$

Thus, collecting Eqs. (20) and (21), we obtain for $F_1[\psi_1{}^t]$

$$F_1[\psi_1{}^t]=\sum_{i,j}{}'\{\langle\chi_{ij}(\mathbf{r},\mathbf{r}')|H_{\mathrm{HF}}(\mathbf{r})+H_{\mathrm{HF}}(\mathbf{r}')-\epsilon_i-\epsilon_j|\chi_{ij}(\mathbf{r},\mathbf{r}')\rangle+2\langle\tfrac{1}{2}\sqrt{2}[\phi_i(\mathbf{r})\phi_j(\mathbf{r}')-\phi_i(\mathbf{r}')\phi_j(\mathbf{r})]|(1/|\mathbf{r}-\mathbf{r}'|)-V_i(\mathbf{r})-V_j(\mathbf{r})-V_i(\mathbf{r}')-V_j(\mathbf{r}')-\epsilon_{ij}{}^{(1)}|\chi_{ij}(\mathbf{r},\mathbf{r}')\rangle\}. \quad (23)$$

We now vary the $\chi_{ij}$ independently and obtain the basic equations for our problem:

$$[H_{\mathrm{HF}}(\mathbf{r})+H_{\mathrm{HF}}(\mathbf{r}')-\epsilon_i-\epsilon_j]\chi_{ij}(\mathbf{r},\mathbf{r}')=-[(1/|\mathbf{r}-\mathbf{r}'|)-V_i(\mathbf{r})-V_j(\mathbf{r})-V_i(\mathbf{r}')-V_j(\mathbf{r}')-\epsilon_{ij}{}^{(1)}]\tfrac{1}{2}\sqrt{2}\times[\phi_i(\mathbf{r})\phi_j(\mathbf{r}')-\phi_i(\mathbf{r}')\phi_j(\mathbf{r})]. \quad (24)$$

Thus, the problem of obtaining $\psi_1$ reduces to a matter of solving a collection of decoupled two-body problems of the type discussed in BJI. The value of $F_1[\psi_1{}^t]$ at the minimizing values of the $\chi_{ij}$ obtained from Eq. (24) is just

$$E_2=\sum_{i,j}{}'\epsilon_{ij}{}^{(2)}, \quad (25a)$$

where

$$\epsilon_{ij}{}^{(2)}=\langle\chi_{ij}(\mathbf{r},\mathbf{r}')|(1/|\mathbf{r}-\mathbf{r}'|)-V_i(\mathbf{r})-V_j(\mathbf{r})-V_i(\mathbf{r}')-V_j(\mathbf{r}')-\epsilon_{ij}{}^{(1)}|\tfrac{1}{2}\sqrt{2}[\phi_i(\mathbf{r})\phi_j(\mathbf{r}')-\phi_i(\mathbf{r}')\phi_j(\mathbf{r})]\rangle. \quad (25b)$$

Returning to our pair equation [Eq. (24)], we note that if we take the inner product of the right-hand side of this equation by $\phi_i(\mathbf{r})\phi_j(\mathbf{r}')$ [or by $\phi_i(\mathbf{r}')\phi_j(\mathbf{r})$] we get zero because of the definitions (1b) of the potentials $V_i$ and (22a)-(22d) for $\epsilon_{ij}{}^{(1)}$. Thus our equation is well posed since, due to the Hermiticity of $H_{\mathrm{HF}}$, the inner product of the left-hand side of Eq. (24) with $\phi_i(\mathbf{r})\phi_j(\mathbf{r}')$ [or $\phi_i(\mathbf{r}')\phi_j(\mathbf{r})$] also vanishes. Let us agree to choose the indeterminate inner product $\langle\phi_i\phi_j|\chi_{ij}\rangle$ to vanish. In solving the pair equations variationally, we will constrain our functions $\chi_{ij}$ to satisfy this condition. Now,

[13] Throughout this paper, we extend the definition of inner products to expressions of the form

$$\langle f(x,y)|g(y)\rangle\equiv\int f^*(x,y)g(y)dy,$$

i.e., such an "inner product" is actually a function of $x$.

[14] We use the notation $\sum_{ij}{}'$ to denote a sum on all *distinct pairs* of indices.

because of the special properties of the Hartree-Fock potentials, if we multiply the right-hand side of Eq. (24) by just, say, $\phi_i(\mathbf{r}')$ and integrate over all space, this integral is identically zero (integration implies, as usual, summation over the appropriate spin indices). Integrating the left-hand side of the equation after multiplying by $\phi_i(\mathbf{r}')$ and using the Hermiticity of $H_{\mathrm{HF}}$, we get the equation

$$[H_{\mathrm{HF}}(\mathbf{r})-\epsilon_j]\langle\phi_i(\mathbf{r}')\,|\,\chi_{ij}(\mathbf{r},\mathbf{r}')\rangle=0\,,$$

which tells us that

$$\langle\phi_i(\mathbf{r}')\,|\,\chi_{ij}(\mathbf{r},\mathbf{r}')\rangle=a\phi_j(\mathbf{r})\,,$$

where $a$ is some constant. Now, taking the inner product of both sides by $\phi_j(\mathbf{r})$, we get

$$\langle\phi_i(\mathbf{r}')\phi_j(\mathbf{r})\,|\,\chi_{ij}(\mathbf{r},\mathbf{r}')\rangle=a.$$

But we have also

$$\langle\phi_i(\mathbf{r}')\phi_j(\mathbf{r})\,|\,\chi_{ij}(\mathbf{r}',\mathbf{r})\rangle=-a$$

because $\chi_{ij}$ is antisymmetric. Since we have already decided to choose this inner product to be zero, we conclude that $a=0$, so that

$$\langle\phi_i(\mathbf{r}')\,|\,\chi_{ij}(\mathbf{r},\mathbf{r}')\rangle=0\,, \tag{26a}$$

and similarly,

$$\langle\phi_j(\mathbf{r}')\,|\,\chi_{ij}(\mathbf{r},\mathbf{r}')\rangle=0. \tag{26b}$$

Hence $\chi_{ij}$ contains no components of $\phi_i$ or $\phi_j$, as we claimed above. Note also that the orthogonality properties of the $\chi_{ij}$ guarantee that $\langle\psi_0|\psi_1\rangle=0$.

One might be tempted to object that although we have succeeded in obtaining a simple set of equations for $\psi_1$, we have not taken $\psi_1{}^t$ to be of the most general form possible, and therefore the true $E_2$ for the four-electron system might be lower than the $E_2$ which we would obtain from our Eq. (24) and Eqs. (25a) and (25b). However, if we substitute into the basic equation [Eq. (13a)] the expression (18) for $\psi_1{}^t$ and make use of the fact that the $\chi_{ij}$ satisfy Eq. (24), a simple calculation shows that the basic equation of first-order wave-function perturbation theory [Eq. (13a)] is actually satisfied exactly. Thus we have found the solution (unique if $\langle\psi_0|\psi_1\rangle=0$) for the first-order wave function.

In concluding this section we note that our above discussion is not at all restricted to four-electron systems, and Eq. (24) which we have derived for the $\chi_{ij}$ applies to atoms with any number of electrons. In Sec. III we point out in what ways the practical matter of actually solving the equations for the $\chi_{ij}$ is expedited by having a four-electron ground state (or more generally, an atom with all angular momentum shells filled).

## III. REDUCTION OF THE PAIR EQUATIONS

We now want to factor out the spin dependence from the pair equations. Let us begin by writing down the six pair equations for the four-electron problem. Using Eq. (24), we get

$$[H_{\mathrm{HF}}(\mathbf{r}_1)+H_{\mathrm{HF}}(\mathbf{r}_2)-2\epsilon_{1s}]\chi_{1s\uparrow1s\downarrow}(\mathbf{r}_1,\mathbf{r}_2)$$
$$=-[1/r_{12}-V_{1s\uparrow}(\mathbf{r}_1)-V_{1s\downarrow}(\mathbf{r}_1)-V_{1s\uparrow}(\mathbf{r}_2)-V_{1s\downarrow}(\mathbf{r}_2)-\epsilon_{1s\uparrow1s\downarrow}{}^{(1)}]\tfrac{1}{2}\sqrt{2}[\phi_{1s\uparrow}(\mathbf{r}_1)\phi_{1s\downarrow}(\mathbf{r}_2)-\phi_{1s\uparrow}(\mathbf{r}_2)\phi_{1s\downarrow}(\mathbf{r}_1)] \tag{27a}$$

$$[H_{\mathrm{HF}}(\mathbf{r}_1)+H_{\mathrm{HF}}(\mathbf{r}_2)-\epsilon_{1s}-\epsilon_{2s}]\chi_{1s\uparrow2s\downarrow}(\mathbf{r}_1,\mathbf{r}_2)$$
$$=-[1/r_{12}-V_{1s\uparrow}(\mathbf{r}_1)-V_{2s\downarrow}(\mathbf{r}_1)-V_{1s\uparrow}(\mathbf{r}_2)-V_{2s\downarrow}(\mathbf{r}_2)-\epsilon_{1s\uparrow2s\downarrow}{}^{(1)}]\tfrac{1}{2}\sqrt{2}[\phi_{1s\uparrow}(\mathbf{r}_1)\phi_{2s\downarrow}(\mathbf{r}_2)-\phi_{1s\uparrow}(\mathbf{r}_2)\phi_{2s\downarrow}(\mathbf{r}_1)] \tag{27b}$$

$$[H_{\mathrm{HF}}(\mathbf{r}_1)+H_{\mathrm{HF}}(\mathbf{r}_2)-\epsilon_{1s}-\epsilon_{2s}]\chi_{1s\uparrow2s\uparrow}(\mathbf{r}_1,\mathbf{r}_2)$$
$$=-[1/r_{12}-V_{1s\uparrow}(\mathbf{r}_1)-V_{2s\uparrow}(\mathbf{r}_1)-V_{1s\uparrow}(\mathbf{r}_2)-V_{2s\uparrow}(\mathbf{r}_2)-\epsilon_{1s\uparrow2s\uparrow}{}^{(1)}]\tfrac{1}{2}\sqrt{2}[\phi_{1s\uparrow}(\mathbf{r}_1)\phi_{2s\uparrow}(\mathbf{r}_2)-\phi_{1s\uparrow}(\mathbf{r}_2)\phi_{2s\uparrow}(\mathbf{r}_1)], \tag{27c}$$

$$[H_{\mathrm{HF}}(\mathbf{r}_1)+H_{\mathrm{HF}}(\mathbf{r}_2)-\epsilon_{1s}-\epsilon_{2s}]\chi_{1s\downarrow2s\uparrow}(\mathbf{r}_1,\mathbf{r}_2)$$
$$=-[1/r_{12}-V_{1s\downarrow}(\mathbf{r}_1)-V_{2s\uparrow}(\mathbf{r}_1)-V_{1s\downarrow}(\mathbf{r}_2)-V_{2s\uparrow}(\mathbf{r}_2)-\epsilon_{1s\downarrow2s\uparrow}{}^{(1)}]\tfrac{1}{2}\sqrt{2}[\phi_{1s\downarrow}(\mathbf{r}_1)\phi_{2s\uparrow}(\mathbf{r}_2)-\phi_{1s\downarrow}(\mathbf{r}_2)\phi_{2s\uparrow}(\mathbf{r}_1)], \tag{27d}$$

$$[H_{\mathrm{HF}}(\mathbf{r}_1)+H_{\mathrm{HF}}(\mathbf{r}_2)-\epsilon_{1s}-\epsilon_{2s}]\chi_{1s\downarrow2s\downarrow}(\mathbf{r}_1,\mathbf{r}_2)$$
$$=-[1/r_{12}-V_{1s\downarrow}(\mathbf{r}_1)-V_{2s\downarrow}(\mathbf{r}_1)-V_{1s\downarrow}(\mathbf{r}_2)-V_{2s\downarrow}(\mathbf{r}_2)-\epsilon_{1s\downarrow2s\downarrow}{}^{(1)}]\tfrac{1}{2}\sqrt{2}[\phi_{1s\downarrow}(\mathbf{r}_1)\phi_{2s\downarrow}(\mathbf{r}_2)-\phi_{1s\downarrow}(\mathbf{r}_2)\phi_{2s\downarrow}(\mathbf{r}_1)], \tag{27e}$$

$$[H_{\mathrm{HF}}(\mathbf{r}_1)+H_{\mathrm{HF}}(\mathbf{r}_2)-2\epsilon_{2s}]\chi_{2s\uparrow2s\downarrow}(\mathbf{r}_1,\mathbf{r}_2)$$
$$=-[1/r_{12}-V_{2s\uparrow}(\mathbf{r}_1)-V_{2s\downarrow}(\mathbf{r}_1)-V_{2s\uparrow}(\mathbf{r}_2)-V_{2s\downarrow}(\mathbf{r}_2)-\epsilon_{2s\uparrow2s\downarrow}{}^{(1)}]\tfrac{1}{2}\sqrt{2}[\phi_{2s\uparrow}(\mathbf{r}_1)\phi_{2s\downarrow}(\mathbf{r}_2)-\phi_{2s\uparrow}(\mathbf{r}_2)\phi_{2s\downarrow}(\mathbf{r}_1)]. \tag{27f}$$

Before proceeding further let us note that these rather messy equations have a very simple physical interpretation. Each equation is precisely what would arise if one performed first-order wave-function perturbation theory on a particular two-body reduced Hamiltonian. For example, Eq. (27a) is obtained from the reduced Hamiltonian $H_{\mathrm{red}}$ given by

$$H_{\mathrm{red}}=T(\mathbf{r}_1)+T(\mathbf{r}_2)+1/r_{12}+V_{2s\uparrow}(\mathbf{r}_1)+V_{2s\downarrow}(\mathbf{r}_1)+V_{2s\uparrow}(\mathbf{r}_2)+V_{2s\downarrow}(\mathbf{r}_2) \tag{28}$$

starting from a zero-order wave function given by

$$\phi_0=\tfrac{1}{2}\sqrt{2}[\phi_{1s\uparrow}(\mathbf{r}_1)\phi_{1s\downarrow}(\mathbf{r}_2)-\phi_{1s\uparrow}(\mathbf{r}_2)\phi_{1s\downarrow}(\mathbf{r}_1)].$$

The same reasoning applies to any pair. This is just the Hamiltonian for two particles moving in a central Coulomb potential of charge $Z$, interacting through the true electron-electron interaction, but seeing also an "average potential" (nonlocal because of exchange) due to the two $2s$ electrons. Unfortunately, because of the interelectronic interactions, this reduced two-body Hamiltonian is not valid to all orders in perturbation theory. If it were, one might envision solving the whole problem by tackling each two-body Hamiltonian of the type given in Eq. (28) by the Rayleigh-Ritz method, employed so successfully in helium by Hylleraas and others.[8,15] Even this program would not be without serious difficulties, however, since in trying to solve the 2s-2s pair problem one would be looking for a very highly excited state of the two-body Hamiltonian.

Returning to our problem of reducing Eqs. (27a)-(27f), we note that the functions $\chi_{ij}$ can be written in general as

$$\chi(\mathbf{r}_1,\mathbf{r}_2)=f(\mathbf{r}_1,\mathbf{r}_2)\Phi_s(1,2)+g(\mathbf{r}_1,\mathbf{r}_2)\Phi_t(1,2),$$

where $f$ and $g$ are functions of position alone, $f$ being a symmetric function and $g$ being antisymmetric. The objects $\Phi_s$ and $\Phi_t$ are spin-wave functions of total spin equal to zero and one, respectively. Now, for the ground state of the four-electron problem, the space parts of $\phi_{1s\uparrow}$ and $\phi_{1s\downarrow}$ are the same, as are the space parts of $\phi_{2s\uparrow}$ and $\phi_{2s\downarrow}$. For this reason, we find that

$$[H_{\mathrm{HF}}(\mathbf{r}_1)+H_{\mathrm{HF}}(\mathbf{r}_2)]\chi(\mathbf{r}_1,\mathbf{r}_2)=[\mathcal{O}(\mathbf{r}_1,\mathbf{r}_2)f(\mathbf{r}_1,\mathbf{r}_2)]\Phi_s(1,2)+[\mathcal{O}(\mathbf{r}_1,\mathbf{r}_2)g((\mathbf{r}_1,\mathbf{r}_2)]\Phi_t(1,2), \quad (29)$$

where $\mathcal{O}(\mathbf{r}_1,\mathbf{r}_2)$ is an operator, symmetric under the interchange of $\mathbf{r}_1$ and $\mathbf{r}_2$, which acts only on space variables;

$$\mathcal{O}(\mathbf{r}_1,\mathbf{r}_2)=T(\mathbf{r}_1)+T(\mathbf{r}_2)+2V_{1s}{}^d(\mathbf{r}_1)+2V_{2s}{}^d(\mathbf{r}_1)+2V_{1s}{}^d(\mathbf{r}_2)+2V_{2s}{}^d(\mathbf{r}_2)-V_{1s}{}^e(\mathbf{r}_1)-V_{2s}{}^e(\mathbf{r}_1)-V_{1s}{}^e(\mathbf{r}_2)-V_{2s}{}^e(\mathbf{r}_2). \quad (30)$$

With these results in hand, the reduction of the basic pair equations is straightforward. For the inner-shell correlation functions $\chi_{1s\uparrow 1s\downarrow}$ and $\chi_{2s\uparrow 2s\downarrow}$, the reduction is particularly simple, since the right-hand side of Eqs. (27a) and (27f) has a pure singlet spin dependence. Hence, we write

$$\chi_{1s\uparrow 1s\downarrow}(\mathbf{r}_1,\mathbf{r}_2)=f(\mathbf{r}_1,\mathbf{r}_2)\tfrac{1}{2}\sqrt{2}[\alpha(1)\beta(2)-\alpha(2)\beta(1)], \quad (31a)$$

$$\chi_{2s\uparrow 2s\downarrow}(\mathbf{r}_1,\mathbf{r}_2)=g(\mathbf{r}_1,\mathbf{r}_2)\tfrac{1}{2}\sqrt{2}[\alpha(1)\beta(2)-\alpha(2)\beta(1)]. \quad (31b)$$

Then from Eqs. (27a) and (27f) we obtain

$$[\mathcal{O}(\mathbf{r}_1,\mathbf{r}_2)-2\epsilon_{1s}]f(\mathbf{r}_1,\mathbf{r}_2)=-[1/r_{12}-V_{1s}{}^d(\mathbf{r}_1)-V_{1s}{}^d(\mathbf{r}_2)-\epsilon_{1s\uparrow 1s\downarrow}{}^{(1)}]\phi_{1s}(\mathbf{r}_1)\phi_{1s}(\mathbf{r}_2), \quad (32)$$

$$[\mathcal{O}(\mathbf{r}_1,\mathbf{r}_2)-2\epsilon_{2s}]g(\mathbf{r}_1,\mathbf{r}_2)=-[1/r_{12}-V_{2s}{}^d(\mathbf{r}_1)-V_{2s}{}^d(\mathbf{r}_2)-\epsilon_{2s\uparrow 2s\downarrow}{}^{(1)}]\phi_{2s}(\mathbf{r}_1)\phi_{2s}(\mathbf{r}_2). \quad (33)$$

Remember that in solving these equations we must constrain $f(\mathbf{r}_1,\mathbf{r}_2)$ to contain no components of $\phi_{2s}$, and $g(\mathbf{r}_1,\mathbf{r}_2)$ must be constrained to contain no components of $\phi_{1s}$. In addition, if we require that

$$\langle\phi_{1s}\phi_{1s}|f\rangle=\langle\phi_{2s}\phi_{2s}|g\rangle=0, \quad (34)$$

then, automatically, $f(\mathbf{r}_1,\mathbf{r}_2)$ will contain no components of $\phi_{1s}$ and $g(\mathbf{r}_1,\mathbf{r}_2)$ will contain no components of $\phi_{2s}$.

In the case of the inter-shell correlation functions, we first note that apart from reversal of spin, Eqs. (27b) and (27d) are equivalent; i.e., we can use the same space function for both equations. The same remark applies to Eqs. (27c) and (27e). In what follows, we will discuss only Eqs. (27b) and (27c). For Eq. (27c), the right-hand side is a pure triplet spin function, so we can write

$$\chi_{1s\uparrow 2s\uparrow}(\mathbf{r}_1,\mathbf{r}_2)=h_{1,1}(\mathbf{r}_1,\mathbf{r}_2)\alpha(1)\alpha(2), \quad (35a)$$

and similarly, from Eq. (27e),

$$\chi_{1s\downarrow 2s\downarrow}(\mathbf{r}_1,\mathbf{r}_2)=h_{1,1}(\mathbf{r}_1,\mathbf{r}_2)\beta(1)\beta(2), \quad (35b)$$

where the subscripts in the function $h_{1,1}$ indicate that we consider a triplet state with total spin $S=1$ and $M_s=\pm1$. A simple calculation yields for $h_{1,1}$ the equation

$$[\mathcal{O}(\mathbf{r}_1,\mathbf{r}_2)-\epsilon_{1s}-\epsilon_{2s}]h_{1,1}(\mathbf{r}_1,\mathbf{r}_2)=-[1/r_{12}-V_{1s}{}^d(\mathbf{r}_1)-V_{2s}{}^d(\mathbf{r}_1)-V_{1s}{}^d(\mathbf{r}_2)-V_{2s}{}^d(\mathbf{r}_2)+V_{1s}{}^e(\mathbf{r}_1)+V_{2s}{}^e(\mathbf{r}_1)+V_{1s}{}^e(\mathbf{r}_2)+V_{2s}{}^e(\mathbf{r}_2)-\epsilon_{1s\uparrow 2s\uparrow}{}^{(1)}]\tfrac{1}{2}\sqrt{2}\times[\phi_{1s}(\mathbf{r}_1)\phi_{2s}(\mathbf{r}_2)-\phi_{1s}(\mathbf{r}_2)\phi_{2s}(\mathbf{r}_1)]. \quad (36)$$

Clearly, because of (35a) and (35b) we have

$$\epsilon_{1s\uparrow 2s\uparrow}{}^{(2)}=\epsilon_{1s\downarrow 2s\downarrow}{}^{(2)}.$$

Finally, we consider Eq. (27b) for $\chi_{1s\uparrow 2s\downarrow}$. In this case, the right-hand side of Eq. (27b) has a combination of singlet ($S=0$, $M_s=0$) and triplet ($S=1$, $M_s=0$) spin dependence, so that we write

$$\chi_{1s\uparrow 2s\downarrow}(\mathbf{r}_1,\mathbf{r}_2)=h_{1,0}(\mathbf{r}_1,\mathbf{r}_2)\tfrac{1}{2}\sqrt{2}[\alpha(1)\beta(2)+\alpha(2)\beta(1)]+h_{0,0}(\mathbf{r}_1,\mathbf{r}_2)\tfrac{1}{2}\sqrt{2}[\alpha(1)\beta(2)-\alpha(2)\beta(1)], \quad (37a)$$

and similarly from Eq. (27d)

$$\chi_{1s\downarrow 2s\uparrow}(\mathbf{r}_1,\mathbf{r}_2)=h_{1,0}(\mathbf{r}_1,\mathbf{r}_2)\tfrac{1}{2}\sqrt{2}[\beta(1)\alpha(2)+\beta(2)\alpha(1)]+h_{0,0}(\mathbf{r}_1,\mathbf{r}_2)\tfrac{1}{2}\sqrt{2}[\beta(1)\alpha(2)-\beta(2)\alpha(1)], \quad (37b)$$

where $h_{1,0}$ is an antisymmetric function and $h_{0,0}$ a symmetric one. A straightforward computation yields

$$[\mathcal{O}(\mathbf{r}_1,\mathbf{r}_2)-\epsilon_{1s}-\epsilon_{2s}]h_{1,0}(\mathbf{r}_1,\mathbf{r}_2)=-\tfrac{1}{2}[1/r_{12}-\epsilon_{1s\uparrow 2s\downarrow}{}^{(1)}]\times[\phi_{1s}(\mathbf{r}_1)\phi_{2s}(\mathbf{r}_2)-\phi_{1s}(\mathbf{r}_2)\phi_{2s}(\mathbf{r}_1)]+\tfrac{1}{2}[V_{2s}{}^d(\mathbf{r}_1)+V_{1s}{}^d(\mathbf{r}_2)]\phi_{1s}(\mathbf{r}_1)\phi_{2s}(\mathbf{r}_2)-\tfrac{1}{2}[V_{1s}{}^d(\mathbf{r}_1)+V_{2s}{}^d(\mathbf{r}_2)]\phi_{1s}(\mathbf{r}_2)\phi_{2s}(\mathbf{r}_1), \quad (38)$$

and

$$[\mathcal{O}(\mathbf{r}_1,\mathbf{r}_2)-\epsilon_{1s}-\epsilon_{2s}]h_{0,0}(\mathbf{r}_1,\mathbf{r}_2)=-\tfrac{1}{2}[1/r_{12}-\epsilon_{1s\uparrow 2s\downarrow}{}^{(1)}]\times[\phi_{1s}(\mathbf{r}_1)\phi_{2s}(\mathbf{r}_2)+\phi_{1s}(\mathbf{r}_2)\phi_{2s}(\mathbf{r}_1)]+\tfrac{1}{2}[V_{2s}{}^d(\mathbf{r}_1)+V_{1s}{}^d(\mathbf{r}_2)]\phi_{1s}(\mathbf{r}_1)\phi_{2s}(\mathbf{r}_2)+\tfrac{1}{2}[V_{1s}{}^d(\mathbf{r}_1)+V_{2s}{}^d(\mathbf{r}_2)]\phi_{1s}(\mathbf{r}_2)\phi_{2s}(\mathbf{r}_1). \quad (39)$$

It is apparent from Eqs. (37a) and (37b) that $\epsilon_{1s\uparrow 2s\downarrow}{}^{(2)}=\epsilon_{1s\downarrow 2s\uparrow}{}^{(2)}$. Apart from a statistical over-all factor $2^{-1/2}$, the difference between Eqs. (36) and (38) is a matter of constraints. We note that Eq. (36) needs no constraints because of the spin factors which multiply

[15] E. A. Hylleraas, Z. Physik 54, 347 (1929).

$h_{1,1}$ in Eq. (35a), i.e., $\chi_{1s\uparrow 2s\uparrow}$ is automatically orthogonal to $\phi_{1s\downarrow}$ and $\phi_{2s\downarrow}$ because of spin considerations. This is further reflected in Eq. (36), where the form of the right-hand side guarantees that $h_{1,1}$ will contain no components of $\phi_{1s}$ or $\phi_{2s}$. On the contrary, in solving Eq. (38) we must constrain $h_{1,0}$ to contain no components of $\phi_{1s}$ or $\phi_{2s}$ (or, alternatively, we must constrain $\chi_{1s\uparrow 2s\downarrow}$ to contain no components of $\phi_{1s\downarrow}$ or $\phi_{2s\uparrow}$) even though the form of Eq. (27c) guarantees that $\chi_{1s\uparrow 2s\downarrow}$ will contain no components of $\phi_{1s\uparrow}$ or $\phi_{2s\downarrow}$. Once the appropriate constraints are applied, it can be shown that Eqs. (36) and (38) turn out to be the same. Therefore, we can omit Eq. (38) in what follows, since on the subspace of the two-particle Hilbert space on which $h_{1,0}$ is different from zero it is equal to a constant multiple of $h_{1,1}$. Finally, in solving Eq. (39), we have to constrain $h_{0,0}$ to contain no components of $\phi_{1s}$ or $\phi_{2s}$.

Thus, we see that the problem comes down to solving Eqs. (32), (33), (36), and (39), the relevant constraints discussed above being taken into account. We now proceed to the solution of these equations.

## IV. SOLUTION OF THE PAIR EQUATIONS

We propose to carry out the solution of the reduced Eqs. (32), (33), (36), and (39) by the variational method discussed in BJI. Each of these equations has the form

$$\mathcal{P}\rho = -Q\phi, \tag{40}$$

where $\mathcal{P}$ and $Q$ are given Hermitian operators, $\rho$ is any of the unknown functions $f$, $g$, $h_{1,1}$, $h_{1,0}$, or $h_{0,0}$ defined in Sec. III, and $\phi$ is a given function. Equation (40) can be derived from the variational expression

$$\epsilon^{(2)}[\rho^t] = \langle\rho^t|\mathcal{P}|\rho^t\rangle + 2\langle\rho^t|Q|\phi\rangle, \tag{41}$$

where we have used the notation $\epsilon^{(2)}[\rho^t]$ because of the fact that each of the variational expressions from which are derived Eqs. (32), (33), (36), (38), and (39) can actually be shown to be a *minimum principle* for the appropriate $\epsilon_{ij}^{(2)}$ referred to in Eq. (25b). We emphasize that the existence of a *minimum* principle for all of the $\epsilon_{ij}^{(2)}$ is directly linked to the constraints imposed upon the functions $f$, $g$, $h_{1,1}$, $h_{1,0}$, and $h_{0,0}$, because the constraints project out of the relevant first-order trial function all components of those states which have lower-zero-order energy than the state in question.

Before proceeding to the solutions it is necessary to decide on a good choice of trial function for $\rho^t$. In BJI, we used for helium the Hylleraas-type function

$$\rho(s,t,u) = \frac{1}{4\pi} e^{-\frac{1}{2}ks} \sum_{l,m,n} C_{lmn} s^l t^m u^n, \tag{42}$$

where $s=r_1+r_2$, $t=r_1-r_2$, $u=r_{12}$, and $k$ is a scale parameter roughly equal to twice the effective charge seen by the two electrons. It is well known from the work of Hylleraas and others[8,15] in the two-electron systems that this set of trial functions is a very good one, and in BJI we found that with a small number of terms in Eq. (42) we could get within 0.5% of the helium correlation energy. However, these functions suffer from one grave defect when applied to a Hartree-Fock–type problem: Because of the existence of the nonlocal exchange potential there occur certain integrals[16] in the variational expression which cannot be done in closed form. This difficulty arises because of the use of the variable $u=r_{12}$ in Eq. (42). As one includes higher and higher powers of $u$, these intergrals require progressively more complicated infinite sums in their evaluation. Now, in BJI it was found that with only 10 parameters in Eq. (42), i.e., only terms linear and quadratic in $u$, a very good value for the correlation energy could be obtained. Clearly, we expect the same to be the case in our four-electron system for the calculation of the $1s$-$1s$ correlation energy, which should be particularly amenable to the treatment of BJI. Indeed, the effective central charge seen by the inner two $1s$ electrons is close to $Z=4$, as opposed to just $Z=2$ in helium. In fact, we employed this method in calculating the $1s$-$1s$ correlation energy and achieved satisfactory results. However, the $2s$-$2s$ correlation energy poses a more serious problem. Here we are dealing with a rather loosely bound *excited* two-body state, and we may anticipate that a wave function with only 10 parameters, even of the Hylleraas type, might not be able to represent adequately such a diffuse structure. In actual calculations of the $2s$-$2s$ correlation energy, just such difficulties were encountered.

At the opposite pole from the Hylleraas-type function is a trial function which is of the "configuration-mixing" type:

$$\rho_\pm(r_1,r_2,\cos\theta_{12}) = \frac{1}{4\pi}\sum_{l,m,n} C_{lmn}(r_1^m r_2^n + r_1^n r_2^m) \times (e^{-\frac{1}{2}\alpha r_1} e^{-\frac{1}{2}\beta r_2} \pm e^{-\frac{1}{2}\alpha r_2} e^{-\frac{1}{2}\beta r_1}) P_l(\cos\theta_{12}), \tag{43}$$

where the choice of the plus sign or of the minus sign gives us a symmetric or antisymmetric function of the space variables. This function, too, has several drawbacks. Calculations by Watson[2] and Weiss[3] utilizing the configuration-mixing method have been able at best to get within 7% of the Be correlation energy. Schwartz[17] has pointed out that one pays dearly for the omission of terms in $r_{12}$, which is in some sense the optimal way of including the angular part of the correlation function. It turns out that the expansion in Legendre polynomials is rather slowly convergent, and the difficulty of getting accurate values for the contribution of the higher terms in $P_l(\cos\rho_{12})$ grows rapidly as $l$ increases.

On the other hand, these functions have some attractive features. All the integrals in the variational princi-

[16] These are the integrals referred to as $g(l,m,n,\alpha,\beta,\gamma,a,b)$ with $a$ and $b$ odd in Appendix III of BJI (Ref. 11).

[17] C. Schwartz, Phys. Rev. **126**, 1015 (1962).

ple are easy to perform, and the constraint requirements on $\rho$ can be seen to be purely on the function of $r_1$ and $r_2$ multiplying the relative $s$-wave ($l=0$) term in $\rho$. Perhaps more important from the computational point of view is the fact that the problem for a particular $\rho$ decouples into a series of relative partial-wave problems, i.e., we can write

$$\epsilon_{ij}^{(2)}=\sum_{l=0}^{\infty}\epsilon_{ij}^{(2)}(l)\,,$$

and obtain trivially minimum principles for each $\epsilon_{ij}^{(2)}(l)$. Also it seems reasonable to suppose that functions of the type (43) might do very nicely in trying to calculate *inter-shell* correlations, since the difficulty in helium, as noted by Schwartz,[17] comes primarily from the problem of representing adequately the cusp behavior of the correlation function at $r_1=r_2$. For inter-shell correlations, we expect the region $r_1=r_2$ to be considerably less important than it would be for inner-shell correlations, where the pair of electrons overlap maximally. We have verified that this is the case in helium, where we have investigated successfully in this way the perturbation theory of the $(1s2s)\ {}^1S_0$ and the $(1s2s)\ {}^3S_1$ states.[18] Therefore, in the case of the four-electron system, we have used trial functions of the type (43) for the inter-shell correlation functions $h_{0,0}(\mathbf{r}_1,\mathbf{r}_2)$ and $h_{1,1}(\mathbf{r}_1,\mathbf{r}_2)$.

Going back to the *inner-shell* correlations, we note that there exists a type of function which retains much of the simplicity of the configuration-mixing type (43), but nevertheless manages to give some cusplike behavior at $r_1=r_2$, namely,

$$\rho(r_>,r_<,\cos\theta_{12})=\frac{1}{4\pi}\sum_{l,m,n}C_{lmn}r_>^{\,m}r_<^{\,n}e^{-(\alpha/2)r_>}e^{-(\beta/2)r_<}\times P_l(\cos\theta_{12})\,,\quad(44)$$

where $r_>$ denotes the greater and $r_<$ the lesser of $r_1$ and $r_2$. Using functions of this type in helium, we were able to get extremely accurate results for the ground-state correlation energy.[18] Therefore, we used the $(r_>r_<)$ basis to calculate the inner-shell correlation functions $f(\mathbf{r}_1,\mathbf{r}_2)$ and $g(\mathbf{r}_1,\mathbf{r}_2)$ in the four-electron system.

To summarize, let us write out our trial functions for the four different cases:

$$f(\mathbf{r}_1,\mathbf{r}_2)=\sum_l F^{(l)}(r_1,r_2)P_l(\cos\theta_{12})\,,\quad(45\text{a})$$

$$g(\mathbf{r}_1,\mathbf{r}_2)=\sum_l G^{(l)}(r_1,r_2)P_l(\cos\theta_{12})\,,\quad(45\text{b})$$

$$h_{1,1}(\mathbf{r}_1,\mathbf{r}_2)=\sum_l H_{1,1}^{(l)}(r_1,r_2)P_l(\cos\theta_{12})\,,\quad(45\text{c})$$

$$h_{0,0}(\mathbf{r}_1,\mathbf{r}_2)=\sum_l H_{0,0}^{(l)}(r_1,r_2)P_l(\cos\theta_{12})\,,\quad(45\text{d})$$

where

$$F^{(l)}(r_1,r_2)=\sum_{m,n}A_{mn}^{(l)}r_>^{\,m}r_<^{\,n}e^{-\frac{1}{2}\alpha r_>}e^{-\frac{1}{2}\beta r_<}\,,\quad(46\text{a})$$

$$G^{(l)}(r_1,r_2)=\sum_{m,n}B_{mn}^{(l)}r_>^{\,m}r_<^{\,n}e^{-\frac{1}{2}\alpha' r_>}e^{-\beta' r_<}\,,\quad(46\text{b})$$

$$H_{1,1}^{(l)}(r_1,r_2)=\sum_{m,n}C_{mn}^{(l)}(r_1^{\,m}r_2^{\,n}+r_1^{\,n}r_2^{\,m})\times(e^{-\frac{1}{2}ar_1}e^{-\frac{1}{2}br_2}-e^{-\frac{1}{2}ar_2}e^{-\frac{1}{2}br_1})\,,\quad(46\text{c})$$

$$H_{0,0}^{(l)}(r_1,r_2)=\sum_{m,n}D_{mn}^{(l)}(r_1^{\,m}r_2^{\,n}+r_1^{\,n}r_2^{\,m})\times(e^{-\frac{1}{2}a'r_1}e^{-\frac{1}{2}b'r_2}+e^{-\frac{1}{2}a'r_2}e^{-\frac{1}{2}b'r_1})\,.\quad(46\text{d})$$

As we remarked before, the constraints all refer to the form of the relative $s$-wave ($l=0$) correlation functions. The simplest constraints, i.e.,

$$\langle\phi_{1s}\phi_{1s}|F^{(0)}\rangle=0\,,\quad(47\text{a})$$

$$\langle\phi_{2s}\phi_{2s}|G^{(0)}\rangle=0\,,\quad(47\text{b})$$

$$\langle\phi_{1s}\phi_{2s}|H_{1,1}^{(0)}\rangle=0\,,\quad(47\text{c})$$

$$\langle\phi_{1s}\phi_{2s}|H_{0,0}^{(0)}\rangle=0\,,\quad(47\text{d})$$

are easily disposed of by the use of Lagrange multipliers. However, for $F^{(0)}$, we must also require that our trial function contain no components of $\phi_{2s}$. Thus, instead of taking $F^{(0)}$ as trial function in the $l=0$ partial wave, we shall use

$$\tilde{F}^{(0)}(r_1,r_2)=F^{(0)}(r_1,r_2)-\phi_{2s}(r_1)\langle\phi_{2s}(r_3)|F^{(0)}(r_3,r_2)\rangle-\phi_{2s}(r_2)\langle\phi_{2s}(r_3)|F^{(0)}(r_1,r_3)\rangle+\phi_{2s}(r_1)\phi_{2s}(r_2)\langle\phi_{2s}\phi_{2s}|F^{(0)}\rangle\,.\quad(48\text{a})$$

Similarly, instead of $G^{(0)}$, we use

$$\tilde{G}^{(0)}(r_1,r_2)=G^{(0)}(r_1,r_2)-\phi_{1s}(r_1)\langle\phi_{1s}(r_3)|G^{(0)}(r_3,r_2)\rangle-\phi_{1s}(r_2)\langle\phi_{1s}(r_3)|G^{(0)}(r_1,r_3)\rangle+\phi_{1s}(r_1)\phi_{1s}(r_2)\langle\phi_{1s}\phi_{1s}|G^{(0)}\rangle\,.\quad(48\text{b})$$

For the inter-shell correlation functions, we have a more complicated expression to guarantee that no components of either $\phi_{1s}$ or $\phi_{2s}$ are present in $H_{0,0}^{(0)}$. Thus we define

$$\begin{aligned}\tilde{H}_{0,0}^{(0)}(r_1,r_2)=H_{0,0}^{(0)}(r_1,r_2)&-\phi_{1s}(r_1)\langle\phi_{1s}(r_3)|H_{0,0}^{(0)}(r_3,r_2)\rangle-\phi_{1s}(r_2)\langle\phi_{1s}(r_3)|H_{0,0}^{(0)}(r_1,r_3)\rangle+\phi_{1s}(r_1)\phi_{1s}(r_2)\\&\times\langle\phi_{1s}\phi_{1s}|H_{0,0}^{(0)}\rangle-\phi_{2s}(r_1)\langle\phi_{2s}(r_3)|H_{0,0}^{(0)}(r_3,r_2)\rangle-\phi_{2s}(r_2)\langle\phi_{2s}(r_3)|H_{0,0}^{(0)}(r_1,r_3)\rangle\\&+\phi_{2s}(r_1)\phi_{2s}(r_2)\langle\phi_{2s}\phi_{2s}|H_{0,0}^{(0)}\rangle\,,\end{aligned}\quad(48\text{c})$$

[18] F. W. Byron, Jr., and C. J. Joachain, preceding paper, Phys. Rev. **157**, 1 (1967), to be referred to as BJII.

where we have omitted two terms in $\phi_{1s}(r_1)\phi_{2s}(r_2)$ and $\phi_{1s}(r_2)\phi_{2s}(r_1)$ which are unnecessary if $H_{0,0}^{(0)}$ already satisfies Eq. (47d).

With this in hand, we substitute our various trial functions into the variational expression for $E_2$, Eq. (16), compute the inner products, and vary the parameters to obtain a minimum. As shown above, the problem breaks up into separate variational problems for the $\epsilon_{ij}^{(2)}$, which in turn split up into still smaller variational problems for the $\epsilon_{ij}^{(2)}(l)$, the energies contributed by each relative partial wave. In practice, we determined reasonable values of the nonlinear parameters $(\alpha,\beta,\alpha',\beta',a,b,a',b')$ in Eqs. (46a)–(46d) by physical considerations connected with effective charges. Thus, our variational expression is a quadratic form in the linear parameters $A_{mn}^{(l)}$, $B_{mn}^{(l)}$, $C_{mn}^{(l)}$, and $D_{mn}^{(l)}$ of Eqs. (46a)–(46d), so that, as in BJI our problem reduces to one of matrix inversion. With the parameters determined in such a way that they minimize the variational expression $F_1[\psi_1^t]$ [Eq. (16)], it is a simple matter to calculate the various $\epsilon_{ij}^{(2)}$ from Eq. (25b). All the integrals appearing in the calculation can be written in terms of finite sums of the basic integrals $V(l,m,\alpha,\beta)$ and $W(l,m,n,\alpha,\beta,\gamma)$ discussed in BJI. These integrals can in turn be written as finite sums of factorials and powers or, more conveniently, may be evaluated by the use of recursion relations.[12] The most complicated expressions encountered are those involved in computing the matrix elements of the exchange potentials $V_i^e$ [see Eq. (5b)]. These actually involve significant computational labor in each relative partial wave.

## V. THE SECOND-ORDER ENERGY

In this section we want to discuss the contributions of the various pairs to $E_2$.

### A. 1s-1s Correlations

For the 1s-1s case, we used 21 terms in Eq. (46a) for each relative partial wave; i.e., we allowed $m+n$ to take on all possible positive values up to $m+n=5$. We evaluated explicitly all terms through $l=6$ in the relative partial-wave expansion (45a). The contribution of the higher partial waves to $\epsilon_{1s\uparrow 1s\downarrow}^{(2)}$ was estimated by extrapolation, using the fact that it drops off like $l^{-4}$ for large $l$.[17] This asymptotic behavior seems to set in quite rapidly after $l=4$. The results are shown in Table II. The most interesting feature concerns the relative $s$-wave contribution $\epsilon_{1s\uparrow 1s\downarrow}^{(2)}(0)$. In doing perturbation calculations for the ground state of helium starting from a hydrogenic Hamiltonian, one finds a very large relative $s$-wave contribution,[18] but here, because of the Hartree-Fock potentials the size of the relative $s$-wave contribution is reduced until it is actually significantly smaller than the relative $p$-wave contribution. Beyond the relative $p$ wave, the terms $\epsilon_{1s\uparrow 1s\downarrow}^{(2)}(l)$ drop off quite rapidly. The same phenomenon occurs in doing Hartree-Fock perturbation theory in helium and is discussed in BJII.

TABLE II. The contribution of each relative partial wave to the $1s\uparrow 1s\downarrow$ correlation energy in second order (in a.u.).

| $l$ | $\epsilon_{1s\uparrow 1s\downarrow}^{(2)}$ Be | $B^+$ |
|---|---|---|
| 0 | −0.01236 | −0.01197 |
| 1 | −0.02242 | −0.02317 |
| 2 | −0.00352 | −0.00360 |
| 3 | −0.00099 | −0.00101 |
| 4 | −0.00038 | −0.00038 |
| 5 | −0.00017 | −0.00017 |
| 6 | −0.00009 | −0.00009 |
| ⩾7 | −0.00015 | −0.00016 |
| Total | −0.04008 | −0.04055 |

Another interesting point about the relative $s$-wave contribution is how it is effected by the constraint which prevents any components of $\phi_{2s}$ from appearing in $\tilde{F}^{(0)}(r_1,r_2)$ [Eq. (48a)]. This constraint must clearly increase $\epsilon_{1s\uparrow 1s\downarrow}^{(2)}(0)$, as may easily be seen by writing

$$\epsilon_{1s\uparrow 1s\downarrow}^{(2)}(0)=\sum_{m\neq 1s,n\neq 1s}\frac{|\langle\phi_{1s}(r_1)\phi_{1s}(r_2)|1/r_{12}|\tfrac{1}{2}\sqrt{2}[\phi_m(r_1)\phi_n(r_2)+\phi_m(r_2)\phi_n(r_1)]\rangle|^2}{2\epsilon_{1s}-\epsilon_m-\epsilon_n}, \tag{49}$$

where the sum runs only over terms in which both $m$ and $n$ refer to $s$ states. If we apply the constraint that no $\phi_{2s}$ components appear in the 1s-1s correlation function, then in the above sum we must exclude all states in which *either* $m$ or $n$ is equal to $2s$. Since $\epsilon_{1s}$ is the lowest single-particle energy, we see that the removal of $\phi_{2s}$ components will remove terms from Eq. (49) with negative energy denominators, thus increasing $\epsilon_{1s\uparrow 1s\downarrow}^{(2)}(0)$. The restriction $m\neq 1s$, $n\neq 1s$ is guaranteed by the form of Eq. (32) for $f(\mathbf{r}_1,\mathbf{r}_2)$. This essentially is the reason that $\epsilon_{1s\uparrow 1s\downarrow}^{(2)}(0)$ is so much larger in hydrogenic than in Hartree-Fock perturbation theory: In first order, the potentials in the Hartree-Fock case act as projection operators to annihilate any component of $\phi_{1s}$. Similar remarks apply, *mutatis mutandis*, to the functions $g$, $h_{1,1}$, and $h_{0,0}$.

If we neglect the constraint imposed in Eq. (48a) and just compute $\epsilon_{1s\uparrow 1s\downarrow}^{(2)}(0)$, including the $\phi_{2s}$ components, we find for Be, a value which is 0.00152 a.u. lower than the number given in Table II, whereas for $B^+$, the value is 0.00204 a.u. lower. This is in accordance with our above discussion. It turns out that this is essentially the only difference between the 1s-1s correlation energy in Be and in $Be^{++}$ or in $B^+$ and in $B^{3+}$. Thus, the outer two electrons in Be and $B^+$ make themselves felt in the $1s$-$1s$ correlation energy *only through the exclusion principle*, not through the electrostatic forces which they exert on the $1s$ electrons.

### B. Inter-shell Correlations

To evaluate the inter-shell correlations, we used 20 parameters in Eqs. (46c) and (46d) for $H_{1,1}^{(l)}$ and $H_{0,0}^{(l)}$, thereby including all terms in the sum with $m+n\leq 7$. Partial-wave contributions through $l=3$ were calculated, and the remaining terms were estimated by extrapolation. We expect the values of $\epsilon_{1s\uparrow 2s\uparrow}^{(2)}$ and $\epsilon_{1s\uparrow 2s\downarrow}^{(2)}$ to be small because of the small overlap between the $1s$-shell and $2s$-shell wave functions. Tables III and IV show the results for these two cases. For the relative $s$-wave contribution to $\epsilon_{1s\uparrow 2s\uparrow}^{(2)}$, the choice of the Hartree-Fock potential, plus the fact that the space function $h_{1,1}(\mathbf{r}_1,\mathbf{r}_2)$, being antisymmetric, already vanishes at $r_1=r_2$, serves to reduce the $s$-wave contribution to insignificance. The fact that the Hartree-Fock function does not vanish properly when $\mathbf{r}_i=\mathbf{r}_j$ can be thought of as giving rise to the correlation effects which we are studying. Of course, the singularity of the Coulomb potential at $\mathbf{r}_i=\mathbf{r}_j$ means that two electrons cannot come arbitrarily close together. For pairs in which both electrons have the same $z$ component of spin, the exclusion principle forces the pair function to be zero at $\mathbf{r}_i=\mathbf{r}_j$, which works in the direction required by the Coulomb force. For this reason, we expect $\epsilon_{1s\uparrow 2s\uparrow}^{(2)}$ to be smaller than $\epsilon_{1s\uparrow 2s\downarrow}^{(2)}$, and we see in Tables III and IV that this is indeed the case, although the difference is only about a factor of 2. Note also that $\epsilon_{1s\uparrow 2s\downarrow}^{(2)}$ contains a contribution from the space-symmetric function $h_{1,1}$ as well as from the space-antisymmetric function $h_{0,0}$, the latter one yielding $\frac{1}{2}\epsilon_{1s\uparrow 2s\uparrow}^{(2)}(l)$ in each partial wave, as pointed out above. Thus, in fact, the *total* contribution of all antisymmetric functions to the inter-shell correlation energy is not significantly smaller than the total arising from all symmetric functions. Only in the $s$ wave is the reduction due to antisymmetry striking. Thus, it is possible to be misled by choosing a "simple" variational function to estimate the relative sizes of the antisymmetric and symmetric contributions,[19] since there is a good chance that the "simple" function will be purely relative $s$ wave in character. Even in the symmetric case, the bulk of the correlation energy comes from relative partial waves with $l\geq 1$.

TABLE III. The contribution of each relative partial wave to the $1s\uparrow 2s\uparrow$ correlation energy in second order (in a.u.).

| $l$ | Be $\epsilon_{1s\uparrow 2s\uparrow}^{(2)}$ | $B^+$ |
|---|---|---|
| 0 | −0.000009 | −0.000014 |
| 1 | −0.000645 | −0.000887 |
| 2 | −0.000041 | −0.000059 |
| 3 | −0.000005 | −0.000008 |
| ⩾4 | −0.000001 | −0.000001 |
| Total | −0.000701 | −0.000969 |

[19] W. R. Conkie, Can. J. Phys. 43, 102 (1965).

TABLE IV. The contribution of each relative partial wave to the $1s\uparrow 2s\downarrow$ correlation energy in second order (in a.u.).

| $l$ | Be $\epsilon_{1s\uparrow 2s\downarrow}^{(2)}$ | $B^+$ |
|---|---|---|
| 0 | −0.000380 | −0.000576 |
| 1 | −0.001065 | −0.001571 |
| 2 | −0.000108 | −0.000167 |
| 3 | −0.000022 | −0.000035 |
| ⩾4 | −0.000012 | −0.000015 |
| Total | −0.001587 | −0.002364 |

### C. 2s-2s Correlations

For the $2s$-$2s$ shell we used the same type of trial function (with the same number of parameters) as was employed in the $1s$-$1s$ calculation. Looking at Table V for the $\epsilon_{2s\uparrow 2s\downarrow}^{(2)}(l)$, we see immediately how different the effects in this shell are from those in the $1s$-$1s$ case. The contribution of the $l=1$ partial wave strongly dominates all the other terms, being an order of magnitude larger than the $l=0$ contribution and a factor of six larger than the $l=2$ contribution. The reason for this is easily seen. In the $2s$-$2s$ case, there is a nearby $2p$-$2p$ state which is expected to have a very large mixing with the $2s$-$2s$ state in second-order perturbation theory because of the smallness of the energy denominator associated with it. (Recall that states like $2s$-$2p$ cannot mix because we have an over-all $s$ state). There is no analogous state in the $1s$-$1s$ case (or $1s$-$2s$ case), so the two shells display very great qualitative as well as quantitative differences. If the above interpretation is correct, we would expect the $\rho$-wave dominance effect to be much more pronounced in $B^+$, where the stronger central field should cause the $2s$-$2s$ and $2p$-$2p$ states to be even more nearly degenerate. Clearly, when electron-electron interactions become "com-

TABLE V. The contribution of each relative partial wave to the $2s\uparrow 2s\downarrow$ correlation energy in second order (in a.u.).

| $l$ | Be $\epsilon_{2s\uparrow 2s\downarrow}^{(2)}$ | $B^+$ |
|---|---|---|
| 0 | −0.00235 | −0.00245 |
| 1 | −0.02213 | −0.03020 |
| 2 | −0.00382 | −0.00449 |
| 3 | −0.00118 | −0.00133 |
| 4 | −0.00048 | −0.00052 |
| 5 | −0.00022 | −0.00024 |
| 6 | −0.00012 | −0.00013 |
| ⩾7 | −0.00020 | −0.00022 |
| Total | −0.03050 | −0.03958 |

TABLE VI. The contribution of each electronic pair to the total second-order correlation energy (in a.u.).

| Electronic pair | $\epsilon_{ij}^{(2)}$ Be | B$^+$ |
|---|---|---|
| $1s\uparrow 1s\downarrow$ | −0.04008 | −0.04055 |
| $1s\uparrow 2s\uparrow$ | −0.00070 | −0.00097 |
| $1s\uparrow 2s\downarrow$ | −0.00159 | −0.00237 |
| $1s\downarrow 2s\uparrow$ | −0.00159 | −0.00237 |
| $1s\downarrow 2s\downarrow$ | −0.00070 | −0.00097 |
| $2s\uparrow 2s\downarrow$ | −0.03050 | −0.03958 |
| Total | −0.07516 | −0.08681 |

pletely negligible" those states would be precisely degenerate if we neglect very small relativistic effects. When we look at B$^+$ in Table V, we find that this is indeed the case. In closing our discussion of the $2s$-$2s$ shell we remark that for $\epsilon_{2s\uparrow 2s\downarrow}^{(2)}(0)$ the subtraction of the $1s$ orbitals gives only a very small effect, changing $\epsilon_{2s\uparrow 2s\downarrow}^{(2)}(0)$ by about 2%.

Table VI summarizes our results for $E_2$. (Note that by our discussion in Sec. IV. $\epsilon_{1s\uparrow 2s\uparrow}^{(2)}=\epsilon_{1s\downarrow 2s\downarrow}^{(2)}$ and $\epsilon_{1s\uparrow 2s\downarrow}^{(2)}=\epsilon_{1s\downarrow 2s\uparrow}^{(2)}$.) We find the total values

$$E_2(\text{Be})=-0.0752 \text{ a.u.},\qquad E_2(\text{B}^+)=-0.0868 \text{ a.u.}, \tag{50}$$

which is about 80% of the correlation energy. We believe that these are very accurate values of $E_2$, probably not differing from the exact $E_2$ by more than one or two in the third significant figure. We should emphasize that these numbers are "exact" in the sense that no approximation other than the use of a finite number of parameters in our trial function has been made in obtaining them.

## VI. THE THIRD-ORDER ENERGY

With the function $\psi_1$ calculated, it is a simple matter to evaluate the third-order energy $E_3$ via the relation

$$E_3=\langle\psi_1|H_1-E_1|\psi_1\rangle. \tag{51}$$

Using Eq. (18), together with the fact that $\mathcal{S}$ is a projection operator, this expression can be written as a sum of 36 terms of which six have the following typical form:

$$\epsilon_{1s\uparrow 1s\downarrow}^{(3)}=\langle[\chi_{1s\uparrow 1s\downarrow}(\mathbf{r}_1,\mathbf{r}_2)\phi_{2s\uparrow}(\mathbf{r}_3)\phi_{2s\downarrow}(\mathbf{r}_4)]|H_1-E_1|\mathcal{S}[\chi_{1s\uparrow 1s\downarrow}(\mathbf{r}_1,\mathbf{r}_2)\phi_{2s\uparrow}(\mathbf{r}_3)\phi_{2s\downarrow}(\mathbf{r}_4)]\rangle. \tag{52}$$

These terms we will refer to as "diagonal," or direct, terms. The remaining ones, which we call "off-diagonal" terms, have typically the form

$$\epsilon_{1s\uparrow 1s\downarrow;1s\uparrow 2s\downarrow}^{(3)}=\langle[\chi_{1s\uparrow 1s\downarrow}(\mathbf{r}_1,\mathbf{r}_2)\phi_{2s\uparrow}(\mathbf{r}_3)\phi_{2s\downarrow}(\mathbf{r}_4)]|H_1-E_1|\mathcal{S}[\chi_{1s\uparrow 2s\downarrow}(\mathbf{r}_1,\mathbf{r}_4)\phi_{1s\downarrow}(\mathbf{r}_2)\phi_{2s\uparrow}(\mathbf{r}_3)]\rangle. \tag{53}$$

Let us focus our attention on terms of the type of Eq. (52). Among the 12 terms which, because of the action of the operator $\mathcal{S}$, appear in Eq. (52), we first look explicitly at

$$\langle\chi_{1s\downarrow 1s\uparrow}(\mathbf{r}_1,\mathbf{r}_2)\phi_{2s\uparrow}(\mathbf{r}_3)\phi_{2s\downarrow}(\mathbf{r}_4)|\frac{1}{r_{12}}+\frac{1}{r_{13}}+\frac{1}{r_{14}}+\frac{1}{r_{23}}+\frac{1}{r_{24}}+\frac{1}{r_{34}}-\mathcal{V}(\mathbf{r}_1)-\mathcal{V}(\mathbf{r}_2)$$
$$-\mathcal{V}(\mathbf{r}_3)-\mathcal{V}(\mathbf{r}_4)-E_1|\chi_{1s\uparrow 1s\downarrow}(\mathbf{r}_1,\mathbf{r}_2)\phi_{2s\uparrow}(\mathbf{r}_3)\phi_{2s\downarrow}(\mathbf{r}_4)\rangle. \tag{54}$$

Performing the integrations on the $\mathbf{r}_3$ and $\mathbf{r}_4$ variables, we get for this term

$$\langle\chi_{1s\uparrow 1s\downarrow}(\mathbf{r}_1,\mathbf{r}_2)|1/r_{12}+2V_{2s}^d(\mathbf{r}_1)+2V_{2s}^d(\mathbf{r}_2)+V_{2s2s}-\mathcal{V}(\mathbf{r}_1)-\mathcal{V}(\mathbf{r}_2)-2V_{2s2s}-4V_{1s2s}-2V_{\text{ex}}-E_1|\chi_{1s\uparrow 1s\downarrow}(\mathbf{r}_1,\mathbf{r}_2)\rangle, \tag{55}$$

where we have used the fact that

$$\langle\phi_{2s\uparrow}|\mathcal{V}|\phi_{2s\uparrow}\rangle=\langle\phi_{2s\downarrow}|\mathcal{V}|\phi_{2s\downarrow}\rangle=V_{2s2s}+2V_{1s2s}-V_{\text{ex}}. \tag{56}$$

The other nonvanishing terms in Eq. (52) contribute precisely the exchange terms which are necessary to reduce Eq. (52) to the form

$$\epsilon_{1s\uparrow 1s\downarrow}^{(3)}=\langle\chi_{1s\uparrow 1s\downarrow}(\mathbf{r}_1,\mathbf{r}_2)|1/r_{12}-V_{1s\uparrow}(\mathbf{r}_1)-V_{1s\downarrow}(\mathbf{r}_1)-V_{1s\uparrow}(\mathbf{r}_2)-V_{1s\downarrow}(\mathbf{r}_2)-\epsilon_{1s\uparrow 1s\downarrow}^{(1)}|\chi_{1s\uparrow 1s\downarrow}(\mathbf{r}_1,\mathbf{r}_2)\rangle. \tag{57}$$

Note that in obtaining this result we have used the fact that the expectation value of the sum $[r_{34}^{-1}-\mathcal{V}(\mathbf{r}_3)-\mathcal{V}(\mathbf{r}_4)]$ precisely cancels all of $E_1$ except $\epsilon_{1s\uparrow 1s\downarrow}^{(1)}$. A similar reduction can be carried out for the remaining diagonal terms. Thus, the six terms of the type of Eq. (52) have the form which we would expect if our two-body model discussed above (Eq. 28) were exact. However, there remain terms like the one written in Eq. (53), which prevent this two-body model from being exact beyond the calculation of $E_2$. Fortunately, it is easily seen by looking at Eq. (53) that the off-diagonal terms make very small contributions to $E_3$. Because of orthogonality, all one-body operators give no contribution, so that we get from Eq. (53).

$$\epsilon_{1s\uparrow 1s\downarrow;1s\uparrow 2s\downarrow}^{(3)}=\langle\chi_{1s\uparrow 1s\downarrow}(\mathbf{r}_1,\mathbf{r}_2)\phi_{2s\uparrow}(\mathbf{r}_3)\phi_{2s\downarrow}(\mathbf{r}_4)|1/r_{12}+1/r_{13}+1/r_{14}+1/r_{23}$$
$$+1/r_{24}+1/r_{34}|\mathcal{S}[\chi_{1s\uparrow 2s\downarrow}(\mathbf{r}_1,\mathbf{r}_4)\phi_{1s\downarrow}(\mathbf{r}_2)\phi_{2s\uparrow}(\mathbf{r}_3)]\rangle.$$

Most of the terms in this inner product vanish, and after a simple calculation one finds

$$\begin{aligned}\epsilon_{1s\uparrow 1s\downarrow;1s\uparrow 2s\downarrow}{}^{(3)}=&\,2\langle\chi_{1s\uparrow 1s\downarrow}(\mathbf{r}_1,\mathbf{r}_2)\phi_{2s\downarrow}(\mathbf{r}_4)|(1/r_{24})|\chi_{1s\uparrow 2s\downarrow}(\mathbf{r}_1,\mathbf{r}_4)\phi_{1s\downarrow}(\mathbf{r}_2)\rangle\\ &-2\langle\chi_{1s\uparrow 1s\downarrow}(\mathbf{r}_1,\mathbf{r}_2)\phi_{2s\downarrow}(\mathbf{r}_4)|(1/r_{24})|\chi_{1s\uparrow 2s\downarrow}(\mathbf{r}_1,\mathbf{r}_2)\phi_{1s\downarrow}(\mathbf{r}_4)\rangle\\ &-\langle\chi_{1s\uparrow 1s\downarrow}(\mathbf{r}_1,\mathbf{r}_2)|\chi_{1s\uparrow 2s\downarrow}(\mathbf{r}_1,\mathbf{r}_2)\rangle\langle\phi_{2s\uparrow}(\mathbf{r}_3)\phi_{2s\downarrow}(\mathbf{r}_4)|(1/r_{34})|\phi_{2s\uparrow}(\mathbf{r}_3)\phi_{1s\downarrow}(\mathbf{r}_4)\rangle. \quad (58)\end{aligned}$$

Expressions analogous to this one are readily obtained for all off-diagonal terms. Clearly, these terms are small compared to the direct terms, since they all involve expectation values of operators between orthogonal functions. We have made estimates of typical terms of the off-diagonal type and found that they are more than an order of magnitude smaller than the diagonal ones. On the basis of these estimates we consider it unlikely that the off-diagonal terms contribute more than 0.001 a.u. to the correlation energy, and in fact they probably contribute less than this amount. Therefore we will neglect them in what follows. It should be mentioned that with functions of the type we have used in calculating the $\chi_{ij}$, it is possible, although tedious, to evaluate the off-diagonal contributions explicitly.

Thus, to obtain $E_3$, we need only evaluate inner products of the type written in Eq. (57). These are not diagonal with respect to the various partial waves (because of the presence of $r_{12}^{-1}$) and therefore the evaluation of the $\epsilon_{ij}^{(3)}$ is more complicated than was the calculation of the $\epsilon_{ij}^{(2)}$. However, this complication is an inessential one, and using the results of our calculation for the $\chi_{ij}$ we obtain the results listed in Table VII for the $\epsilon_{ij}^{(3)}$. Adding the six $\epsilon_{ij}^{(3)}$ together, we get

$$\begin{aligned}E_3(\mathrm{Be})&=-0.01029\text{ a.u.},\\ E_3(\mathrm{B}^+)&=-0.01337\text{ a.u.}.\end{aligned} \quad (59)$$

In looking at the results displayed in Table VII, one is struck by the fact that $\epsilon_{2s\uparrow 2s\downarrow}{}^{(3)}$ is much larger than the other $\epsilon_{ij}^{(3)}$. Of course, we expect both $\epsilon_{1s\uparrow 1s\downarrow}{}^{(3)}$ and $\epsilon_{2s\uparrow 2s\downarrow}{}^{(3)}$ to be much larger than the contribution from inter-shell terms, but $\epsilon_{2s\uparrow 2s\downarrow}{}^{(3)}$ is more than three times larger than $\epsilon_{1s\uparrow 1s\downarrow}{}^{(3)}$, even though $\epsilon_{1s\uparrow 1s\downarrow}{}^{(2)}$ and $\epsilon_{2s\uparrow 2s\downarrow}{}^{(2)}$ differ only by 25%, with $\epsilon_{1s\uparrow 1s\downarrow}{}^{(2)}$ being even larger than $\epsilon_{2s\uparrow 2s\downarrow}{}^{(2)}$. The reason for this phenomenon is easily seen. The outer pair of 2*s* electrons is loosely bound, and we may expect that the convergence of perturbation theory for such a pair would be rather slow, compared, for example, with the very tightly bound pair of 1*s* electrons. In fact, we see from Tables VI and VII that the convergence of the perturbation expansion for the 1*s*-1*s* correlation energy is very rapid. In a similar manner, one also sees that the inter-shell correlation energies are also converging rapidly. However, it is clear that the results for $E_{\mathrm{corr}}{}^{2s\text{-}2s}$, i.e., the 2*s*-2*s* correlation energy, are not converging rapidly, and it would certainly not be surprising if higher-order terms in perturbation theory made significant contributions to $E_{\mathrm{corr}}{}^{2s\text{-}2s}$. We will consider this problem in Sec. VII.

TABLE VII. The contribution of each electronic pair to the total third-order correlation energy (in a.u.).

| Electronic pair | $\epsilon_{ij}^{(3)}$ Be | B$^+$ |
|---|---|---|
| 1*s*↑1*s*↓ | −0.00210 | −0.00174 |
| 1*s*↑2*s*↑ | −0.00011 | −0.00012 |
| 1*s*↑2*s*↓ | −0.00022 | −0.00027 |
| 1*s*↓2*s*↑ | −0.00022 | −0.00027 |
| 1*s*↓2*s*↓ | −0.00011 | −0.00012 |
| 2*s*↑2*s*↓ | −0.00753 | −0.01085 |
| Total | −0.01029 | −0.01337 |

## VII. HIGHER-ORDER EFFECTS

Our results so far are very satisfactory as far as the 1*s*-shell and inter-shell correlations are concerned. However, for Be the sum of $E_2$ and $E_3$ is equal to $-0.0855$ a.u., which is approximately 0.009 a.u. greater than the experimentally observed correlation energy. For B$^+$, $E_2+E_3=-0.100$ a.u., approximately 0.012 greater than the experimental correlation energy. On the basis of what appears to be a very rapid convergence of the 1*s*-1*s* and inter-shell contributions, we conclude that the discrepancy must be due to higher-order effects of pair correlations, primarily in the 2*s* shell and probably also to three- and four-body correlations. Thus, the question arises of how to describe higher-order effects. The inclusion of three- and four-body effects is beyond the scope of this paper, although, as we have already remarked, with $\psi_1$ determined exactly, it is possible to evaluate the off-diagonal contributions to $E_3$, thereby obtaining the lowest-order contribution to the three- and four-body effects.

In order to deal with higher-order effects, we simply move from our equation for $\psi_1$ to a similar equation for $\psi_2$ which is readily obtained from Eq. (13b), namely,

$$(H_0-E_0)\psi_2+(H_1-E_1)\psi_1-E_2\psi_0=0. \quad (60)$$

As noted in BJI, this equation can be derived from a variational expression

$$\begin{aligned}F_2[\psi_2{}^t]=\langle\psi_2{}^t|H_0-E_0|\psi_2{}^t\rangle+2\langle\psi_2{}^t|H_1-E_1|\psi_1\rangle\\ -2E_2\langle\psi_2{}^t|\psi_0\rangle. \quad (61)\end{aligned}$$

Because we have already calculated $\psi_1$, it would appear reasonable to construct a function $\psi_2{}^t$ just as we did for $\psi_1{}^t$ and then form $F_2[\psi_2{}^t]$ which will yield $\psi_2$ upon variation of the parameters in $\psi_2{}^t$ .With $\psi_2$ obtained we then find $E_4$ and $E_5$ by using Eqs. (14d) and (14e). Since we have chosen $\psi_1$ to be orthogonal to $\psi_0$ and since we are free to make the same choice for $\psi_2$ [because of the form of Eq. (60)], Eqs. (14d) and (14e) simplify to

$$E_4=-\langle\psi_2|H_0-E_0|\psi_2\rangle-E_2\langle\psi_1|\psi_1\rangle, \quad (62)$$

$$E_5=\langle\psi_2|H_1-E_1|\psi_2\rangle-2E_2\langle\psi_1|\psi_2\rangle-E_3\langle\psi_1|\psi_1\rangle. \quad (63)$$

Comparing Eqs. (60), (61), and (62) we see that the value taken on by $F_2$ when $\psi_2{}^t$ is equal to the true $\psi_2$ is just $E_4$, apart from a term in $\langle\psi_1|\psi_1\rangle$ which does not depend on $\psi_2$. Thus, Eq. (61) is a variational principle for $E_4$, and in fact it is easy to show that since we are dealing with a ground state we actually have a *minimum* principle for $E_4$.

However, we note that $F_2[\psi_2{}^t]$, because it contains the inner product $\langle\psi_2{}^t|H_1-E_1|\psi_1\rangle$, will not decouple exactly into a set of pair problems as did $F_1[\psi_1{}^t]$. This inner product is exactly of the form of the inner product occuring in the expression for $E_3$ [Eq. (51)], so since $E_3$ did not decouple exactly into a sum of $\epsilon_{ij}{}^{(3)}$, the term $\langle\psi_2{}^t|H_1-E_1|\psi_1\rangle$ will also not decouple in this manner. It should be noted that this is not the only reason that decoupling does not occur. Another, slightly more obscure, reason comes from the term $\langle\psi_2{}^t|H_0-E_0|\psi_2{}^t\rangle$. In discussing a similar term in $F_1[\psi_1{}^t]$ we concluded that it did decouple into a set of pair terms. This decoupling depended critically on the fact that the $\chi_{ij}$ contained no component of *any* filled Hartree-Fock orbital in $\psi_0$. However, for $\psi_2{}^t$, we will want to replace the $\chi_{ij}$ of Eq. (18) for $\psi_1{}^t$ with $\chi_{ij}{}'$, and although it will still be true, for example, that $\chi_{2s\uparrow2s\downarrow}{}'$ will contain no components of $\phi_{1s\uparrow}$ or $\phi_{1s\downarrow}$ because of the exclusion principle, nevertheless, components of $\phi_{2s\uparrow}$ and $\phi_{2s\downarrow}$ are no longer excluded because of the form of the equation for $\psi_2$ as they were in the case of $\psi_1$. This is again a consequence of the term $(H_1-E_1)\psi_1$ in the equation for $\psi_2$.

With these difficulties in mind, let us be guided by the physics of the situation. Since our work on $E_2$ and $E_3$ has shown that the contribution to the correlation energy from inter-shell correlations is small and is mostly exhausted by $\psi_1$, we conjecture that the dominant terms in $\psi_2$ are of the 1s-1s or 2s-2s correlation type. Thus we take our trial function $\psi_2{}^t$ to be of the form

$$\psi_2{}^t=[1/\sqrt{(12)}]\mathcal{S}[\chi_{1s\uparrow1s\downarrow}{}'(\mathbf{r}_1,\mathbf{r}_2)\phi_{2s\uparrow}(\mathbf{r}_3)\phi_{2s\downarrow}(\mathbf{r}_4) +\phi_{1s\uparrow}(\mathbf{r}_1)\phi_{1s\downarrow}(\mathbf{r}_2)\chi_{2s\uparrow2s\downarrow}{}'(\mathbf{r}_3,\mathbf{r}_4) +a\chi_{1s\uparrow1s\downarrow}(\mathbf{r}_1,\mathbf{r}_2)\chi_{2s\uparrow2s\downarrow}(\mathbf{r}_3,\mathbf{r}_4)], \quad (64)$$

where $\chi_{1s\uparrow1s\downarrow}{}'$ and $\chi_{2s\uparrow2s\downarrow}{}'$ contain a large number of parameters to be varied, and $a$ is also a variational parameter. The functions $\chi_{1s\uparrow1s\downarrow}$ and $\chi_{2s\uparrow2s\downarrow}$ are given in Sec. V. Note that in Eq. (64), because of the operator $\mathcal{S}$, $\chi_{2s\uparrow2s\downarrow}{}'$ can be taken to contain no components of $\phi_{1s\uparrow}$ on $\phi_{1s\downarrow}$. A similar comment applies to $\chi_{1s\uparrow1s\downarrow}{}'$.

Now, a term involving $\chi_{1s\uparrow1s\downarrow}{}'$ (or $\chi_{2s\uparrow2s\downarrow}{}'$) can be thought as arising in perturbation theory from two successive interactions between the 1s (or 2s) electrons. Our assumption concerning the form of $\psi_2{}^t$ is essentially that terms in which there are two successive inter-shell interactions between the same electrons are unimportant. However, there is another type of term to be considered in second order, namely, an interaction within one pair followed by an interaction within a different pair. Situations in which the two pairs have an element in common correspond, roughly speaking, to triple correlations in which one electron is in a different shell than the other two. Contributions of this type are expected to be small compared to those coming from terms in which there are successive interactions between two electrons in the same shell. But in an atom with more than three electrons, there exists one more possibility, namely, an interaction within a pair of electrons followed by an interaction within another pair having no element in common with the first. Terms like this correspond to *disconnected* diagrams in perturbation theory. An example of this case is an interaction between the 1s electrons followed by an interaction between the 2s electrons. There are two other possibilities corresponding to inter-shell pairs which, because of the smallness of inter-shell effects, one expects to be negligible. Making a reasonably simple model of correlations, we assume that a term representing two independent correlations in different shells should arise in $\psi_2{}^t$ through a term proportional to the product $\chi_{1s\uparrow1s\downarrow}(\mathbf{r}_1,\mathbf{r}_2)\times\chi_{2s\uparrow2s\downarrow}(\mathbf{r}_3,\mathbf{r}_4)$ of the appropriate first-order correlation functions. This explains the appearance of the last term on the right-hand side of Eq. (64).

If we now insert our trial function (64) into the minimum principle (61) for $E_4$, we get

$$F_2[\psi_2{}^t]=A+2B-2E_2C, \quad (65a)$$

where we have defined

$$A=\langle\psi_2{}^t|H_0-E_0|\psi_2{}^t\rangle, \quad (65b)$$

$$B=\langle\psi_2{}^t|H_1-E_1|\psi_1\rangle, \quad (65c)$$

$$C=\langle\psi_2{}^t|\psi_0\rangle. \quad (65d)$$

The discussion of the term $C$ is trivial. Since we may choose $\langle\psi_2|\psi_0\rangle=0$, we will consider our trial function $\psi_2{}^t$ to be orthogonal to $\psi_0$, or in terms of the functions $\chi_{1s\uparrow1s\downarrow}{}'$ and $\chi_{2s\uparrow2s\downarrow}{}'$, we will constrain them to satisfy

$$\langle\phi_{1s\uparrow}\phi_{1s\downarrow}|\chi_{1s\uparrow1s\downarrow}{}'\rangle=0 \quad (66a)$$

and

$$\langle\phi_{2s\uparrow}\phi_{2s\downarrow}|\chi_{2s\uparrow2s\downarrow}{}'\rangle=0. \quad (66b)$$

Thus the term $C$ enters essentially as a Lagrange multiplier in our problem.

The term $A$ is also straightforward. It breaks up into a simple sum of three terms. There will be no interference between the second-order correlation function terms and the term in $\chi_{1s\uparrow1s\downarrow}\chi_{2s\uparrow2s\downarrow}$ because both $\chi_{1s\uparrow1s\downarrow}$ and $\chi_{2s\uparrow2s\downarrow}$ are orthogonal to *all* Hartree-Fock orbitals (this is also why the term in $\chi_{1s\uparrow1s\downarrow}\chi_{2s\uparrow2s\downarrow}$ does not contribute to $C$.) There will be no interference between the two second-order correlation functions since all but two terms in the overlap expression vanish because $\langle\phi_{1s}|\phi_{2s}\rangle=0$ and the remaining two terms vanish because of the action of $H_0-\epsilon_0$. Thus, in a fairly straightforward manner we get

$$A=\langle\chi_{1s\uparrow1s\downarrow}'|H_{\mathrm{HF}}(\mathbf{r})+H_{\mathrm{HF}}(\mathbf{r}')-2\epsilon_{1s}|\chi_{1s\uparrow1s\downarrow}'\rangle+\langle\chi_{2s\uparrow2s\downarrow}'|H_{\mathrm{HF}}(\mathbf{r})+H_{\mathrm{HF}}(\mathbf{r}')-2\epsilon_{2s}|\chi_{2s\uparrow2s\downarrow}'\rangle$$
$$+a^2\langle\chi_{1s\uparrow1s\downarrow}(\mathbf{r}_1,\mathbf{r}_2)\chi_{2s\uparrow2s\downarrow}(\mathbf{r}_3,\mathbf{r}_4)|H_0-E_0|\mathcal{S}[\chi_{1s\uparrow1s\downarrow}(\mathbf{r}_1,\mathbf{r}_2)\chi_{2s\uparrow2s\downarrow}(\mathbf{r}_3,\mathbf{r}_4)]\rangle. \quad (67)$$

The term $B$ poses the only significant problem in the evaluation of $F_2[\psi_2^t]$. On the right-hand side of the inner product [Eq. (65b)] $\psi_1$ is the sum of six first-order pair-correlation terms, and on the left-hand side we have a sum of three terms given by Eq. (64). The reduction of this expression will involve the neglect of certain terms, and therefore, since we neglect them in the variational principle (i.e., a functional), we are on less firm ground than when we neglect terms in a scalar like $E_3$. However, the 12 terms involving overlaps between inner-shell functions like $\chi_{1s\uparrow1s\downarrow}'$, $\chi_{1s\uparrow1s\downarrow}$, etc., and the inter-shell functions $\chi_{1s\uparrow2s\downarrow}$, etc., are expected to be extremely small, so that neglecting them is probably not a serious error. Thus we are left with six terms in $B$ which should be considered carefully. The two direct terms between the inner-shell correlation functions reduce to

$$\langle\chi_{1s\uparrow1s\downarrow}'(\mathbf{r}_1,\mathbf{r}_2)|(1/r_{12})-V_{1s\uparrow}(\mathbf{r}_1)-V_{1s\downarrow}(\mathbf{r}_1)-V_{1s\uparrow}(\mathbf{r}_2)-V_{1s\downarrow}(\mathbf{r}_2)-\epsilon_{1s\uparrow1s\downarrow}^{(1)}|\chi_{1s\uparrow1s\downarrow}(\mathbf{r}_1,\mathbf{r}_2)\rangle \quad (68a)$$

and

$$\langle\chi_{2s\uparrow2s\downarrow}'(\mathbf{r}_3,\mathbf{r}_4)|(1/r_{34})-V_{2s\uparrow}(\mathbf{r}_3)-V_{2s\downarrow}(\mathbf{r}_3)-V_{2s\uparrow}(\mathbf{r}_4)-V_{2s\downarrow}(\mathbf{r}_4)-\epsilon_{2s\uparrow2s\downarrow}^{(1)}|\chi_{2s\uparrow2s\downarrow}(\mathbf{r}_3,\mathbf{r}_4)\rangle \quad (68b)$$

after a simple calculation completely analogous to the evaluation of similar ("diagonal") terms in $E_3$.

Next we investigate the contributions involving overlaps between the two different inner-shell terms. These are

$$\langle\phi_{1s\uparrow}(\mathbf{r}_1)\phi_{1s\downarrow}(\mathbf{r}_2)\chi_{2s\uparrow2s\downarrow}'(\mathbf{r}_3,\mathbf{r}_4)|(1/r_{12})|\mathcal{S}[\chi_{1s\uparrow1s\downarrow}(\mathbf{r}_1,\mathbf{r}_2)\phi_{2s\uparrow}(\mathbf{r}_3)\phi_{2s\downarrow}(\mathbf{r}_4)]\rangle \quad (69a)$$

and

$$\langle\chi_{1s\uparrow1s\downarrow}'(\mathbf{r}_1,\mathbf{r}_2)\phi_{2s\uparrow}(\mathbf{r}_3)\phi_{2s\downarrow}(\mathbf{r}_4)|(1/r_{34})|\mathcal{S}[\phi_{1s\uparrow}(\mathbf{r}_1)\phi_{1s\downarrow}(\mathbf{r}_2)\chi_{2s\uparrow2s\downarrow}(\mathbf{r}_3,\mathbf{r}_4)]\rangle. \quad (69b)$$

The two terms contributing to each of these expressions which do not involve overlaps between different shells are identically zero because of the constraint equations, Eqs. (66a) and (66b). Again, the remaining terms involve overlap integrals between different shells which we neglect. We note that except for one part of Eq. (69a) and also of Eq. (69b) which is very much smaller (by probably two orders of magnitude) than a typical term of $B$, the remaining terms which are discarded contribute only to the relative $s$-wave part of $\chi_{1s\uparrow1s\downarrow}'$ and $\chi_{2s\uparrow2s\downarrow}'$. We expect the $s$-wave contribution from $\chi_{2s\uparrow2s\downarrow}'$ to be very small by analogy with the $s$-wave contribution from $\chi_{2s\uparrow2s\downarrow}$ (see Table V), so that errors in this partial wave will be relatively unimportant.

Finally, the last contributions to $B$ will come from the overlap of the term in $\chi_{1s\uparrow1s\downarrow}\chi_{2s\uparrow2s\downarrow}$ with the first-order, inner-shell correlation terms in $\psi_1$. These terms are

$$a\langle\chi_{1s\uparrow1s\downarrow}(\mathbf{r}_1,\mathbf{r}_2)\chi_{2s\uparrow2s\downarrow}(\mathbf{r}_3,\mathbf{r}_4)|H_1-E_1|\mathcal{S}[\chi_{1s\uparrow1s\downarrow}(\mathbf{r}_1,\mathbf{r}_2)\phi_{2s\uparrow}(\mathbf{r}_3)\phi_{2s\downarrow}(\mathbf{r}_4)] \quad (70a)$$

and

$$a\langle\chi_{1s\uparrow1s\downarrow}(\mathbf{r}_1,\mathbf{r}_2)\chi_{2s\uparrow2s\downarrow}(\mathbf{r}_3,\mathbf{r}_4)|H_1-E_1|\mathcal{S}[\phi_{1s\uparrow}(\mathbf{r}_1)\phi_{1s\downarrow}(\mathbf{r}_2)\chi_{2s\uparrow2s\downarrow}(\mathbf{r}_3,\mathbf{r}_4)]\rangle, \quad (70b)$$

and they involve only functions which are not varied.

Combining the results expressed in Eqs. (68a), (68b), (70a), and (70b), we get for $B$

$$B=\langle\chi_{1s\uparrow1s\downarrow}'|(1/r_{12})-V_{1s\uparrow}(\mathbf{r}_1)-V_{1s\downarrow}(\mathbf{r}_1)-V_{1s\uparrow}(\mathbf{r}_2)-V_{1s\downarrow}(\mathbf{r}_2)-\epsilon_{1s\uparrow1s\downarrow}^{(1)}|\chi_{1s\uparrow1s\downarrow}\rangle$$
$$+\langle\chi_{2s\uparrow2s\downarrow}'|(1/r_{34})-V_{2s\uparrow}(\mathbf{r}_3)-V_{2s\downarrow}(\mathbf{r}_3)-V_{2s\uparrow}(\mathbf{r}_4)-V_{2s\downarrow}(\mathbf{r}_4)-\epsilon_{2s\uparrow2s\downarrow}^{(1)}|\chi_{2s\uparrow2s\downarrow}\rangle$$
$$+a\langle\chi_{1s\uparrow1s\downarrow}(\mathbf{r}_1,\mathbf{r}_2)\chi_{2s\uparrow2s\downarrow}(\mathbf{r}_3,\mathbf{r}_4)|H_1-E_1|\mathcal{S}[\phi_{1s\uparrow}(\mathbf{r}_1)\phi_{1s\downarrow}(\mathbf{r}_2)\chi_{2s\uparrow2s\downarrow}(\mathbf{r}_3,\mathbf{r}_4)]\rangle$$
$$+a\langle\chi_{1s\uparrow1s\downarrow}(\mathbf{r}_1,\mathbf{r}_2)\chi_{2s\uparrow2s\downarrow}(\mathbf{r}_3,\mathbf{r}_4)|H_1-E_1|\mathcal{S}[\chi_{1s\uparrow1s\downarrow}(\mathbf{r}_1,\mathbf{r}_2)\phi_{2s\uparrow}(\mathbf{r}_3)\phi_{2s\downarrow}(\mathbf{r}_4)]\rangle. \quad (71)$$

Following Eq. (65a), we now combine Eq. (67) for $A$ with Eq. (71) for $B$ to obtain $F_2[\psi_2^t]$. We constrain $\chi_{1s\uparrow1s\downarrow}'$ and $\chi_{2s\uparrow2s\downarrow}'$ according to Eqs. (66a) and (66b) and vary $F_2[\psi_2^t]$ with respect to $\chi_{1s\uparrow1s\downarrow}'$, $\chi_{2s\uparrow2s\downarrow}'$, and $a$, so that we obtain the relations

$$[H_{\mathrm{HF}}(\mathbf{r}_1)+H_{\mathrm{HF}}(\mathbf{r}_2)-2\epsilon_{1s}]\chi_{1s\uparrow1s\downarrow}'(\mathbf{r}_1,\mathbf{r}_2)$$
$$+[(1/r_{12})-V_{1s\uparrow}(\mathbf{r}_1)-V_{1s\downarrow}(\mathbf{r}_1)-V_{1s\uparrow}(\mathbf{r}_2)-V_{1s\downarrow}(\mathbf{r}_2)-\epsilon_{1s\uparrow1s\downarrow}^{(1)}]\chi_{1s\uparrow1s\downarrow}=0, \quad (72a)$$

$$[H_{\mathrm{HF}}(\mathbf{r}_3)+H_{\mathrm{HF}}(\mathbf{r}_4)-2\epsilon_{2s}]\chi_{2s\uparrow2s\downarrow}'+[(1/r_{34})-V_{2s\uparrow}(\mathbf{r}_3)-V_{2s\downarrow}(\mathbf{r}_3)-V_{2s\uparrow}(\mathbf{r}_4)-V_{2s\downarrow}(\mathbf{r}_4)-\epsilon_{2s\uparrow2s\downarrow}^{(1)}]\chi_{2s\uparrow2s\downarrow}=0, \quad (72b)$$

$$a\langle\chi_{1s\uparrow1s\downarrow}(\mathbf{r}_1,\mathbf{r}_2)\chi_{2s\uparrow2s\downarrow}(\mathbf{r}_3,\mathbf{r}_4)|H_0-E_0|\mathcal{S}[\chi_{1s\uparrow1s\downarrow}(\mathbf{r}_1,\mathbf{r}_2)\chi_{2s\uparrow2s\downarrow}(\mathbf{r}_3,\mathbf{r}_4)]\rangle$$
$$+\langle\chi_{1s\uparrow1s\downarrow}(\mathbf{r}_1,\mathbf{r}_2)\chi_{2s\uparrow2s\downarrow}(\mathbf{r}_3,\mathbf{r}_4)|H_1-E_1|\mathcal{S}[\phi_{1s\uparrow}(\mathbf{r}_1)\phi_{1s\downarrow}(\mathbf{r}_2)\chi_{2s\uparrow2s\downarrow}(\mathbf{r}_3,\mathbf{r}_4)]\rangle$$
$$+\langle\chi_{1s\uparrow1s\downarrow}(\mathbf{r}_1,\mathbf{r}_2)\chi_{2s\uparrow2s\downarrow}(\mathbf{r}_3,\mathbf{r}_4)|H_1-E_1)|\mathcal{S}[\chi_{1s\uparrow1s\downarrow}(\mathbf{r}_1,\mathbf{r}_2)\phi_{2s\uparrow}(\mathbf{r}_3)\phi_{2s\downarrow}(\mathbf{r}_4)]\rangle. \quad (72c)$$

We make our usual approximation concerning the neglect of integrals involving overlaps between different shells. Hence, we can easily reduce Eq. (72c) to the form

$$-2a[\epsilon_{1s\uparrow1s\downarrow}^{(2)}\langle\chi_{1s\uparrow1s\downarrow}|\chi_{2s\uparrow2s\downarrow}\rangle+\epsilon_{2s\uparrow2s\downarrow}^{(2)}\langle\chi_{1s\uparrow1s\downarrow}|\chi_{1s\uparrow1s\downarrow}\rangle] +\sqrt{2}[\epsilon_{1s\uparrow1s\downarrow}^{(2)}\langle\chi_{2s\uparrow2s\downarrow}|\chi_{2s\uparrow2s\downarrow}\rangle+\epsilon_{2s\uparrow2s\downarrow}^{(2)}\langle\chi_{1s\uparrow1s\downarrow}|\chi_{1s\uparrow1s\downarrow}\rangle]=0,$$

from which we conclude that $a=2^{-1/2}$.

Equations (72a) and (72b) are just the equations which we would get from a model Hamiltonian of the form of Eq. (28). They can be solved by the methods of BJI, which we have already used in solving for the first-order correlation functions. If we insert the solutions of Eqs. (72a) and (72b), along with the value $a=2^{-1/2}$, into Eq. (64), $\psi_2{}^t$ becomes the true second-order wave function which we will use in calculating $E_4$ and $E_5$. We thus see that when we come to second order in wave-function perturbation theory, our wave function is not strictly speaking what one would obtain from a simple-minded solution of the two-body model Hamiltonians of the type of Eq. (28). However, the terms which must be added are of a very simple and intuitively reasonable kind $(\chi_{1s\uparrow1s\downarrow}\chi_{2s\uparrow2s\downarrow})$ and they can also be obtained in a straightforward manner from the perturbation solutions to the pair equations. Another way of looking at these results is, of course, by examining ordinary perturbation theory in terms of sums over intermediate states and picking out subsums which can be argued to dominate in various orders. This approach lends strong support to the approximations made in this section.

Having determined $\chi_{1s\uparrow1s\downarrow}'$ and $\chi_{2s\uparrow2s\downarrow}'$ by the variational method (we used trial functions of the same functional form as we employed in obtaining $\chi_{1s\uparrow1s\downarrow}$ and $\chi_{2s\uparrow2s\downarrow}$), we now evaluate $E_4$ and $E_5$. The first term in $E_4$ [see Eq. (62)] is $-\langle\psi_2|H_0-E_0|\psi_2\rangle$. In the same way in which we arrived at Eq. (67) for $A$, we find

$$-\langle\psi_2|H_0-E_0|\psi_2\rangle=-\langle\chi_{1s\uparrow1s\downarrow}'|H_{\mathrm{HF}}(\mathbf{r}_1)+H_{\mathrm{HF}}(\mathbf{r}_2)-2\epsilon_{1s}|\chi_{1s\uparrow1s\downarrow}'\rangle-\langle\chi_{2s\uparrow2s\downarrow}'|H_{\mathrm{HF}}(\mathbf{r}_3)+H_{\mathrm{HF}}(\mathbf{r}_4)-2\epsilon_{2s}|\chi_{2s\uparrow2s\downarrow}'\rangle +\epsilon_{1s\uparrow1s\downarrow}^{(2)}\langle\chi_{2s\uparrow2s\downarrow}|\chi_{2s\uparrow2s\downarrow}\rangle+\epsilon_{2s\uparrow2s\downarrow}^{(2)}\langle\chi_{1s\uparrow1s\downarrow}|\chi_{1s\uparrow1s\downarrow}\rangle,$$

where we have reduced the term involving $\chi_{1s\uparrow1s\downarrow}\chi_{2s\uparrow2s\downarrow}$ in a manner similar to that used in solving Eq. (72c) for $a$. The second term in $E_4$ is $-E_2\langle\psi_1|\psi_1\rangle$, which we write as

$$-E_2\langle\psi_1|\psi_1\rangle=-(\epsilon_{1s\uparrow1s\downarrow}^{(2)}+\epsilon_{2s\uparrow2s\downarrow}^{(2)})[\langle\chi_{1s\uparrow1s\downarrow}|\chi_{1s\uparrow1s\downarrow}\rangle+\langle\chi_{2s\uparrow2s\downarrow}|\chi_{2s\uparrow2s\downarrow}\rangle]+\text{remainder}.$$

The terms appearing in the remainder are readily evaluated and are found to be very small compared with the term written out in detail. We neglect these terms since they are analogous to inter-shell terms and terms involving overlaps between different shells, which we have always neglected. An an example, for Be the omitted terms contribute 0.0002 a.u. to $E_2\langle\psi_1|\psi_1\rangle$, while the terms retained contribute 0.0021 a.u. The size of the neglected terms gives a rough idea of the order of magnitude of the overlap effects which we have neglected and which correspond in part to triple and quadruple correlations and in part to pair correlations between electrons in different shells. We see that the term $\chi_{1s\uparrow1s\downarrow}\chi_{2s\uparrow2s\downarrow}$ in $\psi_2$ has the effect of removing terms like $\epsilon_{1s\uparrow1s\downarrow}^{(2)}\langle\chi_{2s\uparrow2s\downarrow}|\chi_{2s\uparrow2s\downarrow}\rangle$ from the final expression for $E_4$, so that finally we have

$$E_4=\sum_{i,j}{}'\epsilon_{ij}^{(4)},\tag{73a}$$

with

$$\epsilon_{ij}^{(4)}=-\langle\chi_{ij}'|H_{\mathrm{HF}}(\mathbf{r})+H_{\mathrm{HF}}(\mathbf{r}')-\epsilon_i-\epsilon_j|\chi_{ij}'\rangle -\epsilon_{ij}^{(2)}\langle\chi_{ij}|\chi_{ij}\rangle,\tag{73b}$$

which is just what we would have obtained if we had calculated the eigenvalue of the appropriate reduced Hamiltonian of the type of Eq. (28) through fourth order in energy. Entirely similar considerations show that, with the omission of terms of the type which we have been neglecting, $E_5$ may be written in a similar manner;

$$E_5=\sum_{i,j}{}'\epsilon_{ij}^{(5)},\tag{74a}$$

with

$$\epsilon_{ij}^{(5)}=\langle\chi_{ij}'|(1/|\mathbf{r}-\mathbf{r}'|)-V_i(\mathbf{r})-V_j(\mathbf{r})-V_i(\mathbf{r}') -V_j(\mathbf{r}')-\epsilon_{ij}^{(1)}|\chi_{ij}'\rangle-2\epsilon_{ij}^{(2)}\langle\chi_{ij}|\chi_{ij}'\rangle -\epsilon_{ij}^{(3)}\langle\chi_{ij}|\chi_{ij}\rangle.\tag{74b}$$

Evaluating $\epsilon_{ij}^{(4)}$ and $\epsilon_{ij}^{(5)}$ for the inner-shell pairs which are of interest, we find for Be:

$\epsilon_{1s\uparrow1s\downarrow}^{(4)}=-0.00026$ a.u., $\epsilon_{2s\uparrow2s\downarrow}^{(4)}=-0.00478$ a.u.,

$\epsilon_{1s\uparrow1s\downarrow}^{(5)}=-0.00003$ a.u., $\epsilon_{2s\uparrow2s\downarrow}^{(5)}=-0.00201$ a.u..

For $B^+$ we get

$\epsilon_{1s\uparrow1s\downarrow}^{(4)}=-0.00018$ a.u., $\epsilon_{2s\uparrow2s\downarrow}^{(4)}=-0.00615$ a.u.,

$\epsilon_{1s\uparrow1s\downarrow}^{(5)}=-0.00002$ a.u., $\epsilon_{2s\uparrow2s\downarrow}^{(5)}=-0.00297$ a.u..

Table VIII summarizes the contributions of various orders of perturbation theory to the quantities $E_{\mathrm{corr}}{}^{1s\text{-}1s}$, $E_{\mathrm{corr}}{}^{1s\text{-}2s}$, and $E_{\mathrm{corr}}{}^{2s\text{-}2s}$ for Be, while Table IX lists the same quantities for $B^+$.

In calculating these results it is found that the $s$-wave contribution to $\epsilon_{2s\uparrow2s\downarrow}^{(4)}$ is smaller than the dominant $p$-wave term by about a factor of 20 or 30. Thus, even errors in the $s$ wave of the order or magnitude of 50% are unimportant for our purposes. This

TABLE VIII. The contributions to $E_{corr}^{1s-1s}$, $E_{corr}^{1s-2s}$, and $E_{corr}^{2s-2s}$ from each order of perturbation theory for Be (in a.u.).

| Order of perturbation theory | 1s-1s | 1s-2s | 2s-2s |
|---|---|---|---|
| Second | −0.04008 | −0.00458 | −0.03050 |
| Third | −0.00210 | −0.00066 | −0.00753 |
| Fourth | −0.00026 | ... | −0.00478 |
| Fifth | −0.00003 | ... | −0.00201 |
| Total | −0.04247 | −0.00524 | −0.04482 |

TABLE IX. The contributions to $E_{corr}^{1s-1s}$, $E_{corr}^{1s-2s}$, and $E_{corr}^{2s-2s}$ from each order of perturbation theory for $B^+$ (in a.u.).

| Order of perturbation theory | 1s-1s | 1s-2s | 2s-2s |
|---|---|---|---|
| Second | −0.04055 | −0.00668 | −0.03958 |
| Third | −0.00174 | −0.00078 | −0.01085 |
| Fourth | −0.00018 | ... | −0.00615 |
| Fifth | −0.00001 | ... | −0.00297 |
| Total | −0.04248 | −0.00746 | −0.05955 |

strengthens the arguments made in reducing the quantity $F_2[\psi_2^t]$. It is also clear that the contributions of the 1s shell to $E_4$ and $E_5$ are insignificant. Thus for the accuracy needed we could have neglected the function $\chi_{1s\uparrow 1s\downarrow}'$ entirely, so that many of our overlap problems would not have existed. We have retained it to illustrate the typical difficulties which one would encounter in a more general situation. It is interesting to see the very significant effect of the term $\chi_{1s\uparrow 1s\downarrow}\chi_{2s\uparrow 2s\downarrow}$ in the trial function $\psi_2^t$. If we had omitted this term, we would have found values of $E_4$ 20 to 30% smaller in magnitude than the results given in Tables VIII and IX. The effect would have been even more drastic in $E_5$, where the magnitude of the results presented in Tables VIII and IX would have been reduced by 35 to 50%. Had we omitted this term we would have found for the total correlation energy $E_{corr}^{total}(Be)=-0.0897$ a.u. and $E_{corr}^{total}(B^+)=-0.1071$ a.u., instead of the results given in Tables VIII and IX. These tables also illustrate the slow rate of convergence of the 2s-2s correlation energy. We see that the final values of $E_{corr}^{2s-2s}$ for Be and $B^+$ will probably be 0.001 to 0.002 a.u. lower than the values given by the tables. We did not pursue our calculation to higher orders, partly because the effects of three- and four-particle correlations will begin to compete seriously with pair correlations and partly because problems relating to loss of precision become more significant in solving for $\psi_3$.

## VIII. CONCLUSIONS

Summarizing our final results in Be, we have

$$E_{corr}^{1s-1s}(Be)=-0.0425 \text{ a.u.},$$
$$E_{corr}^{1s-2s}(Be)=-0.0052 \text{ a.u.},$$
$$E_{corr}^{2s-2s}(Be)=-0.0448 \text{ a.u.},$$
$$E_{corr}^{total}(Be)=-0.0925 \text{ a.u.},$$

while for $B^+$ we have

$$E_{corr}^{1s-1s}(B^+)=-0.0425 \text{ a.u.},$$
$$E_{corr}^{1s-2s}(B^+)=-0.0075 \text{ a.u.},$$
$$E_{corr}^{2s-2s}(B^+)=-0.0596 \text{ a.u.},$$
$$E_{corr}^{total}(B^+)=-0.1096 \text{ a.u.},$$

Thus, our final results in both Be and $B^+$ are in good agreement with the experimental results:

$$E_{corr}^{exp}(Be)=-0.0944 \text{ a.u.},\quad E_{corr}^{exp}(B^+)=-0.1116 \text{ a.u.}$$

There have been several investigations of the binding energy of Be in the past few years, and it is of interest to compare our results with those of other authors. Kelly[6] has used the Bethe-Goldstone version of perturbation theory and has obtained results which are in very close agreement with ours. He finds

$$E_{corr}^{1s-1s}(Be)=-0.0421 \text{ a.u.},$$
$$E_{corr}^{1s-2s}(Be)=-0.0050 \text{ a.u.},$$
$$E_{corr}^{2s-2s}(Be)=-0.0449 \text{ a.u.},$$
$$E_{corr}^{total}(Be)=-0.0920 \text{ a.u.}$$

We see that our result for the 1s-1s contribution is 1% lower than Kelly's, and for the 1s-2s contribution our result is 4% lower. The difference between our figures and those of Kelly in both these cases is about the same as the estimated numerical uncertainty in our results. In our case, one can see explicitly the very rapid convergence of perturbation theory which makes it seem very unlikely that the 1s-1s and 1s-2s correlation energies which we have calculated differ from the actual energies by more than the above stated uncertainties. In particular the results of Sinanoglu *et al.*,[7] $E_{corr}^{1s-1s}=-0.0440$ a.u. and $E_{corr}^{1s-2s}=-0.0065$ a.u., definitely disagree with our results. Sinanoglu's result for the 1s-1s shell is rather close to what we would have obtained if we had not removed the components of $\phi_{2s}$ from our trial function for $\chi_{1s\uparrow 1s\downarrow}$. In the case of the 2s-2s energy, Kelly's result and ours are essentially identical, although we can see from Table VIII that if we carried out pair model to higher orders we would probably obtain a result about 0.001 to 0.002 a.u. *lower* than the result we have found. Again, we are in disagreement with Sinanoglu, who finds $E_{corr}^{2s-2s}=-0.0438$ a.u. a result 2% *higher* than ours.

Basically, we feel that the close agreement between our work and that of Kelly, which are after all just two different approaches to perturbation theory, tends to suggest that the results presented here reflect faithfully the content of the lowest orders of perturbation theory. Furthermore, the rate of convergence of perturbation theory seems to be quite rapid for the 1s-1s and 1s-2s correlation energies, so that for them contributions from higher orders are in all probability not significant. The main uncertaintly in our results comes from the slow convergence of the 2s-2s contributions, but even here the indications are quite good that higher orders will contribute no more than $-0.002$ a.u. to our result.

Regarding $B^+$, there does not seem to be any theoretical work on this ion. We have included it here to point up some of the more interesting features of the four-body system. We see that the $1s$-$2s$ correlation energy increases by about 50% as we go from Be to $B^+$, which illustrates the physically reasonable fact that the inter-shell correlation energy depends critically on the distance between the shells in question. It is for this reason that we expect intershell effects to be much more important in systems with more than four electrons, where there in strong spacial overlap between the $2s$ and $2p$ shells. Also in $B^+$ we see a very large increase in the $2s$-$2s$ contribution to the correlation energy. This occurs because of the fact that the $2p$ state of the Hartree-Fock Hamiltonian gets progressively closer in energy to the $2s$ state as the central field becomes larger and larger, thus contributing smaller and smaller energy denominators in the sums over intermediates states occuring in perturbation theory. Also of interest is the great stability of the $1s$-$1s$ correlation energy, which is to be expected from similar studies of the two-body systems He, $Li^+$, $Be^{++}$, $B^{3+}$, etc.[20]

In closing this discussion, we emphasize again that the results obtained can be viewed in lowest order of perturbation theory as coming out of a simple two-body generalization of Hartree-Fock theory [see Eq. (28)], where each pair of particle is thought of as interacting though the true Coulomb potential, but also seeing an average potential due to the other electrons which remain in this approximation in their unperturbed Hartree-Fock orbitals. However, if we go higher in perturbation theory, correlations within disconnected pairs of electrons modify the perturbation wave functions in a significant way, although it is still possible to take these effects into account in a straightforward manner by using the pair-correlation functions already calculated in the two-body Hartree-Fock model. The effect of this change in the character of the *wave function* is just to give an expression for the perturbation theory *energy* which is precisely what one would obtain by applying the two-body model in a simple-minded way to compute the total energy.

We believe that the method described here can be extended in a trivial way to the lithium atom. There will be coupling between singlet and triplet parts of the two-body correlation functions in this case, but this is a minor complication. In calculating the correlation energy in lithium, one will also need analytic solutions to the true Hartree-Fock equations, i.e., solutions in which the $1s\uparrow$ orbital has a different space wave function than the $1s\downarrow$ orbital because the $2s$ shell is not closed. (Thus, $\chi_{1s\uparrow 1s\downarrow}$ will not be a pure singlet function as it was in Be and $B^+$.) Again, this represents an inessential complication. We believe that one should be able to get extremely accurate results in lithium by going through fifth order in energy. Extending this method into the $2p$ shell (boron, carbon, etc.) will entail some vector-coupling modifications, but, more important, the approximations concerning inter-shell effects ($2s$-$2p$) will have to be examined carefully. A further problem is that of determining the effect of approximate solutions to the Hartree-Fock equation in this method, since the solutions to the true Hartree-Fock equation, involving potentials which are not spherically symmetric, are not readily obtainable, except in very special cases (e.g., closed angular-momentum shells). This is probably the most serious difficulty which will occur in generalizing this method to more complicated systems.

**ACKNOWLEDGMENTS**

We would like to thank Professor Kenneth M. Watson for his interest in this work and Dr. David Judd of the Lawrence Radiation Laboratory for the hospitality which he extended to one of us (C.J.J.) during his stay in Berkeley. We are particularly grateful to the staff of the Lawrence Radiation Laboratory Computer Center for its generous assistance in solving the numerical problems connected with this work.

[20] C. C. J. Roothaan and A. W. Weiss, Rev. Mod. Phys. 32, 194 (1960).

# Chapter VI

# *Atom Viewed as a Nonuniform Electron Gas (Approach II)*

## VI-1

Chapter III included a discussion of the exchange terms of Hartree–Fock theory in their relation to the behavior of electrons in an electron gas. The present chapter deals with electron correlation in atoms viewed as a (strongly) nonuniform electron gas. The local electron density in an atom is high. One is therefore led to the use of methods applicable in the high-density uniform electron gas as treated by Gell-Mann and Brueckner (Chapter VI-2). A recent review of the uniform electron gas theory has been given by Lars Onsager (in *Modern Quantum Chemistry*, Vol. 2, O. Sinanoğlu, ed., Academic Press, New York, 1965, p. 271).

The calculation of correlation energy in the uniform electron gas was the first problem in the quantum theory of interacting many-body systems given careful attention. The problem presented considerable difficulty because of the divergence of perturbation theory applied in a straightforward way, the singularity, which appears first in second order, resulting from the large matrix elements of the Coulomb interaction for small momentum transfer. Although Wigner (1932) was able to obtain an estimate of the correlation energy, the physical origin of the difficulty was not clearly described until Bohm and Pines (1950) recognized the collective properties of the electron gas resulting from the infinite range of the Coulomb force, which gives rise both to screening of the force and to plasma oscillations. Their treatment clarified the problem but was mathematically complicated and the approximations somewhat obscure. The exact solution for the high-uniform-density electron gas was given by Gell-Mann and Brueckner (1957), who showed that an expansion in powers of the density could be obtained by suitable summation of the divergent terms of the perturbation expansion.

The principal problem of the correlation-energy determination, which was recognized by Wigner and by Bohm and Pines, arises from the many-body screening of the electron–electron interaction by the adjustment of many electrons moving between the interacting pair. The collective nature of this phenomenon appears in the perturbation treatment through the difficulty of treatment of the small momentum transfers. As shown by Gell-Mann and Brueckner, the perturbations giving rise to screening result from ring diagrams in which the maximum possible number of electrons in a given order of perturbation contribute. The structure of the perturbation series for the energy then is given by a series of the form

$$\epsilon_{\text{corr}} \sim \int \frac{dq}{q} \sum_n (-1)^n C_n \left(\frac{r_s}{q^2}\right)^n \qquad \text{rydbergs} \tag{1}$$

with the momentum transfer q measured in units of the Fermi momentum, $r_s$ the average electron spacing measured in atomic units, and the coefficients $C_n$ dimensionless constants of the order of unity. The sum of this series gives the Gell-Mann–Brueckner result, which can also be derived in closed form by other methods.

The important contribution of the higher-order diagrams in the electron gas suggests that the corresponding transitions in the atomic case will also be important, although the divergences are of course absent because of the finite atomic size and the discrete energy-level spacing. In the above formula, for example, the effect of the finite size may be estimated by using the Fermi–Thomas radius to determine the average density and the minimum momentum transfer, which gives

$$q_{min} \approx \frac{1}{R_{FT}\ k_F} \approx Z^{1/3} \qquad r_s \approx 1/Z^{2/3} \qquad (r_s/q^2)_{max} \approx 1 \tag{2}$$

Thus the series no longer diverges but the convergence appears to be slow. This problem can, of course, be resolved by computation using, for example, the methods of Kelly. The absence of significant contribution of these many-electron terms to the correlation energy of atoms and molecules will, of course, markedly simplify the correlation energy problem and further justify the methods summarized in Chapters V and VII.

Several attempts have been made to determine the correlation energy of atoms, using the uniform electron gas results, interpolated between the exact high- and low-density results. These calculations give a result approximately double the experimental result, showing that the electrons correlate less effectively in the finite atomic system than in the uniform gas. The importance of resolving this anomaly was increased by the general theory of Hohenberg and Kohn, which gave an improved basis for analysis of the energy starting from the energy density functional and from increasing data on correlation energy which confirmed the early results on the overestimate given by the gas results. This problem was resolved by Ma and Brueckner, who were able, using an extension of the Gell-Mann–Brueckner results and the Kohn density-functional formalism, to obtain the next term in the density gradient expansion. They showed that the correction to the uniform gas energy was large, and that this was to be expected as a consequence of the rapid change in electron density over the screening length estimated from the local electron density. Their result also may be interpreted as showing the very strongly perturbing effect of the strong central Coulomb field on the electron motion, which allows less perturbation of the electron motion by the electron–electron interaction. The reduction of the many-body screening also indicates that much more rapid convergence of the perturbation expansion is to be expected in atomic calculations. Other collective aspects of the structure reflected in modes of excitation, for example, may also be expected to be absent or reduced.

Although Ma and Brueckner were able to obtain a rather accurate fit to the correlation energy of the light atoms by making an *ad hoc* modification of their gradient expansion to obtain convergence, their results indicate that a priori methods starting from the electron gas analysis have little utility. This result also implies that the Hohenberg–Kohn methods can be applied to atoms only with great difficulty because of the requirement for extensive knowledge of the density functional. It is probably safe to conclude that calculations which take account of the finite atomic size in first approximation are required.

It is also worth noting that the special features of the atomic problem further simplify the use of approximation methods. In particular, the use of variational methods may be effective for large atoms under conditions of electron number for which the linked-cluster† expansion is usually required. This may be seen by consideration of the first-order correction to the wave function,

$$\psi = \left(1 + \frac{1}{\Delta E}\, v\right) \psi_0 \tag{3}$$

†The words "cluster" and "linked" and "unlinked" are used in different senses mathematically in the present context of diagrammatic many-body perturbation theory than in the MET work of Sinanoğlu. The reader may consult the precise definitions in respective papers to avoid possible confusion.

The normalization shift of $\psi$ relative to $\psi_0$ is given by

$$N^2 = 1 + \Delta \tag{4}$$

with

$$\Delta = \left(\psi_0, v \frac{1}{(\Delta E)^2} v\psi_0\right)$$

$$= \frac{1}{2} \sum_{iji'j'} v^2_{ij,i'j'} \frac{1}{(\Delta E_{ij,i'j'})^2} \tag{5}$$

This is readily computed, using the methods of Gell-Mann and Brueckner, and is, except for numerical factors of the order of unity,

$$\Delta = \frac{Z}{a_0^2 k_F^2} \int \frac{dq}{q^2} \tag{6}$$

The screening of the interaction cuts off the integral at $q_{min} = r_s^{1/2}$ which is also, as noted above, at the Fermi-Thomas radius. Thus, since $k_F a_0 = (q/4\pi)^{1/3}(1/r_s)$,

$$\Delta \approx Z r_s^{3/2} \tag{7}$$

We therefore note that at fixed density or fixed $r_s$, the normalization correction becomes large as $Z \to \infty$, and $\psi$ becomes orthogonal to $\psi_0$. This is the usual condition in uniform quantum many-body systems and requires the use of the linked-cluster expansion for the energy, the wave function itself being difficult to construct and essentially impossible to use.

In the atomic problem, the increase in the number of electrons is combined with an increase in the nuclear charge and hence in a contraction of the volume with $r_s \sim Z^{-2/3}$. Thus $\Delta$ is independent of Z and the probability of pair excitation does not increase with the number of pairs. Conversely, the probability that the average pair is excited by the electron–electron interaction varies as $1/Z^2$.

The considerations above are only rough order-of-magnitude estimates valid only on the average for the atom. Large variations are to be expected for the last bound electrons for which special features and degeneracies of the configuration become important. The general conclusions, however, show the special features of the atomic problem, which sharply differentiate it from other many-body systems.

 Reprinted from *Physical Review* **106**, 364 (1957)

# Correlation Energy of an Electron Gas at High Density*

MURRAY GELL-MANN, *Department of Physics, California Institute of Technology, Pasadena, California*

AND

KEITH A. BRUECKNER, *Department of Physics, University of Pennsylvania, Philadelphia, Pennsylvania*

(Received December 14, 1956)

The quantity $\epsilon_c$ is defined as the correlation energy per particle of an electron gas expressed in rydbergs. It is a function of the conventional dimensionless parameter $r_s$, where $r_s^{-3}$ is proportional to the electron density. Here $\epsilon_c$ is computed for small values of $r_s$ (high density) and found to be given by $\epsilon_c = A \ln r_s + C + O(r_s)$. The value of $A$ is found to be 0.0622, a result that could be deduced from previous work of Wigner, Macke, and Pines. An exact formula for the constant $C$ is given here for the first time; earlier workers had made only approximate calculations of $C$. Further, it is shown how the next correction in $r_s$ can be computed. The method is based on summing the most highly divergent terms of the perturbation series under the integral sign to give a convergent result. The summation is performed by a technique similar to Feynman's methods in field theory.

WE consider the idealized problem of the ground-state energy of a gas of electrons in the presence of a uniform background of positive charge that makes the system neutral. For most practical problems, of course, the uniform positive charge must be replaced by a lattice of positive ions, but we shall not treat this more realistic case.

We have, then, a fully degenerate Fermi-Dirac system with Coulomb interactions. Let us employ the conventional notation for the problem. The inverse density or volume per electron is set equal to $\frac{4}{3}\pi r_0^3$. The dimensionless parameter $r_s$ is defined as $r_0$ divided by the Bohr radius. The ground state energy per particle in rydbergs is called $\epsilon$ and is a function of $r_s$ only, since there are no other dimensionless quantities involved.

We shall compute $\epsilon$ in the case of high density or small $r_s$. Since $r_s$ is proportional to $e^2$, an expansion in powers of $r_s$ is essentially an expansion in powers of $e^2$, that is, the perturbation expansion. Unfortunately, a straightforward perturbation expansion leads to divergences, but let us ignore that difficulty for a moment.

The leading term in the perturbation series is evidently the Fermi energy, the kinetic energy of the degenerate free electron gas. The maximum electron momentum $P$, the radius of the Fermi sphere in momentum space, is given by $P = (9\pi/4)^{\frac{1}{3}} \hbar r_0^{-1}$. The Fermi energy per particle in rydbergs is thus

$$\epsilon_F = \frac{3}{5}\left(\frac{P^2}{2m}\right) \bigg/ \frac{e^4 m}{2\hbar^2} = \frac{3}{5}\left(\frac{9\pi}{4}\right)^{\frac{2}{3}} \frac{1}{r_s^2} \approx \frac{2.21}{r_s^2}. \quad (1)$$

The next term in the series is the exchange energy, the expectation value of the potential energy in the ground state of a free electron gas. It is one higher order in $e^2$ or $r_s$ than the Fermi energy and so is proportional to $1/r_s$. It is easily evaluated to be

$$\epsilon_x = -\frac{3}{4}\left(\frac{e^2}{\pi}\right) P \bigg/ \frac{e^4 m}{2\hbar^2} = -\frac{3}{2\pi}\left(\frac{9\pi}{4}\right)^{\frac{1}{3}} \frac{1}{r_s} \approx -\frac{0.916}{r_s}. \quad (2)$$

* This study was performed by the authors as consultants to the RAND Corporation, Santa Monica, California, and was sponsored entirely by the U. S. Atomic Energy Commission.

If we now calculate the effect of the potential in second-order perturbation theory, we should expect a term of one higher order in $e^2$ or $r_s$ than (2), that is, a constant independent of $r_s$. However, the second-order perturbation formula diverges logarithmically at small momentum transfers on account of the long-range character of the Coulomb force. Thus some refinement of perturbation theory is necessary in order to carry the computation further.

The terms in the energy beyond (1) and (2) are called collectively the "correlation energy,"

$$\epsilon_c \equiv \epsilon - \epsilon_F - \epsilon_x. \quad (3)$$

This name was introduced by Wigner,[1] who called attention to the importance of the correlation energy in solid-state problems. Following Wigner's lead, calculations were made by Macke[2] and by Pines[3] that led essentially to expressions of the form

$$\begin{aligned}\epsilon_c &= (2/\pi^2)(1-\ln 2)\ln r_s + C \\ &\qquad + \text{terms that vanish as } r_s \to 0, \quad (4)\\ &\approx 0.0622 \ln r_s + C + \cdots,\end{aligned}$$

where in each case the constant $C$ was calculated approximately.[4]

We give here an *exact* evaluation of the constant $C$ by a method that should permit also the calculation of higher corrections in $r_s$. The basic idea of the method is to examine the increasingly divergent terms of the perturbation series and to notice that they fall into

[1] E. P. Wigner, Phys. Rev. **46**, 1002 (1934).

[2] W. Macke, Z. Naturforsch. **5a**, 192 (1950).

[3] D. Pines, Phys. Rev. **92**, 626 (1953); D. Pines, in *Solid State Physics*, edited by F. Seitz and D. Turnbull (Academic Press, Inc., New York, 1955), Vol. 1, p. 367.

[4] Pines' result is actually not of the form of Eq. (4). However, he has neglected a term which he calls the exchange correlation energy and which adds to his result for $\epsilon$ the quantity $0.0311 \ln r_s - 0.0905 + \epsilon_b^{(2)} + O(r_s)$, where $\epsilon_b^{(2)}$ is defined in our Eq. (9). When we supply this term, his final answer takes on the form of Eq. (4). When we quote Pines' result later, we mention this modification. Macke's result does agree with Eq. (4). However, he seems to have made certain unnecessary approximations in his calculation of $C$, and we have therefore recomputed $C$, using his method, in Appendix I.

subseries that can be summed under the integral sign to give convergent results. The logarithmic divergence in second order is then automatically replaced by a logarithmic dependence on the expansion parameter $r_s$, as in (4). In this respect, our method is similar to Macke's.[2] However, Macke fails to sum *all* of the processes that contribute to the constant $C$. In our work, we are able to exhibit all of them and then to sum them by a procedure similar to Feynman's methods in field theory.

Let us discuss, then, the behavior of the formal perturbation series for $\epsilon$. The coefficient of each power of $r_s$ can be written as an integral over various dimensionless vectors $\mathbf{q}_i$, which are virtual momentum transfers divided by $P$, the Fermi momentum. The term independent of $r_s$ then diverges logarithmically, as we have said. The next term, formally linear in $r_s$, diverges quadratically, the succeeding one quartically, etc. Now, since the correlation energy is finite, these integrals, when summed, must cut themselves off at some characteristic value of the dimensionless momentum transfers $q$. Moreover, the nature of the cutoff is clear from the results of work on the plasma vibrations of an electron gas, especially that of Bohm and Pines.[5] It has been shown that collective electron motions effectively screen the Coulomb field at a distance of the order of

$$r_{\max}\sim(\text{const.})\, r_0^{\frac{1}{2}}a^{\frac{1}{2}}+\text{higher terms in } r_0, \tag{5}$$

where $a$ is the Bohr radius. The effective cutoff for $q$ is then

$$q_{\min}\sim(\text{const.})\, r_s^{\frac{1}{2}}+\text{higher terms in } r_s. \tag{6}$$

We use this estimate in conjunction with our estimate of the correlation energy

$$\epsilon_c\sim(\text{log divergence})+r_s\,(\text{quadratic div.})+r_s^2\,(\text{quartic div.})+\cdots, \tag{7}$$

and we deduce the following results:

(i) $\epsilon_c=A\ln r_s+C+$terms that vanish as $r_s\to 0$.

(ii) The coefficient $A$ can be found merely from the strength of the logarithmic divergence in second-order perturbation theory. This leads to the value of 0.0622 quoted above.

(iii) The only virtual processes contributing to $C$ beyond the second order are those which contribute the highest divergence in each order of perturbation theory. Those processes leading to lower divergences will give higher powers of $r_s$ in the final expression for $\epsilon$.

Now the processes that lead to the highest divergences are easily identified. The divergences are caused by the piling-up of factors $1/q^2$ coming from Coulomb interactions in momentum space. Evidently the greatest piling-up occurs when only a single momentum transfer is involved and this single $q$ is handed from electron to electron, contributing a factor $1/q^2$ each time. Moreover, we may distinguish momentum transfers with and without exchange; when exchange occurs the factor is no longer $1/q^2$ but $1/(\mathbf{p}_1-\mathbf{p}_2+\mathbf{q})^2$, where $\mathbf{p}_1$ and $\mathbf{p}_2$ are the initial electron momenta. In the case of exchange, then, the singularity at $q=0$ is not enhanced. We may therefore ignore exchange entirely beyond the second order in computing $C$.

We may now list the processes that contribute to $C$ in the first few orders of perturbation theory. In second order we must include everything; we have two terms, the logarithmically divergent one we have mentioned and a finite one coming from exchange. They may be written as follows:

$$\epsilon_a^{(2)}=-\frac{3}{8\pi^5}\int\frac{d^3q}{q^4}\int_{\substack{p_1<1\\|\mathbf{p}_1+\mathbf{q}|>1}}d^3p_1\int_{\substack{p_2<1\\|\mathbf{p}_2+\mathbf{q}|>1}}d^3p_2\times\frac{1}{q^2+\mathbf{q}\cdot(\mathbf{p}_1+\mathbf{p}_2)}, \tag{8}$$

and

$$\epsilon_b^{(2)}=\frac{3}{16\pi^5}\int\frac{d^3q}{q^2}\int_{\substack{p_1<1\\|\mathbf{p}_1+\mathbf{q}|>1}}d^3p_1\int_{\substack{p_2<1\\|\mathbf{p}_2+\mathbf{q}|>1}}d^3p_2\times\frac{1}{(\mathbf{q}+\mathbf{p}_1+\mathbf{p}_2)^2}\frac{1}{q^2+\mathbf{q}\cdot(\mathbf{p}_1+\mathbf{p}_2)}. \tag{9}$$

In these processes two electrons in the Fermi sea with initial momenta $\mathbf{p}_1P$ and $-\mathbf{p}_2P$ have undergone a collision with momentum transfer $\mathbf{q}P$, emerging into unoccupied states with momenta $(\mathbf{p}_1+\mathbf{q})P$ and $(-\mathbf{p}_2-\mathbf{q})P$ and then returning to their original states. The factor $1/[q^2+\mathbf{q}\cdot(\mathbf{p}_1+\mathbf{p}_2)]$ comes from the energy denominator. In (8) there is a factor $1/q^2$ for each collision, while in the exchange correlation energy (9) one factor is $1/q^2$ and the other $1/(\mathbf{q}+\mathbf{p}_1+\mathbf{p}_2)^2$.

In writing higher order integrals, we shall not indicate explicitly the conditions $|\mathbf{p}+\mathbf{q}|>1$ and $p<1$, but these must still be obeyed by all vectors $\mathbf{p}$ to insure that initial states are occupied and others unoccupied. In third order the processes involving a single momentum transfer are these: Two electrons with momenta $\mathbf{p}_1P$ and $-\mathbf{p}_2P$ emerge from the sea into states with momenta $(\mathbf{p}_1+\mathbf{q})P$ and $(-\mathbf{p}_2-\mathbf{q})P$, as before. One of them now returns to its original state and transfers its excess momentum $\mathbf{q}P$ or $-\mathbf{q}P$ to a third electron, which emerges from a state with momentum $\mathbf{p}_3P$. This third one and the one still outstanding now return to their original states. Both third-order processes contribute equal amounts to the energy, since the first and second electrons are quite equivalent and it does not matter which one first interacts with the third electron.

The processes involved may be represented diagrammatically as in Fig. 1. In second order, two electrons called 1 and 2 are excited and then de-excited. In third order, electrons 1 and 2 are excited, one of them is de-

[5] D. Bohm and D. Pines, Phys. Rev. 92, 609 (1953).

excited while exciting a third electron 3, and then the outstanding electrons are de-excited. And so forth.

The third-order contribution is given by

$$\epsilon^{(3)}=2\left(\frac{\alpha r_s}{\pi^2}\right)\left(\frac{3}{8\pi^5}\right)\int\frac{d^3q}{q^6}\int d^3p_1\int d^3p_2\int d^3p_3$$

$$\times\frac{1}{q^2+\mathbf{q}\cdot(\mathbf{p}_1+\mathbf{p}_2)}\,\frac{1}{q^2+\mathbf{q}\cdot(\mathbf{p}_1+\mathbf{p}_3)}, \qquad (10)$$

where $\alpha$ is $(4/9\pi)^{\frac{1}{3}}$.

In fourth order, rather complicated processes begin to appear, as one may see from Fig. 1. In the first diagram in fourth order, electron 1 is twice replaced before de-exciting with electron 2. In the next, electron 1 is replaced, then electron 2 is replaced, and then the de-excitation takes place. The following two diagrams are similar with electrons 1 and 2 exchanging roles. In the last two diagrams the excitation of 1 and 2 is followed by the excitation of another pair 3 and 4; then electrons 1 and 3 de-excite together and electrons 2 and 4 de-excite together, in either order.

The fourth-order contributions are given by

$$\epsilon^{(4)}=-2\left(\frac{\alpha r_s}{\pi^2}\right)^2\left(\frac{3}{8\pi^5}\right)\int\frac{d^3q}{q^8}\int d^3p_1\int d^3p_2\int d^3p_3\int d^3p_4$$

$$\times\left\{\frac{1}{q^2+\mathbf{q}\cdot(\mathbf{p}_1+\mathbf{p}_2)}\,\frac{1}{q^2+\mathbf{q}\cdot(\mathbf{p}_1+\mathbf{p}_3)}\,\frac{1}{q^2+\mathbf{q}\cdot(\mathbf{p}_1+\mathbf{p}_4)}\right.$$

$$+\frac{1}{q^2+\mathbf{q}\cdot(\mathbf{p}_1+\mathbf{p}_2)}\,\frac{1}{q^2+\mathbf{q}\cdot(\mathbf{p}_1+\mathbf{p}_3)}\,\frac{1}{q^2+\mathbf{q}\cdot(\mathbf{p}_3+\mathbf{p}_4)}$$

$$+\frac{1}{q^2+\mathbf{q}\cdot(\mathbf{p}_1+\mathbf{p}_2)}\,\frac{1}{2q^2+\mathbf{q}\cdot(\mathbf{p}_1+\mathbf{p}_2+\mathbf{p}_3+\mathbf{p}_4)}$$

$$\left.\times\frac{1}{q^2+\mathbf{q}\cdot(\mathbf{p}_1+\mathbf{p}_3)}\right\}. \qquad (11)$$

It is easy now to write down the contributions from any order. The problem then is to sum all these contributions before performing the integral over $q$. In order to see how to sum them, we note the similarity to diagrams in field theory. There, of course, pairs of *electrons* and *holes* (positrons) are under consideration, interacting with an external field. Here we consider pairs of *electrons* in interaction with the Fermi sea and ignore the holes. Nevertheless the similarity is sufficient for the application of Feynman's artifice of considering a pair as composed of one electron traveling forward in time and one backward in time. The creation or annihilation of a pair is interpreted as the turning-around of a particle in time. If we look back at Fig. 1 with Feynman's point of view in mind, we see that in each order all the diagrams are merely versions of one single diagram in which a single electron starts out, is replaced over and over again, and finally returns to its starting point. The various forms of the diagram come from the choice that faces the electron each time it is replaced, a choice of being replaced by an electron going forward in time or one going backward in time.

If we introduce "time variables," then, and let the electrons propagate either forward or backward in time with suitable propagators, we should be able to represent the sum of all diagrams in each order by a single integral. The integrals over the time-variables should give us the various energy denominators we need. We try as the propagator the function

$$F_q(t)=\int d^3p\,\exp[-|t|(\tfrac{1}{2}q^2+\mathbf{q}\cdot\mathbf{p})], \qquad (12)$$

which is arranged so that integration over positive or negative time will introduce into the energy denominator plus or minus $(\frac{1}{2}q^2+\mathbf{q}\cdot\mathbf{p})$, respectively. Now we integrate around a loop using this propagator. In second order we look at

$$A_2\equiv\frac{1}{2}\int_{-\infty}^{\infty}dt_1\int_{-\infty}^{\infty}dt_2F_q(t_1)F_q(t_2)\delta(t_1+t_2), \qquad (13)$$

where the $\delta$ function insures that the electron comes

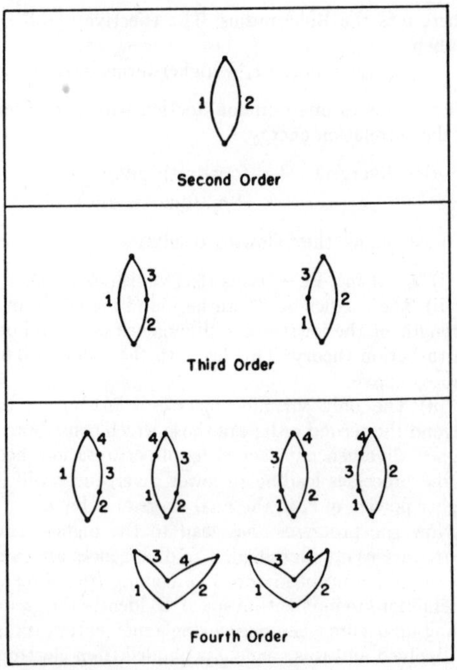


FIG. 1. The relevant second-, third-, and fourth-order processes represented diagrammatically. In second order the process may take place with or without exchange. In higher orders, exchange is neglected.

back to its starting point. Then we have

$$A_2=\int d^3p_1\int d^3p_2\int_0^\infty dt$$
$$\times\exp\{-(\tfrac{1}{2}q^2+\mathbf{q}\cdot\mathbf{p}_1)t+(\tfrac{1}{2}q^2+\mathbf{q}\cdot\mathbf{p}_2)(-t)\} \quad (14)$$
$$=\int d^3p_1\int d^3p_2\frac{1}{q^2+\mathbf{q}\cdot(\mathbf{p}_1+\mathbf{p}_2)},$$

just what we need for Eq. (8).

In third order, we look at

$$A_3\equiv\frac{1}{3}\int_{-\infty}^{\infty}dt_1\int_{-\infty}^{\infty}dt_2\int_{-\infty}^{\infty}dt_3$$
$$\times F_q(t_1)F_q(t_2)F_q(t_3)\delta(t_1+t_2+t_3). \quad (15)$$

We find

$$A_3=2\int d^3p_1\int d^3p_2\int d^3p_3\int_0^\infty dt_1\int_{-t_1}^0 dt_2$$
$$\times\exp\{-(\tfrac{1}{2}q^2+\mathbf{q}\cdot\mathbf{p}_1)t_1+(\tfrac{1}{2}q^2+\mathbf{q}\cdot\mathbf{p}_2)t_2$$
$$+(\tfrac{1}{2}q^2+\mathbf{q}\cdot\mathbf{p}_3)(-t_1-t_2)\}=2\int d^3p_1\int d^3p_2\int d^3p_3$$
$$\times\frac{1}{q^2+\mathbf{q}\cdot(\mathbf{p}_1+\mathbf{p}_2)}\,\frac{1}{q^2+\mathbf{q}\cdot(\mathbf{p}_1+\mathbf{p}_3)}, \quad (16)$$

which is what we need for Eq. (10).

The agreement evidently extends to all orders. It should be noted that there is a direct correspondence between the various forms of the diagram in a given order and the various time-orderings in the corresponding integral $A_n$.

Now let us perform the Fourier transform of the $\delta$ function in the expression for $A_n$. We obtain

$$A_n=\frac{q}{2\pi n}\int_{-\infty}^{\infty}du[Q_q(u)]^n, \quad (17)$$

where

$$Q_q(u)=\int d^3p\int_{-\infty}^{\infty}e^{ituq}\exp\{-|t|[\tfrac{1}{2}q^2+\mathbf{q}\cdot\mathbf{p}]\}dt. \quad (18)$$

The terms in the correlation energy contributing to the constant $C$ may then be summed, leaving aside the exchange term $\epsilon_b^{(2)}$. We have

$$\epsilon'\equiv\epsilon_a^{(2)}+\epsilon^{(3)}+\epsilon^{(4)}+\cdots=-\frac{3}{8\pi^5}\int\frac{d^3q}{q^3}\frac{1}{2\pi}$$
$$\times\int_{-\infty}^{\infty}du\sum_{n=2}^{\infty}\frac{(-1)^n}{n}[Q_q(u)]^n\left(\frac{\alpha r_s}{\pi^2q^2}\right)^{n-2}. \quad (19)$$

This expression may now be enormously simplified when we realize that beyond the second order we are interested only in the most highly divergent part of the $q$ integral, that is, the leading term near $q=0$. We may thus take, beyond the second order, just this leading term and put the upper limit of the $q$ integral equal to some arbitrary number, which we take to be 1.

At small values of $q$, we may approximate $Q_q(u)$ as follows: We apply the restrictions on $\mathbf{p}$ that $p<1$ and $|\mathbf{p}+\mathbf{q}|>1$. If $x$ is the direction cosine between $\mathbf{p}$ and $\mathbf{q}$ we see that at small $q$ the variable $x$ is restricted to the range $0\leq x\leq 1$ and $p$ to the range $1-qx\leq p\leq 1$. Thus we have, for $q\ll 1$,

$$Q_q(u)\approx 2\pi q\int_0^1 xdx\int_{-\infty}^{\infty}dte^{ituq}e^{-|t|qx}$$
$$=2\pi\int_0^1 xdx\int_{-\infty}^{\infty}d^qe^{isu}e^{-|s|x}=4\pi R(u), \quad (20)$$

independent of $q$. We have put

$$R(u)=1-u\arctan u^{-1}. \quad (21)$$

We shall thus approximate $Q_q(u)$ by a function which is equal to $4\pi R(u)$ when $0\leq q\leq 1$ and which vanishes for $q>1$. In second order, we must supply a correction term that restores $\epsilon_a^{(2)}$ to its exact value, but in higher orders the approximation is sufficient. We have, then,

$$\epsilon'\approx-\frac{12}{\pi^3}\int_{-\infty}^{\infty}du\int_0^1\frac{dq}{q}\sum_{n=2}^{\infty}\frac{(-1)^n}{n}$$
$$\times[R(u)]^n\left(\frac{4\alpha r_s}{\pi q^2}\right)^{n-2}+\delta, \quad (22)$$

where we have put

$$\delta\equiv\epsilon_a^{(2)}-\left[-\frac{12}{\pi^3}\int_{-\infty}^{\infty}du\int_0^1\frac{dq}{q}(\tfrac{1}{2}R^2)\right]$$
$$=\lim_{\beta\to 0}\left\{-\frac{3}{8\pi^5}\int_\beta^\infty\frac{d^3q}{q^4}\int_{\substack{p_1<1\\|\mathbf{p}_1+\mathbf{q}|>1}}d^3p_1\int_{\substack{p_2<1\\|\mathbf{p}_2+\mathbf{q}|>1}}d^3p_2\right.$$
$$\left.\times\frac{1}{q^2+\mathbf{q}\cdot(\mathbf{p}_1+\mathbf{p}_2)}+\frac{6}{\pi^2}\int_\beta^1\frac{dq}{q}\int_0^1 xdx\int_0^1 ydy\frac{1}{x+y}\right\}, \quad (23)$$

which is a finite number, the logarithmic divergences cancelling.

We comment here on a difficulty which occurs in carrying out the summation over $n$. For large values of $q$, the series converges and the summation is straightforward. For small $q$, however, the series diverges, with large contributions arising from large $n$. We shall, however, *assume* that the result valid for large $q$ may be continued into the region of divergence. This procedure cannot be justified without a detailed investigation of the behavior of the series for large $n$; we argue, however, that any corrections must either vanish with large $N$ or contribute to higher powers of $r_s$.

We may now perform the sum over $n$ and the integral over $q$ in Eq. (22), remarking that the integral really does cut itself off at a value of $q$ proportional to $r_s^{\frac{1}{2}}$, as predicted in Eq. (6). Dropping terms that vanish as $r_s \to 0$, we have

$$\epsilon' = \frac{3}{\pi^3}\int_{-\infty}^{\infty} du[R(u)]^2\left[\ln\left(\frac{4\alpha r_s}{\pi}\right) + \ln R(u) - \tfrac{1}{2}\right] + \delta$$
$$= \frac{2}{\pi^2}(1-\ln 2)\left[\ln\left(\frac{4\alpha r_s}{\pi}\right) + \langle \ln R\rangle_{\text{Av}} - \tfrac{1}{2}\right] + \delta \tag{24}$$

where

$$\langle \ln R\rangle_{\text{Av}} = \int_{-\infty}^{\infty} duR^2 \ln R \Big/ \int_{-\infty}^{\infty} duR^2. \tag{25}$$

We see that Eq. (24) confirms our value of $(2/\pi^2)\times(1-\ln 2)$ for the constant $A$. For the constant $C$, we have

$$C = \frac{2}{\pi^2}(1-\ln 2)\left\{\ln\left[\frac{4}{\pi}\left(\frac{4}{9\pi}\right)^{\frac{1}{3}}\right] - \tfrac{1}{2} + \langle \ln R\rangle_{\text{Av}}\right\} + \delta. \tag{26}$$

Now Pines[3] has found the value $-0.0508$ for $\delta$, and numerical integration yields the value $-0.551$ for $\langle \ln R\rangle_{\text{Av}}$. The multiple integral (9) for $\epsilon_b^{(2)}$ has been evaluated by the Monte Carlo method, with the result $\epsilon_b^{(2)} = 0.046 \pm 0.002$. Substituting these numbers into (26), we find

$$C = -0.096 \pm 0.002, \tag{27}$$

$$\epsilon_c = 0.0622 \ln r_s - 0.096 + O(r_s). \tag{28}$$

The expression given by Pines[3] is $0.0311 \ln r_s - 0.114 + O(r_s)$, although if the correction mentioned in footnote 4 is taken into account one gets by Pines' method the result $0.0622 \ln r_s - 0.158 + O(r_s)$.

In the Appendix we evaluate $\epsilon_c$ by Macke's method and obtain $0.0622 \ln r_s - 0.128 + O(r_s)$.

We see that the approximations of Macke and Pines tend to overestimate the magnitude of the constant term $C$ in the correlation energy.

In conclusion, let us discuss the calculation of the next correlation to $\epsilon$. In order to include all terms that are genuinely of order $r_s$ or $r_s \ln r_s$, we must improve the present calculation in three ways;

(i) The contribution to $\epsilon'$ from Eq. (19) must be treated more carefully than in Eq. (22), so that terms of order $r_s$ are retained.

(ii) We must calculate the contribution from the diagrams in Fig. 1 beyond the second order when *one* exchange is permitted in each process. Beyond the third order we may employ the crudest approximation that preserves the leading divergence.

(iii) The remaining third-order processes are the following:

$1+2 \to 1'+2'$
$1'+2' \to 1''+2''$ } rescattering.
$1''+2'' \to 1+2$

$1+2 \to 1'+2'$
$1'+3 \to 1'+3$ } direct and exchange scattering with unexcited particles.
$1'+2' \to 1+2$

Those are both logarithmically divergent and must be combined with a sequence of terms similar to those summed to remove the second-order divergence. The methods presented above are easily generalized to this case.

The authors would like to thank Dr. Richard Latter for many valuable discussions and Mr. J. I. Marcum and Mr. H. Kahn for the Monte Carlo computation of $\epsilon_b^{(2)}$.

### APPENDIX. APPROXIMATION OF MACKE

The method of Macke[2] is suggested by the earlier work of Wigner.[1] It consists of summing, instead of the complete set of diagrams indicated in Fig. 1, just the first diagram in each order. Under the integral sign these form a simple geometric series. We obtain, in place of Eq. (19), the following:

$$\epsilon' \approx \frac{3}{8\pi^5}\int\frac{d^3q}{q^4}\int_{\substack{p_1<1\\|\mathbf{p}_1+\mathbf{q}|>1}} d^3p_1 \sum_{n=1}^{\infty}(-1)^n \times\left(\int_{\substack{p_2<1\\|\mathbf{p}_2+\mathbf{q}|>1}} d^3p_2 \frac{1}{q^2+\mathbf{q}\cdot(\mathbf{p}_1+\mathbf{p}_2)}\right)^n\left(\frac{\alpha r_s}{\pi^2 q^2}\right)^{n-1}. \tag{A.1}$$

Making an approximation analogous to (20) in the orders beyond the second, we find instead of (22) the expression

$$\epsilon' \approx \delta + \frac{6}{\pi^2}\int_0^1 \frac{dq}{q}\int_0^1 xdx \sum_{n=1}^{\infty}(-1)^n \times\left(\int_0^1 \frac{ydy}{x+y}\right)^n\left(\frac{2\alpha r_s}{\pi q^2}\right)^{n-1}. \tag{A.2}$$

We may now, as before, perform the sum over $n$ and the integral over $q$, dropping terms that vanish as $r_s \to 0$. Putting $I = \int_0^1 ydy/(x+y)$, we find

$$\epsilon' \approx \delta + \frac{3}{\pi^2}\int_0^1 xdxI \ln\left(\frac{2\alpha r_s I}{\pi}\right). \tag{A.3}$$

The approximate value of $C$ is then

$$C \approx \epsilon_b^{(2)} + \delta + \frac{3}{\pi^2}\int_0^1 xdxI \ln\left(\frac{2\alpha I}{\pi}\right) \approx \epsilon_b^{(2)} - 0.174. \tag{A.4}$$

VI-3 Reprinted from *Physical Review* **165**, 18 (1968)

# Correlation Energy of an Electron Gas with a Slowly Varying High Density*

SHANG-KENG MA AND KEITH A. BRUECKNER
*Institute for Radiation Physics and Aerodynamics and Department of Physics, University of Calfironia, San Diego, La Jolla, California*
(Received 15 June 1967)

The correlation energy (the exact energy minus the Hartree-Fock energy) of an electron gas with a high and slowly varying density is examined. The term proportional to the square of the density gradient is evaluated by the application of perturbation theory to the external field and of the random-phase (or high-density) approximation to the Coulomb interaction. This term has the form $\Delta E_c[\rho]=\int d^3x\, B(\rho(x))|\nabla\rho(x)|^2$, where $\rho(x)$ is the electron density. $B(\rho)$ is found, by summing the leading divergent diagrams, to be $[8.470\times10^{-3}+O(\rho^{-1/3}\ln\rho)+O(\rho^{-1/3})]\rho^{-4/3}$ Ry, with the length measured in units of the Bohr radius. The role of the density gradient in the correlation energy problem of atoms is discussed.

## I. INTRODUCTION

THE correlation energy problem of a many-electron system has not been extensively investigated except for a uniform electron gas at high density, where the random-phase approximation is applicable and the translational symmetry greatly simplifies the problem. The calculation of the correlation energy for a nonuniform electron gas or for systems with a finite number of electrons has not been carried out except for the lightest atoms,[1] although the Hartree-Fock energy for the most important nonuniform electron gas, namely, the many-electron atoms, can be calculated with high accuracy.[2] Because of its complexity, the calculation of the correlation energy of a general nonuniform electron gas seems to be beyond the technique available at present. However, the problem becomes tractable under the idealized condition that the density is high and slowly varying. When the density is slowly varying, the correlation energy may be expanded in powers of the density gradient. Such an expansion can be defined unambiguously with the aid of the density-functional formalism recently developed by Kohn and his coworkers.[3–5] When the density is high, the coefficients of the density gradient expansion may be calculated by the standard diagrammatic perturbation theory.

The objective of this paper is to examine the correlation energy in the limit of high and slowly varying density. The correction term to the correlation energy of a uniform electron gas which is proportional to the square of the density gradient is evaluated by analyzing the second-order contribution of the perturbation expansion in powers of the external field. Qualitative features of the correlation energy problem of atoms are also examined. In Sec. II, we describe the basic physical features of the problem and give a qualitative estimate of the effect of the density gradient on the correlation energy. Since the small density nonuniformity must be caused by an external field, the problem reduces to a study of the response of a uniform electron gas to an external field. Section III is devoted to a derivation of the expression of the correlation energy in terms of the long wavelength values of the density response function of a uniform electron gas. To separate the correlation energy from the total energy in a transparent and unambiguous way, we make moderate use of the basic language of the density-functional formalism[3]. The response function is studied in terms of diagrams, and the subset of diagrams contributing to the correlation energy are identified. The mathematical procedures for evaluating these diagrams are carried out in Sec. IV, and the explicit formula for the correlation energy is determined. In Sec. V, the role of the density gradient in atoms is considered. Because of the large density gradient in atoms, the density gradient expansion is shown to be a divergent series. It is also shown that the collective motion of electrons plays a much less important role in atoms than in the uniform or nearly uniform gas and the notion of the dielectric constant is not very useful in describing electrons in atoms.

The appendices are devoted to the mathematical details involving the derivatives of the Green's function and a few singular integrals needed in the text. Some useful formulas are listed. These appendices should serve as a convenient reference for other calculations involving the high-density electron gas.

## II. QUALITATIVE FEATURES

Let us first review some of the basic characteristics of a uniform electron gas at high density. The average Coulomb repulsion among the electrons is canceled by the field of a uniform positive external charge background and thus, a uniform average density is maintained. By high density, we mean that the average interparticle distance $r_0$ is small compared to the Bohr radius, i.e.,

$$r_0 \ll (e^2m)^{-1},$$

* This research was supported by the Advanced Research Projects Agency (Project DEFENDER) and was monitored by the U. S. Army Research Office (Durham) under Contract No. DA-31-124-ARO-D-257.

[1] H. P. Kelly, Phys. Rev. **144**, 39 (1966).
[2] E. Clementi, J. Chem. Phys. **41**, 303 (1964).
[3] P. Hohenberg and W. Kohn, Phys. Rev. **136**, B864 (1964).
[4] W. Kohn and L. J. Sham, Phys. Rev. **140**, A1133 (1965).
[5] B. Y. Tong and L. J. Sham, Phys. Rev. **144**, 1 (1966).

or

$$\frac{e^2}{r_0} \ll \frac{p_0^2}{2m}, \tag{2.1}$$

where $p_0 \sim r_0^{-1}$ is the Fermi momentum. The energy of the system is the sum of the kinetic energy and the interaction energy due to the correlation of density fluctuations. The correlation over a distance comparable to $r_0$ can be approximately described in terms of the motion of free electrons since, by (2.1), the Coulomb potential becomes much less than the kinetic energy of the electron states which contribute to the correlation. The correlation over a distance large compared to $r_0$ is mainly due to the long-range Coulomb force. On the other hand, the motion of the electrons tends to screen the Coulomb force produced by the density fluctuations. The screening may be characterized by the screening length $L_s$. The density fluctuations separated by a distance much larger than $L_s$ become uncorrelated. The leading terms of the perturbation expansion in powers of $e^2$ and $\ln e^2$ based on the above physical picture has been extensively studied.[6,7,8] The screening length is given by

$$L_s \sim (e^2 m p_0)^{-1/2}. \tag{2.2}$$

Now suppose that the external charge distribution becomes nonuniform. Then, the average electron density will be nonuniform. There will be a change in the electrostatic energy of the charge distribution as a whole, a change in the kinetic energy $E_{\text{kin}}$, and a change in the interaction energy $E_{\text{int}}$ due to the modification of the correlation of density fluctuations. Explicitly, the total energy is

$$E = \int d^3x\, \varphi(\mathbf{x})\rho(\mathbf{x}) + \tfrac{1}{2}\int d^3x d^3r\, \rho(\mathbf{x})\rho(\mathbf{x}+\mathbf{r})\frac{e^2}{r} + E_{\text{kin}} + E_{\text{int}}, \tag{2.3}$$

where $\varphi(\mathbf{x})$ is the potential due to the external charge and $\rho(\mathbf{x})$ is the average density. From the above discussion of a uniform electron gas, we see that $E_{\text{int}}$ involves only the correlation of density fluctuations, which is small beyond a distance $L_s$. More explicitly,

$$E_{\text{int}} \approx \tfrac{1}{2}\int d^3x d^3r\, \langle \rho'(\mathbf{x})\rho'(\mathbf{x}+\mathbf{r})\rangle \frac{e^2}{r},$$

$$\langle \rho'(\mathbf{x})\rho'(\mathbf{x}+\mathbf{r})\rangle \to 0, \quad \text{for } r \gg L_s,$$

where $\rho'(\mathbf{x})$ is the density fluctuation. Therefore, if $\rho(\mathbf{x})$ varies very little over a distance of a screening length, i.e., if

$$L_s^2 |\nabla\rho/\rho|^2 \equiv g \ll 1, \tag{2.4}$$

we expect that

$$E_{\text{int}} \approx \int d^3x\, E_{\text{int}}^0(\rho(x)), \tag{2.5}$$

where $E_{\text{int}}^0(\rho)$ is the interaction energy per unit volume of a uniform gas with density $\rho$. To obtain better approximations for $E_{\text{int}}$, one naturally expects an expansion in powers of $g$ to be helpful. Such an expansion is the "gradient expansion."[3]

We are only interested in the correlation energy of a nonuniform electron gas. The Hartree-Fock energy includes the first three terms of (2.3) and a part of $E_{\text{int}}$ because of the correlation arising from the Pauli principle. If the condition (2.4) is satisfied, we expect that the correlation energy $E_c$ may be written as

$$E_c = \int d^3x\, E_c^0(\rho(x)) + \Delta E_c,$$
$$\Delta E_c \equiv \int d^3x\, B(\rho(x))|\nabla\rho(x)|^2 + O(g^2), \tag{2.6}$$

where $E_c^0(\rho)$ is the correlation energy per unit volume of a uniform electron gas of density $\rho$ and $B(\rho)$ is a function to be determined in the next two sections. Let us give a crude estimate of $B$ here.

In calculating the HF energy of a uniform gas, no screening is taken into account. The correlation energy is roughly the correction due to the effect of screening

$$E_c \approx \tfrac{1}{2}\int d^3x d^3r\, \{\langle \rho'(\mathbf{x})\rho'(\mathbf{x}+\mathbf{r})\rangle_{\text{screened}} - \langle \rho'(\mathbf{x})\rho'(\mathbf{x}+\mathbf{r})\rangle_{\text{unscreened}}\}\frac{e^2}{r}. \tag{2.7}$$

The unscreened density-fluctuation correlation function is well known to be[9]

$$-\tfrac{9}{2}\rho^2[(\sin p_0 r - p_0 r \cos p_0 r)/p_0^3 r^3]^2.$$

Thus, the screened correlation function is, for $r \gg p_0^{-1}$,

$$\langle \rho'(\mathbf{x})\rho'(\mathbf{x}+\mathbf{r})\rangle_{\text{screened}} \approx -9/4(3\pi)^{-4/3}[\rho^{2/3}/r^4]e^{-r/L_s}, \tag{2.8}$$

where we have substituted $(3\pi^2\rho)^{1/3}$ for $p_0$ and $\frac{1}{2}$ for $\cos^2 p_0 r$. We need not consider the small $r$ behavior of the correlation function because the effect of the density gradient is appreciable only for large $r$. The constants $\rho$ and $L_s$ in (2.8) are no longer defined for a nonuniform gas. However, if the density is nearly uniform, one may use the density at $\mathbf{x}$ to evaluate these constants. This approximation is good for $r$ much less than $L_s$. As $r$ becomes larger, one may use the density at $\mathbf{x}+\frac{1}{2}\mathbf{r}$ to evaluate these constants. As a function of $\mathbf{r}$, (2.8) be-

[6] M. Gell-Mann and K. A. Brueckner, Phys. Rev. **106**, 364 (1957).

[7] K. Sawada, Phys. Rev. **106**, 372 (1957).

[8] K. Sawada, K. A. Brueckner, N. Fukuda, and R. Brout, Phys. Rev. **108**, 507 (1957).

[9] See, for example, C. Kittel, *Quantum Theory of Solids* (John Wiley & Sons, Inc., New York, 1963), Chap. V.

comes distorted from the density correlation function for a uniform gas as a result of the slight variation of the constants. An estimate of the effect of the density gradient on the correlation energy may be deduced from such a distortion, which is given by the difference

$$\Delta(\mathbf{x},\mathbf{r})\equiv\rho^{2/3}(\mathbf{x}+\mathbf{r}/2)\{\exp[-r/L_s(\rho(\mathbf{x}+\mathbf{r}/2))]-1\} -\rho^{2/3}(\mathbf{x})\{\exp[-r/L_s(\rho(\mathbf{x}))]-1\}. \quad (2.9)$$

Expanding (2.9) in powers of $\nabla\rho(x)$ assuming

$$\rho\left(\mathbf{x}+\frac{\mathbf{r}}{2}\right)=\rho(\mathbf{x})+\frac{\mathbf{r}}{2}\cdot\nabla\rho(\mathbf{x}),$$

we find, keeping only the second order in $\nabla\rho(\mathbf{x})$

$$\Delta(\mathbf{x},\mathbf{r})=\left\{-\frac{r^3}{12}\nabla\rho^{2/3}\cdot\nabla L_s^{-1}+\frac{r^4}{24}\rho^{2/3}(\nabla L_s^{-1})^2\right\}e^{-r/L_s}. \quad (2.10)$$

A crude estimate of the effect of the density gradient on the correlation energy is given by

$$\Delta E_c\sim\tfrac{1}{2}\int d^3x d^3r[-9(3\pi^2)^{-4/3}/4r^4]\Delta(\mathbf{x},\mathbf{r})\frac{e^2}{r}. \quad (2.11)$$

The $r$ integrals are easily performed. One finds

$$\Delta E_c\sim 8^{-1}3^{-1/3}\pi^{-5/3}e^2\int d^3x\{L_s\nabla\rho^{2/3}\cdot\nabla L_s^{-1} -\tfrac{1}{2}\rho^{2/3}(L_s\nabla L_s^{-1})^2\}. \quad (2.12)$$

Using the fact that $L_s\sim(e^2mp_0)^{-1/2}$ is proportional to $\rho^{-1/6}$, (2.12) reduces to

$$\Delta E_c\sim 2.7\times10^{-3}e^2\int d^3x\,\rho^{-4/3}(\mathbf{x})|\nabla\rho(\mathbf{x})|^2. \quad (2.13)$$

Although (2.13) is a very crude estimate, it does show that $B(\rho)$ [see (2.6)] is of the order $O(e^2)$. Since $E_c^0$, the correlation energy of a uniform gas, is of $O(e^4)+O(e^4\ln e^2)$ [see (3.16)] and $L_s^2=O(e^{-2})$, one might expect, by (2.4), that $\Delta E_c=O(gE_c^0)=O(e^2)+O(e^2\ln e^2)$. Equation (2.13) and the calculation in the next sections show that the $O(e^2\ln e^2)$ term is absent.

The above estimate is based on the picture of a distorted density correlation function and a distorted screening length. The distortion is described by the average density $\rho(x)$, which is regarded as a parameter slowly varying in space. However, nothing has been said about the physical processes which give rise to the distortions. Since the nonuniform density is due to an external field, we have to study the response of the many-electron system to an external field. When the density variation is small, the analysis reduces to the analysis of the density response of a uniform electron gas. The next two sections are devoted to the determination of $B(\rho)$ based on such an analysis.

## III. FORMULATION

In this section, we search for a formula of the form of (2.6) for the correlation energy of a nearly uniform electron gas. It is clear, from the qualitative discussion in the previous section, that the problem should reduce to an analysis of the response of a uniform gas to a weak external field, which causes the small density nonuniformity. From such an analysis, one can determine the energy and the density in terms of the external field. One then looks for a way to eliminate the explicit dependence of the correlation energy on the external field and obtain an expression in terms of the density only. We find it convenient to start our analysis with the density-functional formulation.[3]

The energy $E$ and the average density $\rho(\mathbf{x})$ may be obtained by minimizing the energy functional $E[\rho]$ keeping the total number of electrons fixed. We break up $E[\rho]$ into two parts:

$$E[\rho]\equiv E_{\mathrm{HF}}[\rho]+E_c[\rho]. \quad (3.1)$$

$E_{\mathrm{HF}}[\rho]$ is defined such that, by minimizing it, one would obtain the energy and the density in the Hartree-Fock approximation. $E_c[\rho]$ is the "correlation energy functional." When the density variation is small, we expect that $E_c[\rho]$ may be expanded in powers of the density gradient $\nabla\rho(\mathbf{x})$. If only the two lowest-order terms are kept, one has

$$E_c[\rho]=\int d^3x\,\{A(\rho(\mathbf{x}))+B(\rho(\mathbf{x}))|\nabla\rho(\mathbf{x})|^2\}, \quad (3.2)$$

where $A(\rho)$, $B(\rho)$ can be determined by applying (3.1) and (3.2) to a weakly perturbed uniform gas in the following way.

Consider an electron gas in a unit volume perturbed by a weak static field:

$$\varphi(\mathbf{x})=\sum_{k\neq0}\varphi_{\mathbf{k}}e^{i\mathbf{k}\cdot\mathbf{x}}.$$

The perturbing Hamiltonian is then

$$\mathcal{H}=\sum_{k\neq0}\varphi_{\mathbf{k}}\hat{\rho}_{-\mathbf{k}},$$

where $\hat{\rho}_{\mathbf{k}}$ is the density operator. The energy to the second order in $\varphi$ is

$$E=E^0(n)+\sum_{k\neq0}\varphi_{\mathbf{k}}\varphi_{-\mathbf{k}}\sum_m\frac{\langle0|\hat{\rho}_{\mathbf{k}}|m\rangle\langle m|\hat{\rho}_{-\mathbf{k}}|0\rangle}{E^0-E^m}$$
$$=E^0(n)+\tfrac{1}{2}\sum_{k\neq0}\varphi_{\mathbf{k}}\varphi_{-\mathbf{k}}\mathfrak{F}(\mathbf{k},0), \quad (3.3)$$

where $n$, $|0\rangle$, $E^0$, $|m\rangle$, $E^m$ are, respectively, the density, the exact ground and excited states and their energies of a uniform electron gas, and $\mathfrak{F}(k,\omega)$ is the density response function of a uniform electron gas given by

$$\mathfrak{F}(\mathbf{k},\omega)\equiv -i\int d^4x\, e^{-i\mathbf{k}\cdot\mathbf{x}+i\omega t}\langle T(\hat{\rho}(\mathbf{x},t)\hat{\rho}(0))\rangle$$
$$=\mathfrak{F}(-\mathbf{k},\,\omega)=\mathfrak{F}(\mathbf{k},\,-\omega). \tag{3.4}$$

The average $\langle\cdots\rangle$ is taken over the ground state of a uniform electron gas. The density, to the first order in $\varphi$, is given by

$$\rho_{\mathbf{k}}=\varphi_{\mathbf{k}}\mathfrak{F}(\mathbf{k},0)\,,$$
$$\rho_0=n. \tag{3.5}$$

Equations (3.3) and (3.5) summarize the effect of the external field. $\mathfrak{F}(\mathbf{k},0)$ contains the full information of the response of the uniform electron gas to the external field. Now we have to separate the correlation energy and Hartree-Fock energy from the total energy. This is accomplished in the following manner.

The results (3.3) and (3.5) may be obtained by minimizing the functional

$$E[\rho]=E^0(n)+\sum_{k\neq 0}\varphi_{-\mathbf{k}}\rho_{\mathbf{k}}-\tfrac{1}{2}\sum_{k\neq 0}\rho_{\mathbf{k}}\rho_{-\mathbf{k}}\mathfrak{F}^{-1}(\mathbf{k},0). \tag{3.6}$$

Similarly, minimizing the functional

$$E_{\mathrm{HF}}[\rho]=E^0{}_{\mathrm{HF}}(n)+\sum_{k\neq 0}\varphi_{-\mathbf{k}}\rho_{\mathbf{k}}-\tfrac{1}{2}\sum_{k\neq 0}\rho_{\mathbf{k}}\rho_{-\mathbf{k}}\mathfrak{F}_{\mathrm{HF}}{}^{-1}(\mathbf{k},0) \tag{3.7}$$

will give the density and energy in the Hartree-Fock approximation. Here $\mathfrak{F}_{\mathrm{HF}}$ is the density-response function in the Hartree-Fock approximation. It is the sum of all the Hartree-Fock energy diagrams perturbed twice by an external field. (See Fig. 1.) Thus, the correlation energy functional for a weakly perturbed uniform gas is, by (3.1), (3.6), (3.7),

$$E_c[\rho]=E[\rho]-E_{\mathrm{HF}}[\rho]$$
$$=E_c{}^0(n)-\tfrac{1}{2}\sum_{k\neq 0}\rho_{\mathbf{k}}\rho_{-\mathbf{k}}(\mathfrak{F}^{-1}(\mathbf{k},0)-\mathfrak{F}^{-1}{}_{\mathrm{HF}}(\mathbf{k},0))\,, \tag{3.8}$$

where

$$E_c{}^0(n)\equiv E^0(n)-E_{\mathrm{HF}}{}^0(n)\,,$$

is the correlation energy of a uniform electron gas. The density-response function $\mathfrak{F}$ may be expressed as a geometric sum (see Fig. 2)

$$\mathfrak{F}=F+FVF+FVFVF+\cdots$$
$$=F/(1-VF)\,,$$

and hence,

$$\mathfrak{F}^{-1}=F^{-1}-V\,,$$

where

$$V\equiv 4\pi e^2/k^2\,,$$

and $F$ contains no isolated interaction line. Equation (3.8) now becomes

$$E_c[\rho]=E_c{}^0(n)-\tfrac{1}{2}\sum_{k\neq 0}\rho_{\mathbf{k}}\rho_{-\mathbf{k}}(F^{-1}(\mathbf{k},0)-F_{\mathrm{HF}}{}^{-1}(\mathbf{k},0)). \tag{3.9}$$

The contribution of the $-V$ term of $\mathfrak{F}^{-1}$

$$\tfrac{1}{2}\sum_{k\neq 0}\rho_{\mathbf{k}}\rho_{-\mathbf{k}}(4\pi e^2/k^2)$$

cancels that in $\mathfrak{F}_{\mathrm{HF}}{}^{-1}$. This cancellation means that the electrostatic energy of the electron charge distribution as a whole is included in the Hartree-Fock energy and plays no part in the correlation energy.

Equation (3.9) is an expression with no explicit dependence on the external field. Furthermore, it is applicable even if the density varies rapidly in space as long as the amplitude of the variation is small compared to $n$. To obtain $A$ and $B$ of (3.2), we only need the long-wavelength part of (3.9). On the other hand, (3.2) makes sense even if the density varies appreciably as long as it is slowly varying, i.e., as long as it varies very little over a screening length, according to the qualitative discussion of the previous section. Therefore, to compare (3.9) and (3.2), we expand the integrand of (3.2) in powers of $(\rho(\mathbf{x})-n)$ and the $(F^{-1}-F_{\mathrm{HF}}{}^{-1})$ of

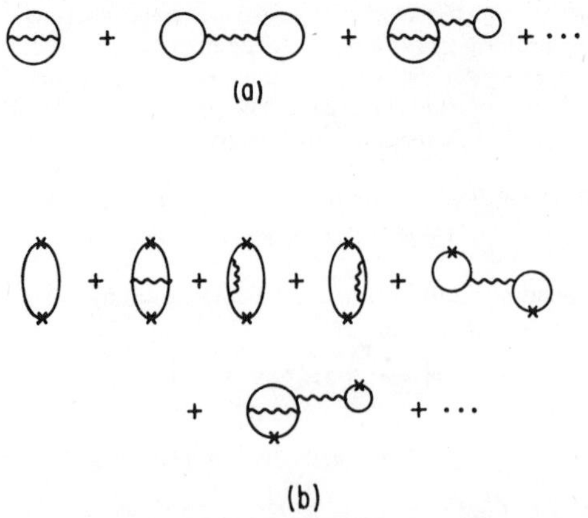


FIG. 1. (a) The Hartree-Fock energy diagrams. Because of the translational symmetry and the uniform charge background, only the first diagram is nonzero for a uniform electron gas. (b) The first diagram gives the second-order perturbation on the kinetic energy. Each cross represents a density vertex with the momentum of the external field.

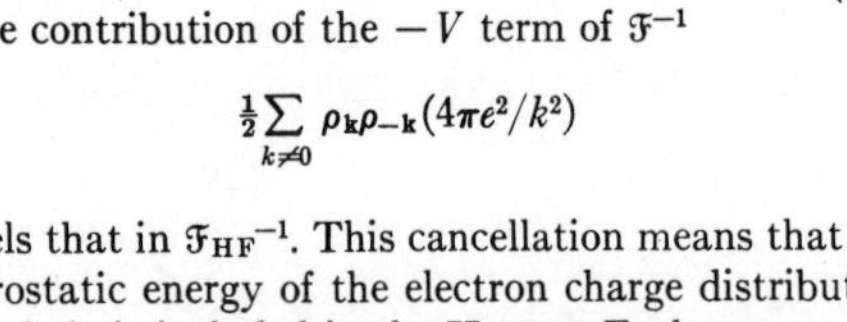


FIG. 2. The full density-response function as a geometric sum of $F$.

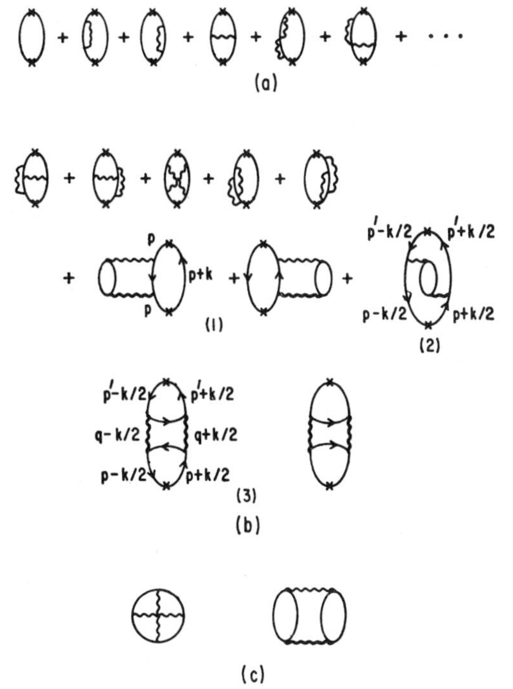


FIG. 3. (a) Diagrams for $F_{\mathrm{HF}}(\mathbf{k},0)$. (b) Additional terms for $F(\mathbf{k},0)$. (c) The correlation energy diagrams calculated by Gell-Mann and Brueckner. The thick wavy line represents an interaction line modified by the dielectric constant in the random-phase approximation.

(3.9) in powers of $k^2$. Let

$$F(\mathbf{k},0)=a^{-1}+bk^2+\cdots, \tag{3.10}$$

then

$$F^{-1}(\mathbf{k},0)=a-a^2bk^2+\cdots. \tag{3.11}$$

$a^{-1}$ is given by the "compressibility sum rule"[10]

$$a^{-1}=\lim_{k\to 0}F(\mathbf{k},0)=-dn/d\mu.$$

Therefore,

$$a=-(d^2/dn^2)E^0(n). \tag{3.12}$$

Expanding (3.2) gives

$$E_c[\rho]=A(n)+\frac{d^2}{dn^2}A(n)\tfrac{1}{2}\sum_{k\neq 0}\rho_{\mathbf{k}}\rho_{-\mathbf{k}}+B(n)\sum_{k\neq 0}k^2\rho_{\mathbf{k}}\rho_{-\mathbf{k}}. \tag{3.13}$$

Substituting (3.11), (3.12) in (3.9) and comparing the result with (3.13), we conclude that

$$A(n)=E_c^0(n),\qquad B(n)=\tfrac{1}{2}(a^2b-a_{\mathrm{HF}}^2b_{\mathrm{HF}}). \tag{3.14}$$

[10] See, for example, D. Pines, in *Lectures on the Many-Body Problem*, edited by E. R. Caianiello (Academic Press Inc., New York, 1964), Vol. II.

The correlation energy can be calculated once $A$ and $B$ are substituted in (3.2). If we regard $E_c[\rho]$ as a small correction to $E_{\mathrm{HF}}[\rho]$, the correlation energy $E_c$ is approximately given by

$$E_c=E_c[\rho_{\mathrm{HF}}], \tag{3.15}$$

where $\rho_{\mathrm{HF}}(\mathbf{x})$ is the density in the Hartree-Fock approximation, and minimizes $E_{\mathrm{HF}}[\rho]$:

$$\left\{\frac{\delta}{\delta\rho(\mathbf{x})}E_{\mathrm{HF}}[\rho]\right\}_{\rho=\rho_{\mathrm{HF}}}=0.$$

The remaining task is to evaluate (3.14), i.e., to find the ground-state energy and the long-wavelength behavior of the density-response function for the uniform electron gas. For such an analysis, we proceed to apply the perturbation theory, which seems to be the only technique available. The perturbation expansion is an expansion in powers of $e^2m/p_0$, which is valid only in the high-density limit. In this limit, the ground-state energy is well known.

$$\begin{aligned}E^0&\equiv E_{\mathrm{HF}}^0+E_c^0,\\ E_{\mathrm{HF}}^0&=n\{\tfrac{3}{5}p_0^2/2m-(3/4\pi)e^2p_0\},\\ E_c^0&=\tfrac{1}{2}e^4mn(0.0622\ln r_s-0.096)+o(e^4),\end{aligned} \tag{3.16}$$

where

$$r_s\equiv\frac{e^2m}{p_0}\left(\frac{9\pi}{4}\right)^{1/3},$$

and $p_0$ is the Fermi momentum. $a$, $a_{\mathrm{HF}}$ can be obtained by differentiating $E^0$ and $E_{\mathrm{HF}}^0$, respectively [see (3.12)]. The function $F(\mathbf{k},0)$, and hence $b$, can be calculated using the perturbation expansion given in Fig. 3. The sum of the diagrams in Fig. 3(a) gives $F_{\mathrm{HF}}$. The sum of all the diagrams in Fig. 3(a) and Fig. 3(b) gives $F$. Let

$$b\equiv b_{\mathrm{HF}}+b_c. \tag{3.17}$$

$b_c$ is then the contribution of Fig. 3(b). The diagrams may seem to indicate that $b_c=O(e^4)$, but the last five diagrams of Fig. 3(b) diverge unless the interaction lines are properly modified by the dielectric constant. As one would expect, after introducing the dielectric constant (in the random-phase approximation), Fig. 3(b) becomes the correlation energy diagrams of Gell-Mann and Brueckner perturbed twice by an external field. [See Fig. 3(c).]

The qualitative argument in the previous section has shown that

$$b_c=O(e^2). \tag{3.18}$$

By (3.14) and (3.17),

$$2B(n)=(a^2-a_{\mathrm{HF}}^2)b_{\mathrm{HF}}+a^2b_c. \tag{3.19}$$

Since $a-a_{\mathrm{HF}}=O(e^4)$, we have

$$B(n)=\left(\frac{\pi^2}{p_0m}\right)^2\tfrac{1}{2}b_c+o(e^2), \tag{3.20}$$

using the fact that

$$a=-\frac{d^2E^0}{dn^2}$$

$$=\frac{-\pi^2}{p_0m}+O(e^2).$$

The formula (3.20) may be understood qualitatively in the following way. $b$ is the long-range part of $F(\mathbf{k},0)$, which may be viewed as the second derivative of the energy with respect to the external potential as is clearly demonstrated by the diagrams. $\pi^2/p_0m$ is the derivative of the chemical potential, or roughly of the external field, with respect to the density. Thus, $B(n)$ is qualitatively the second derivative of the correlation energy with respect to the long-range variation of the density.

The fact that $b_c$ diverges without the modification of the dielectric constant shows that one must have the screening of the Coulomb force for the local density and the expansion in the density gradient to be meaningful. This is obvious in view of the qualitative discussion in Sec. II. By using the dielectric constant, one effectively sums the leading divergent diagrams of all orders and defines a lower cutoff for the divergent integral over the momentum transfer of the Coulomb interaction.

## IV. EVALUATION OF $b_c$

To calculate $b_c$ to $O(e^2)$, we only need to consider the last five diagrams [labeled by (1), (2), and (3)] in Fig. 3(b). It is not practical to evaluate the diagrams for an arbitrary $k$ and then identify $b_c$ with the coefficient of $k^2$ in the $k$ expansion, because multidimensional integrals are involved in the evaluation. We shall instead obtain an expression for $b_c$ by expanding the integrand and extract the coefficient of $k^2$. Then, the integral to be performed, involving no $k$, becomes much simpler. Consider the diagrams (1) first.

### A. Diagrams (1)

These two diagrams have the same value. Their contribution to $F(\mathbf{k},0)$ is

$$F^{(1)}(\mathbf{k},0)=2\times2\int dp\, G(p)^2\Sigma(p)G(p+\mathbf{k}), \quad (4.1)$$

where the symbol $\int dp$ always denotes the four-dimensional integral

$$\int\frac{d^4p}{(2\pi)^4i}.$$

The Green's function is given by

$$G(p)\equiv[\epsilon-\epsilon_p+\mu+i\eta\,\mathrm{sgn}(\epsilon_p-\mu)]^{-1}, \quad (4.2)$$

$$\epsilon_p\equiv p^2/2m,$$

$$\mu=p_0^2/2m,$$

and

$$\Sigma(p)\equiv-\int dqV^-(q)G(p+q), \quad (4.3)$$

with

$$V^-(q)\equiv\frac{4\pi e^2}{q^2}\left(\frac{1}{\epsilon(q)}-1\right). \quad (4.4)$$

$\epsilon(\mathbf{q},\omega)$ is the dielectric constant in the random-phase approximation. The extra factor of 2 in (4.1) is due to the spin multiplicity. The precise meaning of $G^2(p)$ in (4.1) is

$$G^2(p)\equiv-\frac{\partial G(p)}{\partial\epsilon}, \quad (4.5)$$

which has a double pole (forced by momentum conservation) at $\epsilon=\epsilon_p-\mu-i\eta\,\mathrm{sgn}(\epsilon_p-\mu)$. The derivatives of $G(p)$ are discussed in more detail in Appendix A. As will be seen shortly, we need to make the substitution

$$-\frac{\partial}{\partial\epsilon}G(p)=-\frac{\partial}{\partial\mu}G(p)+2\pi i\delta(\epsilon)\delta(\mu-\epsilon_p), \quad (4.6)$$

for $G^2(p)$ in (4.1), which becomes, after integrating the $\delta(\epsilon)\delta(\mu-\epsilon_p)$ term

$$F^{(1)}(\mathbf{k},0)=-4\int dp\,\Sigma(p)G(p+\mathbf{k})\frac{\partial}{\partial\mu}G(p)$$

$$+\Sigma(p_0,0)\frac{\partial}{\partial\mu}F^0(\mathbf{k},0), \quad (4.7)$$

where

$$F^0(\mathbf{k},\omega)\equiv-2\int\frac{d^3p}{(2\pi)^3}\frac{n_{\mathbf{p+k}}-n_{\mathbf{p}}}{\omega-\epsilon_{\mathbf{p+k}}+\epsilon_p+i\eta\,\mathrm{sgn}\omega}, \quad (4.8)$$

$$n_p\equiv\theta(\mu-\epsilon_p).$$

To find the $k^2$ term of $F^{(1)}(\mathbf{k},0)$, we expand $(\partial/\partial\mu)F^0(\mathbf{k},0)$ and $G(p+\mathbf{k})$:

$$\frac{\partial}{\partial\mu}F^0(\mathbf{k},0)=-\frac{m}{\pi^2v_0}+\frac{k^2}{p_0^2}\frac{m}{12\pi^2v_0}+\cdots, \quad (4.9)$$

$$v_0\equiv p_0/m,$$

$$G(p+\mathbf{k})=G(p)+\mathbf{k}\cdot\nabla_pG(p)+\tfrac{1}{2}(\mathbf{k}\cdot\nabla_p)^2G(p)+\cdots$$

$$=G(p)-(\mathbf{k}\cdot\mathbf{p}/m+k^2/2m)\frac{\partial}{\partial\mu}G(p)$$

$$+\frac{(\mathbf{k}\cdot\mathbf{p})^2}{2m^2}\frac{\partial^2}{\partial\mu^2}G(p)+\cdots. \quad (4.10)$$

By symmetry, $(\mathbf{k}\cdot\mathbf{p})^2$ may be replaced by $k^2p^2/3$.

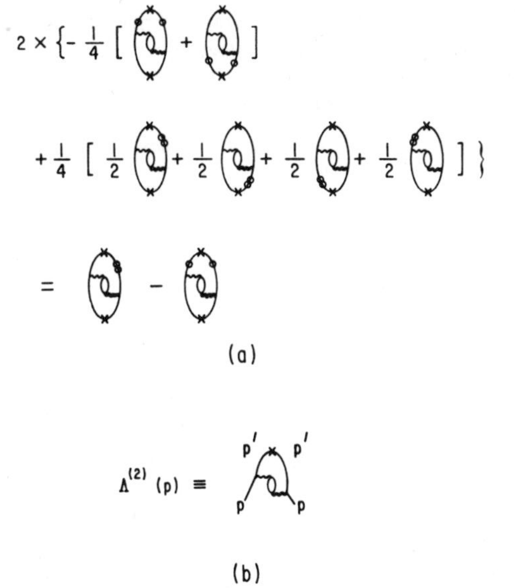


FIG. 4. (a) The $k^2$ term of the diagram (2) in Fig. 3(b). Each circle represents a factor $\mathbf{k}\cdot\nabla$ operating on the line on which the circle is drawn. (b) The zero-momentum-transfer vertex function which constitutes the lower parts of the last two diagrams in (a).

Therefore, the $k^2$ term of $-G(p+\mathbf{k})\partial G(p)/\partial\mu$ is

$$\lim_{\delta\to 0}\left\{-\frac{k^2}{2m}\frac{\partial}{\partial\mu}G(p+\boldsymbol{\delta})+\frac{k^2p^2}{6m^2}\frac{\partial^2}{\partial\mu^2}G(p+\boldsymbol{\delta})\right\}\left\{-\frac{\partial}{\partial\mu}G(p)\right\}$$

$$=-\frac{k^2}{2m}\times\tfrac{1}{6}\frac{\partial^3}{\partial\mu^3}G(p)+\frac{2k^2}{3m}\epsilon_p\times\tfrac{1}{24}\frac{\partial^4}{\partial\mu^4}G(p).\quad(4.11)$$

We have used the identity (see Appendix A)

$$\lim_{\delta\to 0}\frac{1}{(m-1)!}\left(-\frac{\partial}{\partial\mu}\right)^{m-1}G(p+\boldsymbol{\delta})\frac{1}{(n-1)!}\left(-\frac{\partial}{\partial\mu}\right)^{n-1}G(p)$$

$$=\frac{1}{(m+n-1)!}\left(-\frac{\partial}{\partial\mu}\right)^{m+n-1}G(p),\quad(4.12)$$

where $m$, $n$ are integers greater than zero. Therefore, by (4.9) and (4.11), the $k^2$ term of $F^{(1)}(\mathbf{k},0)$ is

$$\frac{k^2}{m}\int dp\,\Sigma(p)\left\{-\tfrac{1}{3}\frac{\partial^3}{\partial\mu^3}G(p)+\tfrac{1}{9}\epsilon_p\frac{\partial^4}{\partial\mu^4}G(p)\right\}$$

$$-k^2\Sigma(p_0,0)/24\pi^2v_0\mu\equiv b^{(1)}k^2.\quad(4.13)$$

### B. Diagrams (2) and (3)

The $k^2$ term of diagram (2) in Fig. 3(b) is given in Fig. 4(a). A circle represents a factor $\mathbf{k}\cdot\nabla$. The $\nabla$ operator operates on the momentum variable of the Green's-function line on which the circle is drawn. The over-all factor of 2 is due to the spin multiplicity.

Figure 4(b) defines the vertex function $\Lambda^{(2)}(p)$, i.e.,

$$\Lambda^{(2)}(p)=-\int dp'\,V(p-p')\left(-\frac{\partial G(p')}{\partial\mu}\right).\quad(4.14)$$

With the aid of the identity (4.12), the analytic expression for the diagrams in Fig. 4(a) can be obtained:

$$\frac{k^2}{m}\int dp\,\Lambda^{(2)}(p)\left(\frac{1}{2}\frac{\partial^2}{\partial\mu^2}G(p)-\tfrac{1}{9}\epsilon_p\frac{\partial^3}{\partial\mu^3}G(p)\right)$$

$$\equiv b^{(2)}k^2.\quad(4.15)$$

Diagrams (3), the last two in Fig. 3(b), contribute equally to the $k^2$ term. Similar to Fig. 4(a), 5(a) gives the $k^2$ term of diagrams (3). The over-all factor of 4 is due to the fact that each vertex has a spin multiplicity 2. The last two terms of Fig. 5(a) give

$$-\tfrac{1}{8}\int dq\left(\frac{\partial}{\partial\mu}F^0(q)\right)^2\{V(q)(\mathbf{k}\cdot\nabla_q)^2V(q)-[\mathbf{k}\cdot\nabla_qV(q)]^2\}$$

$$\equiv b_1^{(3)}k^2,\quad(4.16)$$

where

$$F^0(\mathbf{q},\omega)\equiv 2\int dp\,G(p+q)G(p)$$

is also given by (4.8). Each triangular loop gives a factor $-\frac{1}{4}\partial F^0/\partial\mu$. $V(q)$ is defined by

$$V(q)\equiv 4\pi e^2/q^2\epsilon(q).$$

Let the vertex function $\Lambda^{(3)}(p)$ be defined by Fig. 5(b). Then the first two terms in Fig. 5(a) are the same as those in Fig. 4(a) except that $\Lambda^{(2)}$ is replaced by $\Lambda^{(3)}$. Thus, their contribution $b_2^{(3)}k^2$ is obtained by substituting $\Lambda^{(3)}$ for $\Lambda^{(2)}$ in (4.15). Now $b_c$ can be obtained from

$$b_c=b^{(1)}+b^{(2)}+b_1^{(3)}+b_2^{(3)}.$$

The sum is, by (4.13), (4.15), and (4.16),

$$b_c=b'+b''+b'''-\Sigma(p_0,0)/24\pi^2v_0\mu,\quad(4.17)$$

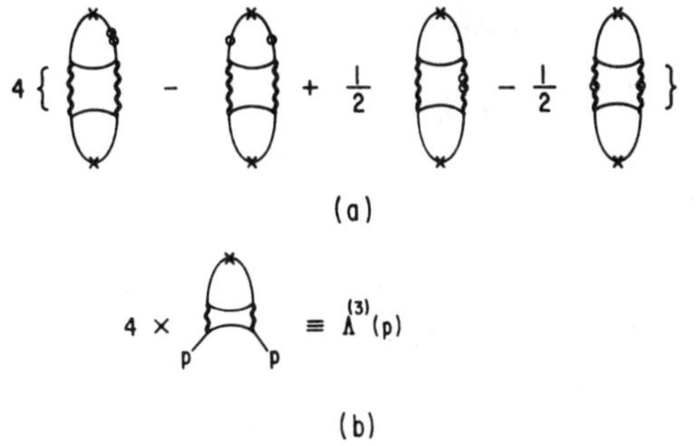


FIG. 5. (a) The $k^2$ term of diagrams (3) in Fig. 3(b). (b) The vertex function constituting the lower parts of the first two diagrams in (a).

where

$$b' \equiv -\frac{1}{m}\frac{\partial}{\partial\mu}\int dp\,\Sigma(p)\left(\tfrac{1}{2}\frac{\partial^2}{\partial\mu^2}G(p)-\tfrac{1}{9}\frac{\partial^3}{\partial\mu^3}G(p)\epsilon_p\right),$$

$$b'' \equiv \frac{1}{m}\int dp\,\Sigma(p)\tfrac{1}{6}\frac{\partial^3}{\partial\mu^3}G(p), \tag{4.18}$$

$$b''' \equiv -\tfrac{1}{24}\int dq\left(\frac{\partial}{\partial\mu}F^0(q)\right)^2 \times\left[V(q)\nabla_q^2V(q)-\left(\frac{d}{dq}V(q)\right)^2\right].$$

In deriving (4.18), we have used the identity

$$\Lambda^{(2)}(p)+\Lambda^{(3)}(p)=-\frac{\partial\Sigma(p)}{\partial\mu}, \tag{4.19}$$

which is easily seen by putting a density vertex of zero momentum on the diagrams for $\Sigma(p)$ in all possible ways.

### C. Evaluation of the Integrals

The next step is to evaluate the integrals in (4.18) and $\Sigma(p_0,0)$ to the order $O(e^2)$. Since $\Sigma(p_0,0)$ is known to be $O(e^4 \ln e^2)$,[11] we need only to consider $b'$, $b''$, and $b'''$.

Recall that [see (4.3), (4.4)]

$$\Sigma(p)=-\int dq\,V^-(q)G(p+q). \tag{4.3}$$

Since the $O(e^2)$ term of $b_c$ is due to the small $q$ part of the Coulomb interaction, and since the Green's function and its derivatives are highly singular, it is more desirable to substitute (4.3) in (4.18) and perform the $p$ integration first. Then the remaining $q$ integral is nonsingular and the $O(e^2)$ term can be extracted easily. Of course, no $p$ integral is needed for $b'''$.

Substituting (4.3) in (4.18), we have

$$b'=-\frac{1}{m}\frac{\partial}{\partial\mu}\int dq\,V^-(q)[\tfrac{1}{2}I_1(q)-\tfrac{1}{9}I_3(q)],$$
$$b''=-\frac{1}{m}\int dq\,V^-(q)\tfrac{1}{6}I_2(q), \tag{4.20}$$

where

$$I_1(q)\equiv\int dp\,G(p+q)\frac{\partial^2}{\partial\mu^2}G(p),$$
$$I_2(q)\equiv\int dp\,G(p+q)\frac{\partial^3}{\partial\mu^3}G(p), \tag{4.21}$$
$$I_3(q)\equiv\int dp\,G(p+q)\frac{\partial^3}{\partial\mu^3}G(p)\epsilon_p.$$

[11] J. Quinn and R. Ferrell, Phys. Rev. **112**, 812 (1958).

For the sake of continuity and clarity, we leave the detailed algebra of these $p$ integrals in Appendix B, and quote the results when they are needed.

The $q$ integral is a four-dimensional integral

$$\int dq\equiv\int\frac{d^3q}{(2\pi)^3}\int_{-\infty}^{\infty}\frac{d\omega}{2\pi i}.$$

It is convenient to turn the $\omega$-integral contour to the imaginary axis and let the dimensionless variable $y$ be defined by

$$\omega=iqv_0y. \tag{4.22}$$

Then, by spherical symmetry and (4.22),

$$\int dq=\frac{4\pi v_0}{(2\pi)^3}\int_0^\infty q^3dq\int_{-\infty}^{\infty}\frac{dy}{2\pi}. \tag{4.23}$$

We consider the $q$ integrals for $b'$, $b''$, and $b'''$ separately.

Since $b_c$ has the dimension of an inverse velocity, and $\partial/\partial\mu=v_0^{-1}\partial/\partial p_0$, $b'$ must be of the form [see Eq. (4.20)]

$$b'=\frac{e^2m}{v_0}\int_0^\infty\frac{dq}{q}\frac{\partial}{\partial p_0}\int_{-\infty}^{\infty}\frac{dy}{2\pi}\left(\frac{1}{\epsilon}-1\right)f\left(\frac{q}{p_0},y\right), \tag{4.24}$$

where $f(x,y)$ is a dimensionless function. Notice that $e^2m$, the inverse Bohr radius, has the dimension of a momentum. Equations (B26) and (B28) show that $I_1(q,y)$ and $I_3(q,y)$ are of $O(q^{-2})$ as $q\to 0$. Therefore, $f(x,y)=O(1)$ as $x\to 0$. The dielectric constant has the form

$$\epsilon=1-\frac{4\pi e^2}{q^2}F^0(q,iqv_0y)$$
$$=1+\frac{e^2mp_0}{q^2}\frac{4}{\pi}R\left(\frac{q}{p_0},y\right), \tag{4.25}$$

where $R(x,y)$ is a dimensionless function, and [see (B16) and (B17)]

$$R(0,y)\equiv R(y)=1-y\tan^{-1}y^{-1}. \tag{4.26}$$

The $O(e^2)$ term of (4.24) can be easily extracted. It comes from the lower limit of the $q$ integral. Let us introduce an arbitrary upper cutoff at $q=p_0$. Above any finite cutoff, the integral is of $O(e^4)$. Explicitly, the $q$ integral is now

$$e^2\int_0^{p_0}\frac{dq}{q}\frac{d}{dp_0}\left[\left(\frac{1}{\epsilon}-1\right)f\left(\frac{q}{p_0},y\right)\right]+o(e^2)$$
$$=e^2\frac{d}{dp_0}\int_0^1\frac{dx}{x}\left[1+\frac{e^2m}{p_0x^2}\frac{4}{\pi}R(y)\right]^{-1}f(0,y)$$
$$-e^2f(0,y)/p_0+o(e^2)$$
$$=-e^2f(0,y)/2p_0+o(e^2). \tag{4.27}$$

Substituting (4.27) in (4.24) and using the expressions for $I_1$, $I_3$ given by (B26) and (B28), we have

$$b'=\frac{4\pi}{(2\pi)^3}4\pi e^2\frac{m}{(2\pi)^2v_0}\int_{-\infty}^{\infty}\frac{dy}{2\pi}\left(-\frac{1}{2p_0}\right)$$

$$\times\left\{\frac{-2y^2}{(1+y^2)^2}+\frac{y^2(5+9y^2)}{9(1+y^2)^3}\right\}+o(e^2)$$

$$=\tfrac{5}{9}\frac{e^2m}{(2\pi)^3v_0p_0}+o(e^2). \quad (4.28)$$

Again, by dimensional argument, $b''$ [see (4.20)] must have the form

$$b''=\frac{e^2m}{p_0v_0}\int_0^{\infty}\frac{dq}{q}\int_{-\infty}^{\infty}\frac{dy}{2\pi}\left(\frac{1}{\epsilon}-1\right)g\left(\frac{q}{p_0},y\right), \quad (4.29)$$

where $g(x,y)$ is a dimensionless function. From (B27), we see that

$$g(0,y)=-\frac{4}{(2\pi)^3}\frac{y^2(3-y^2)}{(1+y^2)^3}. \quad (4.30)$$

Since

$$\int_{-\infty}^{\infty}dy\,g(0,y)=0,$$

(4.29) becomes, after the $q$ integral is performed,

$$b''=\frac{e^2m}{p_0v_0}\int_{-\infty}^{\infty}\frac{dy}{2\pi}[-\tfrac{1}{2}\ln R(y)]g(0,y)+o(e^2)$$

$$=\frac{e^2m}{p_0v_0}\frac{1}{(2\pi)^3}0.82872+o(e^2). \quad (4.31)$$

The $y$ integral was performed numerically.

$b'''$ [see (4.18)] may be obtained in the same manner although the algebra is more tedious. With the aid of the formulas given in Appendix B, $b'''$ reduces to, after some algebra,

$$b'''=\frac{e^2m}{p_0v_0(2\pi)^3}\times\tfrac{2}{9}\int_{-\infty}^{\infty}\frac{dy}{2\pi}\left(2+\frac{1}{1+y^2}\right)\frac{1}{(1+y^2)^2R(y)}$$

$$=\frac{e^2m}{p_0v_0(2\pi)^3}\times 0.59136. \quad (4.32)$$

Combining (4.28), (4.31), and (4.32), we have

$$b_c=b'+b''+b'''+o(e^2)$$

$$=\frac{e^2m}{p_0v_0(2\pi)^3}\times 1.97563+o(e^2). \quad (4.33)$$

The evaluation of $b_c$ to $O(e^2)$ is thus completed. One easily verifies that the small term $o(e^2)$ in (4.33) is of the order $O(e^4)+O(e^4\ln e^2)$.

Substituting (4.33) in (3.20), one obtains $B(\rho)$. Then (3.2) reads

$$E_c[\rho]=\int d^3x\,E_c^0(\rho(x))+\Delta E_c[\rho],$$

$$\Delta E_c[\rho]=4.235\times10^{-3}e^2\int d^3x\,\rho^{-4/3}|\nabla\rho(x)|^2$$

$$\times[1+O(e^2\ln e^2)+O(e^2)]. \quad (4.34)$$

If we measure lengths in units of the Bohr radius $(e^2m)^{-1}$, we have

$$\Delta E_c[\rho]=(8.470\times10^{-3}\,\text{Ry})\int d^3x\,\rho^{-4/3}|\nabla\rho(x)|^2$$

$$\times[1+O(\rho^{-1/3}\ln\rho)+O(\rho^{-1/3})]. \quad (4.35)$$

We see that the results here agree in order of magnitude with the estimate given in Sec. II. [See (2.13).]

The correlation energy is given by (3.15), i.e., by substituting the electron density in the Hartree-Fock approximation in (4.34) or (4.35).

In the above evaluation of $b_c$, there was no approximation made until we performed the $q$ integration, where only the contribution of the small $q$ is extracted. The dielectric constant, which describes the collective motion of the electrons, behaved like a lower cutoff of the otherwise linearly divergent $q$ integral. The lower $q$ cutoff by the dielectric constant near $(e^2mp_0)^{1/2}$ is equivalent to the cutoff of the density correlation outside a sphere of radius $L_s\sim(e^2mp_0)^{-1/2}$. The number of electrons inside this sphere is of the order $(p_0/e^2m)^{3/2}$, which is large in the high-density limit. Thus, the long range correlation and the collective motion of many electrons contribute substantially to the correlation energy.

The above mathematical procedure is very straightforward and involves no numerical work except for two one-dimensional integrals. The calculation of the next-order term, i.e., $O(e^4)+O(e^4\ln e^2)$, would involve more complicated numerical work for the evaluation of the exchange diagrams [the first five of Fig. 3(b)]. We have demonstrated the power of the many-body diagram technique combined with the density-functional formalism. Such a technique may be generalized to study other many-body systems with slowly varying densities.

## V. DENSITY GRADIENT IN ATOMS

For a many-electron atom, a large fraction of the electrons is concentrated within one Bohr radius around the nucleus and the density there becomes very high. In spite of the qualitative difference between the high-density uniform electron gas and the atoms, which we shall discuss in detail later, it is instructive to apply to the atoms the correlation energy formula for a high-

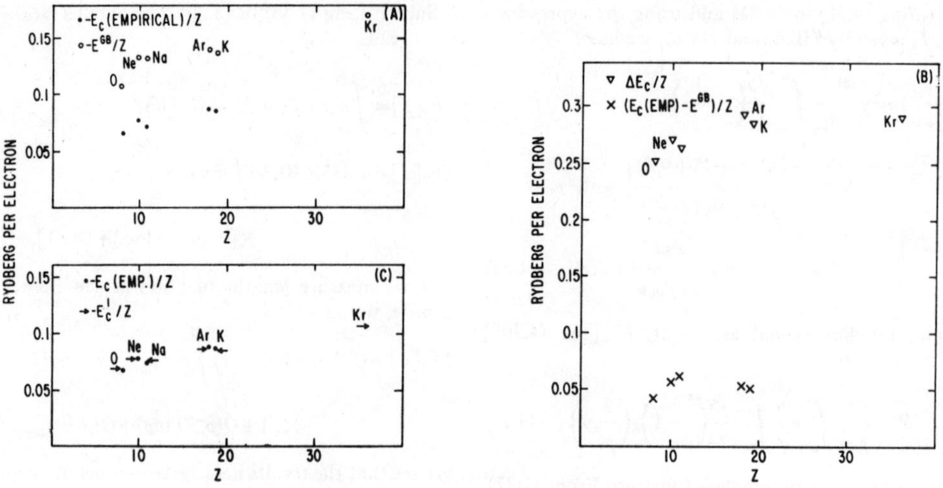


FIG. 6. $-E_c$ (empirical) is taken from the last column of Table III of Ref. 5. $E^{GB}$ is given by (5.1) and (3.16), $\Delta E_c$ from (4.36) and $E_c'$ from (5.2).

density uniform electron gas. The integral

$$E^{GB}=\int d^3x\, E_c^0(\rho(\mathbf{x})), \qquad (5.1)$$

using the Hartree-Fock electron density[12] for $\rho(\mathbf{x})$ and $E_c^0$ given by (3.16), can be easily evaluated numerically. Figure 6(a) shows $E^{GB}$ and the empirical correlation energy[13] for a few atoms. We see that $E^{GB}$ over-estimates the correlation energy by about a factor of 2 in every case. The obvious reason for the discrepancy is the fact that the density is not uniform. To estimate the effect of the nonuniformity, the formula (4.35) can be used to compute the correction to (5.1). The results are shown in Fig. 6(b). We see that (4.35) overestimates in every case by about a factor of 5 the correction needed to reproduce the empirical correlation energy. These overestimates can be easily understood as follows.

If one uses the Thomas-Fermi density instead of the Hartree-Fock density [although, strictly, the latter should be used according to (3.15)], (4.35) would be a logarithmically divergent integral due to the $r^{-3/2}$ behavior of the density near the nucleus, i.e.,

$$4\pi\int r^2dr\,[\rho(r)^{-4/3}|\nabla\rho(r)|^2]_{TF} \sim \int r^2dr\,[(r^{-3/2})^{-4/3}|r^{-5/2}|^2]\sim\int\frac{dr}{r}.$$

This divergence is weak, however, and can be cutoff at a very small $r$ to match the result given by the Hartree-Fock density, which is very high (although finite) at $r=0$. This suggests that the density gradient is too high for the gradient expansion to be useful. Recall that the physical picture behind the correlation energy formula for a uniform electron gas is the long-range correlation due to the Coulomb force. This long range is made finite by the effect of screening owing to the collective motion of electrons. Thus, qualitatively, the correlation energy involves an integral over a sphere of radius $L_s$, the screening length. The same physical picture is behind the formula for the density-gradient square correction term with more emphasis on the outer region of the sphere since $b_c$ is essentially the second moment of the density fluctuation correlation function. One encounters the same situation, when evaluating higher-order terms in the gradient expansion, with the outer region of the sphere contributing even more substantially to the integrals. In an atom, however, the electrons are concentrated around the nucleus (see Fig. 7, for example), so concentrated that a sphere of radius $L_s$ becomes larger than the high-density core, i.e., the outer region of the sphere is essentially empty.[14] In view of the above discussion, it is clear that the gradient expansion tends to overestimate and that the higher the order, the worse the overestimate. In short, the density gradient in atoms is so large that the gradient expansion becomes a divergent series.

The divergent series must be summed formally before it is used for computation. The formal sum is unknown, but a crude estimate can be made as follows. We look for a function $f(\rho,|\nabla\rho|^2)$ which is well behaved for

[12] T. G. Strand and R. A. Bonham, J. Chem. Phys. **40**, 1686 (1964). We are grateful to Dr. Jerry Peacher for pointing out this and related references to us.

[13] The empirical correlation energy has been obtained for many atoms by Clementi (see Ref. 2) by subtracting the HF energy from the experimental total energy.

[14] This point was made by C. Herring during discussion at the Slater Symposium held on Sanibal Island, January, 1967.

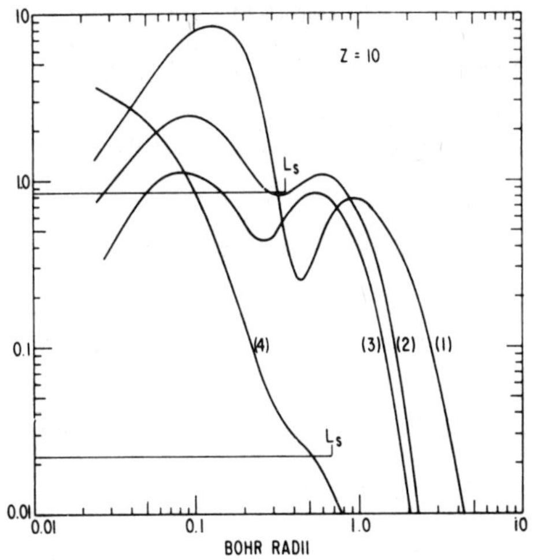


FIG. 7. (1) $4\pi r^2 B(\rho(r))|\nabla\rho(r)|^2$, see (4.36). (2) $4\pi r^2 E_c^0(\rho(r))$, see (3.16). (3) $4\pi r^2 f(\rho(r),\nabla\rho(r))$, see (5.3). (4) $\rho(r)/100$, the Hartree-Fock density in units of hundred electrons per cubic Bohr radius (Ref. 12). The screening length $L_s \sim r_s^{1/2}$ is shown for $r=0.1$ ($L_s \sim 0.37$) and for $r=0.5$ ($L_s \sim 0.68$).

infinite $|\nabla\rho|^2$ and which reduces to the integrand of (4.34) for small $|\nabla\rho|^2$. Then the difficulty arising from the very high density gradient may be avoided, and we expect

$$E_c' \equiv \int_0^\infty 4\pi r^2 dr\, f(\rho(r), |\nabla\rho(r)|^2) \qquad (5.2)$$

to be a reasonable estimate of the correlation energy. Since, if the density gradient is high, a higher density gradient would imply a smaller region in space for the distribution of the same number of electrons, and since the correlation energy is mainly a long range effect, we expect the correlation energy per electron to decrease as the density gradient increases in the region of high density gradient. Since $E_c^0(\rho)$ is roughly proportional to the density, the function

$$f(\rho, |\nabla\rho|^2) = E_c^0(\rho)/[1 - B(\rho)|\nabla\rho|^2/(yE_c^0(\rho))]^y \qquad (5.3)$$

where $y$ is an adjustable constant greater than zero, satisfies the above requirements. We found, if $y=0.32$, then $E_c'$ given by (5.2), agrees well with the empirical value. [See Fig. 6(c).] The integrand of (5.2) is shown in Fig. 7 together with the integrands of (4.34) and the density for $Z=10$. We see that the integral is mainly contributed by the electrons in the middle and outer shells where the density is high.

From the above discussion, we arrive at the following qualitative conclusions on the calculation of the atomic correlation energy. First, the gradient expansion cannot be used unless it is formally summed to infinite orders. Second, the screening due to the collective motion of electrons is not as important as it is in a uniform electron gas because the small size of the atom effectively cuts off the long-range correlation between electrons.

## ACKNOWLEDGMENT

This investigation was initiated as a result of the remarks on the problem of correlation energy in atoms made by Professor J. C. Slater at the Slater Symposium, January, 1967.

## APPENDIX A

In this Appendix, we review a few of the basic mathematical properties of the Green's function:

$$G(p) = [\epsilon - \epsilon_p + \mu + i\eta\, \mathrm{sgn}(\epsilon_p - \mu)]^{-1}$$

$$= P\frac{1}{\epsilon - \epsilon_p + \mu} - \pi i \delta(\epsilon - \epsilon_p + \mu)\, \mathrm{sgn}(\epsilon_p - \mu). \qquad (A1)$$

Since

$$\delta(\epsilon - \epsilon_p + \mu)\, \mathrm{sgn}(\epsilon_p - \mu) = \delta(\epsilon - \epsilon_p + \mu)\, \mathrm{sgn}\epsilon,$$

we have

$$G(p) = [\epsilon - \epsilon_p + \mu + i\eta\, \mathrm{sgn}\epsilon]^{-1}. \qquad (A2)$$

Equation (A2) is a convenient form if $G$ is regarded as a function of $\mu$ or $\epsilon_p$. Because the infinitesimal imaginary number $i\eta$ has a variable coefficient, extra care is needed in manipulating the Green's function. For example, differentiating (A1) with respect to $\epsilon$ and (A2) with respect to $\mu$, we have

$$-\frac{\partial}{\partial\epsilon} G(p) = [\epsilon - \epsilon_p + \mu + i\eta\, \mathrm{sgn}(\epsilon_p - \mu)]^{-2}, \qquad (A3)$$

$$-\frac{\partial}{\partial\mu} G(p) = [\epsilon - \epsilon_p + \mu + i\eta\, \mathrm{sgn}\epsilon]^{-2}. \qquad (A4)$$

Although the right-hand sides of (A3) and (A4) both appear as the square of $G$, they are not the same. The Green's function is a *distribution*. The square of a distribution is in general not defined without additional specifications. Equation (A3) represents a double pole, or the square of the simple pole $G(p)$, only if we regard it as a function of $\epsilon$. Similarly, (A4) is the square of $G(p)$ only if it is regarded as a function of $\mu$ or $\epsilon_p$. Equations (A3) and (A4) are simply related. Differentiating the last line of (A1) shows that

$$-\frac{\partial}{\partial\mu} G(p) = -\frac{\partial}{\partial\epsilon} G(p) - 2\pi i \delta(\epsilon)\delta(\mu - \epsilon_p). \qquad (A5)$$

Similarly, higher powers of $G(p)$ in the variable $\epsilon$ are different from those in the variable $\epsilon_p$ or $\mu$. Their relationship can be found by direct differentiation. For example,

$$\frac{\partial^2}{\partial\mu^2}G(p)=-\frac{\partial}{\partial\epsilon}\frac{\partial}{\partial\mu}G(p)+2\pi i\delta(\epsilon)\delta'(\mu-\epsilon_p),$$

$$=\frac{\partial^2}{\partial\epsilon^2}G(p)+2\pi i\delta'(\epsilon)\delta(\mu-\epsilon_p)+2\pi i\delta(\epsilon)\delta'(\mu-\epsilon_p), \quad \text{(A6)}$$

and

$$-\frac{\partial^3}{\partial\mu^3}G(p)=-\frac{\partial^3}{\partial\epsilon^3}G(p)-2\pi i\delta''(\epsilon)\delta(\mu-\epsilon_p)-2\pi i\delta'(\epsilon)\delta'(\mu-\epsilon_p)-2\pi i\delta(\epsilon)\delta''(\mu-\epsilon_p). \quad \text{(A7)}$$

At a static density vertex of the infinitesimal momentum $\boldsymbol{\delta}$, one encounters the limit

$$\lim_{\delta\to 0}G(p+\boldsymbol{\delta})G(p)\equiv\lim_{\delta\to 0}G(\epsilon,\mathbf{p}+\boldsymbol{\delta})G(\epsilon,\mathbf{p}). \quad \text{(A8)}$$

Regarding the $G$'s as simple poles in the $\mu$ plane, we see that (A8) reduces to a double pole in the $\mu$ plane, i.e.,

$$\lim_{\delta\to 0}G(p+\boldsymbol{\delta})G(p)$$

$$=\lim_{\delta\to 0}(\epsilon-\epsilon_{p+\delta}+\mu+i\eta\,\mathrm{sgn}\epsilon)^{-1}(\epsilon-\epsilon_p+\mu+i\eta'\,\mathrm{sgn}\epsilon)^{-1}$$

$$=(\epsilon-\epsilon_p+\mu+i\eta\,\mathrm{sgn}\epsilon)^{-2}$$

$$=-\frac{\partial}{\partial\mu}G(p). \quad \text{(A9)}$$

More generally, when an $m$th-order pole approaches an $n$th-order pole, the result is an $(m+n)$th-order pole. Thus, we have the identity

$$\lim_{\delta\to 0}\frac{1}{(m-1)!}\left(-\frac{\partial}{\partial\mu}\right)^{m-1}G(p+\boldsymbol{\delta})\frac{1}{(n-1)!}\left(-\frac{\partial}{\partial\mu}\right)^{n-1}G(p)$$

$$=\frac{1}{(m+n-1)!}\left(-\frac{\partial}{\partial\mu}\right)^{m+n-1}G(p),$$

$$m,n=1,2,3,\cdots. \quad \text{(A10)}$$

## APPENDIX B

We perform the integrals

$$I_1(q)=\int dp\,G(p+q)\frac{\partial^2}{\partial\mu^2}G(p),$$

$$I_2(q)=\int dp\,G(p+q)\frac{\partial^3}{\partial\mu^3}G(p), \quad \text{(4.19)}$$

$$I_3(q)=\int dp\,G(p+q)\frac{\partial^3}{\partial\mu^3}G(p)\epsilon_p,$$

and give explicit expressions of the functions needed in the text.

Consider $I_1(q)$. By (A6),

$$I_1(q)=\int\frac{d^3p}{(2\pi)^3}\left\{\int\frac{d\epsilon}{2\pi i}G(p+q)\frac{\partial^2}{\partial\epsilon^2}G(p)+\delta(\mu-\epsilon_p)\left[-\frac{\partial}{\partial\epsilon}G(p+q)\right]_{\epsilon=0}+\delta'(\mu-\epsilon_p)G(\omega,\mathbf{p}+\mathbf{q})\right\}. \quad \text{(B1)}$$

The first term on the right-hand side of (B1) is

$$\int\frac{d^3p}{(2\pi)^3}\left[n_{\mathbf{p}+\mathbf{q}}(1-n_{\mathbf{p}})\frac{\partial^2}{\partial\epsilon^2}G(p)-n_{\mathbf{p}}(1-n_{\mathbf{p}+\mathbf{q}})\frac{\partial^2}{\partial\epsilon^2}G(p)\right]_{\epsilon=\epsilon_{\mathbf{p}+\mathbf{q}}}$$

$$=-\frac{\partial^2}{\partial\omega^2}\int\frac{d^3p}{(2\pi)^3}\frac{n_{\mathbf{p}+\mathbf{q}}-n_{\mathbf{p}}}{D}$$

$$=\frac{\partial^2}{\partial\omega^2}\tfrac{1}{2}F^0(\mathbf{q},\omega), \quad \text{(B2)}$$

where

$$D\equiv[G^{-1}(\omega,\mathbf{p}+\mathbf{q})]_{\mu=\epsilon_p}=\omega-\epsilon_{\mathbf{p}+\mathbf{q}}+\epsilon_{\mathbf{p}}+i\eta\,\mathrm{sgn}(\epsilon_{\mathbf{p}+\mathbf{q}}-\epsilon_{\mathbf{p}}), \quad \text{(B3)}$$

and $F^0$ has been defined by (4.8). The second term on the right-hand side of (B1) is

$$\int\frac{d^3p}{(2\pi)^3}\delta(\mu-\epsilon_p)\left(-\frac{\partial}{\partial\omega}D^{-1}\right). \quad \text{(B4)}$$

Since we are interested eventually only in evaluating the $q$, $\omega$ integral, to which only the part which is even in $\omega$ will contribute, we symmetrize (B4) with respect to $\omega$ and obtain

$$-\tfrac{1}{4}\frac{\partial^2}{\partial\omega\partial\mu}J(\mathbf{q},\omega), \tag{B5}$$

where

$$J(\mathbf{q},\omega)\equiv 2\int\frac{d^3p}{(2\pi)^3}\frac{n_{\mathbf{p+q}}+n_{\mathbf{p}}}{D}. \tag{B6}$$

Henceforth in our formulas we only keep the even $\omega$ part. The third term on the right-hand side of (B1) is

$$\int\frac{d^3p}{(2\pi)^3}\delta'(\mu-\epsilon_p)G(\omega,\mathbf{p+q})$$

$$=\int\frac{d^3p}{(2\pi)^3}\left\{\delta'(\mu-\epsilon_p)[G(\omega,\mathbf{p+q})]_{\mu=\epsilon_p}+\delta(\mu-\epsilon_p)\left[-\frac{\partial}{\partial\mu}G(\omega,\mathbf{p+q})\right]_{\mu=\epsilon_p}\right\}$$

$$=\tfrac{1}{4}\frac{\partial^2}{\partial\mu^2}F^0-\tfrac{1}{4}\frac{\partial^2}{\partial\omega\partial\mu}J-2\pi i\delta(\omega)S(q), \tag{B7}$$

where

$$S(q)\equiv\int\frac{d^3p}{(2\pi)^3}\delta(\mu-\epsilon_p)\delta(\mu-\epsilon_{p+q})$$

$$=\frac{m^2}{(2\pi)^2q}\theta(2p_0-q). \tag{B8}$$

Combining (B2), (B5), and (B7), we have

$$I_1(q)=\tfrac{1}{2}F_{\omega\omega}{}^0-\tfrac{1}{2}J_{\mu\omega}+\tfrac{1}{4}F_{\mu\mu}{}^0-2\pi i\delta(\omega)S(q). \tag{B9}$$

We have used subscripts to denote partial differentiations.

In the same manner, we can express $I_2(q)$ and $I_3(q)$ in terms of $F^0$, $J$, and $S$. By (A7),

$$I_2(q)=\int\frac{d^3p}{(2\pi)^3}\left\{\int\frac{d\epsilon}{2\pi i}G(p+q)\frac{\partial^3}{\partial\epsilon^3}G(p)+\delta(\mu-\epsilon_p)\left[\frac{\partial^2}{\partial\epsilon^2}G(p+q)\right]_{\epsilon=0}+\delta'(\mu-\epsilon_p)\left[-\frac{\partial}{\partial\epsilon}G(p+q)\right]_{\epsilon=0}+\delta''(\mu-\epsilon_p)G(\omega,\mathbf{p+q})\right\} \tag{B10}$$

$$=\tfrac{1}{2}F_{\omega\omega\omega}{}^0+\tfrac{1}{4}F_{\omega\omega}{}^0{}_\mu+\int\frac{d^3p}{(2\pi)^3}\left\{\left[-\frac{\partial}{\partial\omega}G(\omega,\mathbf{p+q})\right]_{\mu=\epsilon_p}\delta'(\mu-\epsilon_p)+\left[\frac{\partial^2}{\partial\omega\partial\mu}G(\omega,\mathbf{p+q})\right]_{\mu=\epsilon_p}\delta(\mu-\epsilon_p)+[G(\omega,\mathbf{p+q})]_{\mu=\epsilon_p}\delta''(\mu-\epsilon_p)+2\left[-\frac{\partial}{\partial\mu}G(\omega,\mathbf{p+q})\right]_{\mu=\epsilon_p}\delta'(\mu-\epsilon_p)+\left[\frac{\partial^2}{\partial\mu^2}G(\omega,\mathbf{p+q})\right]_{\mu=\epsilon_p}\delta(\mu-\epsilon_p)\right\}$$

$$=\frac{\partial}{\partial\mu}[\tfrac{3}{4}F_{\omega\omega}{}^0-\tfrac{3}{4}J_{\mu\omega}+\tfrac{1}{4}F_{\mu\mu}{}^0-3\pi i\delta(\omega)S(q)]. \tag{B11}$$

Remember that only even parts in $\omega$ are kept.

The factor $\epsilon_p$ in $I_3(q)$ slightly complicates the integration. Again, by (A7),

$$I_3(q)=\int\frac{d^3p}{(2\pi)^3}\left\{\int\frac{d\epsilon}{2\pi i}G(p+q)\frac{\partial^3}{\partial\epsilon^3}G(p)\epsilon_p+\left[\frac{\partial^2}{\partial\epsilon^2}G(p+q)\right]_{\epsilon=0}\delta(\mu-\epsilon_p)\mu+\left[-\frac{\partial}{\partial\epsilon}G(p+q)\right]_{\epsilon=0}(\delta'(\mu-\epsilon_p)\mu+\delta(\mu-\epsilon_p))+G(\omega,\mathbf{p+q})[\delta''(\mu-\epsilon_p)\mu+2\delta'(\mu-\epsilon_p)\right\}. \tag{B12}$$

The (even part of) first term is

$$\int\frac{d^3p}{(2\pi)^3}(n_{\mathbf{p+q}}-n_{\mathbf{p}})\frac{\partial^3}{\partial\omega^3}\frac{1}{D}\times\tfrac{1}{2}(\epsilon_{\mathbf{p}}-\epsilon_{\mathbf{p+q}})$$

$$=\tfrac{1}{2}\int\frac{d^3p}{(2\pi)^3}(n_{\mathbf{p}}-n_{\mathbf{p+q}})\frac{\partial^3}{\partial\omega^3}\left(-1+\frac{\omega}{D}\right)$$

$$=\tfrac{3}{4}F_{\omega\omega}{}^0+\tfrac{1}{4}\omega F_{\omega\omega\omega}{}^0.$$

All the other terms in (B12) have been evaluated before. Thus,

$$I_3(q)=\tfrac{1}{4}\omega F_{\omega\omega\omega}{}^0+\tfrac{3}{4}(F_{\omega\omega}{}^0-J_{\mu\omega})+\tfrac{1}{2}F_{\mu\mu}{}^0-4\pi i\delta(\omega)S(q)+\mu\frac{\partial}{\partial\mu}[\tfrac{3}{4}F_{\omega\omega}{}^0-\tfrac{3}{4}J_{\mu\omega}+\tfrac{1}{4}F_{\mu\mu}{}^0-3\pi i\delta(\omega)S(q)]. \tag{B13}$$

$F^0$ and $J$ can be obtained from the definitions (4.8) and (B6). Explicitly,

$$F^0(q,\omega)=\frac{2mp_0}{(2\pi)^2}\left\{-1+\frac{m}{2qv_0}\left[(\alpha_-^2-v_0^2)\ln\left(\frac{\alpha_--v_0}{\alpha_-+v_0}\right)+(\alpha_+^2-v_0^2)\ln\left(\frac{\alpha_++v_0}{\alpha_+-v_0}\right)\right]\right\}, \quad \text{(B14)}$$

$$\frac{\partial}{\partial\mu}J(q,\omega)=\frac{m^2}{(2\pi)^2q}\left[\ln\left(\frac{\alpha_-+v_0}{\alpha_--v_0}\right)-\ln\left(\frac{\alpha_+-v_0}{\alpha_++v_0}\right)\right],$$

where

$$\alpha_\pm\equiv\frac{\omega}{q}\pm\frac{q}{2m}.$$

$\omega$ is regarded as a complex number. Explicit expressions for $I_1$, $I_2$, and $I_3$ can then be obtained by differentiating (B14). The $\omega$ integration contour is below the real axis for $\omega<0$ and above for $\omega>0$. Thus, when differentiating with respect to $\omega$, one must take into account the discontinuities of the functions at $\omega=0$ across the real axis. Let

$$\omega\equiv iqv_0y,$$
$$R(y)\equiv 1-y\tan^{-1}y^{-1}. \quad \text{(B15)}$$

The following formulas are useful. For $q\to 0$, and $y$ fixed,

$$F^0(q,iqv_0y)=\frac{-4mp_0}{(2\pi)^2}R(y)+O(q^2), \quad \text{(B16)}$$

$$\frac{\partial}{\partial\mu}F^0=-\frac{4m}{(2\pi)^2v_0}(1+y^2)^{-1}+O(q^2), \quad \text{(B17)}$$

$$\frac{\partial^2}{\partial\mu^2}F^0=\frac{4}{(2\pi)^2v_0^3}(1-y^2)(1+y^2)^{-2}+O(q^2), \quad \text{(B18)}$$

$$\frac{\partial^2}{\partial\omega^2}F^0=-4\pi i\delta(\omega)S(q)+\frac{8m^2}{(2\pi)^2v_0q^2}(1+y^2)^{-2}+O(1), \quad \text{(B19)}$$

$$\frac{\partial^2}{\partial\mu\partial\omega}J=-8\pi i\delta(\omega)S(q)+\frac{8m^2}{(2\pi)^2v_0q^2}(1+y^2)^{-1}+O(1), \quad \text{(B20)}$$

$$\omega\frac{\partial^3}{\partial\omega^3}F^0=4\pi i\delta(\omega)S(q)-\frac{32m^2}{(2\pi)^2v_0q^2}y^2(1+y^2)^{-3}+O(1), \quad \text{(B21)}$$

$$\frac{\partial}{\partial\mu}(F_{\omega\omega}{}^0-J_{\mu\omega})=\frac{8m}{(2\pi)^2q^2v_0^3}y^2(3-y^2)(1+y^2)^{-3}+O(1), \quad \text{(B22)}$$

$$\frac{\partial}{\partial q}F^0=\frac{4m^2v_0}{(2\pi)^2q}[-(1+y^2)^{-1}+R(y)]+O(q), \quad \text{(B23)}$$

$$\frac{\partial^2}{\partial q^2}F^0=\frac{8m^2v_0}{(2\pi)^2q^2}[(1+y^2)^{-2}-R(y)]+O(1), \quad \text{(B24)}$$

$$I_1(q)=\frac{-4m^2}{(2\pi)^2v_0q^2}y^2(1+y^2)^{-2}+O(1), \quad \text{(B25)}$$

$$I_2(q)=\frac{6m}{(2\pi)^2v_0^3q^2}y^2(3-y^2)(1+y^2)^{-3}+O(1), \quad \text{(B26)}$$

$$I_3(q)=\frac{-m^2}{(2\pi)^2v_0q^2}y^2(5+9y^2)(1+y^2)^{-3}+O(1). \quad \text{(B27)}$$

# Chapter VII

# *Higher Correlations of Hartree-Fock Electrons—Variational Theory MET for Atomic Ground States (Approach I, Part 2)*

## VII-1

Some of the physical aspects of electron correlation in atoms indicated by the theory MET† were mentioned in Chapter V. For ground states an essential prediction was that the $E_{corr}$ should be predominantly given by the sum of N(N − 1)/2 decoupled pair-correlation energies. Methods derived in Approach I for calculating these pairs individually were outlined in that chapter. The pair-correlation theory requires the examination and systematic study of the other effects in the wave function and energy of an N-electron system. This study forms the main topic of the MET for ground states, especially in the papers MET I (Chapter VII-2) and MET III (Chapter VII-3). In the present chapter, we confine ourselves to some of the essentials of the methodology developed in the theory. Several steps are involved. Excited states are discussed in Chapter IX.

## 1. Form of the Exact Wave Function

A detailed form of the exact wave function of an N-electron ground state is derived. The form consists of a finite number of terms, each involving an n-electron *correlation function* or "cluster" ($1 \leqslant n \leqslant N$), $U'_{ij\ldots k}$. The simplest derivation is by the projection out of the N-electron Hilbert space, successively, of the products of 1, 2, 3, . . . , N one-electron Fock-Dirac subspaces. This derivation, which is useful also in analyzing an arbitrary N-electron trial function, is referred to as "method of successive partial orthogonalizations" [O. Sinanoğlu, *Rev. Mod. Phys.* **35**, 517 (1963)]. The exact wave function becomes, in

$$\psi(\mathbf{x}_1, \mathbf{x}_2, \ldots, \mathbf{x}_N) = \phi_0 + X \tag{1}$$

$$X = \sum_{i \geqslant 1}^{N} \{\hat{f}_i\} + \sum_{i>j}^{N} \{\hat{U}_{ij}\} + \sum_{i>j>k}^{N} \{\hat{U}'_{ijk}\} + \cdots + \{\hat{U}'_{123\ldots N}\} \tag{2}$$

†A title used in a series of papers, "*M*any-*E*lectron *T*heory of Atoms and Molecules" communicated in summary form in O. Sinanoğlu, *Proc. U. S. Natl. Acad. Sciences* (1961), followed by a number of detailed papers (1961–1964). An overall presentation of the theory through 1963 is given in *Advan. Chem. Phys.*, **6**, 315–412 (1964). For later developments and applications see in *Atomic Physics* (Plenum Press, New York, 1969).

where the derivation also yields the "orbital-orthogonality" properties, e.g.,

$$\begin{aligned} \langle \hat{f}_i | k \rangle &= 0 \\ \langle \hat{U}'_{ij} | k \rangle &= 0 \qquad (\text{for } k = 1, 2, \ldots, i, \ldots, j, \ldots, N) \\ \langle \hat{U}'_{ijk} | k \rangle &= 0 \end{aligned} \tag{3}$$

The notation is, e.g.,

$$\{\hat{U}'_{ij}\} \equiv \frac{\mathcal{Q}}{\sqrt{2}} \left[ \frac{(123 \cdots N)}{(ij)} \hat{U}'_{ij} \right]$$

## 2. Simultaneous Cluster Correlation Products

Further, each "cluster"† function $\hat{U}'_{ijk \ldots n}$ may be written as the sum of products of smaller "clusters" and a remaining undecomposable, "linked"† n-cluster $\hat{U}'_{ijk \ldots n}$ [cf., e.g., Eq. (10) of Paper VII-3]. Products such as $(\hat{u}_{ij} u_{k\ell} \hat{f}_m)$, etc. correspond to simultaneous but independent correlations involving different H. F. electrons. Sinanoğlu calls these "unlinked clusters," a term more analogous to that in imperfect gas theory but not to be confused with the same words, which have a different meaning in diagrammatic Brueckner-Goldstone perturbation theory (cf. Chapters VI and VIII).

## 3. "Subvariational" Principles—The "Varied-Portions" Approach

A method that may be called "the varied-portions approach" (VPA) is introduced and forms an essential method of calculation in MET. The approach has its roots in a paper on the derivation of different perturbation theories from variational principles [O. Sinanoğlu, *J. Chem. Phys.* **34**, 1237 (1961)]. The method, however, is not confined to perturbation theory; hence rather than the original term used for it ("generalized variation-perturbation approach") in *Advan. Chem. Phys.* **6** (1964), the one above (VPA) is more appropriate. An example actually used in MET will demonstrate the method.

Say one wished to examine the role of each of the terms in Eq. (2) and to calculate the $E_{corr}$ to progressively higher accuracy. The steps become the following:

1. We select first from the full exact $\psi$ in Eq. (2) a portion which should be a fairly significant portion if few later stages are to be required, e.g.,

$$X \approx X_s \equiv \sum_{i>j}^{N} \{\hat{u}_{ij}\}$$

This is substituted in

$$E \leqslant \frac{\langle \tilde{\psi} | H | \tilde{\psi} \rangle}{\langle \tilde{\psi} | \tilde{\psi} \rangle} = E_{HF} + \frac{2\langle \phi_0 | H - E_{HF} | \tilde{X} \rangle + \langle \tilde{X} | H - E_{HF} | \tilde{X} \rangle}{1 + \langle \tilde{X} | \tilde{X} \rangle} \tag{4}$$

where the tilde ($\tilde{\ }$) on $\tilde{\psi}$ and $\tilde{X}$ denote trial functions ($\tilde{X}$ here taken as $X_s$). Written out, Eq. (4) involves energy clusters involving only one kind of $\tilde{u}_{ij}$ or several, as in $\langle \hat{u}_{ij} k | g_{ik} | j \hat{u}_{ik} \rangle$, which is

†See the footnote on page 180.

a *three-electron* correlation term, although resulting only from pair functions. Such terms are analyzed in Chapter VII-2, and the form of the energy, Eq. (4), becomes

$$E \leqslant E_{HF} + \frac{\sum_{i>j}^{N} \tilde{\epsilon}_{ij}}{D} + \frac{R}{D} \tag{5}$$

although Eq. (5) has resulted only from $\sum_{i>j}^{N} \{\hat{u}_{ij}\}$, the pairs are coupled in it through $D = 1 + \langle X_s | X_s \rangle$ and through R, $n \geqslant 3$ energy clusters.

The coupling through D is not physical; it is almost completely eliminated by noting that $X_s$ is inadequate. To be physically more complete, all pair products, "unlinked clusters," should be added to $X_s$. Thus

$$\tilde{X} = \tilde{X}_s' = \sum_{i>j}^{N} \{\hat{u}_{ij}\} + \sum_{\{i>j\}=\{k\}}^{N} \{\hat{u}_{ij}\hat{u}_k\} + \cdots \tag{6}$$

is used in Eq. (4) and yields an energy

$$E \leqslant E_{HF} + \sum_{i>j}^{N} \tilde{\epsilon}_{ij} + R \tag{7}$$

instead of Eq. (5).

The point of step 1 as far as VPA is concerned is that a major portion of $\psi$ has been selected, and a full variational energy expression and upper bound to total E are obtained with it. But now comes the role of VPA: Expression (7) is not very convenient for a calculation. All $\hat{u}_{ij}$'s must be varied in it and we still have almost as major a variational problem as in any of the conventional methods such as C. I., "correlation factor," etc. With C. I., for example, a very large matrix will have to be diagonalized. Instead, VPA leads to a number of independent and considerably smaller "subvariational" calculations (small C. I.'s, for example), each managable even for an atom of sizable N.

2. Instead of the full Eq. (7), *a smaller, but significant, portion of it is varied.* The portion has to be a stationary one. This can be judged, for instance [cf. O. Sinanoğlu, *J. Chem. Phys.* **34**, 1237 (1961)], from the existence theorems for the solutions of the corresponding Euler equation. One such significant portion in Eq. (7) is

$$E \cong E_{MET} \equiv E_{HF} + \sum_{i>j}^{N} \tilde{\epsilon}_{ij}' \tag{8}$$

which is also stationary with respect to variations of each $\hat{u}_{ij}$, as shown in Eq. (24) of Chapter V.

Variation of the portion [Eq. (8)] of the full $\langle \tilde{\psi} | H | \tilde{\psi} \rangle / \langle \tilde{\psi} | \tilde{\psi} \rangle$ [Eq. (7)] thus leads to a number of simpler *decoupled* calculations $\epsilon_{ij} \leqslant \tilde{\epsilon}_{ij}'$ rather than one large variational calculation [Eq. (7)].

3. Each decoupled pair calculation yields one $\hat{u}_{ij}$. In the next step, we substitute all these in the overall $X_s'$ and the full Eq. (7) to get an overall upper bound to the total E. The difference of the expectation values [Eq. (7) minus Eq. (8)] is mainly the "remainder" R, i.e., $n \geqslant 3$ correlations in the energy. These are the dominant $n \geqslant 3$ correlations. If $R'$ is small, others coming from higher clusters, $\hat{U}_{ijk} \cdots$ in Eq. (2) should be smaller still. An estimate of $R'$ gives us an idea of the error involved in using Eq. (8).

The procedure may be continued to higher accuracy but with rapidly increasing difficulty. Now, for example, we take a larger portion of the exact X, Eq. (2). We may add, say, all the $\hat{U}_{ijk}$, three-electron clusters to $X_s'$, write out and examine the new $\langle\tilde{\psi}|H|\tilde{\psi}\rangle / \langle\tilde{\psi}|\tilde{\psi}\rangle$, pick out a significant (and stationary) *portion to be varied*, obtain the $\hat{U}_{ijk}$'s, and then augment again by the calculation of the full expectation value to get another upper bound to the total E.

*The VPA, then, in general consists of obtaining successively larger portions of a wave function from "subvariational principles" which minimize portions of the energy. The resulting functions used in the full variational energy yield full upper bounds.* In addition, the recent work of Goscinski and Sinanoğlu [*Intern. J. Quantum Chem.* 2, 397 (1968)] shows that some subvariational portions constitute lower bounds to the energy.

By the use of VPA, Sinanoğlu has shown in the MET papers that a great many approximations to the N-electron system starting with different portions of Eq. (2) may be derived. Therefore, the working equations of MET, Eqs. (6), (7), and (8) (for ground states only; see Chapter IX for excited states), are based on physical considerations and are designed to include what appear to be the major effects.

The $n \geqslant 3$ electron correlation coupling effects are examined in Chapter VII-2, the one-electron correlation effects on orbitals $\hat{f}_i$ in Chapter VII-3. In arriving at Eq. (6), Chapter VII-2 shows that the coefficients of multiple electron excitations in large C.I. calculations of the conventional type are reproduced well by the products of pair excitation coefficients, showing that in simpler atoms the "linked" $n \geqslant 3$ correlation functions such as $\hat{U}_{1234}$ are indeed much smaller. This has been found to be true recently even with the delocalized, "metallic-like" pi electrons of benzene molecules by Koutecky, Čížek, Paldus, et al. Further tests of this type on other systems would, of course, be very illuminating.

We include the paper in Chapter VII-4 by Clementi (1963), who gives "experimental" correlation values $E_{corr}$ ("exp") for many atoms by subtracting H.F. energies and relativistic estimates from experimental total energies. These values have been useful in examining the predictions mentioned above and made before the data became available on the additivity of decoupled correlations and the transferability of some of them from system to system. From these $E_{corr}$ values of a set of atoms and ions alone, however, it is not possible to obtain the pair correlations (as discussed in Chapter V) correctly, because some pairs change greatly with N due to "exclusion effects." Combined with theoretical calculations of the pairs, however, McKoy and Sinanoğlu (Chapter VII-5) have obtained average total pair values such as $\epsilon(2p^2)$.

Chapter VII-5 was first to note that interorbital pair correlations like $\epsilon(2s2p)$ are as important as intraorbital in contributing to $E_{corr}$. In MET all $N(N-1)/2$ pairs occur from the start in an a priori way, and are naturally included. Previous calculations and discussions had neglected interorbital pairs as in the "geminal" wave functions assumed by Hurley, Lennard-Jones, and Pople (Chapter IV-4); Ruedenberg; Krauss and co-workers; Ebbing; and others.

Chapter VII-8 shows that the interorbital correlations continue to contribute at least as much as the intrapair correlations even when the H.F. orbital set $k = 1, 2, \ldots, N$ in the $\phi_0$ of a closed-shell system is transformed to localized orbitals (LO's). Equations like Eqs. (2) and (6) and many of the procedures of MET apply nearly the same way with LO's, as shown by Sinanoğlu [*Advan. Chem. Phys.* **6** (1964); see pp. 327ff.], although further formal development may be necessary here.

# 4. Configuration-Interaction Procedures for the Estimation of the Ground-State Pair Correlations Individually and of the More-Electron Remainders

The estimation of the $n \geqslant 3$ electron remainders, $R'$, in Eq. (7) is one of the most difficult parts of an N-electron calculation. If $R'$ is small, the calculation reduces to the decoupled ones of Eq. (8), but then an upper bound to E is lost.

To recover an upper bound to E, Krauss and Weiss suggest a modified procedure and a trial function (Chapter VII-6). They obtain an upper bound while eliminating the calculation of the more difficult terms of $R'$. The trial function, however, achieves this simplicity at the expense of omitting the very important interorbital pair correlations from Eq. (6) as well as the unlinked pair products. Taken as a first step for the calculation of yet a different portion of the total $E_{corr}$ of Eq. (7) and of Eq. (6) in VPA, however, it provides an alternative. In a next VPA step, other $\hat{u}_{ij}$'s and parts of Eq. (6) would then have to be added to the result.

Especially for molecules at present, the simplest way to estimate the relative magnitudes of $R'$ with respect to pairs in Eq. (7) is by C.I. For this purpose and for the calculation of separate $\tilde{\epsilon}_{ij}'$'s, a stepwise C.I. procedure was suggested in MET I (Chapter VII-2; see the first section on p. 715 and the second section on p. 716). With recent developments in C.I. technique which improve convergence, achieved especially by E. R. Davidson, M. Krauss, and A. W. Weiss and by A. C. Wahl, the stepwise C.I. procedure now becomes of more quantitative value.

Again for the ground states only, the steps that had been given in MET I are as follows:

1. Take a C.I. trial function containing only double excitation from *a single spin-orbital pair*, (ij), and carry out the C.I. This yields directly the decoupled $\tilde{\epsilon}_{ij}'$ of that particular (ij).

2. Repeat such a "pair C.I." for another (ij) by itself. Obtain the decoupled pair C.I.'s of all the distinct pairs. For example, in the neon atom, this would have to be done for the 11 distinct pairs mentioned in Chapter VI. Adding all these individual C.I. shifts from the H.F. energy, one gets Eq. (8).

3. Now to test the many-electron corrections R in Eq. (5), one can combine all the $\hat{u}_{ij} = \sum_{k>\ell>N}^{\infty} c_{ij}^{k\ell} B(k\ell)$ pair functions into a single trial function, $X_s = \sum_{i>j}^{N} \{\hat{u}_{ij}\}$, and calculate the full *expectation value*, $\langle \phi_0 + X_s | H | \phi_0 + X_s \rangle / [1 + \langle X_s | X_s \rangle]$, i.e., Eq. (5). Comparison of the results [Eqs. (5) and (8)] gives the main $n \geqslant 3$ energy correlations and the D.

4. To test the further $n \geqslant 3$ correlations resulting from the more complete wave function $X_s'$, Eq. (6), the pair C.I. coefficients are multiplied in appropriate fashion to get the $X_s'$. Again a fuller expectation value is calculated to yield the $R'$ [and the small differences $(\tilde{\epsilon}_{ij}'' - \tilde{\epsilon}_{ij}')$].

The first stages of this calculation may involve about N different pair C.I. calculations, each diagonalizing, say, a 100 by 100 matrix to get a fairly good $\tilde{\epsilon}_{ij}'$ value. The procedure proves feasible on the computer. The conventional C.I. calculation, on the other hand, would require at least a 10,000 by 10,000 matrix and becomes more difficult. Yet such calculations are still needed to check the magnitudes of the various effects above.

Various stages of the procedure above have been used recently by Bender, Davidson, and co-workers; Koutecky, Čížek, and Paldus; Harris, Schaefer, and Nesbet; Shavitt; and others (see the footnote on page 119 on confusing matters of misguided terminology).

The success of a C.I. calculation, even for the above much simpler version, depends on the choice of efficient virtual orbitals. Ordinary H.F. virtual orbitals are far removed spatially from the occupied ones and mix poorly. Several powerful methods of determining strongly mixing ghost orbitals are now available.†

1. Davidson and Bender (see, e.g., Chapter VII-7) start with trial orbitals, then obtain approximate natural spin orbitals (NSO's), which by definition converge faster, and use these at the next step. Davidson has obtained the $\tilde{\epsilon}'_{ij}$'s of a whole sequence of hydride molecules such as CH, NH, OH, etc., which should constitute a valuable guide to semiempirical $E_{corr}$ methods for molecules.

2. Krauss, Edmiston, and Weiss obtain the NSO's of an individual pair [since Eq. (24) of Chapter V for a pair is like that of an actual two-electron system and NSO's could be obtained hitherto only on two-electron systems] and then use these "pseudo-NSO"'s as a C.I. basis. This promising approach, which is turning out useful in molecular problems, is incompatible with the Hurley, Lennard-Jones, Pople approach (Chapter IV-4), which requires the use of mutually exclusive subsets for different geminals. A non-closed-shell use of the method by Weiss on the carbon atom is to be found in Chapter IX-3.

3. Kutzellnigg, Ahlrichs, and Bingel in a series of recent papers [see, e.g., *Theoret. Chim. Acta* 5, 289 (1966)] obtain NSO's for geminals directly. Their calculations include chemically interesting systems such as the $BeH_2$ molecule.

4. Another approach to efficient ghost orbitals for C.I. is the solution of "multiconfigurational SCF" (MCSCF) equations. Wahl and co-workers have recently obtained a priori molecular properties and potential curves this way. Some molecular pair correlations are also evaluated.

It is hoped that these "new-generation" C.I. techniques will be used to obtain additional correlation effects for further tests along with the pair correlations and corrections for atomic and molecular properties.

†We thank Dr. Carl Trindle for his helpful comments regarding the recent literature on this subject.

VII-2 Reprinted from *Journal of Chemical Physics* **36**, 706 (1962)

# Many-Electron Theory of Atoms and Molecules.* I. Shells, Electron Pairs vs Many-Electron Correlations

OKTAY SINANOĞLU
*Sterling Chemistry Laboratory, Yale University, New Haven, Connecticut*
(Received July 24, 1961)

A theory is developed (a) to see what the physically important features of correlation in atoms and molecules are; (b) based on this to obtain a quantitative scheme for $N$-electron systems as in He and $H_2$; (c) to see what happens to the "chemical" picture, to semiempirical theories, and to shell structure, when correlation is brought in. It is shown why, unlike in an electron gas, in many atoms and in molecules mainly pair correlations are significant. In configuration-interaction, multiple excitations arise not as three or more electron "collisions," but as several binary "collisions" taking place separately but at the same time. The validity of theory is shown on Be, LiH, and boron. The theory does *not* depend on any perturbation or series expansion and the $r_{12}$-coordinate method can now be used for an $N$-electron system as in He and $H_2$.

## INTRODUCTION

SHELL structure in atoms and molecules is due to nuclear wells and the exclusion principle. Long-range Coulomb repulsions between electrons have a disrupting influence on this structure.

The best one-electron orbits are obtained by letting each electron move in a potential averaged over the motions of all others (Hartree-Fock method). Based on orbitals such as these, much chemical and spectroscopic behavior of atoms and molecules is explained qualitatively or semiempirically.

Yet, full calculations of energy end up with so much error that often even molecule formation is not accounted for. This is usually attributed to electron correlation, the importance of which has been stressed in recent reviews.[1,2]

As the large errors show, electrons affect one another through their instantaneous potentials, not just by their average potentials as in the Hartree-Fock method.[1] Thus, one is faced with a many-electron problem. But what sort of a many-body problem? Does the long range of Coulomb repulsions cause all electrons to correlate with one another in a complicated way? Were this the case to the extent of overcoming the effects of the exclusion principle, shell structure would be wiped out; an atom or molecule would be more like a drop of electron liquid. The actual shell structure and radial electron densities, e.g., from x-ray results[3] are close to those of Hartree-Fock method.

This and *semiempirical success of orbitals must be reconciled with large correlation errors.*

The only accurate calculations on atoms and molecules so far[4] are still those on two-electron systems based on the classical work[1] of Hylleraas on He, and James and Coolidge on $H_2$, using trial functions containing the interelectron coordinate $r_{12}$ directly. This method has not been extended to $N$-electron system. For such systems configuration-interaction[1] (C.I.) is used. This is slowly convergent[5] except when used to remove degeneracy or "resonance." As $N$ and/or the finite orbital basis set used is increased, the number of multiply-excited configurations increases very rapidly into thousands, soon making an atom or molecule too large even for a computer. Best C.I. on small molecules so far has given only about 1/3 to 1/2 of the binding energies.

A method is needed whose difficulty does not increase rapidly with the number of electrons.

## OBJECTIVES OF THE THEORY

The objectives of the theory developed here are:

(a) to examine the physically important features of correlation in atoms and molecules;

(b) to try to obtain a quantitative scheme based on this picture for the $N$-electron system as in the He, $H_2$ case;

(c) to see what happens to the "chemical picture," to semiempirical theories, and to shell structure when correlation is brought in.

Previously[6,7] we used perturbation theory, solving the first-order equations rigorously and starting from Hartree-Fock solutions as unperturbed. The first-order wave function contains correlations in an electron pair at a time. Here we shall go beyond perturbation theory and complete the picture in several directions along objectives (a), (b), and (c). The new results and the over-all scheme that comes out were sum-

* For an over-all view of the scope and main results of this series of articles see reference 8.

[1] P. O. Löwdin, in *Advances in Chemical Physics* (Interscience Publishers, Inc., New York, 1959), Vol. II.

[2] K. S. Pitzer, in *Advances in Chemical Physics* (Interscience Publishers, Inc., New York, 1959), Vol. II.

[3] See, e.g., R. Daudel, R. LeFebvre, and C. Moser *Quantum Chemistry* (Interscience Publishers, Inc., New York, 1959).

[4] The next best calculation is the recent one on Be (reference 19), where $(2s)^2-(2p)^2$ "resonance" is of utmost importance and already corrects for much of the correlation.

[5] For a discussion of the $r_{12}$ method vs C. I. and status of the correlation problem see J. C Slater, *Quantum Theory of Atomic Structure* (McGraw-Hill Book Company, Inc , New York, 1961), Vol. II.

[6] O. Sinanoğlu, Proc. Roy. Soc. (London) **A260**, 379 (1961).

[7] O. Sinanoğlu, Phys. Rev. **122**, 491, 493 (1961).

marized in a communication.[8] This series of articles presents the detailed theory.

Here we look at the physical characteristics of correlation in atoms and molecules as compared to electron gas and nuclear "matter." We show why significant correlations occur only in electron pairs despite long-range Coulomb forces. There are separate correlations in many pairs at the same time, but correlations due to the "collisions" of three or more electrons are unlikely, as shown on Be, LiH, boron, etc. We obtain a variational wave function exact in pairs, and find the correlation energy. The pairs are uncoupled if the molecule does not contain many nearly "metallic" electrons delocalized in the same region. For objective (b), the theory is expressed in terms of correlation functions which are in closed form and not limited to C.I., although C.I. may be used in short steps to estimate three or more electron correlation effects.

In article II we give several methods for obtaining pair correlations exactly. A variation method is applied to electron pairs in the Hartree-Fock "sea" just as in the He or $H_2$ problem. *The $r_{12}$-coordinate can now be used directly in the N-electron system.* Only about $N$ two-electron problems need be solved. There is no "nightmare of innershells." One looks at each pair physically and then selects the appropriate method, $r_{12}$ coordinate, or C.I. "open shell," etc. for it. Perturbation theory comes out as a special case. We derive the Schrödinger equations of electron pairs in the Hartree-Fock "sea" which (for example in ethylene) have precisely the form of the "Π-electron Hamiltonian." By simple transformations, correlation energies are converted into those in bonds and into accurately defined *intra*molecular Van der Waals' interactions. Both quantitative and semiempirical uses of the theory will be discussed.

## COMPARISON WITH OTHER MANY-BODY PROBLEMS

To bring out the main physical features of the many-electron problem in atoms and molecules, we first discuss some related many body problems. Another reason for doing this is to show that it is neither appropriate nor necessary to apply the theories devised for these to most atoms or molecules.

### A. Nuclear Matter

The forces between nucleons are short range. This and the exclusion principle cause only local nucleon pair correlation to dominate inside nuclei. The strong short-range repulsions are described by a hard-core potential. With this type of potential one cannot start with the Hartree-Fock method because orbital energies become infinite.

In the Brueckner theory,[9] nucleons, while correlating pairwise, move in a medium that they constantly polarize. This is analogous to a charged particle moving in a dielectric medium with its polarization cloud around it, whereas in the Hartree-Fock method a particle moves in the potential of the undisturbed charge distribution.

Brueckner theory is applied mainly to infinite nuclear "matter" in which single nucleon orbitals are plane waves. In finite systems[10] difficulties arise. The polarized medium potential is strongly dependent on the orbital upon which it acts, so that orthogonal ground-state orbitals cannot be obtained easily. Also, the basis set is now discrete, and the simple momenta integrations with plane waves must be replaced by slowly convergent infinite sums.[11] As in ordinary perturbation theory,[1] these are difficult to evaluate. Moreover, the self-consistency procedure in getting orbitals, medium potential, pair correlations, back to orbitals, ··· becomes much too complicated and impractical.

### B. Infinite Electron Gas

This idealized model of a metal replaces the periodic lattice (hardly seen by electrons, e.g., in Na) by a smeared out, uniform positive charge distribution. Then Hartree-Fock orbitals are plane waves.

Electron correlation is determined by the deviation of the instantaneous Coulomb potential $g_{ij}=1/r_{ij}$ between two electrons, say $i$ and $j$, from the average [i.e., Hartree-Fock (H.F.)] potential they exert on one another. This deviation will be referred to as "fluctuation potential" (see below).

Let two electrons with opposite spins be in plane waves, $\phi_i$ and $\phi_j$, normalized in a large spherical box of radius $L$. The fluctuation potential is

$$g_{ij}-S_i(j)-S_j(i), \tag{1}$$

with the H.F. part of the potential of, for example, $i$ acting on $j$

$$S_i(j)=\langle\phi_i(\mathbf{r}_i),\, g_{ij}\phi_i(\mathbf{r}_i)\rangle_{r_i}$$
$$=\frac{3}{4\pi L^3}\int(\exp i\mathbf{k}_i\cdot\mathbf{r}_i)\exp(-i\mathbf{k}_i\cdot\mathbf{r}_i)\frac{1}{|\mathbf{r}_{ij}|}d\tau_i. \tag{2}$$

Changing the origin to $j$, so that $\mathbf{r}_i=\mathbf{r}_{ij}$, the simple integral above gives

$$S_i(j)=S_j(i)=3/2L, \tag{3}$$

which is uniform and tends to zero for $L\rightarrow\infty$.

Thus, in the electron gas, the fluctuation potential is the full $g_{ij}$ itself with its long range. Moreover, each electron is totally delocalized and equally likely to be found anywhere. Therefore, *many* electrons "see" each other with their fluctuation potentials at the same time and thus correlate all at once, giving rise to collective screening and oscillation effects.

[8] O. Sinanoglu, Proc. Natl. Acad. Sci. **47,** 1217 (1961).
[9] K. A. Brueckner, in *The Many Body Problem, Grenoble, École d'été de physiqué théoriqué, Les Houches* (John Wiley & Sons, Inc., New York, 1959).
[10] H. A. Bethe, Phys. Rev. **103,** 1353 (1956).
[11] In article II we shall obtain some results that eliminate this difficulty also for the finite nucleus problem.

## PHYSICAL ASPECTS OF ELECTRON CORRELATION IN ATOMS AND MOLECULES

Fortunately, in atoms and molecules (except in organic dyes, etc., where metallic behavior is approached) the problem is very different from that of an electron gas. The symmetry, nuclear wells, and exclusion principle tend to localize electrons (often in pairs) in different regions already in the orbital approximation.

The Hartree-Fock wave function is an ideal starting point for atoms and molecules. It takes care of the long-range effects of Coulomb repulsions, as we shall see; that is why it gives good radial densities. It has other advantages too, which will become apparent. Fortunately, even for molecules, H.F. solutions are rapidly becoming available.

Before examining how electrons correlate, we introduce some definitions. Let

$$\Psi=\phi_0+\chi; \qquad \langle\phi_0,\chi\rangle=0 \tag{4}$$

be the exact wave function of an $N$-electron system with a single-determinant H.F. solution $\phi_0$, and with $\chi$ correcting for correlation. (This article discusses, mainly, closed shells or systems with a single determinant $\phi_0$. For nonclosed shells, degeneracies must be removed first; the qualitative conclusions of this article then apply to what is left over. See II.) Then energy is separated into a H.F. and a correlation part.

$$E=\frac{\langle\psi,H\psi\rangle}{\langle\psi,\psi\rangle}$$

$$=E_{\mathrm{H.F.}}+\frac{2\langle\phi_0,(H-E_{\mathrm{H.F.}})\chi\rangle+\langle\chi,(H-E_{\mathrm{H.F.}})\chi\rangle}{1+2\langle\phi_0,\chi\rangle+\langle\chi,\chi\rangle},$$

$$E_{\mathrm{H.F.}}=\langle\phi_0,H\phi_0\rangle, \tag{5}$$

and

$$H=\sum_{i=1}^{N}h_i^0+\sum_{i>j}^{N}g_{ij}=H_0+H_1. \tag{6}$$

$g_{ij}$ is the bare-nuclei Hamiltonian of electron $i$ and $h_i^0=1/r_{ij}$. Atomic units (1 a.u. = 27.2 ev) are used. We separate $H$ into a part

$$H_0=\sum_{i=1}^{N}(h_i^0+V_i) \tag{7}$$

for independent electrons in the total H.F. potential $V_i$ of the "sea" $\phi_0$ and the residual *fluctuation potential* part[6]

$$H_1=\sum_{i>j}^{N}g_{ij}-\sum_{k=1}^{N}V_k(\mathbf{x}_k)=\sum_{i>j}^{N}[g_{ij}-\bar{S}_i(j)-\bar{S}_j(i)]. \tag{8}$$

In Eq. (7), $V_i$ acts on electron $i$; $V_i=V_i(\mathbf{x}_i)$.

We denote spin-orbitals by numerals 1, 2, $\cdots$, $N$ starting with the lowest orbital with spin $\alpha$; e.g., $1\equiv(1s\alpha)$, $2\equiv(1s\beta)$, $3\equiv(2s\alpha)$, etc. Then

$$\phi_0=\mathcal{A}(123\cdots N); \qquad \langle\phi_0,\phi_0\rangle=1, \tag{9}$$

where $\mathcal{A}$ is the antisymmetrizer

$$\mathcal{A}=\frac{1}{(N!)^{\frac{1}{2}}}\sum_P(-1)^P P. \tag{10}$$

In Eq. (9), before $\mathcal{A}$ is applied, *electron* $i$ with space, spin coordinates $\mathbf{x}_i$ occupies *orbital* $i$; e.g., $1(\mathbf{x}_1)$, $2(\mathbf{x}_2)$, etc. *Thus in Eq.* (8), $i$, $j$ *can refer equally to orbitals or to electrons*, where we separated the total H.F. potentials into those of orbitals.[6]

$$\sum_{k=1}^{N}V_k(\mathbf{x}_k)=\sum_{i>j=1}^{N}[\bar{S}_i(j)+\bar{S}_j(i)]. \tag{11}$$

$\bar{S}_i(j)$ is the Coulomb ($S_i$) plus the exchange ($R_i''$) potential of orbital $i$ acting on electron $j$.

$$\bar{S}_i(j)=S_i(j)-R_i''(j), \tag{12}$$

$$S_i(j)\equiv S_i(\mathbf{x}_j)=\langle i(\mathbf{x}_i),g_{ij}i(\mathbf{x}_i)\rangle_{\mathbf{x}_i}, \tag{13a}$$

$$R_i''(\mathbf{x}_j)j(\mathbf{x}_j)=\langle i(\mathbf{x}_i),g_{ij}j(\mathbf{x}_i)\rangle_{\mathbf{x}_i}i(\mathbf{x}_j). \tag{13b}$$

$\langle\ \ \rangle_{\mathbf{x}_i}$ means integration over coordinates $\mathbf{x}_i$ only. It is important to include "self-potentials" ($i=j$) in $V_i$ to make it the same acting on any electron in $\phi_0$.

From

$$H_0\phi_0=E_0\phi_0, \tag{14a}$$

H.F. spin-orbitals satisfy

$$e_i i=0,$$

where

$$e_i=h_i^0+V_i-\epsilon_i \tag{14b}$$

and $\epsilon_i$ are orbital energies. Then

$$E_{\mathrm{H.F.}}=E_0+E_1,$$

$$E_0\equiv\langle\phi_0,H_0\phi_0\rangle=\sum_{i=1}^{N}\epsilon_i. \tag{15a}$$

and

$$E_1\equiv\langle\phi_0,H_1\phi_0\rangle=-\sum_{i>j}^{N}(J_{ij}-K_{ij}'') \tag{15b}$$

$J_{ij}$ and $K_{ij}$ are the Coulomb and exchange integrals.

$$J_{ij}=\langle ij,g_{ij}ij\rangle; \qquad K_{ij}''=\langle ij,g_{ij}ji\rangle.$$

Writing

$$H-E_{\mathrm{H.F.}}=\sum_{i=1}^{N}e_i+\sum_{i>j}^{N}m_{ij}. \tag{16}$$

and defining[12]

$$m_{ij}=g_{ij}-\bar{S}_i(j)-\bar{S}_j(i)+J_{ij}-K_{ij}'', \tag{17}$$

Eq. (5) becomes

$$E-E_{\mathrm{H.F.}}=\frac{2\langle\phi_0,\sum_{i>j}m_{ij}\chi\rangle+\langle\chi,(\sum_i e_i+\sum_{i>j}m_{ij})\chi\rangle}{1+2\langle\phi_0,\chi\rangle+\langle\chi,\chi\rangle}. \tag{18}$$

[12] This notation is different from that used in reference 6. There $m_{ij}$ denoted only part of Eq. (17).

$m_{ij}$ is the complete fluctuation potential since Eq. (15b) is the expectation value of Eq. (8). Our potential must not be confused with the "*correlation potential*" $t_{ij}$ of the Brueckner theory which need not be introduced for atoms and molecules (see below). It would unnecessarily complicate this problem.

Two things in Eq. (18) determine the nature of electron correlation in atoms and molecules: (i) charge distribution in the "sea" $\phi_0$, (ii) pairwise fluctuation potentials $m_{ij}$.

### i. Charge Distribution

Although Hartree-Fock orbitals (in molecules SCF MO's) are suitable in describing electronic excitation and ionization, they do not necessarily reflect the probabilities of finding electrons at various places relative to one another. The exclusion principle, $\mathcal{A}$ in Eq. (9), tends to arrange electrons with like spin in $\phi_0$ to make them as far apart as possible. Linnett and Pöe[13] looked at $|\phi_0|^2$ as a function of $3N$ coordinates and determined the positions of electrons that make it a maximum. For example, in Ne eight electrons are found arranged on two tetrahedra with arbitrary orientation (since Ne is $^1S$). *Localization of electron pairs due to antisymmetry goes beyond the rule: two electrons per orbital.*

Directional bonding and shapes of molecules are mainly due to these electron distributions and are better represented by transforming the H.F. orbitals into "*equivalent orbitals,*" leaving $\phi_0$ unchanged.[14] The new orbitals are often localized and look like bonds or inner shells except in some $\Pi$ systems containing electrons that are too free. Our theory will show that qualitative conclusions drawn on these orbitals should be valid in spite of correlation.

### ii. Fluctuation Potential

Coulomb wells of nuclei and resulting geometric features of H.F. orbitals make also the fluctuation potentials, $m_{ij}$ in Eq. (17), very different from those in an electron gas. In atoms and molecules with no strong delocalization, $m_{ij}$ is usually of short range in directions going from orbital to orbital.

This is shown in Fig. 1 for the Be atom. Take a $(1s)$ electron and place a second arbitrary electron with opposite spin ar $r_2$. The $(1s)$ electron [taking a $(1s)$ Slater orbital with exponent 3.70 as sufficiently close to H.F.] is most likely to be found at $r_1=0.27$ a.u., the Bohr orbit of $Be^{2+}$. Were it there instantaneously, electron-two would see the full repulsion $g_{12}=r_{12}^{-1}$, not just the average Coulomb potential $S_1(r_2)$. Both potentials are shown in Fig. 1. The $g_{12}$ is for the two electrons and the nucleus placed on the same line.

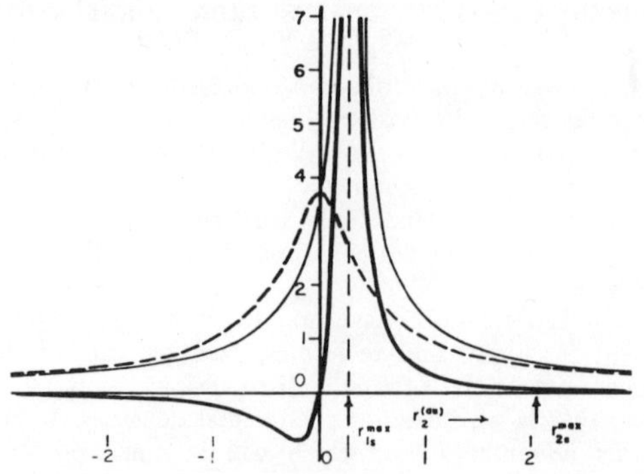


FIG. 1. Beryllium atom. Fluctuation potential as seen by an electron at $r_2$ due to a $(1s)$ electron ($\delta_1=3.70$) with opposite spin instantaneously at its Bohr radius ($r_1^{max}$). The nucleus, $r_1$ and $r_2$ are on the same line. ——— $=g_{12}=1/r_{12}$ (a.u.); - - - $=S_1(r_2)=$ H.F. potential of electron 1 acting on 2; ——— $=g_{12}-S_1(2)=$ fluctuation potential (a.u.).

The deviation, $g_{12}-S_1(2)$ differing from $m_{12}$ by the constant value, 0.39 a.u. is the fluctuation potential (also in Fig. 1), which shows where electron-two ($r_2$) wants to be if electron-one is at $r_1=0.27$ a.u. Both $g_{12}$ and $S_1(2)$ are of long range, but $m_1(2)$ has a short range. It has died off before reaching a distance of 2.1 a.u., about the Bohr radius of $(2s)$ in Be. The singularity of $g_{12}$ at $r_{12}=0$ is the main feature left in $m_{12}$. Note also the minimum in $m_1(2)$ on the other side of the nucleus. Similar curves and potential surfaces can be drawn by varying $r_1$ and $\theta_{12}$ (the angle between $r_1$ and $r_2$) in addition to $r_2$. These would aid the selection of trial functions to give maximum charge density at the dips and zero at the singularities.

The short range of $m_{12}$ in the radial direction explains, of course, why correlation energy between a $(1s)$ and a $(2s)$ electron is much smaller than that in the $(1s)^2$ shell even though Coulomb force is long range. More important, this also makes three- and four-electron correlations negligible. While correlating with one another, two $(1s)$ electrons do not see the third and fourth electrons that the exclusion principle has put in a different shell, *a safe distance away from the fluctuation potential* of $(1s)$.

The exclusion principle and fluctuation potentials operate this way generally. Even in many-electron shells (single determinant) pairs stay apart and cannot quite see other's $m_{ij}$ potential. Out of any three electrons in a system, at least two will have the same spin and will not come together.

Thus in atoms and molecules with a single determinant $\phi_0$, (except large and highly conjugated ones with nearly free electrons) we expect local pair correlations to dominate strongly over three or more electron "collisions." This is substantiated with more evidence below.

---

[13] J. W. Linnett and A. J. Pöe, Trans. Faraday Soc. 47, 1033 (1951).

[14] See, e.g., J. A. Pople, Quart. Revs. 11, 291 (1957). These orbitals are due to Lennard-Jones.

## PREVIOUS PERTURBATION THEORY RESULTS

The first-order wave function $\chi_1$ of the perturbation theory contains just these pair correlations if we start with the H.F. "sea" $\phi_0$ as unperturbed. Previously[6] we solved the first-order equation with operator techniques and got

$$\chi_1=\sum_{i>j}^{N}\frac{\mathfrak{a}}{\sqrt{2}}\left\{123\cdots N\frac{\hat{u}_{ij}^{(1)}}{(ij)}\right\}, \tag{19}$$

which gives for second-order energy (the first term in correlation energy)

$$E_2=\sum_{i>j}^{N}\langle\mathfrak{B}(ij), g_{ij}\hat{u}_{ij}^{(1)}\rangle. \tag{20}$$

The pair correlation in $(ij)$ is given by $\hat{u}_{ij}^{(1)}$, which satisfies the first-order part[15] of the Schrödinger equation[6] of two electrons in the H.F. "sea." $\mathfrak{B}$ is the two electron antisymmetrizer

$$\mathfrak{B}=(1/\sqrt{2})(1-P_{ij}). \tag{21}$$

One crucial effect of the H.F. "sea" on a pair is that two electrons, say $i$ and $j$ with opposite spins, correlating with each other, keep away from any other electron $k$ which will have the same spin as either $i$ or $j$. Thus $i$ and $j$, not only in H.F. orbital motion, but also while correlating, are prevented from going into spin-orbitals, $k\neq i, j$, already occupied in $\phi_0$.

Hartree-Fock orbitals $i$ and $j$ are solutions of Eq. (14b), and so are already adjusted to the polarizing influence of the average "sea" potential $V_i$ of $\phi_0$. Therefore, in the language of C.I., single excitations which would have otherwise resulted from this "orbital average polarization" of individual electrons[16] are also missing from $\hat{u}_{ij}^{(1)}$.

The rigorous solution[6] reflects these effects in the *orthogonality* of $\hat{u}_{ij}^{(1)}$ to the *orbitals k*.

$$\langle\hat{u}_{ij}^{(1)}, k\rangle=\langle\hat{u}_{ij}^{(1)}(\mathbf{x}_i, \mathbf{x}_j), k(\mathbf{x}_i)\rangle_{\mathbf{x}_i}=0, \tag{22}$$

where $k=1, 2, 3, \cdots N$ includes $i$ and $j$. Integration is again over coordinates of only one electron. How pair functions are made orthogonal to orbitals was shown in detail in reference 6 (see also article II). The procedure is quite similar to ordinary Schmidt orthogonalization, except that we have scalar products over only one electron space, as in Eq. (22).

$\chi_1$ and $E_2$ are often sufficient approximations (see II), as expected physically from the importance of pair correlations. But sometimes the strength of $m_{ij}$ may require the pair functions to be determined to greater accuracy than first order.

Here we shall go beyond perturbation theory in two respects and (i) examine many electron correlations; (ii) use in part a variational approach to take pair correlations into account exactly, and not just to first order.

[15] In reference 6 we denoted the first-order pairs, $\hat{u}_{ij}^{(1)}$ by $\hat{u}_{ij}$. From here on $\hat{u}_{ij}$ will be used to denote a general pair function not restricted to any order or to perturbation theory.

[16] See, e.g., O. Sinanoğlu, J. Chem. Phys. **33**, 1212 (1960).

## CLUSTERS OF ELECTRONS IN THE HARTREE-FOCK "SEA"

The exact wave function $\chi$ will contain not just binary "collisions" in the Hartree-Fock "sea" as does $\chi_1$, Eq. (19), but also the "collisions" of successive 3, 4, $\cdots$, and $N$ electrons at a time. By "collisions" we mean, of course, electrons coming within reach of the fluctuation potentials $m_{ij}$ of each other. $\chi$ may be written *exactly* as

$$\chi=\mathfrak{a}\left\{(1234\cdots N)\left[\sum_{i=1}^{N}\frac{\hat{f}_i}{(i)}+\frac{2}{(2!)^{\frac{1}{2}}}\sum_{i>j}^{N}\frac{\hat{u}_{ij}}{(ij)}+\frac{1}{(3!)^{\frac{1}{2}}}\sum_{i,j,k}^{N}\frac{\hat{U}'_{ijk}}{(ijk)}+\cdots+\frac{1}{(N!)^{\frac{1}{2}}}\frac{\hat{U}'_{123\cdots N}}{(123\cdots N)}\right]\right. \tag{23}$$

Each term replaces some of the initial H.F. orbitals by a function involving only the corresponding electrons. These functions are antisymmetric with respect to odd permutations; i.e.,

$$\hat{u}_{ij}(\mathbf{x}_i, \mathbf{x}_j)=-\hat{u}_{ij}(\mathbf{x}_j, \mathbf{x}_i), \tag{24}$$

$$\hat{U}'_{ijk}(\mathbf{x}_i, \mathbf{x}_j, \mathbf{x}_k)=-\hat{U}'_{ijk}(\mathbf{x}_j, \mathbf{x}_i, \mathbf{x}_k), \text{ etc.} \tag{25}$$

The caret indicates orthogonality to all the orbitals of the H.F. "sea" as in Eq. (22).

Equation (23) is *not* an infinite series for a system with a finite number $N$ of electrons. But it can be obtained from the configuration-interaction (C.I.) series by summations. The infinite C.I. expansion[1] contains all unique Slater determinants formed from a complete one-electron basis set. Equation (23) replaces all single, double, etc., excitations from different H.F. orbitals by closed, as yet undetermined, functions. One could substitute this *exact* $\chi$ into the variational expression, Equation (18), and obtain a finite set of coupled integro-differential equations. But this would only amount to rewriting the original Schrödinger equation. Instead we look at each term and pick out the important effects.

The $\hat{f}_i$ in Eq. (23) adjusts the orbital $i$ to the field of all other electrons to an extent beyond the H.F. potential. $\hat{u}_{ij}$ are pair correlations as in Eq. (19), but exact to all orders. $\hat{U}'_{ijk}$ contains all triple excitations from the orbitals $i$, $j$, and $k$ in $\phi_0$. In $\hat{u}_{ij}$, electrons $i$ and $j$ "collide" (via $m_{ij}$) while each one is moving in the H.F. potential $V_i=V_j$. Part of $\hat{U}'_{ijk}$, on the other hand, changes this $V_i$ so that while $i$ and $j$ "collide," $i$, say, will be moving in a "sea" in which $k$ is being polarized.[9,17] This effect and $\hat{f}_i$ would come in if we went from the H.F. "sea" to the Brueckner "sea." However, in terms of perturbation theory, which is rapidly convergent if one starts with H.F., both effects alter the energy only in the fourth and higher orders, and so are both negligible. That electron densities come out well

[17] R. Brout, Phys. Rev. **111**, 1324 (1958).

with H.F. orbitals is also evidence that $\hat{f}_i$ may be neglected.

*Parts* of $\hat{U}'_{123}$, $\hat{U}'_{1234}\cdots$ correct for three, four, $\cdots$, $n$, $\cdots$ electron "collisions." Such $n$ electron correlations will be large only if the probability of finding all the $n$ electrons within reach of each other's fluctuation potential is appreciable. Above, the nature of these potentials and exclusion effects led us to believe that this probability is small. Then we expect $\chi$ to be well approximated (see, however, below) just by

$$\chi_s = \frac{\mathcal{A}}{\sqrt{2}} \sum_{i>j}^{N} (123\cdots N)\frac{\hat{u}_{ij}}{(ij)}, \tag{26a}$$

$$\langle \hat{u}_{ij}(\mathbf{x}_i, \mathbf{x}_j), k(\mathbf{x}_i)\rangle_{\mathbf{x}_i} = 0, \qquad (k=1, 2, 3, \cdots N)$$

simple generalization of $\chi_1$, Eq. (19), to *exact pairs* still in the H.F. "sea."

Terms in Eq. (23) involving progressively larger numbers of electrons will be referred to as "clusters," just as in imperfect gas theory.[18] We shall show below that there is still an important physical effect missing from Eq. (26). In this problem, too, we will find two types of clusters not apparent from C.I.

## RECENT CONFIGURATION-INTERACTION RESULTS ON Be AND LiH

Recently Watson[4,19] studied the Be atom by C.I. by starting with the Hartree-Fock solution, and examining the effect of adding single, double, triple, and quadruple excitations. Single ($f_i$) and triple ($U_{ijk}'$) excitations were found totally negligible, as we anticipated. ***But appreciable quadruple excitations appeared.*** At first sight this is a surprising result; for, if three-electron correlations are negligible, there ought to be even less chance for four electrons to "collide."

Ebbing[20] too, did such a study on the LiH molecule, transforming his orbitals to H.F. SCF MO's with similar results. (There are also some single excitations. These still ought to vanish in energy to second order so long as the approximate H.F. molecular orbitals are obtained from Roothaan's self-consistent field procedure.[3] Nevertheless, if there is a nonzero effect on energy—no separate energy estimate was available—it must be due to the relative crudeness of molecular Hartree-Fock orbitals compared to atomic H.F.)

We show below that important four-excitations in Be and LiH found in these studies are *not* due to four-electron "collisions."

## "UNLINKED CLUSTERS"—SIMULTANEOUS CORRELATIONS IN SEPARATE PAIRS

According to $\chi_s$, Eq. (26), only one pair of electrons correlate at any one time. Actually, it is very likely that when two electrons are "colliding" in some part of the atom or molecule, other binary "collisions" should be taking place in other regions separately but at the same time.[17] We can talk about "time" here since the theory could have been based equally on a time-dependent formalism.[21] To account for these simultaneous pairs, all possible products of $\hat{u}_{ij}$'s which have no electrons in common must be added to $\chi_s$

$$\chi \cong \chi_s + \chi_{ss} + \chi_{sss} + \cdots$$

$$\equiv \mathcal{A}\left\{\frac{1}{\sqrt{2}}\sum_{i>j}(123\cdots N)\frac{\hat{u}_{ij}}{(ij)} + \sum_{\substack{i>j \\ (i,\, j\neq k,\, l)}}^{N}\sum_{k>l}^{N}\frac{\hat{u}_{ij}\hat{u}_{kl}}{(ijkl)} + \frac{1}{2\sqrt{2}}\sum_{\substack{i>j \\ i,\, j\neq k,\, l\neq m,\, n}}^{N}\sum_{k>l}^{N}\sum_{m>n}^{N}(123\cdots N)\frac{\hat{u}_{ij}\hat{u}_{kl}\hat{u}_{mn}}{(ijklmn)}\cdots\right\}. \tag{27}$$

For instance, in a 4-electron system (e.g., Be or LiH)

$$\chi_{ss} = (\mathcal{A}/\sqrt{2})(\hat{u}_{12}\hat{u}_{34} + \hat{u}_{13}\hat{u}_{24} + \hat{u}_{14}\hat{u}_{23}). \tag{28}$$

Such products[22] are "unlinked clusters" as in imperfect gas theory[18] and also play an important role in other many-body problems.[9,17,21] They are automatically taken care of in Rayleigh-Schrödinger perturbation theory,[23] but in, for example, variational methods,[17] must be specially introduced.

Then in Eq. (23), "clusters" should be separated into "linked" and "unlinked" parts; e.g.,

$$\frac{2}{(3!)^{\frac{1}{2}}}\hat{U}'_{1234} = \hat{u}_{12}\hat{u}_{34} + \hat{u}_{13}\hat{u}_{24} + \hat{u}_{14}\hat{u}_{23} + \hat{U}_{1234}, \tag{29}$$

where now $\hat{U}_{1234}$ is the true four-electron correlation ("collision"), the one expected to be small. Drawing straight lines between each pair joined by a "$U$,"

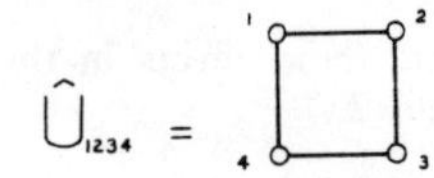


(Linked cluster)

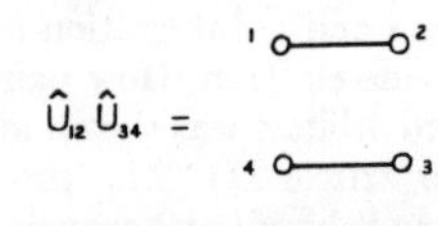


(Unlinked cluster)

## FOUR-ELECTRON EXCITATIONS IN Be AND LiH ARE MAINLY UNLINKED CLUSTERS

Watson's[19] final wave function for Be contains 37 selected configurations with coefficients obtained by solving the 37×37 secular equation. We renormalize his function so as to have $\langle\phi_0, \Psi\rangle = \langle\phi_0, \phi_0\rangle = 1$ as in

[18] See, e.g., T. L. Hill, *Statistical Mechanics* (McGraw-Hill Book Company, Inc., New York, 1956).

[19] R. E. Watson, Phys. Rev. **119**, 170 (1960).

[20] D. D. Ebbing, Ph.D. thesis, Department of Chemistry, Indiana University, June, 1960.

[21] Such as used by J. Goldstone, Proc. Roy. Soc. (London) **A239**, 267 (1957).

[22] Just the first term in Eq. (28) is similar to a product of group functions in the "Atoms and molecules" method of M. Moffitt [Proc. Roy. Soc. (London) **A210**, 245 (1951)]; but Eq. (28) is much more accurate, containing all the other pairs, and has many advantages such as a direct relation to C. I. Other group function methods will be discussed in II.

[23] K. A. Brueckner, Phys. Rev. **100**, 36 (1955).

TABLE I. Unlinked clusters vs four-electron excitations in Be atom.

| Quadruply excited configuration[a] | Coefficient from 37-configuration wave function[a] | Coefficient calculated from double excitations as unlinked clusters [Eq. (28)] |
|---|---|---|
| $p_1^2(^1S)p_{11}^2(^1S)$ | 0.007063 | 0.0073 |
| $p_1^2(^1S)s_1^2(^1S)$ | 0.005651 | 0.00647 |
| $p_1^2(^1S)d_{11}^2(^1S)$ | 0.001585 | 0.00168 |
| $p_{11}^2(^1S)d_1^2(^1S)$ | 0.0004639 | 0.000478 |
| Energy contribution (Table III) | −0.075 ev | −0.074 ev |

[a] R. E. Watson, Phys. Rev. **119**, 170 (1960).

Eq. (4) instead of $\langle\Psi, \Psi\rangle = 1$, and use his notation for his orbitals.[19] Only configuration $1s^2\ p_1^2$ has an appreciable coefficient, −0.29706, compared to unity. This results from the near-degeneracy[19] of $2s^2$ with $2p^2$ and accounts for most of the correlation error in the outer shell. Such an important "resonance" is to be expected since Be is not an inert gas, but is capable of forming a metal. Most of the other configurations have coefficients less than | −0.027 | and contribute to the inner shell.

The C.I. wave function has the form

$$\Psi = \mathfrak{A}(1234) + \sum_{k>l>4}^{\infty} a_{kl}\mathfrak{A}(kl34) + \sum_{m>n>4}^{\infty} a_{mn}\mathfrak{A}(12mn) + \cdots + \sum_{k>l>m>n>4}^{\infty} \gamma_{klmn}\mathfrak{A}(klmn). \quad (30)$$

If the linked cluster $\hat{C}_{1234}$ in Eq. (29) is negligible, then coefficients of quadruple excitations must be simply obtainable from Eq. (28), i.e., by multiplying together already determined coefficients, $a_{kl}$, $a_{mn}$, etc. of double excitations ($\hat{u}_{ij}$'s in C.I. form).

Table I compares coefficients calculated from Eqs. (28) and (30) taking all contributing pair products, with the complete[19] $\gamma_{klmn}$ from the full C.I. secular equation. The agreement is indeed satisfactory. Thirty-seven configuration results lie somewhat lower, indicating, as expected, small *negative* coefficients for $\hat{C}_{1234}$. One can now estimate coefficients for many more four-excited configurations from the available double excitations and Eq. (28) without augmenting the C.I. secular equation any further.

We found similar results (Table II) upon examining the Hartree-Fock C.I. coefficients by Ebbing[20] on LiH. Unlinked cluster coefficients are again calculated by multiplying coefficients of appropriate pair excitations and taking the right spin components. In Ebbing's notation ([11] is the inner, [22] the outer shell and 3, 4, 5, ⋯ are some excited orbitals), the main contribution to the coefficient of [3377] comes from the product of those of [1133] and [2277]; similarly [3357] comes from [1133] and [2257] (see calculation "*D*" in his[20] Table 6). There is no appreciable contribution from unlinked clusters to [3456] and [3457]. Their C.I. values and the difference between full C.I. and Eq. (28) values for [3357] and [3355] which are of the same magnitude, must be due to the linked cluster $\hat{C}_{1234}$. They are all small.

## THREE OR MORE ELECTRON CORRELATIONS ARISE IN ENERGY EVEN WITH JUST PAIR FUNCTIONS

From physical considerations we arrived at the hypothesis that in atoms and in molecules with no near-metallic electrons, the wave function needs to correct only for pair correlations. Supported also by perturbation theory, this led us to Eq. (27), which we confirmed on Be and LiH.

Even with just pair functions, however, the energy will contain three or more electron correlations. The same thing happens in perturbation theory, since the first-order wave function $\chi_1$ [Eq. (19)] determines the energy to third order.[6]

In the following sections, we obtain the complete energy given by $\chi_s$, Eq. (26), and then by Eq. (27) using the variation method. We show that many electron correlation terms arising this way should be small too. Thus we end up with uncoupled pair energies reducing the problem to several He or $H_2$-like problems [objective (b)].

## CORRELATION IN ONE PAIR AT A TIME

Take first $\chi_s$, Eq. (26), which represents only one binary "collision" ($\hat{u}_{ij}$) in the entire system at a time. Use it as a trial function for $\chi$ in Eq. (5) or (18). In article II we show that, depending on what portion of this energy is minimized, one obtains different approximations for $\hat{u}_{ij}$. In particular, if just the

$$2\langle\phi_0, (H_1 - E_1)\chi\rangle + \langle\chi, (H_0 - E_0)\chi\rangle \quad (31)$$

part of Eq. (5) is minimized one gets $\chi_1$. Then $\langle\chi_1, (H_1 - E_1)\chi_1\rangle$ becomes $E_3$, the third-order energy. Thus the analysis in this section includes perturbation

TABLE II. Unlinked clusters vs four-electron correlations in LiH.

| Configuration[a] | Coefficient[a] from full C. I. | Coefficient calculated from unlinked clusters; [Eq. (28)] |
|---|---|---|
| [3377][b] | 0.00113 | 0.00113 |
| [3357] | −0.00086 | −0.00078 |
| [3355] | 0.00062 | 0.00052 |
| [3456][c] | {−0.00023, 0.00006 | ~0 |
| [3457][c] | {0.00014, 0.00007 | ~0 |

[a] From calculation "D", Table 6 of D. D. Ebbing, Ph.D. thesis, Department of Chemistry, Indiana University, June, 1960.

[b] These are excited orbitals without spin in Ebbing's nomenclature, not our spin orbitals.

[c] These give two independent "codetors."

theory as a special case, but due to the generality of $\chi_s$ is by no means limited to it.

To analyze $E-E_{\text{H.F.}}$ substitute $\chi_s$, Eq. (26), and $\phi_0=\mathcal{A}(123\cdots N)$ into Eq. (18) and note that $\mathcal{A}$ leaves

$$\sum_{i=1}^{N} e_i \quad \text{and} \quad \sum_{i>j}^{N} m_{ij}$$

unchanged. Also $\mathcal{A}^2=(N!)^{\frac{1}{2}}\mathcal{A}$. Thus each term is reduced to a form having only one $\mathcal{A}$ in it on the left side; e.g.,

$$\langle\phi_0, \sum_{i>j} m_{ij}\chi_s\rangle$$

$$=\left(\frac{N!}{2}\right)^{\frac{1}{2}}\left\langle \mathcal{A}(123\cdots N), \left(\sum_{i>j} m_{ij}\right)\sum_{k>l}^{N}(12\cdots N)\frac{\hat{u}_{kl}}{(kl)}\right\rangle. \tag{32}$$

**DIAGRAM TECHNIQUE**

To proceed further, a diagram technique is very convenient. In a matrix element, such as in Eq. (32), we look at the direct term (without the $\mathcal{A}$), draw a solid line for each $\hat{u}_{ij}$ and a dotted line for an $m_{ij}$. Thus, e.g.,

$\langle \mathcal{B}(12), m_{12}\,\hat{u}_{12}\rangle$ = 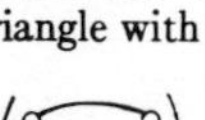 (33)



Instead of doing complicated algebra, one now determines all possible diagrams arising from a matrix element. The exchange parts are then obtained by applying $\mathcal{A}$ in the submatrix element corresponding to each diagram. This is done easily since a diagram is nonzero only for a subgroup of permutations that does not put electrons into orbitals so as to come out with $\langle\hat{u}_{ij}, k\rangle=0$. One first finds this small subgroup, then applies only those permutations.

**ENERGY WITH $\chi_s$**

With $\chi_s$ alone, the energy has a particularly simple form when $\phi_0$ is the Hartree-Fock wave function. The first term in Eq. (18) gives

$$\langle\phi_0, (H_1-E_1)\chi_s\rangle=\sum_{i>j}^{N}\langle\mathcal{B}(ij), m_{ij}\hat{u}_{ij}\rangle. \tag{34}$$

This has the same *form* as $E_2$, Eq. (20), but now it is general. One point should be emphasized. When $\mathcal{A}$ is applied in Eq. (32) to get Eq. (34), all the exchange terms (except $\mathcal{B}$) drop out because of $\langle\hat{u}_{ij}, k\rangle=0$, Eq. (26b). This does not mean however, that these exchange terms have been dropped by some artificial orthogonality condition. If we had had a $u_{ij}$ in Eq. (32), nonorthogonal to $k\neq i, j$, after applying $\mathcal{A}$ the new exchange terms would have been just the ones to Schmidt-orthogonalize $u_{ij}$ to $k$ to make it $\hat{u}_{ij}$. Thus just as in the perturbation result,[6,7] Eq. (34), is a shorthand form containing all the exchange effects.

In the other part of Eq. (18),

$$\langle\chi_s, (\sum_i e_i+\sum_{i>j} m_{ij})\chi_s\rangle,$$

the only nonvanishing diagrams, including their exchange parts, are

 $\equiv \langle \hat{u}_{ij}, (e_i+e_j+m_{ij})\hat{u}_{ij}\rangle$ (35a)



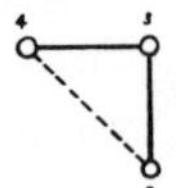 $= 2\left\{\langle 2_2\,\hat{u}_{34}(3,4), m_{24}\,\hat{u}_{23}(2,3)\,4_4\rangle - \langle 2_4\,\hat{u}_{34}(3,2), m_{24}\,\hat{u}_{23}(2,3)\,4_4\rangle\right\}$ (35b)



Also:

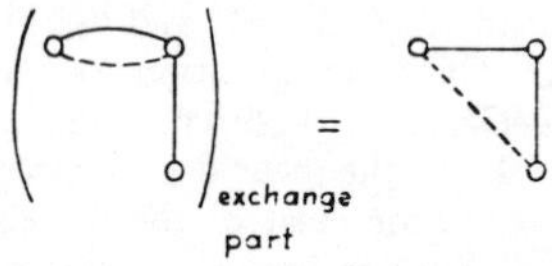 (35c)



In Eq. (35b), e.g., $2_4$ means $2(\mathbf{x}_4)$ and $\hat{u}_{34}(3, 4)\equiv\hat{u}_{34}(\mathbf{x}_3, \mathbf{x}_4)$. Only the triplet (234) is shown. There is one such term for every $(ijk)$ and for each orientation of the triangle with numbering fixed on paper.

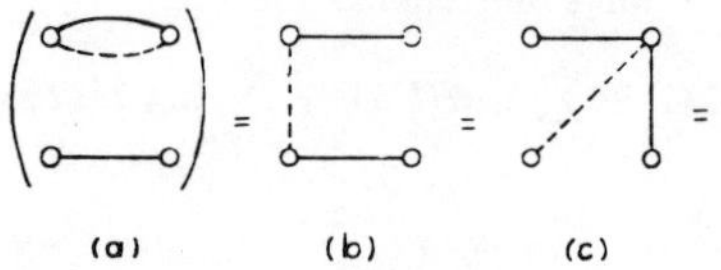 (36)



including their exchange parts vanish due to $\langle\hat{u}_{ij}, k\rangle=0$.

The only other diagrams left have a dotted line with an open end.

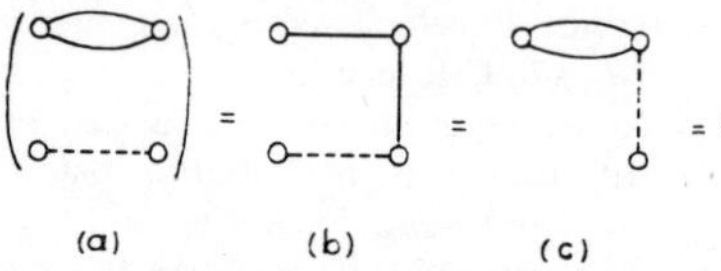 (37)



All these vanish, because $(---\circ\ j)$ corresponds to the average of the fluctuation potential $m_{ij}$ over orbital $j$. For example, Eq. (37c) has in it

$$\langle(1-P_{13})1\hat{u}_{134}(3, 4), m_{13}1_1\hat{u}_{34}(3, 4)\rangle=0, \tag{38}$$

which vanishes because $\langle(1-P_{13})1_1, m_{13}1_1\rangle_{\mathbf{x}_1}=0$ from Eqs. (12), (13), and (17). Such open-end dotted lines indicate polarization by the average potential of orbitals. The H.F. choice for $\phi_0$ and $V_s$ eliminates the effects of this average polarization not only on single excitations, but also on diagrams such as Eq. (37).

Finally, using Eqs. (34) to (37), Eq. (18) becomes

 (39a)

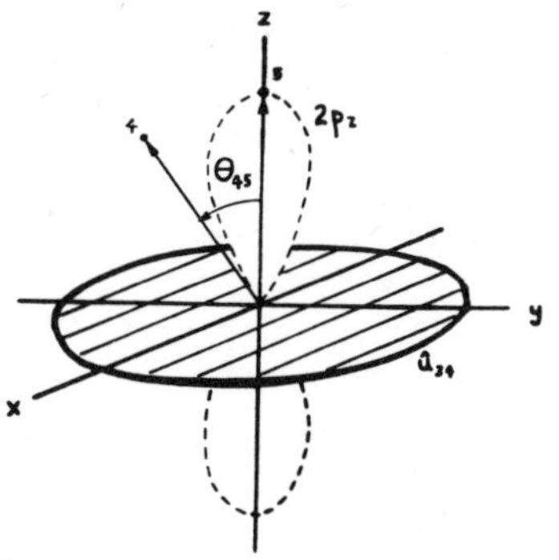


FIG. 2. The exclusion effect in boron atom ($1s^2 2s^2 2p_z$). Due to ($2p_z$), the $(2s)^2$ correlation $\hat{u}_{34}$ is confined to the vicinity of the $xy$ plane. Electron 5 is shown at a most probable position.

where

$$\bar{\epsilon}_{ij} \equiv 2\langle B(ij), m_{ij}\hat{u}_{ij}\rangle + \langle \hat{u}_{ij}, (e_i+e_j+m_{ij})\hat{u}_{ij}\rangle \quad (39b)$$

and

$$D \equiv 1+\langle \chi_s, \chi_s\rangle = 1+\sum_{i>j}\langle \hat{u}_{ij}, \hat{u}_{ij}\rangle. \quad (39c)$$

The energy contains the pair correlations, but also the normalization change, discussed in the next section, and three-body correlations. The $(ijk)$ triangle in Eq. (39a) means that the fluctuation potential of $j$ causes simultaneous fluctuations in the orbital charge distributions $i$ and $k$. Instantaneously induced distortions in the two charge clouds than interact electrostatically.

Now we see more specifically why such three-electron correlations should be small. Because of the exclusion principle $j$ cannot be close both to $i$ and $k$ at the same time. When its fluctuation disturbs $i$, in general it will not reach $k$ [see Fig. 1]. Another way of putting it: in

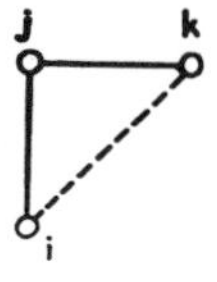


[see also Eq. (35b)],

$$\langle i\hat{u}_{kj}, g_{ki}\hat{u}_{ij}k\rangle, \quad (40)$$

the extensions of $\hat{u}_{kj}$ and $\hat{u}_{ij}$ in space are determined by the ranges of $m_{kj}$ and $m_{ij}$ and by where orbitals $(ijk)$ are. Because of the exclusion principle, at least one of the electrons is in a different orbital. This, supplemented by Eq. (26b), means $\hat{u}_{kj}$ and $\hat{u}_{ij}$ will not both be large in the same region of space; therefore their product will be small everywhere.

This is easily seen in Be or LiH, where two pairs are localized in radially different regions. But what about cases with three or more electrons radially in close proximity?

## FLUCTUATION POTENTIAL AND EXCLUSION EFFECT IN BORON

The boron atom ($1s^2 2s^2 2p_z$) is the simplest example of such a case. Consider the three-electron correlation in $(2s_\alpha 2s_\beta 2p_z\alpha) \equiv (345)$. Almost all of $(2s)^2$ correlation, $\hat{u}_{34}$ in the Be atom[19] and presumably also in Boron is due to strong mixing ("resonance") with $(2p)^2$. In Be and $B^+$, $\hat{u}_{34}$ is $^1S$, a combination of three $(2p)^2$ determinants containing $p_x$, $p_y$, and $p_z$. When $(2p_z\alpha)$, i.e., spin-orbital 5 is placed on top of the $(2s)^2$ shell to construct the boron atom; however, $\hat{u}_{34}$ gets changed to become orthogonal to $p_z$; $\langle \hat{u}_{34}, 5\rangle = 0$. The new $\hat{u}_{34}$ therefore is now a combination of only $p_x$ and $p_y$ and instead of over the whole sphere is large only on the $xy$ plane as shown in Fig. 2.

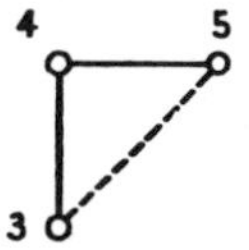


will be appreciable if the product $\hat{u}_{34}\hat{u}_{45}$ is large. But $\hat{u}_{45}$ is large only where electrons 4 and 5 [orbitals $\mathcal{B}(45)$] come together, i.e., around $p_z$ and where $m_{45}$ is large. The fluctuation potential $m_{45} \sim [g_{45} - S_5(4)]$ that electron 4 sees along the fixed radius as a function of angle $\theta_{45}$ from $z$ when 5 is placed at its most probable place is shown in Fig. 3. Note that $m_{45}$, hence $\hat{u}_{45}$ is very small just where ($\sim 90°$) $\hat{u}_{34}$ is large. Thus, their product and the three-electron correlation will be small everywhere. The same arguments apply to other triplets; e.g., $\hat{u}_{35}\hat{u}_{45}$ is small because, for the $\alpha\alpha$ spins pair, $\hat{u}_{35}$ is small.

In *neon*, $\hat{u}_{34}$ for $(2s)^2$ should practically vanish, since no $2p$ is left available. There is only *one* triplet in Ne, $(2s_\alpha 2s_\beta 2p_z\alpha)$, about which little may be said *a priori*. We expect to make direct calculations on Ne comparing many electron correlations with just pairs.

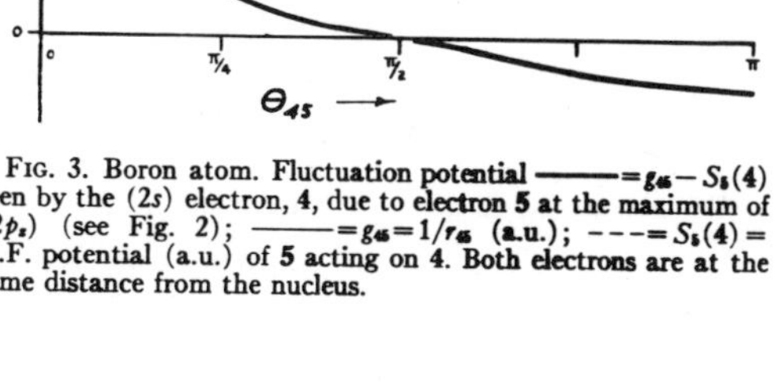

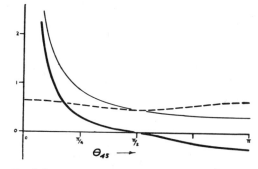

FIG. 3. Boron atom. Fluctuation potential ——— $= g_{45} - S_5(4)$ seen by the $(2s)$ electron, 4, due to electron 5 at the maximum of $(2p_z)$ (see Fig. 2); ——— $= g_{45} = 1/r_{45}$ (a.u.); - - - $= S_5(4) =$ H.F. potential (a.u.) of 5 acting on 4. Both electrons are at the same distance from the nucleus.

## ESTIMATE OF TRIANGLES BY C.I.

In doubtful cases triangles may be estimated by C.I. (see II for other methods). Calculations in three stages would be illuminating:

(a) Take a trial function like $\chi_s$, Eq. (26a), but with just one pair, say $\hat{u}_{34}$ in it. Then the entire energy is

$$E \leq E_0 + E_1 + \frac{\tilde{\epsilon}_{34}}{1+\langle \hat{u}_{34}, \hat{u}_{34}\rangle}. \tag{41}$$

Hitherto C.I. has been done mainly this way on molecules (e.g., $Li_2$).[24] Take another pair, say $\hat{u}_{23}$ and calculate it separately this way.

(b) Use the pairs from (a) and calculate the triangles they enter into.

(c) Take a $\chi_s$ with both $\hat{u}_{34}$ and $\hat{u}_{23}$ in. Determine them together by solving the big secular equation. Compare the C.I. coefficients from (a) and (c); also energies from (a), (b), and (c).

From Ebbing's C.I. (his calculations[20] "$A$", "$B$" and "$C$", and Table 5) on LiH, we estimate total triangle energies less than 0.01 ev. Coefficients obtained minimizing first Eq. (41), then Eq. (39a), (cf. Ebbing,[20] Table 6, "$B$" and "$C$") differ only by about 0.2%.

In general, we expect that Eq. (39a) can be approximated by

$$\tilde{E} = E_0 + E_1 + \frac{1}{D}\sum_{i>j}^{N} \tilde{\epsilon}_{ij}. \tag{42}$$

## DIAGRAMS DUE TO UNLINKED CLUSTERS

To see the effect of unlinked clusters on energy, we substitute all of $\chi$ from Eq. (27) into Eq. (18). Terms coming from $\chi_{ss}$, $\chi_{sss}$, etc., will then be added to Eq. (39a).

All of $\langle \phi_0, \sum m_{ij}(\chi_s+\chi_{ss}+\cdots\rangle$ vanishes with diagrams like Eqs. (36a) and (36b). For simplicity consider a $4e^-$ case so that $\chi=\chi_s+\chi_{ss}$ only, where $\chi_{ss}$ is given by Eq. (28). This is a much more realistic wave function than $\chi_s$ alone since it allows for several separate binary "collisions" at the same time.

$\chi_{ss}$ has two effects on $E$; (i) it tends[25] to cancel the normalization change, $(1+\langle\chi, \chi\rangle)$ although with a finite number of electrons this cancellation is incomplete; (ii) it introduces some new linked clusters, many electron correlations analogous to the effect of $\hat{U}_{1234}$ etc.

For the $4e^-$ case, Eq. (28), we get in Eq. (18)

$$\langle \chi_s, (\sum e_i + \sum m_{ij})\chi_{ss}\rangle$$
$$= \sum_{\substack{i>j\ k>l \\ i,j\neq k,l}}^{N}\sum^{N} \langle \mathcal{B}(ij), n_{ij}\hat{u}_{ij}\rangle\langle \hat{u}_{kl}, \hat{u}_{kl}\rangle + R'. \tag{43}$$

The first terms in product form are unlinked clusters, Eq. (44a). $R'$ contains the nonzero four-particle effects, Eqs. (44b), (44c), and (44d).

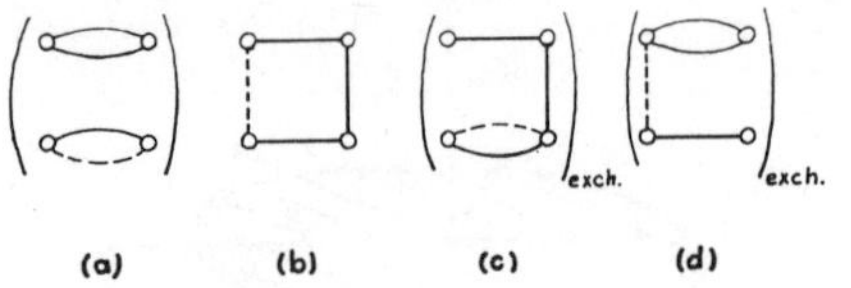

(44)

also

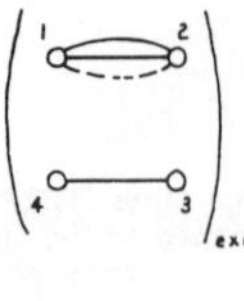

$$= \langle 34, m_{12}\hat{u}_{12}\rangle\langle \hat{u}_{12}, \hat{u}_{34}\rangle \tag{45}$$

and

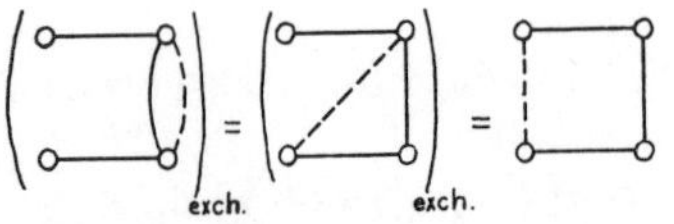

(46)

For example, Eq. (44b) is $\langle \mathcal{B}(12)\hat{u}_{34}, m_{12}\hat{u}_{14}\hat{u}_{23}\rangle$.

Equation (45) represents an exclusion effect not considered by $\langle \hat{u}_{ij}, k\rangle = 0$. Both pairs of electrons in $\hat{u}_{12}$ and $\hat{u}_{34}$ make excitations to the same states while correlating. However, electrons of one pair will be excluded from an excited state if that state happens to be occupied instantaneously by an electron of the other pair. This effect should be very small, since the fraction of time spent by each pair in any one excited state is already very slight. It is also dropped in the Brueckner method and also with generalized orthogonality conditions devised previously for many-electron group functions.[26]

The other term in $E$, Eq. (18), is

$$\langle \chi_{ss}, (\sum e_i + \sum m_{ij})\chi_{ss}\rangle$$
$$= \sum_{i>j}^{N}\sum_{k>l}^{N} \langle \hat{u}_{ij}, (e_i+e_j+m_{ij})\hat{u}_{ij}\rangle\langle \hat{u}_{kl}, \hat{u}_{kl}\rangle + 2R'', \tag{47}$$

where

$R'' =$ 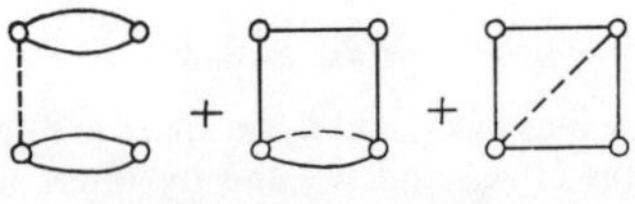

(48)

For a system with more than 4 electrons $R'$ and $R''$ in Eqs. (43) and (47) will contain larger linked clusters in addition to Eqs. (44), (45), (46), and (48). All these represent many electron effects much smaller than the triangles discussed after Eq. (39a) and are expected to be negligible.

---

[24] E. Ishiguro, K. Kayama, M. Kotani, and Y. Mizuno, J. Phys. Soc. (Japan) **12**, 1355 (1957).

[25] With the exact $\chi$, Eq. (4), $(1+\langle\chi, \chi\rangle)$ in Eq. (5) would be canceled completely giving just $E = E_{\text{H.F.}} + \langle \phi_0, H_1\chi\rangle$.

[26] A. C. Hurley, J. E. Lennard-Jones, and J. A. Pople, Proc. Roy. Soc. (London) **A220**, 446 (1953).

TABLE III. Unlinked cluster effects in Be and LiH.

| | |
|---|---|
| Be | |
| $D=1+\sum_{i>j}^{4}\langle \hat{u}_{ij}, \hat{u}_{ij}\rangle$ | 1.0905 |
| $D'$ [Eq. (50)] | 1.0906 |
| $D_{12}=1+\langle \hat{u}_{34}, \hat{u}_{34}\rangle$ | 1.08861 |
| $D_{34}=1+\langle \hat{u}_{12}, \hat{u}_{12}\rangle$ | 1.00172 |
| $D_{13}=1+\langle \hat{u}_{24}, \hat{u}_{24}\rangle$ | 1.0000 |
| $D_{14}=1+\langle \hat{u}_{23}, \hat{u}_{23}\rangle$ | 1.0001 |
| $(\tilde{\epsilon}_{12}/D)'$ (Configurations[a] "3" to "5" only) | −0.831 ev |
| Energy of four-excited configurations (6, 9, 19, 32)[a] from 37×37 C.I. | −0.075 ev |
| $(\tilde{\epsilon}_{12}/D)'\times(D_{12}-1)$ = unlinked cluster energy. | −0.0074 ev |
| Unlinked cluster energy of $(p_1^2 d_{11}^2)$ | −0.00019 ev |
| Full C.I. energy[a] of $(p_1^2 d_{11}^2)$ | −0.00019 ev |
| LiH[b] | |
| $D'=D$ | 1.020 |
| $D_{12}$ | 1.018 |
| $D_{34}$ | 1.001 |
| $D_{13}$, $D_{24}$, $D_{23}$, $D_{14}$ | 1.000 |

[a] R. E. Watson, Phys. Rev. **119**, 170 (1960); (configuration nos. refer to his Table III).
[b] Calculation "D", Table 6 in reference 13.

## ENERGY WITH UNLINKED CLUSTERS

The total energy resulting from a $\chi$ with all pairs and their unlinked clusters, Eq. (27), is now obtained from Eqs. (18), (39a), (43), and (47) as

$$E\leq E_0+E_1+\sum_{i>j}^{N}\tilde{\epsilon}_{ij}(D_{ij}/D')+\frac{R}{D'}. \quad (49)$$

$R$ is the sum of all true many electron correlation effects, i.e., linked clusters, e.g., triangles, squares, ⋯, Eqs. (44) to (46), (48) and larger ones for $N\geq 5$. The pair correlation energy $\tilde{\epsilon}_{ij}$ was defined in Eq. (39b)

$$D'=1+\sum_{i>j}^{N}\langle \hat{u}_{ij}, \hat{u}_{ij}\rangle+\langle \chi_{ss}, \chi_{ss}\rangle+\cdots. \quad (50)$$

$D_{ij}$'s come from factoring out the $\langle \hat{u}_{kl}, \hat{u}_{kl}\rangle$ in Eqs. (43) and (47).

$$D_{ij}=1+\sum_{\substack{k>l\\ k,l\neq i,j}}\langle \hat{u}_{kl}, \hat{u}_{kl}\rangle+\cdots. \quad (51)$$

For $N>4$ terms like

$$\sum\sum_{k>l\neq m>n}\langle \hat{u}_{kl}, \hat{u}_{kl}\rangle\langle \hat{u}_{mn}, \hat{u}_{mn}\rangle$$

are added to $D_{ij}$ corresponding to $\chi_{ss}$, $\chi_{sss}$, etc. $D_{ij}$ is essentially $D'$ from which the normalization effect of all pairs, which involve electrons $i$ or $j$ or both are missing. Since $\langle \hat{u}_{ij}, \hat{u}_{ij}\rangle$'s are already small (see Table III) their products, $\langle \chi_{ss}, \chi_{ss}\rangle$, etc. are entirely negligible and so using Eq. (39c)

$$D'\cong D \quad\text{and}\quad D_{ij}\cong 1+\sum_{k>l\neq i,j}^{N}\langle \hat{u}_{kl}, \hat{u}_{kl}\rangle. \quad (52)$$

The last equation is exact for $N=4$.

For reasons given in previous sections, $R/D$ in Eq. (49) is expected to be negligible. Then in going from simple pairs, Eq. (42), to Eq. (49) with unlinked clusters the main difference will be just the replacement of $(1/D)$ by $(D_{ij}/D)$. That unlinked clusters affect the energy only in this way is confirmed below on Be and LiH.

With $N$ electrons, the number of pairs missing from $D$ to give $D_{ij}$ is $(2N+1)$, whereas $D$ itself contains $N(N-1)/2$ pairs. Thus $D_{ij}/D$ is of the order

$$D_{ij}/D\sim 1-2(2N+1)/N(N-1) \quad (53)$$

so that as $N\to\infty$, $(D_{ij}/D)\to 1$, giving[9,17,21]

$$E\sim E_0+E_1+\sum_{i>j}^{N}\tilde{\epsilon}_{ij}.$$

In atoms and molecules $N$ is not large enough for such a complete cancellation of $D$'s in Eq. (49).

## UNLINKED CLUSTERS AND C.I.

As in the estimate of triangle energies, various effects of unlinked clusters can easily be tested by doing C.I. in several steps:

(i) Obtain the coefficient of double excitations for one electron pair at a time [see Eq. (41)].

(ii) Do all the pairs together with full C.I. This is the step in the estimation of triangle energies [Eq. (39a)]. Presumably, the coefficients will not change appreciably as compared to (i).

(iii) As Tables I and II showed, e.g., quadruple excitation coefficients can be obtained just from those of double excitations assuming only unlinked clusters are important. This is checked by adding these multiple excitations to C.I. and doing the full C.I. as in Be. The new energy is Eq. (49).

(iv) Calculate the $D_{ij}/D$'s from the pair coefficients. The main unlinked cluster contribution should be [compare Eqs. (49) and (42) or (39a)]

$$\sum_{i>j}^{N}\frac{\tilde{\epsilon}_{ij}}{D}(D_{ij}-1) \quad (54)$$

which can be checked by comparing it to the full C.I. energy in (iii). The difference is due to linked clusters $R$.

## MAGNITUDES IN Be AND LiH

Watson's[19] 37-configuration wave function for Be gives a total correlation energy −2.30 ev vs experimental −2.57 ev. In Table III we show the $D_{ij}$'s calculated from his coefficients renormalized to $\langle\phi_0, \phi_0\rangle=1$. The main contribution to $D$ comes from the coefficient of $(1s^2p_1^2)$, the $(2s^2-2p^2)$ mixing term in $\hat{u}_{34}$. The largest $D_{ij}$ too is $D_{12}\doteq 1+\langle\hat{u}_{34},\hat{u}_{34}\rangle$, which therefore is close to $D$. Thus $D_{12}/D\sim 1$, making Eq. (49)

$$E_{Be}\cong E_0+E_1+\bar{\epsilon}_{12}+(\bar{\epsilon}_{34}+\bar{\epsilon}_{13}+\bar{\epsilon}_{24}+\bar{\epsilon}_{14}+\bar{\epsilon}_{23}+R)/D. \quad (55)$$

When four excitations (Table I) are added to a wave function with just double excitations and full C.I. repeated,[19] the energy is lowered by 0.075 ev (sum of $D_{ij}/D$ and $R$ effects).

On the other hand, just from renormalization, $D_{ij}/D$ [Eq. (54)], and by looking at double excitations (Watson's[19] configurations "3" to "5") which when multiplied together give the four-excitations, we obtain 0.074 ev. Thus in their energy effect too, four excitations represent not four electron "collisions," but two binary ones. Especially[27] the agreement for $p_1^2d_{11}^2$ indicates the negligibility of $R$.

Similar results are found upon examining LiH (Table III), although the C.I. here was less complete and gave only a third of the correlation energy. In Ebbing's calculations we observe that

(i) Double excitation coefficients are hardly changed after four-excitations are added to C.I. (cf. "$C$" and "$D$" in his Table 6) indicating the smallness of $D_{ij}$ and $R$.

(ii) His four excitations lower the energy by 0.001 ev. This is only part of $\bar{\epsilon}_{12}(D_{12}/D-1)$; very few terms were taken in $\hat{u}_{12}$ (inner shell) so that $\bar{\epsilon}_{12}/D$ came out only about 0.3 ev. With $(\bar{\epsilon}_{12}/D)\cong 1$ ev and $D_{12}\cong 1.02$ (Table III), the unlinked cluster energy will be increased to $\sim$0.02 ev.

## CONCLUSION

The results and physical arguments given indicate that in single determinant atoms and molecules with no nearly "metallic" electrons, mainly pair correlations are important. The fluctuation potential acts locally; in three or more electron "collisions" its effect is overshadowed by the Fermi holes of pairs. A realistic wave function, Eq. (27), includes many separate binary "collisions" at the same time. Multiple excitations of significance arise in C.I. this way. The energy given by Eqs. (27), (39b) and (51) neglecting the three or more electron collisions in Eq. (49) is

$$E\cong E_0+E_1+\sum_{i>j}^{N}\bar{\epsilon}_{ij}(D_{ij}/D'). \quad (56)$$

Neglected "linked clusters" $R$ can be estimated too, and an upper limit to $E$ obtained once the pair functions, $\hat{u}_{ij}$'s, are known (see II). This is especially easy with C.I. done stepwise, as indicated above.

Equation (56), where the last sum is close to just

$$\sum_{i>j}^{N}\bar{\epsilon}_{ij},$$

has significance for the following reasons:

(1) The energy is the sum of uncoupled pair (except trivially by $D_{ij}/D$) energies. Thus by methods developed in II, each pair correlation $\hat{u}_{ij}$ can be obtained separately. Now the $r_{12}$ coordinate can be introduced into trial functions directly as in He and $H_2$.

(2) For an $N$-electron problem, there are just $\sim N$ pair functions to be determined, not $N(N-1)/2$ (see II). Thus even a large atom or molecule can be calculated quite easily.

(3) Had multiple excitations other than "unlinked clusters," or many-electron collisions been important, triplets, quadruplets, etc. of electrons would have had to be calculated. The number of these increase as $\sim N^3$, $N^4$ etc. with $N$ so that the problem would have become huge very soon.

(4) Equation (56) has already the "chemical picture" and the reconcilation of the qualitative vs. quantitative situation of orbital theories in it. This picture to be given later will in particular (a) justify the "Π-electron" theory; (b) give a precise way of defining and calculating intramolecular van der Waals forces and bond energies.

(5) Adding the correlation energy $\bar{\epsilon}_{ij}$ to $(J_{ij}-K_{ij}'')$, Eq. (15b), we find the same *form* for the energy as given by the Hartree-Fock method. Correlation does not disturb the H.F. radial density much either, since it acts locally and has no long-range effects. Thus it is expected that, e.g., the shapes of molecules are determined[13,14] mainly by the orbitals and the exclusion principle (equivalent orbitals).

(6) The results of this article are easily generalized to, say, an outer shell which has either degeneracy (see II) or significant $n$-electron correlations in it just by introducing some $\hat{U}_{123\cdots n}$ in Eq. (27) in addition to $\hat{u}_{ij}$'s. With, say, organic dye molecules one then gets the cores in terms of $u_{ij}$'s and also a separate, "metallic" shell which can be dealt with in a number of special ways such as C.I. or free-electron or plasma models.

I wish to thank Professor Raymond Fuoss and Professor Lars Onsager for the kind interest they showed in this work. This research was supported partly by a grant from the National Science Foundation.

[27] Configurations "6" and "9" do not give any $R$'s. Their effect on $R$ can be checked by mixing them in towards the end of 37 C.I.

 Reprinted from *Journal of Chemical Physics* **38**, 1740 (1963)

# Many-Electron Theory of Atoms and Molecules. III. Effect of Correlation on Orbitals*

OKTAY SINANOĞLU† AND DEBBIE FU-TAI TUAN

*Sterling Chemistry Laboratory, Yale University, New Haven, Connecticut*

(Received 5 November 1962)

The exact wavefunction of an $N$-electron atom or molecule contains, after the Hartree–Fock (HF) part, correlation terms involving successively one, two, ... $N$ electrons at a time. Particularly in closed shells, one-electron terms $f_i$ result mainly from pair correlations. The $f_i$ were previously neglected in the many-electron theory. Reasons for the smallness of $f_i$ are summarized. Different types of correlation effects are classified, and methods for estimating each type of $f_i$ are given. $f_i$ in closed form, i.e., including infinitely many single excitations is less than 2.8% of the Hartree–Fock orbital in He with an energy contribution 0.0001 a.u. (63 cal/mole). In the $H_2$ molecule $f_i$ is negligible for $(R/R_e)<2$. At larger $R$, as $(1\sigma_g)^2$ becomes degenerate with $(1\sigma_u)^2$ the $f_i$ effect increases to $\sim$0.4 eV at dissociation. However, in such cases and in actual nonclosed shells, these nondynamical $f_i$ are removed if HF orbitals are obtained after the removal of degeneracies. Dynamical correlation effects give negligible $f_i$ and so generalized SCF methods are not necessary. Qualitative quantum chemistry can be based on just HF orbitals or approximations to them, though energies include localized pair correlations.

## INTRODUCTION

THIS series of papers has been developing a theory of many-electron atoms and molecules including electron correlation effects.[1,2] Paper I showed that, starting from a Hartree–Fock solution, an exact wavefunction (wf) in closed form can be formulated in which correlation effects are systematically separated into parts: $\hat{f}_i$, corrections to the Hartree–Fock (HF) orbitals, $\hat{U}_{ij}$, electron-pair correlations and $\hat{U}_{ijk}$, $\hat{U}_{ijkl}\cdots$, etc. many-electron correlations. From an examination of the physical and mathematical aspects of correlation, many-electron terms were shown to be small in most atoms and molecules. Some comments were also made as to why $\hat{f}_i$ is expected to be small too, particularly in closed-shell systems. Then the energy of a single determinant atom or molecule reduces to the sum of variational expressions for the energies of two-electron systems in the Hartree–Fock medium. This allows the calculations to become much simpler; it also provides a basis for the separation of shells and for semiempirical theories including correlation.

Paper II[3] dealt with two electron correlations showing how for different types of pairs various two-electron methods can be used.

This article deals with $\hat{f}_i$, any corrections to the Hartree–Fock SCF orbitals due to effects, mainly correlation, not already included in the HF SCF method. We shall actually calculate $\hat{f}_i$ for some typical cases and show that they are indeed small where they were expected to be. In addition, further physical aspects of $\hat{f}_i$ will be discussed. The relation between Hartree–Fock and other orbitals such as those of Brueckner method, generalized SCF method, etc., will be shown to emphasize again why the Hartree-Fock method is to be preferred as a starting point.

## GENERAL FORMULATION

For a closed-shell system there is a unique Hartree–Fock method: that based on the single Slater determinant. Thus "correlation energy," too, is uniquely defined. Our many-electron theory has so far dealt mainly with closed shells. The correlation here is mostly of the short-range type. The long-range effects of the Coulomb repulsions are taken care of in the Hartree–Fock wf, so that the residual correlation effects are not expected to affect the HF orbitals much; thus the $\hat{f}_i$ will be small.

For a nonclosed shell system, however, many variants of the HF method are in use.[1,4] For example, orbitals may be obtained from a single determinant, though it may not be a symmetry eigenfunction, and then a projection operator may be applied. Alternatively, the proper symmetry eigenfunction as a linear combination of several determinants may be varied as a whole to get the orbitals ("extended" HF).[5] The "correlation energy" of a nonclosed shell system then depends on which HF is used. If HF is done before all the degeneracies are removed, some of the "correlation" error will be due to the change of orbitals that would be observed in going from one HF method to a more complete one. Thus it is possible to find non-negligible $\hat{f}_i$ with nonclosed shells, unless the most complete HF has been used.

* This work was supported by a research grant from the National Science Foundation.

† Alfred P. Sloan Fellow.

[1] For perspective and a nonmathematical discussion of the overall theory, including semiempirical implications see: O. Sinanoğlu, J. Phys. Chem. **66**, 2283 (1962) (ACS Symposium on Molecular Quantum Mechanics issue).

[2] O. Sinanoğlu, J. Chem. Phys. **36**, 706 (1962); hereafter referred to as I.

[3] O. Sinanoğlu, J. Chem. Phys. **36**, 3198 (1962); hereafter referred to as II.

[4] R. K. Nesbet, Rev. Mod. Phys. **33**, 28 (1961).

[5] P.-O. Löwdin, Phys. Rev. **97**, 1509 (1955).

There are also intermediate cases of near degeneracy, systems that are becoming nonclosed. A most interesting case is diatomic molecules near dissociation ($R\rightarrow\infty$). For instance, the $H_2$ molecule behaves as a closed-shell system $(1\sigma_g)^2$ near its equilibrium distance, with a correlation behavior similar to that of the He atom. But at dissociation, $(1\sigma_g)^2$ becomes degenerate with $(1\sigma_u)^2$. Then its HF treatment can no longer be based on the single determinant $(1\sigma_g)^2$ that was appropriate near the equilibrium distance. We shall return below to a quantitative discussion of $f_i$ in such cases and to nonclosed shells. The main part of this paper deals, nevertheless with, $\hat{f}_i$ in closed shells. We begin with a general formulation for the closed-shell case.

The exact wavefunction of the many-electron system can be written as[2,3]

$$\Psi=\phi_0+\chi; \qquad \langle\phi_0,\phi_0\rangle=1; \qquad \langle\phi_0,\chi\rangle=0. \quad (1)$$

$\phi_0$ is the HF determinant

$$\phi_0=\mathcal{A}(123\cdots N); \qquad 1\equiv 1(\mathbf{x}_1)=1s_\alpha, \quad \text{etc.}, \quad (2)$$

$$\chi=\sum_{i=j}^{N}\{\hat{f}_i\}+\sum_{i>j}^{N}\{\hat{U}_{ij}'\}+\sum_{i>j>k}^{N}\{U_{ijk}'\}\cdots+U_{123\cdots N}', \quad (3a)$$

$$\{\hat{f}_i\}=\mathcal{A}\{(123\cdots N)[\hat{f}_i/(i)]\};$$

$$\{\hat{u}_{ij}'\}=(\mathcal{A}/\sqrt{2})\{(123\cdots N)[\hat{U}_{ij}'/(ij)]\}, \quad (3b)$$

i.e., in Eq. (2), spin-orbitals $i$, $ij$, $\cdots$ are replaced by $f_i$, $\hat{U}_{ij}'$, $\cdots$, respectively. Also[2]

$$\langle\hat{f}_i,k\rangle=0, \qquad \langle\hat{U}_{ij}',k\rangle=0\cdots, \qquad \langle U_{ijk\cdots N}',k\rangle=0, \quad (4)$$

for $k=1, 2, 3, \cdots, i, j, \cdots N$, where again,[2,3] e.g.,

$$\langle\hat{U}_{ij}',k\rangle=\int\hat{U}_{ij}'(\mathbf{x}_i,\mathbf{x}_j)k(\mathbf{x}_i)d\mathbf{x}_i.$$

Equation (3a) has a finite number of terms, but if the functions $\hat{f}_i$, $\hat{u}_{ij}'$, etc., in closed form, are each expanded in a complete one-electron basis set, one obtains the infinite configuration-interaction (CI) series.

Two kinds of "orbital orthogonality" are implied in Eq. (4): (i) A given $U$, say $U_{ijl}'$, is orthogonal to each $k$ of $\phi_0$ except $i$, $j$, and $l$. This is the "exclusion effect," discussed before,[2,6] due only to $\mathcal{A}$. (ii) A given $\hat{U}$ is also orthogonal to the orbitals which it replaced in Eq. (3); e.g.,

$$\langle\hat{U}_{ij}',i\rangle=\langle\hat{U}_{ij}',j\rangle=0. \quad (5)$$

This is because in comparison with CI, $(\hat{f}_i)$ in Eq. (3) contains all the single excitations, $(\hat{U}_{ij}')$ only the double excitations, and so on with $(\hat{U}_{ijk}')$. An arbitrary function can be put in this form by first Schmidt-orthogonalizing it to orbitals $i, j, \cdots$ in the sense of Eq. (5),[6] then adding the subtracted terms—which are the $(f_i)$—back again.

Consider, for example, the wavefunction of the helium atom written as[3]

$$\Psi=\mathcal{B}(12)+u_{12},$$

where

$$\phi_0=\mathcal{B}(12),$$

and

$$\langle u_{12},\phi_0\rangle=0. \quad (6)$$

In general, $\langle u_{12},1\rangle$ or $\langle u_{12},2\rangle\neq 0$ for some trial function $\Psi$. But

$$u_{12}=\hat{U}_{12}+\mathcal{B}\{\hat{f}_12\}+\mathcal{B}\{1\hat{f}_2\},$$

and

$$\langle\hat{U}_{12},1\rangle=\langle\hat{U}_{12},2\rangle=0, \quad (7)$$

if

$$\hat{f}_1(\mathbf{x}_1)\equiv(2!)^{\frac{1}{2}}\int u_{12}(\mathbf{x}_1,\mathbf{x}_2)2(\mathbf{x}_2)d\mathbf{x}_2=(2!)^{\frac{1}{2}}\langle u_{12},2\rangle,$$

and

$$\hat{f}_2(\mathbf{x}_2)\equiv(2!)^{\frac{1}{2}}\langle u_{12},1\rangle. \quad (8)$$

Similarly in Eq. (3),

$$\hat{f}_i=(2!)^{\frac{1}{2}}\sum_{j\neq i}^{N}\langle\hat{U}_{ij},j\rangle+(3!)^{\frac{1}{2}}\sum_{j,l\neq i}^{N}\langle U_{ijl},jl\rangle\cdots$$
$$=(N!)^{\frac{1}{2}}\langle\chi,[1,2,3(i-1)(i+1)\cdots N]\rangle \quad (9)$$

with $\hat{U}_{ij}$, $U_{ijl}$, etc., satisfying only Eq. (1) and the "exclusion effect" parts of Eqs. (4). Equations similar to Eq. (7) give Eq. (3) and the complete Eqs. (4).

By a continuation of this type of procedure, Eq. (3) can be written in the much more detailed form which separates "linked" and "unlinked" clusters[1,2]

$$\hat{U}_{ij}'=\mathcal{B}(\hat{f}_i\hat{f}_j)+\hat{U}_{ij},$$

$$\hat{U}_{ijk}'=\mathcal{A}\{\hat{f}_i\hat{f}_j\hat{f}_k+\hat{f}_i[\hat{U}_{jk}/(2!)^{\frac{1}{2}}]+\cdots\hat{U}_{ijk}\},$$

$$\hat{U}_{ijkl}'=\mathcal{A}\{\hat{f}_i\hat{f}_j\hat{f}_k\hat{f}_l+\hat{f}_i\hat{f}_j[\hat{U}_{kl}/(2!)^{\frac{1}{2}}]\cdots\hat{f}_i[\hat{U}_{jkl}/(3!)^{\frac{1}{2}}]\cdots$$
$$+[\hat{U}_{ij}\hat{U}_{kl}/(2!)^{\frac{1}{2}}(2!)^{\frac{1}{2}}]\cdots+\hat{U}_{ijkl}\}, \quad (10)$$

and so on.[7]

The exact correlation energy, too, is separated into the effects of various pairs, shells, etc., by the substitution of Eqs. (10) and (3) in the variational expression.[1,2] The product terms (unlinked clusters) serve only to cancel similar terms that show up in the denominator, $1+\langle\chi,\chi\rangle$.

According to Eq. (9), $\hat{f}_i$ is the result of averaging all the two, three, $\cdots$ electron correlations over all the electrons involved except $i$. The unlinked clusters in $\hat{U}$'s do not contribute to $\hat{f}_i$ as can be seen by substituting Eqs. (10) in Eq. (9) and noting that $\langle\hat{f}_i,k\rangle=0$. Moreover, in *I* many-electron correlations were found

[6] (a) O. Sinanoğlu, Proc. Roy. Soc. (London) **A260**, 379 (1961). See also L. Szasz, J. Chem. Phys. **37**, 193 (1962). (b) O. Sinanoğlu, J. Chem. Phys. **33**, 1212 (1960).

[7] In paper I, $\hat{f}_i$ were assumed small and dropped from the beginning; so their unlinked clusters did not come in.

small (by an order of magnitude or better) compared to pair correlations.[8] Thus Eq. (9) reduces to

$$\hat{f}_i \cong \sum_{j\neq 1}^{N} \langle u_{ij}, j \rangle. \qquad (11)$$

It now remains to examine the various types of pair correlations[3] occurring in a many-electron system and to find their effect on $\hat{f}_i$. This will show whether the HF orbitals $i$ should be replaced by new ones, $(i+\hat{f}_i)$, or whether we can really drop $f_i$ from Eqs. (3) and (10). Before doing this, we discuss the nature and general properties of $\hat{f}_i$ further.

## PHYSICAL SIGNIFICANCE OF $f_i$ —EFFECT OF CORRELATIONS ON ORBITALS

Consider again the closed-shell system. If one took as the starting point not the Hartree–Fock $\phi_0$, but for instance, hydrogenlike orbitals (i.e., for the bare nuclei), there would be large $f_i$'s due to the long-range Coulomb repulsions between electrons. But, these $f_i$'s would be just the terms that would try to turn the hydrogenlike $\phi_0$ into a Hartree–Fock one. The SCF method calculates these $f_i$ that come from the long-range Coulombic potentials and incorporates them into $\phi_0$ completely.[9]

Since our theory starts with the Hartree–Fock $\phi_0$, our $f_i$ can be due only to the residual, "fluctuation" potentials.[2] These are the potentials that remain between electrons after the average, long-range Coulomb repulsions are taken out by HF. They are responsible for "correlation," and on account of Eq. (11), for $f_i$.

How $\hat{f}_i$ comes about can be seen also from the Schrödinger equation. Consider, for example, the helium atom. After writing the wf of helium as $\Psi = \phi_0 + u_{12}$, with $\langle \phi_0, u_{12} \rangle = 0$, and $\phi_0$ not necessarily HF, the Schrödinger equation corresponding just to the $u_{12}$ part becomes[3] as in Paper II:

$$(e_1+e_2)u_{12} + m_{12}\mathcal{B}(12) + m_{12}u_{12} = \epsilon_{12}u_{12}, \qquad (12)$$

where again[3,10]

$$m_{12} = g_{12} - \bar{S}_1(2) - \bar{S}_2(1) + J_{12} - K_{12}'' \qquad (13)$$

is the fluctuation potential

$$\epsilon_{12} = E - \langle \phi_0, H\phi_0 \rangle,$$

and $E$ the exact ground-state energy of He.

If $\phi_0$ refers to bare-nuclei orbitals $i^0$, the entire Coulomb repulsions remain, and

$$m_{12} \to g_{12} = 1/r_{12}. \qquad (14)$$

If now Eq. (12) with Eq. (14) substituted in it is multiplied by orbital $i^0$, say, and integrated over $\mathbf{x}_1$, using $e_1 1^0 = 0$, one gets[11]

$$e_2\hat{f}_2^0 + \langle 1(\mathbf{x}_1), g_{12}1(\mathbf{x}_1)\rangle_{\mathbf{x}_1} 2^0(\mathbf{x}_2) + \langle 1^0(\mathbf{x}_1), g_{12}u_{12}\rangle_{\mathbf{x}_1} = \epsilon_{12}(2^0 + \hat{f}_2^0), \qquad (15)$$

where again

$$\hat{f}_2^0(\mathbf{x}_2) \equiv (2!)^{\frac{1}{2}} \langle 1^0(\mathbf{x}_1), u_{12}(\mathbf{x}_1, \mathbf{x}_2)\rangle_{\mathbf{x}_2}. \qquad (16)$$

The second term in Eq. (15) is the ordinary Coulomb potential, $S_1^0(\mathbf{x}_2)$, of orbital $1^0$ acting on $\mathbf{x}_2$. The next term multiplied by $(2^0/2^0)$ gives the "correlation potential" acting on $\mathbf{x}_2$. This is the potential considered in Brueckner method after turning $g_{12}$ into pair-dependent interaction

$$t_{12} = g_{12} + g_{12}[u_{12}/\mathcal{B}(12)]. \qquad (17)$$

The Coulomb potential $S_1^0$ in Eq. (15) determines the part of $\hat{f}_2$ which turns $2^0$ into the Hartree–Fock orbital, 2. The remaining "correlation potential," on the other hand, is the one that would give our $f_i$. If $\phi_0$ is a Hartree–Fock wavefunctio , then $g_{12}$ in Eq. (15) is replaced by $m_{12}$ (Eq. 13) nd the Coulomb potential disappears. Thus, the on' contribution of $f_i$ then comes from the correlation potential. The correlation potential is a small and short-range potential, as can be seen[2] from the behavior of $u_{12}$ and $m_{12}$. So, in the case of atoms and molecules, its effect is expected to be small.

The Brueckner method puts the emphasis on this potential, however, because in nuclear matter this potential is due to the hard-core interactions. It cancels the preceding potential in Eq. (15), which in this case would be infinite. Other generalized self-consistent field (SCF) methods[12–14] also aim at calculating this $f_i$ due to the Brueckner-type potentials, but including many-electron correlations. Since many-electron correlations are quite unimportant,[2] for closed-shell atoms and molecules, the complications introduced by these methods are unnecessary. The resulting double SCF procedures that come in are anyhow highly impracticable.[15]

[8] *Ab initio* calculations based on pair correlations along with estimates of many electron terms are in progress in the writers' laboratory.

[9] Hartree–Fock SCF method amounts to the replacement of some initial orbitals $i^c$ by $(i^c + \hat{f}_i^c)$ in each iteration and then renormalizing the new orbitals. This means "unlinked cluster" of $f_i$'s, i.e., $f_i^c f_j^c f_k^c \cdots$ products are also included. They come from the expansion of $(i^c+f_i^c)(j^c+f_j^c)(k^c+f_k^c)\cdots$. Approximations to Hartree–Fock orbitals can also be obtained by perturbation theory, starting, say, with bare-nuclei orbitals (hydrogenlike) $i^c$, as done by J. Linderberg ["Study of atoms by Perturbation Theory," Tech. Note 74, Quantum Chemistry Group, Uppsala University, Uppsala, Sweden; (December 15, 1961)]. Then, the first iteration in SCF corresponds to the first-order $f_i$.

[10] For the rest of notation see Eqs. (5) through (14) of I.

[11] Similar successive partial integrations on the many-electron Schrödinger equation lead to a hierarchy of such coupled equations for $u_{12}$, $u_{123}$, etc., analogous to the set of equations in the Kirkwood–Bogolubov theory of fluids. As pointed out in I, these only amount to rewriting the original Schrödinger equation. [See also W. Brenig, Nuclear Physics **4**, 363 (1957) for the nuclear case.]

[12] J. C. Slater, Phys. Rev. **91**, 528 (1953).

[13] V. Fock, M. Veselov, and M. Petrashen, Soviet Phys.—JETP (U.S.S.R.) **10**, 723 (1940).

[14] P.-O. Löwdin, Preprint No. 48, Quantum Chemistry Group, Uppsala University, Uppsala, Sweden (1 August 1960).

[15] There are many other difficulties with Brueckner-type methods since their potentials depend on which electrons are involved. See O. Sinanoğlu, Proc. Natl. Acad. Sci. U. S. **47**, 1217 (1962); also R. K. Nesbet, Phys. Rev. **109**, 1632 (1958).

### REASONS FOR EXPECTING $\hat{f}_i$ TO BE SMALL

Summarized below are the reasons why we expect $\hat{f}_i$ to be small in closed shells:

(1) Starting from Hartree–Fock, the first-order wavefunction $\chi_1$ in perturbation theory contains no $\hat{f}_i$ [see reference (6) and those cited in (6b) for the CI form]. Since each order of wavefunction determines the energy to two new orders, this means any effect of $\hat{f}_i$ would show up only in the fourth order of energy. But starting with Hartree–Fock the perturbation series is rapidly convergent (a measure of this convergence is given by $\langle \chi_1, \chi_1 \rangle / 1$), so that these higher-order terms ought to be small.

(2) In CI language, $\hat{f}_i$ corresponds to the sum of single excitations. Calculations, for instance those of R. E. Watson[16] on Be using CI, show entirely negligible effects from single excitations.

(3) As we discussed above, once the long-range effects of Coulombic repulsion are taken out in $\phi_0$, the remaining correlation effects are only important when two electrons come quite close. The effects of such collisions ($\hat{U}_{ij}$) are smeared out over an entire orbital in getting $\hat{f}_i$ [see Eqs. (11) and (15)]; so that the resulting $\hat{f}_i$ should be small.

(4) Observed x-ray electron densities are clcse to those calculated from Hartree–Fock orbitals (see reference 3 of I). Any appreciable $\hat{f}_i$ would modify the density.

(5) In addition to these arguments we can also calculate $\hat{f}_i$ directly for typical cases. We shall do this below in a much more efficient way than CI which would need infinitely many single excitations. Actual means of calculating $f_i$ are important even if $f_i$ may turn out small, because in a satisfactory theory one should be able to estimate the magnitude of neglected terms.

In Eq. (11) we showed that the main $\hat{f}_i$ come from all possible pair correlations within a many-electron atom or molecule. Therefore we classify below the different types of pair correlations that exist in many-electron atoms and molecules. Then it remains to obtain the $\hat{u}_{ij}$ for each type of correlation as discussed in II, and then to obtain the corresponding $\hat{f}_i$ by Eq. (8).

### DIFFERENT TYPES OF CORRELATION EFFECTS

The following main types of pair correlations occur in atoms and molecules:

(1) "Tight pairs": For example, the $(1s)^2$ in He, Li, $Li_2$, etc., the $(2p_z)^2$ in Ne, or $(\sigma_{1g})^2$ in $H_2$, etc., $\cdots$. Here two electrons are tightly packed into one orbital. These are the cases where CI converges most slowly, and the $r_{12}$ or "different orbitals for different spins" methods are suitable. To distinguish it from effects of degeneracy, this type of correlation is sometimes referred to as "dynamical correlation."

(2) Pairs with near degeneracy: The near degeneracy might arise: (a) from cases that would actually be degenerate in the limit of infinite $Z$ as in the Be atom[16,17]; (b) in dissociating diatomic molecules, as in the well-known case of $H_2$. In extreme cases of this (b) type of near degeneracy, the Hartree–Fock method must be based not on the single determinant appropriate near the equilibrium configuration, but on the suitable linear combination of determinants.[4] The "correlation" problem then becomes the same as in actual nonclosed shells. In the near-degeneracy type of correlation, electrons are not tightly packed together; however, greater "correlation" errors in energy result.

(3) Intershell effects: These can be between concentric or nonconcentric shells. The $1s2s$ correlations[18] in Li and $Li_2$, the correlations between the *sigma* and *pi* shells of a conjugated system are examples of the concentric type. London dispersion forces between two helium atoms, or between different bonds of a saturated hydrocarbon[19] are of the same nonconcentric type.

Both types may be viewed as generalized by Van der Waals forces.[20]

The above cases referred mainly to closed shells. The main difference in nonclosed shells is in the dependence of "correlation error" on the HF method used[1,4] as the starting point. Some "Hartree–Fock" variants remove degeneracies *after* SCF orbitals are obtained, say, from a single determinant which is not a pure symmetry eigenfunction. Thus, in nonclosed shells, "correlation error" may include some effects of degeneracies. These effects are of the order of the energy differences observed from one HF variant to another and are often very small,[1,4] though they may not be negligible as in dissociating diatomic molecules (see below). In addition, there are the *bona fide* correlation effects that remain after all the degeneracies have been removed. These resemble those in closed shells (dynamical correlation) and exhibit the same pair types as listed above.

### CALCULATION OF $\hat{f}_i$

We examine now the $\hat{f}_i$ in these different types of correlations.

#### Type I. "Tight Pairs"

We shall demonstrate that the $\hat{f}_i$ of "tight pairs" are small by calculating it directly for the helium atom $(1s)^2$.

From Eqs. (6) and (8) the trial $\hat{f}_i$, $\bar{f}_i$, calculated

[16] R. E. Watson, Phys. Rev. **119**, 170 (1960).

[17] J. Linderberg and H. Shull, J. Mol. Spectry. **5**, 1 (1960).

[18] O. Sinanoğlu and E. M. Mortensen, J. Chem. Phys. **34**, 1968 (1961).

[19] K. S. Pitzer, Advan. Chem. Phys. **2**, 59 (1959).

[20] O. Sinanoğlu, J. Chem. Phys. **33**, 1212 (1960).

TABLE I. He atom. Values of $(\tilde{f}_i/i)$, percentage change in the Hartree–Fock orbital $i$ due to correlation.[a]

| $r_i$ (a.u.) | $\lvert\tilde{f}_i/i\rvert\times100$ (from Löwdin and Redei's wf) | $\lvert\tilde{f}_i/i\rvert\times100$ (from Hylleraas' wf) |
|---|---|---|
| 0.0 | 4.1 | 2.8 |
| 0.1 | 2.4 | 2.1 |
| 0.2 | 1.0 | 1.5 |
| 0.3 | 0.07 | 0.90 |
| 0.4 | 0.60 | 0.60 |
| 0.5 | 1.1 | 0.30 |
| 0.6 | 1.3 | 0.01 |
| 0.7 | 1.4 | 0.10 |
| 0.8 | 1.3 | 0.30 |
| 0.9 | 1.2 | 0.30 |
| 1.0 | 0.20 | 0.40 |
| 1.5 | 1.2 | 0.30 |
| 2.0 | 3.7 | 0.40 |
| 2.5 | ... | 0.90 |
| 3.0 | 7.1 | 2.00 |

[a] Beyond 3.0 a.u. both $i$ and $\tilde{f}_i$ become very small.

from any trial function $\Psi_{tr}$, depends on the original $\Psi_{tr}$ we chose. When $\Psi_{tr}$ is a good approximation to the exact wavefunction of the two-electron system, $\tilde{f}_i$ is a good approximation to $\hat{f}_i$, which includes *only* the effects of the sum of all virtual single excitations. The criteria for testing whether the magnitude of $\hat{f}_i$ is negligible is as follows: (1) By comparing different $\tilde{f}_i$'s obtained from different $\Psi_{tr}$'s. If $\tilde{f}_i$ approaches a small value as $\Psi_{tr}$ approaches the exact wavefunction (using energy as a criterion), then $\hat{f}_i$ is negligible. (2) By comparing the part of the correlation energy contributed by $\tilde{f}_i$'s: We set up a new trial function by taking out the $\tilde{f}_i$ part of the original one; i.e.,

$$\begin{aligned}\tilde{\Psi}_{tr}&=\phi_0+\tilde{\tilde{u}}_{12},\\&=\phi_0+\tilde{u}_{12}-1\tilde{f}_2-2\tilde{f}_1,\\&=\Psi_{tr}/(1+c)-1\tilde{f}_2-2\tilde{f}_1,\end{aligned}\tag{18}$$

where

$$\tilde{u}_{12}=[\chi_{12}-\langle\chi_{12},\phi_0\rangle\phi_0],$$

$$c=\langle\chi_{12},\phi_0\rangle,\qquad \chi_{12}=\Psi_{tr}-\phi_0,$$

and the relationships between $\tilde{\Psi}_{tr}$, $\tilde{f}_1$, $\tilde{f}_2$ and $\tilde{u}_{12}$ fulfill the condition given by Eqs. (6), (7), and (8). $\tilde{\Psi}_{tr}$ can be split into two parts:

$$\tilde{\Psi}_{tr}=\Psi_{tr}^c+\Delta,\tag{19}$$

where $\Psi_{tr}^c=\Psi_{tr}/(1+c)$, and $\Delta=-(1\hat{f}_2+2\tilde{f}_1)$. Then, according to the variational method,

$$\begin{aligned}\tilde{E}_{tr}&=\langle\tilde{\Psi}_{tr}H\tilde{\Psi}_{tr}\rangle/\langle\tilde{\Psi}_{tr}^2\rangle\\&=\frac{\langle\Psi_{tr}^cH\Psi_{tr}^c\rangle+2\langle\Psi_{tr}^cH\Delta\rangle+\langle\Delta H\Delta\rangle}{\langle(\Psi_{tr}^c)^2\rangle+2\langle\Psi_{tr}^c,\Delta\rangle+\langle\Delta^2\rangle}\\&=E_{tr}\\&\quad+\frac{(-E_{tr})(2\langle\Psi_{tr}^c,\Delta\rangle+\langle\Delta^2\rangle)+2\langle\Psi_{tr}^cH\Delta\rangle+\langle\Delta H\Delta\rangle}{\langle(\Psi_{tr}^c)^2\rangle+2\langle\Psi_{tr}^c,\Delta\rangle+\langle\Delta^2\rangle}\\&=E_{tr}+\Delta E.\end{aligned}\tag{20}$$

$\Delta E=\tilde{E}_{tr}-E_{tr}$ is the energy contributed from the terms that involve $\tilde{f}_i$, i.e., $-1\tilde{f}_2-2\tilde{f}_1$. If $\Delta E$ approaches a small value as $\Psi_{tr}$ approaches the true wavefunction, it implies the $\hat{f}_i$ contribution to correlation energy is small, and the $\hat{f}_i$ contained in the true wf is small. *In other words, if the exact wf $\Psi$ contains negligible $\hat{f}_i$, we can improve a trial function by taking out its $\tilde{f}_i$ part.* The trial functions that we choose are: (1) The scaled wavefunction[21]

$$\Psi(r_1,r_2)=N(1+\eta\alpha r_{12})u(\eta r_1)u(\eta r_2),\tag{21}$$

where $u(\eta r_i)$'s are single-electron two-term exponentials. They are scaled analytic SCF functions. $\eta$ is the scaling parameter; $\alpha$ is a variational one. The wf Eq. (21) gives 80% of the correlation energy. (2) Hylleraas' six-parameter wavefunction[22]

$$\phi(s,t,u)=\exp(-\tfrac{1}{2}s)\sum_{l,m,n=0}c_{n,2l,m}s^nt^{2l}u^m,\tag{22}$$

where $s=r_1+r_2$, $t=r_2-r_1$, $u=r_{12}$ and each $C_{n,2l,m}$ is a parameter. This wavefunction gives 98% of the correlation energy. The analytic Hartree–Fock SCF $1s$ wavefunction chosen is the two-term exponential form given by Löwdin.[21,23]

The magnitude of $\tilde{f}_i$ expressed as $\tilde{f}_i/i$ is given in Table I, and plotted in Fig. 1 as a function of $r_i$. For the simple trial function of Löwdin and Redei, the maximum effect of $\tilde{f}_i$ is about 4% of the Hartree–Fock orbital at the origin (where the Hartree–Fock orbital is also the largest). $\tilde{f}_i$ from the Hylleraas trial function is considerably smaller. It is 2.8% of the Hartree–Fock orbital at the origin. It is much smaller elsewhere, except at very large distances where the actual values of both $\tilde{f}_i$ and $i$ become very small.

We should also remark that in using a trial function of the form Eq. (21), it is very important that it be scaled. The same trial function, if unscaled, gives only 45% of the correlation energy. The $\tilde{f}_i$ calculated from it shows, even qualitatively, the wrong behavior. This $\tilde{f}_i$ becomes negative at small $r_i$, instead of turning up as in Fig. 1.

The $\Delta E$ obtained from different trial functions are

[21] P.-O. Löwdin and L. Redei, Phys. Rev. **114**, 752 (1959).
[22] E. A. Hylleraas, Z. Physik **54**, 347 (1929), also reference 6.
[23] P.-O. Löwdin, Phys. Rev. **90**, 120 (1953).

given in Table II. It is found that the value $\Delta E/E_{corr}$, the fraction of the correlation energy contributed by $\bar{f}_i$, decreases as the wavefunction improves from the simple trial wavefunction of Löwdin and Redei ($\Delta E/E_{corr}=1.7\%$) to the Hylleraas' wavefunction ($\Delta E/E_{corr}=0.24\%$). The trends in the change of $\Delta E/E_{corr}$ and $\bar{f}_i$ are the same: as the trial wavefunction approaches the exact eigenfunction, $\bar{f}_i$ and the correlation energy contributed by $\bar{f}_i$ become very small. Hylleraas' wavefunction already gives 98% of the correlation energy, and so it is a very good one. Better wavefunctions, such as Kinoshita's, may be expected to give even smaller $\bar{f}_i$ effects. In other words, $\hat{f}_i$ is small.

It is clear from the above results on helium with its tightly bound electrons that the effects of $\hat{f}_i$ can be neglected for innermost shells like $(1s)^2$. The errors in the energy by neglecting $\hat{f}_i$ in the case of Hylleraas' six-term wavefunction would be only 63 cal/mole, negligible even for thermochemical calculations.

The correlation behavior of the two electrons of the $H_2$ molecule near its equilibrium distance is very similar to that in the He atom. A good, yet still simple, wf for $H_2$ is obtained by multiplying the HF wavefunction of $H_2$ by a correlation factor as in the Löwdin and Redei wf for helium. Kolos and Roothaan[24a] have obtained 86% of the correlation energy of $H_2$ at the equilibrium distance $R=R_e$ by such a closed-shell three-term correlation function (see their Table V, p. 209). From this, by exactly the same procedure as used above for He, one can calculate the $f_i$ in closed form. This is not be done here,[24b] however, since this case is very similar to He. Instead, below we discuss $H_2$ and its $f_i$ as $R$ varies from $R_e$ to $\infty$ in connection with recent CI results (see below) which show $f_i$ to be negligible near $R_e$, but not at large $R$ where the system becomes a "nonclosed shell."

TABLE II. Ground-state energy of He—contribution of $\bar{f}_i$ to energy.

| $\Psi_{tr}$ | Löwdin and Redei's wavefunction | Hylleraas' wavefunction (six terms) |
|---|---|---|
| $E_{exact}$[a,c] (a.u.) | −2.90374 | −2.90374 |
| $E_{HF}$[c] (a.u.) | −2.8617 | −2.8617 |
| $E_{corr}$ (a.u.) $=E_{exact}-E_{HF}$ | −0.0421 | −0.0421 |
| $E_{tr}$ (a.u.) | −2.8954[c] | −2.90324[b] |
| $\Delta E$ (a.u.) $=E_{tr}-E_{tr}$ | −0.0007[d] | +0.0001 |
| $\Delta E/E_{corr}$ | 1.7% | 0.24% |
| $\bar{f}_i/i$(max) | 4.0% | 2.8% |

[a] See T. Kinoshita, Phys. Rev. **105**, 1490 (1957).
[b] Reference 23.
[c] Reference 22.
[d] The value of $E_{tr}$ used in this case is −2.8959 a.u. (calculated by us) instead of Löwdin and Redei's value (−2.8954 a.u.).

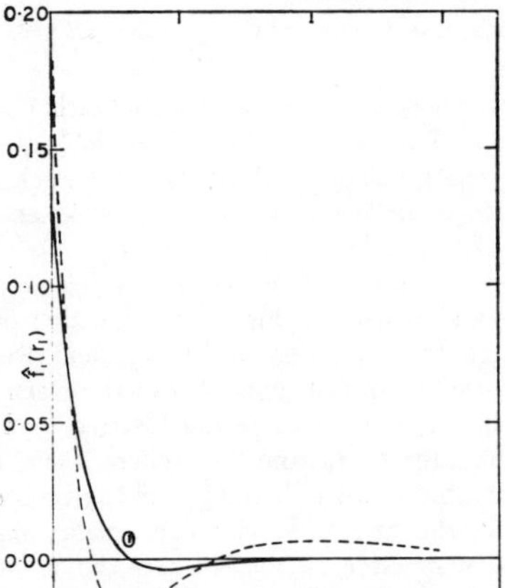


FIG. 1. Calculated $\bar{f}_i$ of He. ① $\bar{f}_i$ from Hylleraas' wavefunction, ② $\bar{f}_i$ from Löwdin's and Redei's wavefunction, $r_i$ is the distance of electron $i$ from the nucleus.

Another important type of "tight-pair" correlation occurs in $(2p_x)^2$, say in Ne, with a correlation energy of about 1.6 eV.[25] The $\hat{f}_i$ of such a pair is probably small, too, since the orbitals are quite diffuse and have a node at the origin where the contribution of $\hat{f}_i$ may have been largest.

We conclude that in "tight pairs" (dynamical correlation), $\hat{f}_i$ should be negligible.

## Type 2. Pairs with Near Degeneracy

### (a) *Ground State of Atoms*

For Be single excitation was found negligible in CI studies.[15] The correlation in the outer shell of Be is mainly due to the near-degeneracy of $(2s)^2$ with $(2p)^2$. But double excitations by definition contain no $\hat{f}_i$. Since other configurations make small contributions to the outer shell of Be, its $\hat{f}_i$ should be negligible.

### (b) *Diatomic Molecules near Dissociation (Interatomic Potentials)*

In the case of $H_2$ molecule, at $R=\infty$, if HF is still based on the single determinant $(1\sigma_g)^2$, which was ap-

[24] (a) W. Kolos and C. C. J. Roothaan, Rev. Mod. Phys. **32**, 205 (1960). (b) *Note added in proof*: E. R. Davidson and L. L. Jones [J. Chem. Phys. **37**, 2966 (1962)] have just calculated the natural spin orbitals of $H_2$ at $R=R_e$ using the Kolos–Roothaan wavefunction. By showing that the first NSO is approximately equal to $(i+\hat{f}_i)(1+\langle \hat{f}_i\hat{f}_i\rangle)^{-\frac{1}{2}}$ and by comparing it with the Hartree–Fock orbital, they too find $\hat{f}_i$ negligible.

[25] E. Clementi, J. Chem. Phys., "Correlation Energy in Atoms" (to be published).

TABLE III. Estimated energy contributions of single excitations or $\hat{f}$ in the $H_2$ molecule (see text for explanation; $R=\infty$ case is calculated by a different method than the other cases).

| $R$ (a.u.) | Energy contribution (eV) of $\hat{f}_i$ or $(1\sigma_g2\sigma_g)$ and $(1\sigma_g3\sigma_g)$ [a] |
|---|---|
| $\infty$ | $-0.4$ |
| 12.0 | $-0.33$ |
| 4.0 | $-0.32$ |
| 3.0 | $-0.13$ |
| 2.0 | $-0.027$ |
| 1.4 ($\sim R_e$) | $-0.0092$ |
| 1.1 | $-0.0046$ |

[a] See text and reference 28.

propriate near $R=R_e$, one gets a "correlation" error 7.74 eV.[26] Will such a large "correlation" error give a large $\hat{f}_i$ effect too? Not as large as might be expected, but not negligible either.

This 7.74 eV is not a short-range dynamical correlation effect, but comes about because of the degeneracy of $(1\sigma_g)^2$ with $(1\sigma_u)^2$ at $R=\infty$. Thus the problem becomes that of a nonclosed shell. Most of this error is eliminated by simply removing the degeneracy by a two-by-two CI between $(1\sigma_g)^2$ and $(1\sigma_u)^2$. This $(1\sigma_u)^2$ is now a double excitation and so has no $\hat{f}_i$. But if HF was done just on $(1\sigma_g)^2$, even at $R=\infty$, there is still an error partly due to $\hat{f}_i$.[27] How much is this $\hat{f}_i$ and how does it become negligible as $R\rightarrow R_e$?

Consider still the HF based on $(1\sigma_g)^2$. Löwdin[26] has given the HF $(1\sigma_g)$ at $R=\infty$ in the form $1\sigma_g=(1/\sqrt{2})(u_a+u_b)$ where $a$ and $b$ are the two nuclei. Taking $1\sigma_u=(1/\sqrt{2})(u_a-u_b)$, we can calculate the energy of $(1\sigma_g)^2$ and $(1\sigma_u)^2$ mixed together to give the symmetrized product $(u_au_b)$, and get $E=2\langle u_a, H_a, u_a\rangle=-26.4$ eV, as compared to the exact energy $2\langle 1s_a, H_a, 1s_a\rangle=-27.2$ eV. The total error remaining at $R=\infty$ after degeneracy is removed is therefore 0.8 eV. Part of this comes from having $u_a$ instead $1s_a$ in $(1\sigma_g)$ and part in $(1\sigma_u)$. Only the former part represents an $\hat{f}_i$ effect. The corrections in $(1\sigma_u)^2$ would be double excitations relative to $(1\sigma_g)^2$. Thus the $\hat{f}_i$ error at $R=\infty$ will be about 0.4 eV.

This "nondynamical" or nonclosed-shell-type $\hat{f}_i$ would be removed if at $R=\infty$, HF were based not on $(1\sigma_g)^2$ but on $(u_au_b)$, i.e., the proper combination of $(1\sigma_g)^2$ and $(1\sigma_u)^2$. Then, of course, the best $u_a$ would come out $1s_a$ and any explicit $\hat{f}_i$ eliminated.

But near $R=R_e$, $H_2$ does behave like an He atom, and HF is based on $(1\sigma_g)^2$ with a negligible $\hat{f}_i$ resulting only from "dynamical correlation." The question important for the theory of interatomic forces at intermediate separations is: "Where do we draw the line?"

Fraga and Ransil have studied[28] $H_2$ by $(1\sigma_g)^2$ HF followed by CI. From their results we can get an idea of how the $\hat{f}_i$ contribution to energy varies with $R$.

They used a limited LCAO–MO–SCF plus CI; the single excitations included were $(1\sigma_g2\sigma_g)$ and $(1\sigma_g3\sigma_g)$. These represent two terms in a CI expansion of $\hat{f}_i$.

The full CI coefficients are very often well approximated by the first-order ones—$\langle\phi_0, H_1\phi_k\rangle/(E_k-E_0)=c_{k(1)}$ where $(E_k-E_0)$ is the excitation energy of the configuration $\phi_k$ relative to $(1\sigma_g)^2$. When this is the case, the energy contribution of a configuration can be approximated by its second-order energy. However, the single excitations we are interested in do not appear in the wf until the second order[2,3,6] and in the energy until the fourth order due to Brillouin's theorem.

The full CI coefficients $c_k$ of single excitations $\phi_k$ should be well approximated by the coefficients in the second order w.f., $c_{k(2)}$. Similarly the energy contribution of $\phi_k$ in full CI should be close to its fourth-order energy contribution. With these assumptions, as shown in the Appendix, the approximate energy lowering by a singly excited configuration $\phi_k$ is $\approx E_4(k)=-(E_k-E_0)c_{k(2)}^2\approx-(E_k-E_0)c_k^2$. The full CI coefficients $c_k\approx c_{k(2)}$ are given in Table III of reference.[28] Further, $(E_k-E_0)$ may be approximated by the differences of HF orbital energies $\epsilon_i$ to within 10% on the average.

The main $\hat{f}_i$-type contribution comes from $(1\sigma_g3\sigma_g)$ with a very slight amount from $(1\sigma_g2\sigma_g)$. The estimated total energy contributions of these configurations are given in Table III.

It will be noted that at $R=12$ a.u., the $\hat{f}_i$ correction to the energy of $(1\sigma_g)^2$ HF is about $-0.33$ eV, close to our previous estimate $-0.4$ eV at $R=\infty$. As $R$ gets smaller and $(1\sigma_g)$, $(1\sigma_u)$ energy separation increases, the $\hat{f}_i$ contribution drops, though slowly at first. It is still about $-0.32$ eV at $R=4.0$ a.u., but at $R=3.0$ a.u. $(R/R_e=2.1)$ has fallen off to $-0.13$ eV. At $R=R_e\cong1.4$ a.u. it is $-0.009$ eV, negligible as expected.

Thus our usual theory for closed shells with the single determinant $(1\sigma_g)^2$ as HF $\phi_0$ will be applicable and $\hat{f}_i$ negligible, i.e., less than $(-0.1$ eV$)$ for $(R/R_e)<2$. At larger $R$'s, either these "nondynamical" $\hat{f}_i$'s can be introduced as corrections to $\phi_0+\sum_{i>j}\{\hat{u}_{ij}\}$ or, better yet, where this is simple enough one may use "extended HF" on the proper linear combination of degenerate determinants for large $R$.

This is easy enough in $H_2$, but what about other molecules and calculation of intermolecular potentials at intermediate $R$? Fortunately nonbonded interactions do not suffer from this $\hat{f}_i$ complication. The He–He system has[29] a unique HF at all $R$, based on the

[26] P.-O. Löwdin, Advan. Chem. Phys. 2, 303 (1959).
[27] We are indebted to E. R. Davidson for calling our attention to this point.
[28] S. Fraga and B. J. Ransil, J. Chem. Phys. 35, 1967 (1961).
[29] B. J. Ransil, J. Chem. Phys. 34, 2109 (1961).

always closed shell $(1\sigma_g)^2(1\sigma_u)^2$. Exclusion effects of outer electrons eliminate the degeneracy of $(1\sigma_g)^2$. Similarly for Ne–Ne, etc., or the nonbonded interactions between the bonds of a saturated hydrocarbon[30] our usual theory with the single determinant HF $\phi_0$ plus pair correlations applies with negligible $\hat{f}_i$'s.

Nondynamical $\hat{f}_i$ complication is at its worst in, e.g., dissociating $N_2$. The behavior of determinental wf's for this molecule as $R\rightarrow\infty$ has been admirably discussed by Nesbet.[31] At $R=\infty$, $\pi_u$, $\sigma_g 2p_z$, $\pi_g$, and $\sigma_u 2p_z$ all become degenerate involving six electrons at once. Here one may have to solve the HF equations for the proper linear combinations of determinants that remove the degeneracy and give the correct dissociation. Since this may get too lengthy, one may use the alternative procedure for getting these combinations in terms of localized orbitals as discussed by Nesbet.[31]

In any case, a new nondegenerate HF will take care of these major, nondynamical (long-range type) $\hat{f}_i$'s. Then, there will still be left dynamical type $\hat{f}_i$'s resulting from the residual short-range type $\hat{u}_{ij}$ correlations to which our theory above is applicable. These dynamical correlation $\hat{f}_i$'s should again be negligible. These comments apply also to the usual nonclosed shells in general.

### Type 3. Intershell Correlations

These van der Waals type correlation effects are already small; therefore, second-order energy gives good approximations to them. This means in the wf, $\chi_1$ is sufficient and there is no $f_i$.[32]

## CONCLUSION

We showed here how $f_i$, i.e., corrections to the Hartree–Fock orbitals due to correlation in many-electron closed-shell atoms and molecules can be calculated to test their smallness. Though in CI the $f_i$ correspond to infinitely many singly excited configurations, we gave methods for the calculation of these in closed form. Once the pair correlation functions are obtained, as discussed elsewhere,[3] the $f_i$'s are obtained from these simply by integrations.

We classified the different types of electron-pair correlations, discussed typical cases and their $f_i$.

The $\hat{f}_i$ of successively better trial functions for helium $(1s)^2$ were shown to become smaller and contribute less and less to the energy. When the Hylleraas six-term wavefunction is chosen as the trial function, the magnitude of $\hat{f}_i$ expressed as $\hat{f}_i/i$ is less than 2.8%, and the part of correlation energy contributed by $\hat{f}_i$ is around 0.0001 a.u.

"Dynamical" correlation in "tight pairs" like He, $Li_2$ core $(1s)^2$, $H_2$ near its equilibrium distance, etc., does not affect the over-all charge distribution or the orbitals; i.e., $\hat{f}_i$ are negligible. So are the $\hat{f}_i$ of intershell correlations or of some near-degeneracy-type correlations as in the outer shell of Be atom. Nonbonded interactions as in He–He, Ne–Ne, etc., can also be calculated from the many-electron theory developed so far; $\hat{f}_i$ may be neglected.

But degeneracies arise as diatomic molecules, e.g., $H_2$, $N_2$, etc., dissociate. If, e.g., in $H_2$, Hartree–Fock method is still based on $(1\sigma_g)^2$ at $R=\infty$, there will be 0.4 eV error due to nondynamical $\hat{f}_i$ resulting from the $(1\sigma_g)^2$, $(1\sigma_u)^2$ degeneracy. This $\hat{f}_i$ contribution becomes less than 0.1 eV for $(R/R_e)<2$. In cases of such degeneracy and in nonclosed shells "extended" HF may be used first. The remaining $\hat{f}_i$ will be of the dynamical type as in, say, He, and small.

Once these typical cases have been examined and the magnitudes estimated, $\hat{f}_i$ can be safely discarded from the wavefunction of appropriate many-electron atoms or molecules. This simplifies the energy and the wavefunction and the methods of calculation greatly, confirming our initial assumptions in Papers I and II. It also shows that there is no need for generalized SCF methods since these are ways for calculating dynamical type $\hat{f}_i$'s. They put all the emphasis on these terms that are small anyway.

The smallness of $\hat{f}_i$, along with the unimportance of many-electron correlations discussed in I means that Hartree–Fock orbitals form an excellent basis for quantum chemistry. Even though correlations change the energy significantly, their effect on orbitals is slight. Therefore all qualitative considerations on the shapes of molecules, bond angles, electron densities, can be safely based on Hartree–Fock orbitals or some good approximations to them.

## ACKNOWLEDGMENT

One of the authors (D. F. T.) wishes to thank Neil R. Kestner for many helpful discussions during the preparation of this paper. Our thanks are due also to W. H. Schmidt who helped in some of the calculations.

## APPENDIX

### Estimate of the Energy Contributions of Singly-Excited Configurations

We wish to estimate the energy contributions of individual configurations without repeating the full CI calculations.

[30] O. Sinanoğlu, J. Chem. Phys. **37**, 191 (1962).
[31] R. K. Nesbet, Phys. Rev. **122**, 1497 (1961).
[32] This is true in both concentric and nonconcentric cases, if $\phi_0$ is the HF wf for the entire atom or molecule. In ordinary London dispersion forces there is still no $f_i$ provided each atom is spherical and nonpolar. However if one starts with the Heitler–London type description and then allows overlap, orbital average polarization effects $f_i^{\circ}$—as with the bare nuclei $\phi_0$—will arise; these are eliminated by going over into MO SCF.

With $\phi_0$ the Hartree–Fock wf and $\chi$ the full CI part

$$\Psi=\phi_0+\chi=\phi_0+\chi_1+\chi_2+\cdots. \tag{A1}$$

In CI form

$$\chi=\sum_L^{\infty} c_L\phi_L. \tag{A2}$$

From ordinary perturbation theory,

$$\chi_1=\sum_L^{\infty} c_{L(1)}\phi_L, \tag{A3}$$

$$c_{L(1)}=-\frac{\langle\phi_0, H_1\phi_L\rangle}{E_L-E_0}. \tag{A4}$$

But

$$E_2(L)=-\frac{\langle\phi_0, H_1\phi_L\rangle^2}{E_L-E_0}, \tag{A5}$$

therefore,

$$E_2(L)=-(E_L-E_0)c_{L(1)}^2. \tag{A6}$$

Thus, when—as is usually the case—$c_L\approx c_{L(1)}$, the energy lowering in CI by $\phi_L$ is $\approx E_2(L)$. But for singly excited configurations, and when $\phi_0$ is HF, both $C_{L(1)}$ and $E_2(L)$ are rigorously zero on account of Brillouin's theorem. Moreover, the first-order wf $\chi_1$ determines the energy to third order. Thus the main contributions of single excitations $\{\hat{f}_i\}$ come in the second-order wf $\chi_2$ and the fourth-order energy $E_4$.

To these orders a formula similar to Eq. (A6) holds for single excitations. The proof given below is due to Kestner of our laboratory.

In Rayleigh–Schrödinger type perturbation theory[33]

$$E_4=\langle\chi_1, q\chi_2\rangle-E_2\langle\chi_1, \chi_1\rangle, \tag{A7}$$

using $\langle\phi_0, \chi_1\rangle=\langle\phi_0, \chi_2\rangle=0$. Also,

$$e\chi_2=-q\chi_1-E_2\phi_0, \tag{A8}$$

where $e=H_0-E_0$ and $q=H_1-E_1$.

Substituting Eq. (A8) in Eq. (A7), one gets

$$E_4=-\langle e\chi_2, \chi_2\rangle-E_2\langle\chi_1, \chi_1\rangle. \tag{A9}$$

Now assume

$$\phi_0+\chi\cong\phi_0+\chi_1+\chi_2, \tag{A10a}$$

$$\phi_0+\chi\approx\phi_0+\sum_{L\neq k}(c_{L(1)}+c_{L(2)})\phi_L+\sum_{k\neq L}c_{k(2)}\phi_k, \tag{A10b}$$

$\phi_k$ are the single excitations, $\phi_L$ all others. Since $\phi_k$ make no contribution to the last term of Eq. (A9), their energy contribution is

$$\sum_k E_4(k)=-\sum_k(E_k-E_0)c_{k(2)}^2, \tag{A11}$$

where again $H_0\phi_k=E_k\phi_k$.

Assuming $c_{k(2)}\approx c_k$, the full CI coefficient, the energy lowering $\Delta E(k)$ due to $\phi_k$ in full CI is

$$\Delta E(k)\approx E_4(k)=-(E_k-E_0)c_{k(2)}^2\approx-(E_k-E_0)c_k^2. \tag{A12}$$

The estimates in Table III, except the $R=\infty$ case, are based on the last part of this equation.

[33] O. Sinanoğlu, J. Chem. Phys. **34**, 1237 (1961).

VII-4 Reprinted from *Journal of Chemical Physics* **38**, 2248 (1963)

# Correlation Energy for Atomic Systems*

Enrico Clementi

*International Business Machines Corporation, Research Laboratory, San Jose, California*

(Received 6 September 1962)

The correlation energies are estimated for the isoelectronic series of atoms with 2 to 10 electrons in the ground state and in the excited states of the lowest electronic configuration. The obtained values of the correlation energy point out clearly the importance of precorrelation energy, i.e., the energy difference between Hartree and Hartree–Fock energies, the quasi-additivity of 1*s*, 2*s*, 2-, ⋯ intra- and intershell correlation, the relative importance of spin-multiplicity correlation and total angular-momentum correlation, and the $Z$ effects on correlation. The estimate of the correlation energy makes use of energy data for the relativistic energy: the accuracy of the relativistic energy computations is discussed. The 2 to 10 electron correlation energies were used as a basis to estimate the 11 to 18 electron correlation energies and to draw general conclusions on the value of the correlation energy for any atomic system.

## I. INTRODUCTION

The electronic correlation energy is a stabilization effect for any electronic system and is associated with the instantaneous motion and the potential energy of the electrons in the system. If one computes the energy of a system, taking into account from the start all the instantaneous effects of the system, then one obtains an expectation value for the energy which is the sum of average and instantaneous effects. On the other hand, if one computes the energy of a system taking into effect only average interactions between electrons, then one obtains an expectation value for the energy which does not include the effect of the instantaneous motion and the potential of each electron. The latter case will give a higher energy than the former case, and the difference between these two energies is the correlation energy.

The correlation energy is commonly defined as the difference between the exact nonrelativistic energy and the Hartree–Fock energy.[1]

It is worthwhile to note that there are several Hartree–Fock schemes, each leading to a somewhat different energy and, consequently, to different values of the correlation energy. For this reason we state from the beginning that in the following, when we refer to the Hartree–Fock energy, we refer to the best energy one can obtain by the analytical self-consistent field method as put forward by Roothaan.[2] The reason for this choice is simply that by now this method has been used to obtain many atomic functions and energies and a large number of molecular functions and energies.

From a conceptual point of view one might prefer to define the correlation energy as the difference between the exact nonrelativistic energy and the Hartree energy. The reason is that the Hartree–Fock method presents an unbalanced situation when we look at the way in which electrons with same spins and those with different spins are considered. The Hartree–Fock method partially correlates electrons with the same spins. This correlation present in the Hartree–Fock method will be hereafter referred to as *precorrelation*, where we define the precorrelation energy as the difference between the Hartree–Fock energy and the Hartree energy. This energy difference is a correlation energy, but in view of the accepted definition of correlation, we might say that it is a correlation energy *ante litteram*.

We note that the emphasis on the nonrelativistic exact energy in the definition of the correlation energy has mainly a practical value. The relativistic energy itself can be partitioned into a correlated and an uncorrelated relativistic energy. We note that for the relativistic energy we mean the expectation values that are associated with the relativistic Hamiltonian as given by Hirschfelder[3] (in general) or by Bethe and Salpeter[4] for the two-electron problem.

It is well known that there are, in principle, several methods available in order to obtain correlated wavefunctions. At present it seems that a common characteristic of these methods is that they are not easily applicable and are often outside of today's computational capabilities. The so-called "many-body problem" has been investigated with two different points of view. There are those who seem to use the word "many" in the restricted sense that "many" is less than a dozen or so. There are others who use the word "many" in the sense that the number of particles

* Partially presented at the Symposium on Molecular Structure and Spectroscopy, June 1962, Columbus, Ohio, and to the International Symposium on Molecular Structure and Spectroscopy, September 1962, Tokyo, Japan.

[1] P.-O. Löwdin, Advan. Chem. Phys. **2**, 207 (1959).

[2] C. C. J. Roothaan, Rev. Mod. Phys. **32**, 179 (1960); C. C. J. Roothaan and P. S. Bagus, "Methods of Computational Physics" (to be published).

[3] J. O. Hirschfelder, C. F. Curtiss, and R. B. Bird, *Molecular Theory of Gases and Liquids* (J. Wiley & Sons, Inc., New York, 1954), p. 1045.

[4] H. A. Bethe and E. E. Salpeter, *Handbuch der Physik*, edited by S. Flügge (Springer-Verlag, Berlin, 1957), Band XXXV.

is as large as the Avogadro number.[5] The statement at the beginning of this paragraph concerns mostly the "*many* compared to one, *but few* compared to two or three dozen" type of available techniques for the many-electron problem. This unsatisfactory situation is easily understandable in view of the very complex nature of the problem. We refer to Slater's work[6] for a review of the "many—but few" methods of computing correlated functions. Finally, mention should be made to the existence of more recent work,[7,8] but it is too early to comment on the practical use of these methods, since numerical applications of these new theoretical approaches are not available at present.

For the above reasons, it might be of interest to give an estimate of the correlation energy for the isoelectronic series of atomic systems with 2 to 18 electrons. The estimate we present varies in accuracy. For low $Z$ atoms ($Z \leq 10$) we estimate a 5% error; for intermediate $Z$ atoms ($10 \leq Z \leq 20$) the maximum error should be around 10%, and for large $Z$ ($20 \leq Z \leq 36$) the values reported give certainly the order of magnitude, but probably not much more than that.

## II. PRECORRELATION ENERGY AND CORRELATION ENERGY

Since the electrons with parallel spins are somewhat correlated in the Hartree–Fock method, one can expect that the correlation energy in the ground state of neutral atoms is not a linear function of the number of electrons. The Hartree–Fock method uses antisymmetrized wavefunctions; this is done to satisfy the Pauli principle and brings about the exchange energy which is the origin of the precorrelation energy. Electrons with the *same* spin find themselves encircled by a Fermi hole which prevents electrons with the same spin from approaching each other.

From the work of Fröman[9] and that of Linderberg and Shull[10] we expect a strong correlation for pairs of electrons of the same shell (intrashell correlation), a smaller correlation between electrons of different shells (intrashell correlation), and a quasi-constancy (it turns out that this expectation is quantitatively incorrect, as shown later) for the correlation of a pair of electrons with opposite spin.

With the Fröman and Linderberg work in mind, one will predict that the correlation energy of the ground-state first-row atoms will behave as follows: There is a given correlation for the pair of electrons in the He atom. For the beryllium atom, the correlation is about twice that of helium. For the neon atom, the correlation energy is given by a linear relation connecting the He and Be correlation energies versus $Z$. The weighting of the energies of the three atoms, He, Be, and Ne should be considered equal since they all have closed shells. Lithium will have an intermediate correlation energy between He and Be. Since the extra electron (compared to He) is a $2s$ electron, which has a maximum radial probability far from the $1s$ electrons, its correlation with the $1s$ electrons is certainly small. In fact, from the Linderberg and Shull[10] values, we know it to be very small (the intershell correlation for $1s$–$2s$ is much smaller than the $1s$ (or $2s$) intrashell correlation). The correlation energy of B, C, and N can be estimated by keeping in mind that the $2p$ electrons have all parallel spins and consequently the precorrelation existing in the Hartree–Fock energy will take care of most of the correlation for the $2p$ electrons. There will certainly be some intershell correlation of $1s$–$2p$ type and $2s$–$2p$ type. Since the $2s$ electrons are in the same spatial neighborhood as the $2p$ electrons, one is tempted to assume that ($1s$–$2p$) intershell correlation $\ll$ ($2s$–$2p$) intershell correlation.

The correlation energy for O, F, and Ne should increase sharply. With those atoms we start building one, two, and respestively, three pairs of unparallel spin electrons in the same shell (the $2p$ shell). The sharp increase is due to the lack of precorrelation for those newly added electrons.

It is fairly simple to be more quantitative about all the above reasoning. Accurate Hartree–Fock energies for 2 to 18 electrons are available.[11] a few relativistic energies were available from Fröman[9]; and the total energy can be obtained experimentally by adding the ionization potentials from Moore.[12] Then the correlation energy is simply the total energy minus the Hartree–Fock energy minus the relativistic energy. We note, however, that the Fröman data are available only for closed-shell atoms. The extrapolation to open shell can be made quite simply since (i) the main contribution of the relativistic energy is in the inner shell electrons; (ii) the spin–orbit and spin–spin correction for open shell is very small up to $Z=10$ (only of the order of few hundreds of $cm^{-1}$); and (iii) the

[5] See for example D. Pines, *The Many Body Problem* (Benjamin, Inc., New York, 1961); C. Dewitt, *The Many Body Problem* (John Wiley & Sons, Inc., New York, 1959).
[6] J. C. Slater, *Quantum Theory of Atomic Structures* McGraw-Hill Book Company, Inc., New York, 1962) Vol. II. See also H. Yoshimine, *Advan. Chem. Phys.* **II**, 323 (1959).
[7] L. Szasz, Z. Naturforsch. **15a**, 909 (1960); and see especially MIT Solid State and Molecular Group Theory Group, Quarterly Progress Report No. 45.
[8] O. Sinanoğlu, Proc. Natl. Acad. Sci. U. S. **47**, 1217 (1961); J. Chem. Phys. **36**, 706, 3198 (1962).
[9] A. Fröman, Phys. Rev. **112**, 807 (1958), Rev. Mod. Phys. **32**, 317 (1960).
[10] J. Linderberg and H. Shull, J. Mol. Spectry. **5**, 1 (1960).
[11] Neutral first-row atoms: E. Clementi, C. C. J. Roothaan, and M. Yoshimine, Phys. Rev. **127**, 1618 (1962), Neutral second-row atoms: E. Clementi and C. C. J. Roothaan (to be published); Isoelectronic Series for 2 to 10 electrons: E. Clementi, J. Chem. Phys. **38**, 996 (1963); Isoelectronic Series for 11 to 18 electrons: E. Clementi, J. Chem. Phys. **38**, 1001 (1963); Isoelectronic Series for 19 to 36 electrons: E. Clementi (to be submitted); First-row negative ions: E. Clementi and A. D. McLean (to be submitted).
[12] C. Moore, *Atomic Energy Levels* (National Bureau of Standards, Washington 25, D. C., 1958).

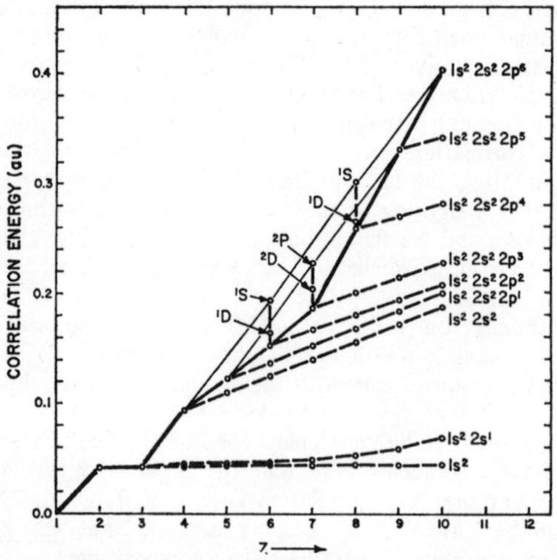


FIG. 1. Correlation energy vs $Z$ for 2 to 10 electrons. The dashed lines connect isoelectronic series in the ground state, the straight solid lines connect ground states or excited states with the same symmetry. For each isoelectronic series the electronic configuration is reported in this figure. The excited states are indicated by their state designation.

closed-shell energy contribution can be partitioned into the $1s$, the $2s$, and the $2p$ relativistic energies and is additive.

With these data we obtain[13] an estimate of the correlation energy for 2 to 10 electrons in neutral and in positive ions with $Z$ from 2 to 10. The results are condensed into a diagram (Fig. 1) where the correlation energy is plotted against $Z$.

In Fig. 1 we have also given the correlation energy for excited states of the ground-state configuration for neutral atoms. The correlation energy for excited states of the positive ions was not indicated, to avoid confusion on the diagram. The solid lines connect states with the same multiplicity and total angular momentum. The linear dependency on the number of electrons for the correlation energy of $^1S$(Be), $^1S$(Ne), excited $^1S$(C), and $^1S$(O) was obtained as expected. The same is true for the $^2P$(B), $^2P$(F), and excited $^2P$(N). We note that for these excited states the relativistic correction was taken equal to that of the corresponding ground state. Clearly this assumption is valid only because of the small energy error introduced in this way.

This diagram, we feel, reveals the essential features of the correlation energy problems.

## III. RELATIVISTIC CORRECTION

From the previous section it is clear that in order to obtain more accurate values for the correlation energy in the computational scheme we have adopted, one needs more accurate values of the relativistic correction. Hartmann and the author[14] have computed the relativistic correction for 2-, 4-, 10-, 12-, and 18-electron isoelectronic series with the method used by Fröman.[9] Our wavefunctions[11] are more accurate than those available to Fröman, and thus we could expect more accurate results for the relativistic correction. On the other hand, one point has to be made clear. A Hartree–Fock function cannot give accurate relativistic correction since in the relativistic energy there is a relativistic correlation energy contribution which clearly cannot be obtained from direct use of Hartree–Fock functions. (Details are given in the Hartmann and Clementi work.[14]) In addition, the relativistic correction depends strongly on the wavefunction behavior near the nucleus (particularly on the $1s$ function). Thus, one can expect to have significant variations in the relativistic energy by using different analytical Hartree–Fock functions, even if these functions give the same total energy. This is one more case which points out that the total energy is a necessary, but not sufficient, criterion for the "goodness" of a function.[11,15,16] (On the variations of the relativistic energy for different self-consistent field functions, see reference 14.)

For closed-shell determinantal functions, there is no intershell contribution to the total relativistic energy, i.e., the matrix elements $\langle \phi_i H(\text{REL}) \phi_j \rangle$ are zero if $i$ and $j$ refer to orbitals of different shells (see reference 14). We note that this is not true for the individual contributions of $E_1$, $E_4$, and $E_5$, the expectation values of the relativistic Hamiltonians $H_1$, $H_4$, and $H_5$ in the Bethe and Salpeter notation (see reference 4, page 267), but is true for $E(\text{REL}) = E_1 + E_4 + E_5$. For closed shells we can alternatively write for the determinantal functions $E(\text{REL}) = E(1s) + E(2s) + E(2p) + \cdots$, where $E(1s)$, $E(2s)$, and $E(2p)$ are the relativistic energy contributions to the total relativistic energy $E(\text{REL})$ from the $1s$, $2s$, and $2p$ orbitals, respectively. This latter way of presenting $E(\text{REL})$ allowed Hartmann and the author[14] to verify systematically the approximate constancy of the $E(1s)$ for different positive ions and for the neutral atom with the same $Z$. In addition, the relativistic computations did show that $E(1s) > E(2s) > E(3s)$ and $E(1s > E(2p) > E(3p)$. The computations did reveal that there are deviations for computations made with different SCF functions regardless of the fact that the different functions gave equally good energies. Again, this is not too important for the estimation of the correla-

[13] E. Clementi, IBM Research Note NJ-7, 3 January 1962.

[14] H. Hartmann and E. Clementi, "Relativistic Correction for Closed Shells Atoms" Phys. Rev. (to be published), see also E. Clementi, J. Mol. Spectry. (to be published).

[15] C. C. J. Roothaan and A. Weiss, Rev. Mod. Phys. **32**, 194 (1960).

[16] See, for example, C. C. J. Roothaan, L. M. Sachs, and A. W. Weiss, Rev. Mod. Phys. **32**, 186 (1960).

TABLE I. Relativistic correction (in a.u.) for closed shells.[a]

| | 2 Electrons | | 4 Electrons | | 10 Electrons | |
|---|---|---|---|---|---|---|
| Z | Computed[b] | S.S.M.[c] | Computed[b] | S.S.M.[c] | Computed[b] | S.S.M.[c] |
| 2 | −7.06663−5 | −5.99324−5 | | | | |
| 3 | −5.50782−4 | −4.99904−4 | | | | |
| 4 | −2.08991−3 | −1.87755−3 | −2.19995−3 | −1.91796−3 | | |
| 5 | −5.65361−3 | −5.09238−3 | −6.16455−3 | −5.35035−3 | | |
| 6 | −1.25252−2 | −1.13454−2 | −1.40783−2 | −1.22509−2 | | |
| 7 | −2.43071−2 | −2.21658−2 | −2.79750−2 | −2.45757−2 | | |
| 8 | −4.29177−2 | −3.73950−2 | −5.02963−2 | −4.24156−2 | | |
| 9 | −7.05914−2 | −6.52018−2 | −8.38850−2 | −7.45716−2 | | |
| 10 | −1.09890−1 | −1.20275−1 | −1.32001−1 | −1.18384−1 | −1.31293−1 | −1.19728−1 |
| 11 | −1.63676−1 | −1.52847−1 | −1.98315−1 | −1.79167−1 | −2.00652−1 | −1.83718−1 |
| 12 | −2.35176−1 | −2.20655−1 | −2.86891−1 | −2.61126−1 | −2.95056−1 | −2.71429−1 |
| 13 | −3.27911−1 | −3.08996−1 | −4.02248−1 | −3.68576−1 | −4.19983−1 | −3.90575−1 |
| 14 | −4.45684−1 | −4.21671−1 | −5.49291−1 | −5.06481−1 | −5.81509−1 | −5.44285−1 |
| 15 | −5.92731−1 | −5.62833−1 | −7.33382−1 | −6.80347−1 | −7.86226−1 | −7.38850−1 |
| 16 | −7.73495−1 | −7.36965−1 | −9.60270−1 | −8.95355−1 | −1.04129+0 | −9.84055−1 |
| 17 | −9.92841−1 | −9.48891−1 | −1.23627+0 | −1.15828+0 | −1.35444+0 | −1.28548+0 |
| 18 | −1.25571+0 | −1.20375+0 | −1.56789+0 | −1.47567+0 | −1.73394+0 | −1.65197+0 |
| 19 | −1.56785+0 | −1.50706+0 | −1.96236+0 | −1.85443+0 | −2.18874+0 | −2.09335+0 |
| 20 | −1.93464+0 | −1.86462+0 | −2.42721+0 | −2.30231+0 | −2.72809+0 | −2.61961+0 |

[a] The last figure preceded by a sign in each entry is the power of 10 by which the relativistic correction has to be multiplied.
[b] Computed by H. Hartmann and E. Clementi[14] (one set of computations).
[c] Empirical data from Scherr, Silverman, and Matsen.[18]

tion energy for low $Z$, but it becomes more important for large $Z$, since there the relativistic effects are larger than the correlation energy.

For open shells the largest additional contribution to $E(\text{REL})$ not present in the closed-shell energy is the spin–orbit energy. The spin–orbit operator affects the energies in a twofold manner. It splits the $J$ degeneracy and in addition it lowers the baricenter of the term manifold by a small amount compared to the splitting of the terms. In the Lande approximation[17] the sum of the term splitting weighted by the statistical factor $(2J+1)$ is zero (the center of gravity of the term is not affected by the spin orbit perturbation). Since the splitting of the terms is available from Moore's tables,[12] we can estimate the spin–orbit correction. By adding the spin–orbit contribution to the closed-shell contribution, one obtains the total relativistic energy for open shells.

If one performs the computations for the relativistic closed and open shells as indicated above, one is still without any check (i) on the value for the relativistic energy one would obtain by using correlated functions; (ii) which analytical SCF computation is to be preferred when one computes the relativistic correction.

Recently Scherr, Silverman, and Matsen[18] made a critical survey of the experimental ionization potentials for the isoelectronic series of the ground state atoms and ions with 2 to 10 electrons and $Z=2$ up to $Z=20$. In the Scherr, Silverman, and Matsen work a semiempirical scheme is used to obtain the relativistic energy. In Table I we compare these empirical data (indicated as S.S.M.) with one set of our computed values. The $Z$ dependency is essentially the same, but the absolute values we computed[14] are constantly larger than those of S.S.M. In view of the careful empirical analysis done in the S.S.M. data, and in view of the ambiguity implicit in our relativistic computation, we have finally decided to use Scherr, Silverman, and Matsen data as an estimate of the relativis-

[17] See, for example, E. U. Condon and G. H. Shortley, *The Theory of Atomic Spectra* (Cambridge University Press, New York, 1957).

[18] C. W. Scherr, J. N. Silverman, and F. A. Matsen, Phys. Rev. **128**, 2675 (1962).

TABLE II. Estimated correlation energy for atomic isoelectronic series with 2 to 10 electrons (in a.u.).[a]

| $Z$ | $2-{}^1S_0$ | $3-{}^2S_{1/2}$ | $4-{}^1S_0$ | $5-{}^2P_{1/2}$ | $6-{}^3P_0$ | $6-{}^1D_2$ | $6-{}^1S_0$ |
|---|---|---|---|---|---|---|---|
| 2 | −0.0421 | | | | | | |
| 3 | −0.0435 | −0.0453 | | | | | |
| 4 | −0.0443 | −0.0475 | −0.0944 | (−0.109) | | | |
| 5 | −0.0448 | −0.0489 | −0.1123 | −0.125 | (−0.146) | | |
| 6 | −0.0451 | −0.0498 | −0.1268 | −0.139 | −0.158 | −0.169 | −0.199 |
| 7 | −0.0453 | −0.0505 | −0.1412 | −0.151 | −0.167 | −0.180 | −0.219 |
| 8 | −0.0455 | −0.0510 | −0.1551 | −0.162 | −0.175 | −0.190 | −0.236 |
| 9 | −0.0456 | −0.0513 | −0.1684 | −0.173 | −0.182 | −0.198 | −0.254 |
| 10 | −0.0457 | −0.0516 | −0.1814 | −0.182 | −0.188 | −0.206 | −0.270 |
| 11 | −0.0458 | −0.0519 | −0.1941 | −0.191 | −0.193 | −0.216 | −0.287 |
| 12 | −0.0459 | −0.0521 | −0.2066 | −0.200 | −0.199 | −0.225 | −0.307 |
| 13 | −0.0459 | −0.0523 | −0.2190 | −0.208 | −0.204 | −0.236 | −0.329 |
| 14 | −0.0460 | −0.0524 | −0.2313 | −0.216 | −0.209 | −0.246 | −0.349 |
| 15 | −0.0461 | −0.0525 | −0.2435 | −0.225 | −0.214 | −0.256 | −0.371 |
| 16 | −0.0461 | −0.0527 | −0.2556 | −0.232 | −0.218 | −0.27 | −0.415 |
| 17 | −0.0462 | −0.0528 | −0.2677 | −0.240 | −0.222 | −0.27 | −0.436 |
| 18 | −0.0463 | −0.0529 | −0.2797 | −0.248 | −0.227 | −0.28 | −0.457 |
| 19 | −0.0463 | −0.0529 | −0.2917 | −0.255 | −0.231 | −0.29 | −0.478 |
| 20 | −0.0463 | −0.0530 | −0.3037 | −0.263 | −0.235 | −0.30 | −0.499 |
| 21 | −0.046 | −0.053 | −0.316 | −0.270 | −0.24 | −0.31 | −0.52 |
| 22 | −0.046 | −0.053 | −0.327 | −0.278 | −0.24 | −0.32 | −0.54 |
| 23 | −0.047 | −0.053 | −0.339 | −0.285 | −0.25 | −0.33 | −0.57 |
| 24 | −0.047 | −0.053 | −0.351 | −0.293 | −0.25 | −0.34 | −0.59 |
| 25 | −0.047 | −0.053 | −0.363 | −0.300 | −0.25 | −0.35 | −0.60 |
| 26 | −0.047 | −0.053 | −0.375 | −0.31 | −0.26 | −0.36 | −0.62 |
| 27 | −0.047 | −0.053 | −0.387 | −0.32 | −0.26 | −0.36 | −0.64 |
| 28 | −0.047 | −0.053 | −0.398 | −0.32 | −0.26 | −0.37 | −0.66 |
| 29 | −0.047 | −0.053 | −0.411 | −0.33 | −0.27 | −0.38 | −0.68 |
| 30 | −0.047 | −0.053 | −0.423 | −0.34 | −0.27 | −0.39 | −0.71 |
| 31 | −0.047 | −0.053 | −0.434 | −0.35 | −0.27 | | |
| 32 | −0.047 | −0.054 | −0.446 | −0.35 | −0.28 | | |
| 33 | −0.047 | −0.054 | −0.458 | −0.36 | −0.28 | | |
| 34 | −0.047 | −0.054 | −0.470 | −0.37 | −0.28 | | |
| 35 | −0.047 | −0.054 | −0.482 | −0.38 | −0.29 | | |
| 36 | −0.047 | −0.054 | −0.494 | −0.38 | −0.29 | | |

| $Z$ | $7-{}^4S_{3/2}$ | $7-{}^2D_{3/2}$ | $7-{}^2P_{1/2}$ | $8-{}^3P_2$ | $8-{}^1D_2$ | $8-{}^1S_0$ | $9-{}^2P_{3/2}$ | $10-{}^1S_0$ |
|---|---|---|---|---|---|---|---|---|
| 7 | −0.188 | −0.208 | −0.232 | | | | | |
| 8 | −0.193 | −0.213 | −0.242 | −0.258 | −0.269 | −0.306 | | |
| 9 | −0.197 | −0.217 | −0.251 | −0.260 | −0.272 | −0.315 | −0.324 | (−0.398)[b] |
| 10 | −0.200 | −0.222 | −0.261 | −0.267 | −0.279 | −0.329 | −0.328 | −0.393 |
| 11 | −0.203 | −0.227 | −0.271 | −0.274 | −0.286 | −0.343 | −0.336 | −0.396 |
| 12 | −0.205 | −0.233 | −0.282 | −0.279 | −0.292 | −0.356 | −0.344 | −0.402 |
| 13 | −0.207 | −0.238 | −0.291 | −0.285 | −0.297 | −0.366 | −0.350 | −0.409 |
| 14 | −0.209 | −0.243 | −0.301 | −0.291 | −0.301 | −0.378 | −0.358 | −0.417 |
| 15 | −0.211 | −0.249 | −0.308 | −0.296 | −0.307 | −0.389 | −0.366 | −0.426 |
| 16 | −0.213 | −0.254 | −0.315 | −0.301 | −0.313 | −0.401 | −0.372 | −0.434 |
| 17 | −0.214 | −0.259 | −0.325 | −0.305 | −0.317 | −0.412 | −0.379 | −0.442 |
| 18 | −0.215 | −0.264 | −0.332 | −0.309 | −0.321 | −0.422 | −0.384 | −0.449 |
| 19 | −0.217 | −0.270 | −0.341 | −0.313 | −0.325 | −0.433 | −0.389 | −0.456 |
| 20 | −0.218 | −0.275 | −0.349 | −0.317 | −0.329 | −0.443 | −0.394 | −0.463 |
| 21 | −0.22 | −0.28 | −0.36 | −0.32 | −0.33 | −0.45 | −0.40 | −0.47 |
| 22 | −0.22 | −0.28 | −0.37 | −0.32 | −0.34 | −0.46 | −0.40 | −0.48 |
| 23 | −0.22 | −0.29 | −0.37 | −0.33 | −0.34 | −0.47 | −0.41 | −0.48 |
| 24 | −0.22 | −0.30 | −0.38 | −0.33 | −0.34 | −0.48 | −0.41 | −0.49 |
| 25 | −0.22 | −0.30 | −0.39 | −0.34 | −0.35 | −0.49 | −0.42 | −0.50 |
| 26 | −0.23 | −0.31 | −0.40 | −0.34 | −0.35 | −0.50 | −0.42 | −0.50 |
| 27 | −0.23 | −0.31 | −0.41 | −0.34 | −0.36 | −0.51 | −0.43 | −0.51 |
| 28 | −0.23 | −0.32 | −0.42 | −0.35 | −0.36 | −0.52 | −0.43 | −0.52 |
| 29 | −0.23 | −0.32 | −0.43 | −0.35 | −0.36 | −0.53 | −0.44 | −0.52 |
| 30 | −0.24 | −0.33 | −0.43 | −0.35 | −0.37 | −0.54 | −0.44 | −0.53 |
| 31 | −0.24 | | | −0.36 | | | −0.45 | −0.53 |
| 32 | −0.24 | | | −0.36 | | | −0.45 | −0.54 |
| 33 | −0.24 | | | −0.37 | | | −0.46 | −0.55 |
| 34 | −0.24 | | | −0.37 | | | −0.46 | −0.55 |
| 35 | −0.24 | | | −0.37 | | | −0.46 | −0.56 |
| 36 | −0.25 | | | −0.38 | | | −0.47 | −0.56 |

[a] The correlation energy for the negative ions is obtained by extrapolation of the corresponding isoelectronic series.

[b] This value should be accurate within ±0.003 a.u. The correlation energy for $O^{--}$ ($^1S$) is estimated at −0.406±0.006 a.u.

tic energy in the computation of the correlation energy. The difference between our computed values and the S.S.M. values was used as a criterion to establish the error in our estimate of the correlation energy.

It is noted that these uncertainties in the relativistic energy present the most difficult block in computing the correlation energy in the method we have adopted. Extrapolations by means of $Z$ power expansions were made for the relativistic data of S.S.M. and used for the correlation energy from $Z=20$ to $Z=36$. For the excited states the relativistic correction is obtained by subtracting from the S.S.M. relativistic energies for the ground states the spin–orbit correction and adding to it the spin–orbit correction for the excited state. The spin–orbit correction is obtained from Moore's data on term splitting or extrapolations of the same up to $Z=36$.

## IV. ESTIMATED VALUES OF THE CORRELATION ENERGY AND DISCUSSION

In the previous section we have indicated how we can obtain information about the relativistic correction. We need, in addition, the total energies for the isoelectronic series of the ground state and excited states. The sum of the ionization potentials gives us the total energy of the ground state. By subtracting the excitation energy from the total ground-state energy we obtain the total energy of the excited states. Again from Moore's tables or from an extrapolation of the data of Moore by means of power expansion in $Z$ we calculate the total energies of the excited state. Thus, we have the total energy, $E$(TOT), the relativistic energy, $E$(REL), and the Hartree–Fock energy, $E$(HF), for the ground and the excited states.[11] We note that the Hartree–Fock energy is computed in the approximation of infinite nuclear mass. Consequently, the Hartree–Fock energies must be mass corrected. (See for example, Bethe and Salpeter, reference 4, page 253.) The correlation energy is simply $E(\mathrm{COR})=E(\mathrm{TOT})-E(\mathrm{REL})-E(\mathrm{HF})$. The results we have obtained are given in Table II (in atomic units; one atomic unit equals 27.2097 eV). The Hartree–Fock energies are computed with seven significant figures and the relativistic and the total energies are computed with the same number of figures. In Table II the correlation is given to the number of figures we feel are correct (plus one not to introduce roundoff error.) One can notice that the number of significant figures we give varies from four to one. As previously stated, this is done in view of the uncertainty in the relativistic energy and the extrapolations of the total energy. To estimate the possible error in the relativistic energy we have looked at the difference between our computed values[14] and those of S.S.M.[18] To estimate the possible error in the extrapolations of Moore's excitation energies, we compared the mean percentage deviation in the values calculated by the power expansion in $Z$ with the known excited-state energies. We feel that these criteria, to establish the error in our estimated correlation energy, are somewhat conservative.

TABLE III. Correlation energies from *ab initio* calculations (in a.u.).

| Case | $E$(R.W.)[a] | Case | $E$(W)[b] | Case | $E$(W) |
|---|---|---|---|---|---|
| He | −0.0424 | | | | |
| $Li^+$ | −0.0435 | Li | −0.0444 | | |
| $Be^{2+}$ | −0.04427 | $Be^+$ | −0.0462 | Be | −0.0869 |
| $B^{3+}$ | −0.04474 | $B^{2+}$ | −0.0472 | $B^+$ | −0.1038 |
| $C^{4+}$ | −0.04506 | $C^{3+}$ | −0.0479 | $C^{2+}$ | −0.1177 |
| $N^{5+}$ | −0.04529 | $N^{4+}$ | −0.0483 | $N^{3+}$ | −0.1305 |
| $O^{6+}$ | −0.04546 | $O^{5+}$ | −0.0486 | $O^{4+}$ | −0.1424 |
| $F^{7+}$ | −0.04558 | | ... | | ... |
| $Ne^{8+}$ | −0.04570 | | ... | | ... |

[a] $E$ (R.W.) refers to the Roothaan and Weiss computed correlation energy.[15]
[b] $E$(W) refers to the Weiss computed correlation energy.[20]

There are cases in which the correlation is directly known from the computations of the Hartree–Fock and correlated functions. For the two electron cases, the data of Roothaan and Weiss[15] and those of Pekeris[19] give us accurate correlation energies up to $Ne^{8+}$. The computations by Weiss[20] of three and four electron systems up to $O^{5+}$ and $O^{4+}$, respectively, give us data for the correlation energies for three and four electrons. There are no other data of comparable accuracy available. In Table III the Roothaan and Weiss data for two electrons and the Weiss data for three and four electrons are given. It is noted that for the three and four electrons the Weiss computations give only a lower limit of the correlation energy since the exact nonrelativistic energy was not fully obtained. Comparing Table II and Table III, our estimated correlation energies are in agreement with the data in Table III.

It is quite interesting to note that not only the $1s^2\,2s^2$ isoelectronic series has a strong $Z$ dependency, but also the $2p(n)$ isoelectronic series shows large $Z$ dependency. One notices that the $Z$ dependency is pronounced in the Be($^1s$) series, and progressively less in B($^2P$), C($^3P$), and N($^4S$). From O($^3P$) to Ne($^1S$) the dependency is about constant. We note that Linderberg and Shull[10] have discussed the $Z$ dependency of the $1s^2\,2s^2$ configuration in terms of $2s$, $2p$ degeneracy.

We can comment on the excited states correlation energy somewhat further than that done in Sec. II. The correlation energy for the multiplets of a given

[19] C. L. Pekeris, Phys. Rev. **112,** 1649 (1958).
[20] A. W. Weiss, Phys. Rev. **122,** 1826 (1961). A. Fröman communicated to us a value of 0.0544 a.u. for the correlation energy in the three-electron problem in the limit of $Z=\infty$.

TABLE IV. Correlation energy for 1 to 18 electrons (in a.u.).[a]

| State | $-IP$ | $EHF(N-1)$ | $-EHF(N)$ | $EC(N-1)$ | $EC(N)$ | Correct |
|---|---|---|---|---|---|---|
| Ne | −0.7923 | −127.81765 | 128.54698 | −0.332 | −0.395 | −0.397 |
| Na | −0.1888 | −161.67676 | 161.85857 | −0.395 | −0.402 | −0.404 |
| Mg | −0.2809 | −199.37132 | 199.61432 | −0.402 | −0.440 | −0.444 |
| Al | −0.2199 | −241.67408 | 241.87630 | −0.440 | −0.458 | −0.480 |
| Si | −0.2995 | −288.57276 | 288.85420 | −0.458 | −0.476 | −0.503 |
| P | −0.3853 | −340.34947 | 340.71857 | −0.476 | −0.492 | −0.540 |
| S | −0.3806 | −397.17285 | 397.50469 | −0.492 | −0.541 | −0.596 |
| Cl | −0.4781 | −459.04839 | 459.48181 | −0.541 | −0.586 | −0.642 |
| Ar | −0.5790 | −526.27990 | 526.81705 | −0.586 | −0.628 | −0.692 |

[a] These correlations are for the ground state.

term is approximately the same. The variation in correlation energy, for example, between $B(^2P_{1/2})$ and $B(^2P_{3/2})$ are very small and within the error of the estimate. For this reason no such data are reported.

For different states of the same electronic configuration the correlation energy has the following characteristics. *First*, the lowest correlation is for the state of higher spin multiplicity. For example, in the $^3P$, $^1D$, and $^1S$ series of C (or the series of O) the $E(\text{COR. }^3P) < E(\text{COR. }^1D)$ and $E(\text{COR. }^3P) < E(\text{COR. }^1S)$. This is, as mentioned previously at length, a consequence of the spin precorrelation in the Hartree–Fock method. *Second*, for states with the same spin multiplicity the smaller correlation is for the states of highest angular momentum. For example, $E(\text{COR. }^1D) < E(\text{COR. }^1S)$ for the carbon and the oxygen series, and $E(\text{COR. }^2D) < E(\text{COR. }^2P)$ for the nitrogen series. Since states with the same spin multiplicity but different angular momenta do not have the same correlation energy (for the same $Z$ and the same number of electrons), one concludes that in the Hartree–Fock method we have not only spin precorrelation but also angular precorrelation. The angular precorrelation being in the sense that the higher the angular momentum (total angular momentum) the higher the angular precorrelation. This is quite interesting because it tells us that we cannot obtain excitation energies of the correct magnitude with the Hartree–Fock method even for states of the same multiplicity. A simple explanation of the differences of the correlation energies between states of the same multiplicity but different total angular momentum is the following: The larger the angular momentum, the more "preferential" is the electron's motion about the nucleus. Talking about preferential motions is tantamount to talking of correlation of the motion. Since the correlation energy is higher for the states of lowest angular momentum, the correlation energy tends to oppose Hund's rules, which are exaggerated by the Hartree–Fock method.

## V. IONIZATION POTENTIALS AND CORRELATION ENERGY

The first ionization potential of a system of $N$ electrons is by definition the difference between the total energy of the ground state for the same system with $N$-1 electrons and the total energy of the ground state for the system with $N$ electrons. By using the notation $ET$, $EHF$, $ER$, and $EC$ for the total, Hartree–Fock, relativistic, and correlation energies, respectively, and $IP$ for the ionization potential, we can write the obvious relation:

$$IP = EHF(N\text{-}1) - EHF(N) + ER(N\text{-}1) - ER(N) + EC(N\text{-}1) - EC(N).$$

We note that if one puts $ER(N\text{-}1)$ or $ER(N)$ equal to zero, one would introduce a serious error, but the assumption $ER(N\text{-}1) - ER(N) = 0$ is not too bad for $N \geq 3$. This is due to the "concentration" of the relativistic energy on the inner shells, previously discussed at length. Hence,

$$IP \cong EHF(N\text{-}1) - EHF(N) + EC(N\text{-}1) - EC(N),$$

or

$$EC(N) \cong EHF(N\text{-}1) - EHF(N) + EC(N\text{-}1) - IP. \tag{1}$$

This last equation is quite useful to obtain the correlation energy when the $EC(N\text{-}1)$ term is known. In Table IV we use this method to estimate the correlation energies for the second group. The entries given in Table IV are in the same order as those of the last above equation. It is noted, however, that this method of estimation introduces an accumulation of errors. The Hartree–Fock energy for the neutral and positive ions were recently computed by us.[11] We have repeated the computation of $EC$ for the Ne atom, and obtained −0.395 a.u. whereas, before we obtained −0.397 a.u.

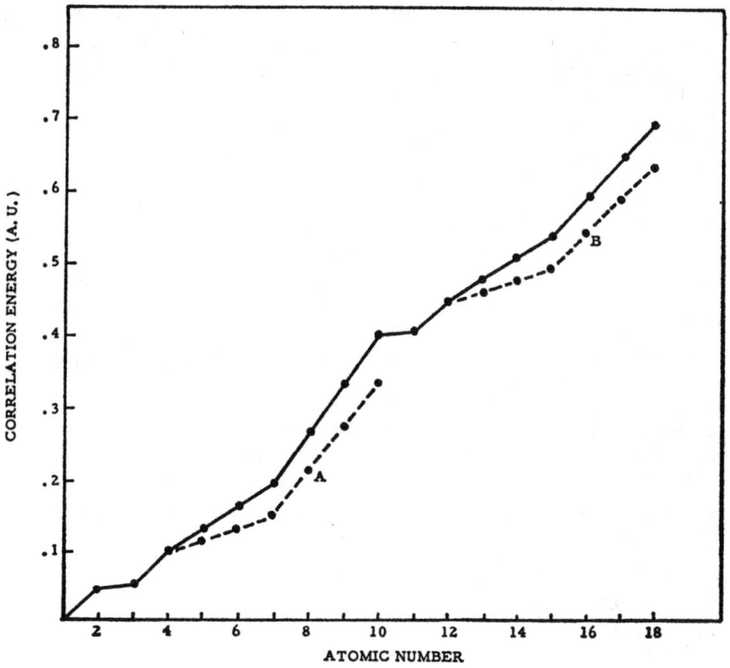


FIG. 2. Correlation energy for the ground states of neutral atoms. The full line refers to the values for the correlation energy when the relativistic energy is carefully considered. The dotted lines refer to the value for the correlation energy one would obtain by use of Eq. (1), where the relativistic effect is approximated.

(see Table II). The essential features of the computation for the correlation energies of the second group reveal what is expected: an essential repetition of the features encountered in the first group.

In order to gain information on the accuracy of this method in computing the correlation energy, we have performed the same computation for the first group, using the correlation $EC(N\text{-}1) = EC(2)$ of $Li^+$ as given in Table II. The correlation energies computed this way for the neutral ground-state atoms are −0.046 (−0.046), −0.093 (−0.095), −0.106 (−0.126), −0.123 (−0.158), −0.144 (−0.190), −0.208 (−0.261), −0.270 (−0.327), and −0.333 (−0.397) a.u. for Li, Be, B, C, N, O, F, and Ne, respectively. (In parentheses we have reported the values of Table II for the corresponding atoms). These new values computed are systematically lower than those in Table II, and the reason is the error due to the neglect of the difference of relativistic energy $ER(N\text{-}1) - ER(N)$.

One way to improve the estimate for the second-row correlation is to add to the values of the correlation energy in Table III the difference between the values of the correlation energies for the first group, estimated above, and the values of the corresponding correlation energies for the first group as given in Table II. The repetition of the physical situations of the first and second group of atoms is the justification for this correction. If we do this we obtain the correlation energy given in the last column of Table IV. In Fig. 2 we give the plot of the correlation energies of the first and second group neutral atoms vs $Z$.

The full line indicates the final estimated values (from Table II and the last column of Table IV). The dotted line $A$ gives the correlation energy obtained by the same method used to obtain the $EC(N)$ of Table IV. Finally, the dotted line $B$ gives the values $EC(N)$ of Table IV.

What about the correlation energies of the remaining atoms of the periodic system? One can expect to a large degree the same behavior as pointed out by Fig. 2, i.e., a somewhat linear increase with $N$, the number of electrons. For the third group the deviation from linearity will be given by the $3p$ as well as $3d$ electrons. Again, in adding $3d$ electrons, the first five will have same spin direction and will be spin precorrelated. The last five electrons of $3d$ type will not be precorrelated and, hence, here the correlation will increase more rapidly. The same expectation should be correct for the $f$ electrons. It is noted that these variations from linearity will always be of less and less percentage importance in the correlation energy of atoms with more and more electrons. The method used in Table IV can be readily adopted for the estimate of the correlation energy of the second group isoelectronic series. More accurate data will be presented in a forthcoming paper.

[21] E. Clementi, "Correlation Energy II" J. Chem. Phys. (to be published).

### VI. CONCLUSIONS

Clearly this present scheme of computing the correlation energy in atomic systems can be improved. When that is done, it should be quite possible to give an empirical rule to predict quantitatively the correlation as a function of the number of electrons, the spin multiplicity and the total angular momentum.

The problem of relating atomic correlation to molecular correlation is expected to become clearer in the near future. The present possibility of obtaining Hartree–Fock energies for the diatomic molecules near future. The present possibility of obtaining Hartree–Fock energies for the diatomic molecules will quantitatively elucidate the role of the molecular extra-correlation energy.[22] It is expected that this energy is a very important feature of the chemical bonds containing hydrogen atoms. Relations between the atomic correlation energy and the correlation energy in diatomic[23] and polyatomic molecules[24] are presented in forthcoming papers.

Meanwhile, more accurate work on atomic systems will provide correct values for atomic correlation energies. Since this goal is not easily obtainable in the near future, we feel justified in presenting our estimate of the correlation energy, although with some embarrassment because of the crude nature of this work.

### ACKNOWLEDGMENTS

We are deeply indebted to Professor C. C. J. Roothaan for having provided us with a program for Hartree–Fock computations, without which this work would not have been done. We are grateful to Dr. C. W. Scherr for his kindness in letting us use his data before publication, and to Dr. Fröman for comments on our work.

[22] E. Clementi, J. Chem. Phys. **35**, 33 (1962).

[23] E. Clementi, "Dissociation Energy in Diatomic Molecules," J. Chem. Phys. (to be published).

[24] E. Clementi, "Correlation Energy in $CH_4$" J. Chem. Phys. (to be published).

VII-5 Reprinted from *Journal of Chemical Physics* **41**, 2689 (1964)

# Many-Electron Theory of Atoms and Molecules. V. First-Row Atoms and Their Ions*,†

V. McKoy‡ and O. Sinanoğlu§

*Sterling Chemistry Laboratory, Yale University, New Haven, Connecticut*

(Received 28 May 1964)

The correlation energies of the $1s^22s^2$ inner cores of the first-row atoms B, C, ···, Ne are found to be very different from those of the corresponding four-electron Be-like ions, $B^+$ to $Ne^{6+}$, due to the exclusion effects of the outer $2p$ electrons. Whereas the $2s^2$ correlation, $\epsilon(2s^2)$, in the $1s^22s^2$ ions increases from $-1.13$ eV in Be to $-3.2$ eV in $Ne^{6+}$, the $2s^2$ correlation in the neutral atoms decreases from $-1.13$ eV in Be to $-0.27$ eV in Ne. The many-electron theory was used for the nonempirical $2s^2$ calculations and included the use of the $r_{12}$ coordinate. With these theoretical $\epsilon(2s^2)$ values the correlation of a $2p$ electron with the $1s^22s^2$ inner core is found to be large, $\sim -1$ eV. Also the $2p^2$ correlation comes out about $-1$ eV. The results show that core energies will, in general, depend strongly on the state and number of the outer-shell electrons and that intershell correlation interactions may be appreciable. Implications for $\pi$-electron systems and the ligand-field theory of inorganic complexes are discussed.

## INTRODUCTION

THIS series of papers has been developing a theory of many-electron atoms and molecules including electron correlation effects.[1] Paper I systematically analyzed the correlation effects in the wavefunction and energy of a many-electron system showing the pair correlations to be by far the most important. Paper II developed methods for obtaining the pair correlations. Paper III obtains the effect of correlation on orbitals and shows them to be negligible on the Hartree–Fock part of the wavefunction. In Paper IV nonempirical calculations are carried out on the Be atom and its ions. The changes in $Be^{2+}$ ion when it is turned into the core of $Be^{1+}$ or Be are obtained quantitatively studying the usual "neglect" of the $1s^2$-type inner shells. The $1s^2$ pair correlation in Be, $\epsilon(1s^2)$, differs very little from $\epsilon(1s^2)$ in $Be^{2+}$ or $Be^+$. Such pair correlations are transferable. This "inner-shell" cancellation is basic to much of quantum chemistry.

Paper V gives a theoretical analysis of the changes of correlation energies in the ground states of the first-row atoms and their ions. This includes changes both with the total number of electrons $N$, and the nuclear charge $Z$. The $1s^2$ core of Be is very much like the $1s^2$ electrons in $Be^{2+}$ or $Be^+$ (Paper IV). Comparing $Be^+$ and Be we can "cancel" out such inner cores,[2] $\epsilon(1s^2)$. But how far can we go with such "inner-shell cancellations?" Is the $1s^22s^2$ core of carbon the same as $1s^22s^2$ of $C^{2+}$? Can we put the $1s^22s^2$ $F^{5+}$ ion into a F atom? Our results show that the answer is no. These cores, the $2s^2$ part, change drastically when we put $2p$ electrons on top of them.[2] The $\epsilon(2s^2)$ will be calculated nonempirically using the many-electron theory. $\epsilon(2s^2)$ drops from $-1.13$ eV in Be to almost zero[3] in nitrogen through to Ne (see Paper I). Contrary to previous empirical results,[4] which show $\epsilon(2s^2)$ increasing from $-1.03$ eV in Be to $-3.2$ eV in Ne, these new $\epsilon(2s^2)$ show that $\epsilon(2p\rightarrow 1s^22s^2)$, the correlation of a $2p$ electron with the $1s^22s^2$ core of the atom, is large $\sim -1.0$ eV. Previous results[4] gave $\epsilon(2p\rightarrow 1s^22s^2)\approx -0.25$ eV. Also $\epsilon(2p^2)$ is about $-1.0$ eV and not $-1.72$ eV.[4]

## "EXPERIMENTAL" CORRELATION ENERGIES OF ATOMS AND THEIR IONS

From total experimental energies, Hartree–Fock (HF) energies ($E_{HF}$) and relativistic estimates (accurate to within 5%),[4] Clementi[4] has obtained the correlation energies of atoms and ions of the first row (see also Fig. 1). His graphs[4] show the change in total $E_{corr.}$ with $N$ number of electrons, and $Z$ nuclear charge [see also Ref. 3(b) for some detailed graphs]. These graphs bring out three main features: (a) In the Be-like ions, Be, $B^+\cdots Ne^{6+}$, the total correlation energy increases linearly with $Z$. The $2s^2$ correlation increases with $Z$, not $\epsilon(1s^2)$. Linderberg and Shull[5] predicted this $Z$ dependence from $2s$–$2p$ degeneracy in the limit of infinite $Z$. (b) When one $2p$ electron is added,

* Research supported by a grant from the National Science Foundation.

† Presented at the International Symposium of Quantum Chemistry, January 1964 at Sanibel Island, Florida.

‡ Yale University Predoctoral Fellow.

§ Alfred P. Sloan Fellow.

[1] Papers are referred to as follows: (I) O. Sinanoğlu (pair versus many-electron correlations), J. Chem. Phys. **36**, 706 (1962); (II) O. Sinanoğlu (methods for pair correlations), J. Chem. Phys. **36**, 3198 (1963); (III) O. Sinanoğlu and D. F. Tuan (effect of correlation or orbitals, $f_i$'s), J. Chem. Phys. **38**, 1740 (1963); (IV) D. F. Tuan and O. Sinanoğlu ($Be^{2+}$, $Be^+$, and Be), J. Chem. Phys. **41**, 2677 (1964).

[2] This paper deals only with changes in the correlation part of a pair of electrons. Changes in the Hartree–Fock part are well understood.

[3] See (a) O. Sinanoğlu, Proc. Natl. Acad. Sci. (U.S.) **47**, 1217 (1961) (b) O. Sinanoğlu, Advan. Chem. Phys. **6**, 315–412 (1964).

[4] E. Clementi, J. Chem. Phys. **38**, 2248 (1963); see also L. C. Allen, E. Clementi, and H. M. Gladney, Rev. Mod. Phys. **34**, 465 (1963).

[5] J. Linderberg and H. Shull, J. Mol. Spectry. **5**, 1 (1960).

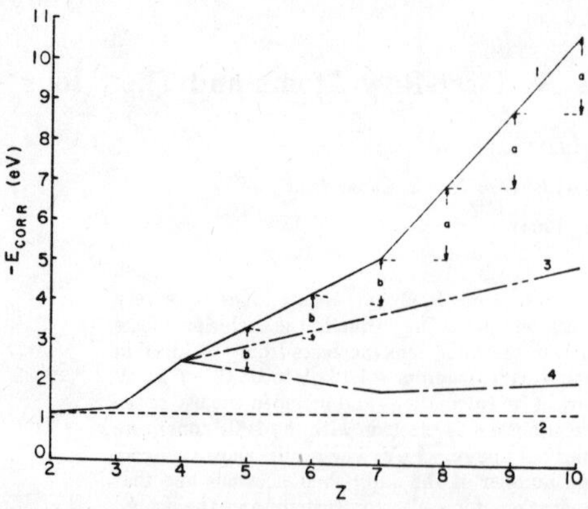


FIG. 1. Correlation energy of first-row atoms and their ions. Curve 1: Total "experimental" correlation energy of the ground states of first-row atoms (Ref. 4). Curve 2: Total correlation energy of He-like ions (Ref. 4). Curve 3: Correlation energy of four-electron $1s^22s^2$ ions (Ref. 4). Curve 4: Correlation energy of $1s^22s^2$ inner cores in neutral atoms (obtained by adding the calculated $\epsilon(2s^2)$ [see text] to the semiempirical $\epsilon(1s^2)$ and $1s$–$2s$ intershell correlations [see Paper IV[1]]. Segment $a$ is the $2p^2$ correlation plus the correlation of a $2p$ electron with the $1s^22s^2$ core (see text). Segment $b$ is the correlation energy of one $2p$ electron with the $1s^22s^2$ core (see text).

turning Be-like ions into B-like ions: total correlation energy increases by $-0.35$ eV in B and $-0.16$ eV in $Ne^{5+}$. Total $E_{corr.}$ still increases with $Z$ and almost parallel to the Be-like ions. (c) There is an increase of about $-1.7$ eV as soon as $2p$ electrons pair up, e.g., $(2p_0\alpha 2p_0\beta)$. We expect such an increase.

The small increase in total $E_{corr.}$ on adding a $2p$ electron, (b) above, naturally leads to two empirical conclusions[4]: (i) This very small increase in total $E_{corr.}$ is the correlation of a $2p$ electron with the inner $1s^22s^2$ core; (ii) the inner core, $1s^22s^2$, has pretty much the same correlation as the $1s^22s^2$ of the free four-electron ion.

But $2s^2$ correlation is a near-degeneracy-type correlation ("nondynamical"),[3b,5] very different from $1s^2$ correlation ("dynamical"). The many-electron theory shows that these "nondynamical" pairs, in contrast to "dynamical" pairs, are very sensitive to the rest of the Hartree–Fock "sea." The total Hartree–Fock potential influences the pair of electrons ("immersion effects") and also while correlating in this sea a "nondynamical" pair is strongly prevented from making excitations to other occupied orbitals[6] ("exclusion" effects).

On this basis we expect the $2s^2$ correlation to decrease from $-1.13$ eV in Be to about zero in nitrogen and remain about zero all the way up to neon.[3b] What is the source of the apparent contradiction between the empirical $2s^2$ and $2s$–$2p$ correlation picture and these theoretical predictions? There are three possible reasons: (a) "Dynamical" correlation may be increasing in $\epsilon(2s^2)$ of B, C, etc., even though it is negligible in Be. (b) Perhaps the relativistic corrections are wrong. This is unlikely. (c) $2s$–$2p$ correlations are large, so that they actually compensate for the decrease in the $2s^2$ correlations and then total $E_{corr.}$'s seem just slightly increased upon the addition of $2p$ electrons.

In the following sections we make detailed calculations with the many-electron theory (a) to get the correct nondynamical effects as influenced by the other electrons and the nuclear charge, (b) to see if dynamical correlation in $2s^2$ subshells remains as small as it is in Be, and (c) to get the correct values for other pair correlations, e.g., $\epsilon(2p\rightarrow 1s^22s^2)$. The results resolve the discrepancy and show that the third alternative (c), mentioned above, is the one that correctly accounts for the total $E_{corr.}$ trends.

## THEORY

All our results are for ground-state atoms of the first row. All are closed-shell systems or have single-determinant Hartree–Fock wavefunctions. For such single-determinant states the total wavefunction[1,3b] $\psi$ is

$$\psi \cong \phi_0 + \chi_s', \tag{1}$$

with $\phi_0$ the HF wavefunction and $\chi_s'$ the correlation part so that $\langle \phi_0, \chi_s' \rangle = 0$. For $\phi_0$ we use Roothaan's SCF orbitals.[7] $\chi_s'$ is

$$\chi_s' = \mathcal{A}\left\{(1\,2\,3\cdots N)\left[\frac{1}{\sqrt{2}}\sum_{i>j}^{N}\frac{\hat{u}_{ij}}{(ij)} + \frac{1}{2}\sum_{i>j}^{N}\sum_{k>l}^{N}\frac{\hat{u}_{ij}\hat{u}_{kl}}{(ijkl)} + \cdots\right]\right\}\cdots, \quad i,j\neq k,l. \tag{2}$$

The second and higher terms in Eq. (2) are unlinked clusters [see Eq. (77) of Ref. 3b]. $\mathcal{A}$ is the $N$-electron antisymmetrizer, 1, $2\cdots i$, $j$ are HF spin orbitals and $\hat{u}_{ij}$'s are pair functions. Closed shell $\hat{f}_i$'s are about zero (Paper III).[1,3b] Nonclosed-shell $\hat{f}_i$'s here are also very small because we have single determinant HF $\phi_0$'s.[3b] With the very small $\hat{f}_i$ effects neglected (estimates given in Paper III),[1] the total energy $E$ of the many-electron atom becomes[3b]

$$E \leqslant E'_s \equiv E_{HF} + \sum_{i>j}^{N} \tilde{\epsilon}'_{ij} + \frac{R'}{D'} \tag{3}$$

with

$$D' = 1 + \langle \chi_s', \chi_s' \rangle. \tag{4}$$

---

[6] O. Sinanoğlu, J. Chem. Phys. **33**, 1212 (1960); see also O. Sinanoğlu, Proc. Roy. Soc. (London) **A260**, 379 (1961).

[7] C. C. J. Roothaan and P. S. Bagus in *Methods in Computational Physics* (Academic Press, Inc., New York, 1963), Vol. 2, p. 47.

$\bar{\epsilon}'_{ij}$ is the pair correlation energy and $R'$ the three or more Coulomb correlations.

$$\bar{\epsilon}'_{ij} \cong \frac{2\langle \mathcal{B}(ij), m_{ij}\hat{u}_{ij}\rangle + \langle \hat{u}_{ij}, (e_i+e_j+m_{ij})\hat{u}_{ij}\rangle}{1+\langle \hat{u}_{ij}, \hat{u}_{ij}\rangle}, \quad (5)$$

$\mathcal{B}$ is the two-electron antisymmetrizer.

$$m_{ij} = (r_{ij})^{-1} - \bar{S}_{ij} + \bar{J}_{ij}, \quad (6)$$

$$\bar{S}_{ij} = \bar{S}_i(\mathbf{x}_j) + \bar{S}_j(\mathbf{x}_i). \quad (7)$$

The self-potentials $\bar{S}_i(\mathbf{x}_i)$ should be included in $\bar{S}_{ij}$, $e_i$, and $m_{ij}$. They are not shown in Eq. (7) since in the exact pair expressions $\epsilon_{ij}$ these self-potentials cancel out. $\bar{S}_j(\mathbf{x}_i)$ is the Coulomb plus exchange potential of spin-orbital $j$ acting on $i$, i.e.,

$$\bar{S}_j(\mathbf{x}_i) i(\mathbf{x}_i) = \langle j(\mathbf{x}_j), (r_{ij})^{-1} j(\mathbf{x}_j)\rangle_{\mathbf{x}_j} i(\mathbf{x}_i) - \langle j(\mathbf{x}_j), (r_{ij})^{-1} i(\mathbf{x}_j)\rangle_{\mathbf{x}_j} j(\mathbf{x}_i), \quad (8)$$

$$\bar{J}_{ij} = J_{ij} - K''_{ij}. \quad (9)$$

$J_{ij}$ and $K''_{ij}$ are Coulomb and exchange integrals. $e_i$ is

$$e_i \equiv h_i^0 + V_i(\mathbf{x}_i) - \epsilon_i. \quad (10)$$

$\epsilon_i$ is the HF orbital energy of $i$ and $V_i$ the total HF potential acting on spin-orbital $i$. $h_i^0$ is the bare nuclei Hamiltonian. For details see Paper I and Ref 3(b). $\hat{u}_{ij}$ is orthogonal to all occupied HF spin-orbitals ("exclusion" effect), i.e.,

$$\langle \hat{u}_{ij}, k(\mathbf{x}_i)\rangle_{\mathbf{x}_i} = 0. \quad (11)$$

The pair energy $\bar{\epsilon}'_{ij}$, Eq. (5), is for a closed-shell single-determinant HF. For single-determinant non-closed-shell HF $\bar{\epsilon}'_{ij}$ must be re-examined because the total HF potential $V_i$ acting on each electron is different. We do this in Appendix A and the new $\bar{\epsilon}'_{ij}$ has almost the same form as Eq. (5). From Eq. (5), pairs can now be calculated one by one just like the helium-type atom. But Eq. (5) contains the total Hartree–Fock potential in $V_i(\mathbf{x}_i)$ ["immersion" effects Eq. (10)], and the exclusion effect [Eq. (11)], keeping a $\hat{u}_{ij}$ orthogonal to occupied HF orbitals.

To use Eq. (5) we start from any two-electron function $\psi_{ij}(\mathbf{x}_i, \mathbf{x}_j)$. To fit this into the many-electron theory $\psi_{ij}$ is written as

$$\psi'_{ij} = \mathcal{B}(ij) + u_{ij}^0 \quad (12)$$

and

$$\langle \mathcal{B}(ij), u_{ij}^0\rangle = 0 \quad (13)$$

by a change of normalization. The projection operator $Q$ (see Paper II)

$$Q = 1 - (2)^{\frac{1}{2}} \sum_{k\geq 1}^{N} \mathcal{B}\{k\rangle\langle k\} + \sum_{l>k\geq 1}^{N} \mathcal{B}(kl)\rangle\langle \mathcal{B}(kl) \quad (14)$$

then turns $u_{ij}^0$ into $\hat{u}_{ij}$. $\hat{u}_{ij}$ can contain $r_{ij}$ or be of the CI type, etc.

These pair functions also give $R'$, Eq. (3), but this is small.[3b]

## CALCULATION OF $2s^2$ CORRELATIONS (NONDYNAMICAL)

In this section we use a "near-degeneracy" type $\hat{u}_{ij}$ to get $2s^2$ correlation in Be, B, C, $\cdots$, Ne. For Be

$$\hat{u}_{34} = c(\mathcal{B}(2p_{+1}\alpha 2p_{-1}\beta) - \mathcal{B}(2p_{+1}\beta 2p_{-1}\alpha) - \mathcal{B}(2p_0\alpha 2p_0\beta)). \quad (15)$$

This gives almost all of the $\epsilon(2s^2)$ in Be (Table I; see also Paper IV[1]) and Be-like ions (Table II). The "virtual" $2p$ functions used in these calculations [$\hat{u}_{34}$, Eq. (15)] were obtained from the $1s^2 2p^2(^1S)$ states using Roothaan's Hartree–Fock programs[7] (two STO's per $2p$).

In boron

$$\phi_0 = \mathcal{A}(1s\alpha 1s\beta 2s\alpha 2s\beta 2p_0\alpha)$$

and

$$\langle \hat{u}_{34}, 2p_0\alpha\rangle = 0. \quad (16)$$

$2p_0\alpha$ is an occupied HF orbital; this exclusion effect,[8] Eq. (16), eliminates the $2p_0$ term of $\hat{u}_{34}$ of $B^+$, Eq. (15), and makes the nondynamical $\hat{u}_{34}$ in boron:

$$\hat{u}_{34}\,(\text{boron}) = c_1[\mathcal{B}(2p_{+1}\alpha 2p_{-1}\beta) - \mathcal{B}(2p_{+1}\beta 2p_{-1}\alpha)]. \quad (17)$$

Adding another $2p$ electron gives, in the $^3P$ term of carbon,

$$\hat{u}_{34}\,(\text{carbon}) = c_2\mathcal{B}(2p_0\alpha 2p_0\beta). \quad (18)$$

In Eqs. (16) to (18) it is assumed that the "*virtual*" $2p$ orbitals taken in the $\hat{u}_{34}$ functions are the same as the *actual* occupied $2p$ orbitals that occur in the ground state $\phi_0$'s. We calculated these nondynamical $\epsilon(2s^2)$'s in boronlike and carbonlike ions with such $2p$ orbitals. All the HF orbitals were generated with Roothaan's programs. The basis set was two STO's per $2p$ HF orbital and three STO's for each $2s$ or $1s$ (the $\bar{\epsilon}'_{34}$ is insensitive to the size of the orbital basis set; e.g., in Be going from a three-STO basis to a six-STO basis changes the $\bar{\epsilon}'_{34}$ only by $\sim$0.01 eV; see Paper IV[1]).

In $N(^4S)$, $O(^3P)$, F, and Ne $\hat{u}_{34}$ (nondyn) = 0 rigorously[8] if the $2p$ function in $\hat{u}_{34}$ (nondyn) is the $2p$ orbital of the ground state of each atom (Paper I, p. 714).[1]

---

[8] Nondynamical correlation remains in $N(^2P)$, $\epsilon(2s^2) \approx -0.94$ eV; $O(^1S)$, $\epsilon(2s) \approx -1.44$ eV; also $\epsilon(2s^2)$ in $^1S$ of carbon $= -1.88$ eV, 4 times larger than nondynamical correlation in carbon $^3P$.

TABLE I. $2s^2$ correlation, $\epsilon(2s^2)$, in first-row atoms (electron volts).

| Atom | Empirical[a] $\epsilon(2s^2)$ assuming transferability of ion cores (no exclusion effect) | Theoretical $\epsilon(2s^2)$[b] (nondynamical with exclusion effect) | Theoretical $\epsilon(2s^2)$ with dynamical ($r_{12}$) and exclusion effects[c] |
|---|---|---|---|
| Be | −1.034[a] | −1.132 | −0.892 |
| B | −1.387 | −0.811 | −0.702 |
| C | −1.741 | −0.457 | −0.408 |
| N | −2.149 | −0.0 | −0.166 |
| O | −2.421 | −0.0 | ··· |
| F | −2.829 | −0.0 | ··· |
| Ne | −3.155 | −0.014[d] | −0.272 |

[a] L. C. Allen, E. Clementi, and H. M. Gladney, Rev. Mod. Phys. **35**, 465 (1963).

[b] This work. Calculations based on many electron theory. "Nondynamical" refers to "near-degeneracy" type, here $2s^2$–$2p^2$ mixing.

[c] This work. Pair functions containing $r_{12}$ to include the "dynamical" (slowly convergent with CI) part of the correlation. Calculations include all the orbital orthogonalizations (exclusion effects, etc.) according to the many-electron theory.

[d] See text.

Tables I and II give these calculated nondynamical $\bar{\epsilon}'_{34}$'s for the ground states of Be-like, B-like, and C-like ions ($Z \leqslant 10$) and those of the neutral atoms Be to Ne. The $\hat{u}_{34}$ (nondyn) in N($^4S$), O($^3P$), F, and Ne will not be rigorously zero if the virtual $2p$ function of $\hat{u}_{34}$ (nondyn) is different from the $2p$ orbital of the neutral atom. The $\hat{u}_{34}$ can contain any $2p$ function to start with but must then be orthogonalized to the $2p$ orbitals *occupied* in $\phi_0$ of the atom according to Eqs. (11) and (14). This will then reduce such a $\hat{u}_{34}$ (nondyn) to almost zero. For example, a Roothaan HF $2p$ function taken from the $1s^22p^2(^1S)$ state of $Ne^{6+}$ gives a nondynamical $2s^2$ correlation of only $-0.014$ eV (Table I) when used in the $\hat{u}_{34}$ of $^1S$ Ne atom but $-3.25$ eV for $2s^2$ correlation in the ground state of $Ne^{6+}$. Thus the use of the occupied $2p$ orbitals in calculating the nondynamical $\hat{u}(2s^2)$ is justified.

Table I also shows the empirical[4] $\epsilon(2s^2)$ in these atoms. These are values carried over from the Be-like ion $2s^2$ correlation and assumed unchanged in the neutral atoms.[4] The empirical and the nonempirical $\epsilon(2s^2)$ have very different trends. But these nonempirical $\hat{u}(2s^2)$ are only nondynamical. Is the dynamical part of $\epsilon(2s^2)$ in B, C, $\cdots$, Ne becoming large enough to account for these different trends? We investigate this in the next section.

## CALCULATION OF $2s^2$ CORRELATION WITH $r_{12}$ (DYNAMICAL EFFECTS)

In this section we get total $2s^2$ correlation, dynamical plus nondynamical, with a pair function containing $r_{12}$. The two-electron trial function $\psi_{34}$ is

$$\psi_{34} = \mathcal{B}[2s(3)\alpha(3)2s(4)\mathcal{B}(4)](1+\gamma r_{34}); \quad (19)$$

$\gamma$ is a variational parameter. From Eq. (12)

$$u_{34}{}^0 = c^{-1}\psi_{34}(\mathbf{x}_3, \mathbf{x}_4) - \mathcal{B}[2s(3)\alpha(3)2s(4)\beta(4)], \quad (20)$$

$$c = \langle \psi_{34}, \mathcal{B}[2s(3)\alpha(3)2s(4)\beta(4)]\rangle. \quad (21)$$

The projection operator, $Q$, of Eq. (14) turns $u_{34}{}^0$ into $\hat{u}_{34}$. This $\hat{u}_{34}$ is not difficult to handle. All integrals were done on an IBM 709 (see Appendix B for details).

Table I gives the $\epsilon(2s^2)$ from these pair functions. $2s^2$ correlation decreases from $-0.892$ eV in Be to $-0.166$ eV in N and increases slightly to $-0.27$ eV in Ne. Optimized values of the variational parameter $\gamma$, Eq. (19), are: 0.974(Be), 0.506(B), 0.313[C($^3P$)], 0.207[N($^4S$)], 0.414(Ne). Thus we find that the dynamical correlation indeed remains small. The trend in $\epsilon(2s^2)$ we found above with the nondynamical pair functions is real. From an analysis of the correlation energies of Ne $2s$ hole and $2p$ hole states Kestner[9] finds that in ground state neon $\epsilon(2s^2) < -0.5$ eV, in good agreement with these results.

It must be pointed out that our $r_{12}$ calculations were not made to get really accurate results but just to see if the total $\epsilon(2s^2)$'s show the same trend as the nondynamical $\epsilon(2s^2)$. Had our purpose been to get very accurate results we would have included more terms, e.g., $b(r_1+r_2)$ terms to help scale our results.[10]

## OTHER PAIRS

With these nonempirical $\epsilon(2s^2)$ above, we can get the other "dynamical" and quite "transferable" (Paper IV) $\epsilon_{ij}$'s semiempirically from the data. These again are very different from the empirical[4] ones.

In IV, $\epsilon(1s^2)$ and $1s^2 \rightarrow 2s^2$ intershell correlations were discussed and analyzed, and correct values for the $1s2s$ pairs were obtained for $Z=3$ to 10. Two important pairs still remain for first row atoms: (i) $\epsilon(2p \rightarrow 1s^22s^2)$, correlation of a $2p$ electron with the $1s^22s^2$ core; (ii) $\epsilon(2p^2)$, tight pair $2p$ correlation. Table III gives $\epsilon(2p^n \rightarrow 1s^22s^2) + \epsilon(2p^n)$ i.e., the correlation of $n$ $2p$ electrons with the $1s^22s^2$ core and among themselves. This is what is left after the correlation energies of the $1s^22s^2$ cores, i.e., the ones in the $N$-electron "medium" with the proper immersion and exclusion effects calculated above, are subtracted from the total $E_{corr.}$. These new $\epsilon(2p^n \rightarrow 1s^22s^2) + \epsilon(2p^n)$ are much larger than the previous non-theoretical ones. Below we obtain the $\epsilon(2p \rightarrow 1s^22s^2)$ and the $\epsilon(2p^2)$ parts of these values separately.

[9] N. R. Kestner, J. Chem. Phys., "$2s$ and $2p$ Hole States of Ne" (to be published).

[10] For a discussion of scaling see P.-O. Löwdin, J. Mol. Spectry. **3**, 46 (1959).; M. Veselov, M. Petrashen, and A. Krichaniga [Zh. Eksperim. i Teor. Fiz. **10**, 857 (1940)] included such a term, i.e., $b(r_1+r_2)$ along with $\alpha r_{12}$ to put in correlation between the $2s$ electrons in Be. They used an equation derived by V. Fock *et al.* [Zh. Eksperim. i Teor. Fiz. **10**, 723 (1940)] and obtained a total energy 0.045 a.u. lower than their H.F. energy 14.529 a.u.

TABLE II. $\epsilon(2s^2)$ in Be, B, C ($^3P$) isoelectronic series. Nondynamical (near-degeneracy) correlation here from $2p^2$ mixing [$Z$ is nuclear charge; $N$ is number of electrons (electron volts)].

| $N \backslash Z$ | 4 | 5 | 6 | 7 | 8 | 9 | 10 |
|---|---|---|---|---|---|---|---|
| 4 | −1.132[a] | −1.417 | −1.828 | −2.309 | −2.554 | −2.900 | −3.248 |
| 5 | | −0.811 | −1.069 | −1.270 | −1.477 | −1.692 | −1.882 |
| 6 | | | −0.457 | −0.563 | −0.656 | −0.745 | −0.832 |

[a] See Paper IV of this series for $2p$ function used in Be calculation, all other $2p$ functions from HF of ($1s^22p^n$) states (see text).

### a. $\epsilon(2p\rightarrow 1s^22s^2)$

In C($^3P$) and N($^4S$) all $2p$ electrons have parallel spins. $\epsilon(2p^n)$ is then small [Ref. 3(b) and Paper I]. Most of $\epsilon(2p^n)+\epsilon(2p^n\rightarrow 1s^22s^2)$ (Table III) is really $\epsilon(2p^n\rightarrow 1s^22s^2)$. From Table III $\epsilon(2p\rightarrow 1s^22s^2)$ is $\sim$1 eV and $\epsilon(2p^n\rightarrow 1s^22s^2)$ is in the ratio 1:2:3 for $n=1, 2, 3$ in B, C, and N, respectively. $\epsilon(2s^2)$ is very different in B, C, and N so the total $E_{corr.}$ behavior which led to the previous[4] fortuitous conclusion that $\epsilon(2p\rightarrow 1s^22s^2)$ is small $\approx -0.25$ eV (also decreasing with $Z$ in boron-like ions)[4] is accounted for by the actually large $2s$–$2p$ correlations. As expected total $2s$–$2p$ correlation is proportional to the number of $2p$ electrons. More results below support this estimate of $\epsilon(2p\rightarrow 1s^22s^2)=-1$ eV.

These large values of $\epsilon(2p\rightarrow 1s^22s^2)$ are also in line with: (a) From CI Donath[11] gives −1.35 to −2.72 eV for the total of six $\epsilon(2p\rightarrow 1s^22s^2)$ in $F^-$, Ne, $Na^+$. This gives an $\epsilon(2p\rightarrow 1s^22s^2)\approx -0.35$ eV in Ne, already twice the previous estimate[4] $\approx -0.16$ eV. This is only with two configurations for $2s$–$2p$ correlation. Because of the slow convergence of CI the final $\epsilon(2p\rightarrow 1s^22s^2)$ should be at least twice Donath's value. (b) The large $2s$–$2p$ radial overlap is

$$\int R_{2s}R_{2p}r^2dr\approx 0.92.$$

The $2s$–$2p$ correlation is not an "intershell effect" like $\epsilon(1s\text{–}2s)$.

TABLE III. $\epsilon(2p^n)+\epsilon(2p^n\rightarrow 1s^22s^2)$ in first-row atoms and ions (eV), i.e., the correlation energy of $n$ $2p$ electrons among themselves, $\epsilon(2p^n)$, and with the $1s^22s^2$ core, $\epsilon(2p^n\rightarrow 1s^22s^2)$. ($n$ is the number of $2p$ electrons; $Z$ is the nuclear charge.)

| $n \backslash Z$ | 5 | 6 | 7 | 8 | 9 | 10 |
|---|---|---|---|---|---|---|
| 1[a] | −1.137 | −1.232 | −1.333 | −1.411 | −1.491 | −1.602 |
| 1[b] | −0.354 | −0.326 | −0.272 | −0.245 | −0.190 | −0.165 |
| 2[a] | | −2.317 | −2.477 | −2.586 | −2.737 | −2.897 |
| 3[a] | | | −3.449 | −3.533 | −3.631 | −3.754 |

[a] This work. Using the strong changes in $1s^22s^2$ ion cores due to exclusion effects on $\epsilon(2s^2)$.

[b] Based on the assumption that $1s^22s^2$ ions remain the same in the neutral atoms [L. C. Allen, E. Clementi, and H. M. Gladney, Revs. Mod. Phys. 35, 465 (1963)].

[11] W. Donath, Rev. Mod. Phys. **35**, 490 (1963). This follows comment by O. Sinanoğlu in Discussion [Rev. Mod. Phys. **35**, 489 (1963)] where this problem with $2s^2$ correlation and discrepancies is pointed out. See also Ref. 3(b) and Paper I.

To get a crude theoretical estimate of $\epsilon(2p\rightarrow 1s^22s^2)$, we mixed in the $^2P$ term of ($1s^22s2p3d$) configuration (double excitation). The $3d$ function is a single Slater orbital, ($r^2e^{-\alpha r}$), with the radial part pulled into the region of the $2s$ and $2p$ orbitals. This gives an $\epsilon(2p\rightarrow 1s^22s^2)$ of −0.44 eV in B and −0.76 eV in $Ne^{5+}$ already larger than previous estimates[4] and we did not optimise the $3d$ orbital exponent nor scale the final wavefunction.[10]

### b. $\epsilon(2p^2)$: $2p^2$ Correlation

Assuming $\epsilon(2p\rightarrow 1s^22s^2, A)=\epsilon(2p\rightarrow 1s^22s^2, A^+)$ for F, O, Ne, and neglecting some small effects which can be estimated from a full theoretical analysis (non-closed-shell many-electron theory; to be published) we get $\epsilon(2p^2)+\epsilon(2p\rightarrow 1s^22s^2)=-1.72$ eV (Fig. 1). $\epsilon(2p\rightarrow 1s^22s^2)$ lies between −0.4 eV and −0.8 eV. This −0.4 eV is Donath's[11] estimate and −0.8 eV is 50% of $\epsilon(2p\rightarrow 1s^22s^2)$ in $Ne^{5+}$. We take $\epsilon(2p\rightarrow 1s^22s^2, \text{Ne})\approx \frac{1}{2}\epsilon(2p\rightarrow 2s^21s^2; Ne^{5+})$ because 50% of $\epsilon(2p\rightarrow 1s^22s^2, Ne^{5+})$ comes from $1s^22s$–$2p3d(2P)$ mixing and this configuration will not mix in Ne($^1S$). With these considerations we find that $\epsilon(2p^2)$ is about $-1\mp 0.3$ eV[12] less than the previous estimate[4] of −1.72 eV.

## CONCLUSIONS

The $1s^22s^2$ cores of B, C, and Ne are very different from the $1s^22s^2$ Be-like ion of these atoms (Fig. 1). $\epsilon(2s^2)$ in the Be-like ion is a nondynamical pair, and unlike the "dynamical" $1s^2$ pair (slowly convergent CI), is drastically reduced when we turn this core into say the $1s^22s^2$ core of the five-electron boron atom. The $\epsilon(2s^2)$ decreases from −1.13 eV in Be to −0.27 eV in Ne whereas the assumption of transferability of $1s^22s^2$ ion into the neutral atom would have led[4] to the large value of $\epsilon(2s^2)$ in Ne=−3.2 eV. These pairs are nontransferable. The $r_{12}$ calculations show that the dynamical part of $\epsilon(2s^2)$ remains negligible. Nonempirical calculation with e.g., $\alpha r_{12}$, $\beta r_{12}^2$, etc., are carried out easily one pair at a time. Putting $r_{12}$ into the wavefunction avoids the slow convergence of CI and does

[12] Three-body effects may not be totally negligible in Ne itself and Kestner (private communication) estimates them to be 1 eV in Ne. We find them to be 0.07 eV in $Ne^{5+}$. Differences between these effects in $Ne^{1+}$ and Ne will be much less and not important for our argument above.

not lead to difficult integrals even with the inclusion of all the orbital orthogonalizations.

$\epsilon(2p\rightarrow 1s^2 2s^2)$, the correlation of a $2p$ electron with the $1s^2 2s^2$ core, is large, $\sim -1$ eV. The nontheoretical comparison of the total $E_{\text{corr.}}$'s of the $1s^2 2s^2$ ions and the neutral atoms would have led, on the other hand,[4] to the conclusion that $2s$–$2p$ correlations are negligible [$\epsilon(2p\rightarrow 1s^2 2s^2) = -0.25$ eV].[4] The $\epsilon(2s^2)$ is not transferable from ions to neutral atoms and ions with more electrons. With the correct $\epsilon(2s^2)$, $\epsilon(2p^2)$ now comes out $\sim -1$ eV and not $-1.72$ eV.[4]

Since $\epsilon(2s^2)$ is so sensitive to the occupation of the $2p$ shell the $1s^2 2s^2$ core correlation will strongly depend on the state of the atom, e.g., $^1S$ or $^3P$ in carbon. The core energy will also depend on ionization and on molecule formation. Thus the usual assumption of the cancellation of inner shells is not valid except for $1s^2$. Correlation errors resulting from this assumption may be several electron volts. Such effects are being applied in this laboratory in connection with the prediction of excitation energies and molecular binding energies.

For heavier atoms, e.g., with $3s$, $3p$, $3d$ and higher $d$ and $f$ electrons we will find similar large exclusion effects making the inner ion cores of these atoms nontransferable. Again the usual assumption of the cancellation of inner cores will not hold. Also, as we found in the $2s$–$2p$ case above various "interorbital" correlations, e.g., between $3s$ and $3p$, $3p$ and $3d$, $3d$ and $4f$, etc., electrons of the larger atoms are expected to be appreciable. A rough idea about the importance of such correlations may be obtained from the magnitudes of the radial overlap integrals, e.g.,

$$\int R_{3d}R_{3p}r^2 dr,$$

etc., between the radial parts of the appropriate Hartree–Fock orbitals. Such correlations may need to be taken into account in the ligand-field theory of inorganic complexes.

The exclusion effects on $2s^2$ correlation and the large $2s$–$2p$ correlation found in this paper should be significant also in $\pi$-electron theory. Because of these effects the $\sigma$ core correlation in, say benzene, will change depending on which of the $\pi$ orbitals are occupied in the particular electronic state. In addition the large $2s$–$2p$ correlations mean that now there should be appreciable $\sigma$–$\pi$ correlation[13] interaction which will also depend on state.

## ACKNOWLEDGMENTS

We thank Professor C. C. J. Roothaan for making his Hartree–Fock programs available and Mr. William Schmidt for computational assistance.

## APPENDIX A. THEORY APPLIED TO SINGLE-DETERMINANT NONCLOSED SHELL STATES

For nonclosed shell single-determinant HF we cannot use Eq. (5) as it stands. The traditional HF potential, $V_i(\mathbf{x}_i)$, is not the same for every spin–orbital $i$. Roothaan's HF orbitals are identical with the traditional HF ones for the configurations dealt with in this paper. We used the operators of the traditional HF method but Roothaan's analytical representation[7] of the orbitals. To see what happens to Eq. (5) for these single determinant nonclosed shell atoms we put, e.g., for boron $\chi=(2)^{-\frac{1}{2}}\mathfrak{A}(1s\alpha 1s\beta\hat{u}_{34}2p_0\alpha)$ and $\phi_0$ into

$$E_{\text{corr.}}=E_{\text{exp.}}-E_{\text{HF}}$$
$$\leqslant\frac{2\langle\phi_0,\,(H-E_{\text{HF}})\chi\rangle+\langle\chi,\,(H-E_{\text{HF}})\chi\rangle}{1+\langle\chi,\,\chi\rangle}. \quad \text{(A1)}$$

We write out in detail the matrix elements of Eq. (A1). The final result after the appropriate cancellations looks very much like Eq. (5), though the variational pair energy expressions are not identical. For example, one has in boron

$$\epsilon_{34}(2s^2)\leqslant\frac{\begin{array}{l}2\langle\mathfrak{B}(2s(3)\alpha(3)2s(4)\beta(4)),\,m_{34}\hat{u}_{34}\rangle+\langle\hat{u}_{34},\,(e'_3+e'_4+m_{34}\\ \quad -\langle 2p_0(3)\alpha(3),\,g_{34}2p_0(3)\alpha(3)\rangle P_{34}+\langle 2p_0(4)\beta(4),\,g_{34}2p_0(4)\beta(4)\rangle P_{34})\hat{u}_{34}\rangle\end{array}}{1+\langle\hat{u}_{34},\,\hat{u}_{34}\rangle}. \quad \text{(A2)}$$

$P_{34}$ permutes Electrons 3 and 4. $m_{34}$ contains the self-potentials but these cancel out in Eq. (A2). $e'_3$ is like $e_3$, Eq. (10), but $V_3$ and $\epsilon_3$ are from the traditional HF operators. The extra terms are differences in exchange potentials.

## APPENDIX B. DETAILS OF $r_{12}$ CALCULATIONS

The $\hat{u}_{34}$ with the orbital orthogonalizations and $r_{12}$ are easily handled. For example, with $2s_\alpha=2s$, $2s_\beta=2\bar{s}$,

$$u_{34}{}^0=\mathfrak{B}[2s(3)2\bar{s}(4)](\omega+\mu r_{34}), \quad \text{(B1)}$$

$$\omega=-\mu\langle 2s(3)2s(3)2s(4)2s(4)r_{34}\rangle, \quad \text{(B2)}$$

$g(3)=\langle 1s(4)2s(4)r_{34}\rangle_{\mathbf{x}_4}$, $h(3)=\langle 2s(4)2s(4)r_{34}\rangle_{\mathbf{x}_4}$, and $f(3)=\langle 2p_0(4)2s(4)r_{34}\rangle_{\mathbf{x}_4}$, we have

$$\begin{aligned}\hat{u}_{34}=u_{34}{}^0&-\mu\mathfrak{B}[1s(3)2\bar{s}(4)g(4)-\mu\mathfrak{B}[2s(3)g(3)1\bar{s}(4)]\\&-\mu\mathfrak{B}[2s(3)2\bar{s}(4)h(4)]-\mu\mathfrak{B}[2s(3)h(3)2s(4)]\\&-2\omega\mathfrak{B}[2s(3)2\bar{s}(4)]-\mu\mathfrak{B}[2p_0(3)2\bar{s}(4)f(4)]\\&+\mu\mathfrak{B}[1\bar{s}(4)1s(3)]\langle 1s(3)1s(4),\,2s(3)2s(4)r_{34}\rangle\\&+\mu\mathfrak{B}[1s(3)2\bar{s}(4)]\langle 1s(3)2s(4),\,2s(3)2s(4)r_{34}\rangle\\&+\mu\mathfrak{B}[2s(3)1\bar{s}(4)]\langle 2s(3)1s(4),\,2s(3)2s(4)r_{34}\rangle.\end{aligned}$$

[13] Sigma–pi correlation energies were discussed previously in terms of the core polarization theory (Ref. 6). They are also included in the many electron formulation of the Σ, Π separation and the $\pi$-electron problem (Papers I, II).

$r_{34}$ occurs explicitly only in $u_{34}^0$. The other terms are products of one electron functions, a convenient result.

All $r_{12}$ integrals, except one, are reduced to a linear combination of integrals of the type[14]:

$$\int f(r_1)\, Y_l^m(\theta_1\phi_1)\, h(r_{12})\, g(r_2)\, Y_\lambda^\mu(\theta_2\phi_2)\, d\tau_1 d\tau_2.$$

Rotating the co-ordinate system and integrating we obtain this integral in closed form. These expressions are then programmed.

The other integral contains three "inseparable" interelectronic coordinates. We programmed the expressions derived by Öhrn and Nordling[15] for the Yale IBM 709 computer. The general form of the integral is $\langle f_1(r_1)\, f_2(r_2)\, f(r_3)\, Y_k^m(1)\, Y_j^\nu(2)\, Y_l^\mu(3)\, r_{12}^\lambda r_{13}^\mu r_{23}^\nu \rangle$. This reduces to an infinite series. For spherically symmetrical integrands ($Y_0^0$) and $\lambda=1$, $\mu=-1$, $\nu=1$, the second term in the series is 3% of the first. For nonspherically symmetrical integrands ($Y_1^0$ and same $\lambda$, $\mu$, $\nu$) the second term is about 8% of the first.

---

[14] J. L. Calais and P.-O. Löwdin, J. Mol. Spectry. **8**, 203 (1962).

[15] Y. Öhrn and J. Nordling, J. Chem. Phys. **39**, 1864 (1963).

VII-6 Reprinted from *Journal of Chemical Physics* **40,** 80 (1964)

# Pair Correlations in Closed-Shell Systems

M. Krauss and A. W. Weiss
*National Bureau of Standards, Washington, D. C. 20234*
(Received 15 July 1963)

Pair function equations have been derived exactly for a restricted class of trial functions containing only closed-orbital pair correlations. In a matrix representation, it is shown that they reduce to a homogeneous pseudo-eigenvalue equation. The results are equally applicable to any set of unitarily transformed set of Hartree–Fock orbitals. Comparisons are made with the Sinanoğlu scheme, and computational problems are discussed.

## INTRODUCTION

THE Hartree–Fock wavefunctions and energies have been determined for a large number of atoms[1] and a growing number of diatomic molecules.[2] The same accuracy has not yet been obtained in the calculation of the energy of polyatomic molecules, but recent calculations in this area have been promising.[3] The analysis of these results has brought about a renewed interest in the correlation problem for such systems.[4] Although consideration has been given to the general $n$-electron correlation function,[5] attention has been focused mainly on the determination of the best two-electron correlation function which minimizes the correlation energy.

The first study of this kind was undertaken by Fock *et al.*[6] who considered the correlation of only the valence electrons for alkaline atoms. Recent investigations have been more general[7]; Sinanoğlu,[8] in particular, has derived the variational equations for any orbital pair, although certain approximations and orthogonality

[1] D. R. Hartree, *The Calculation of Atomic Structures* (John Wiley & Sons, Inc., New York, 1957).
[2] See, for example, E. Clementi, J. Chem. Phys. **36,** 33 (1962); R. K. Nesbet, *ibid.*, p. 1518.
[3] See, for example, A. D. McLean, J. Chem. Phys. **32,** 1595 (1960); M. Krauss, *ibid.* **38,** 564 (1963).
[4] E. Clementi, J. Chem. Phys. **38,** 2248 (1963).
[5] L. Szasz, Phys. Rev. **126,** 169 (1962).
[6] V. Fock, M. Wesselow, and M. Petrashen, Zh. Eksperim. i Teor. Fiz. **10,** 723 (1940).
[7] A. C. Hurley, J. E. Lennard-Jones, and J. A. Pople, Proc. Roy. Soc. (London) **A220,** 446 (1953); L. Szasz, J. Math. Phys. **3,** 1147 (1962); W. Brenig, Nucl. Phys. **4,** 363 (1957).
[8] O. Sinanoğlu, J. Chem. Phys. **36,** 706, 3198 (1962).

constraints were used to simplify the equations. He chose for the trial function of the correlation function[8]

$$\chi = A\left\{\frac{1}{\sqrt{2}}\sum_{i>j}(\varphi_1\cdots\varphi_N)\frac{\hat{u}_{ij}}{\varphi_i\varphi_j} + \tfrac{1}{2}\sum_{\substack{i>j\\ i,j\neq k,l}}\sum_{k>l}(\varphi_1\cdots\varphi_N)\frac{\hat{u}_{ij}\hat{u}_{kl}}{\varphi_i\varphi_j\varphi_k\varphi_l}+\cdots\right\}, \quad (1)$$

where the pair correlation function $\hat{u}_{ij}$ replaces the spin orbitals $\varphi_i$ and $\varphi_j$ in the product $(\varphi_i\cdots\varphi_N)$ and $A$ is the antisymmetrizing operator. He found it necessary to ignore many so-called linked-cluster terms in the energy expression. In addition, a significant simplification was obtained by assuming that the remaining integrals arising from the so-called unlinked clusters would approximately cancel the normalization terms. The variational equation for each orbital pair is then decoupled from the equation for any other orbital pair. The sum of the pair energies, however, need not be an upper bound to the exact energy because of the approximations in the expression for the total energy. We should also note that the trial function as it is written need not be an eigenfunction of the operator for the total electron spin if all possible pairs are considered. The effect of explicitly introducing symmetry conditions into the pair equations was not considered by Sinanoğlu.

With these thoughts in mind then, our basic attitude has been to derive the pair function equations exactly for a special class of trial functions, with no approximations except for those inherent in the structure of the original trial function. The trial functions are of the form, Hartree–Fock plus a correction function, where the correction function allows only the two electrons in doubly occupied orbitals (closed orbital pairs) to correlate one pair at a time. The interorbital correlation effects are not considered in this paper.

We require, as Sinanoğlu did, that the pair functions be one-electron orthogonal to *all* the occupied Hartree–Fock orbitals. This is equivalent, in a configuration interaction representation, to disregarding single excitations from the zero-order Hartree–Fock function.[9] Neglecting these terms in the trial function is estimated to have little effect on the energy[10] although for certain molecular situations this is not at all certain.

The pair equations are derived in such a manner that they apply equally to any unitarily transformed set of Hartree–Fock functions. The spin dependence are also separated from the total energy expression before the variation principle is applied. If we choose orbitals that are optimally localized[11] from each other, the intraorbital correlation should be much greater than the interorbital correlation since the electrons in one orbital are almost completely out of the range of the dynamically screened potential of the other orbital's electrons. This should serve to ameliorate the errors due to the neglect of the interorbital effects.

## DERIVATION OF THE CLOSED-SHELL PAIR

We restrict this entire discussion to closed shell systems, for which the Hartree–Fock wavefunction is a single determinant of doubly occupied orbitals, $\varphi_p$, i.e.,

$$\Phi = A\varphi_1\alpha(1)\varphi_1\beta(2)\varphi_2\alpha(3)\cdots\varphi_n\beta(2n)$$
$$=[(2n)!]^{-\frac{1}{2}}\,|\,\varphi_1\alpha(1)\varphi_1\beta(2)\varphi_2\alpha(3)\cdots\varphi_n\beta(2n)\,|, \quad (2)$$

where $A$ is the antisymmetrizing operator. The $\varphi_p$ need not be eigenfunctions of the Fock operator but may be any unitary linear combination of such eigenfunctions. The exact wavefunction is then written as

$$\Psi = \Phi + \chi, \quad (3)$$

where $\chi$ is the correlation part of the wavefunction, which can, without loss of generality, be taken orthogonal to the Hartree–Fock function. The total energy is given by

$$E = E_{\mathrm{HF}} + 2\langle\Phi\,|\,\mathcal{H}\,|\,\chi\rangle + \langle\chi\,|\,\mathcal{H}\,|\,\chi\rangle/(1+\langle\chi\,|\,\chi\rangle). \quad (4)$$

Here, $\mathcal{H}$ is the total nonrelativistic $N$-electron Hamiltonian,

$$\mathcal{H} = \sum_i h(i) + \sum_{i<j} g_{ij}, \quad (5)$$

where $g_{ij} = r_{ij}^{-1}$ is the interelectronic repulsion and $h(i)$ the "bare nucleus" Hamiltonian of the $i$th electron. The total energy can also be written as

$$E = E_{\mathrm{HF}} + \mathcal{E}, \quad (6)$$

with the correlation energy $\mathcal{E}$ given by,

$$\mathcal{E} = 2\langle\Phi\,|\,\mathcal{H}-E_{\mathrm{HF}}\,|\,\chi\rangle + \langle\chi\,|\,\mathcal{H}-E_{\mathrm{HF}}\,|\,\chi\rangle/(1+\langle\chi\,|\,\chi\rangle). \quad (7)$$

We now assign a specific approximation form to $\chi$ and concentrate our attention on the structure of $\mathcal{E}$ and the results of applying to it the variational principle. The trial function $\chi$, is obtained from the Hartree–Fock function by replacing each closed shell pair, *one at a time*, by an unspecified pair function $\hat{u}_p(1, 2)$, i.e.,

$$\chi = \sum_{p=1}^{n} \chi_p,$$
$$\chi_p = (1/\sqrt{2})\,A\varphi_1\alpha(1)\varphi_1\beta(2)\cdots\varphi_{p-1} \times\beta(2p-2)\hat{u}_p(2p-1, 2p)\cdots\varphi_n\beta(2n). \quad (8)$$

Because of the restricted choice of the trial function we have considered, $\chi$, in general, is not invariant under the same unitary transformation applied to the Hartree–Fock orbitals. Therefore, the choice of $\chi$ is dependent on the orbital basis we use. Furthermore,

[9] L. Szasz, J. Chem. Phys. **37**, 193 (1962).
[10] E. R. Davidson and L. L. Jones, J. Chem. Phys. **37**, 2966 (1962); O. Sinanoğlu and D. F. Tuan, *ibid.* **38**, 1740 (1963).
[11] C. Edmiston and K. Ruedenberg, Rev. Mod. Phys. **35** (to be published); G. G. Hall, Rept. Progr. Phys. **22**, 1 (1959); J. M. Foster and S. F. Boys, Rev. Mod. Phys. **32**, 300 (1960).

we take[12]

$$\hat{u}_p^{12}=u_p^{12}[(\alpha^1\beta^2-\beta^1\alpha^2)/\sqrt{2}], \tag{9}$$

where $u_p$ is a purely spatial function, symmetrical in the interchange of the electron coordinates. Following Sinanoğlu, we further assume that every pair function is strongly orthogonal, in the following sense, to all the Hartree–Fock orbitals.

$$\langle u_p^{12} \mid \varphi_q^2\rangle_2=\langle u_p^{12} \mid \varphi_q^1\rangle_1=0 \qquad \text{for all } p \text{ and } q.$$

It should be noted that this represents a real restriction on the wavefunction, as mentioned in the introduction. This particular form of the trial function (8) is referred to as the "closed-orbital pair" approximation.

The denominator of Eq. (7) reduces to

$$D=1+\langle\chi \mid \chi\rangle=1+\sum_p\langle u_p^{12} \mid u_p^{12}\rangle. \tag{10}$$

The first part of the numerator of (7), by virtue of the strong orthogonality constraint, is nonzero only for the two-electron part of the Hamiltonian.

$$\langle\Phi \mid \mathcal{H} \mid \chi\rangle=\sum_p\langle\varphi_p^1\varphi_p^2 \mid g_{12} \mid u_p^{12}\rangle. \tag{11}$$

The one-electron part of the last term in the numerator is a sum over pair function terms of the form

$$\langle\chi_p \mid h \mid \chi_p\rangle=2\langle u_p^{12} \mid h^1 \mid u_p^{12}\rangle +2\langle u_p^{12} \mid u_p^{12}\rangle\sum_{q\neq p}\langle\varphi_q \mid h \mid \varphi_q\rangle. \tag{12}$$

When one considers the direct and exchange effects, the part of $\langle\chi \mid \sum g_{ij} \mid \chi\rangle$ diagonal in $p$ is seen to be given by

$$\begin{aligned}\langle\chi_p \mid \textstyle\sum g_{ij} \mid \chi_p\rangle=&\langle u_p^{12} \mid g_{12} \mid u_p^{12}\rangle\\&+\langle u_p^{12} \mid u_p^{12}\rangle\sum_{q\neq p}\sum_{k\neq p}(2J_{qk}-K_{qk})\\&+\sum_{q\neq p}\{4\langle\varphi_q^1u_p^{23} \mid g_{12} \mid \varphi_q^1u_p^{23}\rangle\\&-2\langle\varphi_q^1u_p^{23} \mid g_{12} \mid \varphi_q^2u_p^{13}\rangle\}.\end{aligned} \tag{13}$$

$J_q$ and $K_q$ are the Coulomb and exchange operators, respectively, and $J_{pq}$ and $K_{pq}$ are the Coulomb and exchange integrals. Combining (12) and (13), one has exactly

$$\langle\chi_p \mid \mathcal{H}-E_{\mathrm{HF}} \mid \chi_p\rangle=\langle u_p^{12} \mid \mathcal{H}_p^{12} \mid u_p^{12}\rangle, \tag{14}$$

where

$$\begin{aligned}\mathcal{H}_p^{12}=&h^1+h^2+\sum_q(2J_q^1-K_q^1+2J_q^2-K_q^2)+g_{12}\\&-(2J_p^1-K_p^1)-(2J_p^2-K_p^2)-2\epsilon_p+J_{pp}\\=&F^1+F^2+g_{12}-(2J_p^1-K_p^1)\\&-(2J_p^2-K_p^2)-2\epsilon_p+J_{pp},\end{aligned} \tag{15}$$

and, where

$$F^i=h^i+\sum_q(2J_q^i-K_q^i) \tag{16}$$

is the Fock operator with the eigenvalue $\epsilon_p$. We have used the definitions of these operators in rewriting the three-electron integrals of Eq. (13), e.g.,

$$\langle\varphi_q^1u_p^{23} \mid g_{12} \mid \varphi_q^1u_p^{23}\rangle=\langle u_p^{12} \mid J_q^1 \mid u_p^{12}\rangle.$$

Furthermore, in putting (15) into the form (14), we have used the expression for the Hartree–Fock total energy and the general property of the Fock operator,

$$E_{\mathrm{HF}}=\sum_p[2\epsilon_p-\sum_q(2J_{pq}-K_{pq})], \tag{17}$$

$$F\varphi_p=\epsilon_p\varphi_p+\sum_{q\neq p}\epsilon_{pq}\varphi_q. \tag{18}$$

The crossed-pair terms only give rise to

$$\langle\chi_p \mid \mathcal{H}-E_{\mathrm{HF}} \mid \chi_q\rangle=\langle\varphi_p^1\varphi_p^2 \mid g_{12} \mid \varphi_q^1\varphi_q^2\rangle\langle u_p^{12} \mid u_q^{12}\rangle. \tag{19}$$

Using the results given by Eqs. (10), (11), (14), and (19), $\mathcal{E}$ is now given exactly by

$$\mathcal{E}=\frac{\sum_p\{2\langle\varphi_p^1\varphi_p^2 \mid \mathcal{H}_p^{12} \mid u_p^{12}\rangle+\langle u_p^{12} \mid \mathcal{H}_p^{12} \mid u_p^{12}\rangle\}+\sum_{p,q\neq p}K_{pq}\langle u_p^{12} \mid u_q^{12}\rangle}{1+\sum_p\langle u_p^{12} \mid u_p^{12}\rangle} \tag{20}$$

The strong orthogonality condition has allowed us to rewrite the first term of the numerator as

$$\langle\varphi_p^1\varphi_p^2 \mid g_{12} \mid u_p^{12}\rangle=\langle\varphi_p^1\varphi_p^2 \mid \mathcal{H}_p^{12} \mid u_p^{12}\rangle. \tag{21}$$

$\mathcal{E}$ is now varied subject to the constraints

$$\left.\begin{aligned}\langle u_p^{12} \mid \varphi_q^1\rangle_1=0\\\langle u_p^{12} \mid \varphi_q^2\rangle_2=0\end{aligned}\right\}, \tag{22}$$

and the variation of $\mathcal{E}$ is given in a straightforward

[12] From this point on, for compactness of notation, the functional dependence on electron coordinates is written as a superscript; i.e., $f_p(i)=f_p^i$. Furthermore, when an integration is not carried out over all available variables, the integrated variables is indicated by a subscript, e.g., $\langle g^{12} \mid f^1\rangle_1\equiv\int dV_1g(1,2)f(1)$.

manner by

$$\delta\mathcal{E}=(2/D)\sum_p\langle\delta u_p^{12}\mid\mathcal{H}_p^{12}(\varphi_p^1\varphi_p^2+u_p^{12})-\mathcal{E}u_p^{12}+\sum_{q\neq p}K_{pq}u_q^{12}\rangle, \quad (23)$$

where the denominator is given by Eq. (10). In order to obtain the Euler equation for our problem we must now add to (23) the sum of the variations of the constraint equations (22), multiplied by undetermined Lagrangian multipliers. Since there is one set of constraints (22) for each value of one of the electron coordinates, the Lagrangian multipliers are also functions of those coordinates, and the sum is an integral; i.e., the following kinds of terms must be considered.

$$\int dv_2\lambda_{pq}^2\langle\delta u_p^{12}\mid\varphi_q^1\rangle_1. \quad (24)$$

When this is done, one obtains as the extremum conditions,

$$\mathcal{H}_p^{12}\psi_p^{12}+\sum_{q\neq p}K_{pq}u_q^{12}=\mathcal{E}u_p^{12}+\sum_q(\lambda_{pq}^1\varphi_q^2+\lambda_{pq}^2\varphi_q^1), \quad (25)$$

where we have defined

$$\psi_p^{12}=\varphi_p^1\varphi_p^2+u_p^{12}. \quad (26)$$

The Lagrangian multipliers can now be eliminated from the above equations. Multiplying (25) by $\varphi_k^1$, integrating over coordinates of Electron 1 and utilizing the strong orthogonality conditions gives

$$\langle\varphi_k^1\mid\mathcal{H}_p^{12}\mid\psi_p^{12}\rangle_1=\lambda_{pk}^2+\sum_q\langle\varphi_k\mid\lambda_{pq}\rangle\varphi_q^2. \quad (27)$$

Multiplying (27) by $\varphi_l^2$ and integrating yields,

$$\langle\varphi_k^1\varphi_l^2\mid\mathcal{H}_p^{12}\mid\psi_p^{12}\rangle=\langle\varphi_l\mid\lambda_{pk}\rangle+\langle\varphi_k\mid\lambda_{pl}\rangle. \quad (28)$$

Combining these results, we find

$$\sum_q(\lambda_{pq}^2\varphi_q^1+\lambda_{pq}^1\varphi_q^2)=\sum_q\{\langle\varphi_q^1\mid\mathcal{H}_p^{12}\mid\psi_p^{12}\rangle_1\varphi_q^1+\langle\varphi_q^2\mid\mathcal{H}_p^{12}\mid\psi_p^{12}\rangle_2\varphi_q^2\}-\sum_{ql}\langle\varphi_l^1\varphi_q^2\mid\mathcal{H}_p^{12}\mid\psi_p^{12}\rangle\varphi_q^1\varphi_l^2. \quad (29)$$

The variation on $\mathcal{E}$ subject to the strong orthogonality constraints can also be carried out in a different way. An arbitrary pair function can be made strongly orthogonal to all the orbitals of the system by applying the following projection operator,[13]

$$Q=1-\sum_q\{\mid\varphi_q^1\rangle\langle\varphi_q^1\mid+\mid\varphi_q^2\rangle\langle\varphi_q^2\mid\}+\sum_{ql}\mid\varphi_q^1\varphi_l^2\rangle\langle\varphi_q^1\varphi_l^2\mid, \quad (30)$$

whence

$$\langle\varphi_k^1\mid Qv^{12}\rangle_1=\langle\varphi_k^2\mid Qv^{12}\rangle_2=0.$$

The operator $Q$ projects the pair functions onto the function space strongly orthogonal to all the orbitals. One can make a completely arbitrary variation of the strongly orthogonal pair functions, providing the Hamiltonian, $\mathcal{H}'$, of the *constrained* system is used in the energy integral, and $\mathcal{H}'$ is obtained just by operating on the original Hamiltonian with $Q$.[14]

$$\mathcal{H}'=Q\mathcal{H}. \quad (31)$$

This leads to the set of equations identical to (25),

$$Q\mathcal{H}_p^{12}\psi_p^{12}+\sum_{q\neq p}K_{pq}u_q^{12}=\mathcal{E}u_p^{12}. \quad (32)$$

For the projection of the Hamiltonian, we have

$$Q\mathcal{H}_p^{12}\psi_p^{12}=\mathcal{H}_p^{12}\psi_p^{12}-\sum_q\{\langle\varphi_q^1\mid\mathcal{H}_p^{12}\mid\psi_p^{12}\rangle_1\varphi_q^1+\langle\varphi_q^2\mid\mathcal{H}_p^{12}\mid\psi_p^{12}\rangle_2\varphi_q^2\}+\sum_{ql}\langle\varphi_q^1\varphi_l^2\mid\mathcal{H}_p^{12}\mid\psi_p^{12}\rangle\varphi_q^1\varphi_l^2. \quad (33)$$

The last two sums are recognized as just the negative of the Lagrangian multiplier terms of Eq. (29).

Before defining the operators necessary to make these equations more tractable, we need to digress briefly for one more result. By multiplying Eqs. (25) or (32) by $u_p^{12}$, integrating over 1 and 2, and summing the resulting equations, one obtains another expression for $\mathcal{E}$, which when combined with the original energy expression (20) gives,

$$\mathcal{E}=\sum_p\langle\varphi_p^1\varphi_p^2\mid\mathcal{H}_p^{12}\mid u_p^{12}\rangle. \quad (34)$$

Furthermore, since

$$\langle\varphi_p^1\varphi_p^2\mid\mathcal{H}_p^{12}\mid\varphi_p^1\varphi_p^2\rangle=\langle\varphi_p^1\varphi_p^2\mid F^1+F^2-2\epsilon_p+g_{12}-(2J_p^1-K_p^1)-(2J_p^2-K_p^2)+J_{pp}\mid\varphi_p^1\varphi_p^2\rangle=0,$$

this can also be written as

$$\mathcal{E}=\sum_p\langle\varphi_p^1\varphi_p^2\mid\mathcal{H}_p^{12}\mid\psi_p^{12}\rangle. \quad (35)$$

We now define a set of Hermitian operators such that Eqs. (25) are brought into as nearly homogeneous form as possible, i.e., the operators should remove the terms arising from the strong orthogonality constraints

[13] This operator notation may warrant some comment. An operator such as $\mathcal{O}^1=\mid\phi^1\rangle\langle\psi^1\mid$ is defined to be shorthand for

$$\mathcal{O}^1\chi^1=\langle\psi^2\mid\chi^2\rangle_2\phi^1.$$

[14] S. H. Gould, *Variational Methods for Eigenvalue Problems* (University of Toronto Press, Toronto, Canada, 1957).

and transform all the $u_p$'s into $\psi_p$'s. These operators are

$$\begin{aligned}\tilde{\mathcal{H}}_p^{12}=&-\sum_q\{|\varphi_q^1\rangle\langle\varphi_q^1\mathcal{H}_p^{12}|+|\varphi_q^2\rangle\langle\varphi_q^2|\mathcal{H}_p^{12}|+|\mathcal{H}_p^{12}\varphi_q^1\rangle\langle\varphi_q^1|+|\mathcal{H}_p^{12}\varphi_q^2\rangle\langle\varphi_q^2|\}\\ &+\sum_{kq}\{|\varphi_k^1\varphi_q^2\rangle\langle\varphi_k^1\varphi_q^2\mathcal{H}_p^{12}|+|\mathcal{H}_p^{12}\varphi_k^1\varphi_q^2\rangle\langle\varphi_k^1\varphi_q^2|\}\\ &-(n-2)\{|\mathcal{H}_p^{12}\varphi_p^1\varphi_p^2\rangle\langle\varphi_p^1\varphi_p^2|+|\varphi_p^1\varphi_p^2\rangle\langle\varphi_p^1\varphi_p^2\mathcal{H}_p^{12}|\}\\ &+(n-1)\langle\varphi_p^1\varphi_p^2|\mathcal{H}_p^{12}|\psi_p^{12}\rangle|\varphi_p^1\varphi_p^2\rangle\langle\varphi_p^1\varphi_p^2|\\ &-\sum_q\langle\psi_q^{12}|\mathcal{H}_p^{12}|\varphi_p^1\varphi_p^2\rangle\{|\varphi_q^1\varphi_q^2\rangle\langle\varphi_p^1\varphi_p^2|+|\varphi_p^1\varphi_p^2\rangle\langle\varphi_q^1\varphi_q^2|\}. \qquad (36)\end{aligned}$$

$$\begin{aligned}\mathcal{L}_{pq}^{12}=&K_{pq}[1-|\varphi_p^1\varphi_p^2\rangle\langle\varphi_p^1\varphi_p^2|-|\varphi_q^1\varphi_q^2\rangle\langle\varphi_q^1\varphi_q^2|]+|\varphi_p^1\varphi_p^2\rangle\langle\varphi_q^1\varphi_q^2\mathcal{H}_q^{12}|\\ &+|\mathcal{H}_q^{12}\varphi_q^1\varphi_q^2\rangle\langle\varphi_p^1\varphi_p^2|+|\varphi_q^1\varphi_q^2\rangle\langle\varphi_p^1\varphi_p^2\mathcal{H}_p^{12}|+|\mathcal{H}_p^{12}\varphi_p^1\varphi_p^2\rangle\langle\varphi_q^1\varphi_q^2|. \qquad (37)\end{aligned}$$

One can now verify, by straightforward calculation, that

$$\begin{aligned}\tilde{\mathcal{H}}_p^{12}\psi_p^{12}+\sum_{q\neq p}\mathcal{L}_{pq}^{12}\psi_q^1=&-\sum_q(\lambda_{pq}^2\varphi_q^1+\lambda_{pq}^1\varphi_q^2)\\ &\sum_q\langle\varphi_q^1\varphi_q^2|\mathcal{H}_p^{12}|\psi_q^{12}\rangle\varphi_p^1\varphi_p^2+\sum_{q\neq p}K_{pq}u_q^{12}. \qquad (38)\end{aligned}$$

In Eqs. (36), $n$ is the total number of closed orbital pair functions.

In view of (35) and (38), the Eq. (25) determining the pair functions can now be written as

$$(\mathcal{H}_p^{12}+\tilde{\mathcal{H}}_p^{12})\psi_p^{12}+\sum_{q\neq p}\mathcal{L}_{pq}^{12}\psi_q^{12}=\mathcal{E}\psi_p^{12}. \qquad (39)$$

The structure of these equations is such that the solutions are automatically assured to be of the form $\varphi_p^1\varphi_p^2+u_p^{12}$, where $u_p^{12}$ is strongly orthogonal to all the orbitals of the system. We should note that the first order perturbation correction to the Hartree–Fock function has the same strong orthogonality property.[15] Thus, with *only* the initial approximation of a "closed-orbital pair" type of trial function, the $n$-electron closed shell problem can be rigorously reduced to a set of coupled inhomogeneous equations, which are of the two-electron type. It is now somewhat illuminating to pass to a matrix representation, where the pair functions are assumed to be expanded in some two-electron basis set—such as a configuration interaction or Hylleraas-like ($r_{12}$) expansion.

$$\psi_p^{12}=\sum_\lambda c_{p\lambda}v_{p\lambda}^{12}. \qquad (40)$$

Equation (39) then becomes a matrix equation,

$$\mathbf{H}_p\mathbf{C}_p+\sum_{q\neq p}\mathbf{L}_{pq}\mathbf{C}_q=\mathcal{E}\mathbf{S}_p\mathbf{C}_p. \qquad (41)$$

where

$$\begin{aligned}(\mathbf{H}_p)_{\lambda\mu}&=\langle v_{p\lambda}^{12}|\mathcal{H}_p^{12}+\tilde{\mathcal{H}}_p^{12}|v_{p\mu}^{12}\rangle,\\ (\mathbf{S}_p)_{\lambda\mu}&=\langle v_{p\lambda}^{12}|v_{p\mu}^{12}\rangle,\\ (\mathbf{L}_{pq})_{\lambda\mu}&=\langle v_{p\lambda}^{12}|\mathcal{L}_{pq}^{12}|v_{q\mu}^{12}\rangle. \qquad (42)\end{aligned}$$

Furthermore, by virtue of the symmetry of the operator $\mathcal{L}_{pq}^{12}$,

$$\mathbf{L}_{pq}=\mathbf{L}_{qp}^{+}, \qquad (43)$$

where one should not confuse the subscripts $p$ and $q$ with matrix indices; $\mathbf{L}_{pq}$ and $\mathbf{L}_{qp}$ are initially two distinct matrices. Equation (41), however, is really just one large matrix eigenvalue equation, where $\mathbf{H}_p$ are the diagonal blocks and $\mathbf{L}_{pq}$ the off-diagonal blocks: i.e., (41) is the same as

$$\mathbf{Hc}=\mathcal{E}\mathbf{Sc},$$

where the structure of the symmetric matrices and the vector is

$$\mathbf{H}=\left[\begin{array}{c|c}\mathbf{H}_1&\mathbf{L}_{12}\\ \hline \mathbf{L}_{21}&\mathbf{H}_2\end{array}\right];\quad \mathbf{S}=\left[\begin{array}{c|c}\mathbf{S}_1&0\\ \hline 0&\mathbf{S}_2\end{array}\right];\quad \mathbf{C}=\left[\begin{array}{c}C_1\\ \hline C_2\end{array}\right]. \qquad (44)$$

It should be noted, however, that this is really only a pseudo-eigenvalue equation, since the solutions are implicit in the matrix itself via the operator $\mathcal{H}_p^{12}$. Presumably the solutions would have to be found by iteration.

## DISCUSSION

The pair function equations, (39) or (41), are exact for the trial function assumed here. While these functions are restricted to closed-orbital correlations, this may not prove to be a significant physical constraint since the equations are applicable to any initial choice of an orbital set obtained by a unitary transformation from the Hartree–Fock set. This allows us to consider the intraorbital correlation for a maximally localized orbital basis.[11] Since electrons in orbitals which do not interpenetrate are effectively screened from one another, they should weakly correlate their motions. The orbital choice would determine the partition of the correlation energy and hence the magnitude of the energy improvement determined by this model. Certainly, the physically different situations involved in atoms and molecules would imply more ease in localizing orbitals for the molecules and, therefore, both a smaller coupling between the various orbital pairs and a corresponding decrease in the energy ascribed to interorbital correlation.

[15] M. Cohen and A. Dalgarno, Proc. Phys. Soc. (London) **77**, 748 (1961); C. Møller and M. S. Plesset, Phys. Rev. **46**, 618 (1934).

At the present time the predominant effort is in the calculation of the correlation energy. Expression (35) would provide a value of the correlation energy which is an upper bound. Within this context the importance of the interorbital correlation could be assessed without resorting immediately to a completely general calculation.

The complexity of the operators, diagonal and nondiagonal, does not permit an *a priori* judgment on the extent the coupling would affect the pair correlation function. Even if the interorbital correlation energy is small, the solutions of the coupled equations may still be significantly perturbed from the results of an uncoupled set of equations, particularly in those regions of configuration space which are less significant for determining the energy but may be important in the calculation of other properties than the energy. In any case the use of optimally localized orbitals may not always be most convenient for the problem at hand and then the coupling would increase together with the interorbital correlation which would have to be treated explicitly.

We note that the equations presented here should be only slightly more difficult to solve than those presented by Sinanoğlu. The Sinanoğlu pair equations differ in two ways from the ones derived here. First, the diagonal operator is simplified and does not implicitly involve the solution. Secondly, the off-diagonal blocks are set equal to zero; i.e., the matrix is partitioned. Some sort of iterative procedure is necessary to solve the equations in this paper. This is the price that must be paid to obtain a rigorous bound.

Calculation of the matrix elements entails the same degree of difficulty in either case. In both the present scheme and the Sinanoğlu scheme, we appear to have reduced the many-electron problem to a set of two-electron equations. It is, therefore, tempting to think of making a very accurate $N$-electron calculation by approximating the pair functions by some Hylleraas-like expansion (containing $r_{12}$). In either of the schemes, one must compute matrix elements of the form,

$$\langle v_{p\lambda}^{12} \mid \mathcal{H}_p^{12} \mid v_{p\mu}^{12} \rangle.$$

Recalling the definition of this operator, Eq. (15), one must compute matrix elements with the Coulomb and exchange operators, as well as the simple two-electron (heliumlike) Hamiltonian. In particular, since the exchange operator is defined by,

$$K_p^1\varphi^1 = \left[ \int dv_3 \varphi_p^3 \varphi^3 g_{13} \right] \varphi_p^1, \tag{45}$$

the matrix elements over it become

$$\langle v_{p\lambda}^{12} \mid K_q^1 \mid v_{p\mu}^{12} \rangle = \langle v_{p\lambda}^{12} \varphi_q^3 \mid g_{13} \mid \varphi_q^1 v_{p\mu}^{23} \rangle \tag{46}$$

which, for functions containing $r_{12}$, is just the difficult three-electron integral that has always plagued any attempt to extend the $r_{12}$ method to more than two-electron systems,[16] and is common to *both* our scheme and Sinanoğlu's. It should be emphasized that these integrals do not arise from "many-electron collision" terms in the energy, *in addition* to the purely pair energy terms, in which case they would be included in one of Sinanoğlu's approximations. They are inherent in the $\mathcal{H}_p^{12}$ operator, and neglecting them in the original energy expression would disrupt the structure of this operator and probably invalidate some of its properties used in deriving the pair equations.

One also gets three-electron integrals from the $\mathcal{H}_p^{12}$ operator. However they are not quite so troublesome, since they do not involve three cyclic $r_{ij}$ pairs. For a two-electron system, both schemes become the same, the exchange operator problem does not arise, but it is interesting to notice these integrals remain—i.e., for helium, one must still compute three-electron integrals.

In addition to the Hylleraas procedure, one can of course solve the pair equations by a configuration interaction approach, which, however, is usually slowly convergent. In any case the configuration interaction or perturbation approach could be applied directly without the need for determining the variational equations given above. While the difficulties in computing the three-electron integrals in the usual approach are well known, there is no special difficulty if one uses a completely Gaussian basis set. Some success has recently been obtained for the hydrogen molecule with such sets,[17] and calculations on four-electron systems have been initiated by one of us (M.K.), utilizing the procedures discussed in this paper.

In conclusion, it should be observed that nowhere in the present treatment was it necessary for the one-electron functions to be eigenfunctions of the Fock operator. The only requirement was that they be some unitary transform of such eigenfunctions—see Eq. (18). Any set of localized orbitals may therefore be used from the outset, and the solutions will automatically be the concomitant localized pair functions.

---

[16] L. Szasz, J. Chem. Phys. **35**, 1072 (1961); H. M. James and A. S. Coolidge, Phys. Rev. **49**, 676 (1936).

[17] J. V. L. Longstaff and K. Singer, Proc. Roy. Soc. (London) **A285**, 412 (1960).

VII-7 Reprinted from *Journal of Chemical Physics* **47**, 360 (1967)

# Correlation Energy and Molecular Properties of Hydrogen Fluoride

CHARLES F. BENDER AND ERNEST R. DAVIDSON

*Department of Chemistry, University of Washington, Seattle, Washington*

(Received 9 March 1967)

A natural-orbital–configuration-interaction calculation has been carried out on the ground state of the hydrogen fluoride molecule at the experimental equilibrium internuclear separation (1.7328 a.u.). A nonrelativistic energy of −100.2577 a.u. was obtained with 39 configurations. Individual pair correlation energies totaling to over 80% of the total correlation energy were obtained by considering double excitations of each pair.

## INTRODUCTION

Although there have been around 25 calculations of the energy of the HF molecule, the best calculation prior to the present one was the SCF calculation reported by Cade and Huo[1] (who also included an extensive bibliography of past calculations). Very little work has been done on trying to obtain the correlation energy for this molecule by *a priori* calculation.

Since it can be shown that, for closed-shell configurations, most of the correlation energy can be obtained with configurations which are doubly excited relative to the SCF configuration,[2] the study of individual pairs was used as a starting point for obtaining the correlated wavefunction. Higher excitations were, by definition, omitted in studying individual pairs. Quadruple excitations were considered only in the final step of constructing the wavefunction.

The method used is basically the same as that outlined in previous work on LiH by Bender and Davidson.[3] In this method approximate natural orbitals (NO) are constructed and configuration interaction (CI) is carried out to obtain a wavefunction which can be analyzed into natural orbitals. This procedure is repeated until convergence to the NO–CI wavefunction is approached. If all configurations are considered in each iteration, the orbitals will approach the natural orbitals of the molecule (within constraints imposed by the incompleteness of the basis set). If only configurations corresponding to a particular type of correlation are included, however, the wavefunction will converge to an optimum representation of that type of correlation. This property greatly facilitates the study of individual pair correlation effects.

## BASIS SET

Elliptical orbitals were used as the basis set;

$$(njm\alpha\beta)=2R^{-3/2}\pi^{-1/2}\lambda^n\eta^j(\lambda^2-1)^{|m|/2}(1-\eta^2)^{|m|/2}\times\exp(-\alpha\lambda+\beta\eta+im\phi),$$

where $\lambda=(r_a+r_b)R^{-1}$, $\eta=(r_a-r_b)R^{-1}$, and $\phi$ is the cylindrical angle of rotation around the internuclear axis. The distances from the nuclei are $r_a$ and $r_b$, $R$ is the internuclear separation and the $z$ axis for molecular properties is along the internuclear axis and measured from the midpoint.

In the final calculation $19\sigma$, $17\pi(m=1)$, $17\bar{\pi}(m=-1)$, $7\delta$, $7\bar{\delta}$, $1\phi$, and $1\bar{\phi}$ functions were used. The values of $\alpha$, with $\beta$ constrained to be $\pm\alpha$, were partially optimized for $m=0$, $\pm1$ to minimize the SCF energy. The values of $\alpha$ for $m=\pm2$ and $\pm3$ were estimated from past work. This basis set gave an SCF energy of −100.0486 compared with Cade and Huo's value of −100.0703. Values of the orbital energies are given in Table I. The dipole moment was 1.65 D compared with the experimental result of 1.74. Since past experience has shown that energy lost at this point cannot be gained later as correlation energy, an error of 0.02 a.u. was introduced at this point. Perhaps more serious is the fact that configuration interaction did not change the molecular properties appreciably, so the error in the dipole moment at this point was carried into the final results. In future calculations more serious attention will be paid to this problem. The basis set also suffered from the defects of being unnecessarily linearly dependent which gave rise to numerical instability trouble later.

## RESULTS

The $2\sigma$, $3\sigma$, and $1\pi$ natural orbitals are shown in Figs. 1, 2, and 3, respectively. The contours in these figures are drawn at 0.1, 0.5, and 0.9 times the maximum absolute value of the molecular orbital. Nodes are shown by dotted lines and nuclei by squares. It is noticed that the bond orbital is nearly pure $2p_F$–$1s_H$ with a slight $2s_F$ hybridization away from the hydrogen. The lone-pair orbital is polarized into the bond in compensation. This is in agreement with results shown by Kern and Karplus.[4] The $\pi$ orbital is visibly polarized toward the hydrogen.

The correlation of each pair was examined by considering all double excitations of the pair into vacant natural orbitals. It should be emphasized that these excitations were double only in the spatial sense. All spin eigenfunctions were used so that some configurations, and some determinants within any one spin-

[1] P. E. Cade and W. M. Huo, J. Chem. Phys. **47**, 614 (1967).
[2] O. Sinanoğlu, J. Chem. Phys. **36**, 706 (1962).
[3] C. F. Bender and E. R. Davidson, J. Phys. Chem. **70**, 2675 (1966).
[4] C. W. Kern and M. Karplus, J. Chem. Phys. **40**, 1374 (1964).

TABLE I. SCF orbital energies at $R=1.7328$ for HF.

| | Present work | Cade and Huo[a] |
|---|---|---|
| $1\sigma$ | $-26.27861$ | $-26.29428$ |
| $2\sigma$ | $-1.58915$ | $-1.60074$ |
| $3\sigma$ | $-0.75246$ | $-0.76810$ |
| $1\pi$ | $-0.63719$ | $-0.65009$ |

[a] Reference 1.

adapted configuration, were often of higher than double spin-orbital excitation. That is, for example, in considering the $2\sigma 3\sigma$ pairs, all spin arrangements for $2\sigma^2 3\sigma^2 \rightarrow 2\sigma 3\sigma 2\pi 2\bar{\pi}$ were considered in constructing spin eigenfunctions. For this reason only the average $2\sigma 3\sigma$ correlation energy can be given for the four pairs of this type and individual results like $2\sigma\alpha 3\sigma\alpha$ and $2\sigma\alpha 3\sigma\beta$ cannot be discussed.

The total correlation energy obtained from individual pairs was 0.309. The estimated total correlation energy of HF based on Cade and Huo's results and Clementi's estimate of the relativistic correction[5] for the F atom is 0.376. If pair results are additive, this indicates that the basis set used can give at least 80% of the correlation energy. Table II gives the results for individual pairs. A "scaled-up" estimate of the actual pair energy is also given. Each pair was scaled by 0.376/0.309 so the total pair energy would be 0.376. (It is shown later that quadruple excitations also contribute. But these are already included in the sum of pair energies[2] so the total pair correlation energy should be the total correlation energy).

The estimated pair correlation energies given in Table II are probably reliable to within ±0.001. These numbers are in disagreement with both the conclusions of Clementi[5] and those of McKoy and Sinanoğlu.[6] The former of these authors held that $-\epsilon(2s^2)$ was increasing from Be to Ne and was nearly +0.15 for Ne. He also felt that $-\epsilon(2p^2)$ was 0.063 for Ne and $\epsilon(2p \rightarrow 1s^2 2s^2)$ was small. The latter authors felt that $-\epsilon(2s^2)$ was decreasing from Be to Ne and was small for Ne. They held that $-\epsilon(2p^2)$ was only ~0.035 and $-\epsilon(2p \rightarrow 1s^2 2s^2)$ was about 0.04. The present results agree well with McKoy and Sinanoğlu concerning $\epsilon(2s^2)$ and $\epsilon(2p^2)$. Table II shows, however, that $-\epsilon(2p \rightarrow 1s^2 2s^2)$ is only about 0.01. The remaining correlation energy is due to cross-shell correlation between $2p$ orbitals which was not considered important by either of the above authors.

The preceding discussion is based on the assumption that the HF molecule has a similar pattern of correlation energy to the united atom (Ne). The recent results of Nesbet for neon are quoted in Table II for comparison. Except for the changes resulting from the mixing of the $2s$ and $2p\sigma$ orbitals, the pair correlation energies of Ne and HF are the same (within the accuracy to which they are known).

From Table II one might hope to find the contribution of correlation energy to the dissociation energy.

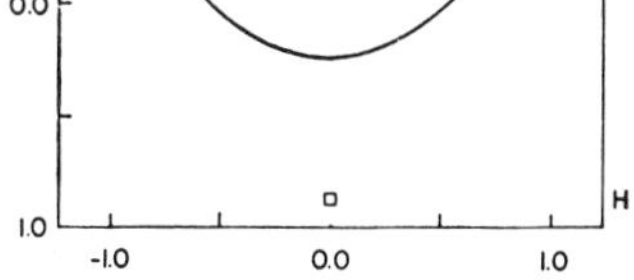

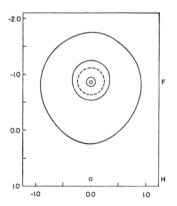

FIG. 1. The $2\sigma$ natural orbital of HF.

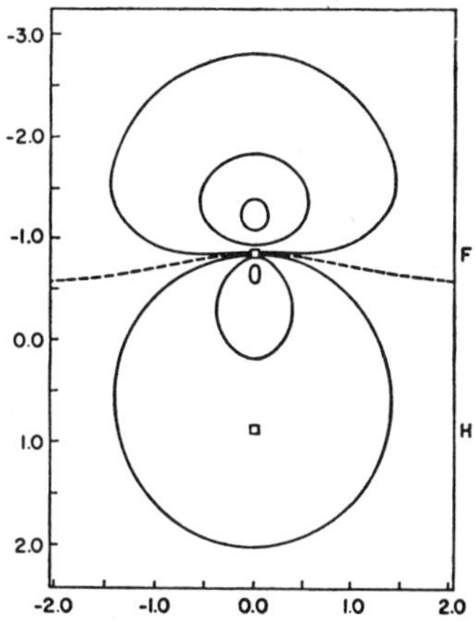


FIG. 2. The $3\sigma$ natural orbital of HF.

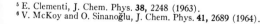

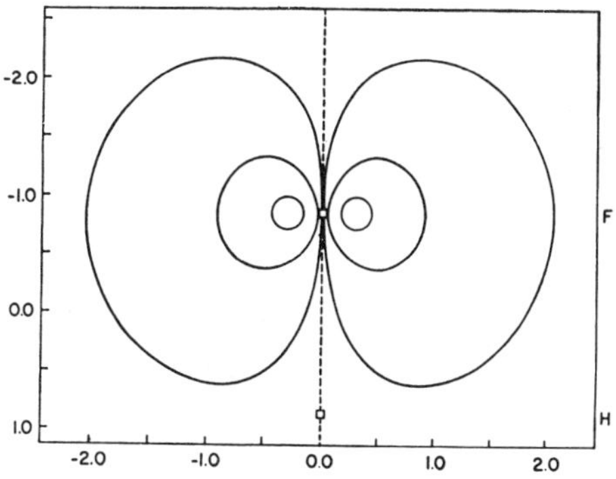


FIG. 3. The $1\pi$ natural orbital of HF.

[5] E. Clementi, J. Chem. Phys. **38,** 2248 (1963).
[6] V. McKoy and O. Sinanoğlu, J. Chem. Phys. **41,** 2689 (1964).

TABLE II. Pair correlation energies.

| Type | No. of pairs of this type | Energy contribution in final 39-configuration calculation | Energy from double excitations of this type only | Scaled-up estimate | Estimated pair correlation energy | Neon pair correction energy[a] |
|---|---|---|---|---|---|---|
| $1\sigma^2$ | 1 | −0.0192 | −0.0316 | −0.0385 | −0.0385 | −0.0399 |
| $2\sigma^2$ | 1 | −0.0109 | −0.0061 | −0.0074 | −0.0074 | −0.0108 |
| $3\sigma^2$ | 1 | −0.0220 | −0.0265 | −0.0323 | −0.0323 | −0.0258 |
| $\pi^2$ | 2 | −0.0266 | −0.0318 | −0.0387 | −0.0193 | −0.0165 |
| $1\sigma 2\sigma$ | 4 | −0.0008 | −0.0035 | −0.0043 | −0.0011 | −0.0013 |
| $1\sigma 3\sigma$ | 4 | 0. | −0.0027 | −0.0033 | −0.0008 | −0.0016 |
| $2\sigma 3\sigma$ | 4 | −0.0119 | −0.0192 | −0.0234 | −0.0058 | −0.0068 |
| $\pi\bar{\pi}$ | 4 | −0.0489 | −0.0580 | −0.0706 | −0.0174 | −0.0168 |
| $1\sigma\pi$ | 8 | 0. | −0.0096 | −0.0117 | −0.0014 | −0.0016 |
| $2\sigma\pi$ | 8 | −0.0254 | −0.0428 | −0.0521 | −0.0065 | −0.0068 |
| $3\sigma\pi$ | 8 | −0.0460 | −0.0768 | −0.0915 | −0.0114 | −0.0123 |
| Total | 45 | −0.2091 | −0.3088 | −0.376[b] | | |

[a] R. K. Nesbet, Phys. Rev. **155**, 56 (1967).

[b] Based on the Cade and Huo[1] result of −100.070, the experimental result of −100.530, and the relativistic correction for the F atom of 0.084 reported by Clementi.[6]

Removing the hydrogen atom to infinity breaks the $3\sigma^2$ pair, two $3\sigma$–$1\sigma$ pairs, two $3\sigma$–$2\sigma$ pairs, and four $3\sigma$–$\pi$ pairs. If the pair correlation energies of the fluorine atom were the same as the HF molecule, the change of correlation energy upon dissociation would be 0.0911 *of which only one-third would be due to breaking the* $3\sigma^2$ *pair*. The actual correlation energy of the F atom is about −0.312, so the actual change in correlation energy is 0.072. This discrepancy is due largely to the increase of correlation energy of other pairs in the F atom because of the availability of a half-filled orbital and to the increased importance of single excitations in the atomic wavefunction.

The natural orbitals for the correlation of each type of pair were compared and merged in such a way that an orthogonal set could be formed which included the main contributors to the energy. This set was then used as a starting set for computing the wavefunction. The best nine configurations were selected and the nine-by-nine secular equation was solved to get a starting $\psi_0$. Using perturbation theory, the energy contribution from all possible single and double excitations (and a few quadruple excitations) was estimated. The best set of configurations was then selected and used in a variational calculation to get a new wavefunction. This function was analyzed into natural orbitals and the entire process was repeated until the energy and natural orbital seemed to have converged.

The final 39-configuration wavefunction gave 50% of the correlation energy. If the total energy obtained, −100.2577, is corrected for relativistic effects it is just barely good enough to predict binding relative to the experimental separated atoms. Table III gives the orbitals used in the final calculation, and Table IV gives some one-electron expectation values for these orbitals. The occupation number $n$ for each orbital in the wavefunction is also included. It should be noted that $2\sigma$ and $3\sigma$ in Table III are not the canonical SCF orbitals because the canonical orbitals mixed slightly

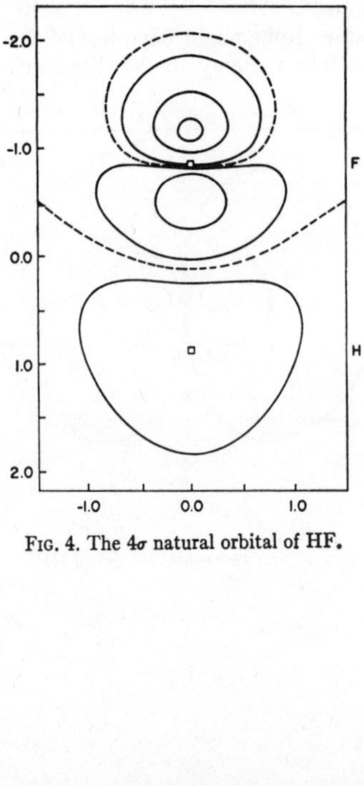


FIG. 4. The $4\sigma$ natural orbital of HF.

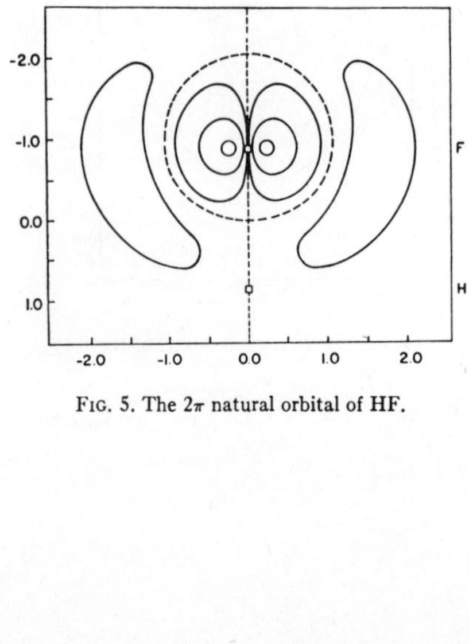


FIG. 5. The $2\pi$ natural orbital of HF.

TABLE III. Natural orbitals for HF.

| $n$ | $j$ | $m$ | $\alpha$ | $\beta$ | $1\sigma$ | $2\sigma$ | $3\sigma$ | $4\sigma$ | $5\sigma$ | $6\sigma$ | $7\sigma$ | $8\sigma$ |
|---|---|---|---|---|---|---|---|---|---|---|---|---|
| 0 | 0 | 0 | 8.6310 | 8.6310 | −29.4101 | 7.8780 | −9.2766 | 10.3189 | 52.4806 | 0.2785 | −21.0433 | 63.3321 |
| 0 | 0 | 0 | 1.0928 | −1.0928 | −0.0094 | −0.2372 | −2.0704 | 1.4676 | 2.4672 | 1.5426 | −1.6043 | 6.5645 |
| 0 | 1 | 0 | 1.0928 | −1.0928 | −0.0101 | 0.1231 | −1.1842 | 2.1713 | 2.3685 | 1.2352 | −1.2712 | 4.6724 |
| 1 | 0 | 0 | 1.0928 | −1.0928 | 0.0046 | −0.0385 | 0.4719 | −0.2095 | −0.5322 | −0.1483 | 0.2385 | −1.2589 |
| 1 | 1 | 0 | 1.0928 | −1.0928 | 0.0050 | −0.1342 | 0.3897 | −0.2785 | −0.4869 | −0.1436 | 0.1138 | −0.8349 |
| 0 | 0 | 0 | 2.3473 | 2.3473 | −0.3558 | −3.3111 | −8.2918 | 12.4647 | −2.9257 | −107.1772 | 0.5448 | 26.2895 |
| 0 | 1 | 0 | 2.3473 | 2.3473 | −0.6472 | 7.4921 | 2.6685 | 1.6126 | −2.4620 | 90.5010 | 14.9535 | −129.9442 |
| 1 | 0 | 0 | 2.3473 | 2.3473 | 0.4132 | −3.0975 | 3.7101 | −8.4120 | −18.3506 | 109.3603 | 6.4318 | −18.0549 |
| 1 | 1 | 0 | 2.3473 | 2.3473 | 0.4710 | 0.3762 | −6.9199 | 17.0764 | 2.1449 | −63.2644 | −20.9394 | 67.9940 |
| 0 | 2 | 0 | 2.3473 | 2.3473 | −0.0706 | −3.3277 | 13.2393 | −27.9149 | 23.4311 | 17.5107 | 28.9344 | 269.6438 |
| 2 | 0 | 0 | 2.3473 | 2.3473 | −0.1236 | −0.9089 | 0.5188 | −5.4196 | 11.0290 | −27.8202 | −0.6966 | −18.3180 |
| 1 | 2 | 0 | 2.3473 | 2.3473 | −0.0486 | 1.6051 | −5.0403 | −17.7374 | −19.9876 | −35.0429 | −15.1188 | −205.1471 |
| 2 | 1 | 0 | 2.3473 | 2.3473 | −0.0615 | −0.2474 | 4.7484 | 12.6531 | 2.7494 | 19.6903 | −0.0789 | 56.1040 |
| 2 | 2 | 0 | 2.3473 | 2.3474 | 0.0185 | −0.0421 | −0.3675 | −1.3690 | −1.3151 | −0.6982 | 5.4771 | −6.0053 |
| 0 | 1 | 0 | 8.6310 | 8.6310 | 21.1960 | −13.8612 | 55.0723 | 65.8548 | 82.2170 | 62.7574 | 205.7311 | 878.0574 |
| 1 | 0 | 0 | 8.6310 | 8.6310 | −21.1039 | 13.6516 | −56.9010 | −72.6674 | −88.7614 | −66.2321 | −178.4751 | −907.9421 |
| 1 | 1 | 0 | 8.6310 | 8.6310 | 66.9872 | −67.8383 | 89.0228 | −311.5294 | −76.7789 | 114.1869 | −723.4498 | 349.6364 |
| 0 | 2 | 0 | 8.6310 | 8.6310 | −32.3842 | 31.8973 | −60.9640 | 89.9696 | −45.4317 | −95.6190 | 682.8322 | −526.7089 |
| 2 | 0 | 0 | 8.6310 | 8.6310 | −34.3359 | 36.3397 | −22.1984 | 234.3816 | 93.2948 | −21.9641 | 105.0600 | 106.7858 |
| | | | | | $1\pi$ | $2\pi$ | $3\pi$ | $4\pi$ | $5\pi$ | | | |
| 0 | 0 | 1 | 2.3050 | 2.3050 | −10.2735 | 10.2160 | 33.5164 | −22.1010 | −1.5367 | | | |
| 0 | 1 | 1 | 2.3050 | 2.3030 | −3.8782 | 6.8809 | 10.9320 | 80.4072 | 6.2065 | | | |
| 1 | 0 | 1 | 2.3050 | 2.3050 | 5.1864 | −5.2137 | −21.2243 | 10.0016 | −2.2934 | | | |
| 1 | 1 | 1 | 2.3050 | 2.3050 | 5.9277 | 8.5177 | −20.8211 | −54.2017 | −9.6652 | | | |
| 0 | 2 | 1 | 2.3050 | 2.3050 | −3.7581 | −1.8546 | −6.2364 | −11.5052 | 5.9724 | | | |
| 2 | 0 | 1 | 2.3050 | 2.3050 | −2.7938 | −4.3577 | 7.2683 | 7.5713 | 3.0987 | | | |
| 0 | 0 | 1 | 1.0231 | −1.0231 | −0.0195 | 0.4412 | 0.1318 | −6.7725 | −0.3129 | | | |
| 0 | 1 | 1 | 1.0231 | −1.0231 | 0.4403 | 2.2475 | −3.7569 | −1.6942 | −1.4503 | | | |
| 1 | 0 | 1 | 1.0231 | −1.0231 | −0.0449 | −0.1371 | 0.0128 | 0.9946 | 0.0603 | | | |
| 1 | 1 | 1 | 1.0231 | −1.0231 | −0.1253 | −0.4039 | 0.5555 | 0.3934 | 0.2189 | | | |
| 0 | 0 | 1 | 13.5870 | 13.5870 | −2.7192 | 0.5977 | 98.8842 | −335.3675 | 1216.2365 | | | |
| 0 | 1 | 1 | 13.5870 | 13.5870 | 28.0031 | −104.9143 | −37.6857 | 333.6567 | 2904.8647 | | | |
| 1 | 0 | 1 | 13.5870 | 13.5870 | −28.0031 | 104.9143 | 37.6857 | −333.6567 | −2904.8647 | | | |
| 1 | 1 | 1 | 13.5870 | 13.5870 | 1.3625 | −2.3705 | −112.1591 | 330.1407 | −60.6498 | | | |
| 0 | 0 | 1 | 1.6500 | −1.6500 | 0.0605 | −1.4501 | 1.6617 | −3.4980 | 0.5154 | | | |
| 0 | 1 | 1 | 1.6500 | −1.6500 | −0.1179 | −2.1618 | 1.9049 | 4.7168 | 1.0769 | | | |
| 1 | 0 | 1 | 1.6500 | −1.6500 | 0.1586 | 0.8890 | −2.5807 | 6.8446 | −0.5860 | | | |
| | | | | | $1\delta$ | $2\delta$ | $3\delta$ | $4\delta$ | | | | |
| 0 | 0 | 2 | 15.5870 | 15.5870 | 22.1645 | 250.2620 | 373.3587 | −6301.3757 | | | | |
| 0 | 0 | 2 | 3.4500 | 3.4500 | −24.6438 | −134.1448 | −189.8953 | 33.8203 | | | | |
| 0 | 1 | 2 | 3.4500 | 3.4500 | 30.9665 | 154.7113 | 59.5995 | −17.0284 | | | | |
| 1 | 0 | 2 | 3.4500 | 3.4500 | 43.8702 | 130.8553 | 101.3420 | −21.4456 | | | | |
| 1 | 1 | 2 | 3.4500 | 3.4500 | −35.9216 | −180.2639 | −17.3529 | 15.8966 | | | | |
| 0 | 0 | 2 | 1.5400 | −1.5400 | −0.0265 | −0.2111 | −0.1229 | 0.0301 | | | | |
| 0 | 0 | 2 | 2.8000 | −2.8000 | 0.1098 | −0.1354 | 0.9882 | 0.0115 | | | | |
| | | | | | $1\Phi$ | | | | | | | |
| 0 | 0 | 3 | 5.8000 | 5.8000 | 321.1883 | | | | | | | |

TABLE IV. Orbital properties.

| | 1σ | 2σ | 3σ | 4σ | 5σ | 6σ | 7σ | 8σ |
|---|---|---|---|---|---|---|---|---|
| $\langle\delta(\mathbf{r}_F)\rangle$ | 212.399 | 10.5748 | 0.3814 | 0.0324 | 33.8592 | 2.5906 | 1976.9082 | 11.1512 |
| $\langle\delta(\mathbf{r}_H)\rangle$ | 0.0000 | 0.0426 | 0.1466 | 0.4552 | 0.0005 | 0.0681 | 0.0016 | 0.9441 |
| $\langle r_F\rangle$ | 0.1757 | 1.0397 | 1.2481 | 1.1372 | 1.1969 | 1.1010 | 0.2314 | 1.3985 |
| $\langle r_H\rangle$ | 1.7406 | 1.8521 | 1.7969 | 1.5441 | 2.2124 | 1.8925 | 1.7660 | 1.6346 |
| $\langle Z\rangle$ | −0.8662 | −0.7248 | −0.5688 | −0.2914 | −0.9845 | −0.8397 | −0.8638 | −0.3755 |
| $\langle r_F r_H\rangle$ | 0.3069 | 1.9792 | 2.1891 | 1.5873 | 2.9958 | 2.2785 | 0.4599 | 2.3538 |
| $\langle X^2+Y^2\rangle$ | 0.0278 | 0.8038 | 0.6209 | 0.4786 | 1.4161 | 0.6246 | 0.1269 | 0.9987 |
| $\langle Z^2\rangle$ | 0.7642 | 1.0186 | 1.6094 | 1.1767 | 1.5459 | 1.5109 | 0.8016 | 1.2462 |
| $\langle r_F^{-1}\rangle$ | 8.6310 | 1.3966 | 1.1160 | 1.4778 | 1.9730 | 1.1025 | 14.9288 | 0.9965 |
| $\langle r_H^{-1}\rangle$ | 0.5772 | 0.6403 | 0.7954 | 0.9484 | 0.5099 | 0.7048 | 0.5743 | 1.1727 |
| $\langle P_2(\cos\theta_F)/r_F^3\rangle$ | −0.0031 | 0.0379 | 2.1630 | 4.9034 | 1.2200 | 0.8204 | 4.5710 | 3.2416 |
| $\langle P_2(\cos\theta_H)/r_H^3\rangle$ | 0.1923 | 0.1819 | 0.1683 | 0.0881 | 0.1080 | 0.5292 | 0.1888 | 1.1329 |
| $\langle Z_F/r_F^3\rangle$ | 0.0526 | 0.0804 | 0.0010 | −0.0099 | −1.6023 | 0.3833 | −12.3840 | 0.5089 |
| $\langle Z_H/r_H^3\rangle$ | −0.3331 | −0.3543 | −0.3366 | −0.1963 | −0.2402 | −0.5849 | −0.3275 | −1.3013 |
| $n$ | 1.9998 | 1.9934 | 1.9843 | 0.0117 | 0.0047 | 0.0028 | 0.0001 | 0.0001 |

| | 1π | 2π | 3π | 4π | 5π | 1δ | 2δ | 3δ | 4δ | 1φ |
|---|---|---|---|---|---|---|---|---|---|---|
| $\delta(\mathbf{r}_F)$ | 0. | 0. | 0. | 0. | 0. | 0. | 0. | 0. | 0. | 0. |
| $\delta(\mathbf{r}_H)$ | 0. | 0. | 0. | 0. | 0. | 0. | 0. | 0. | 0. | 0. |
| $\langle r_F\rangle$ | 1.1408 | 1.3468 | 1.1113 | 1.6339 | 0.1577 | 1.0932 | 1.1716 | 1.0677 | 0.2037 | 0.6722 |
| $\langle r_H\rangle$ | 2.0526 | 2.2170 | 2.0259 | 1.3698 | 1.7499 | 2.0133 | 2.0703 | 2.0541 | 1.7459 | 1.8575 |
| $\langle Z\rangle$ | −0.8150 | −0.8200 | −0.9279 | 0.0701 | −0.8664 | −0.8233 | −0.8141 | −0.9274 | −0.8644 | −0.8664 |
| $\langle r_F r_H\rangle$ | 2.5658 | 3.4049 | 2.3606 | 2.3259 | 0.3055 | 2.2711 | 2.4546 | 2.4351 | 0.3608 | 1.2667 |
| $\langle X^2+Y^2\rangle$ | 1.3630 | 2.0042 | 0.7807 | 1.1769 | 0.0661 | 1.1788 | 1.2832 | 1.0909 | 0.0533 | 0.4463 |
| $\langle Z^2\rangle$ | 1.0255 | 1.2651 | 1.4967 | 0.9276 | 0.7714 | 0.8494 | 1.0174 | 1.1613 | 0.7565 | 0.8064 |
| $\langle r_F^{-1}\rangle$ | 1.2244 | 1.3499 | 1.0566 | 0.6937 | 9.8586 | 1.0587 | 1.0774 | 1.2006 | 6.0626 | 1.6736 |
| $\langle r_H^{-1}\rangle$ | 0.5265 | 0.5002 | 0.5796 | 1.1141 | 0.5746 | 0.5162 | 0.5045 | 0.5351 | 0.5745 | 0.5470 |
| $\langle P_2(\cos\theta_F)/r_F^3\rangle$ | −1.3847 | −2.6926 | 2.7236 | 0.0956 | −507.705 | −0.6418 | −0.5946 | −0.8374 | −115.4610 | −2.3810 |
| $\langle P_2(\cos\theta_H)/r_H^3\rangle$ | 0.0933 | 0.0604 | 0.1109 | −0.7630 | 0.1882 | 0.0842 | 0.0584 | 0.1353 | 0.1867 | 0.1378 |
| $\langle Z_F/r_F^3\rangle$ | 0.0243 | −0.1115 | 0.0981 | 0.2151 | 0.3141 | 0.0172 | −0.3315 | 0.4286 | −0.0128 | 0.0000 |
| $\langle Z_H/r_H^3\rangle$ | −0.2410 | −0.2005 | −0.3069 | −0.03767 | −0.3291 | −0.2300 | −0.2039 | −0.2769 | −0.3283 | −0.2832 |
| $n$ | 3.9669 | 0.0213 | 0.0067 | 0.0002 | 0.0001 | 0.0073 | 0.0003 | 0.0001 | $1\times10^{-6}$ | 0.0001 |

TABLE V. Wavefunction and energy convergence for HF.

| | Configuration | Square of coefficient | −Energy | $-\Delta E$ |
|---|---|---|---|---|
| 1 | $1\sigma^2 2\sigma^2 3\sigma^2 1\pi^2 1\bar{\pi}^2$ | 0.97226 | 100.04861 | |
| 2 | $1\sigma^2 2\sigma^2 3\sigma^2 1\pi 1\bar{\pi} 2\pi 2\bar{\pi}$ | 0.00370 | 100.07575 | 0.0271 |
| 3 | $1\sigma^2 2\sigma^2 3\sigma^2 1\pi 1\bar{\pi} 1\delta 1\bar{\delta}$ | 0.00217 | 100.09383 | 0.0181 |
| 4 | $1\sigma^2 2\sigma^2 3\sigma 4\sigma (1\pi^2 1\bar{\pi} 2\bar{\pi}+\text{c.c.})$ | 0.00514 | 100.12706 | 0.0332 |
| 5 | $1\sigma^2 2\sigma^2 1\pi^2 1\bar{\pi}^2 4\sigma^2$ | 0.00232 | 100.13761 | 0.0106 |
| 6 | $1\sigma^2 3\sigma^2 2\sigma 5\sigma (1\pi^2 1\bar{\pi} 2\bar{\pi}+\text{c.c.})$ | 0.00267 | 100.15669 | 0.0190 |
| 7 | $2\sigma^2 3\sigma^2 1\pi^2 1\bar{\pi}^2 5\pi 5\bar{\pi}$ | 0.00003 | 100.16498 | 0.0083 |
| 8 | $1\sigma^2 1\pi^2 1\bar{\pi}^2 2\sigma 3\sigma 4\sigma 5\sigma$ | 0.00116 | 100.17227 | 0.0073 |
| 9 | $2\sigma^2 3\sigma^2 1\pi^2 1\bar{\pi}^2 7\sigma^2$ | 0.00003 | 100.17933 | 0.0071 |
| 10 | $1\sigma^2 2\sigma^2 3\sigma 6\sigma (1\pi^2 1\bar{\pi} 3\bar{\pi}+\text{c.c.})$ | 0.00140 | 100.19156 | 0.0124 |
| 11 | $1\sigma^2 2\sigma^2 3\sigma (1\pi^2 1\bar{\pi} 3\pi 1\bar{\delta}+\text{c.c.})$ | 0.00156 | 100.20309 | 0.0116 |
| 12 | $1\sigma^2 2\sigma^2 3\sigma^2 1\pi 1\bar{\pi} 3\pi 3\bar{\pi}$ | 0.00068 | 100.20827 | 0.0052 |
| 13 | $1\sigma^2 2\sigma^2 3\sigma^2 (1\pi^2 2\bar{\pi}^2+\text{c.c.})$ | 0.00276 | 100.21936 | 0.0111 |
| 14 | $1\sigma^2 2\sigma^2 1\pi^2 1\bar{\pi}^2 3\pi 3\bar{\pi}$ | 0.00041 | 100.22240 | 0.0030 |
| 15 | $1\sigma^2 2\sigma^2 6\sigma^2 1\pi^2 1\bar{\pi}^2$ | 0.00029 | 100.22483 | 0.0024 |
| 16 | $2\sigma^2 3\sigma^2 1\pi^2 1\bar{\pi}^2 5\sigma 7\sigma$ | 0.00002 | 100.22718 | 0.0024 |
| 17 | $1\sigma^2 3\sigma^2 1\pi^2 1\bar{\pi}^2 5\sigma^2$ | 0.00025 | 100.22943 | 0.0023 |
| 18 | $1\sigma^2 2\sigma^2 3\sigma^2 6\sigma (1\pi^2 1\bar{\delta}+\text{c.c.})$ | 0.00058 | 100.23343 | 0.0040 |
| 19 | $1\sigma^2 2\sigma^2 3\sigma^2 (1\bar{\pi}^2 3\pi^2+\text{c.c.})$ | 0.00048 | 100.23663 | 0.0032 |
| 20 | $1\sigma^2 3\sigma^2 1\pi^2 1\bar{\pi}^2 3\pi 3\bar{\pi}$ | 0.00016 | 100.23798 | 0.0014 |
| 21 | $1\sigma^2 3\sigma^2 1\pi^2 1\bar{\pi}^2 1\delta 1\bar{\delta}$ | 0.00014 | 100.23925 | 0.0013 |
| 22 | $1\sigma^2 2\sigma^2 3\sigma^2 1\pi 1\bar{\pi} 5\sigma^2$ | 0.00017 | 100.24029 | 0.0010 |
| 23 | $2\sigma^2 3\sigma^2 1\pi^2 1\bar{\pi}^2 4\delta 4\bar{\delta}$ | 0.00000 | 100.24136 | 0.0011 |
| 24 | $1\sigma^2 3\sigma^2 1\pi^2 1\bar{\pi}^2 4\sigma^2$ | 0.00017 | 100.24237 | 0.0010 |
| 25 | $1\sigma^2 2\sigma^2 3\sigma^2 1\pi 1\bar{\pi} 3\delta 3\bar{\delta}$ | 0.00006 | 100.24335 | 0.0010 |
| 26 | $1\sigma^2 2\sigma^2 3\sigma^2 1\pi 1\bar{\pi} 2\delta 2\bar{\delta}$ | 0.00007 | 100.24432 | 0.0010 |
| 27 | $1\sigma^2 3\sigma^2 2\sigma (1\pi^2 1\bar{\pi} 1\delta 1\bar{\phi}+\text{c.c.})$ | 0.00009 | 100.24632 | 0.0020 |
| 28 | $1\sigma^2 3\sigma^2 2\sigma 4\sigma (1\pi^2 1\bar{\pi} 3\bar{\pi}+\text{c.c.})$ | 0.00028 | 100.24817 | 0.0018 |
| 29 | $1\sigma^2 3\sigma^2 1\pi^2 1\bar{\pi}^2 2\pi 2\bar{\pi}$ | 0.00014 | 100.24896 | 0.0008 |
| 30 | $1\sigma^2 1\pi^2 1\bar{\pi}^2 2\sigma 3\sigma 4\sigma 6\sigma$ | 0.00009 | 100.24971 | 0.0008 |
| 31 | $1\sigma^2 2\sigma^2 1\pi^2 1\bar{\pi}^2 8\sigma^2$ | 0.00002 | 100.25040 | 0.0007 |
| 32 | $1\sigma^2 2\sigma^2 3\sigma^2 1\pi 1\bar{\pi} 1\delta 1\bar{\delta}$ | 0.00013 | 100.25128 | 0.0009 |
| 33 | $1\sigma^2 3\sigma^2 1\pi^2 1\bar{\pi}^2 6\sigma^2$ | 0.00008 | 100.25201 | 0.0007 |
| 34 | $1\sigma^2 3\sigma^2 2\sigma (1\pi^2 1\bar{\pi} 2\pi 1\bar{\delta}+\text{c.c.})$ | 0.00020 | 100.25364 | 0.0016 |
| 35 | $1\sigma^2 2\sigma^2 1\pi^2 1\bar{\pi}^2 4\pi 4\bar{\pi}$ | 0.00008 | 100.25437 | 0.0007 |
| 36 | $1\sigma^2 2\sigma^2 3\sigma^2 1\pi 1\bar{\pi} 1\phi 1\bar{\phi}$ | 0.00002 | 100.25515 | 0.0008 |
| 37 | $1\sigma^2 1\pi^2 1\bar{\pi}^2 2\sigma 3\sigma 4\sigma 8\sigma$ | 0.00007 | 100.25578 | 0.0006 |
| 38 | $1\sigma^2 3\sigma^2 2\sigma (1\pi^2 1\bar{\pi} 2\pi 2\bar{\delta}+\text{c.c.})$ | 0.00012 | 100.25693 | 0.0012 |
| 39 | $3\sigma^2 1\pi^2 1\bar{\pi}^2 1\sigma 2\sigma 7\sigma^2$ | 0.00000 | 100.25771 | 0.0008 |

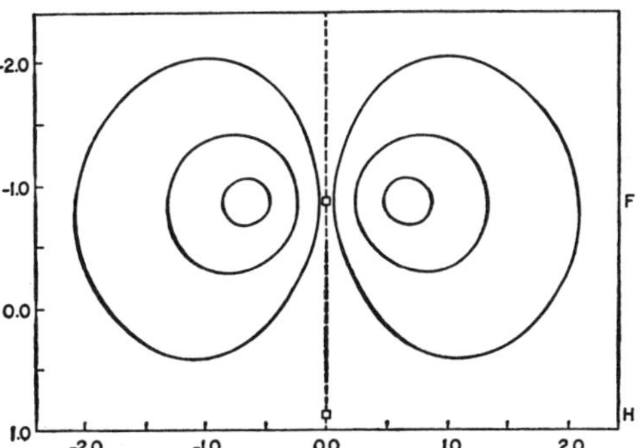


FIG. 6. The $1\delta$ natural orbital of HF.

during the NO transformation. Consequently the correlation for each type of pair in Table II refers to a slightly different pair for the final wavefunction than it did for the pair-by-pair solution.

Table V gives the configurations included in the final calculation and the square of the coefficients obtained by the variational method. The energies obtained by solving all secular determinants from 1×1 to 39×39 is also included for comparison. The orbitals involved in the first few configurations are shown in Figs. 4–6. The second configuration is mainly in–out correlation between $\pi$ and $\bar{\pi}$. The third configuration (and also the 32 which involves a different spin eigenfunction) gives angular correlation between $\pi$ and $\bar{\pi}$. The fourth configuration is cross-shell correlation between $3\sigma$ and $\bar{\pi}$

TABLE VI. Molecular properties of the hydrogen fluoride molecule[a] (atomic units).

| Property | SCF | NO–CI | Exptl |
|---|---|---|---|
| $R_e$ (not calculated) | 1.7330 | 1.7330 | 1.7330 |
| $E$ | −100.0486 | −100.2577 | −100.530 |
| $T$ | 100.1079 | 100.3262 | |
| $V/T$ | −1.9994 | −1.9993 | −2.0000 |
| Force on F | 0.446 | 0.480 | 0. |
| Force on H | −0.014 | −0.012 | 0. |
| $\langle r_F \rangle$ | 0.9491 | 0.9493 | |
| $\langle r_H \rangle$ | 1.8991 | 1.8992 | |
| $\langle Z \rangle$ | −0.7581 | −0.7580 | |
| $\langle r_H r_F \rangle$ | 1.9218 | 1.9227 | |
| $\langle X^2+Y^2 \rangle$ | 0.8359 | 0.8368 | |
| $\langle Z^2 \rangle$ | 1.0889 | 1.0890 | |
| $\langle r_F^{-1} \rangle$ | 2.7182 | 2.7193 | |
| $\langle r_H^{-1} \rangle$ | 0.6131 | 0.6132 | |
| $\langle P_2(\cos\theta_F)/r_F^3 \rangle$ | −0.1143 | −0.1153 | |
| $\langle P_2(\cos\theta_H)/r_H^3 \rangle$ | 0.1458 | 0.1457 | |
| $\langle Z_F/r_F^3 \rangle$ | 0.0383 | 0.0386 | |
| $\langle Z_H/r_H^3 \rangle$ | −0.3011 | −0.3010 | |
| $\langle r_F^2 \rangle$ | 1.3619 | 1.3631 | |
| $\langle r_H^2 \rangle$ | 3.9893 | 3.9900 | |
| $\rho(F)$ | 446.4376 | 446.8381 | |
| $\rho$(H) | 0.3783 | 0.3815 | |
| Electric field gradient at F | 2.6693 | 2.6901 | |
| Electric field gradient at H | 0.5430 | 0.5456 | |
| Dipole moment (Debyes) | −1.6484 | −1.6485 | −1.74 |

[a] $\langle A \rangle = \langle \Sigma_i A(i) \rangle$ The positive $z$ axis points from F toward H for all expectation values.

(and $\pi$ in the complex conjugate term). The fifth configuration is left–right correlation in the valence orbital. This is the term needed to make the SCF function approach the correct limit at large $R$. The sixth configuration involves cross-shell correlation between $2\sigma$ and $\bar{\pi}$ (or $\pi$). The very small coefficient for the seventh configuration is the result of the fact that coefficients go like the reciprocal of effective charge even though the energy contribution is nearly independent of $z$ in the inner shell. Clearly, cross-shell correlation is an important feature of this wavefunction and *it is not at all a simple antisymmetrized product of geminals*. A transformation to localized orbitals in the sense of Edmiston and Ruedenberg[7] might reduce the amount of cross-shell correlation (particularly $\pi\bar{\pi}$). No transformation can make it really small, however, since orbital distributions around the F atom must have considerable penetration. No single or quadruple excitations gave energy contributions as large as the smallest in Table V. The estimated contribution from all quadruple excitations is 0.005.

Table VI gives the molecular properties calculated with the SCF wavefunction and with the full 39-configuration function. The results are in good agreement, as was found previously for LiH.[3] This is further confirmation of the fact that the charge density derived from an SCF wavefunction is nearly correct for a closed-shell molecule. The corrections to the SCF wavefunction changes only those properties which cannot be calculated from the charge density. This emphasizes the necessity for getting the SCF answer correct before proceeding with configuration interaction.

## ACKNOWLEDGMENTS

The authors would like to acknowledge the generous grant of time by the University of Washington Computer Center and the financial support of the National Science Foundation.

[7] C. Edmiston and K. Ruedenberg, J. Chem. Phys. **43**, S97 (1965).

VII-8 Reprinted from *Chemical Physics Letters* **1**, 699 (1968)

# CORRELATIONS BETWEEN TETRAHEDRALLY LOCALIZED ORBITALS

Oktay SINANOĞLU and Bolesh SKUTNIK *
*Sterling Chemistry Laboratory, Yale University, New Haven, Connecticut, USA*

Received 11 March 1968

It is shown that correlations between neighboring tetrahedral orbitals contribute about twice as much to the valence shell correlation energies of neon and methane, as the correlations within the four doubly occupied ("bonds") tetrahedral pairs. "Non-bonded attractions" are thus not negligible but may be an important part of the correlation energies of saturated molecules.

The empirical bond additivity of thermochemistry leads one to expect that the valence shell correlation energies of molecules like $CH_4$, $C_2H_6$, ... would consist mainly of bond correlations. This assumption is built into "geminal" or "separated-pair" type approaches [1]. To consider that correlation energy resides mainly in doubly occupied orbitals is an assumption also in some recent empirical discussions of correlation [2]. On this basis an $N$-electron atom or molecule contains $(N/2)$-"geminal" correlations with additional effects as small corrections.

In the rigorous wave function and energy of an $N$-electron ground state on the other hand, written out in a cluster form [3], intra- and inter-orbital correlations occur on pretty much the same footing. It has been found first by perturbation theory [4], next by a more general, non-perturbative treatment [5,3a] (referred to as "M.E.T.", "many electron theory of atoms and molecules") that the correlation energy of a ground state is given to a good approximation as the sum of $\frac{1}{2}N(N-1)$ spin-orbital pair correlations each of whoch may be evaluated separately non-empirically or by a combination of computational and semi-empirical ways [6]. In refs. [3], [4] and [5], three types of pair correlations, related to each other via unitary transformations, were introduced: (i) Reducible ("B($ij$)-type"), i.e. Hartree-Fock spin-orbital pairs, $\epsilon_{ij}$, (ii) "Irreducible-pairs" $\epsilon_K^{\text{irr}}$, and (iii) localized orbital (LO)-pairs, $\epsilon_{\rho\nu}$. Thus

$$E_{\text{CORR}}(N\text{-electronic ground state}) \cong \sum_{i>j}^{N} \epsilon_{ij} = \sum_{K}^{\frac{1}{2}N(N-1)} \epsilon_K^{\text{irr.}} = \sum_{\rho>\nu}^{N} \epsilon_{\rho\nu}. \qquad (1)$$

The significant magnitudes of inter-orbital correlations were first noted in [6] where sizable 2s-2p correlations, $\epsilon$(2s-2p), were found in first row atoms, also in He-He with $\epsilon(1\sigma_g^2) \approx \epsilon(1\sigma_u^2) \approx \epsilon(1\sigma_g^\alpha 1\sigma_u^\beta)$ [7], and in the computations of Kelly [8], Grimaldi [9], Davidson [10] and others. The sizable role of inter-orbital correlations remains even after transformation to the LO-description, $\epsilon_{\rho\nu}$'s [11].

In the present note, the correlations between tetrahedral orbitals are evaluated. In M.E.T. the valence shell $E_{\text{CORR}}$ of neon atom and of $CH_4$ may be writen rigorously as the sum of four "bond" correlations ("intra-bond") plus the correlations between the bonds ("inter-bod") (cf. also [4a]). It is shown below that the sum of "inter-bond" correlations contribute roughly two times the four ($sp^3$ or C-H) "bond" correlations in Ne and $CH_4$.

Let $\eta_1$, $\eta_3$, $\eta_5$, $\eta_7$ be the tetrahedral, ($sp^3$)-hybrid spin-LO's of neon with $\alpha$-spins. The $\eta_2$, $\eta_4$, $\eta_6$, $\eta_8$ have $\beta$-spins. The tetrahedron is oriented such that $\eta_1$ and $\eta_2$ are along the $Z$-axis and $\eta_3$ and $\eta_4$ (the same spatial LO) are in the $XZ$-plane. The LO-transformation matrix $t$ is

* Present address: Department of Chemistry, Brandeis University, Waltham, Massachusetts, USA.

$$\begin{pmatrix}\eta_1\\ \eta_3\\ \eta_5\\ \eta_7\end{pmatrix} = \frac{1}{2}\begin{pmatrix}1 & \sqrt{3} & 0 & 0\\ 1 & -\frac{1}{\sqrt{3}} & 2\sqrt{\frac{2}{3}} & 0\\ 1 & -\frac{1}{\sqrt{3}} & -\sqrt{\frac{2}{3}} & \sqrt{2}\\ 1 & -\frac{1}{\sqrt{3}} & -\sqrt{\frac{2}{3}} & -\sqrt{2}\end{pmatrix}\begin{pmatrix}2s\alpha\\ 2p_z\alpha\\ 2p_x\alpha\\ 2p_y\alpha\end{pmatrix} \tag{2}$$

The pair functions $\hat{u}_{ij}$ are then transformed by the direct product matrix $\boldsymbol{T} = \boldsymbol{t} \times \boldsymbol{t}$ (cf. [5b,4a]). Each LO-pair correlation $\epsilon_{\eta_i\eta_j}$ may then be expressed in terms of the "reducible" pairs $\epsilon_{kl}$ and the "cross pairs" $\epsilon_{kl;mn}$ [12], where

$$\epsilon_{kl;mn} = \langle B(kl)|\frac{1}{r_{12}}|\hat{u}_{mn}\rangle . \tag{3}$$

We have for an $(LO)^2$-correlation,

$$\epsilon_{\eta_1\eta_2} = \frac{1}{16}\left[\epsilon(2s^2) + 6\epsilon(2s\overline{2p}_z) + 9\epsilon(2p_z^2) + 6\epsilon(2s^2;2p_z^2) + 6\epsilon(2s\overline{2p}_z;2p_z\overline{2s})\right] \tag{4}$$

where ($\overline{2p}_z \equiv 2p_z\beta$; $2p_z \equiv 2p_z\alpha$). For an inter-orbital correlation $(LO)_\alpha$-$(LO')_\beta$:

$$\epsilon_{\eta_1\eta_4} = \frac{1}{16}\left[\epsilon(2s^2) + 6\epsilon(2s\overline{2p}_z) + \epsilon(2p_z^2) + 8\epsilon(2p_z\overline{2p}_x) - 2\epsilon(2s^2;2p_z^2) - 2\epsilon(2s\overline{2p}_z;2p_z\overline{2s})\right] \tag{5}$$

and for a $(LO)_\alpha$-$(LO')_\alpha$:

$$\epsilon_{\eta_1\eta_3} = \frac{1}{2}\left[\epsilon(\overline{2s2p}_z) + \epsilon(\overline{2p_z2p_x})\right] . \tag{6}$$

To get just the total $E_{CORR}$, we would not need the "cross-pairs". However here we ask for the individual $\epsilon_{\eta_i\eta_j}$, hence we must know the values of the $\epsilon_{kl;mn}$ as well as the $\epsilon_{ij}$. All "cross-pairs", except $\epsilon(2s^2;2p_z^2)$ are evaluated by group theory in terms of "irreducible" (symmetry) pairs $\epsilon_K^{irr}$. One obtains

$$\epsilon_{\eta_1\eta_2} = \frac{1}{16}\left[\epsilon(2s^2) + 6\epsilon(2s2p;{}^1P) + 3\epsilon(2p^2;{}^1S) + 6\epsilon(2p^2;{}^1D) + \frac{6}{\sqrt{3}}\epsilon(2s^2\,{}^1S;2p^2\,{}^1S)\right] \tag{7a}$$

$$\epsilon_{\eta_1\eta_4} = \frac{1}{16}\left[\epsilon(2s^2) + 2\epsilon(2s2p;{}^1P) + 4\epsilon(2s2p;{}^3P) + \frac{1}{3}\epsilon(2p^2;{}^1S) + \frac{14}{3}\epsilon(2p^2;{}^1D) + 4\epsilon(2p^2;{}^3P) - \frac{2}{\sqrt{3}}\epsilon(2s^2\,{}^1S;2p^2\,{}^1S)\right] \tag{7b}$$

$$\epsilon_{\eta_2\eta_3} = \frac{1}{2}\left[\epsilon(2s2p;{}^3P) + \epsilon(2p^2;{}^3P)\right] . \tag{7c}$$

The total $(2s^22p^6)$ L-shell correlation energy of Ne in the three different descriptions [5,4] is:
In "reducible pairs" ($s,x,y,z$ form):

$$E_{CORR}(\text{L-shell;Ne}) = \epsilon(2s^2) + 6\epsilon(\overline{2s2p}_z) + 6\epsilon(2s\overline{2p}_z) + 3\epsilon(2p_x^2) + 6\epsilon(2p_x2p_y) + 6\epsilon(p_x\overline{p}_y) . \tag{8}$$

In "irreducible pairs":

$$E_{CORR}(\text{L-shell;Ne}) = \epsilon(2s^2) + 3\epsilon(2s2p;{}^1P) + 9\epsilon(2s2p;{}^3P) + 9\epsilon(2p^2;{}^3P) + 5\epsilon(2p^2;{}^1D) + \epsilon(2p^2;{}^1S) . \tag{9}$$

In tetrahedral "LO-pairs":

$$E_{CORR}(\text{L-shell;Ne}) = 4\epsilon_{\eta_1\eta_2} + 12\epsilon_{\eta_1\eta_4} + 12\epsilon_{\eta_1\eta_3} . \tag{10}$$

Also

$$E_{CORR}(\text{Ne}) = \epsilon(1s^2) + 2\epsilon(1s^2 \to 2s) + 6\epsilon(1s^2 \to 2p) + E_{CORR}(\text{L-shell}) . \tag{11}$$

Very similar equations apply to $CH_4$ with s, $p_x$, $p_y$, $p_z$ replaced by a, $t_x$, $t_y$, $t_z$ MO's, and irreducible pairs defined with respect to the point group $T_d$ instead of the rotation group O(3).

We have for Ne and $CH_4$:

$$E_{CORR}(\text{L-shell}) = 4\epsilon_{\eta_1\eta_2} \times (1 + \text{Rt}). \qquad (12)$$

The $4\epsilon_{\eta_1\eta_4}$ would be just the "bonds only" correlation. Rt is the ratio indicating the deviation from this concept, the ratio of the total inter-"bond" correlations to the "bonds" part:

$$\text{Rt} \equiv \frac{E^{\text{inter}}_{\text{CORR}}}{E^{\text{intra}}_{\text{CORR}}} = \frac{3(\epsilon_{\eta_1\eta_4} + \epsilon_{\eta_1\eta_3})}{\epsilon_{\eta_1\eta_2}}. \qquad (13)$$

The $\epsilon_{ij}$ and $\epsilon^{\text{irr}}_K$ for first row atoms have been evaluated using a non-closed shell version of M.E.T. by a combination of computation and experimental $E_{CORR}$'s on $1s^2 2s^n 2p^m$ configurations [13]. For Ne the values are (in eV): $\epsilon(2s^2) = -0.275$; $\epsilon(2s2p, {}^3P) = -0.123$; $\epsilon(2s2p; {}^1P) = -0.585$; $\epsilon(2p^2; {}^3P) = -0.174$; $\epsilon(2p^2; {}^1D) = -0.545$; $\epsilon(2p^2; {}^1S) = -1.202$. Also, $\epsilon(1s^2 \rightarrow 2p) = -0.102$; $\epsilon(1s^2 \rightarrow 2s) = -0.075$, and $\epsilon(1s^2) =$ $= -1.272$ eV. These values not only reproduce the ground state $E_{CORR}$ of Ne, but also those of many Ne-ions and excited states after the addition of non-closed shell effects [12,13].

In eqs. (7), we also need the $\epsilon(2s^2; 2p_z^2)$ cross term. The individual LO-pair $\epsilon_{\eta_i\eta_j}$ values are shown in table 1 for two estimates of $\epsilon(2s^2; 2p_z)$. To second order in energy one has $\epsilon^{(2)}(2s^2; 2p_z^2) = \epsilon^{(2)}(2p^2; 2s_z^2)$. Taking this as valid for the "exact pairs", and neglecting the small dynamical $\hat{u}_{2s^2}$ [6], we have

$$\text{(A)} \qquad \epsilon(2s^2; 2p_z^2) \approx \epsilon(2p_z^2; 2s^2)_{\text{non-dyn.}} = 0. \qquad (14)$$

On the other hand, a rough estimate, (likely to be on the large side) is obtained as

$$\text{(B)} \qquad \frac{1}{\sqrt{3}}\epsilon(2s^2; 2p^2\,{}^1S) = \epsilon(2s^2; 2p_z^2) \approx -0.313 \text{ eV}. \qquad (15)$$

This estimate is based on applying the Cauchy-Schwartz inequality twice to $\langle B(2s^2)|g_{12}|\hat{u}_{2p_z^2}\rangle$. Using also the rough rule $\langle \hat{u}_{ij}|\hat{u}_{ij}\rangle \approx \epsilon_{ij}/E_{\text{H.F.}}$, one gets

$$|\langle B(2s^2)|g_{12}|\hat{u}_{2p_z^2}\rangle| \approx J_{2s^2}\sqrt{|\epsilon(2p_z^2)|/E_{\text{H.F.}}} \qquad (16)$$

where $J_{2s^2}$ is the 2s-2s Coulomb integral. Three different equations of this type are obtained yielding 0.411, 0.298 and 0.250 eV with the geometric mean 0.313 eV in eq. (15).

Table 1 indicates that the $\epsilon_{\eta_i\eta_j}$ and the Rt ("inter-bonds"/"intra-bonds" ratio) are not affected in their general features by (A) or (B). The inter-LO correlations contribute to the valence shell $E_{CORR}$ of Ne, as much as 1.8 to 2.2 times the 4-tetrahedrally localized pairs.

Does the same drastic picture hold true in $CH_4$? How large are the non-bonded (C-H)--(C-H) attractions within the same molecule [14], compared to the four (C-H) bond correlations?

Table 1

Correlations between and within tetrahedral orbitals in the valence shell of neon[a] (in eV). Even with such localized orbitals, correlations between "bonds" are found to contribute more than the four "bonds" (Rt > 1).

| | A[b] | B[b] |
|---|---|---|
| $\epsilon_{\eta_1\eta_2}$ (a "bond" or "lone pair" type) | -0.666 eV | -0.784 eV |
| $\epsilon_{\eta_1\eta_4}$ ("inter-bond" opposite spin) | -0.349 eV | -0.310 eV |
| $\epsilon_{\eta_1\eta_3}$ ("inter-bond" same spin) | -0.149 eV | -0.149 eV |
| $\sum_{i>j}^{N} \epsilon_{\eta\,\eta} = E^{(\text{Ne;L-shell})}_{\text{CORR}}$ | -8.64 eV | -8.64 eV |
| $\text{Rt} \equiv \frac{E^{\text{inter}}_{\text{CORR}}}{E^{\text{intra}}_{\text{CORR}}} =$ | 2.24 | 1.75 |

[a] The irreducible pair correlation values of İ. Öksüz and O. Sinanoğlu [13] for neon are used.
[b] Calculation A uses eq. (14); B uses eq. (15) in text.

The $CH_4$, LO-pair correlations have been obtained using a semi-empirical MO-pairs (or LO-pairs) correlation theory [15] by Pamuk. These $CH_4$ values are: $\epsilon_{\eta_1\eta_2}$ (C-H bond) = -0.64 eV, $(\epsilon_{\eta_1\eta_4}+\epsilon_{\eta_1\eta_3})$ = = -0.40 eV, and Rt = 1.9. Thus the tentative $CH_4$ results are in line with those in table 1 for Ne. The tetrahedral (C-H) bond correlation value as well as the Ne values in table 1 may be compared with the bonding MO-pair $\epsilon_{ij}$ = -0.87 eV in the $2\pi$ C-H radical, calculated non-empirically by Davidson [10]. The latter value however is for a different hybridization as well as being for a non-closed shell system in which total pair correlation (including the "semi-internal" correlation effect which is zero in $CH_4$ and Ne (cf. refs. [12,13])) are expected to be larger.

In conclusion: the inter-bond correlations above and in other molecules studied in this laboratory are found to contribute at least as much to valence shell correlation energies as the "bond" correlations. Both effects however naturally occur and are predicted in the perturbation theory of correlation [4] and in M.E.T. [5]. A semi-empirical MO-pairs version of M.E.T. [15] is being applied to classes of molecules whose LO's have been obtained by a recently developed localization method [16].

This work was supported by grants from the U.S. National Science Foundation and the Alfred P. Sloan Foundation. It is a pleasure to thank Mr. İskender Öksüz and Mr. Ö. Pamuk for helpful discussions.

*References*


[1] a) A.C.Hurley, J.E.Lennard-Jones and J.A.Pople, Proc.Roy.Soc. (London) A220 (1953) 446.
b) R.McWeeny, Rev.Mod.Phys. 32 (1960) 335.
c) C.Edmiston and Ruedenberg, Rev.Mod.Phys. 35 (1963) 457.
d) R.K.Nesbet, Advs.Chem.Phys. 9 (1965) 321, cf. p. 355.
e) cf. also remarks on this point by E.Kapey, J.Chem.Phys. 44 (1966) 956; and by C.Edmiston, J.Chem.Phys. 39 (1963) 2394.

[2] a) D.C.Pan and L.C.Allen, J.Chem.Phys. 46 (1967) 1797; cf. also p. 49 of L.C.Allen, in: Quantum Theory of Atoms, Molecules and the Solid State, ed. P.O.Löwdin (Acad.Press, New York, 1966).
b) E.Clementi, J.Chem.Phys. 38 (1963) 2248; and L.C.Allen, E.Clementi and H.M.Gladney, Rev.Mod.Phys. 35 (1963) 465.

[3] a) O.Sinanoğlu, Proc.Nat'l.Acad.Sci.U.S. 47 (1961) 1217;
b) Rev.Mod.Phys. 35 (1963) 517.

[4] a) O.Sinanoğlu, J.Chem.Phys. 33 (1960) 1212;
Proc.Roy.Soc. (London) A260 (1961) 379.

[5] a) O.Sinanoğlu, J.Chem.Phys. 36 (1962) 706 [M.E.T. 1]
b) Advs.Chem.Phys. 6 (1964) and later papers of M.E.T. series.

[6] V.McKoy and O.Sinanoğlu, J.Chem.Phys. 41 (1964) 2689.

[7] O.Sinanoğlu, in: Modern Quantum Chemistry, 1964 Istanbul Lectures, Vol. II (Acad.Press, New York, 1965) p. 231.

[8] H.P.Kelly, Phys.Rev. 144 (1966) 39.

[9] F.Grimaldi, J.Chem.Phys. 43 (1965) S59.

[10] C.F.Bender and E.R.Davidson, J.Phys.Chem. 70 (1966) 2675; and other papers. E.R.Davidson, to be published. That the correlations between different 2p-orbitals are also important is pointed out by C.F.Bender and E.R.Davidson, J.Chem.Phys. 47 (1967) 360.

[11] As presented and emphasized by one of us (O.S.), in a number of recent symposia (e.g. Canadian Qu.Chem. Symp., Montreal 1967; Qu.Chem.Symp., Kutna Hora, Czechoslovakia 1967; Nato Frascati Summer School on Electron Correlation 1967; Yale Seminar on Sigma MO Theories 1967).

[12] See e.g. ref. [7], also O.Sinanoğlu, in: Advances in Chemical Physics (1968), (Frascati volume; R.Lefebvre and C.Moser editors). In closed shells, the sum of all cross-pairs is zero. They occur only in transforming individual $\epsilon_{ij}$'s. They do not enter into the separate variational calculation of each $\epsilon_{ij}$. The "cross-pairs" are particularly significant in non-closed shell states where they do not cancel out.

[13] İ.Öksüz and O.Sinanoğlu, to be published (cf. also ref. [12]). These values use many excited states of different $N$'s and $Z$'s to extract $\epsilon_{ij}$'s, by a generalization of the methods used in ref. [6]. Ref. [6] has used only ground states, hence less detail and accuracy were attainable. Some preliminary calculations of the above detailed type have also been made by B.Skutnik (Ph.D.Thesis, 1967; Yale University).

[14] Intra-molecular $C/R^6$ London dispersion forces were invoked by K.S.Pitzer (Advs.Chem.Phys. 2 (1959) 59), in estimating heats of isomerization. Inter-bond LO-correlations are a more fundamental formulation of the intramolecular attraction concept [4a].

[15] Ö.Pamuk and O.Sinanoğlu, to be published.

[16] C.Trindle and O.Sinanoğlu, J.Chem.Phys., in press.

# Chapter VIII

# *Diagrammatic Many-Body Perturbation Theory (Approach III) and Some Field Theoretic Techniques*

## VIII-1

Several methods for treating atomic and molecular many-electron problems have been discussed in the earlier chapters. The Hartree–Fock and related methods do not include correlation; the methods starting from the electron gas and including gradient corrections converge badly in the steep density gradients characteristic of atomic systems. The pair and cluster correlation approach (I) gives useful results, particularly when supplemented by semiempirical methods. The latter uses mostly variational techniques in a way that approximations can be improved by considering successively larger numbers of electrons correlating at a time. The analysis of higher effects and some observables, however, becomes particularly convenient in the diagrammatic many-body perturbation theory, which we shall now discuss.

The techniques developed specifically for atomic and molecular problems have special features pertinent to such problems. Thus they differ considerably from methods which were developed for the study of other types of quantum many-body systems (such as nuclear matter, liquid helium, etc.) and which have experienced such rapid growth in the 1950s and 1960s. The developments given in this chapter concentrate on the more general aspects of the methods of quantum many-body theory, of particular significance because they provide powerful techniques new to atomic and molecular problems. They should be particularly useful in the study of properties requiring the inclusion of higher-order effects, as in hyperfine structure and polarizability.

The development of quantum many-body theory in its present form was initiated by Brueckner (1953) in the study of the structure of many-body systems of infinite extent, in particular, infinite nuclear matter. The principal formal difficulty, divergences in the structure of the perturbation expansion in higher orders for large numbers of particles, was analyzed by Brueckner (1956), who introduced for their elimination the "linked cluster expansion."† The full resolution of this problem was given by Goldstone (1957). It was immediately recognized by the quantum field theorists that the many-body problem offered a very fertile field for application of their techniques, and in a series of basic papers the full formal development of quantum many-body theory was quickly given. This is the diagrammatic many-body perturbation theory, which removes the difficulties of conventional perturbation expansions when there are infinitely many particles. The particular problem of nuclear matter, on the other hand, treated extensively by Brueckner and co-workers and by Bethe and Goldstone,

†See the footnotes on pages 119 and 180 regarding questions of terminology and different usage of these words.

involves, as its main concern, the nucleon–nucleon hard core, in addition to the difficulties of infinitely many particles.

The atomic (and molecular) problem presents few of the formal problems which have drawn the concentrated attention of the theorist in studying the uniform quantum liquids. The problems of superconductivity, superfluidity, and phase transitions are absent; the collective phenomena peculiar to the plasma are also suppressed or totally inhibited by the strong central field. The problems of the nucleus as a many-body system, resulting from the collective excitations, the cooperative pairing, the close admixture of collective and single-particle excitation energies, and the complications due to the very complicated, strong short-range forces are absent in the atom. The problem for the quantum chemist lies perhaps more in the enormous variety of structures which he must consider, the many observables to be predicted, the accuracy required, and the need to consider many excited states. In the more complicated molecular structures, the close admixture of vibrational, rotational, and electronic states presents new problems. Finally, it should be emphasized that the relativistic corrections in heavy atoms are very poorly understood and present a major obstacle to quantitative predictions of structure, at least in purely a priori calculations.

The calculations by Kelly are an application of the Brueckner–Goldstone diagrammatic many-body perturbation theory, and do not make use of the more powerful methods of quantum field theory and of Green's functions. The latter do not seem to be necessary, at least for the problems of the relative simplicity considered by Kelly.

To apply perturbation theory, two principal problems to be resolved are in the optimum relation of a basis set and in the treatment of intermediate sums over bound states with large quantum numbers and over the continuum. The latter problem, in particular, caused much difficulty before Kelly showed that a straightforward treatment is possible.

In the choice of a basis set, the optimum choice should give most rapid convergence of the perturbation expansion while maintaining orthogonality and completeness. The requirement of convergence arises from the fact that the Hamiltonian is separated into a single-particle part, $H_0 = \Sigma_i(T_i + V_i)$, plus a perturbation, $H_1 = \frac{1}{2}\Sigma_{ij}v_{ij} - \Sigma_i V_i$, and the matrix elements of $V_i$ in excited states will not usually vanish. Thus the choice of $V_i$ should be to minimize matrix elements of $H_1$, as far as possible. The choice of the H. F. potential, for example, gives zero diagonal matrix elements of $H_1$ in the ground state. For this reason the single-particle term $V_i$ has often been taken to be the H. F. potential for excited as well as ground states. It is, however, immediately apparent that this is a poor choice for excited states, because of the inclusion of spurious self-interaction terms which cancel in the ground state but not in excited states. These appear as follows: The H. F. potential is

$$V_{HF}(1) = \sum_j (v_{ij,\,ij} - v_{ij,\,ji}) \tag{1}$$

with the sum over filled states. The term j = i then cancels if i is any filled state. For excited states, however, formed by removing an electron from state i and exciting it to i′, the H.F. potential may be written

$$\begin{aligned} V_{HF}(i') &= \sum_j (v_{i'j,\,i'j} - v_{i'j,\,ji'}) \\ &= \sum_{j \neq i} (v_{i'j,\,i'j} - v_{i'j,\,ji'}) + (v_{i'i,\,i'i} - v_{i'i,\,ii'}) \end{aligned} \tag{2}$$

The second term is the interaction of the excited electron with a fictitious electron left in the state it vacated. This term is repulsive, being dominated by the electrostatic repulsion of the spurious electron. The potential so constructed is in part the interaction of an added electron with the neutral atom and not with the positive ion left upon single-electron excitation. The convergence of this choice, as pointed out by Kelly, is the loss of the bound excited states in the light atoms, the neutral atom not binding an additional electron.

The obvious modification to make in the single-particle potential is to omit the self-interaction term, i.e., to use

$$V'_{HF}(i'i) = \sum_{j \neq i} (v_{i'j,\, i'j} - v_{i'j,\, ji'}) \tag{3}$$

with the potential labeled by the holes and particles present. This choice leads to considerable complication, however, particularly in the loss of hermiticity of the single-particle Hamiltonian and hence wave-function orthogonality. Kelly showed, however, that a reasonable compromise for each angular mementum (or symmetry) state was to omit in the filled-state sum the last filled state and to use this potential for all states. The potential then becomes $\ell$ (or symmetry)-dependent but remains Hermitian. The wave functions of the last filled states then are still the original Hartree–Fock, the excited states are correctly a positive ion, but the lower filled states are now incorrect. This adds small correction terms to the energy expansion.

The remaining problem of inclusion of high excited states and the continuum presents no conceptual difficulty but is primarily one of computational convenience. Kelly showed how this may be done in a straightforward way, when replacing the integrals by sums, as is required in the numerical computation.

The range of applications already made by Kelly reflects the power and conceptual simplicity of the methods, although the numerical difficulties are, of course, nontrivial. He has analyzed the contributions of different diagrams to the ground state and several low excited states of beryllium and oxygen in detail. The relative importance of various terms observed is in agreement so far with the importance of effects noted in the Sinanoglu theory and which must be included in other approximation methods. Kelly has not, however, analyzed the third- and higher-order ring diagrams, so that the convergence of the method still remains to be verified, and the actual magnitudes of higher electron effects in both Approaches I and III are still unknown.

Kelly also considered other atomic observables (i.e., f numbers and polarizabilities) and has shown the importance of correlation corrections. The use of perturbation theory in these problems is particularly effective because the systematic nature of the procedures and the powerful conceptual clarification of the diagrammatic techniques ensures that all effects are included. This becomes particularly important in the higher orders of correction.

Presently there is some interest among theoretical chemists and physicists in seeing also whether other field theoretic techniques, such as Green's functions, will be useful in atomic and molecular problems. We include in this chapter two such early papers, by Tolmachev and by Čížek, in this direction.

VIII-2 Reprinted from *Proceedings of the Royal Society* (London) A239, 267 (1957)

# Derivation of the Brueckner many-body theory

By J. Goldstone
*Trinity College, University of Cambridge*‡

(*Communicated by N. F. Mott, F.R.S.—Received 24 August* 1956)

An exact formal solution is obtained to the problem of a system of fermions in interaction. This solution is expressed in a form which avoids the problem of unlinked clusters in many-body theory. The technique of Feynman graphs is used to derive the series and to define linked terms. The graphs are those appropriate to a system of many fermions and are used to give a new derivation of the Hartree–Fock and Brueckner methods for this problem.

## 1. Introduction

The Hartree–Fock approximation for the many-body problem uses a wave function which is a determinant of single-particle wave functions—that is, an independent-particle model. The single-particle states are eigenstates of a particle in a potential $V$, which is determined from the two-body interaction $v$ by a self-consistent calculation. The Brueckner theory (Brueckner & Levinson 1955; Bethe 1956; Eden 1956) gives an improved method of defining $V$ and shows why the residual effects of $v$ not allowed for by $V$ can be small. In particular, in the nuclear problem the corrections to the energy are small, even though the corrections to the wave function are large. The theory thus gives a reconciliation of the shell model, the strong two-nucleon interactions, and the observed two-body correlations in the nucleus. The smallness of the corrections is due to the operation of the exclusion principle. Bethe (1956) has shown that this same exclusion effect makes even the Hartree–Fock approximation good for quite strong interactions, such as an exponential potential fitted to low-energy nucleon-nucleon scattering.

The first problem on which calculations have been made is that of 'nuclear matter', that is, a very large nucleus with surface effects neglected (Brueckner 1955*a*; Wada & Brueckner 1956). In this problem the aim is to show that at a fixed density the energy is proportional to the number of particles, and that as the density is varied the energy per particle has a minimum at the observed density of large nuclei, and that this minimum value gives the observed volume energy of large nuclei. The single-particle wave functions are plane waves, and the potential $V$ is diagonal in momentum space (in contrast to the ordinary Hartree potential which is diagonal in configuration space). The independent-particle model state is a 'Fermi gas' state with all the one-particle states filled up to the Fermi momentum $k_F$ which depends only on the density.

Brueckner & Levinson's derivation, and that of Eden, is based on the multiple scattering formalism of Watson (Watson 1953). The proportionality of the energy of nuclear matter of a given density to the number of particles follows at once from the theory provided certain terms which represent several interactions occurring

‡ Author's present address: Institut for Teoretisk Fysik, Blegdamevej 17, Copenhagen.

independently are not present. There is no satisfactory proof of this in the usual presentation of the theory. It has been shown (Brueckner 1955*b*) that the usual perturbation theory for bound states can be recast so that these terms disappear from the first few orders. The present paper proves a new perturbation formula in which these terms are absent and so completely solves this problem of 'unlinked clusters'.

The method of Feynman graphs (Feynman 1949) is used to enumerate the terms of the perturbation series. To derive the 'linked cluster' result it is essential to describe states in a particular way explained later, which is equivalent to treating the independent-particle ground state as a 'vacuum' state. This description then emphasizes the important exclusion effects, and is used to give a derivation of the Hartree–Fock approximation which seems very natural in this context. The ideas of the Brueckner method for dealing with strong potentials are then introduced and are shown to fit naturally into the Feynman graph treatment.

## 2. Time-dependent perturbation theory and Feynman graph analysis

Consider $A$ particles with the Hamiltonian

$$H = \sum_{i=1}^{A} T_i + \sum_{i<j} v_{ij}. \tag{2·1}$$

$T_i$ is the kinetic energy of the $i$th particle and $v_{ij}$ the interaction potential between particles $i$ and $j$. Introduce the one-body potential $V$ which is to be chosen later to give a reasonable independent-particle model of the system. Let $V_i$ be this potential acting on particle $i$. Define

$$H_0 = \sum_i (T_i + V_i), \tag{2·2}$$

$$H_1 = \sum_{i<j} v_{ij} - \sum_i V_i, \tag{2·3}$$

so that

$$H = H_0 + H_1. \tag{2·4}$$

Expansions will be in powers of $H_1$, but the complete series obtained will finally be rearranged so that higher-order terms represent small effects when $V$ is suitably defined. Let the solutions of the one-particle Schrödinger equation

$$(T+V)\psi = E\psi \tag{2·5}$$

be a series of one-particle eigenstates $\psi_n$ with eigenvalues $E_n$. $V$ must be a potential which gives a discrete series of bound eigenstates $\psi_n$. (From now on suffixes $m$, $n$, etc., will refer to these states, not to particles.)

The second-quantized formalism will be used. Let $\eta_n^\dagger$, $\eta_n$ be creation and destruction operations for the state $\psi_n$ with the usual anti-commutation relations. Define matrix elements of $v$ and $V$ by

$$\langle rs \mid v \mid mn \rangle = \int \psi_r^*(1)\,\psi_s^*(2)\,v_{12}\,\psi_m(1)\,\psi_n(2)\,d\tau_1 d\tau_2, \tag{2·6}$$

$$\langle r \mid V \mid m \rangle = \int \psi_r^*(1)\,V_1\,\psi_m(1)\,d\tau_1. \tag{2·7}$$

The matrix element of $v$ defined by (2·6) is not antisymmetrized and corresponds to an interaction in which one particle goes from state $\psi_m$ to state $\psi_r$, while the other goes from state $\psi_n$ to state $\psi_s$. With these definitions

$$H_0 = \sum_n E_n \eta_n^\dagger \eta_n, \tag{2·8}$$

$$\begin{aligned} H_1 &= \Sigma \langle rs \mid v \mid mn \rangle \eta_r^\dagger \eta_s^\dagger \eta_n \eta_m \\ &\quad - \Sigma \langle r \mid V \mid m \rangle \eta_r^\dagger \eta_m. \end{aligned} \tag{2·9}$$

The first sum in (2·9) is over all distinct matrix elements, a matrix element $\langle rs \mid v \mid mn \rangle$ being characterized by the pair of transitions ($\psi_m$ to $\psi_r$) and ($\psi_n$ to $\psi_s$). Thus $\langle sr \mid v \mid nm \rangle$ is not distinct from $\langle rs \mid v \mid mn \rangle$, but $\langle sr \mid v \mid mn \rangle$ is distinct. This way of introducing antisymmetry is the most suitable for graphical representation.

An eigenstate $\Phi$ of $H_0$ is a determinant formed from $A$ of the $\psi_n$ and can be described by enumerating these $A$ one-particle states. A different description is necessary to obtain the results of this paper. It is supposed that $H_0$ has a non-degenerate ground state $\Phi_0$ formed from the lowest $A$ of the $\psi_n$. The proofs of this paper only apply to this case, that is, only to the ground state of a closed-shell nucleus or the ground state of 'nuclear matter'. The states $\psi_n$ occupied in $\Phi_0$ will be called unexcited states, and all the higher states $\psi_n$ will be called excited states. Thus for 'nuclear matter' with a Fermi momentum $k_F$, an unexcited state means one with momentum $k < k_F$, an excited state one with $k > k_F$. An eigenstate $\Phi$ of $H_0$ can now be described by enumerating all the excited states which are occupied, and all the unexcited states which are not occupied. An unoccupied unexcited state is regarded as a 'hole', and the theory will deal with particles in excited states and holes in unexcited states. This treatment is analogous to the theory of positrons, with $\Phi_0$ as the 'vacuum' state. An unexcited state is automatically regarded as occupied and so excluded for other particles, unless a hole in that state is introduced explicitly. Thus the chief effect of the exclusion principle is emphasized by this description. This is the essential difference from the theory of positrons, in which there is symmetry between particles and holes. In this theory the asymmetry between particles and holes is emphasized. To introduce this method formally equations (2·8) and (2·9) are retained, but the interpretation of $\eta_n^\dagger$, $\eta_n$ for unexcited $\psi_n$ is altered. $\eta_n$ will now be the operator creating a hole in state $\psi_n$, $\eta_n^\dagger$, the operator destroying a hole.

The following derivation of the perturbation formula uses time-dependent perturbation theory in the interaction representation. In this way certain of the results needed appear more naturally than in a completely time-independent presentation. Let $\Phi_0$ be the ground state of $H_0$ as described above, assumed to be non-degenerate, and let $\Psi_0$ be the lowest eigenstate of $H$. $\Psi_0$ will be derived from $\Phi_0$ by adiabatically switching on the interaction $H_1$ over the time interval $-\infty$ to 0. For this case of a discrete series of eigenstates with a unique ground state the adiabatic theorem can be proved in the following form (Gell-Mann & Low 1951).

Define
$$H_1(t) = e^{iH_0 t} H_1 e^{-iH_0 t} e^{\alpha t}, \tag{2·10}$$

and let
$$U_\alpha = \sum_{n=0}^{\infty} (-i)^n \int_{0 > t_1 > t_2 \ldots > t_n} H_1(t_1) H_1(t_2) \ldots H_1(t_n) \, dt_1 \ldots dt_n. \tag{2·11}$$

As $\alpha \to 0$ the unitary operator $U_\alpha$ describes the adiabatic process.

Let
$$\Psi_0 = \lim_{\alpha\to 0} \frac{U_\alpha \Phi_0}{\langle \Phi_0 | U_\alpha | \Phi_0 \rangle}. \tag{2·12}$$

By using Feynman graphs this limit will be shown to exist and an explicit expression derived for it. Then the adiabatic theorem states that

$$H\Psi_0 = (E_0 + \Delta E)\Psi_0, \tag{2·13}$$

where
$$H_0 \Phi_0 = E_0 \Phi_0 \tag{2·14}$$

and
$$\Delta E = \langle \Phi_0 | H_1 | \Psi_0 \rangle = \lim_{\alpha \to 0} \frac{\langle \Phi_0 | H_1 U_\alpha | \Phi_0 \rangle}{\langle \Phi_0 | U_\alpha | \Phi_0 \rangle}. \tag{2·15}$$

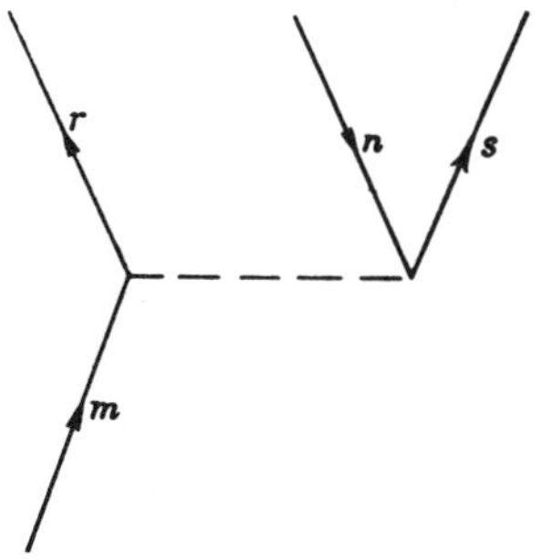


FIGURE 1. In all the graphs the direction of increasing time is upwards.

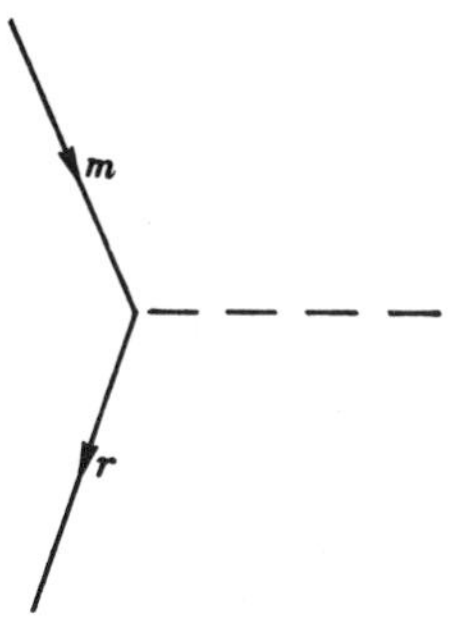


FIGURE 2

The required perturbation formulae for $\Psi_0$ and $\Delta E$ will be obtained on carrying out the time integrations in the expression for the limits in equations (2·12) and (2·15).

$H_1(t)$ is derived from equation (2·9) for $H_1$ by substituting $\eta_n(t)$ for $\eta_n$, where

$$\eta_n(t) = \eta_n \mathrm{e}^{-\mathrm{i}E_n t} \tag{2·16}$$

and then multiplying by $\mathrm{e}^{\alpha t}$. The expression (2·11) for $U_\alpha$ then becomes a sum of products of $v$ and $V$ matrix elements, $\mathrm{e}^{\mathrm{i}Et}$ and $\mathrm{e}^{\alpha t}$ factors and operators $\eta^\dagger$ and $\eta$. Analysis of the products of operators by the same algebra as is used in proving Wick's theorem (Wick 1950) leads to the following expression for $U_\alpha \Phi_0$ as a sum of terms represented by Feynman graphs. Each graph represents a series of $H_1(t)$ interactions. A particle in an excited state is represented by a line in the direction of increasing time. A hole in an unexcited state is represented by a line in the opposite direction. A matrix element $\langle rs | v | mn \rangle$ in $H_1(t)$ is represented as in figure 1. This is for the case in which $\psi_m, \psi_r, \psi_s$ are excited states and $\psi_n$ an unexcited state. It represents an interaction between two particles in which one is scattered from $\psi_m$ to $\psi_r$ while the other jumps from $\psi_n$ into $\psi_s$ leaving a hole in $\psi_n$. With this graph is associated a time factor $\mathrm{e}^{\mathrm{i}(E_r+E_s-E_m-E_n)t}\mathrm{e}^{\alpha t}$. The other combinations of excited and unexcited states $\psi_m \psi_n \psi_r \psi_s$ are represented similarly.

A matrix element $\langle r | V | m \rangle$ is represented as in figure 2. This shows a particle scattered from state $\psi_m$ to $\psi_r$ by $V$, both states unexcited. Initially there was a hole in state $\psi_r$, otherwise the interaction is excluded, and finally there is a hole in state $\psi_m$.

There are further possibilities which do not occur in positron theory. Here the unexcited states are occupied by real particles not explicitly represented in the graphs, but interacting with each other and with the particles represented in the graphs. These particles will be called passive unexcited particles. Their interactions are the most important ones present, and it is these interactions which must be

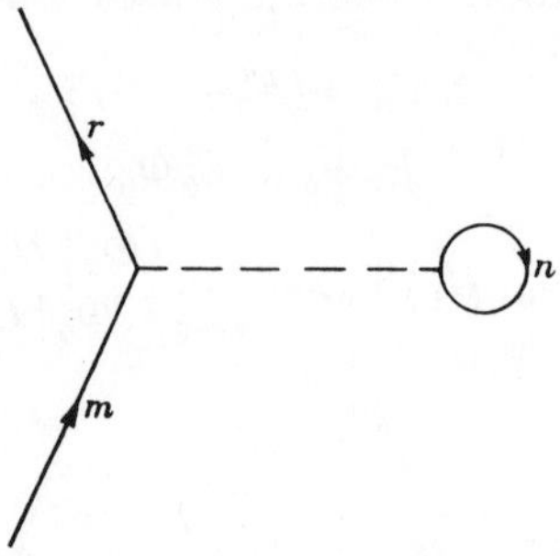


FIGURE 3

allowed for in the choice of $V$. They are represented as in figure 3. This shows a particle scattered from excited state $\psi_m$ to excited state $\psi_r$ by the particle in the unexcited state $\psi_n$ which remains in the same state after the interaction. (In 'nuclear matter' this is 'forward' scattering.) Figure 3 corresponds to a factor

$$\langle rn \mid v \mid mn \rangle \, e^{i(E_r - E_m)t}.$$

The 'exchange' term corresponding to this contains the matrix element $\langle rn \mid v \mid nm \rangle$ and is represented as in figure 4. Finally, figure 5 shows the graphs representing interactions in which only passive unexcited particles take part. The matrix elements are for figure 5 (*a*), $\langle mn \mid v \mid mn \rangle$; for figure 5 (*b*) $\langle mn \mid v \mid nm \rangle$; for figure 5 (*c*) $\langle n \mid V \mid n \rangle$.

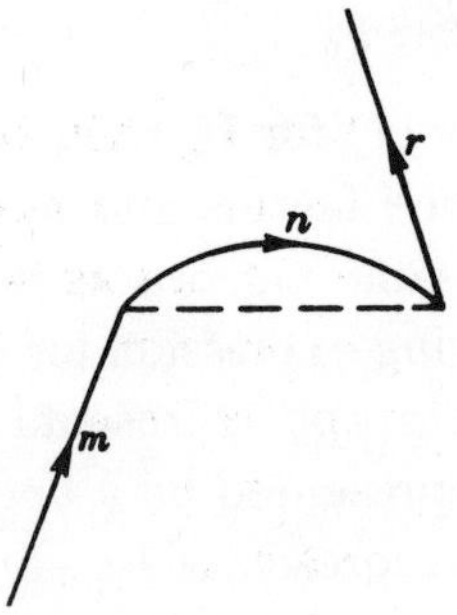


FIGURE 4

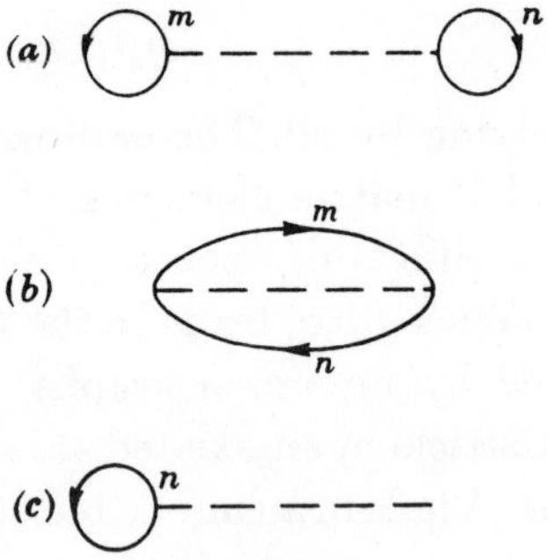


FIGURE 5

The algebra of Wick's theorem now gives the following rule for $U_\alpha \Phi_0$. All distinct graphs starting with no free lines at the bottom, that is, with $\Phi_0$, are drawn. Each such graph consists of a number of open loops of nucleon lines and a number of closed loops. For example, figure 6 contains one open loop and two closed loops. For each graph multiply the $v$ and $V$ matrix elements and the $e^{iEt}$ and $e^{\alpha t}$ factors and a factor $(-1)^{h+l}$, where $h$ is the number of internal hole lines (four in figure 6; the line labelled $m$ is an external line) and $l$ the number of closed loops. A passive unexcited particle loop as in figure 5 (*c*) contributes a plus sign, counting as one hole line and one closed loop, while figure 5(*b*) has a minus sign having two hole lines and one closed loop. Each $V$ matrix element has a minus sign attached since it occurs with a minus sign in $H_1$. Attach the pairs of creation operators corresponding to the external lines at the ends of each open loop with the hole operator to the right ($\eta_r^\dagger \eta_m$ for figure 6). Finally, carry out the time integrations. Then $U_\alpha \Phi_0$ is the sum of all these terms acting on $\Phi_0$. It is important to note that

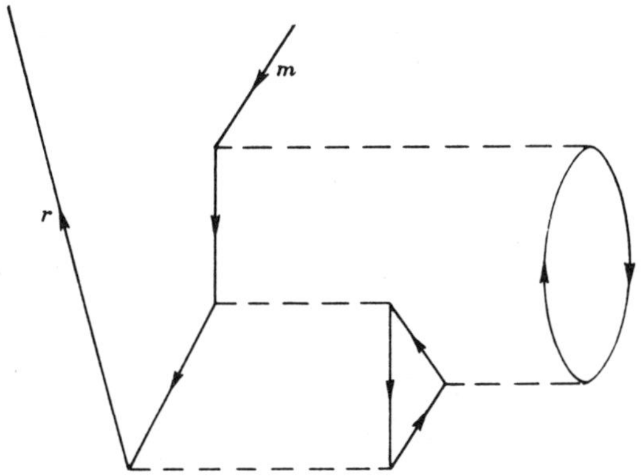


FIGURE 6

the exclusion principle is to be ignored in labelling the graphs. The major effects of exclusion are already taken into account by the 'hole' picture as described above. The rest must not be included if the results of § 3 are to be derived. The algebra of the $\eta$ operators does give this result, which is merely a careful application to this case of the principle that intermediate states need not be anti-symmetrized. In fact all graphs which contradict the exclusion principle are exactly cancelled by the corresponding 'exchange' graphs. However, in § 3 certain graphs will be removed and then this cancellation will no longer occur and the graphs contradicting exclusion will represent important physical effects. This representation is essential for the derivation of the 'linked cluster' result.

## 3. The linked-cluster perturbation formula

Any part of a graph which is completely disconnected from the rest of the graph and which has no external lines attached will be called an unlinked part. In the expression for $U_\alpha \Phi_0$ before the time integrations are carried out the lines of a graph can be labelled independently of each other (this is where it is essential not to have to take exclusion into account), and the factors attached to the interaction lines are independent of each other. Now consider a graph containing unlinked parts and take together with it all the graphs which differ only by having the interactions in the unlinked parts in different positions relative to those in the rest of the graph. The order of the interactions in the two parts separately is kept fixed. Let the times of the interactions in the unlinked part be $t_1, t_2, \ldots, t_n$ and in the rest be $t_1' t_2' \ldots t_m'$, where the order of the two parts separately is given by $0 > t_1 > t_2 > \ldots > t_n$ and $0 > t_1' > t_2' > \ldots > t_m'$. The sum over all the different relative positions of the two parts is obtained by carrying out the time integrations with only these restrictions on the order of the times and so is the product of the expressions obtained from the two parts separately. A graph containing no unlinked parts will be called a linked graph. It follows that $U_\alpha \Phi_0$ is given by the rules of § 2 applied to the sum of linked graphs only, multiplied by a factor given by the sum of all graphs consisting only of unlinked parts. This factor is just what the rules of § 2 give for $\langle \Phi_0 | U_\alpha | \Phi_0 \rangle$. Thus $\Psi_0$ as defined by (2·12) is given by taking the limit $\alpha \to 0$ in the sum of linked graphs only.

The result of carrying out the time integrations in this sum may be written as

$$\Psi_0 = \lim_{\alpha \to 0} \sum_L \frac{1}{E_0 - H_0 + in\alpha} H_1 \cdots \frac{1}{E_0 - H_0 + 2i\alpha} H_1 \frac{1}{E_0 - H_0 + i\alpha} H_1 \Phi_0. \tag{3·1}$$

$\sum_L$ means that the terms are to be enumerated by the linked graphs described above. $\Phi_0$ cannot occur as an intermediate state in a linked graph as the part of the graph below that intermediate state would be an unlinked part. Since all other

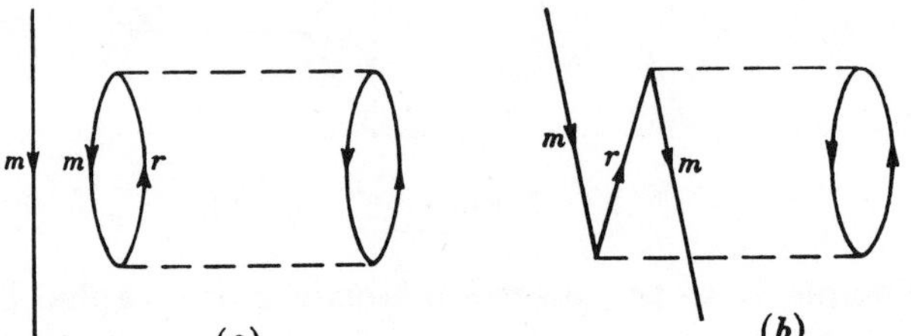


FIGURE 7

intermediate states have energies greater than $E_0$ (this is where the limitation to non-degenerate ground states is useful), the limit in (3·1) can be taken by putting $\alpha = 0$ as no zero energy denominators can occur. The final result can then be written

$$\Psi_0 = \sum_L \left( \frac{1}{E_0 - H_0} H_1 \right)^n \Phi_0. \tag{3·2}$$

The energy shift $\Delta E$ is given by (2·15), and using the same arguments as for $U_\alpha \Phi_0$,

$$\Delta E = \sum_L \left\langle \Phi_0 \middle| H_1 \left( \frac{1}{E_0 - H_0} H_1 \right)^n \middle| \Phi_0 \right\rangle, \tag{3·3}$$

where now $\sum_L$ means summed over all connected graphs leading from $\Phi_0$ to $\Phi_0$, that is, with no external lines. (3·2) and (3·3) are the linked-cluster perturbation formulae. They differ from the usual bound state perturbation formula by having $E_0$ in the denominator instead of the usual $E_0 + \Delta E$. This difference is compensated by the different enumeration of terms, that is, by summing only over linked graphs and by ignoring exclusion as described in § 2.

A typical graph contradicting the exclusion principle is figure 7 (*b*). Before the unlinked parts were removed this was cancelled by figure 7 (*a*) which has the same matrix elements and an extra minus sign. Figure 7 (*a*) represents an interaction of the passive particle in the unexcited state $\psi_m$. Many repetitions of figure 7 (*b*) combine to give the modification of the energy of state $\psi_m$ due to this interaction.

These linked-cluster expansions can be derived without using time-dependent theory. $\Psi_0$ can be defined to be given by (3·2). It then follows that

$$(E_0 - H_0)\Psi_0 = \sum_L H_1 \left(\frac{1}{E_0 - H_0} H_1\right)^n \Phi_0 \tag{3·4}$$

and

$$H_1 \Psi_0 = H_1 \sum_L \left(\frac{1}{E_0 - H_0} H_1\right)^n \Phi_0. \tag{3·5}$$

The right-hand side of (3·5) is given by those graphs which are linked when the last $H_1$ is removed. Some care is needed to prove this, since Wick's theorem does not immediately apply to the time-integrated expression (3·4). Subtracting (3·4) from (3·5) gives

$$(H - E_0)\Psi_0 = \Sigma' H_1 \left(\frac{1}{E_0 - H_0} H_1\right)^n \Phi_0, \tag{3·6}$$

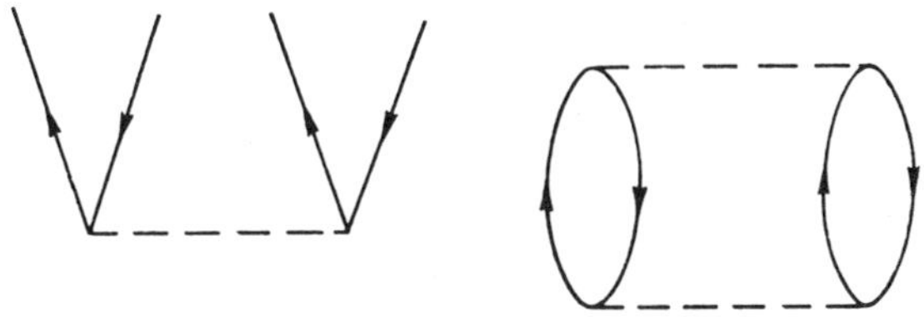

FIGURE 8

where $\Sigma'$ means summed over all graphs containing an unlinked part but which are linked when the last $H_1$ line is removed. Such graphs must be of the type shown in figure 8. Now the last $H_1$ line in the unlinked part may be kept fixed and a sum taken over the different positions of the rest of the unlinked part relative to the rest of the graph. By using algebraic identities on the energy denominators which are equivalent to the separation of the time integrations in the time-dependent proof it can be shown that the right-hand side of (3·6) is equal to the product of $\Psi_0$ with the sum of all connected closed graphs, that is, with $\Delta E$ as defined by (3·3). Then (3·6) gives

$$H\Psi_0 = (E_0 + \Delta E)\Psi_0, \tag{3·7}$$

the required result.

This method of proof has one advantage over the other in that it does not use time-dependent methods to prove a time-independent result. However, the time-dependent proof gives the easiest way of enumerating the terms correctly and of combining the contributions of different positions of unlinked parts. The adiabatic theorem used can be strictly proved under the conditions of this paper. The time-independent method has been used by the author to extend the results to excited and degenerate states.

## 4. Choice of $V$: the Hartree–Fock method

The simplest way to choose $V$ is to make it allow for the first-order interactions with passive unexcited particles. This is done by making the graph parts in figure 9 cancel, that is, by defining

$$\langle r \mid V \mid m\rangle = \sum_n \{\langle rn \mid v \mid mn\rangle - \langle rn \mid v \mid nm\rangle\}. \tag{4·1}$$

The sum is over all unexcited states $\psi_n$. The states $\psi_n$ are determined by

$$(T+V)\,\psi_n = E_n \psi_n. \tag{4·2}$$

(4·1) and (4·2) are the Hartree–Fock self-consistent equations.

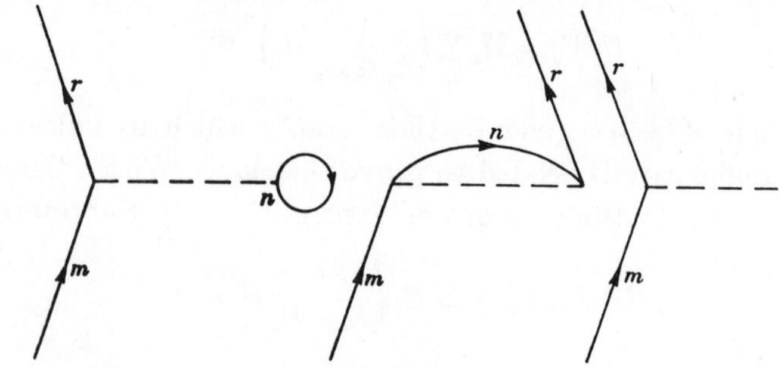


FIGURE 9

This definition ensures the complete disappearance of the $V$ interaction and the interactions with passive unexcited states from all graphs except the connected closed parts in figure 5. These represent the first-order terms in $\Delta E$. Figure 5($c$) contributes $-\sum_n \langle n \mid V \mid n \rangle$, while figures 5($a$) and ($b$) contribute

$$\tfrac{1}{2}\sum_{m,n}\{\langle mn \mid v \mid mn\rangle - \langle mn \mid v \mid nm\rangle\} = \tfrac{1}{2}\sum_n \langle n \mid V \mid n\rangle \tag{4·3}$$

when summed over all distinct possibilities. Also,

$$E_0 = \sum_n \{\langle n \mid T \mid n\rangle + \langle n \mid V \mid n\rangle\}. \tag{4·4}$$

Thus, to the first order in $v$,

$$E = E_0 + \Delta E = \sum_n \langle n \mid T \mid n\rangle + \tfrac{1}{2}\sum_n \langle n \mid V \mid n\rangle. \tag{4·5}$$

This factor of $\frac{1}{2}$ is familiar in the Hartree–Fock method.

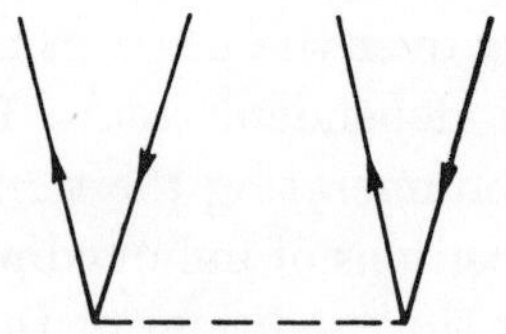

FIGURE 10

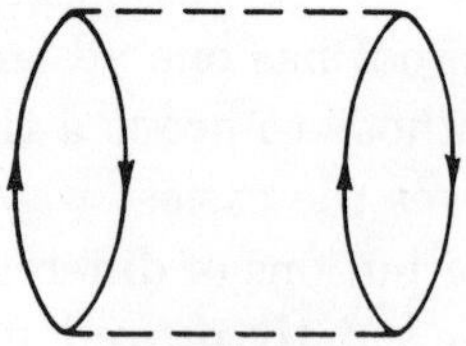

FIGURE 11

The higher-order corrections to $E$ are given by the sum (3·3) over all connected graphs with no external lines and with no $V$ interactions and no interactions with passive unexcited particles. The wave function $\Psi_0$ is given by the sum (3·2) again without the above interactions. The expression for $\Psi_0$ contains terms which are the product of many factors represented by graphs like figure 10. The result is that $\Phi_0$ is only a very small component of $\Psi_0$. (Note that $\Psi_0$ is normalized to $\langle \Phi_0 \mid \Psi_0 \rangle = 1$.) However, the corresponding correction to the energy can contain the factor represented by figure 11 once only. Bethe (1956) has shown that the exclusion principle which limits the particles in excited states to states with momentum $> k_F$ can make this correction fairly small even for strong potentials for the values of $k_F$ of interest. Thus the Hartree–Fock method can give the energy quite well even for strong potentials. This is a quantitative version of the old argument that strong interactions would be inhibited by the exclusion principle. It applies to the energy but not to the wave function. There are certainly strong correlations between nucleons in a nucleus and the Brueckner theory can be used to explain them (Brueckner, Eden & Francis 1955).

## 5. The Brueckner theory

The nucleon-nucleon potential very probably has a steep repulsive core at small distances. (This will certainly ensure saturation but a proper theory is needed to obtain an energy minimum at the observed nuclear density.) For this $v$ it is clearly impossible to choose $V$ by the Hartree–Fock method, as the matrix elements of

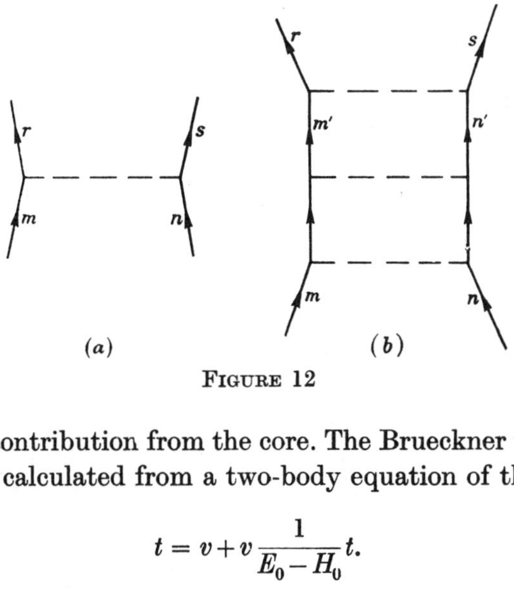


(a) (b)

Figure 12

$v$ will have a large contribution from the core. The Brueckner theory replaces $v$ by a reaction matrix $t$ calculated from a two-body equation of the type

$$t = v + v\frac{1}{E_0 - H_0}t. \tag{5·1}$$

The idea is to derive $V$ from $t$ instead of from $v$. Since $H_0$ contains $V$, $V$ occurs in the energy denominator so that there is a further self-consistency requirement in addition to the Hartree–Fock condition on the wave functions. In fact, for 'nuclear matter' the Hartree–Fock self-consistency disappears since the wave functions must be plane waves. Brueckner (1955*a*) has shown that this new self-consistency is important.

The procedure in terms of graphs is as follows. Corresponding to any graph with a single $v$ line in a certain position as in figure 12 (*a*), there are more complicated ones in which figure 12 (*a*) is replaced by the 'ladder' graph of figure 12 (*b*). In the intermediate states of figure 12 (*b*) both particles are in excited states. The sum of all such parts is given by an integral equation of the type (5·1). When figure 12 (*b*) occurs as part of a larger graph the energy denominator for the intermediate state containing $\psi_{m'}\psi_{n'}$ is

$$-E_{m'} - E_{n'} - E_R = E_m + E_n - E_{m'} - E_{n'} - \delta E, \tag{5·2}$$

where $E_R$ is the excitation energy of the other particles present while the interaction represented in 12 (*b*) occurs and $\delta E$ is the excitation energy of the complete intermediate state at the beginning of the interaction. The excitation energy of a state is the sum of the energies $E_n$ of occupied excited states minus the sum of the energies of unexcited states in which there are holes. The integral equation for the sum of the terms represented by figure 12 is then

$$\langle rs \mid t \mid mn \rangle = \langle rs \mid v \mid mn \rangle + \sum_{m'n'} \frac{\langle rs \mid v \mid m'n' \rangle \langle m'n' \mid t \mid mn \rangle}{E_m + E_n - E_{m'} - E_{n'} - \delta E}, \qquad (5 \cdot 3)$$

where the sum is over $\psi_{m'}, \psi_{n'}$ excited states only. The solution is a matrix $t(\delta E)$ which can be used to replace $v$ and which is finite even for a repulsive core potential.

A graph will be called irreducible if it contains no 'ladders' of the type of figure 12 (*b*). A sequence of $v$ interactions as in figure 12 (*b*) only forms a 'ladder' if all the intermediate states are excited and if there are no other interactions in other

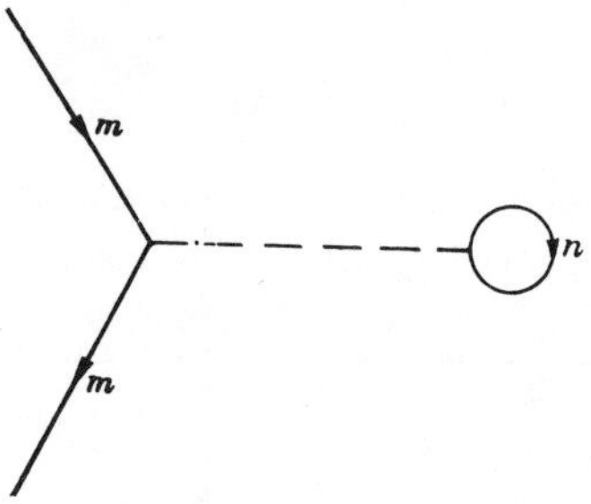


FIGURE 13

parts of the graph between the ends of the ladder. All graphs can be obtained by substituting independently 'ladders' for each $v$ line in the irreducible graphs. The terms of the linked cluster expansion can thus be grouped together so that each $v$ matrix element is replaced by a matrix element of $t(\delta E)$. $\delta E$ is the excitation energy of the intermediate state to the right of the matrix element in the series (below it in the graph). The sums must now be taken over linked irreducible graphs only.

Figure 7 (*b*) is an important type of ladder graph and is absorbed into the $t$ matrix element $\langle mn \mid t \mid mn \rangle$ represented in figure 13. Note that (5·3) is not antisymmetrized. The ladder graph in which the lines of figure 12 (*b*) cross over is counted in $\langle sr \mid t \mid mn \rangle$.

$V$ can now be defined to cancel the $t$-interactions with passive unexcited states, that is, by (4·1) with $v$ replaced by $t$. However, the cancellation cannot be complete because of the dependence of $t$ on $\delta E$. (The procedure in this problem contrasts with that in field theory in which the time ordering and the dependence of one part of a graph on another are completely removed by introducing an extra energy variable for each particle. This does not seem appropriate here.) The best that can be done is to choose some average value of $\delta E$ appropriate to the matrix element of $t$ being evaluated.

Equation (4·5) is replaced by the following expression for the energy to first order in $t$:

$$E = \sum_n \langle n \mid T \mid n \rangle + \tfrac{1}{2} \sum_{m,n} \{\langle mn \mid t(0) \mid mn \rangle - \langle nm \mid t(0) \mid mn \rangle\}. \qquad (5 \cdot 4)$$

The $\langle n \mid V \mid n \rangle$ in $E_0$ is cancelled by the term represented by figure 5 (*c*) whatever the definition of $V$. The second term in (5·4) will equal $\frac{1}{2} \sum_n \langle n \mid V \mid n \rangle$ only if $\langle n \mid V \mid n \rangle$ is derived from $t$ with $\delta E = 0$. This is the most straightforward choice for the diagonal elements of $V$ between unexcited states.

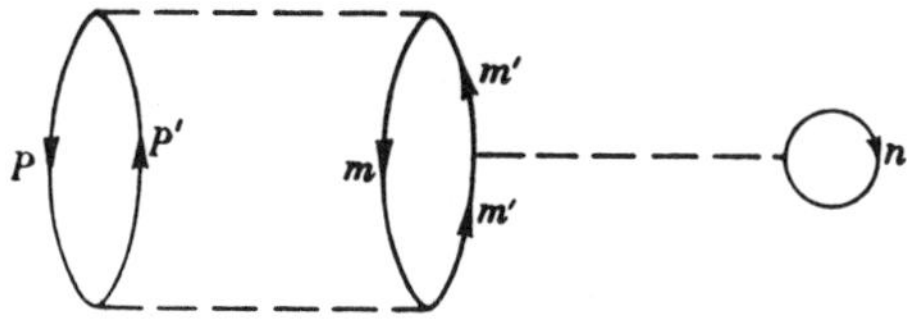


FIGURE 14

Figure 14 represents a term in the energy given by

$$\langle mp \mid t(\delta E) \mid m'p' \rangle \left( -\frac{1}{\delta E} \right) \langle m'n \mid t(\delta E) \mid m'n \rangle \left( -\frac{1}{\delta E} \right) \langle m'p' \mid t(0) \mid mp \rangle, \tag{5·5}$$

where

$$\delta E = E_{p'} - E_p + E_{m'} - E_m. \tag{5·6}$$

An average of this $\delta E$ used in the definition of $\langle m' \mid V \mid m' \rangle$ will ensure as much cancellation of this term as possible. For 'nuclear matter' the conservation of momentum limits the possible values of $\psi_p \psi_{p'}$ and $\psi_m$ given $\psi_{m'}$.

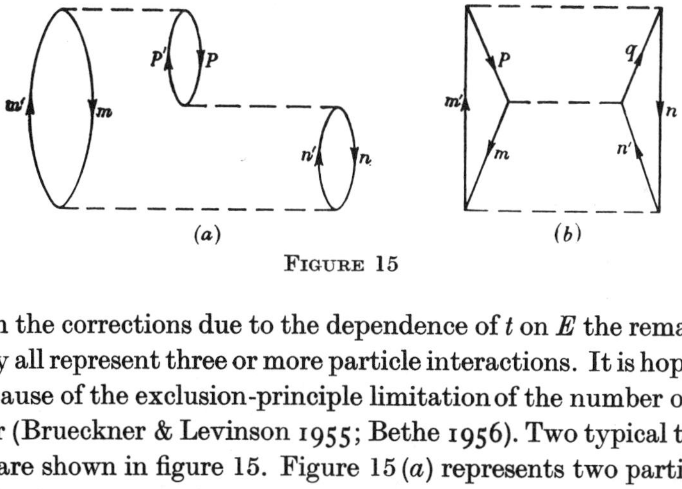


FIGURE 15

Apart from the corrections due to the dependence of $t$ on $E$ the remaining graphs for the energy all represent three or more particle interactions. It is hoped that these are small because of the exclusion-principle limitation of the number of states to be summed over (Brueckner & Levinson 1955; Bethe 1956). Two typical three-particle interactions are shown in figure 15. Figure 15 (*a*) represents two particles jumping from states $\psi_m \psi_n$ into excited states $\psi_{m'} \psi_{n'}$. Then the particle in $\psi_{n'}$ falls back into $\psi_n$ while another particle jumps from $\psi_p$ into $\psi_{p'}$. Finally the particles in $\psi_{p'}$ and $\psi_{m'}$ interact and fall back. The corresponding matrix elements are

$$\langle pm \mid t \mid p'm' \rangle \langle np' \mid t \mid n'p \rangle \langle m'n' \mid t \mid mn \rangle. \tag{5·7}$$

Figure 15 (*b*) represents two particles jumping from states $\psi_m \psi_n$ into states $\psi_{m'} \psi_{n'}$. The particle in $\psi_{n'}$ then interacts with the particle in the occupied state $\psi_p$. The particle in $\psi_{n'}$ is scattered into $\psi_q$ while that in $\psi_p$ jumps into the hole in $\psi_m$ leaving a hole in $\psi_p$. Finally the particles in $\psi_{m'}$, $\psi_q$ fall back into the holes in $\psi_p$, $\psi_n$. The corresponding matrix elements are

$$\langle np \mid t \mid qm' \rangle \langle qm \mid t \mid n'p \rangle \langle m'n' \mid t \mid mn \rangle. \tag{5·8}$$

I am very grateful to Professor H. A. Bethe and Dr R. J. Eden for much advice and discussion on this problem and for very helpful criticism of the first draft of this paper. I also thank the Department of Scientific and Industrial Research for a maintenance grant.

## References

Bethe, H. A. 1956 *Phys. Rev.* **103**, 1353.
Brueckner, K. A. 1955*a* *Phys. Rev.* **97**, 1353.
Brueckner, K. A. 1955*b* *Phys. Rev.* **100**, 36.
Brueckner, K. A., Eden, R. J. & Francis, N. C. 1955 *Phys. Rev.* **98**, 1445.
Brueckner, K. A. & Levinson, C. A. 1955 *Phys. Rev.* **97**, 1344.
Eden, R. J. 1956 *Proc. Roy. Soc.* A, **235**, 408.
Feynman, R. P. 1949 *Phys. Rev.* **76**, 749.
Gell-Mann, M. & Low, F. 1951 *Phys. Rev.* **84**, 350 (Appendix).
Wada, W. W. & Brueckner, K. A. 1956 *Phys. Rev.* **103**, 1008.
Watson, K. M. 1953 *Phys. Rev.* **89**, 575.
Wick, G. C. 1950 *Phys. Rev.* **80**, 268.

VIII-3 Reprinted from *Physical Review* **131**, 684 (1963)

# Correlation Effects in Atoms

HUGH P. KELLY
*Lawrence Radiation Laboratory, University of California, Berkeley, California*
(Received 5 November 1962)

The applicability of many-body perturbation theory in calculating electron correlations in atoms has been investigated. It was found that, in order for results to agree well with experiment, certain classes of diagrams, representing terms of the perturbation series, must be included to many orders. The higher order diagrams were found to be more important in calculating correlations among the outer electrons than among inner electrons. Methods for summing various classes of these diagrams were found. The total correlation energy in beryllium was calculated by perturbation theory, by the use of an IBM 7090 computer, to be $-2.48\pm0.11$ eV. The value obtained from experiment is $-2.59$ eV. The calculated correlation energies for the different electronic shells were found to be $-1.19$ eV among $2s$ electrons, $-1.15$ eV among $1s$ electrons, and $-0.135$ eV between the $1s$ and $2s$ shells.

## I. INTRODUCTION

THE Rayleigh-Schrödinger perturbation theory for many-body systems has been described by Goldstone[1] and is often referred to as Goldstone's perturbation theory. Goldstone has shown that in this theory certain terms containing "unlinked clusters" may be omitted and that the terms for the energy beyond first order are then proportional to the number of particles present. In this description of perturbation theory, annihilation and creation operators are used to account for Fermi-Dirac statistics, and the terms of the expansion are represented by diagrams or graphs similar to the Feynman diagrams of field theory.

Although this perturbation method has been used mainly in solving problems with an infinite number of identical fermions, it is, in principle, quite applicable to other problems and, in particular, to the atomic problem, where a finite number of identical fermions move in the potential field of a nucleus.

This problem is, of course, different from that of an infinite number of fermions, and it is not surprising that terms of the perturbation series which are important in the infinite electron gas problem, for example, are of less importance in the atomic problem, and that other terms assume greater importance. One of the greatest differences in the atomic problem is that the basis states for the perturbation expansion are no longer plane-wave states but are solutions of the Hartree-Fock equations for the atom and are eigenstates of orbital angular momentum. One consequence of this is that the infinite diagrams which were summed by Gell-Mann and Brueckner[2] in the problem of the (infinite) dense electron gas are now finite and are, in many cases, extremely small.

Another major difference is the importance in the atomic problem of terms in which the exclusion principle is violated in intermediate states. Goldstone[1] has stressed that such terms exist. In the following discussion these terms are denoted as exclusion-principle-violating (EPV) terms.

[1] J. Goldstone, Proc. Roy. Soc. (London) **A239**, 267 (1957).
[2] M. Gell-Mann and K. Brueckner, Phys. Rev. **106**, 364 (1957).

The purpose of this investigation is to determine the corrections to the Hartree-Fock wave functions and energies for atoms by means of Goldstone's theory. Errors in the Hartree-Fock (HF) wave functions arise from the smearing out or averaging of electronic wave functions in the HF equation. As a result the repulsion between electrons is not accounted for properly. The correlation corrections rectify this deficiency; and the correlation energy is negative because of electron repulsion. This repulsion is not so important for electrons with parallel spins because they are already kept apart by the exclusion principle incorporated into the HF equation by the antisymmetrization of the total wave function. In the perturbation expansion there are exchange diagrams that tend to cancel direct diagrams when the spins of the two interacting electrons are parallel.

The correction for the energy is called the correlation energy and is the difference between the exact eigenvalue $E$ of the Schrödinger equation for $N$ electrons,

$$\left[\sum_{i=1}^{N}\left(-\frac{\hbar^2\nabla_i^2}{2m}-\frac{Ze^2}{r_i}\right)+\sum_{i<j}^{N}\frac{e^2}{r_{ij}}\right]\psi=E\psi, \tag{1.1}$$

and the exact HF energy, $E_{\mathrm{HF}}$. For atoms of low atomic number $Z$ the difference between $E$ and the total experimental energy $E_{\mathrm{exp}}$ is quite small and is due to relativistic effects, which may be estimated. This enables one to compare the calculated correlation energy with the experimental value, as is done for the beryllium atom in Sec. V.

Section II is devoted to a review of many-body perturbation theory. In Sec. III the exclusion-principle-violating diagrams are discussed in detail, and formulas summing these diagrams are derived. Section IV deals with the explicit application of the perturbation theory to atomic systems. The basis states for the expansion are considered in detail, and formulas for the sums over intermediate states are given. In Sec. V the total correlation energy for Be is calculated and found to be in good agreement with the experimental result. The total correlation energy is the sum of the correlations among the various electrons of the atom; correlations for both the 1s and 2s shells of Be and for the 1s-2s shell interactions are given.

Correlation energies in boron were calculated by use of a screened Coulomb potential, rather than the correct HF potential, to obtain the set of basis states. These results are described in Sec. VI along with an approximate formula for the total correlation energy as a function of the atomic number. The last two sections contain the conclusions (Sec. VII) and Appendix.

## II. REVIEW OF GOLDSTONE'S PERTURBATION THEORY

### A. Time-Dependent Perturbation Theory

The problem now considered is that of a system of $N$ identical fermions interacting through two-body potentials. The potential between particles $i$ and $j$ is written $v_{ij}$. There may be also one-body potentials present. The total Hamiltonian for the system is

$$H=\sum_{i=1}^{N} T_i+\sum_{i<j}^{N} v_{ij}. \tag{2.1}$$

The symbol $T_i$ is the sum of the kinetic energy for the $i$th particle and all one-body potentials acting on it. For example, in the atomic problem the one-body potential for the $i$th electron is its potential energy due to the presence of the nucleus.

The true ground state of the system is $\psi_0$, given by

$$H\psi_0=(E_0+\Delta E)\psi_0. \tag{2.2}$$

The effect of the $N$ interacting particles may be approximated by a single-particle potential $V$, and the true system is approximated by an unperturbed system $\Phi_0$ with a total Hamiltonian

$$H_0=\sum_{i=1}^{N} (T_i+V_i), \tag{2.3}$$

and the eigenvalue equation[3]

$$H_0\Phi_0=E_0\Phi_0. \tag{2.4}$$

The potential $V$ is required to be Hermitian so that the single-particle wave functions $\phi_n$, which are solutions of

$$(T+V)\phi_n=\epsilon_n\phi_n, \tag{2.5}$$

constitute an orthonormal set. The state $\Phi_0$ is a determinant formed from the $N$ solutions of Eq. (2.5) that are lowest in energy. It is assumed that this state is nondegenerate. The states occupied in $\Phi_0$ are called unexcited states. An unoccupied unexcited state is called a hole, and an occupied excited state is called a particle.

[3] There are essentially three energies in this problem: the unperturbed energy $E_0$, the exact energy of the system $E_0+\Delta E$, and the Hartree-Fock energy $E_{\text{HF}}$, which is $E_0$ plus the first-order energy term when the potential $V$ is chosen as the Hartree-Fock potential of Eq. (2.14).

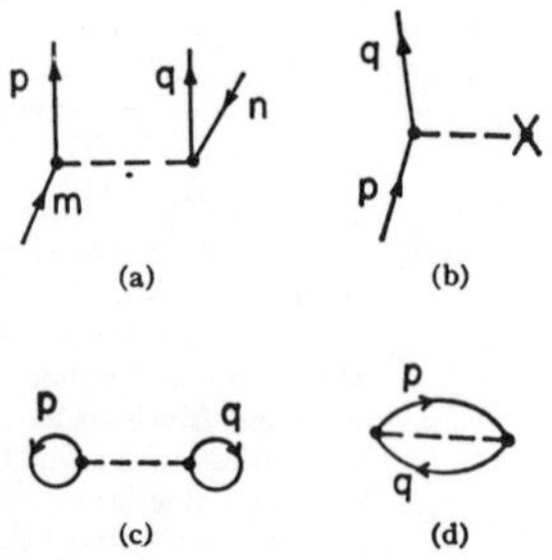


FIG. 1. Diagrams associated with matrix elements: (a) $\langle pq|v|mn\rangle$. Particles in the excited state $\phi_m$ and the unexcited state $\phi_n$ interact through $v$ and scatter into excited states $\phi_p$ and $\phi_q$, leaving a hole in $\phi_n$. (b) $-\langle q|V|p\rangle$. A particle in an excited state $\phi_p$ is scattered by the potential $V$ to the excited state $\phi_q$. (c) $\langle pq|v|pq\rangle$. (d) $\langle pq|v|qp\rangle$. Diagrams (c) and (d) represent the interactions of passive unexcited particles.

In the matrix element representation, one has

$$H_0=\sum_n \epsilon_n\eta_n^\dagger\eta_n \tag{2.6}$$

and

$$H'=H-H_0=\sum_{p,q,m,n} \langle pq|v|mn\rangle\eta_p^\dagger\eta_q^\dagger\eta_n\eta_m - \sum_{p,m} \langle p|V|m\rangle\eta_p^\dagger\eta_m. \tag{2.7}$$

The $\eta^\dagger$ and $\eta$ operators satisfy the usual Fermi-Dirac anticommutation relations. The sums run over all states, unexcited and excited. In the summation, $\sum_{p,q,m,n}$, only distinct matrix elements are included. For example, $\langle pq|v|mn\rangle$ is not distinct from $\langle qp|v|nm\rangle$; however, $\langle pq|v|nm\rangle$ is distinct.

The true ground state is

$$\psi_0=\lim_{\alpha\to 0}\frac{U_\alpha(0)\Phi_0}{\langle\Phi_0|U_\alpha(0)|\phi_0\rangle}, \tag{2.8}$$

with

$$\Delta E=\langle\Phi_0|H'|\psi_0\rangle, \tag{2.9}$$

where

$$U_\alpha(t)=\sum_{n=0}^{\infty}(-i)^n\int_{t>t_1>t_2\cdots>t_n} H'(t_1)\cdots \times H'(t_n)dt_1\cdots dt_n \tag{2.10}$$

and

$$H'(t)=e^{iH_0t}H'e^{-iH_0t}e^{\alpha t}. \tag{2.11}$$

By use of Wick's theorem,[4] $U_\alpha(0)\Phi_0$ becomes a sum of terms that may be represented by Feynman diagrams or graphs. A particle in an excited state is represented by a line directed upwards, and a hole in an unexcited state is represented by a line directed downwards. The direction of increasing time is upwards. See Fig. 1.

The rules for obtaining $U_\alpha(0)\Phi_0$ are:

(i) Draw all distinct diagrams with no free lines at the bottom.

(ii) For each diagram multiply the $v$ and $V$ matrix elements and the appropriate $e^{iEt}$ and $e^{\alpha t}$ factors and

[4] G. C. Wick, Phys. Rev. 80, 268 (1950).

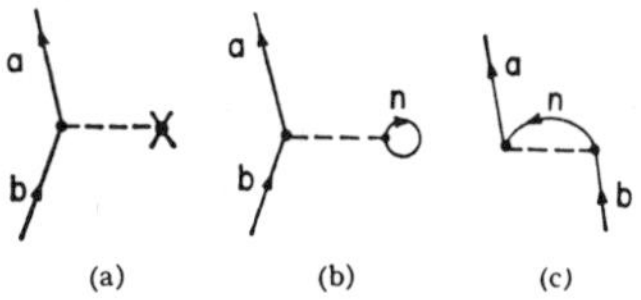


FIG. 2. Diagrams that sum to zero if the Hartree-Fock potential is used to obtain the single-particle states.

include a factor $(-1)^{h+l}$, where $h$ is the number of internal-hole lines and $l$ is the number of closed loops.

(iii) Note that each $V$ has a minus sign with it.

(iv) Attach the pairs of $\eta^\dagger$ and $\eta$ operators corresponding to the free (external) lines of each open loop with $\eta^\dagger$ to the left of $\eta$.

(v) Carry out the time integrations and note that the diagrams are time ordered. Goldstone stresses that the exclusion principle is to be ignored in labeling the diagrams.

### B. Factorization of Diagrams

Goldstone defines the unlinked part of a diagram as any part completely disconnected from the rest and with no external lines attached. A diagram with no unlinked parts is called linked.

For a given diagram with an unlinked part there are other diagrams in which the interactions of the unlinked parts are in different positions in time relative to the interactions of the linked part. If all these diagrams are summed, the result is that the time integrations of the linked and unlinked parts may be carried out independently. This results in $U_\alpha(0)\Phi_0$ expressed as the product of a sum of linked terms and a sum of unlinked terms. The expression $\langle\Phi_0|U_\alpha(0)|\Phi_0\rangle$ is the sum of all possible unlinked terms and cancels the same factor in $U_\alpha(0)\Phi_0$.

After carrying out the time integrations, one obtains

$$\psi_0=\sum_L\left(\frac{1}{E_0-H_0}H'\right)^n\Phi_0, \tag{2.12}$$

where $\sum_L$ means that only linked diagrams are to be included. Also,

$$\Delta E=\sum_{L'}\left\langle\Phi_0\middle|H'\left(\frac{1}{E_0-H_0}H'\right)^n\middle|\Phi_0\right\rangle, \tag{2.13}$$

where the indicated sum is over connected diagrams with no external lines. The potential $V$ is now chosen to be the Hartree-Fock potential. Thus,

$$\langle a|V|b\rangle=\sum_{n=1}^{N}\left(\langle an|v|bn\rangle-\langle an|v|nb\rangle\right). \tag{2.14}$$

The sum is over the $N$ unexcited states. The sum of the diagrams of Fig. 2 is now zero except for unexcited states $a=b$. When $a=b$, unexcited, the diagrams of Fig. 2 represent the first-order energy term

$$\langle\Phi_0|H'|\Phi_0\rangle=-\tfrac{1}{2}\sum_{n=1}^{N}\langle n|V|n\rangle. \tag{2.15}$$

To first order, then, the energy is given by

$$E_{\mathrm{HF}}=\sum_{n=1}^{N}\left(\langle n|T|n\rangle+\tfrac{1}{2}\langle n|V|n\rangle\right), \tag{2.16}$$

which is the Hartree-Fock result.

## III. EXCLUSION-PRINCIPLE-VIOLATING DIAGRAMS

### A. EPV Diagrams Arising from the Linked-Cluster Formula

In applying the perturbation theory to the beryllium atom it was found that certain terms that violate the exclusion principle in intermediate states are numerically significant. In fact, they are the most important terms in third- and higher order energy diagrams and must be considered if good numerical values for the correlation energies are desired. These exclusion-principle-violating (EPV) diagrams are important in finite systems, but in extended systems where the basis states are plane waves these diagrams may be neglected as the volume and density approach infinity.

The EPV diagrams arise from two sources. The first source to be discussed is the factorization of the $U_\alpha(0)\Phi_0$ diagrams into products of linked and unlinked diagrams. The second source of EPV diagrams is the use of the Hartree-Fock potential, and this is discussed in Sec. IIIB.

Goldstone[1] has shown that in order to factor a given diagram with an unlinked part into the product of the linked part times the unlinked part, it is necessary that the interactions of the unlinked part have all possible time orderings relative to the linked part. For example, consider the diagrams shown in Fig. 3. In order to make the factorization shown in Fig. 3(c), it is necessary to include the diagram shown in Fig. 3(b) as well as that of Fig. 3(a). The diagram of Fig. 3(a) vanishes unless the states $p$, $q$, $r$, and $s$ are all included in the unperturbed state $\Phi_0$. If, however, $p$, $q$, $r$, and $s$ are not all different, then the diagram of Fig. 3(b) vanishes because of the exclusion principle. If $p=q$ and $r=s$, then the diagram of Fig. 3(b) may be added and subtracted and the factorization may still be made, but a negative diagram remains after the factorization has taken place. Goldstone states that such diagrams arise from the application of Wick's theorem and that after the factorization, exclusion-principle-violating diagrams that represent important physical effects remain.

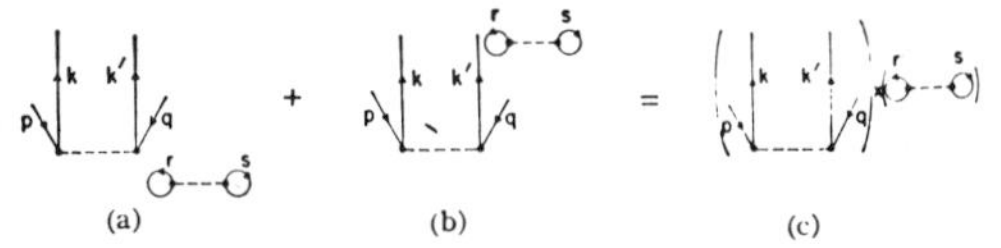


FIG. 3. Factorization of diagrams into the product of a linked and an unlinked diagram. Diagrams (a) and (b) combine to give the product shown in (c).

At this point it is instructive to consider the contraction of the annihilation and creation operators associated with the diagram of Fig. 3(b) when $p=r$ and $q=s$:

$$\langle pq|v|pq\rangle\langle kk'|v|pq\rangle\eta_p{}^\dagger\eta_q{}^\dagger\eta_q\eta_p\eta_k{}^\dagger\eta_{k'}{}^\dagger\eta_q\eta_p. \qquad (3.1)$$

These operators may be contracted in four different ways and the resulting diagrams are shown in Fig. 4. Since each internal-hole line contributes a factor $(-1)$ and each closed loop a factor $(-1)$, the diagrams of Figs. 4(a) and 4(d) are positive and those of Figs. 4(b) and 4(c) are negative. It should be remembered that in this theory contractions are defined as follows[1]:

$$C(\eta_n\eta_n{}^\dagger)=1;\quad \phi_n \text{ excited}$$

and

$$C(\eta_n{}^\dagger\eta_n)=1;\quad \phi_n \text{ unexcited.}$$

All other contractions are zero.

Diagrams (b) and (c) of Fig. 4 are cancelled because of the Hartree-Fock definition of the potential $V$. That is, there are diagrams resulting from factorizations similar to that of Fig. 3 that cancel Figs. 4(b) and 4(c). These additional factorizations have unlinked parts with interactions of states $\phi_p$ and $\phi_q$ with the potential $V$ and with the other unexcited states. The diagram of Fig. 4(a) is used in the factorization of the unlinked part and only the diagram of Fig. 4(d) remains. It is a linked diagram with one hole-hole interaction. It is readily seen that this diagram violates the exclusion principle because holes are created in the states $\phi_p$ and $\phi_q$, then one more hole is created in each of these states, and then the first two holes are filled. The exclusion principle is violated when the second two holes are created.

A similar examination of the next order in the perturbation expansion yields a diagram with two hole-hole interactions. The expression for the first-order linked diagram with unexcited states $\phi_p$ and $\phi_q$ is

$$\sum_{k,k'}\frac{1}{(E_0-H_0)}\langle kk'|v|pq\rangle\eta_k{}^\dagger\eta_p\eta_{k'}{}^\dagger\eta_q. \qquad (3.2)$$

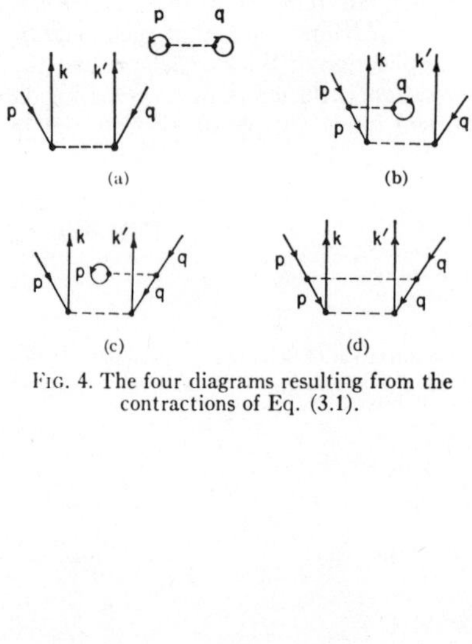


FIG. 4. The four diagrams resulting from the contractions of Eq. (3.1).

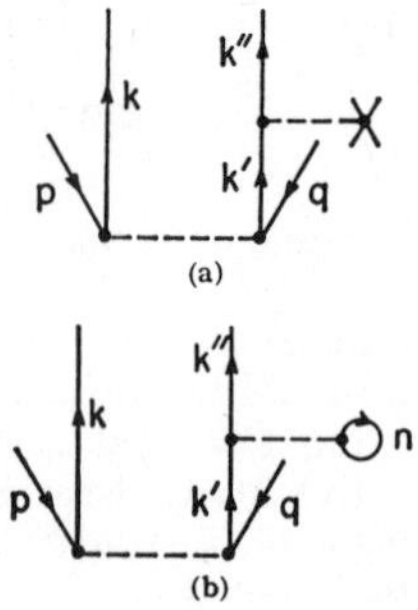


FIG. 5. (a) Scattering by the Hartree-Fock potential. (b) Diagram causing partial cancellation of the diagram shown in (a).

The expression for Fig. 4(d) is

$$\sum_{k,k'}\frac{1}{(E_0-H_0)}\langle pq|v|pq\rangle \times\frac{1}{(E_0-H_0)}\langle kk'|v|pq\rangle\eta_k{}^\dagger\eta_p\eta_{k'}{}^\dagger\eta_q. \qquad (3.3)$$

In both the above expressions, the denominator is

$$E_0-H_0=\epsilon_p+\epsilon_q-\epsilon_k-\epsilon_{k'},$$

where $\epsilon_n$ is the energy for the state $\phi_n$. The next higher order expression due to hole-hole interactions is obtained by multiplying by the factor

$$\frac{1}{E_0-H_0}\langle pq|v|pq\rangle.$$

Since, for these diagrams, $(E_0-H_0)$ always has the same value, the sum of all such diagrams is a geometric sum that converges provided

$$\langle pq|v|pq\rangle(E_0-H_0)^{-1}<1.$$

The result is that the sum to all orders of EPV diagrams due to hole-hole interactions is given by

$$\sum_{k,k'}\frac{1}{(E_0-H_0-\langle pq|v|pq\rangle)}\langle kk'|v|pq\rangle\eta_k{}^\dagger\eta_p\eta_{k'}{}^\dagger\eta_q. \qquad (3.4)$$

Since there may be also hole-hole EPV diagrams involving exchange, the denominator of Eq. (3.4) should be replaced by

$$(E_0-H_0-\langle pq|v|pq\rangle+\langle pq|v|qp\rangle). \qquad (3.5)$$

In the applications of this theory discussed in later sections, the quantity

$$(\langle pq|v|pq\rangle-\langle pq|v|qp\rangle)(E_0-H_0)^{-1}$$

is always less than one and so the summation is valid. The quantity $(E_0-H_0)$ is always negative, since $\Phi_0$ is the unperturbed ground state of the system. The direct matrix element $\langle pq|v|pq\rangle$ is positive and is, in general, much larger than the exchange term for the

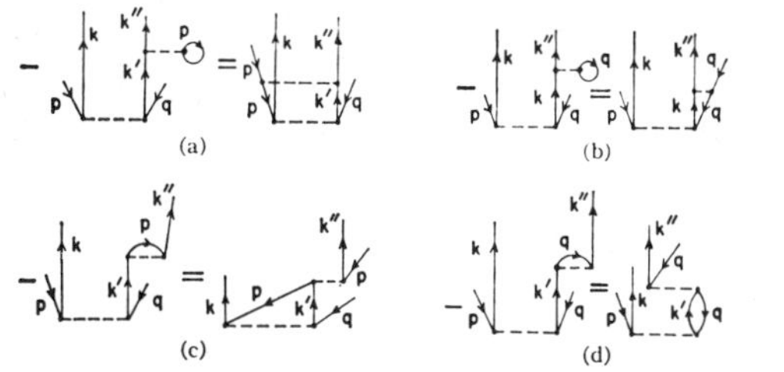

FIG. 6. Diagrams resulting from the interaction of an excited particle with the Hartree-Fock potential after the cancellation of all diagrams with passive unexcited particles. Diagrams (a) and (b) are hole-particle EPV diagrams. Diagrams (c) and (d) are the corresponding exchange terms.

applications to be considered. The resulting denominators are then always negative and never zero in these cases. In all further discussions of diagrams the energy denominators are assumed shifted to account for EPV hole-hole interactions to all orders. The origin of the EPV hole-hole diagrams may be demonstrated also by invoking the Hartree-Fock cancellation before the factorizations. All EPV diagrams may be factored into products of linked EPV diagrams and unlinked diagrams when higher orders are considered. When third-order terms, for example, are investigated, there will be factorizations similar to that shown in Fig. 3(c) but with the linked part of Fig. 3(c) replaced by the linked diagram of Fig. 4(d).

### B. EPV Diagrams Arising from the Hartree-Fock Cancellation

In using this perturbation theory it seems natural to ask whether all states should be determined with a fixed potential, since the effective potential may be changed by excitations. In fact, the effect of the changed potential is accounted for by the existence of a certain class of diagrams now described.

Consider the diagram of Fig. 5(a). This diagram is cancelled by diagrams of the type shown in Fig. 5(b) and the corresponding exchange diagrams, provided the state $\phi_n$ ranges over all the unexcited states. This, of course, does not happen because the states $\phi_p$ and $\phi_q$ are no longer occupied and so the cancellation is incomplete. This difficulty is removed by adding and subtracting the necessary diagrams to complete the cancellation. The remaining subtracted diagrams are shown in Fig. 6. Note that in diagrams (a), (b), and (d) of Fig. 6 the minus sign on the left is accounted for by the hole line on the right. In Fig. 6(c) the minus sign is accounted for by the rule that the $\eta^\dagger$ and $\eta$ operators for the free lines of each open loop go together with $\eta^\dagger$ to the left of $\eta$. The diagram of Fig. 6(c) can exist only when states $\phi_p$ and $\phi_q$ have parallel spins. In this order of perturbation theory there are also four additional diagrams resulting from the interaction of the particle in the state $\phi_k$ with the potential $V$. These results may be obtained also by consideration of the contractions of the annihilation and creation operators. Diagrams of the type described have been found to be very important in calculating correlations in Be, particularly among the outer electrons.

### C. Higher-Order EPV Diagrams

Another class of EPV diagrams arising from the factorization into products of linked and unlinked diagrams is illustrated in Fig. 7. After the factorization of the leftmost diagram, there are two remaining diagrams shown on the right. Ultimately, of course, these too will acquire similar factors when higher orders are considered, as did the simple diagram on the left. As Goldstone showed, all these unlinked diagrams are cancelled by the factor $\langle\Phi_0|U_\alpha(0)|\Phi_0\rangle$ in the denominator of $\psi_0$. The drawing of these two subtracted diagrams as shown does not mean that they may be considered as having unlinked parts as defined in Ref. 1. The method of subtracting diagrams in order to achieve factorizations often gives simpler results than the contraction of operators and it also enables one to keep in mind orders of magnitudes of higher order diagrams when the sizes of the component parts are known. If the contraction of operators is employed, then $-Y1$ and $-Y2$ of Fig. 7 are each replaced by three complicated linked diagrams of equal magnitudes but with different signs.

The expression for the first of the subtracted diagrams of Fig. 7 is (omitting the $\eta^\dagger$ and $\eta$ operators)

$$Y1=\sum_{kk'k''k'''}\frac{1}{D(k'',k''')}\langle pq|v|kk'\rangle\frac{1}{D(k,k')+D(k'',k''')}\times\langle k''k'''|v|pq\rangle\frac{1}{D(k,k')}\langle kk'|v|pq\rangle, \quad (3.6)$$

where

$$D(k,k')=\epsilon_p+\epsilon_q-\epsilon_k-\epsilon_{k'}-\langle pq|v|pq\rangle+\langle pq|v|qp\rangle. \quad (3.7)$$

The sum of all hole-hole EPV diagrams is implied in Fig. 7 and is reflected in the shifted denominators of Eq. (3.7).

The expression for $Y2$ of Fig. 7 is

$$Y2=\sum_{kk'k''k'''}\frac{1}{D(k'',k''')}\langle pq|v|kk'\rangle\frac{1}{D(k,k')+D(k'',k''')}\times\langle kk'|v|pq\rangle\frac{1}{D(k'',k''')}\langle k''k'''|v|pq\rangle. \quad (3.8)$$

Again, the annihilation and creation operators are implied. The matrix elements $\langle kk'|v|pq\rangle$ are, in general, significant only over a limited range of the excited states $k$ and $k'$. For many problems it is a good approximation to replace Eqs. (3.6) and (3.8) by

$$Y1\approx Y2\approx\tfrac{1}{2}\gamma\sum_{k''k'''}\frac{1}{D(k'',k''')}\langle k''k'''|v|pq\rangle E_2(pq), \quad (3.9)$$

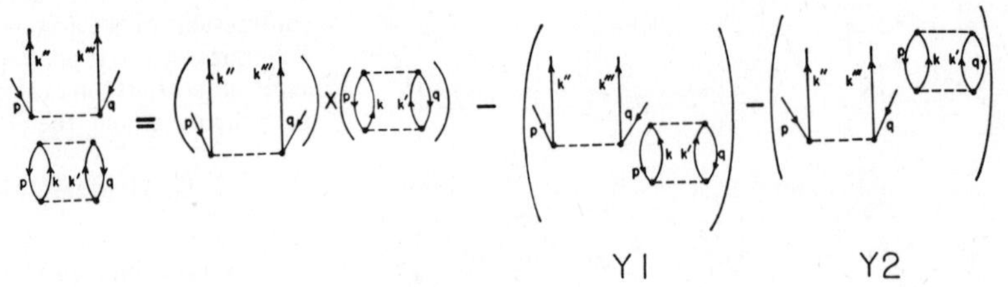


FIG. 7. Factorization of an unlinked diagram resulting in two EPV diagrams on the right. The two subtracted diagrams may not be considered as unlinked in this case.

where

$$\gamma = E_2^{(2)}(p,q)/E_2(p,q), \tag{3.10}$$

$$E_2(p,q) = \sum_{kk'} \langle pq|v|kk'\rangle \frac{1}{D(k,k')} \langle kk'|v|pq\rangle, \tag{3.11}$$

and

$$E_2^{(2)}(p,q) = \sum_{kk'} \langle pq|v|kk'\rangle \frac{1}{[D(k,k')]^2} \langle kk'|v|pq\rangle. \tag{3.12}$$

In the applications to Be such terms as Eq. (3.9) are negligible for correlations among the inner electrons but are significant (of order 10%) for the outer-electron correlations. For the outer electrons the energy denominators are dominated by $\epsilon_p+\epsilon_q-\langle pq|v|pq\rangle$ and the approximations in Eq. (3.9) are valid.

The quantity $L_n(p,q)$ is now defined as any linked diagram of $n$th order with only two external-hole lines, referring to states $\phi_p$ and $\phi_q$. There may be many internal-hole lines but these also are restricted to states $\phi_p$ and $\phi_q$. The quantity $E_n(p,q)$ is the sum of all energy diagrams of $n$th order in which the hole lines are restricted to states $\phi_p$ and $\phi_q$. Hole-hole EPV diagrams are considered summed, and this discussion does not include them. For example, $L_1(p,q)$ is given by Eq. (3.4) with the denominator of (3.4) replaced by (3.5); $E_2(p,q)$ (excluding exchange) is given by Eq. (3.11).

By use of Eq. (3.9), the two subtracted diagrams on the right of Fig. 7 give $-\gamma L_1(p,q)E_2(p,q)$. The term $L_1(p,q)$ is replaced now by

$$L_1(p,q)[1-\gamma E_2(p,q)]. \tag{3.13}$$

Further investigation shows that $L_1(p,q)$ is modified, to a good approximation, by the factor $[1-\gamma S(p,q)]$, where

$$S(p,q) = \sum_{m=2}^{\infty} E_m(p,q) \times\{1-\gamma S(p,q)[1-\gamma S(p,q)]^2\}^{m+1}. \tag{3.14}$$

Although Eq. (3.14) appears to be a difficult expression for $S(p,q)$, the quantity $S(p,q)$ is very nearly the correlation energy $E_{\text{corr}}(p,q)$ between two particles in the state $\phi_p$ and $\phi_q$ and is generally small. Equation (3.14) may be replaced by

$$S(p,q) \approx E_{\text{corr}}(p,q)[1-\gamma E_{\text{corr}}(p,q)]^2. \tag{3.14a}$$

Often $E_{\text{corr}}(p,q)$ may be estimated by physical arguments or on the basis of previous calculations. For the $2s$ states of Be, $E_{\text{corr}}(2s,2s) \approx -0.043$ a.u. and $\gamma = -0.789$ a.u.$^{-1}$.

The linked term $L_m(p,q)$ is modified to

$$L_m(p,q)[1-\gamma S(p,q)]^m. \tag{3.15}$$

These results are due to EPV diagrams resulting from factorizations in which the unlinked part is second order or higher. The corresponding modification for energy diagrams is

$$E_{m,i}(p,q) \to E_{m,i}(p,q)[1-\gamma S(p,q)]^{m-1}, \tag{3.16}$$

where $E_{m,i}(p,q)$ is the $i$th energy diagram of order $m$ with hole lines referring to $\phi_p$ and $\phi_q$. Equations (3.14), (3.15), and (3.16) are justified in the Appendix.

## IV. USE OF PERTURBATION THEORY IN ATOMIC SYSTEMS

### A. The Hartree-Fock Potential and Single-Particle States

In making use of Goldstone's perturbation theory[1] it is necessary first to obtain a complete set of single-particle states determined by a potential $V$. Because of the cancellation of the diagrams of Fig. 2, a great simplification in the number and types of diagrams to be considered results when $V$ is chosen to be the Hartree-Fock (HF) potential. In the matrix-element representation the self-consistent HF potential $V$ is defined by Eq. (2.14). In Eq. (2.14) states $|a\rangle$ and $|b\rangle$ may be any two states of the basis set. The potential $v$ is the Coulomb potential between any two of the interacting particles:

$$v_{ij} = e^2/r_{ij}. \tag{4.1}$$

The electron states $\phi_n$ are determined by the eigenvalue Eq. (2.5). In Eq. (2.5), one has

$$T = -\frac{\hbar^2\nabla^2}{2m} - \frac{Ze^2}{r}. \tag{4.2}$$

The second term of Eq. (4.2) is the electron's potential energy with the nucleus of atomic number $Z$. The distance from the electron to the nucleus is $r$, and $e$ and $m$ are the electronic charge and mass, respectively. Atomic units[5] are now chosen in which $e=1$, $m=1$, and

[5] D. R. Hartree, *The Calculation of Atomic Structures* (John Wiley & Sons, Inc., New York, 1957), p. 5.

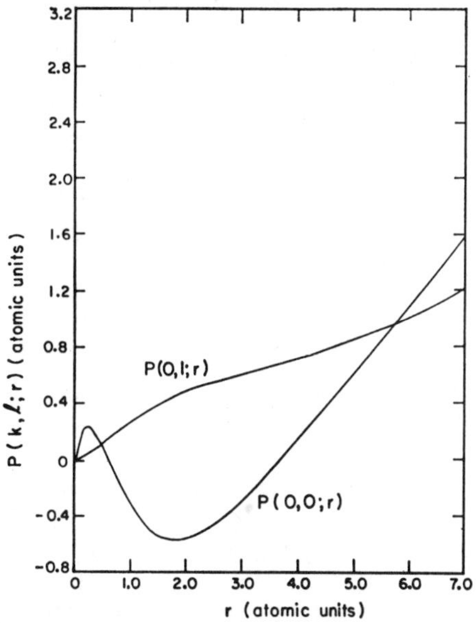


FIG. 8. Behavior of $P(k,l;r)$ for $\epsilon=0$ in the potential field of a neutral beryllium atom. There are no bound $l=1$ states, since $P(0,1;r)=0$ only at $r=0$. There are two bound $l=0$ states since $P(0,0;r)$ equals zero at two values of $r\neq 0$.

$\hbar=1$. Energies are expressed in atomic units (a.u.): 1 a.u. $=27.210$ eV.

Equations (2.5) and (2.14) constitute the self-consistent Hartree-Fock equations. The usual form for Eq. (2.5) in configuration space is

$$\frac{-\nabla^2}{2}\phi_n(\mathbf{r})-\frac{Z}{r}\phi_n(\mathbf{r})+\left(\sum_{j=1}^{N}\int\frac{\phi_j^*(\mathbf{r}')\phi_j(\mathbf{r}')d\mathbf{r}'}{|\mathbf{r}-\mathbf{r}'|}\right)\phi_n(\mathbf{r})$$
$$-\sum_{j=1}^{N}\left(\delta(m_{sn},m_{sj})\int\frac{\phi_j^*(\mathbf{r}')\phi_n(\mathbf{r}')d\mathbf{r}'}{|\mathbf{r}-\mathbf{r}'|}\phi_j(\mathbf{r})\right)$$
$$=\epsilon_n\phi_n(\mathbf{r}), \quad (4.3)$$

where $m_{sn}$ is the spin projection of the electron in state $\phi_n$.[6] The $N$ Hartree-Fock solutions $\phi_n$ of lowest energy are determined by solving a set of $N$-coupled integro-differential equations, Eq. (4.3). These $N$ states determine the potential. Additional states $\phi_n$ are obtained from Eq. (4.3), and it is not necessary to consider a set of coupled equations for them because they are calculated in the potential field of the $N$ states of lowest energy. Since the HF potential $V$ as written in Eq. (2.14) is Hermitian, an infinite set of orthonormal states comes from Eq. (4.3).[6] The $N$ lowest in energy constitute the ground state and are called "unexcited states." All other states are called "excited." There is a tendency for excited states of the perturbation expansion to be confused with true excited states of the total system. It should be remembered that the excited states in this theory are the solutions to the HF equation that are orthonormal to the $N$ unexcited states. These excited states are not true physical states of any system, although they may be given a certain physical interpretation.

### B. The Excited States

Consider now the expectation values of the HF potential $V$ as given by Eq. (2.14). If $|a\rangle=|b\rangle$, there are two possibilities. First, $|a\rangle$ may be one of the $N$ unexcited states. In this case when the summation over unexcited states is made in Eq. (2.14), there will be a term for which $|n\rangle=|a\rangle$. The exchange term then cancels the direct term and the resulting sum gives only $N-1$ direct terms in addition to the exchange terms. If $|a\rangle$ is one of the excited states, however, there is no cancellation due to the equality of a direct and an exchange term, and there are $N$ direct terms included in the expectation value of the potential. The physical interpretation is that the potential for an unexcited state is due to the $N-1$ other electrons of the system. The potential for an excited state contains $N$ direct terms and represents interactions with all $N$ particles present in the ground state. One consequence is that in a neutral atom the excited states are determined by a total potential in which the net charge is zero. This, in turn, presents the possibility that for such a potential there may be no bound excited states for the perturbation theory.

The problem as to whether bound, excited, perturbation theory states will occur for neutral atoms is similar to that of the formation of negative ions. For neutral atoms the excited states are determined in the potential field of a nucleus of charge $+N$ and $N$ electrons. The $N$ states that determine this potential are fixed and not affected by the excited state. In the formation of a negative ion, however, the original $N$ states are changed by the presence of the additional electron and are rearranged so as to minimize the energy. Values of the electron affinities in eV are listed by Massey for neutral atoms.[7] He states that hydrogen has a relatively high affinity of 0.74 eV. Beryllium, for example, has an affinity of 0.2 eV. However, calculations by Hylleraas show that the undisturbed potential of the neutral hydrogen atom is too weak for the existence of a bound state.[8] It is expected that many other neutral atoms, especially those with lower electron affinities, will behave similarly. If bound states do exist for a neutral atom, there should be very few.

Whenever calculations are made on a particular atom, there still should be a search for bound excited

[6] J. C. Slater, *Quantum Theory of Atomic Structure* (McGraw-Hill Book Company, Inc., New York, 1960), Vol. II, Chap. 17, p. 6.

[7] H. S. W. Massey, *Negative Ions* (Cambridge University Press, New York, 1950), 2nd ed., p. 19.

[8] E. A. Hylleraas, Z. Physik **60**, 624 (1930).

states. One way is to consider the Schrödinger equation as written in Eq. (4.4):

$$\frac{d^2}{dr^2}P(k,l;r)=\frac{l(l+1)}{r^2}P(k,l;r)$$
$$+2V(r)P(k,l;r)-2\epsilon P(k,l;r). \quad (4.4)$$

The function $P(k,l;r)$ is $rR(k,l;r)$, where $R(k,l;r)$ is the radial part of the wave function with angular momentum $l$ and energy $\epsilon=k^2/2$. The usual separation of the wave function into angular and radial parts is assumed. Attractive potentials $V(r)$ pull the wave function toward the $r$ axis. The attraction is enhanced by the energy term for $\epsilon>0$. For $\epsilon=0$, the behavior of $P(k,l;r)$ indicates the number of bound states of angular momentum $l$ for the potential $V(r)$. The functions $P(0,0;r)$ and $P(0,1;r)$ for Be are shown in Fig. 8. Since $P(0,0;r)$ fails to turn over a third time, the potential cannot support a $3s$ bound state and the only bound states for $l=0$ are the $P_{1s}$ and $P_{2s}$ unexcited states. There are no bound $l=1$ states, since $P(0,1;r)$ does not turn over.

It may be argued that the exchange part of the potential is different for different energies and that the failure of the wave function to turn over at zero energy does not guarantee the absence of a bound state. This is true, but it seems most unlikely that the difference in exchange contributions for zero energy and near-zero energy is sufficient to cause a bound state.

Another way to look for bound states is to make use of Levinson's theorem,[9] which relates the number of bound states for a potential $V(r)$ to the phase shifts $\delta_l(k)$ at zero and infinite energies. According to Levinson's theorem, if the potential $V(r)$ satisfies

$$\int_0^\infty dr\, r|V(r)|<\infty$$

and (4.5)

$$\int_0^\infty dr\, r^2|V(r)|<\infty,$$

then $n_l$, the number of bound states of angular momentum $l$, is given by

$$\pi n_l=\delta_l(0)-\delta_l(\infty). \quad (4.6)$$

Conditions (4.5) are clearly satisfied by the potentials of neutral atoms. For Be, $\delta_0(0)=2\pi$, and all other phase shifts at zero and infinite energies are zero. The value $2\pi$ for $\delta_0(0)$ is seen in Fig. 8, where $P(0,0;r)$ undergoes a complete cycle. Jauch has shown[10] that Eq. (4.6) is a simple consequence of the orthogonality and completeness relation for the set of eigenfunctions of the total-energy operator $H=H_0+V$. He states that this is a property that must be assumed to hold for any reasonable quantum-mechanical system.

[9] R. G. Newton, J. Math. Phys. 1, 319 (1960).
[10] J. M. Jauch, Helv. Phys. Acta 30, 143 (1957).

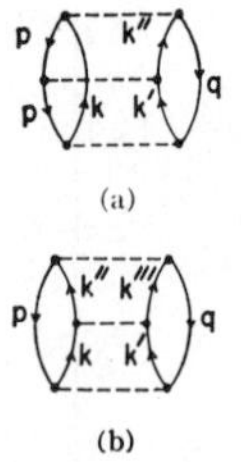

FIG. 9. Third-order energy diagrams with transitions between excited states. (a) EPV hole-particle diagram. (b) Ladder diagram. Two particles in excited states interact and scatter into two new excited states.

In Goldstone's theory the limitation to nondegenerate ground states is a restriction to systems in which the energy denominators $(E_0-H_0)$ cannot vanish for excited states of the perturbation theory. This requirement is certainly met in atomic systems even though the ground state may be degenerate in the usual sense.

### C. Wave Functions for Continuum States

In determining the form of the continuum ($\epsilon>0$) states it is necessary to consider the Hartree-Fock potential of Eq. (2.14). For those atoms that do not have spherical symmetry in their electron distributions, this HF potential is not a central potential, and the eigenstates of Eq. (4.3) are not eigenstates of orbital angular momentum $l$. It is expected, however, that it is a good approximation to make a spherical average so that the excited states are eigenstates of $l$. This may be accomplished by multiplying Eq. (4.3) by the appropriate spherical harmonic and integrating over all angles. For Be, which has spherical symmetry, no approximation is necessary. It was chosen for the numerical calculations in order that any discrepancies between calculated and experimental results might be linked directly to the perturbation series.

An additional complication arises from the fact that the potential of Eq. (2.14) is the unrestricted Hartree-Fock (UHF) potential in which states with identical quantum numbers except for spin may have different radial wave functions and energies. An example is the lithium atom with two $1s$ electrons and one $2s$ electron. The $1s$ electron with spin parallel to that of the $2s$ electron has an exchange potential term with the $2s$ electron. The other $1s$ electron does not have this exchange term and so the two $1s$ radial orbitals differ. The situation is similar for excited states, and so it may be necessary to determine excited states of spin up and spin down. Again, Be does not have this complication.

The continuum states $|k,l,m,m_s\rangle$ are determined by Eq. (4.3) with $\epsilon=k^2/2$. Spherical symmetry is assumed so that

$$\langle \mathbf{r}|k,l,m,m_s\rangle=R(kl;r)Y_{lm}(\theta,\phi)X_s(m_s). \quad (4.7)$$

In Eq. (4.7), $r$, $\theta$, $\phi$ are spherical coordinates, $l$ and $m$ are the orbital angular momentum and azimuthal quantum number, respectively, and $m_s$ is the spin projection. The spin eigenfunction is $X_s(m_s)$. The state

$|k,l,m,m_s\rangle$ is now written as $|k\rangle$, where the remaining quantum numbers are implied.

Consider the atom enclosed in a large spherical volume of radius $R_0$ that tends to infinity. At distances far from the atom the potential is effectively zero (for neutral atoms), and the radial solutions $R(kl;r)$ to Eq. (4.3) are linear combinations of $j_l(kr)$ and $n_l(kr)$, the spherical Bessel and Neumann functions.[11] As $r\rightarrow\infty$,

$$R(kl;r)\sim\cos\frac{[kr+\delta_l-\frac{1}{2}(l+1)\pi]}{kr}. \tag{4.8}$$

Since $\int_0^\infty dr\, r^2R^2(kl;r)=1$, the normalization is $(2/R_0)^{1/2}k$, and

$$\underset{r\rightarrow\infty}{R(kl;r)}=\left(\frac{2}{R_0}\right)^{1/2}\cos\frac{[kr+\delta_l-\frac{1}{2}(l+1)\pi]}{r}. \tag{4.9}$$

Since the wave function must vanish on the boundary $R_0$,

$$kR_0+\delta_l-\tfrac{1}{2}(l+1)\pi=n\pi, \tag{4.10}$$

where $n$ is an integer. The number of eigenstates $\Delta n$ for fixed $l$ in the range $\Delta k$ is determined by

$$\Delta kR_0+\Delta\delta_l=\Delta n\pi. \tag{4.11}$$

Since $(\Delta\delta)R_0^{-1}\rightarrow 0$ for finite $\Delta k$, it follows that

$$\Delta n=(R_0/\pi)\Delta k \tag{4.12}$$

and

$$\sum_k=\frac{R_0}{\pi}\int_0^\infty dk. \tag{4.13}$$

In calculating energy diagrams, every state that is excited is in turn de-excited. It is possible then to replace $\sum_k$ by $(2/\pi)\int_0^\infty dk$, provided the normalization factor $(2/R_0)^{1/2}$ is dropped from $R(kl;r)$ in Eq. (4.9). Sums over intermediate states involve sums over $l$, $m$, and $m_s$ in addition to $\int_0^\infty dk$. When sums over $m$ values are made for excitations, it is found that the resulting intermediate state has the same orbital momentum as the initial state. This is expected since the total angular-momentum operator commutes with the perturbation $(r_{ij})^{-1}$.

Consider the diagram of Fig. 9(a) which is a hole-particle EPV energy diagram. In calculating diagrams of this type in the atomic problem, the intermediate matrix element $\langle pk''|v|pk'\rangle$ can become infinite for $k'=k''$ as $R_0\rightarrow\infty$, but the integrations over $k'$ and $k''$ remove the infinity. In making numerical calculations it is a practical necessity to truncate the integrations at some finite $R_0$, and the resulting error $\epsilon 1$ is given approximately by

$$\epsilon 1\approx\left(\int_{R_0}^\infty dr\frac{\sin\mu r}{r^2}\right)E_2^{(2)}(p,q), \tag{4.14}$$

where $E_2^{(2)}(p,q)$ is given by Eq. (3.12). The symbol $\mu$ is the value of $|k''-k'|$ for which the matrix elements begin to show significant errors due to the finite $R_0$. The truncation radius $R_0$ was chosen as 120 a.u. for the calculations of Sec. V. In the diagram of Fig. 9(b), infinities as in Fig. 9(a) arise when $k''=k$ or $k'''=k'$. When both equalities occur, the infinity is stronger but is again removed by integrations. This error, $\epsilon 2$, is given roughly by

$$\epsilon 2\approx\int_{R_0}^{R_F} dr\frac{\sin\mu r}{r^2}(\mu r-\mu R_0)E_2^{(2)}(p,q), \tag{4.15}$$

where $R_F\approx 2/\mu$.

## V. THE CORRELATION ENERGY OF BERYLLIUM

### A. Observed Correlation Energies

The procedures described are equally applicable in obtaining many-body corrections to atomic wave functions and energies. The many-body property most readily determined experimentally is the correlation energy, defined as the difference between the eigenvalue $E$ of the Schrödinger Eq. (1.1) and the exact Hartree-Fock energy $E_{\mathrm{HF}}$:

$$E_{\mathrm{corr}}=E-E_{\mathrm{HF}}. \tag{5.1}$$

The eigenvalue $E$ is the difference between the experimental energy of the atom and the relativistic contributions to the energy, and so

$$E_{\mathrm{corr}}=E_{\mathrm{exp}}-E_{\mathrm{rel}}-E_{\mathrm{HF}}. \tag{5.2}$$

By use of spectroscopic data,[12] $E_{\mathrm{exp}}=-14.6682$ a.u. Fröman[13] has estimated the relativistic corrections to the Be energy as $-0.002$ a.u. Calculations by Watson give $E_{\mathrm{HF}}=-14.57299$ a.u. for Be.[14] The value for $E_{\mathrm{HF}}$ should be multiplied by the reduced electron mass for Be divided by the electron mass to give $E_{\mathrm{HF}}=-14.57211$ a.u. With the use of these numbers, Eq. (5.2) gives $E_{\mathrm{corr}}=-0.0941$ a.u. A more accurate result may be obtained by use of the spectroscopic data to obtain the first two ionization potentials for Be and then by use of Pekeris' calculations[15] for the nonrelativistic energy of $Be^{++}$. The result is $E_{\mathrm{corr}}=-0.0953$ a.u.[16] The results of calculations reported in this investigation give $E_{\mathrm{corr}}=-0.091$ a.u.

[11] L. I. Schiff, *Quantum Mechanics* (McGraw-Hill Book Company, Inc., New York, 1955), 2nd ed., pp. 77, 78.

[12] *Atomic Energy Levels*, edited by C. E. Moore, National Bureau of Standards Circular No. 467 (U. S. Government Printing Office, Washington, D. C., 1949), Vol. I.

[13] A. Fröman, Phys. Rev. **112**, 870 (1958).

[14] R. E. Watson, Phys. Rev. **119**, 170 (1960).

[15] C. L. Pekeris, Phys. Rev. **112**, 1649 (1958).

[16] This procedure is that used by Watson in Ref. 14, where he reports $E_{\mathrm{corr}}=-0.0944$ a.u. The small discrepancy between Watson's result and that of this investigation is believed to be due to Watson's conversion factor between atomic units and eV.

### B. Second-Order Energy Calculations

The HF ground-state wave functions $P_{1s}(r)$ and $P_{2s}(r)$ of Kibartas and Yutsis[17] were used in the numerical calculations. As explained in Sec. IVB, there are no bound excited states (of the perturbation theory) for Be. The continuum states were obtained by solving Eq. (4.3) on the IBM 7090 computer, using Noumerov's method[18] for the integration. Matrix elements were calculated, and then the appropriate double integrals were performed to obtain second-order energy terms. The results of these calculations are listed in Table I. The unmodified second-order energy term is

$$E_2{}^0(l;p,q)=\left(\frac{2}{\pi}\right)^2\int_0^\infty dk \times\int_0^\infty dk'\frac{\langle pq|v|kk';l\rangle\langle kk';l|v|pq\rangle}{\epsilon_p+\epsilon_q-\frac{1}{2}k^2-\frac{1}{2}k'^2}. \quad (5.3)$$

The orbital angular momentum $l$ is identical for each of the states $k$ and $k'$. A consideration of the expansion $1/|\mathbf{r}_i-\mathbf{r}_j|$ in spherical harmonics[19] shows that both $k$ and $k'$ must have the same $l$ because all unexcited states of Be are $s(l=0)$ states. In addition, the azimuthal quantum numbers $m$ and $m'$ of states $k$ and $k'$ must add to zero. In Eq. (5.3) the sum over $l$ is omitted but the sums over $m$ and $m_s$ are still implied. The second-order term with the modified denominator because of hole-hole EPV diagrams [see Eq. (3.5)] is

$$E_2(l;p,q)=\left(\frac{2}{\pi}\right)^2\int_0^\infty dk\int_0^\infty dk' \times\frac{\langle pq|v|kk';l\rangle\langle kk';l|v|pq\rangle}{\epsilon_p+\epsilon_q-\frac{1}{2}k^2-\frac{1}{2}k'^2-\langle pq|v|pq\rangle+\langle pq|v|qp\rangle}. \quad (5.4)$$

For the exchange terms, the first matrix element in both Eqs. (5.3) and (5.4) is changed to $\langle pq|v|k'k;l\rangle$, and the usual minus sign for exchange multiplies the terms. In the numerical calculations for $|pq\rangle=|1s1s\rangle$, the values of $k$ and $k'$ for significant contributions ranged from 0.0 to 10.0 a.u. The significant contributions for $|pq\rangle=|2s2s\rangle$ ranged from $k=0.0$ to $k=1.0$ a.u. It is seen from Table I that the second-order contributions give less than 70% of $E_{\text{corr}}$. Even if the true second-order terms $E_2{}^0$ are used, they still fail to give $E_{\text{corr}}$, which is approximately $-0.095$ a.u., and so it is necessary to consider higher order terms in the perturbation expansion.

### C. Correlations among 2s Electrons

An investigation of the correlation energies in atoms by means of configuration interaction has been made by Linderberg and Shull.[20] They conclude that in many-electron atoms the inner pair of electrons is very similar to the $1s$ pair in the He-like series. For $Z=4$, they report $E_{\text{corr}}=-0.0439$ a.u. for the configuration $1s^2$. Among the outer pair of electrons in Be they found $-0.043$ a.u. from $l=1$ correlations. Reference to Table I indicates that the second-order results are worse for $2s$ than for $1s$ correlations. Higher order terms for $2s$ correlations are now examined, and it is found that the series converges rather slowly. Convergence is more rapid for the $1s$ terms.

Consider the EPV diagrams of Sec. III B. There are four possible energy diagrams arising from linked diagrams of the types shown in Figs. 6(a) and (b). They arise from interactions of each excited state with each of the unexcited states of the diagram, and they are called hole-particle EPV diagrams. There are two exchange energy diagrams, one corresponding to the linked exchange diagram of Fig. 6(d) and the other to the diagram of Fig. 6(d) in which $k$ replaces $k'$ in the second interaction. All four hole-particle EPV diagrams are numerically equal, since both $2s^+$ and $2s^-$ states have identical wave functions except for spin. Both exchange diagrams are also numerically equal.

The third-order hole-particle EPV diagram is denoted by

$$E_{3\text{hp}}(l;2s,2s) =-\left(\frac{2}{\pi}\right)^3\int_0^\infty dk\int_0^\infty dk'\int_0^\infty dk''\langle 2s2s|v|kk''\rangle \times\frac{1}{D(k,k'')}\langle 2sk''|v|2sk'\rangle\frac{1}{D(k,k')}\langle kk'|v|2s2s\rangle, \quad (5.5)$$

where

$$D(k,k')=\epsilon_{2s}+\epsilon_{2s}-\tfrac{1}{2}k^2-\tfrac{1}{2}k'^2-\langle 2s2s|v|2s2s\rangle. \quad (5.6)$$

The exchange diagram is denoted

$$E_{3\text{ex}}(l;2s,2s) =\left(\frac{2}{\pi}\right)^3\int_0^\infty dk\int_0^\infty dk'\int_0^\infty dk''\langle 2s2s|v|kk''\rangle \times\frac{1}{D(k,k'')}\langle k''2s|v|2sk'\rangle\frac{1}{D(k,k')}\langle kk'|v|2s2s\rangle. \quad (5.7)$$

In Eqs. (5.5) and (5.7), the orbital momentum $l$ is implied for intermediate states. The quantity $a(l;2s,2s)$ is defined:

$$a(l;2s,2s)=E_{3\text{hp}}(l;2s,2s)/E_2(l;2s,2s), \quad (5.8)$$

[17] V. V. Kibartas and A. P. Yutsis, Zh. Eksperim. i Teor. Fiz. **25**, 264 (1953).

[18] M. G. Salvadori and M. L. Baron, *Numerical Methods in Engineering* (Prentice-Hall, Inc., Englewood Cliffs, New Jersey, 1961), 2nd ed., Chap. 3, p. 137.

[19] E. U. Condon and G. H. Shortley, *The Theory of Atomic Spectra* (Cambridge University Press, New York, 1957), p. 174.

[20] J. Linderberg and H. Shull, J. Mol. Spectry. **5**, 1 (1960).

TABLE I. Second-order energies for Be calculated with normal and shifted denominators.[a]

| $pq$ | $l$ | $E_2^0(l;\,p,q)$ | $E_2(l;\,p,q)$ | $\sum_{l=0}^{2} E_2(l;\,p,q)$ |
|---|---|---|---|---|
| $1s^+1s^-$ | 0 | | $-0.01152$ | |
| $1s^+1s^-$ | 1 | $-0.02255$ | $-0.02077$ | |
| $1s^+1s^-$ | 2 | | $-0.00345$ | $-0.03574$ |
| $2s^+2s^-$ | 0 | $-0.00241$ | $-0.00192$ | |
| $2s^+2s^-$ | 1 | $-0.02228$ | $-0.01605$ | |
| $2s^+2s^-$ | 2 | $-0.00383$ | $-0.00326$ | $-0.02123$ |
| $1s2s$[b] | 0 | | $-0.002033$ | |
| $1s2s$[b] | 1 | $-0.004991$ | $-0.004719$ | |
| $1s2s$[b] | 2 | | $-0.000612$ | |
| $1s2s$[c] | 0 | | 0.0009899 | |
| $1s2s$[c] | 1 | | 0.001108 | |
| $1s2s$[c] | 2 | | 0.00030[d] | $-0.004966$ |
| Total | | | | $-0.06194$ |

[a] All energies listed are in atomic units. According to Ref. 17, $\epsilon_{1s}=-4.7353$, $\epsilon_{2s}=-0.3092$, $\langle 1s1s|v|1s1s\rangle=2.2731$, $\langle 2s2s|v|2s2s\rangle=0.3420$, $\langle 1s2s|v|1s2s\rangle=0.4805$, and $\langle 1s2s|v|2s1s\rangle=0.0250$.
[b] Results listed include a factor of 4 due to four possible electron pairs.
[c] These are exchange terms and include a factor of 2 due to the two possible exchange pairs.
[d] This number was estimated and not calculated.

where $E_2$ is defined by Eq. (5.4). The quantity $a(l;\,2s,2s)$ is sometimes written simply $a$ when the arguments of $a(l;\,2s,2s)$ are understood. Now, one defines

$$b(l;\,2s,2s)=E_{3\text{ex}}(l;\,2s,2s)/E_2(l;\,2s,2s), \qquad (5.9)$$

and it is written as $b$ when the arguments are understood.

A calculation of $a$ and $b$ gave the values

$$a(1;\,2s,2s)=0.28504$$

and

$$b(1;\,2s,2s)=-0.06158.$$

The ratio of all third-order hole-particle and exchange EPV diagrams to the second-order energy diagram is then $4a+2b=1.0170$ for $l=1$. An examination of the numerical calculations of $E_{3\text{hp}}(1;\,2s,2s)$ showed that the effect of the intermediate interaction may be accurately approximated by a constant. That is, the relative magnitude of different $k''$ excitations after the interaction are very nearly the same as the relative $k'$ magnitudes. This leads to the following approximate equation for $a(l;\,2s,2s)$:

$$a_{\text{app}}(l;\,2s,2s)=\left(-\frac{2}{\pi}\int_0^\infty dk'\langle 2sk''|v|2sk'\rangle\frac{1}{D(k,k')}\langle kk'|v|2s2s\rangle\right)\Big/\langle kk''|v|2s2s\rangle. \qquad (5.10)$$

Again, the value $l$ is implied for all intermediate states. The values for $k$ and $k''$ should be chosen as representative values found in the calculation of $E_2$. An indication of the validity of Eq. (5.10) may be obtained from Table II.

Since $a_{\text{app}}$ differs very little from $a$ over a wide range of excitations, Eq. (5.10) is a good approximation for $a$. The relatively large value for $4a+2b$ indicates the necessity of investigating higher order terms and of looking for other third-order diagrams that will effect at least partial cancellations. For every third-order diagram there are four corresponding fourth-order hole-particle $E_{4\text{hp}}$ diagrams and two $E_{4\text{ex}}$ diagrams. By invoking the relative constancy of these intermediate interactions, the sum of fourth-order terms of the two types considered is $E_2(4a+2b)^2$ to a good approximation. Higher order sums simply involve higher powers of $(4a+2b)$. The cancellations needed for convergence come from the so-called ladder diagrams, which are important in calculating the $t$ matrix of nuclear physics. Additional terms that help convergence come from the EPV diagrams of Sec. IIIC, but the effects of ladder diagrams are greater.

A third-order ladder diagram is shown in Fig. 9(b). The sign of the diagram is positive, and there are two energy denominators that are always negative. The contribution to the correlation energy is positive and tends to cancel $E_2$. In all ladder diagrams considered it is assumed that all excited states have the same $l$ unless it is explicitly stated otherwise. Scattering processes as in Fig. 9(b), where $k$ and $k'$ have one value of $l$ and $k''$ and $k'''$ have a different value of $l$, are considered later. Calculations of third-order ladder diagrams are lengthy since four integrations are involved. However, a simplification is achieved by assuming that the relative magnitudes of the different $k$ and $k'$ excitations remain unchanged under such an interaction. The effect of the ladder interaction is then to multiply the lower order diagram by a constant factor. This assumption is similar to that which led to Eq. (5.10), and it is justified by the numerical calculations. The third-order ladder term is written as $E_{3t}(l;\,2s,2s)$ and

$$t(l;\,2s,2s)=E_{3t}(l;\,2s,2s)/E_2(l;\,2s,2s). \qquad (5.11)$$

To a good approximation,

$$t(l;\,2s,2s)=\left[\left(\frac{2}{\pi}\right)^2\int_0^\infty dk\int_0^\infty dk'\langle k''k''';l|v|kk';l\rangle\frac{1}{D(k,k')}\times\langle kk';l|v|2s2s\rangle\right]\Big/\langle k''k''';l|v|2s2s\rangle, \qquad (5.12)$$

TABLE II. The effects of hole-particle EPV interactions on excited states.

| $k$ | $k''$ (atomic units) | $a_{\text{app}}(1;\,2s,2s)/a(1;\,2s2s)$ |
|---|---|---|
| 0.2 | 0.4 | 1.060 |
| 0.4 | 0.2 | 0.995 |
| 0.4 | 0.4 | 1.017 |
| 0.4 | 0.8 | 1.165[a] |
| 0.6 | 0.2 | 0.957 |
| 0.6 | 0.4 | 0.960 |

[a] The most important contributions come from $k$, $k''$ between 0.2 and 0.6, with $k=k''=0.4$ most nearly in the center of the range of important excitations.

where the excited states all have orbital momentum, $l$, and the states $k''$, $k'''$ are chosen as typical excited states contributing to the diagram. Sums over $m$ and $m'$, the azimuthal quantum numbers of $k$ and $k'$, are implied in Eq. (5.12). By use of the assumption of Eq. (5.12), the ladder diagrams may be summed, including second order, to give $E_2(1-t)^{-1}$, where $t$ is negative. When the reduction of energy terms caused by the third class of EPV diagrams, given by Eq. (3.16), is included, the ladder diagrams sum geometrically to

$$E_2(1-\gamma S)[1-t(1-\gamma S)]^{-1}. \quad (5.13)$$

The factor

$$t_r=[1-t(1-\gamma S)]^{-1} \quad (5.14)$$

is an effective reduction of the Coulomb interaction due to summing ladder diagrams. If a nonladder interaction is multiplied by $t_r$, the effect is a summation of all ladder terms of fixed $l$ between it and the next higher nonladder or $l$-changing ladder interaction. The total correlation energy among $2s$ electrons for intermediate states of fixed $l$ is written

$$E_{corr}(l;2s,2s) = E_2t_r(1-\gamma S)[1+(4a+2b)t_r(1-\gamma S) + (4a+2b)^2t_r^2(1-\gamma S)^2+\cdots]=E_2t_r(1-\gamma S) \times[1-(4a+2b)t_r(1-\gamma S)]^{-1}, \quad (5.15)$$

where the arguments $l$ and $2s$, $2s$ are implied in the quantities on the right.

It is convenient to introduce still another symbol

$$\mathrm{Ce}(l;2s,2s) =t_r(1-\gamma S)[1-(4a+2b)t_r(1-\gamma S)]^{-1}. \quad (5.16)$$

The quantity $\mathrm{Ce}(l;2s,2s)$ is now called the coefficient of enhancement and represents the factor by which a basic diagram is enhanced owing to repeated interactions in which the value of $l$ remains unchanged. Reference to Eq. (5.15) and then to Eq. (3.14) for $S$ shows that there is a problem of self-consistency. It is necessary to know $S$ in order to calculate $E_{corr}(l;2s,2s)$, but $S$ is determined from the sum of energy diagrams for all $l$ values. By use of Eq. (3.10), $\gamma$ was calculated as $-0.7887$ a.u.$^{-1}$ In this calculation $S(2s,2s)=-0.0395$ a.u.; this value may be checked for consistency after calculation of all energy terms. Results of the numerical calculations of $t(1;2s,2s)$ and $t_r(1;2s,2s)$ are presented in Table III for different $k''=k'''$. The excitations $k''=k'''=0.40$ are expected to give the most accurate result of those listed because they are most nearly in the middle of the range of important excitations. By use of this result for $t_r$ and the other numerical results already reported for $S$, $\gamma$, $(4a+2b)$, and $E_2$, it was found from Eq. (5.15) that $E_{corr}(1;2s,2s)=-0.04256$ a.u. From Eq. (5.16), $\mathrm{Ce}(1;2s,2s)=2.6515$. If an average of the three $t$ values of Table III is made, then $E_{corr}(1;2s,2s)=-0.0402$ a.u. and $\mathrm{Ce}(1;2s,2s)=2.5042$. The first calculation of $E_{corr}(1;2s,2s)$ is believed to be more accurate, but the second calculation gives some indication of the error. Results for the correlation energies among the $2s$ electrons are listed in Table IV for three values of $l$. Most of the correlation energy comes from $l=1$ contributions. The sum for $l=0$, 1, and 2 is $-0.0516$ a.u.

There are still more diagrams involving only $2s$ unexcited states. These are ladder diagrams in which the interaction changes the value of $l$ for the excited states. The third-order energy diagram of this type is written $E_3(l,l';2s,2s)$, where $l\neq l'$. The enhanced diagram

$$E_{3Ce}(l,l';2s,2s)=E_3(l,l';2s,2s)\,\mathrm{Ce}(l;2s,2s) \times\mathrm{Ce}(l';2s,2s) \quad (5.17)$$

corresponds to the summation of all possible diagrams in which there is only one change of orbital angular

TABLE III. Calculations of ladder effects using different intermediate excitations.

| $k''$ (atomic units) | $k'''$ | $t(1;2s,2s)$[a] | $t_r(1;2s,2s)$[b] |
|---|---|---|---|
| 0.4 | 0.4 | −0.3620 | 0.7404 |
| 0.4 | 0.6 | −0.3785 | 0.7317 |
| 0.4 | 0.8 | −0.4120[c] | 0.7147 |

[a] See Eq. (5.12).
[b] See Eq. (5.14).
[c] The most important contributions to diagrams are from $k=0.20$ to $k=0.60$.

TABLE IV. Correlation energies among $2s$ electrons in Be for different $l$.[a]

| $l$ | $a$ | $b$ | $t$ | $t_r$ | Ce$(l;2s,2s)$ | $E_{corr}(l;2s,2s)$ |
|---|---|---|---|---|---|---|
| 0 | 0.18330 | −0.02778 | −0.1650 | 0.8622 | 1.925 | −0.003704 |
| 1 | 0.28504 | −0.06158 | −0.3620 | 0.7404 | 2.652 | −0.04256 |
| 2 | 0.16480 | −0.01423 | −0.2113 | 0.8301 | 1.632 | −0.005321 |

[a] All energies are in atomic units.

TABLE V. Contributions to the correlation energy among $2s$ electrons in Be.

| Term[a] | Energy (a.u.) |
|---|---|
| $E_{corr}(0)$ | −0.003704 |
| $E_{corr}(1)$ | −0.04256 |
| $E_{corr}(2)$ | −0.005321 |
| $E_{3Ce}(0,1)$ | 0.00203 |
| $E_{3Ce}(1,0)$ | 0.00203 |
| $E_{3Ce}(1,2)$ | 0.00282 |
| $E_{3Ce}(2,1)$ | 0.00282 |
| $E_{4Ce}(0,1,0)$ | −0.000096 |
| $E_{4Ce}(0,1,2)$ | −0.000107 |
| $E_{4Ce}(1,0,1)$ | −0.001109 |
| $E_{4Ce}(1,2,1)$ | −0.000959 |
| $E_{4Ce}(2,1,0)$ | −0.000107 |
| $E_{4Ce}(2,1,2)$ | −0.000120 |
| $\sum_{n=5}^{\infty} E_{nCe}$ | 0.00051 |
| Total | −0.04387 |

[a] The arguments $2s$, $2s$ are implied.

TABLE VI. Correlation energies among 1s electrons in Be for different $l$.[a]

| $l$ | $a$ | $b$ | $t$ | $t_r$ | Ce($l$; 1s,1s) | $E_{corr}$($l$; 1s,1s) |
|---|---|---|---|---|---|---|
| 0 | 0.07672 | −0.01823 | −0.08784 | 0.9193 | 1.222 | −0.01408 |
| 1 | 0.07614 | −0.01302 | −0.10205 | 0.9075 | 1.213 | −0.02518 |
| 2 | | | | | 1.20[b] | −0.00414 |

[a] Energies are in atomic units.
[b] This value was estimated.

momentum, from $l$ to $l'$, in the excited states. The corresponding enhanced fourth-order energy diagram is $E_{4Ce}(l,l',l''; 2s,2s)$, where $l\neq l'\neq l''$. As might be expected from the previous numerical results, the most important diagrams of this type are those in which $l=l''=1$. The approximation of Eq. (5.12) was used in calculating diagrams in which $l$ changes. The results are listed in Table V along with previous calculations of $E_{corr}$ from Table IV. The sum of all these terms is the total correlation energy among the 2s electrons of Be and is −0.0439 a.u. It is believed that all diagrams that can be of importance have been included in this calculation. It is interesting to note that the total calculated correlation energy among 2s electrons is very nearly that obtained from a consideration of $l=1$ terms alone. Correlations involving $l=0$ and $l=2$ states are very nearly cancelled by the ladder diagrams that involve changes from $l=1$ states.

### D. Correlations among 1s Electrons

Correlations among the 1s electrons are considered now for $l=0$, 1, and 2 intermediate states. Equations (5.10) through (5.17) are used in the calculations with the arguments 2s, 2s replaced by 1s, 1s. Results are listed in Table VI. The quantity $\gamma$ in Eq. (3.10) equals −0.0286 a.u.$^{-1}$ and is much smaller than that for the 2s calculations because of the larger energy denominators for the 1s diagrams. In these calculations $S=-0.040$ a.u. The coefficient of enhancement Ce(2; 1s,1s) was estimated. As in the case of 2s correlations, there are terms in which the excited electrons change orbital momentum after the interactions. All calculated contributions to the 1s correlations are listed in Table VII. The terms $E_{3Ce}(l,l'; 1s,1s)$ are less important for 1s than for 2s correlations.

TABLE VII. Contributions to the correlation energy among 1s electrons in Be.

| Term[a] | Energy (a.u.) |
|---|---|
| $E_{corr}(0)$ | −0.01408 |
| $E_{corr}(1)$ | −0.02518 |
| $E_{corr}(2)$ | −0.00414 |
| $E_{3Ce}(1,0)$ | 0.000412 |
| $E_{3Ce}(0,1)$ | 0.000412 |
| $E_{3Ce}(1,2)$ | 0.000226 |
| $E_{3Ce}(2,1)$ | 0.000226 |
| Total | −0.04212 |

[a] The arguments 1s, 1s are implied.

### E. Correlations between 1s and 2s Electrons

The results of the second-order energy correlations among 1s and 2s electrons are listed in Table I, where the total 1s−2s correlation energy is −0.00497 a.u. The coefficients of enhancement Ce($l$; 1s,2s) have not been calculated, but are expected to be small because the energy denominators are much larger than for the 2s calculations. A rough estimate for the coefficients of enhancement for the 1s−2s diagrams is 1.10 to 1.20. Third-order diagrams that tend to reduce the 1s−2s correlation energy are the ring diagrams, which were found to be of great importance in calculating the correlation energy of a dense electron gas.[2] A typical third-order ring diagram is shown in Fig. 10(a). This diagram was calculated and found to be 0.0000136 a.u. There are eight such diagrams, each of which should be multiplied by the appropriate coefficient of enhancement. The result is approximately 0.000288 a.u. There are additional diagrams in which the relative positions of the 1s and 2s states differ from those shown in Fig. 10(a). The approximate sum of all third-order ring diagrams, including enhancement, is 0.0007 a.u. The ring diagrams are reduced by the hole-particle diagrams of Fig. 10(b). These diagrams are quite small, like the ring diagrams. However, there are not so many hole-particle diagrams because spins must be parallel in scattering from one unexcited state to another. A very rough estimate indicates that these diagrams reduce the ring diagrams by one-third. It seems then that the combined effects of enhancement, ring diagrams, and hole-particle diagrams roughly cancel for 1s−2s correlations and that the second-order result may be used. It should be remembered, however, that certain higher order diagrams are included, since the second-order calculations were made with shifted energy denominators.

By use of the results of the 1s- and 2s-correlation energies, physical arguments may be used to estimate the 1s−2s correlation energy. Reference to the Be wave functions of Kibartas and Yutsis shows that the radial distribution for 1s electrons, $P_{1s}^2$, is peaked sharply at $r=0.26$ a.u.[17] The normalization of $P_{1s}$ and $P_{2s}$ is

$$\int_0^\infty P_{1s}^2 dr=\int_0^\infty P_{2s}^2 dr=1. \tag{5.18}$$

The ratio $P_{2s}^2/P_{1s}^2$ represents the relative probability of finding a 2s electron. Neglecting exchange, the 1s−2s correlation energy should be given roughly by the product of $P_{2s}^2/P_{1s}^2$ and the total correlation energy for the 1s electrons. At $r=0.26$ a.u., $P_{2s}^2/P_{1s}^2=0.0259$, and the resulting 1s−2s correlation energy is −0.00109 a.u. Since there are four 1s−2s pairs, the energy becomes −0.00436 a.u. It is also necessary to consider the ratio $P_{1s}^2/P_{2s}^2$, where $P_{2s}^2$ peaks. This is more difficult because $P_{2s}^2$ is not so sharply peaked as $P_{1s}^2$, which is a rapidly decaying exponential in the region where $P_{2s}^2$

is large. At best, it may be estimated that the correlations due to $P_{1s}^2$ contributions, where $P_{2s}^2$ is large, are a little smaller than those due to $P_{2s}^2$ contributions, where $P_{1s}^2$ peaks. Since the exchange contributions may be expected to be somewhat less than half the total $1s-2s$ correlation energy, the exchange energy should approximately cancel the contributions from considerations of large $P_{2s}^2$. The result of these rough physical considerations is then $-0.00436$ a.u., which is in reasonable agreement with the result $-0.00497$ a.u. calculated from perturbation theory.

### F. Discussion of Results

The sum of all correlation energies in Be, obtained from Tables V and VII and Sec. V E, is $-0.09096$ a.u. It is difficult to assess the accuracy of the calculations, but limits of $\pm 0.004$ a.u. seem reasonable. In these calculations intermediate states were restricted to orbital angular momenta $l=0$, 1, and 2. Additional contributions to the correlation energy may come from higher angular-momentum states, and these will probably affect the $1s$ more than the $2s$-correlation results. The result for the total correlation energy compares favorably with the experimental value $-0.0953$ a.u., which was obtained from spectroscopic data[12] and Pekeris' energy for $Be^{++}$.[15] Watson's calculation of the Be correlation energy by configuration interactions resulted in $-0.0844$ a.u. when he used thirty-seven configurations.[14] However, he was able to obtain 74% of the correlation energy by use of only four configurations. Watson also found that about 5% of the correlation energy was caused by interactions between the $1s$ and $2s$ shells. This result also agrees favorably with the calculations of Sec. V E that gave $-0.004966$ a.u. for the $1s-2s$ correlation energy. It is interesting to note that correlations among $1s$ and $2s$ electrons, though small, must be considered if accurate values for the correlation energy are desired. The fact that four such pairs must be considered increases their effect.

The calculations reported here may also be compared with those of Linderberg and Shull.[20] For $Z=4$ in the He-like series, they find by use of configuration interaction that the total correlation energy is $-0.0439$ a.u. and that $l=1$ terms contribute $-0.0217$ a.u. and $l=2$ terms contribute $-0.0025$ a.u. The corresponding results of Table VII agree reasonably well with these figures. Linderberg and Shull also report that in Be, the $2s$ correlations are almost entirely due to $l=1$ contributions that give $-0.043$ a.u. This value agrees with the results of Table V when the $l$-changing ladder-diagram results are included.

## VI. APPROXIMATE METHODS FOR CORRELATION ENERGIES

At the beginning of this investigation the correlation energy of boron was calculated by use of second-order

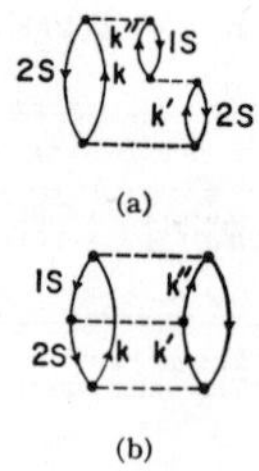

FIG. 10. (a) A typical ring diagram. (b) A hole-particle diagram.

perturbation theory.[21] In order to simplify the numerical work, the Hartree-Fock potential was approximated by the screened Coulomb potential in which the excited states were calculated. The bound states were solutions of the Hartree-Fock equations. The results for the correlations among the $1s$ electrons were in good agreement with the energies reported by Linderberg and Shull.[20] However, the correlations among the $2s$ electrons were much too large. The poor results for the $2s$ correlations may be attributed to the fact that the important excited states for $2s$ matrix elements are those of low energy ($k\sim 0.3$ a.u.), whereas the important excitations for $1s$ matrix elements are in the range $k\sim 4.0$ a.u. On physical grounds, it is expected that wave functions for high energies will not be affected by the details of the potential so much as will low-energy wave functions. Since the approximate solutions are not orthogonal to the ground state HF wave functions, it is not surprising that the matrix elements are too large. These results indicate the importance of using the correct Hartree-Fock excited states in the perturbation theory.

An estimate of the total correlation energy as a function of the atomic number $Z$ was obtained from the Gell-Mann and Brueckner[2] formula for the correlation energy of a dense electron gas:

$$\epsilon_c = -0.096 + 0.0622 \ln r_s \text{ Ry per particle.} \quad (6.1)$$

The dimensionless spacing parameter $r_s$ depends upon the particle density. Assuming that the density of electrons, $\rho$, within the atom is given by a Fermi-Thomas distribution,[22] the resulting total correlation energy is

$$E_{\text{corr}} = \int \epsilon_c(\rho)\rho d\tau = -Z[0.0062 + 0.0207 \ln Z] \text{ a.u.} \quad (6.2)$$

This formula gives a rough estimate of $E_{\text{corr}}$ and is too large by a factor of approximately 2 for small $Z$. The Fermi-Thomas electron distribution fulfills the requirements of a dense electron gas only for the inner part of the atom for low $Z$.

---

[21] H. Kelly, in *Proceedings of the Second International Conference on the Physics of Electronic and Atomic Collisions, Abstracts of Papers* (W. A. Benjamin, Inc., New York, 1961), p. 136.

[22] E. U. Condon and G. H. Shortley, *The Theory of Atomic Spectra* (Cambridge University Press, New York, 1957), p. 335.

## VII. CONCLUSIONS

By use of Hartree-Fock wave functions, accurate perturbation-theory calculations were made of the correlation energies among the electrons of Be. The total calculated correlation energy was found to be $-0.0910$ a.u. as compared with $-0.0953$ a.u., which is believed to be the correct result. The correlations in the 1*s* and 2*s* shells were found to be each approximately one-half of the total, and the intershell correlations contributed approximately 6% of the total correlation energy. In calculating the 2*s* correlations it was found that the exclusion-principle-violating (EPV) diagrams are extremely important and that these must be summed to high orders. It is expected that EPV diagrams will be important in calculating outer electron correlations in all atoms. If accurate results are desired, it is necessary to include the intershell correlations. These may become more important as the atomic number increases beyond four because the 2*s* and 2*p* wave functions are expected to overlap more than 1*s* and 2*s* wave functions. When the number of 2*p* electrons equals three or more, it is possible to have third-order ring diagrams in which all three unexcited-state wave functions are the same, and the higher-order ring diagrams may be more important for such atoms than for those of fewer electrons. The third-order ring diagrams in Be were found to be very small. The calculations on boron indicate the importance of using Hartree-Fock wave functions for the excited states.

It has been shown in this investigation that many-body perturbation theory is very useful in correcting atomic energies. Although the complexity of the calculation increases with the number of electrons, this method should still be quite useful in calculating correlations in many-electron atoms. When there are open shells, the calculation may be less accurate because of the lack of spherical symmetry in the true atomic potentials. There may also be the difficulty due to the necessity of distinguishing between single-particle states with different spins. This theory is also applicable in obtaining correlation corrections for atomic wave functions and therefore for all atomic properties. Future investigations will examine the correlation corrections for transition probabilities and nuclear quadrupole coupling constants.

## ACKNOWLEDGMENTS

I am extremely grateful to Professor Kenneth M. Watson, who suggested this problem, for his advice and encouragement; to Valerie M. Burke, who assisted with the initial phases of this problem; and to Dr. Philip G. Burke for several helpful discussions.

This work was done under the auspices of the U. S. Atomic Energy Commission.

## APPENDIX

In Sec. III C formulas were given to account for the effects of the third class of EPV diagrams. These diagrams arise from factorizations in which the unlinked part is of second order or higher. It was shown that $L_1$ is modified by the factor $(1-\gamma E_2)$. In a similar way it may be shown that, to a good approximation, the linked diagram $L_m$ is modified by $(1-\gamma S)^m$, where $S$ is given by Eq. (3.14). Since these formulas are concerned with correlations between two electrons, each diagram has the same unexcited states, although there is no limit to the number of hole lines. Other cases may be handled similarly.

Consider the diagrams of Fig. 11. The results must be the same for any unlinked third-order diagram. After factorization, three EPV diagrams remain and sum to $-\gamma L_1(p,q)E_{3t}(p,q)$ in the approximation of Eq. (3.9). The fraction under each diagram gives the fraction of $-\gamma L_1E_{3t}$ and is determined from the increased energy denominators. A consideration of higher order unlinked diagrams with $L_1$ shows that the reduction is still the same and so $L_1(p,q)$ is modified by $[1-\gamma \sum_{m=2}^{\infty} E_m(p,q)]$. If factorizations like that of Fig. 11 are made with a linked part $L_2(p,q)$ and an unlinked

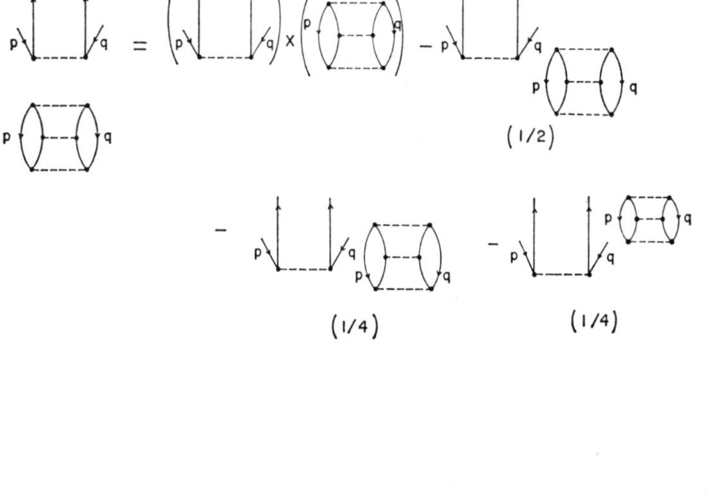


FIG. 11. Factorization of $L_1(p,q)$ and $E_{3t}(p,q)$. The three subtracted diagrams on the right are EPV diagrams that sum to $-\gamma L_1(p,q)E_{3t}(p,q)$ in the approximation of Eq. (3.9). Fractions of this sum are shown below each diagram. The arrows and labels are implied for excited states when not written explicitly.

part of second order or higher, the result is that $L_2(p,q)$ is modified by $[1-2\gamma\sum_{m=2}^{\infty}E_m(p,q)]$. The general result is that the linked diagram $L_n(p,q)$ is multiplied by the factor $[1-n\gamma\sum_{m=2}^{\infty}E_m(p,q)]$, and this may be verified by an examination of higher order linked diagrams. The form $[1-\gamma\sum_{m=2}^{\infty}E_m(p,q)]^n$ is now suggested as the modification of $L_n(p,q)$, and this form may be checked as approximately correct by enumeration of the EPV diagrams resulting from factorizations of diagrams like Fig. 12.

Before the factorizations of Fig. 12 are actually made, the factorizations of Fig. 13 should be investigated. These lead to further modifications of the factor $[1-\gamma\sum_{m=2}^{\infty}E_m(p,q)]^n$ for $L_n(p,q)$. After the factorizations of Fig. 13, there are three factored diagrams $G0$, $G1F$, and $G2F$ and a sum of EPV diagrams that add to $L_1 3(\gamma E_2)^2$. The diagram $L_1$ is now modified to

$$L_1[1-\gamma E_2(1-3\gamma E_2)]. \tag{A1}$$

The form

$$L_1[1-\gamma E_2(1-\gamma E_2)^3] \tag{A2}$$

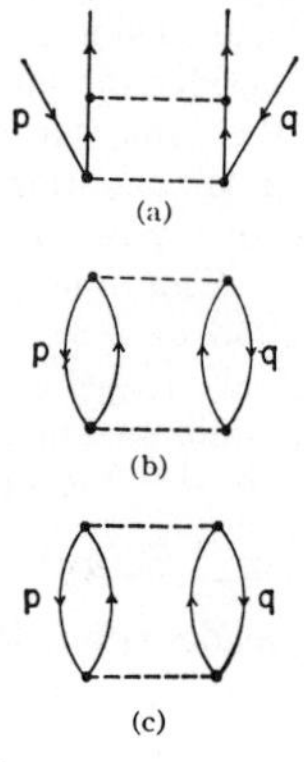


FIG. 12. Factorization of this diagram leads to the modification of $L_2(p,q)$ given by Eq. (A3).

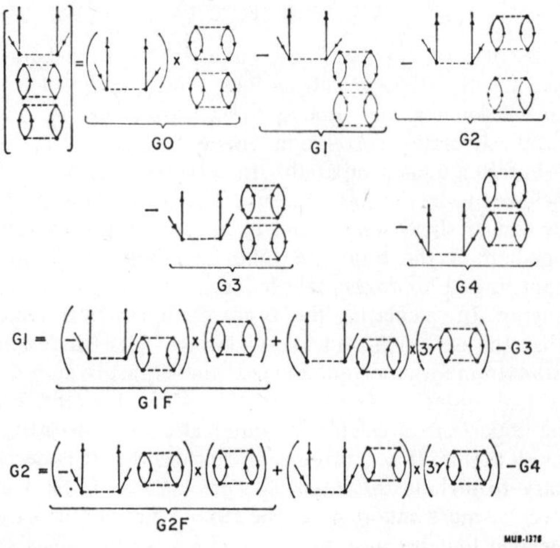


FIG. 13. EPV diagrams resulting from factorization of the diagram shown at the top left. After all factorizations there are three factored diagrams $GO$, $G1F$, and $G2F$ and a sum of EPV diagrams that add to modify $L_1$ to $L_1[1-\gamma E_2(1-3\gamma E_2)]$. The hole line on the left of each disconnected part refers to the state $\phi_p$ and the hole line on the right refers to the state $\phi_q$.

is then suggested and may be checked. Further factorizations of diagrams like Fig. 12 indicate that a good approximation to the modifications is given by

$$L_n(p,q)(1-\gamma S)^n, \tag{A3}$$

where

$$S(p,q)=\sum_{m=2}^{\infty}E_m(p,q)[1-\gamma S(p,q)]^{m+1}. \tag{A4}$$

Factorizations of $L_1$ and three unlinked parts indicate that $S$ on the right of Eq. (A4) should be modified approximately by $(1-\gamma S)^2$. This last modification leads to Eq. (3.14). The modifications for the energy diagrams are readily obtained from (A3), since $E_m$ is related to $L_{m-1}$. The validity of considering all hole-hole EPV diagrams summed in making this analysis may be checked by a consideration of the appropriate factorizations.

VIII-4 Reprinted from *Physical Review* **136**, B896 (1964)

# Many-Body Perturbation Theory Applied to Atoms*

HUGH P. KELLY
*Department of Physics, University of California, San Diego, La Jolla, California*
(Received 29 June 1964)

Many-body perturbation theory as formulated by Brueckner and Goldstone is applied to atoms to obtain corrections to Hartree-Fock wave functions and energies. Calculations are made using a complete set of single-particle Hartree-Fock wave functions which includes both the continuum and an infinite number of bound states. It is shown how one may readily perform the sums over an infinite number of bound excited states. In order to demonstrate the usefulness of many-body perturbation theory in atomic problems, calculations are made for a wide variety of properties of the neutral beryllium atom. The calculated $2s$-$2s$ correlation energy is $-0.0436$ atomic unit for $l=1$ excitations. The calculated dipole and quadrupole polarizabilities are $6.93\times10^{-24}$ cm$^3$ and $14.1\times10^{-40}$ cm$^5$, respectively. The calculated dipole and quadrupole shielding factors are 0.972 and 0.75. Results are given for oscillator strengths, photoionization cross sections, and the Thomas-Reiche-Kuhn sum rule, which is 4.14 as compared with 4.00, the theoretical value.

## I. INTRODUCTION

MANY-BODY perturbation theory as developed by Brueckner[1] and Goldstone[2] has proven very useful in the study of many-particle systems. As shown by Brueckner, the appropriate form of perturbation theory as the number of particles becomes large is Rayleigh-Schrödinger theory modified so as to eliminate the "unlinked clusters." The principal applications of the Brueckner-Goldstone linked cluster expansion (BG expansion) to many-fermion systems have thus far been investigations of nuclear structure[3] and of the electron gas.[4] However, the BG theory, which corrects both wave functions and energies, should also prove very useful in calculations of atomic structure and in other fields. In applying this theory to atoms, where the interparticle forces are well known, one also gains information as to its general applicability to finite systems.

A previous application of BG theory to the calculation of correlation energies in the neutral beryllium atom yielded excellent results.[5] However, it was found necessary to calculate high orders in the expansion. This difficulty was related to the set of single-particle Hartree-Fock states which were used. The purpose of this paper is to investigate the use of a different basis set for the expansion and to show the usefulness of perturbation calculations using this set. The states used are the ground-state Hartree-Fock orbitals and single-particle excitations calculated in the Hartree-Fock potential field of the nucleus and $N-1$ of the $N$ ground-state orbitals. The use of this set is justified in Sec. II. In Sec. III it is shown how sums over an infinite number of bound excited states may be carried out. In Sec. IV the $l=1$ correlation energy among the two $2s$ electrons of Be is calculated. In Sec. V calculations are given for the dipole and quadrupole polarizabilities and shielding factors for Be. In Sec. VI many oscillator strengths and the photoionization cross section curve are calculated. Section VII contains the conclusions.

---

* Work supported in part by the U. S. Atomic Energy Commission.

[1] K. A. Brueckner, Phys. Rev. **97**, 1353 (1955); **100**, 36 (1955); *The Many-Body Problem* (John Wiley & Sons, Inc., New York, 1959).

[2] J. Goldstone, Proc. Roy. Soc. (London) **A239**, 267 (1957).

[3] K. A. Brueckner and J. L. Gammel, Phys. Rev. **109**, 1023 (1958); K. A. Brueckner and K. S. Masterson, Jr., *ibid.* **128**, 2267 (1962).

[4] M. Gell Mann and K. Brueckner, Phys. Rev. **106**, 364 (1957).

[5] H. P. Kelly, Phys. Rev. **131**, 684 (1963), hereafter referred to as K. Correlation energies are defined in K.

## II. PERTURBATION THEORY

### A. Review of the Brueckner-Goldstone Expansion

The total Hamiltonian for a system of $N$ identical fermions interacting through two-body ptoentials $v_{ij}$ is

$$H=\sum_{i=1}^{N} T_i+\sum_{i<j}^{N} v_{ij}. \quad (1)$$

The kinetic energy for the $i$th particle and the sum of all one-body potentials acting on it are given by $T_i$. For atoms $T_i$ includes the interaction of the $i$th electron with the nucleus. The true ground state of the system is $\Psi_0$ given by

$$H\Psi_0=(E_0+\Delta E)\Psi_0. \quad (2)$$

The effect of the $N$ interacting particles may be approximated by a single-particle potential $V$ and then $\Psi_0$ is approximated by $\Phi_0$ where

$$H_0\Phi_0=E_0\Phi_0 \quad (3)$$

and

$$H_0=\sum_{i=1}^{N}(T_i+V_i). \quad (4)$$

The single-particle wave functions, $\varphi_n$ which are solutions of

$$(T+V)\varphi_n=\epsilon_n\varphi_n, \quad (5)$$

constitute an orthonormal set, provided $V$ is Hermitian. The state $\Phi_0$ is a determinant formed from the $N$ solutions of Eq. (5) which are lowest in energy. The states occupied in $\Phi_0$ are called unexcited states. An unoccupied unexcited state is called a hole and an occupied excited state is called a particle.

The BG result is that

$$\Psi_0=\sum_L\left(\frac{1}{E_0-H_0}H'\right)^n\Phi_0, \quad (6)$$

where $\sum_L$ means that only "linked" terms are to be included[2] and

$$H'=\sum_{i<j}^{N} v_{ij}-\sum_{i=1}^{N} V_i. \quad (7)$$

Also,

$$\Delta E=\sum_{L'}\langle\Phi_0|H'\left(\frac{1}{E_0-H_0}H'\right)^n|\Phi_0\rangle, \quad (8)$$

where $L'$ indicates that the sum is only over those terms which are "linked" when the leftmost $H'$ interaction is removed.

To first order, the energy is

$$E^{(1)}=E_0+\langle\Phi_0|H'|\Phi_0\rangle, \quad (9)$$

and when a Hartree-Fock basis is used

$$E^{(1)}=E_{\mathrm{HF}}=\sum_{n=1}^{N}(\langle n|T|n\rangle+\tfrac{1}{2}\langle n|V|n\rangle). \quad (10)$$

### B. Choice of the Single-Particle Potential $V$

The Hartree-Fock potential is defined by[2]

$$\langle a|V_{\mathrm{HF}}|b\rangle=\sum_{n=1}^{N}(\langle an|v|bn\rangle-\langle an|v|nb\rangle), \quad (11)$$

where $a$ and $b$ are arbitrary. This potential was used in $K$ to obtain the complete set of single-particle Hartree-Fock states which were used in calculating the correlation energy for Be. When $V_{\mathrm{HF}}$ is used, Eq. (5) for the single-particle states $\varphi_n$ becomes

$$-\tfrac{1}{2}\nabla^2\varphi_n(\mathbf{r})-Z/r\varphi_n(\mathbf{r})+\left(\sum_{j=1}^{N}\int d\mathbf{r}'\frac{\varphi_j^*(\mathbf{r}')\varphi_j(\mathbf{r}')}{|\mathbf{r}-\mathbf{r}'|}\right)\varphi_n(\mathbf{r})$$
$$-\sum_{j=1}^{N}\left(\delta(m_{sn},m_{sj})\int d\mathbf{r}'\frac{\varphi_j^*(\mathbf{r}')\varphi_n(\mathbf{r}')}{|\mathbf{r}-\mathbf{r}'|}\varphi_j(\mathbf{r})\right)$$
$$=\epsilon_n\varphi_n(\mathbf{r}). \quad (12)$$

This is the usual Hartree-Fock equation considered, for example, by Slater.[6] Atomic units are used throughout this paper except where specified otherwise. Once the $N$ unexcited states have been calculated self-consistently by solving the coupled Eqs. (12), $V_{\mathrm{HF}}$ is determined and the remaining states of the orthonormal set are obtained by solving the single Eq. (12). As pointed out in K, excited states are calculated in the potential of $N$ particles but unexcited states are calculated in the potential of $N-1$ particles because of cancellation of direct and exchange terms when $\varphi_j=\varphi_n$. In K this led to the surprising result that all excited single-particle Hartree-Fock states for Be were in the continuum and it was conjectured that excited states of Eq. (12) would all lie in the continuum for most if not all neutral atoms.

There are two advantages in dealing with only continuum excited states. First, it is much simpler to solve Eq. (12) for continuum states than for bound states for which it becomes an eigenvalue equation. Second, sums over excited states are more readily performed when only continuum excited states need be considered. However, in the numerical work reported in the later sections of this paper it was found quite feasible both to solve Eq. (12) for bound states and to sum over excited bound and continuum states. There is also a disadvantage in using $V_{\mathrm{HF}}$ as defined by Eqs. (11) or (12) for excited states. In K it was found that the perturbation expansion converged slowly for the correlation energy among $2s$ electrons; this was due to large effects from certain hole-particle interactions referred to as second-type EPV (exclusion-principle-violating) diagrams. They were shown to arise from the fact that interactions of excited particles with the occupied unexcited states do not cancel the interaction with $V_{\mathrm{HF}}$ as shown in Fig. 1.

[6] J. C. Slater, *Quantum Theory of Atomic Structure* (McGraw-Hill Book Company, Inc., New York, 1960), Vol. II, Chap. 17, p. 6.

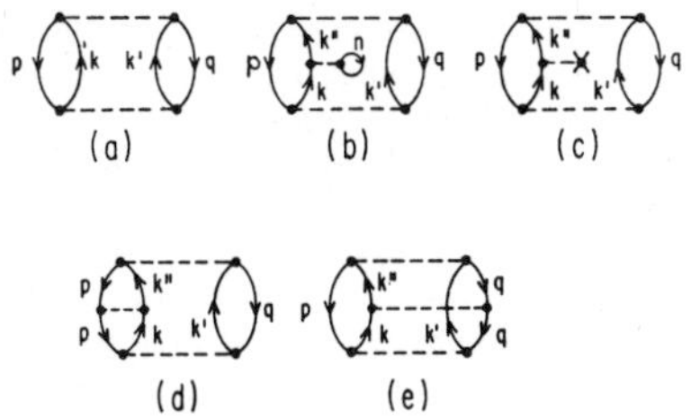


FIG. 1. (a) Second-order energy diagram. (b) Particle in excited state $\varphi_k$ interacts with all occupied unexcited states $\varphi_n$. Note that $\varphi_p$ and $\varphi_q$ are no longer occupied. (c) Particle in state $\varphi_k$ interacts with the potential $V$. When the Hartree-Fock potential $V_{HF}$ of Eq. (11) is used, diagram (b) does not fully cancel (c) because $V_{HF}$ includes interactions with all unexcited states. Diagrams (d) and its exchange and (e) result. If $\varphi_p$ and $\varphi_q$ have parallel spins there is also the exchange of diagram (e). There are similar diagrams for the interactions of $\varphi_{k'}$.

Since the excited particles actually interact with $N-1$ other particles, it appears desirable to choose the potential accordingly.[7] In the calculations of this paper for Be the excited states were calculated in the potential field of the nucleus and $(1s)^2(2s)$ where $1s$ and $2s$ are the Hartree-Fock orbitals of the neutral beryllium atom. In this potential the excited states have direct interactions with two $1s$ electrons and one $2s$ electron and also an exchange interaction with one $1s$ electron. The excited states then correspond closely to the physical single-particle excitations of beryllium. The $2s$ state for this potential coincides with the usual Hartree-Fock $2s$ state. The new $1s$ state differs from the Hartree-Fock solution but this difference is expected to be small as the $1s$ potential depends strongly on the interaction with the nucleus. However, there are now first-order corrections to the wave function as shown in Fig. 2. When these terms and similar terms in higher orders are added to $\Phi_0$ the wave function becomes the usual Hartree-Fock $\Phi_0$. The appropriate procedure is to omit the diagrams of Fig. 2 in calculations and to use the Hartree-Fock solutions for both $1s$ and $2s$ states. Bound and continuum states were calculated for $l=0$, 1, and 2. The $l=1$ and $l=2$ states are orthonormal because they were all calculated with the same Hermitian potential and they are automatically orthonormal with respect to all $l=0$ states. The $2s$ Hartree-Fock (HF) state and all excited $l=0$ states were calculated with the same Hermitian potential and are orthonormal. The only deviations from orthonormality in the basis set arise from the nonorthogonality of the HF $1s$ state with excited $l=0$ states. This nonorthogonality is not an error or an approximation as may be seen from the diagrams of Fig. 2. In actual calculations the nonorthogonality of the $1s$ state and excited $l=0$ states is expected

TABLE I. Orthogonality of $s$ states.[a]

| | | | |
|---|---|---|---|
| $\langle 1s\|1s\rangle$ | 1.0000012 | $\langle 2s\|2s\rangle$ | 1.0000001 |
| $\langle 1s\|2s\rangle$ | 0.0000357 | $\langle 2s\|3s\rangle$ | 0.0000012 |
| $\langle 1s\|3s\rangle$ | 0.0000418 | $\langle 2s\|4s\rangle$ | −0.0000037 |
| $\langle 1s\|4s\rangle$ | 0.0000272 | $\langle 2s\|5s\rangle$ | −0.0000013 |
| $\langle 1s\|5s\rangle$ | 0.0000191 | $\langle 2s\|0.1s\rangle$ | 0.0000010 |
| $\langle 1s\|0.1s\rangle$ | 0.0000723 | $\langle 2s\|0.2s\rangle$ | 0.0000041 |
| $\langle 1s\|0.2s\rangle$ | 0.0001049 | $\langle 2s\|0.3s\rangle$ | 0.0000075 |
| $\langle 1s\|0.6s\rangle$ | 0.0002253 | $\langle 2s\|0.4s\rangle$ | 0.0000067 |
| $\langle 1s\|1.0s\rangle$ | 0.0003638 | $\langle 2s\|0.6s\rangle$ | −0.0000066 |
| $\langle 1s\|1.6s\rangle$ | 0.0005253 | $\langle 2s\|1.0s\rangle$ | 0.0000007 |
| $\langle 1s\|2.0s\rangle$ | 0.0005626 | $\langle 2s\|2.0s\rangle$ | −0.0000014 |
| $\langle 1s\|3.0s\rangle$ | 0.0005056 | $\langle 2s\|3.0s\rangle$ | 0.0000008 |
| $\langle 1s\|4.0s\rangle$ | 0.0003611 | $\langle 2s\|4.0s\rangle$ | 0.0000002 |
| $\langle 1s\|5.0s\rangle$ | 0.0002009 | $\langle 2s\|5.0s\rangle$ | −0.0000001 |
| $\langle 1s\|6.0s\rangle$ | 0.0001000 | $\langle 2s\|6.0s\rangle$ | −0.0000015 |
| $\langle 1s\|8.0s\rangle$ | 0.0000254 | $\langle 2s\|8.0s\rangle$ | 0.0000009 |

[a] Continuum states are normalized so that asymptotically $P_k(r)=\sin(kr+(1/k)\ln 2kr+\delta)$.

TABLE II. Radial functions $p_{2s}(r)$ for Be.

| $r$ | $P_{2s}{}^{a}(r)$ | $P_{2s}{}^{b}(r)$ | $r$ | $P_{2s}{}^{a}(r)$ | $P_{2s}{}^{b}(r)$ |
|---|---|---|---|---|---|
| 0.01 | 0.02569 | 0.02569 | 1.60 | −0.58549 | −0.58534 |
| 0.02 | 0.04936 | 0.04935 | 1.80 | −0.61697 | −0.61683 |
| 0.03 | 0.07110 | 0.07109 | 2.00 | −0.62889 | −0.62878 |
| 0.04 | 0.09104 | 0.09101 | 2.20 | −0.62566 | −0.62559 |
| 0.05 | 0.10926 | 0.10922 | 2.40 | −0.61110 | −0.61106 |
| 0.10 | 0.17771 | 0.17765 | 2.60 | −0.58835 | −0.58833 |
| 0.15 | 0.21536 | 0.21530 | 2.80 | −0.55996 | −0.55997 |
| 0.20 | 0.22985 | 0.22980 | 3.00 | −0.52797 | −0.52799 |
| 0.25 | 0.22700 | 0.22694 | 3.40 | −0.45905 | −0.45911 |
| 0.30 | 0.21123 | 0.21118 | 3.80 | −0.39016 | −0.39027 |
| 0.35 | 0.18596 | 0.18592 | 4.20 | −0.32597 | −0.32611 |
| 0.40 | 0.15381 | 0.15377 | 4.60 | −0.26870 | −0.26884 |
| 0.45 | 0.11681 | 0.11679 | 5.00 | −0.21910 | −0.21923 |
| 0.50 | 0.07655 | 0.07654 | 6.00 | −0.12683 | −0.12692 |
| 0.55 | 0.03428 | 0.03428 | 7.00 | −0.07065 | −0.07071 |
| 0.60 | −0.00904 | −0.00901 | 8.00 | −0.03828 | −0.03834 |
| 0.65 | −0.05261 | −0.05257 | 9.00 | −0.02031 | −0.02037 |
| 0.70 | −0.09585 | −0.09580 | 10.00 | −0.01060 | −0.01065 |
| 0.75 | −0.13827 | −0.13821 | 11.00 | −0.00546 | −0.00550 |
| 0.80 | −0.17951 | −0.17944 | 12.00 | −0.00278 | −0.00281 |
| 0.85 | −0.21929 | −0.21921 | 14.00 | −0.00070 | −0.00071 |
| 0.90 | −0.25739 | −0.25730 | 16.00 | −0.00017 | −0.00017 |
| 0.95 | −0.29366 | −0.29356 | 18.00 | −0.00004 | −0.00004 |
| 1.00 | −0.32798 | −0.32788 | | | |
| 1.20 | −0.44492 | −0.44479 | $-\epsilon_{2s}$(a.u.) | 0.30942 | 0.30927 |
| 1.40 | −0.52968 | −0.52953 | | | |

[a] Calculated for this investigation. 1 a.u. = 27.21 eV.
[b] Calculated by Roothaan, Sachs, and Weiss, Ref. 9.

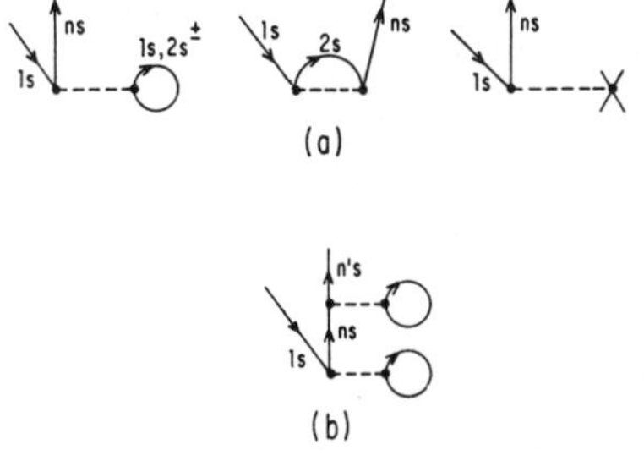


FIG. 2. (a) Single-particle corrections to the wave function $\Phi_0$ in first order. When $\varphi_{1s}$ is a Hartree-Fock orbital, these corrections vanish. When $\varphi_{1s}$ is not determined by the Hartree-Fock potential, these terms and similar higher order terms as shown in (b) added to $\Phi_0$ give effectively the Hartree-Fock result.

[7] I would like to thank Professor K. A. Brueckner for stressing the desirability of using a potential for excited states which has a physical basis and yields rapid convergence of the perturbation expansion.

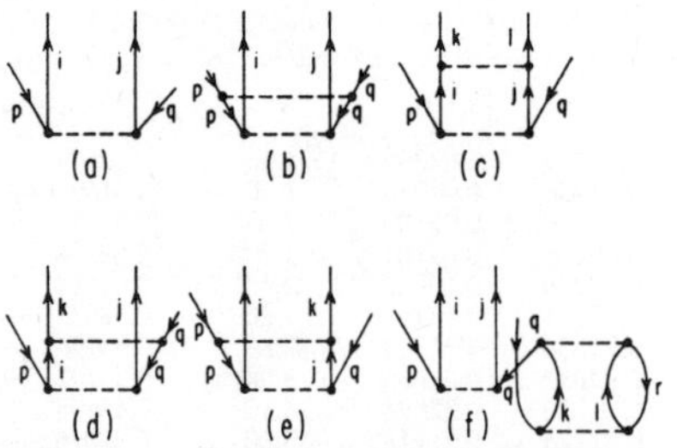


FIG. 3. Corrections to $\Phi_0$ due to correlations among electrons in states $p$ and $q$. (a) First-order term. (b) Diagonal hole-hole interaction. (c) Particle-particle interaction or ladder diagram. (d) and (e) are hole-particle interactions which represent the net effect of interactions of the particles in states $i$ and $j$ with all occupied unexcited states and with the potential $V$. It is assumed that $i$ is calculated in the field of all unexcited states except $p$ and $j$ in the field of all but $q$. (f) Exclusion principle violating diagram arising from the linked cluster factorization.

to be very small. Overlap intergrals of $l=0$ states are given in Table I. All integrals were calculated by Simpson's rule. Orthogonality for the $l=0$ states is quite good, even for the 1s state with excited states for which, in principle, exact orthogonality is not expected. The lack of exact orthogonality between 1s and 2s states may be attributed to limits on the numerical accuracy of the HF 1s state taken from the Kibartas and Yutsis solution.[8] All other states were calculated by the author's Hartree-Fock program which solves the following equation for $\varphi_n$.

$$-\tfrac{1}{2}\nabla^2\varphi_n(\mathbf{r})-Z/r\varphi_n(\mathbf{r})$$
$$+\left(2\int d\mathbf{r}'\frac{\varphi_{1s}^*(\mathbf{r}')\varphi_{1s}(\mathbf{r}')}{|\mathbf{r}-\mathbf{r}'|}+\int d\mathbf{r}'\frac{\varphi_{2s}^*(\mathbf{r}')\varphi_{2s}(\mathbf{r}')}{|\mathbf{r}-\mathbf{r}'|}\right)\varphi_n(\mathbf{r})$$
$$-\int d\mathbf{r}'\frac{\varphi_{1s}^*(\mathbf{r}')\varphi_n(\mathbf{r}')}{|\mathbf{r}-\mathbf{r}'|}\varphi_{1s}(\mathbf{r})=\epsilon_n\varphi_n(\mathbf{r}). \quad (13)$$

For Be, $Z=4$. The 1s and 2s states used in Eq. (13) were taken from Kibartas and Yutsis.[8] The 2s state calculated by the author's program was in good agreement with the Kibartas solution. Comparison with the 2s Be solution of Roothaan[9] is given in Table II. The very small disagreement in the fourth decimal place may not be attributed to a limit of accuracy for the author's program but is due to use of the Kibartas 1s and 2s states rather than Roothaan's in Eq. (13). In the calculations of this paper the most important perturbation terms are those involving 2s states and excited $l=1$ and $l=2$ states.

### C. Summation of Diagrams

In order to obtain the corrections to $\Phi_0$ due to correlations among a pair of electrons in states $p$ and $q$, the diagrams of Fig. 3 are considered. For simplicity, states $p$ and $q$ have opposite spins. Figure 3(b) shows the diagonal part of the hole-hole interaction which violates the exclusion principle as discussed in K. In general, hole-hole nondiagonal interactions are quite small. The diagram of Fig. 3(c) shows a particle-particle interaction. It is assumed that the excited state $i$ of Figs. 3(a) and 3(d) was calculated in the potential field of all unexcited states except for the state $p$. Interactions with the occupied unexcited states and with the potential combine to give the diagram (d). This diagram is analogous to that of Fig. 1(e). Similarly, it is assumed that the state $j$ was calculated in the potential field of all unexcited states except for $q$ and this gives Fig. 3(e). Diagrams shown in Fig. 1(d) are not included because excited states are now calculated in the field of $N-1$ unexcited states. When $i$ and $j$ are bound-states diagrams (b), (c), (d), and (e) of Fig. 3 are largest for diagonal matrix elements. Diagram (c) is largest for $i=k$, $j=l$ and (d) and (e) are largest for $i=k$ and $j=k$, respectively. The expression for the diagram of Fig. 3(a) is

$$(\epsilon_p+\epsilon_q-\epsilon_i-\epsilon_j)^{-1}\langle ij|v|pq\rangle. \quad (14)$$

For the diagonal interactions just described, the sum of diagrams 3(b) through 3(e) is given by

$$[(\epsilon_p+\epsilon_q-\epsilon_i-\epsilon_j)^{-1}(\langle pq|v|pq\rangle+\langle ij|v|ij\rangle-\langle iq|v|iq\rangle - \langle pj|v|pj\rangle)](\epsilon_p+\epsilon_q-\epsilon_i-\epsilon_j)^{-1}\langle ij|v|pq\rangle. \quad (15)$$

When these diagonal interactions occur in the next order of perturbation theory the factor in brackets in Eq. (15) is repeated, and so the diagonal interactions give a geometric series which is readily summed to

$$D^{-1}\langle ij|v|pq\rangle, \quad (16)$$

where

$$D=(\epsilon_p+\epsilon_q-\langle pq|v|pq\rangle)-(\epsilon_i+\epsilon_j+\langle ij|v|ij\rangle - \langle iq|v|iq\rangle-\langle pj|v|pj\rangle). \quad (17)$$

When diagrams of the type shown in Fig. 3(f) are considered, $D$ of Eq. (17) is further modified to[10]

$$D=[\epsilon_p+\epsilon_q-\langle pq|v|pq\rangle+E_{\text{corr}}(p,q)+E_{\text{corr}}(p,r\neq q) + E_{\text{corr}}(r\neq p,q)]-[\epsilon_i+\epsilon_j+\langle ij|v|ij\rangle - \langle iq|v|iq\rangle-\langle pj|v|pj\rangle-E_{\text{corr}}'(i,j)]. \quad (18)$$

The term $E_{\text{corr}}(p,q)$ is the correlation energy among the two electrons in states $p$ and $q$. The term $E_{\text{corr}}(p,r\neq q)$ is the total correlation energy of an electron in state $p$ from interactions with all unexcited states except for $q$ and similarly for $E_{\text{corr}}(r\neq p,q)$. The term $E_{\text{corr}}'(i,j)$ is the sum of all terms contributing to the total correlation energy in which either of the excited states $i$ or $j$ occurs and in which the hole states differ from $p$ and $q$. Equa-

[8] V. V. Kibartas and A. P. Yutsis, Zh. Eksperim. i Teor. Fiz. 25, 264 (1953).

[9] C. C. J. Roothaan, L. M. Sachs, and A. W. Weiss, Rev. Mod. Phys. 32, 186 (1960).

[10] In Ref. 5 diagrams of this type were shown to arise from the factorization of diagrams and they were labeled third class exclusion-principle-violating (EPV) diagrams. A more detailed analysis of such terms and their effects on Eq. (18) is given by H. Kelly, Phys. Rev. 134, A1450 (1964).

TABLE III. Dependence of matrix elements on $n$.[a]

| $n$ | $\lvert\langle 2pnp\lvert v\rvert 2s2s\rangle\rvert^2$ | $n^3\lvert\langle 2pnp\lvert v\rvert 2s2s\rangle\rvert^2$ |
|---|---|---|
| 2 | $3.644\times10^{-2}$ | 0.2915 |
| 3 | $2.569\times10^{-3}$ | 0.0694 |
| 4 | $7.371\times10^{-4}$ | 0.0472 |
| 5 | $3.171\times10^{-4}$ | 0.0396 |
| 6 | $1.667\times10^{-4}$ | 0.0360 |
| 7 | $9.889\times10^{-5}$ | 0.0339 |
| 8 | $6.364\times10^{-5}$ | 0.0326 |

| $n$ | $\lvert\langle np\lvert r\rvert 2s\rangle\rvert^2$ | $n^3\lvert\langle np\lvert r\rvert 2s\rangle\rvert^2$ |
|---|---|---|
| 2 | $7.849\times10^{0}$ | 62.794 |
| 3 | $1.480\times10^{-2}$ | 0.400 |
| 4 | $1.299\times10^{-2}$ | 0.831 |
| 5 | $7.446\times10^{-3}$ | 0.931 |
| 6 | $4.443\times10^{-3}$ | 0.960 |
| 7 | $2.823\times10^{-3}$ | 0.968 |
| 8 | $1.894\times10^{-3}$ | 0.970 |

| $n$ | $\lvert\langle np\lvert r^{-2}\rvert 2s\rangle\rvert^2$ | $n^3\lvert\langle np\lvert r^{-2}\rvert 2s\rangle\rvert^2$ |
|---|---|---|
| 2 | $1.830\times10^{-2}$ | 0.1464 |
| 3 | $1.711\times10^{-3}$ | 0.0462 |
| 4 | $5.260\times10^{-4}$ | 0.0337 |
| 5 | $2.327\times10^{-4}$ | 0.0291 |
| 6 | $1.241\times10^{-4}$ | 0.0268 |
| 7 | $7.424\times10^{-5}$ | 0.0255 |
| 8 | $4.804\times10^{-5}$ | 0.0246 |

| $n$ | $\lvert\langle np\lvert r\rvert 1s\rangle\rvert^2$ | $n^3\lvert\langle np\lvert r\rvert 1s\rangle\rvert^2$ |
|---|---|---|
| 2 | $2.515\times10^{-2}$ | 0.2012 |
| 3 | $4.774\times10^{-3}$ | 0.1289 |
| 4 | $1.791\times10^{-3}$ | 0.1146 |
| 5 | $8.637\times10^{-4}$ | 0.1080 |
| 6 | $4.819\times10^{-4}$ | 0.1041 |
| 7 | $2.960\times10^{-4}$ | 0.1015 |
| 8 | $1.947\times10^{-4}$ | 0.0997 |

| $n$ | $\lvert\langle nd\lvert r^2\rvert 2s\rangle\rvert^2$ | $n^3\lvert\langle nd\lvert r^2\rvert 2s\rangle\rvert^2$ |
|---|---|---|
| 3 | $5.169\times10^{1}$ | 1395.6 |
| 4 | $1.604\times10^{1}$ | 1026.3 |
| 5 | $7.022\times10^{0}$ | 877.8 |
| 6 | $3.724\times10^{0}$ | 804.3 |
| 7 | $2.223\times10^{0}$ | 762.5 |
| 8 | $1.438\times10^{0}$ | 736.2 |
| 9 | $9.857\times10^{-1}$ | 718.6 |
| 10 | $7.062\times10^{-1}$ | 706.2 |

| $n$ | $\lvert\langle nd\lvert r^{-3}\rvert 2s\rangle\rvert^2$ | $n^3\lvert\langle nd\lvert r^{-3}\rvert 2s\rangle\rvert^2$ |
|---|---|---|
| 3 | $7.317\times10^{-5}$ | $1.976\times10^{-3}$ |
| 4 | $4.099\times10^{-5}$ | $2.623\times10^{-3}$ |
| 5 | $2.328\times10^{-5}$ | $2.910\times10^{-3}$ |
| 6 | $1.419\times10^{-5}$ | $3.064\times10^{-3}$ |
| 7 | $9.176\times10^{-6}$ | $3.147\times10^{-3}$ |
| 8 | $6.259\times10^{-6}$ | $3.204\times10^{-3}$ |
| 9 | $4.437\times10^{-6}$ | $3.235\times10^{-3}$ |
| 10 | $3.284\times10^{-6}$ | $3.284\times10^{-3}$ |

| $n$ | $\lvert\langle nd\lvert r^2\rvert 1s\rangle^2$ | $n^3\lvert\langle nd\lvert r^2\rvert 1s\rangle\rvert^2$ |
|---|---|---|
| 3 | $9.711\times10^{-5}$ | $2.622\times10^{-3}$ |
| 4 | $5.707\times10^{-5}$ | $3.653\times10^{-3}$ |
| 5 | $3.304\times10^{-5}$ | $4.130\times10^{-3}$ |
| 6 | $2.030\times10^{-5}$ | $4.385\times10^{-3}$ |
| 7 | $1.323\times10^{-5}$ | $4.537\times10^{-3}$ |
| 8 | $9.051\times10^{-6}$ | $4.634\times10^{-3}$ |
| 9 | $6.446\times10^{-6}$ | $4.699\times10^{-3}$ |
| 10 | $4.746\times10^{-6}$ | $4.746\times10^{-3}$ |

| $n$ | $\langle 2s\lvert r^2\rvert nd\rangle\langle nd\lvert r^{-3}\rvert 2s\rangle$ | $n^3\langle 2s\lvert r^2\rvert nd\rangle\langle nd\lvert r^{-3}\rvert 2s\rangle$ |
|---|---|---|
| 3 | $6.150\times10^{-2}$ | 1.660 |
| 4 | $2.564\times10^{-2}$ | 1.641 |
| 5 | $1.279\times10^{-2}$ | 1.598 |
| 6 | $7.268\times10^{-3}$ | 1.570 |
| 7 | $4.517\times10^{-3}$ | 1.549 |
| 8 | $3.000\times10^{-3}$ | 1.536 |
| 9 | $2.091\times10^{-3}$ | 1.525 |
| 10 | $1.523\times10^{-3}$ | 1.523 |

[a] Only radial parts of matrix elements are given.

tion (18) is the two-particle energy for states $p$ and $q$ minus the approximate two-particle energy for states $i$ and $j$. In the first bracket, subtraction of $\langle pq|v|pq\rangle$ from $\epsilon_p+\epsilon_q$ corrects for the fact that each single-particle state was calculated in the potential field of the other and so the interaction of $p$ with $q$ was counted twice. Then $\epsilon_p+\epsilon_q-\langle pq|v|pq\rangle$ is the Hartree-Fock energy for the pair $pq$. $E_{\mathrm{corr}}(p,q)$ accounts for higher order interactions of $p$ with $q$ and $E_{\mathrm{corr}}(p,r\neq q)$ and $E_{\mathrm{corr}}(r\neq p,q)$ account for the higher order interactions of $p$ and $q$ with the other unexcited states since these interactions are not included in the HF calculation of $\epsilon_p$ and $\epsilon_q$. In the second bracket, $\langle ij|v|ij\rangle$ accounts for the interaction of $i$ with $j$ and $-\langle iq|v|iq\rangle$ corrects for the fact that $i$, which was calculated with interactions with $q$, does not interact with $q$ which is now unoccupied. The term $-\langle pj|v|pj\rangle$ corrects similarly for state $j$. The first five terms of the second bracket of Eq. (18) then give essentially the HF energy of the excited pair $ij$. The term $-E_{\mathrm{corr}}'(i,j)$ does not give correlations between $i$ and $j$ but accounts for the fact that all correlations among unexcited pairs which involve excitations into states $i$ or $j$ are eliminated by the Pauli principle when $p$ and $q$ are excited into $i$ and $j$. Although the discussion of this section has treated bound excited states, it is readily extended to continuum states as shown previously.[5,11] In numerical applications the nondiagonal higher order terms are calculated but they converge rapidly.

### III. SUMS OVER BOUND EXCITED STATES

In later sections the BG theory is used to calculate the correlation energy among $2s$ electrons and other properties for Be. In using BG perturbation theory, it is necessary to sum over all excited states. When the continuum is considered, the sums are readily evaluated by numerical integrations as shown in K. It is not obvious that the sums over bound states may be handled so simply, however. One often includes just the first few bound states and assumes that the remaining contributions are small. This is probably reasonable in many cases; however, it is preferable to sum over all bound excited states and this is now shown to be feasible. In the numerical work reported here it was found that matrix elements such as $\langle mpnp|v|2s2s\rangle$ are proportional to $n^{-3/2}$ for fixed $m$ as $n$ becomes large. This behavior

[11] H. P. Kelly and A. M. Sessler, Phys. Rev. 132, 2091 (1963).

is expected to hold true for other operators and for other atoms when the asymptotic potential is Coulombic as in this case. The explanation lies in the fact that $\varphi_{2s}$ lies much closer to the nucleus than $\varphi_{np}$ for $n$ large. When we compute $\varphi_{n+1p}$ there is very little change in the single-particle energy and the behavior of $\varphi_{n+1p}$ is very close to that of $\varphi_{np}$ (except for normalization) in the region of space where $\varphi_{2s}$ is substantially nonzero. The principal change in the matrix element in going from $n$ to $n+1$ then is due to the change in the normalization factors.

It is shown, for example, by Bethe and Salpeter[12] that for hydrogen-like atoms the behavior of the eigenfunctions for large principal quantum number $n$ is

$$R_{nl}\approx 2\left(\frac{Z}{n}\right)^{3/2}\frac{(2Zr)^l}{(2l+1)!}\left[1-\frac{2rZ}{2l+2}+\cdots\right]. \tag{19}$$

The potential used in Eq. (13) is asymptotically Coulombic and since $\varphi_{np}$ for large $n$ is located mainly in the asymptotic region of the potential, it is expected that the normalization of $\varphi_{np}$ should contain the factor $n^{-3/2}$ as does that of hydrogen. The numerical checks on this rule are given in Table III. When the product $n^3$ times matrix element squared has not completely reached its asymptotic value, a curve may be drawn to estimate the higher values and the limit. In perturbation theory calculations we consider terms of the form

$$\sum_{m=2}^{\infty}\sum_{n=2}^{\infty}|\langle mpnp|v|2s2s\rangle|^2D^{-1}, \tag{20}$$

where $D$ is given by Eq. (18). The double summation presents no essential complication in the following discussion. As $n$ in Eq. (20) becomes large, $D$ approaches a constant value. This allows us to carry out the summations of Eq. (20). For example, for fixed $m$ we might carry out the sum from $n=2$ to $n=8$ by explicit calculation of the terms. Then from $n=9$ to approximately $n=15$ we would calculate terms by using the $n^{-3}$ rule for the matrix elements squared and we would make the necessary extrapolations to obtain accurate denominators. For example, $\epsilon_{np}\propto n^{-2}$. The remainder of the sum is obtained to a good approximation from

$$C\int_{N_f+1}^{\infty} n^{-3}dn=C/[2(N_f+1)^2], \tag{21}$$

where $N_f$ is the last $n$ value calculated by discrete summation and

$$C=\lim_{n\to\infty} n^3|\langle mpnp|v|2s2s\rangle|^2D^{-1}. \tag{22}$$

For greater accuracy we may also use $\zeta(3)$ where $\zeta(s)$ is the zeta function of Riemann[13] given by

$$\zeta(s)=\sum_{n=1}^{\infty} 1/n^s.$$

In the calculations of the next sections $N_f$ is typically 15. This procedure may be carried out to any desired accuracy by calculation of a sufficient number of excited states. The sums of Eq. (20) must be repeated for different values of $m$ and an extrapolation made for $m\to\infty$ just as for $n$.

TABLE IV. Numerical values in a.u. for $D$ of Eq. (23) and for excitation matrix element.

| | |
|---|---|
| $\epsilon_{2s}$ | $-0.30942$ |
| $\epsilon_{2p}$ | $-0.17951$ |
| $E_{corr}(2s,2s)$ | $-0.0439$ |
| $4E_{corr}(2s;1s)$ | $-0.00497$ |
| $\langle 2s2s\|v\|2s2s\rangle$ | 0.34331 |
| $\langle 2p2p\|v\|2p2p\rangle$[a] $(m_l=\pm1)$ | 0.28652 |
| $\langle 2p2p\|v\|2p2p\rangle$ $(m_l=0)$ | 0.30180 |
| $\langle 2p2s\|v\|2p2s\rangle$ | 0.30867 |
| $\langle 2p2p\|v\|2s2s\rangle$[b] | 0.0636315 |

[a] One $2p$ state has $m_l=+1$ and the other $m_l=-1$.
[b] This term is negative for $m_l=\pm1$ and positive for $m_l=0$.

## IV. Be CORRELATION ENERGY FOR 2s ELECTRONS

In K it was found that almost all the contribution to $2s-2s$ correlations in Be came from excitations into $l=1$ states. This calculation has been again made using BG theory but with the set of single-particle states of Eq. (13). Diagonal terms beyond second order are included in the "second-order" calculation by using the denominator $D$ of Eq. (18). The nondiagonal third-order and higher terms are also calculated. The states $p$ and $q$ in Eq. (18) are now the Hartree-Fock $2s$ states of Be. The term $E_{corr}(2s,2s)$ was found to be $-0.0439$ atomic units (a.u.) in K. The terms $E_{corr}(p,r\neq q)+E_{corr}(r\neq p,q)$ give the total correlation energy between the $2s$ and $1s$ shells which was calculated to be $-0.00497$ a.u. in K. One a.u. $=27.21$ eV. Most of the contribution to the $2s$ correlation energy will be shown to come from excitations into $2p$ states. The term $E_{corr}'(i,j)$ is the contribution to the correlation energy among $1s$ electrons when at least one of the excited states coincides with $i$ or $j$. For $2p$ excitations $E_{corr}'$ was calculated to be $-0.00027$ a.u. This is quite small relative to the other terms in Eq. (18) so $E_{corr}'(i,j)$ is omitted. When both $2s$ electrons are excited into $2p$ states

$$D=\epsilon_{2s}+\epsilon_{2s}-\epsilon_{2p}-\epsilon_{2p}+E_{corr}(2s,2s)+4E_{corr}(2s,1s)-\langle 2s2s|v|2s2s\rangle-\langle 2p2p|v|2p2p\rangle+2\langle 2p2s|v|2p2s\rangle. \tag{23}$$

The numerical values are given in Table IV. Although the notation has so far been suppressed, excited states

[12] H. A. Bethe and E. E. Salpeter, *Quantum Mechanics of One- and Two-Electron Systems* (Academic Press Inc., New York, 1957), p. 18.

[13] E. T. Whittaker and G. N. Watson, *Modern Analysis* (Cambridge University Press, Cambridge, 1927), Chap. XIII, p. 265.

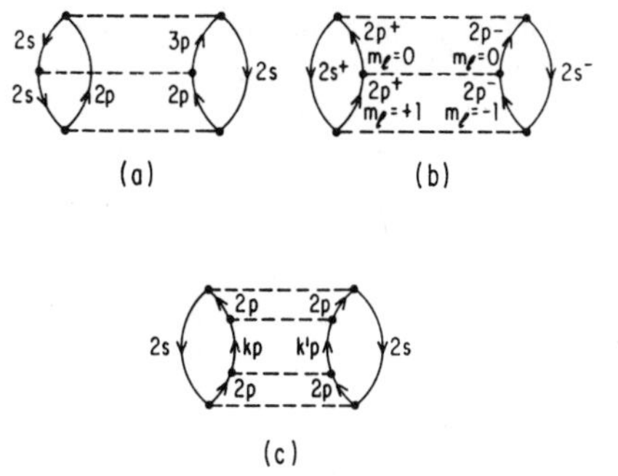


FIG. 4. Changes in excited states due to nondiagonal interactions. (a) Hole-particle interaction. (b) Particle-particle interaction. (c) Two nondiagonal interactions.

are labeled by $m_l$ and $m_s$ in addition to $n$ and $l$. The angular factors for matrix elements, which may be obtained from Condon and Shortley,[14] affect $\langle 2p2p|v|2p2p\rangle$ differently according to the $m_l$ values of the $2p$ excited states. When we write the sums over $m_l$ explicitly,

$$E_{\text{corr}}(2s,\,2s\to mp,\,np)=2|\langle mpnp|v|2s2s\rangle|^2/D(m_l=\pm1)+|\langle mpnp|v|2s2s\rangle|^2/D(m_l=0).\quad(24)$$

The results of Eq. (24) are given in Table V for $m=2$ and $n$ variable.

The sums over continuum states are readily performed as described in K. In the following term one excitation is into a bound state and one into the continuum

$$\sum_k E_{\text{corr}}(2s,\,2s\to np,\,kp)=\frac{2}{\pi}\int_0^\infty dk|\langle 2s2s|v|npkp\rangle|^2/D(k),\quad(25)$$

where

$$D(k)=\epsilon_{2s}+\epsilon_{2s}-\epsilon_{np}-k^2/2+E_{\text{corr}}(2_s,2_s)+4E_{\text{corr}}(2s,1s)-\langle 2s2s|v|2s2s\rangle+\langle np2s|v|np2s\rangle.\quad(26)$$

TABLE V. Bound-state contributions to $E_{\text{corr}}$ involving $2p$ states.

| $n$ | $E_{\text{corr}}(2s2s\to 2pnp)$ in a.u. |
|---|---|
| 2 | −0.037244 |
| 3 | −0.003781 |
| 4 | −0.001011 |
| 5 | −0.000423 |
| 6 | −0.000219 |
| 7 | −0.000129 |
| 8 | −0.000082 |
| 9 | −0.000056 |
| 10 | −0.000040 |
| $\sum_{11}^{\infty}$ | −0.000170 |
| Total | −0.04316 |

[14] E. U. Condon and G. H. Shortley, *The Theory of Atomic Spectra* (Cambridge University Press, Cambridge, England, 1957), p. 178.

TABLE VI. Diagonal contributions to $E_{\text{corr}}(2s,2s)$ from $p$ states.[a]

| Lowest state[b] | Sum of bound states[c] | Continuum states[d] | Bound +continuum |
|---|---|---|---|
| $2p$ | $-4.3155\times10^{-2}$ | $-3.804\times10^{-3}$ | $-4.6959\times10^{-2}$ |
| $3p$ | $-3.003\times10^{-4}$ | $-5.225\times10^{-4}$ | $-8.228\times10^{-4}$ |
| $4p$ | $-3.89\times10^{-5}$ | $-1.794\times10^{-4}$ | $-2.183\times10^{-4}$ |
| $5p$ | $-9.61\times10^{-6}$ | $-8.210\times10^{-5}$ | $-9.171\times10^{-5}$ |
| $6p$ | $-3.36\times10^{-6}$ | $-4.456\times10^{-5}$ | $-4.792\times10^{-5}$ |
| $7p$ | $-1.44\times10^{-6}$ | $-2.693\times10^{-5}$ | $-2.837\times10^{-5}$ |
| $8p$ | $-7.10\times10^{-7}$ | $-1.754\times10^{-5}$ | $-1.825\times10^{-5}$ |
| $9p$ | $-3.75\times10^{-7}$ | $-1.209\times10^{-5}$ | $-1.247\times10^{-5}$ |
| $10p$ | $-2.16\times10^{-7}$ | $-8.680\times10^{-6}$ | $-8.900\times10^{-6}$ |
| $11p$ | $-1.33\times10^{-7}$ | $-6.440\times10^{-6}$ | $-6.570\times10^{-6}$ |
| $12p$ | $-9.0\times10^{-8}$ | $-4.910\times10^{-6}$ | $-5.000\times10^{-6}$ |
| $\sum_{n=13}^{\infty} np$ | $-4.14\times10^{-7}$ | $-2.470\times10^{-5}$ | $-2.510\times10^{-5}$ |
| | | Bound+continuum total: | $-4.8245\times10^{-2}$ |
| | | Continuum−continuum states: | $-1.6827\times10^{-3}$ |
| | | Diagonal total | $-4.9928\times10^{-2}$ |

[a] All energies are in a.u. Second-order and only diagonal higher order bound-state contributions are included. See Eq. (24). Sums over $m_l$ have been made.
[b] One of the two excited states has this quantum number. The other state has a principal quantum number greater than or equal to this.
[c] The sum is over all bound excited states with principal quantum number greater than or equal to that at the left. In the first row the sum runs from $2p$ to $\infty$. In second row the sum runs from $3p$ to $\infty$, etc.
[d] Hole-particle and particle-particle interactions are included.

Note that diagonal bound-state contributions are included. The continuum particle interaction with the $2s$ hole is treated as in K and similarly for the particle-particle interaction. That is, we consider

$$a_{np}(k)=-\frac{2}{\pi}\int_0^\infty dk'\langle 2sk|v|2sk'\rangle D^{-1}(k')\times\langle npk'|v|2s2s\rangle\langle npk|v|2s2s\rangle^{-1}\quad(27)$$

and

$$t_{np}(k)=-\frac{2}{\pi}\int_0^\infty dk'\langle npk|v|npk'\rangle D(k')^{-1}\times\langle npk'|v|2s2s\rangle\langle npk|v|2s2s\rangle^{-1}.\quad(28)$$

In Eq. (28) there is also a sum over $m_l$. As found previously in K for two continuum excitations, both $a_{np}(k)$ and $t_{np}(k)$ were almost constant, with a very small dependence on $k$. The ladder or particle-particle interactions and the continuum particle-hole interactions are then summed by multiplying Eq. (25) by the factor

$$(1-a_{np}(k)-t_{np}(k))^{-1},\quad(29)$$

where an average value for $k$ is used. The term $t_{np}$ differs by 4% for $np(m_l=\pm1)$ and $np(m_l=0)$.

For $n=2$, $a_{2p}(0.4)=0.3385$, $t_{2p}(0.4)=0.323(m_l=\pm1)$, and $t_{2p}(0.4)=0.337(m_l=0)$. The factor of Eq. (29) is then 1.0109, a small correction.

Contributions to the $2s$ correlation energy from the various excited $l=1$ states are given in Table VI. The principal contribution is seen to come from $2p$ excitations.

Some typical nondiagonal terms are given in Fig. 4. Changes from one continuum state to another have already been included in Table VI. All calculations are made with the shifted denominator of Eq. (18). All nondiagonal terms with a significant contribution to $E_{corr}(2s,2s)$ are listed in Table VII. The total contributions of Tables VI and VII are added to give $E_{corr}(2s,2s)\times(l=1)=-0.04357$ a.u. In the previous calculations of K, $E_{corr}(2s,2s)(l=1)$ was found to be $-0.04256$ a.u. The value calculated here is believed to be a much more accurate result than the calculation of K. If the new value for $E_{corr}(2s,2s)(l=1)$ is added to the other contributions to the total correlation energy calculated in K, the total correlation energy of Be is changed from $-0.091$ to $-0.092$ a.u. which improves agreement with the value $-0.0953$ a.u. deduced from experiment.

## V. POLARIZABILITIES AND SHIELDING FACTORS

### A. Dipole Polarizability for Be

An atom perturbed by an external charge $Z'$ becomes polarized; and the effect of the external electric field on the atom depends upon the atomic dipole polarizability $\alpha_d$. An extensive discussion of atomic polarizabilities and shielding factors has been given by Dalgarno[15] and his notation is used in this section.

The unperturbed Hamiltonian is

$$H=-\sum_{i=1}^{N}\left(\frac{\nabla_i^2}{2}+\frac{Z}{r_i}\right)+\sum_{i<j}^{N}|\mathbf{r}_i-\mathbf{r}_j|^{-1}. \tag{30}$$

Atomic units are used in all formulas. The ground-state wave function $\Psi_0$ satisfies[16]

$$(H-E)\Psi_0=0. \tag{31}$$

TABLE VII. Nondiagonal terms in $E_{corr}(2s,2s)$ for $l=1$.[a]

| | |
|---|---|
| $2p2p\rightarrow 2p2p$[b] | $4.058\times10^{-3}$ |
| $\sum_{n=3}^{\infty}(2p2p\rightarrow 2pnp)$ | $2.016\times10^{-4}$ |
| $2p2p\rightarrow 2pkp$ | $-2.655\times10^{-4}$ |
| $\sum_{n=3}^{\infty}(2p2p\rightarrow 3pnp)$ | $2.45\ \times10^{-4}$ |
| $2p2p\rightarrow 3pkp$ | $4.11\ \times10^{-4}$ |
| $\sum_{n=4}^{\infty}\left(2p2p\rightarrow np\begin{Bmatrix}mp\\kp\end{Bmatrix}\right)^{c}$ | $3.66\ \times10^{-4}$ |
| $2p2p\rightarrow kpkp$ | $1.399\times10^{-3}$ |
| Nondiagonal total | $6.355\times10^{-3}$ |

[a] Hole-particle diagrams and ladder diagrams are included.
[b] This includes $2p^+(m_l=\pm1)2p^-(m_l=\mp1)\rightarrow 2p^+(m_l=\pm1)2p^-(m_l=\pm1)$ $2p^+(m_l=\pm1)2p^-(m_l=\mp1)\leftrightarrow 2p^+(m_l=0)2p^-(m_l=0)$.
[c] This includes sum $m\geq n$.

[15] A. Dalgarno, Advan. Phys. 11, 281 (1962).
[16] Note that $\Psi_0$ is unperturbed with respect to interactions with the external charge $Z'$. However, $\Psi_0$ includes correlation effects and is perturbed with respect to the Hartree-Fock solution.

The interaction potential between the external charge $Z'$ at $\mathbf{r}'$ and the atom is given by

$$V_{ext}=-Z'\sum_{i=1}^{N}\sum_{k=1}^{\infty}(r_i^k/r'^{k+1})P_k(\cos\theta_i), \tag{32}$$

where the polar axis has been chosen along the lien between the nucleus and $\mathbf{r}'$ and the contant, spherically symmetric part of $V_{ext}$ has been omitted. The perturbed wave function

$$\Psi=\Psi_0+Z'\sum_{k=1}^{\infty}\Psi_1^{(k)}/r'^{k+1}+O(Z'^2). \tag{33}$$

The dipole polarizability is

$$\alpha_d=2\langle\Psi_0|\sum_{i=1}^{N}r_iP_1(\cos\theta_i)|\Psi_1^{(1)}\rangle/\langle\Psi_0|\Psi_0\rangle, \tag{34}$$

where $Z'$ is assumed small. The wave function $\Psi_1^{(1)}$ then is the function $\Psi_0$ perturbed once by the term $-U_1$ where

$$U_k=\sum_{i=1}^{N}r_i^kP_k(\cos\theta_i). \tag{35}$$

In our case the function $\Psi_0$ is not known at the outset; so we start from the Hartree-Fock $\Phi_0$ and use BG theory to calculate $\Psi$.

The perturbation is

$$H'=\sum_{i<j}^{N}|\mathbf{r}_i-\mathbf{r}_j|^{-1}-\sum_{i=1}^{N}V_i-Z'\sum_{k=1}^{\infty}U_k/r'^{k+1}. \tag{36}$$

The BG linked cluster result is derived as usual[2] and

$$\Psi=\sum_L\left(\frac{1}{E_0-H_0}H'\right)^n\Phi_0, \tag{37}$$

where $\sum_L$ indicates that we sum over all linked terms.[2] The function $\Psi_0$ is given by the sum of all terms of $\Psi$ in which there are no interactions involving $U_k$. The term $\Psi_1^{(1)}$ is the sum of all terms of $\Psi$ in which $-U_1$ acts once and only once, $Z'/r'^2$ being factored out. $\Psi_0$ obtained from Eq. (37) is not normalized to unity. However, in the numerator of Eq. (34) we may factor the disconnected terms into a product of terms involving $U_1$ times all other terms. If we neglect the exclusion principle, the second factor is $\langle\Psi_0|\Psi_0\rangle$ and cancels the denominator. This factorization proceeds as in the derivation of the linked cluster result. However, the exclusion principle must be considered in this factorization and it will be shown to have a significant effect for small systems.

The terms contributing to Eq. (34) may be represented by diagrams as in Fig. 5. The ordering of interactions from the bottom to the top of the diagram corresponds to interactions proceeding from right to left in Eq. (34). The interaction lines labeled DP for dipole

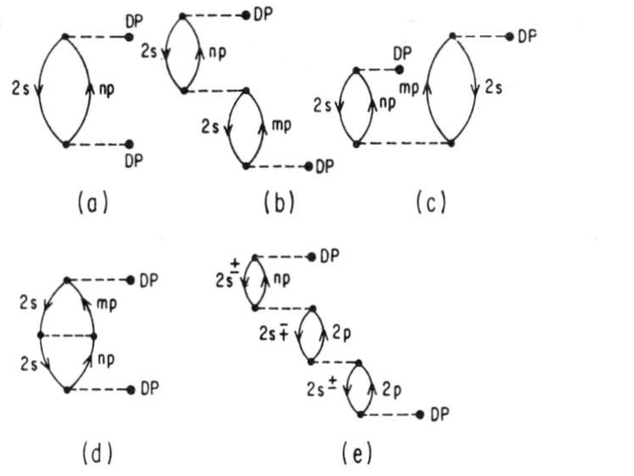


FIG. 5. Lowest order terms contributing to the dipole polarizability $\alpha_d$. The interaction labeled DP refers to the dipole polarizability operator $r\cos\theta$. (a) Second-order term. (b) and (c) are third-order terms with one correlation interaction. There is also the term obtained by inverting (c). (d) Hole-particle interaction diagram which does not occur when the single-particle states are calculated as in this paper. (e) Fourth-order diagram which is an iteration of the diagram (b).

polarizability indicate interactions through the operator $U_1$. The diagrams are calculated by the usual perturbation theory rules[2] and multiplied by $-2$ to give $\alpha_d$. The factor 2 comes from Eq. (34) and $(-1)$ from Eq. (36). The lowest order contribution to $\alpha_d$ is positive; it is shown in Fig. 5(a) and has the value

$$\alpha_d(2s\rightarrow np)=-4|\langle 2s|r\cos\theta|np\rangle|^2/(\epsilon_{2s}-\epsilon_{np}). \quad (38)$$

Equation (38) includes a factor 2 because there are two $2s$ electrons. Transitions to continuum states are given by

$$\sum_k \alpha_d(2s\rightarrow kp)=-\frac{8}{\pi}\int_0^\infty dk|\langle 2s|r\cos\theta|kp\rangle|^2/(\epsilon_{2s}-k^2/2), \quad (39)$$

where the continuum states are normalized so that

$$P_k(r)=\sin(kr+(1/k)\ln 2kr+\delta) \quad (40)$$

as $r\rightarrow\infty$ and $\sum_k=(2/\pi)\int_0^\infty dk$ as shown in K. The

TABLE VIII. Dipole polarizability in second order.[a]

| $n$ | $\alpha_d(2s\rightarrow np)$ Å$^3$ | $n$ | $\alpha_d(1s\rightarrow np)$ Å$^3$ |
|---|---|---|---|
| 2 | 11.9371 | 2 | 0.001091 |
| 3 | 0.01213 | 3 | 0.000202 |
| 4 | 0.00939 | 4 | 0.0000753 |
| 5 | 0.00513 | 5 | 0.0000362 |
| 6 | 0[illegible]298 | 6 | 0.0000202 |
| 7 | .00187 | 7 | 0.0000124 |
| 8 | [illegible]00124 | 8 | 0.0000081 |
| $\sum_{n=9}^{\infty}$ | 0.00393 | $\sum_{n=9}^{\infty}$ | 0.0000257 |
| continuum | 0.17049 | continuum | 0.005848 |
| $2s$ total | 12.1443 Å$^3$ | $1s$ total | 0.007319 Å$^3$ |

[a] 1 Å$^3$ = $10^{-24}$ cm$^3$.

second-order contribution to $\alpha_d$ from $1s$ electrons is obtained by replacing $2s$ by $1s$ in Eqs. (38) and (39). Numerical results for $\alpha_d$ in second order are given in Table VIII. The validity of the $n^{-3}$ rule for $|\langle 2s|r\cos\theta|np\rangle|^2$ may be checked in Table III. It is interesting to note that almost the entire contribution to $\alpha_d$ comes from $2p$ excited states and that excitations of $1s$ electrons contribute negligibly. The second-order $\alpha_d$ is 12.15 Å$^3$ as compared with 4.54 Å$^3$ obtained by Kelly and Taylor[17] in a second-order calculation using the set of Hartree-Fock states described in K in which all excited states are in the continuum. It was pointed out[17] that this second-order calculation is equivalent to the uncoupled Hartree-Fock approximation of Dalgarno.[15] In higher order calculations using this continuum set it is necessary to calculate the diagram of Fig. 5(d) and higher iterations. This type of diagram was called a second class EPV diagram in K. It arises from the fact that for this set the interactions of an excited particle with the Hartree-Fock potential do not cancel the interactions with the occupied unexcited states. This is the analog of Fig. 1(d).

In K these diagrams were found to be comparable in size to the second-order term and of the same sign. In the calculations of this paper the single-particle states were calculated so that interactions of excited states with the potential are canceled by interactions with the occupied unexcited states when there is only a single $2s$ excitation. Another way to look at this problem is to note that all terms of the type of Fig. 5(d) have been summed and need no longer be considered when the states of this paper are used. Since these terms are all of the same sign as the second-order term, it is understandable that the second order result of this paper should be much larger than that reported in KT.

The third order terms of Figs. 5(b) and (c) which reduce the second-order result are found to be large. This is expected since they differ from the second-order term by one $l=1$ correlation interaction, and in Sec. IV

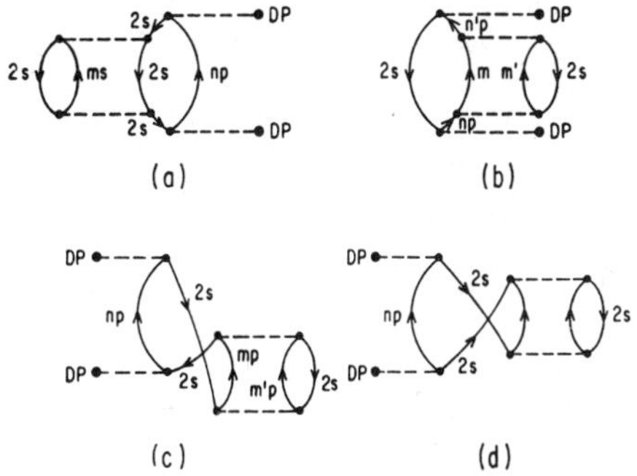


FIG. 6. Fourth-order diagrams which modify single-particle excitations. Diagrams (a), (c), and (d) are rearrangement diagrams discussed in Ref. 18.

[17] H. P. Kelly and H. S. Taylor, J. Chem. Phys. 40, 1478 (1964), hereafter referred to as KT.

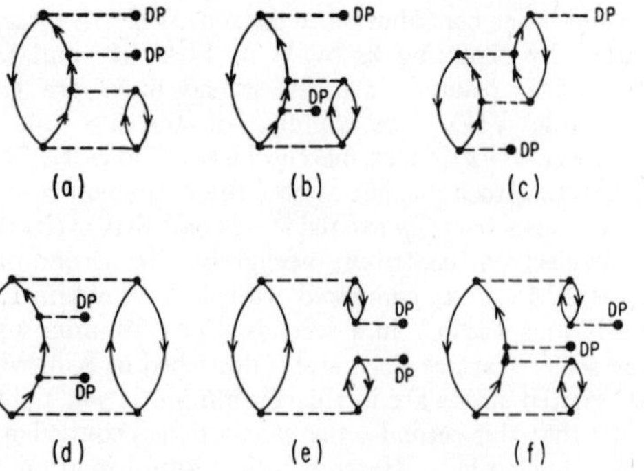


FIG. 7. Basic fourth-order diagrams contributing to the dipole polarizability $\alpha_d$ are given by diagrams (a), (b), (c), (d), and (e). Diagrams (a), (b), and (f) may be inverted. In diagrams (b) and (d) the first DP interaction may also occur on the other particle line. A fifth-order diagram modifying (a) is given by (f). Similar diagrams modify (b) and (c).

these correlations were large. It is desirable to include as many higher order effects as possible when the basic third-order diagrams are calculated. This is achieved by first considering modifications to the single-particle excitations shown in Figs. 6(a), (b), (c), and (d). These modifications will be shown to be included by an effective shift in the energy denominators for single-particle excitations and they are found to be small compared to other effects. The modification which is numerically largest is shown in Fig. 5(e). That is, whenever we have a single-particle excitation we include all correlation interactions which interchange the excited and unexcited 2*s* electrons. Due to the dominance of excitations into 2*p* states as seen from Tables II and VI, it is possible to sum exactly the principal part of this modification by considering the geometrical series

$$\begin{aligned} MF &= 1+(\langle 2p2s|v|2s2p\rangle/(\epsilon_{2s}-\epsilon_{2p}+d))+\cdots \\ &= [1-\langle 2p2s|v|2s2p\rangle/(\epsilon_{2s}-\epsilon_{2p}+d)]^{-1}. \end{aligned} \tag{41}$$

The term $d$ is due to the modifications of Fig. 6. Whenever there is a single excitation into the state 2*p*, the term is multiplied by $MF$. In Fig. 5(b) for $m=2$ and $n\neq 2$, we multiply by $MF$ and have included all diagrams like that of Fig. 5(e) to all orders. If we multiply the diagram of Fig. 5(a) by $MF$ then we must be careful not to include $n=m=2$ in Fig. 5(b) as this is already included in $MF$. In these calculations, $MF=0.696$. When the diagram of Fig. 5(c) is calculated, the denominator $D$ of Eq. (18) is used to account for higher correlation effects. The nondiagonal terms discussed in Sec. III must also be included. When $m=2$, this diagram is multiplied by $MF$. The inverted diagram is numerically identical.

Modifications due to the diagrams of Fig. 6 are now considered. Diagrams (a), (c), and (d) are "rearrangement" diagrams discussed by Brueckner and Goldman.[18]

[18] K. A. Brueckner and D. T. Goldman, Phys. Rev. **117**, 207 (1960).

Diagrams (c) and (d) result from the linked cluster factorization and may be added to give the negative product of the second-order term for $\alpha_d$ and the second-order correlation energy term. Higher order diagrams like (c) and (d) give additional factors of correlation energy terms and the result is a geometrical series which may be summed to give the second-order term for $\alpha_d$ with the shifted denominator[10]

$$D=\epsilon_{2s}-\epsilon_{2p}+E_{\text{corr}}(2s,2s)+2E_{\text{corr}}(2s,1s). \tag{42}$$

Calculations of 6(a) and (b) were made for $n=n'=2$ as 2*p* excitations are dominant. The ratio of diagrams (a) and (b) to the second-order diagram of $\alpha_d$ establishes the ratio of terms in a geometric series which is also summed to give a shifted denominator. In calculating (b), $m$ and $m'$ are the possible excited bound and continuum states consistent with the rules for allowed angular momenta of single-particle states in Coulomb matrix elements.[14] That is, we may have $mp\,m's$, $mp\,m'd$, $md\,m'p$, etc. In calculating 6(b), only $s$, $p$, and $d$ states were considered. The shift in Eq. (42) due to diagrams like 6(c) and (d) is $-0.0465$ a.u. However, this number is largely canceled by the shift due to diagrams like 6(a) and (b) and the net shift is only $-0.0156$ a.u. For comparison, $\epsilon_{2s}-\epsilon_{2p}=-0.1299$ a.u. Also, this effect for third-order terms in this case tends to cancel that for second-order terms. The second- and third-order diagrams were calculated with the modifications just described to account for iterations of certain terms beyond third order. The result was $\alpha_d=5.569$ Å$^3$.

In fourth order, new types of diagrams enter and examples are shown in Fig. 7. We may note that we have now included terms in which the correlations and DP interactions have assumed all possible relative positions. Diagrams 7(a), (b), and (f) may also be inverted and in 7(b) and (d) the first DP interaction may occur on the other particle line. When we have a single 2*p* excitation, we multiply by the factor $MF$ of Eq. (41). When the diagram begins or ends with a single excitation as for 7(a), (b), and (c), there is also the modification of the type shown in 7(f) which modifies 7(a). Numerical calculations of the diagrams of Fig. 7 are given in Table IX. Excited bound and continuum states were included for $l=0$, 1, and 2.

The factor which results from the normalization $\langle\Psi_0|\Psi_0\rangle$ in the denominator of Eq. (34) for $\alpha_d$ must be

TABLE IX. Contributions to $\alpha_d$ from diagrams of Fig. 7.[a]

| Diagram | Value in Å$^3$ |
|---|---|
| a | 0.8484 |
| b | 0.4257 |
| c | 0.5957 |
| d | 0.3383 |
| e | 0.3389 |
| f | −0.4232 |
| Total | 2.1238 |

[a] 1 Å = $10^{-8}$ cm.

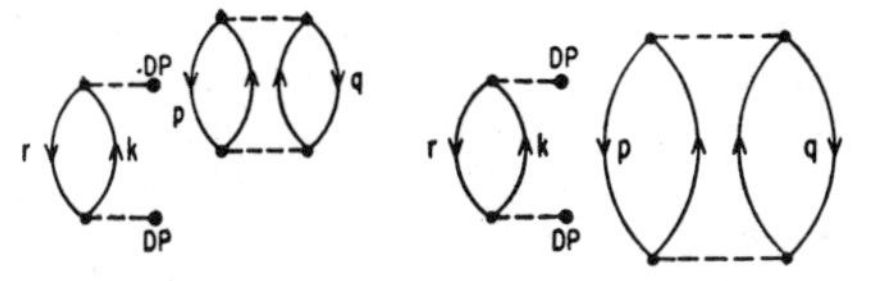

(a) (b)

FIG. 8. Disconnected diagrams which factor when added. States $p$ and $q$ must be different from $r$ and the excitations of $p$ and $q$ must differ from $k$ because of the exclusion principle. Diagrams of this type give the factor of Eq. (45).

included. This effect of this factor is reduced when we consider the higher order terms of Fig. 8. When 8(a) and 8(b) are added, the disconnected parts factor into the product $\alpha_d$ in second order ($\alpha_d^{(2)}$) times the lowest order contribution of $p$, $q$ to the correlation part of the normalization which is

$$Nm(p,q)=\sum_{kk'}|\langle pq|v|kk'\rangle|^2/D^2, \tag{43}$$

$$\langle\Psi_0|\Psi_0\rangle=1+\sum_{pq} Nm(p,q). \tag{44}$$

Although the nondiagonal terms have not been explicitly written in Eq. (43), they are assumed included in $Nm(p,q)$. The diagrams like those of Fig. 8 and the normalization $\langle\Psi_0|\Psi_0\rangle^{-1}$ combine to multiply $\alpha_d^{(2)}(r\rightarrow k)$ by the factor

$$(1+\sum_{pq\neq r} Nm'(p,q))/(1+\sum_{pq} Nm(p,q)). \tag{45}$$

The sums $pq$ extend over all unexcited states except in the numerator where $p$ and $q$ must not equal $r$ because of the exclusion principle. The prime in the numerator indicates that the excited state $k$ is not to be included in calculating $Nm(p,q)$. The higher order diagrams for $\alpha_d$ are treated similarly. Only connected terms are then retained, the corrections from disconnected terms being contained in Eq. (45). For a large system, Eq. (45) becomes effectively one. However, for a small system such as Be, the restrictions on the sum in the numerator of Eq. (45) may have an important effect. Calculations of the normalization terms $Nm$ resulted in $Nm(2s,2s)=0.1107$, $Nm(1s,1s)=0.0028$, and $Nm(1s,2s)=0.000125$. In calculating $\langle\Psi_0|\Psi_0\rangle$, $4Nm(1s,2s)$ is included to account for the four $1s-2s$ pairs. The total value for $\alpha_d$ before normalization is 7.69 Å$^3$ and after normalization becomes 6.93 Å$^3$. For Be the normalization effect is approximately 10%. This correction may be especially large for Be due to the low-lying $2p$ excitation which enhances the effect of $D^2$ in Eq. (43).

## B. The Dipole Shielding Factor

When an external charge $Z'$ is placed at $\mathbf{r}'$, the electric field at the nucleus due to the electrons divided by the electric field at the nucleus due to $Z'$ is defined as the dipole shielding factor $\beta_\infty$.

$$\beta_\infty=(\langle\Psi|\sum_{i=1}^{N} P_1(\cos\theta_i)/r_i^2|\Psi\rangle/\langle\Psi|\Psi\rangle)(Z'/r'^2)^{-1}$$

$$=2\langle\Psi_0|\sum_{i=1}^{N} P_1(\cos\theta_i)/r_i^2|\Psi_1^{(1)}\rangle/\langle\Psi_0|\Psi_0\rangle, \tag{46}$$

where $Z'$ is assumed small and terms of second and higher powers of $Z'$ are neglected. The formula for $\beta_\infty$ differs from Eq. (34) for $\alpha_d$ only by the replacement of the second interaction $U_1$ by the shielding term

$$S_1=\sum_{i=1}^{N} P_1(\cos\theta_i)/r_i^2. \tag{47}$$

Therefore, $\beta_\infty$ may be calculated analogously to $\alpha_d$, the second one-body interaction being $S_1$.

The formulas for $\beta_\infty$ in second order are

$$\sum_{n=2}^{\infty}\beta_\infty(2s\rightarrow np)=-4\sum_{n=2}^{\infty}\langle 2s|r^{-2}\cos\theta|np\rangle$$
$$\times(\epsilon_{2s}-\epsilon_{np})^{-1}\langle np|r\cos\theta|2s\rangle \tag{48}$$

and

$$\sum_k\beta_\infty(2s\rightarrow kp)=-(8/\pi)\int_0^\infty dk\langle 2s|r^{-2}\cos\theta|kp\rangle$$
$$\times(\epsilon_{2s}-k^2/2)^{-1}\langle kp|r\cos\theta|2s\rangle. \tag{49}$$

The $1s$ contributions are obtained by replacing $2s$ by $1s$ in Eqs. (48) and (49). These equations are analogous to Eqs. (38) and (39) for $\alpha_d$. The second-order contributions to $\beta_\infty$ given in Table X add to 4.296 which is in poor agreement with the theoretical value 1.00,[19] and so it is necessary to consider the higher order terms. There are terms of the same type as considered for $\alpha_d$ and shown in Figs. 5, 6, and 7. In addition there are now important contributions from the $1s$ electrons. This is not surprising as $\beta_\infty$ involves matrix elements of $r^{-2}$

TABLE X. Second-order contributions to the dipole shielding factor $\beta_\infty$.

| $n$ | $\beta_\infty(2s\rightarrow np)$ | $n$ | $\beta_\infty(1s\rightarrow np)$ |
|---|---|---|---|
| 2 | 3.88977 | 2 | 0.03267 |
| 3 | −0.02783 | 3 | 0.00635 |
| 4 | −0.01276 | 4 | 0.00240 |
| 5 | −0.00612 | 5 | 0.00116 |
| 6 | −0.00337 | 6 | 0.00065 |
| 7 | −0.00205 | 7 | 0.00040 |
| 8 | −0.00134 | 8 | 0.00026 |
| $\sum_{n=9}^{\infty}$ | −0.00422 | $\sum_{n=9}^{\infty}$ | 0.000829 |
| continuum | −0.01766 | continuum | 0.4373 |
| $2s$ total | 3.814 | $1s$ total | 0.482 |

[19] R. M. Sternheimer, Phys. Rev. **96**, 951 (1954).

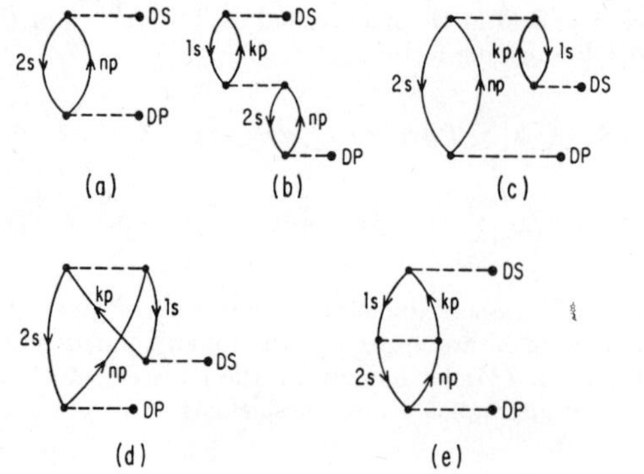


FIG. 9. (a) Second-order diagram for the dipole shielding factor $\beta_\infty$. (b), (c), (d), and (e) are third-order diagrams involving 1*s* and 2*s* states. In diagrams (d) and (e) the two hole states must have parallel spins.

which emphasizes the inner atomic regions. In the most important of these $1s-2s$ terms shown in Fig. 9 the 2*s* electron contributes to the $r$ matrix element and the 1*s* electron to $r^{-2}$. The line labeled DS represents the interaction with the shielding operator $S_1$. Diagram (c) may be inverted except that the DS interaction must appear above DP. Diagrams (b) and (c) are similar to $2s-2s$ third-order diagrams except that one of the hole states is labeled 1*s*. In diagrams (d) and (e) the hole states must have parallel spins.

When diagram 9(b) was calculated with the factor *MF* of Eq. (41) the result was $-0.592$. The sum of 9(c) and its inverted form was also $-0.592$. Diagram (d) was calculated to be $-0.0489$ and (e) was $-0.126$. Additional modifications to these diagrams come from the fact that when there is a single 1*s* excitation, the interactions with the potential do not exactly cancel interactions with the occupied unexcited states. This was found to give approximately a 10% increase to diagrams like (b) and (e). This increase was approximately canceled by inclusion of higher order $1s-1s$ interactions which modify a single 1*s* excitation in the same way that the single 2*s* excitation in Fig. 5(a) is modified by 5(b) and (c).

The result of calculating the modified second- and third-order diagrams involving two 2*s* electrons was found to be 1.248. The basic diagrams are shown in Figs. 5(a), (b), and (c) except that the topmost interaction is now DS rather than DP. The appropriate modifications to the basic diagrams were discussed in Sec. IV A. Similar calculations involving the two 1*s* electrons gave the result 0.438. Calculation of the $1s-2s$ diagrams of Figs. 9(b), (c), (d), and (e) gave the result $-1.189$. This number includes the modifications discussed above and a very small contribution from interations of the basic diagrams. The modified total result through third order is then 0.498.

The final result was obtained by considering higher order diagrams of the type shown in Fig. 7 with DS replacing the upper DP and by including the normalization factor of Eq. (45). The higher order diagrams included not only the basic structures of Fig. 7 but also the possible additions to them by adding on $1s-2s$ interactions in the same way that diagrams in Figs. 9(b), (c), (d), and (e) may be considered as additions to the basic diagram of Fig. 9(a). Before normalization the higher order terms contributed 0.532, giving a total 1.030. After normalization the total dipole shielding result was 0.972. The normalization factors significantly affected (by approximately 10%) only the terms in which there was at least one 2*s* electron excited.

### C. Quadrupole Polarizability

The quadrupole polarizability is defined by

$$\alpha_q \equiv 2\langle\Psi_0|\sum_{i=1}^{N} r_i^2 P_2(\cos\theta_i)|\Psi_1^{(2)}\rangle/\langle\Psi_0|\Psi_0\rangle. \quad (50)$$

The term $\Psi_1^{(2)}$ is defined by Eq. (33) and is obtained from BG theory by collecting all the terms of $\Psi$ given by Eq. (37) in which the perturbation $-U_2$ acts once and only once. The calculation of $\alpha_q$ proceeds in the same manner as that of $\alpha_d$ except that the operator $U_1$ is replaced by $U_2$.

The second-order formulas are

$$\sum_{n=2}^{\infty} \alpha_q(2s\rightarrow nd) = -4\sum_{n=2}^{\infty} |\langle 2s|r^2P_2(\cos\theta)|nd\rangle|^2 \times(\epsilon_{2s}-\epsilon_{nd})^{-1}, \quad (51)$$

$$\sum_k \alpha_q(2s\rightarrow kd) = -\frac{8}{\pi}\int_0^\infty dk|\langle 2s|r^2P_2(\cos\theta)|kd\rangle|^2 \times(\epsilon_{2s}-k^2/2)^{-1}. \quad (52)$$

The angular integrations contribute a factor $\frac{1}{5}$ on the right-hand side of Eqs. (51) and (52). Both equations contain a factor 2 for two 2*s* electrons. The 1*s* terms are obtained by substituting 1*s* for 2*s* in Eqs. (51) and (52).

The results of the second-order calculations are given in Table XI and the total second-order result is 15.09 Å$^5$.

TABLE XI. Second-order contributions to the quadrupole polarizability.[a]

| $n$ | $\alpha_q(2s\rightarrow nd)$ Å$^5$ | $n$ | $\alpha_q(1s\rightarrow nd)$ Å$^5$ |
|---|---|---|---|
| 3 | 6.7760 | 3 | $6.889\times10^{-7}$ |
| 4 | 1.9161 | 4 | $4.028\times10^{-7}$ |
| 5 | 0.8059 | 5 | $2.326\times10^{-7}$ |
| 6 | 0.4184 | 6 | $1.427\times10^{-7}$ |
| 7 | 0.2467 | 7 | $9.293\times10^{-8}$ |
| 8 | 0.1583 | 8 | $6.355\times10^{-8}$ |
| 9 | 0.1079 | 9 | $4.525\times10^{-8}$ |
| 10 | 0.07701 | 10 | $3.330\times10^{-8}$ |
| $\sum_{n=11}^{\infty}$ | 0.3218 | $\sum_{n=11}^{\infty}$ | $1.444\times10^{-7}$ |
| continuum | 4.2620 | continuum | 0.000609 |
| 1*s* total | 15.0901 Å$^5$ | 1*s* total | 0.000611 Å$^5$ |

[a] 1 Å$^5$ = $10^{-40}$ cm$^5$.

TABLE XII. Second-order contributions to the quadrupole shielding factor $\gamma_\infty$.

| $n$ | $\gamma_\infty(2s\to nd)$ | $n$ | $\gamma_\infty(1s\to nd)$ |
|---|---|---|---|
| 3 | 0.19430 | 3 | $2.546\times10^{-5}$ |
| 4 | 0.07383 | 4 | $1.514\times10^{-5}$ |
| 5 | 0.03537 | 5 | $8.778\times10^{-6}$ |
| 6 | 0.01968 | 6 | $5.385\times10^{-6}$ |
| 7 | 0.01208 | 7 | $3.517\times10^{-6}$ |
| 8 | 0.00796 | 8 | $2.405\times10^{-6}$ |
| 9 | 0.00552 | 9 | $1.719\times10^{-6}$ |
| 10 | 0.00399 | 10 | $1.252\times10^{-6}$ |
| $\sum_{n=11}^{\infty}$ | 0.01928 | $\sum_{n=11}^{\infty}$ | $6.888\times10^{-6}$ |
| continuum | 0.51589 | continuum | 0.16987 |
| $2s$ total | 0.88790 | $1s$ total | 0.16994 |

The contribution from $1s$ terms is negligible. The third-order terms correspond to those in Figs. 5(b) and (c) for $\alpha_d$ including the inverted form of 5(c). The interaction lines labeled DP now are labeled QP and correspond to the interaction $U_2$. The excited states are $nd$ and $kd$. Since the correlations between two $2s$ electrons are given almost entirely by excitations into $p$ states,[5] the third-order terms for $\alpha_q$ were expected to be much less important than they were for $\alpha_d$. This was found to be true; the third-order terms contributed $-1.034$ Å$^5$, giving a total result 14.06 Å$^5$ for second- and third-order terms. Fourth-order terms were not calculated and the normalization factors were omitted for consistency as they correspond to fourth-order and higher terms. Most fourth-order terms are expected to increase the value for $\alpha_q$, and $\alpha_q$ is reduced by the normalization factors so there should be some cancellation between these two effects. However, since $l=1$ states may enter into the fourth-order terms, the fact that third-order terms are small does not necessarily imply fourth-order terms are also small.

### D. The Quadrupole Shielding Factor

The quadrupole shielding factor $\gamma_\infty$ is defined as the change in the gradient of electric field at the nucleus due to the electrons divided by the gradient of electric field at the nucleus due to the external charge $Z'$.

$$\gamma_\infty \equiv 2\langle\Psi_0|\sum_{i=1}^{N} r_i^{-3}P_2(\cos\theta_i)|\Psi_1^{(2)}\rangle/\langle\Psi_0|\Psi_0\rangle. \quad (53)$$

The second-order terms are

$$\sum_{n=2}^{\infty}\gamma_\infty(2s\to nd)=\sum_{n=2}^{\infty}-4\langle 2s|r^{-3}P_2(\cos\theta)|nd\rangle(\epsilon_{2s}-\epsilon_{nd})^{-1} \times\langle nd|r^2P_2(\cos\theta)|2s\rangle, \quad (54)$$

$$\sum_k \gamma_\infty(2s\to kd)=-\frac{8}{\pi}\int_0^\infty dk\langle 2s|r^{-3}P_2(\cos\theta)|kd\rangle \times(\epsilon_{2s}-k^2/2)^{-1}\langle kd|r^2P_2(\cos\theta)|2s\rangle, \quad (54a)$$

and similarly for excitations of $1s$ electrons. The second order contributions to $\gamma_\infty$ are given in Table XII. Both $1s$ and $2s$ contributions are now significant. More than half of the total result 1.058 comes from continuum states; this is not surprising since the wave function for $nd$ states is generally far from the origin due to the centrifugal barrier and so the $r^{-3}$ matrix elements are small. However, the continuum states have sufficient energy to overcome much of the barrier.

The third-order $2s-2s$ terms which were calculated correspond to the third-order correction terms for $\alpha_d$ shown in Fig. 5. However, the bottom interaction is now QP and the top interaction is through the quadrupole shielding (QS) term $r^{-3}P_2(\cos\theta)$. Excited states have $l=2$. Third-order terms involving $1s-2s$ interactions of the types shown in Fig. 9 but with changes to QP, QS, and $l=2$ excitations were also calculated. Corrections were made which account for the fact that for $1s$ excitations the interactions with the potential do not cancel interactions with occupied unexcited states. This effect gave a 7% increase to the terms with $1s$ excitations. The contributions from third-order $2s-2s$ terms was calculated to be $-0.144$ and from $1s-2s$ terms $-0.163$. The $1s-1s$ terms were small. The final result of these calculations is 0.751. The fourth-order and higher terms were not considered in calculating $\gamma_\infty$ and it is possible that they might contribute significantly since excitations into $l=1$ states are now possible. Very rough calculations of some fourth-order terms indicated that $\gamma_\infty$ might increase by as much as 20% and $\alpha_q$ change less. Any increase, of course, would be partly offset by the normalization factor. It is possible that the second-order result could turn out to be in better agreement with experiment than the result including third-order terms.

(*Note added in proof*. The results of this section are in good agreement with those of Professor A. Dalgarno who has used the coupled Hartree-Fock approximation for Be. An analysis of the coupled method indicates that it includes the second- and third-order diagrams of this section and higher iterations of these basic diagrams, and so the coupled method actually includes some of the correlation effects. I am grateful to Professor Dalgarno for forwarding his results prior to publication.)

## VI. OSCILLATOR STRENGTHS

The oscillator strength $f_{ni}^z$ for a transition between an initial state $i$ and an excited state $n$ is given in atomic units by[20]

$$f_{ni}^z=2\omega_{ni}|\langle n|Z_{Op}|i\rangle|^2, \quad (55)$$

where $Z_{Op}=\sum_{j=1}^{N}z_j$. The energy difference between states $n$ and $i$ is given by $\omega_{ni}$ in a.u. The ground state of Be is $(1s)^2(2s)^2\,{}^1S$ and transitions are calculated to excited states $(1s)^2(2s)(np)\,{}^1P$ which may be written in

[20] H. A. Bethe, *Intermediate Quantum Mechanics* (W. A. Benjamin, Inc., New York, 1964), Chap. 13, p. 147.

TABLE XIII. Excitation energies in a.u. from $(2s)^2\,^1S$ to $(2s)(np)\,^1P$.

| $n$ | $\omega_{ni}$[a] | $\omega_{ni}$[b] |
|---|---|---|
| 2 | 0.1935 | 0.1939 |
| 3 | 0.2467 | 0.2742 |
| 4 | 0.2749 | 0.3063 |
| 5 | 0.2877 | |
| 6 | 0.2946 | |
| 7 | 0.2986 | |
| 8 | 0.3012 | |
| Ionization limit | 0.3094 | 0.3426 |

[a] Calculated from Eq. (58).
[b] Experimental values obtained from Ref. 21. Only values up to $n=4$ were listed.

second quantized notation

$$|2snp\,^1P\rangle = 2^{-1/2}(\eta_{2s^+}{}^+\eta_{np^-}{}^+ - \eta_{2s^-}{}^+\eta_{np^+}{}^+)|0\rangle, \quad (56)$$

when correlations in these states are neglected. The notation $2s^+$, $np^-$ indicates $2s$ electron with spin up, etc. The state $|0\rangle$ is the "core" state $(1s)^2\,^1S$. The operators $\eta^+$ satisfy the usual Fermi-Dirac anticommutation relations.[2] The excitation energy

$$\omega_{ni} = \langle 2snp\,^1P|H|2snp\,^1P\rangle - \langle (2s)^2\,^1S|H|(2s)^2\,^1S\rangle. \quad (57)$$

If we use the ground-state Hartree-Fock solution for $(2s)^2\,^1S$ and Eq. (56) for $(2s)(np)\,^1P$ with the $np$ single-particle orbitals determined by Eq. (13),

$$\omega_{ni} = \epsilon_{2p} - \epsilon_{2s} + \langle 2snp|v|np2s\rangle. \quad (58)$$

The $1s$ and $2s$ orbitals used in $(2s)(np)\,^1P$ are the same as for $(2s)^2\,^1S$.

Excitation energies calculated from Eq. (58) are given in Table XIII and are compared with the ob-observed energies obtained from the Charlotte Moore

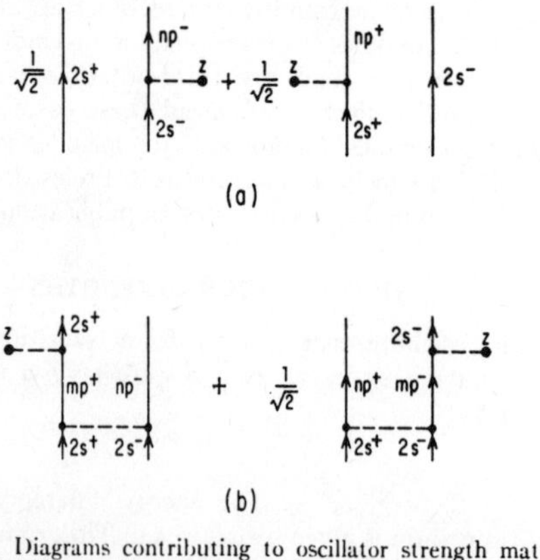

FIG. 10. Diagrams contributing to oscillator strength matrix elements. (a) Hartree-Fock approximation. (b) Correlation terms. These diagrams differ from the usual diagrams of Ref. 2 in that both hole and particle lines are directed upwards. The lines at the bottom of the diagrams represent single-particle states occupied in the initial state $i$. Lines at the top of the diagram represent single-particle states occupied in the final state $n$.

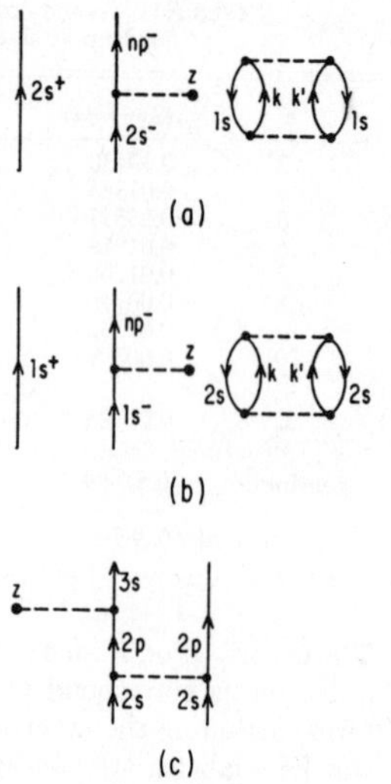

FIG. 11. Higher order diagrams contributing to oscillator strength matrix elements. (a) and (b) are disconnected diagrams. (c) Transitions $(2s)^2\,^1S \rightarrow (ns)\times(2p)\,^1P$ are possible when correlations among the two $2s$ electrons are included.

Tables.[21] The discrepancy between calculated and observed values increases with the excitations and is mostly due to omission of correlation corrections for the calculated energies. When a $2s$ electron has been raised to a highly excited state it is expected to have little correlation energy with the remaining $2s$ electron. If the $2s-2s$ correlation energy of the ground state is included with the ionization limit calculated from Eq. (58), the result is 0.353 a.u. which improves agreement with the observed value 0.343 a.u.

When the oscillator strengths are calculated in the first approximation,

$$f_{ni}{}^z = 4\omega_{ni}|\langle np|z|2s\rangle|^2. \quad (59)$$

This approximation for the matrix elements $\langle n|Z_{op}|i\rangle$ is illustrated by diagram (a) of Fig. 10. Unlike the usual many-body diagrams, both particle and hole lines point upwards; the lines at the bottom of the diagram correspond to the initial unexcited states of $i$ and the lines at the top correspond to the unexcited single-particle states of the final state $n$. The ground-state Hartree-Fock determinant $i$ is connected to each of the excited-state determinants in the linear combinations (56) through the matrix element $\langle np|z|2s\rangle$. It is also desirable to include the effects of correlations among the two $2s$ electrons in the ground state and these are represented by Fig. 10(b). The contribution to $\langle n|G_{op}|i\rangle$ from diagrams (a) and (b) together is

$$\langle n|Z_{op}|i\rangle = \sqrt{2}\Big(\langle np|z|2s\rangle + \sum_m \langle 2s|z|mp\rangle D^{-1} \times \langle mpnp|v|2s2s\rangle\Big). \quad (60)$$

[21] *Atomic Energy Levels*, edited by C. E. Moore, Natl. Bur. Std. Circ. No. 467 (U. S. Government Printing Office, Washington, D. C., 1949), Vol. I.

TABLE XIV. Oscillator strengths for transitions $(2s)^2\,{}^1S \to 2snp\,{}^1P$.

| $n$ | $f_{ni}^z(\mathrm{HF})$[a] | $f_{ni}^z(\mathrm{corr})$[b] |
|---|---|---|
| 2 | 2.0293 | 1.2540 |
| 3 | 0.00541 | 0.01676 |
| 4 | 0.00530 | 0.01013 |
| 5 | 0.00318 | 0.00549 |
| 6 | 0.00194 | 0.00321 |
| 7 | 0.00125 | 0.00202 |
| 8 | 0.00084 | 0.00135 |
| $\sum_{n=9}^{\infty}$ | 0.00295 | 0.00467 |

[a] Calculated from Eq. (59) using Hartree-Fock single-particle states.
[b] Ground-state correlations are included in these values.

The denominator $D$ is calculated by Eq. (18) to account for higher order terms and account is also taken of nondiagonal terms as explained in Sec. IV. There is still the normalization correction to be considered because the correlated ground state is obtained by BG theory and in Eq. (55) it is assumed that states $n$ and $i$ are normalized to unity. Most of the normalization corrections come from $2s$ electrons and the very small effects from $1s$ contributions to the normalization are essentially canceled by terms of the form shown in Fig. 11(a). A more detailed analysis of this point is given later in this section. Oscillator strengths calculated by Eq. (59), using single-particle Hartree-Fock states of Eq. (13), are compared in Table XIV with those calculated with correlation and normalization corrections. The observed excitation energies listed in Table XIII were used and extrapolations were made to obtain higher excitation energies. The values in Table XIV may be compared with the results of Altick and Glassgold[22] who used Hartree wave functions and employed the methods of the random-phase approximation.[23] Their oscillator strength for the $n=2$ transition is 2.34 using Hartree single-particle states and 1.71 including correlations by the random-phase approximation. For the higher levels, their Hartree oscillator strengths are larger than the Hartree-Fock values listed here but their correlated values are lower than the Hartree-Fock values of the present calculation.

It is of interest also to compute the oscillator strengths of $1s$ excitations to bound excited states and those for $1s$ and $2s$ transitions to the continuum so that we may evaluate the Thomas-Reiche-Kuhn sum rule.[20]

$$\sum_n f_{ni}^z = N. \tag{61}$$

When $1s \to np$ or $1s \to kp$ transitions are calculated, terms of the form shown in Fig. 11(b) should be included. These account for correlations among the two unexcited $2s$ electrons. These correlations should also be included in the normalization for both the ground state $i$ and excited state $n$. Because the particles in excited states are propagating in the presence of different unexcited states above and below the $z$ interaction, the denominators as given by Eq. (18) are different in these two cases. The correlation part of Fig. 11(b) is

$$|\langle 2s2s|v|kk'\rangle|^2/D_nD_i\,, \tag{62}$$

where $D_i$ is the denominator for $(1s)^2$ occupied and $D_n$ is the appropriate denominator when $1s^+$ and $np^-$ are occupied. For $1s$ transitions, Eq. (55) then becomes

$$f_{ni}^z = 2\omega_{ni}|\langle n|Z_{Op}|i\rangle|_c^2\Big(1+\sum_{kk'}{}'|\langle 2s2s|v|kk'\rangle|^2/D_nD_i\Big)^2$$
$$\times\Big[\Big(1+\sum_{kk'}|\langle 2s2s|v|kk'\rangle|^2/D_n^2\Big)$$
$$\times\Big(1+\sum_{kk'}|\langle 2s2s|v|kk'\rangle|^2/D_i^2\Big)\Big]^{-1}. \tag{63}$$

The prime in the first sum indicates $k,k'$ do not equal the excited single-particle state in $n$. The subscript $c$ after $|\langle n|Z_{Op}|i\rangle|^2$ indicates that only connected terms are included. This means that terms as shown in Fig. 11(b) are not to be included as they are accounted for by the factor following $|\langle n|Z_{Op}|i\rangle|_c^2$. All normalization corrections are accounted for by the last factor of Eq. (63). The sums over $k$ and $k'$ include all excited states although for Be the $(2p)(2p)$ excitation dominates. The terms $|\langle 2s2s|v|kk'\rangle|^2$ should include appropriate factors to account for the small nondiagonal correlation terms discussed in Sec. V. For example, we may multiply $\langle kk'|v|2s2s\rangle$ by the factor

$$1+\Big(\sum_{k''k'''\neq kk'}\langle kk'|v|k''k'''\rangle D^{-1}\langle k''k'''|v|2s2s\rangle\Big)$$
$$\times\langle kk'|v|2s2s\rangle^{-1}, \tag{64}$$

which accounts for nondiagonal particle-particle interactions, and similarly for particle-hole interactions. $D_i$ is given by Eq. (18) with $p$, $q$ replaced by $2s$, $2s$ and $D_n$ becomes

$$\begin{aligned}D_n = {} & (\epsilon_{2s}-\langle 2s1s|v|2s1s\rangle+\langle 2snp|v|2snp\rangle)\\ & +(\epsilon_{2s}-\langle 2s1s|v|2s1s\rangle+\langle 2s1s|v|1s2s\rangle\\ & +\langle 2snp|v|2snp\rangle-\langle 2snp|v|np2s\rangle\\ & -(\epsilon_k-\langle k1s|v|k1s\rangle+\langle knp|v|knp\rangle)\\ & -(\epsilon_k{}'-\langle k'1s|v|k'1s\rangle+\langle k'1s|v|1sk'\rangle\\ & +\langle k'np|v|k'np\rangle-\langle k'np|v|npk'\rangle)-\langle 2s2s|v|2s2s\rangle\\ & -\langle kk'|v|kk'\rangle+\langle k2s|v|k2s\rangle+\langle k'2s|v|k'2s\rangle\\ & +E_{\mathrm{corr}}(2s,2s;1snp\text{ unex})+2E_{\mathrm{corr}}(2s,1s;1snp\text{ unex})\\ & +2E_{\mathrm{corr}}(2s,np;1snp\text{ unex}).\end{aligned} \tag{65}$$

The correlation terms in Eq. (65) are written so as to emphasize the fact that the correlations are computed for the state $(1s)(np)(2s)^2$. The terms added to the single-particle energies $\epsilon$ account for the fact that the single-particle states were computed in the potential field of the nucleus and $(1s)^2(2s)$ but one $1s$ electron is now in the state $np$.

[22] P. L. Altick and A. E. Glassgold, Phys. Rev. **133**, A632 (1964).
[23] K. Sawada, Phys. Rev. **106**, 372 (1957).

TABLE XV. Sum-rule evaluation for oscillator strengths in Be.

| Transitions | Hartree-Fock | Correlated |
|---|---|---|
| $2s\rightarrow np$ | 2.0499 | 1.2977 |
| $2s\rightarrow kp$ | 0.6420 | 0.5969 |
| $1s\rightarrow np$ | 0.2091 | 0.2029 |
| $1s\rightarrow kp$ | 2.0663 | 1.9744 |
| $(2s)^2\rightarrow(ns)(2p)$ | | 0.0690 |
| Total | 4.967 | 4.141 |

When the state $np$ is in the continuum the matrix elements of Eq. (65) which involve this state are zero because of the continuum states' normalization $(2/R_0)^{1/2}$, where $R_0$ is the radius of a large sphere tending to infinity.[5] For $np$ in the continuum, $D_n=0.438$ for $k$, $k'=2p$, $2p$ and $m_l=\pm1$ from Eq. (65). For comparison, $D_i=0.3212$ for two $2p$ excitations and $m_l=\pm1$. Calculation of the factors on the right of Eq. (63) which multiply $|\langle n|Z_{op}|i\rangle|_c^2$ gave the result 0.993 for $np$ in the continuum.

The sum rule of Eq. (61) was evaluated and the results are given in Table XV. The transitions $(2s)^2\,{}^1S\rightarrow(ns)(2p)\,{}^1P$ are possible only when correlations of the two $2s$ electrons are included. The process is shown in Fig. 11(c). The most important transition is $(2s)^2\rightarrow(3s)(2p)\,{}^1P$ for which the oscillator strength was found to be 0.0642. The correlated sum rule result 4.14 is in considerably better agreement with the theoretical value 4.00 than is the Hartree-Fock result 4.97.[24]

The Be photoionization cross section $\sigma_k$ for transitions of a $2s$ electron into a continuum $p$ state is

$$\sigma_k=4\pi^2\frac{\omega_E\alpha}{k}\sum_{M_L=-L}^{+L}|\langle 2snp\,{}^1P(M_L)|Z_{op}|(2s)^2\,{}^1S\rangle|^2, \quad (66)$$

where the continuum electron has energy $k^2/2$, $\omega_E$ is the ionization potential plus $k^2/2$, and $\alpha$ is the fine structure constant.[25] The continuum function in this case has the normalization factor $(2/\pi)^{1/2}$. The cross section $\sigma_k$ is plotted in Fig. 12. The curve labeled "corr" includes correlation effects from the two $2s$ electrons in the ground state and also the normalization factor. The excited state $(2s)(kp)\,{}^1P$ is obtained from the Hartree-Fock single-particle states of Eq. (13) and does not include correlations. The curve labeled $\sigma_{\rm HF}$ omits the correlation and normalization corrections of the ground state.

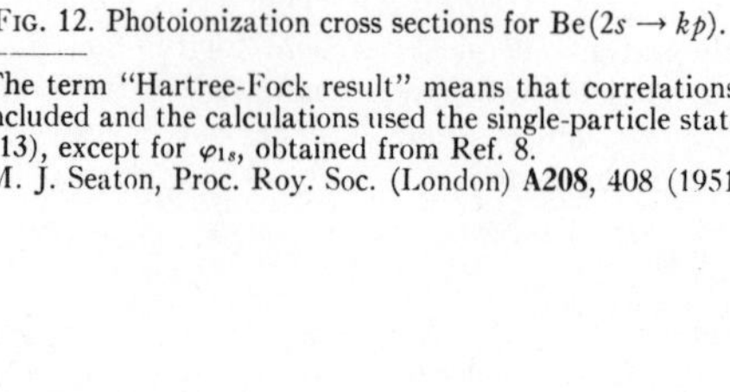

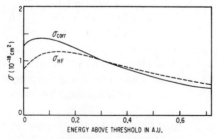

FIG. 12. Photoionization cross sections for Be$(2s\rightarrow kp)$.

[24] The term "Hartree-Fock result" means that correlations are not included and the calculations used the single-particle states of Eq. (13), except for $\varphi_{1s}$, obtained from Ref. 8.

[25] M. J. Seaton, Proc. Roy. Soc. (London) **A208**, 408 (1951).

## VII. DISCUSSION AND CONCLUSIONS

In the previous sections it was shown that many-body perturbation theory may be used to obtain many varied atomic properties from correlation energies to oscillator strengths. Much of the value of this approach lies in the fact that once the set of single-particle states for the perturbation theory has been calculated and used for one property, it is relatively easy to use the same states and many of the matrix elements from the first calculation to obtain additional atomic properties. Also, from the evaluation of diagrams for one calculation, one often develops a physical feeling as to which diagrams will be important in other calculations since many of the matrix elements and denominators are equal in different diagrams. In many cases it is possible to relate quantitatively the diagrams for different calculations.

The convergence of the perturbation expansion is strongly dependent on the choice of the basis set of single-particle states, and in Sec. II it was pointed out that it is desirable to choose the potential in such a way that the excited states correspond essentially to the physical single-particle excitations. This approach, which is a departure from the previous use of the Hartree-Fock potential in Ref. 5, was justified in Sec. II. For all atoms there is now an infinite number of excited bound states and the continuum to be included in the perturbation expansion. However, it was shown in the calculations of Secs. III, IV, V, and VI that perturbation calculations are readily made using this set of states and the convergence of the expansion is much more rapid than in Ref. 5. The sums over the infinite number of bound excited states were easily carried out by use of the $n^{-3}$ rule which was demonstrated in Sec. III.

The correlation energy for Be $2s$ electrons excited into $l=1$ states was found to be $-0.0436$ a.u. as compared with $-0.0426$ a.u. calculated in K. This particular calculation was repeated in this paper since in K the perturbation expansion converged very slowly for $2s$ correlations and the accuracy of the calculation was estimated to be approximately 5%. However, with the basis set of this paper the convergence was quite rapid and the accuracy of the result is estimated at better than 1%.

The perturbation theory may also be used to calculate other quantities such as polarizabilities and shielding factors as shown in Sec. V. The perturbation is more complicated than in the correlation energy calculation because there is now an additional perturbation due to the presence of an external charge. The second-order result for the dipole polarizability $\alpha_d$ was found to be 12.15 Å$^3$ but when the higher order terms were included the result was changed to 6.93 Å$^3$ with an expected accuracy of a few percent. The higher order terms are particularly important in this case because the two $2s$ electrons have strong correlations into $l=1$ states. This result may be compared with the second-order calculations of Kelly and Taylor[17] which gave 4.54 Å$^3$. As pointed out by KT,[17] their approach is equivalent to the uncoupled Hartree-Fock approximation of Dalgarno[15] which also yielded 4.5 Å$^3$. The second-order result of KT differs greatly from the second-order result of this paper because of the different basis sets of single-particle states. There are propagation corrections to the second-order calculations of KT because the interactions of excited states with the occupied unexcited states do not cancel the interaction with the Hartree-Fock potential. Since there is partial cancellation between the propagation corrections shown in Fig. 5(d) and the correlation terms of Figs. 5(b) and (c), the approach of KT can give reasonable results in second order when the correlations are strong. In this case the second-order results of this paper are poor and higher order terms must be included. However, when the correlations become small the propagation corrections do not necessarily also become small; and the second-order calculations of KT and of Dalgarno's uncoupled Hartree-Fock method may give results which are less than the correct solution. In general, it should be preferable to use either the coupled Hartree-Fock method of Dalgarno[15] or the perturbation theory approach presented in this paper for calculating polarizabilities and shielding factors. Dalgarno has previously pointed out that the coupled Hartree-Fock method is much more accurate than the uncoupled method.[15]

The dipole shielding calculations gave the second-order result 4.296 which is considerably higher than the 1.77 result of KT. This second-order difference has the same explanation as for $\alpha_d$. After the higher order terms were included, the final result was 0.972, with an estimated accuracy of approximately 5%. This value is in good agreement with the theoretical value 1.00. In order to obtain the value 0.972, it was necessary to consider all types of diagrams and the calculations were more complicated than for $\alpha_d$ because of large effects from diagrams involving $1s-2s$ correlations. The calculation of higher order diagrams may also be carried out with the basis set used by KT as shown in K. However, it is then necessary to sum diagrams like that of Fig. 5(d). The basis set of this paper sums these diagrams exactly and seems to be both more accurate and more convenient when higher order terms are to be included.

The calculated quadrupole polarizability was 15.09 Å$^5$ in second order and was changed to 14.06 Å$^5$ by inclusion of third-order terms. Again the second-order result is much higher than 9.26 Å$^5$ as calculated by KT. The accuracy of the second- and third-order calculations is expected to be within 2%. However, there is no assurance that the fourth-order terms which have been omitted are small. The calculated quadrupole shielding factor $\gamma_\infty$ was 1.06 in second order as compared with 0.67 computed by KT. After inclusion of third-order terms $\gamma_\infty$ was reduced to 0.75. The $1s-2s$ correlations contributed significantly to the third-order terms for $\gamma_\infty$ just as for $\beta_\infty$.

The methods of Sec. V may be readily applied to the calculation of higher order polarizabilities and shielding factors and it is probably a good approximation to limit these calculations to second order because correlations in higher $l$ states are expected to be quite small.

In Sec. VI it was shown that the basis set of single-particle states of this paper is also useful in calculating quantities such as excitation energies, oscillator strengths, and photionization cross sections. Correlations generally were included only for the ground state where they are expected to be most important. However, in a more detailed calculation the correlations in the excited states could also be included. The accuracy of these calculations is indicated by the evaluation of the sum rule which is theoretically 4.00 and was calculated as 4.14 including ground-state correlations and 4.97 without correlations.

The numerical calculations of this paper were for Be which has only four electrons and a simple closed shell structure. However, the perturbation theory is applicable to other atoms and may be particularly useful for atoms with a large number of electrons. In addition, many of the features of the perturbation theory which were used in the previous sections may be applied not only to other atoms but to other types of finite systems.

**ACKNOWLEDGMENTS**

I would like to thank Professor Keith A. Brueckner for his support and for very helpful suggestions, Dr. Howard S. Taylor for stimulating my interest in polarizability and shielding calculations, and Dr. Robert H. Traxler for helpful discussions.

VIII-5 Reprinted from *Physical Review* **144**, 39 (1966)

# Many-Body Perturbation Theory Applied to Open-Shell Atoms*

HUGH P. KELLY†
*Institute for Radiation Physics and Aerodynamics, University of California, San Diego, La Jolla, California*
(Received 9 September 1965)

It is shown how many-body perturbation theory may be applied to the problem of correcting Hartree-Fock energies and wave functions for the degenerate ground states of open-shell atoms. The choice of an appropriate potential for the calculation of the single-particle states of the perturbation expansion is discussed in detail. As an example of these methods, the correlation energies among all pairs of electrons in the neutral oxygen atom are calculated for excitations into $l=0$, 1, 2, and 3 states. The total calculated pair correlation energy is $-0.274$ atomic units (a.u.) as compared with the total correlation energy $-0.258$ a.u. which is deduced from experiment. The correlation energy among $2p$-$2p$ electron pairs is $-0.0906$ a.u.; among $2s$-$2p$ pairs, $-0.1004$ a.u.; $2s$-$2s$, $-0.0150$ a.u.; $1s$-$2p$, $-0.014996$ a.u.; $1s$-$2s$, $-0.00629$ a.u.; $1s$-$1s$, $-0.0438$ a.u. There is also a contribution $-0.0028$ a.u. which corrects for the use of a restricted Hartree-Fock solution.

## I. INTRODUCTION

THE many-body perturbation theory developed by Brueckner[1] and Goldstone[2] has been successfully used in calculations of nuclear structure[3] and of correlations in the high-density electron gas.[4] In addition, this theory has recently been applied to atoms to obtain corrections to Hartree-Fock (HF) solutions and to calculate other atomic properties.[5,6] Results have been obtained for the correlation energies among the different electron pairs in the neutral beryllium atom which are in good agreement with experiment.[5] Calculations have also been made of dipole and quadrupole polarizabilities and shielding factors and of transition probabilities, all for neutral beryllium.[6]

Although the many-body perturbation theory is quite applicable to atoms with few electrons, its greatest value may be in calculations for atoms with many electrons. The purpose of this paper is to consider the application of many-body perturbation theory to atoms which are considerably more complicated than those of beryllium. As an example, the correlations among all electron pairs of the neutral oxygen atom are studied in detail. Oxygen is an eight-electron atom with a nonspherically symmetric ground state.

The perturbation theory of Brueckner and Goldstone is reviewed in Sec. II. In Sec. III the choice of an appropriate unperturbed ground-state wave function is considered and it is shown how the perturbation theory may be applied in cases where the electrons do not form closed shells. The selection of the potential to be used in calculating the single-particle states is discussed in Sec. IV. In Sec. V the correlation energies among all pairs of electrons in oxygen are calculated. Section VI contains the discussion and conclusions.

* This research was supported by the Advanced Research Projects Agency (Project DEFENDER) and was monitored by the U. S. Army Research Office—Durham under Contract DA-31-124-ARO-D-257.

† Present address: Physics Department, University of Virginia, Charlottesville, Virginia.

[1] K. A. Brueckner, Phys. Rev. **97**, 1353 (1955); **100**, 36 (1955); *The Many-Body Problem* (John Wiley & Sons, Inc., New York, 1959).

[2] J. Goldstone, Proc. Roy. Soc. (London) **A239**, 267 (1957).

[3] K. A. Brueckner and J. L. Gammel, Phys. Rev. **109**, 1023 (1958); K. A. Brueckner and K. S. Masterson, Jr., *ibid.* **128**, 2267 (1962).

[4] M. Gell-Mann and K. A. Brueckner, Phys. Rev. **106**, 364 (1957).

[5] H. P. Kelly, Phys. Rev. **131**, 684 (1963).

[6] H. P. Kelly, Phys. Rev. **136**, B896 (1964).

## I. REVIEW OF PERTURBATION THEORY

The total Hamiltonian $H$ for a system of $N$ identical fermions interacting through two-body potentials $v_{ij}$ is given by

$$H=\sum_{i=1}^{N} T_i+\sum_{i<j}^{N} v_{ij}, \quad (1)$$

where $T_i$ is the sum of the kinetic-energy operator for the $i$th particle and all one-body potentials for the $i$th particle. For atoms,

$$T_i=-\nabla_i^2/2-Z/r_i, \quad (2)$$

where the term $-Z/r_i$ is the interaction of the $i$th electron with the nucleus of charge $Z$. Atomic units are used throughout this paper.[7] The correct ground-state wave function $\Psi_0$ satisfies the eigenvalue equation

$$H\Psi_0=E\Psi_0, \quad (3)$$

where $E$ is the exact nonrelativistic ground-state energy.

A great simplification is achieved by approximating the effect of the $N$ interacting particles by a single-particle potential $V$. The Hamiltonian $H$ is then replaced by

$$H_0=\sum_{i=1}^{N}(T_i+V_i), \quad (4)$$

and $\Psi_0$ is approximated by $\Phi_0$ which satisfies the eigenvalue equation

$$H_0\Phi_0=E_0\Phi_0. \quad (5)$$

The correct ground-state energy $E$ is now approximated by $E_0$. The potential $V$ must be Hermitian so that the

[7] D. R. Hartree, *The Calculation of Atomic Structures* (John Wiley & Sons, Inc., New York, 1957), p. 5.

single-particle wave functions $\varphi_n$, which are solutions of

$$(T+V)\varphi_n=\epsilon_n\varphi_n, \qquad (6)$$

constitute an orthonormal set. The wave function $\Phi_0$ is a determinant of the $N$ solutions of Eq. (6) which are lowest in energy. The states occupied in $\Phi_0$ are called unexcited states. The remaining single-particle states of the orthonormal set are called excited states. The unoccupied unexcited states are called holes and the occupied excited states are called particles. The correct ground-state wave function $\Psi_0$ and energy $E$ are obtained by perturbing the approximate $\Phi_0$ by

$$H'=H-H_0=\sum_{i<j}^{N} v_{ij}-\sum_{i=1}^{N} V_i, \qquad (7)$$

and we obtain[2]

$$\Psi_0=\lim_{\alpha\to 0} U_\alpha(0)\Phi_0/\langle\Phi_0|U_\alpha(0)|\Phi_0\rangle, \qquad (8)$$

and

$$E=E_0+\langle\Phi_0|H'|\Psi_0\rangle, \qquad (9)$$

where

$$U_\alpha(t)=\sum_{n=0}^{\infty}(-i)^n \times\int_{t>t_1>t_2\cdots>t_n} H'(t_1)\cdots H'(t_n)dt_1\cdots dt_n, \qquad (10)$$

and

$$H'(t)=e^{iH_0t}H'e^{-iH_0t}e^{\alpha t}. \qquad (11)$$

In the matrix element representation,

$$H_0=\textstyle\sum_n \epsilon_n\eta_n{}^\dagger\eta_n, \qquad (12)$$

and

$$H'=\tfrac{1}{2}\sum_{pqmn}\langle pq|v|mn\rangle\eta_p{}^\dagger\eta_q{}^\dagger\eta_n\eta_m -\sum_{pm}\langle p|V|m\rangle\eta_p{}^\dagger\eta_m. \qquad (13)$$

The operators $\eta^\dagger$ and $\eta$ satisfy the usual Fermi-Dirac anticommutation relations.[2] Wick's theorem may be used to express $U_\alpha(0)$ by sums of terms, each of which may be expressed by a Feynman-type diagram.[2]

The term $U_\alpha(0)$ may be expressed as a product of a sum of "linked" terms and a sum of "unlinked" terms equal to $\langle\Phi_0|U_\alpha(0)|\Phi_0\rangle$. After the time integrations in Eq. (10) have been carried out, one obtains the result

$$\Psi_0=\sum_L\left(\frac{1}{E_0-H_0}H'\right)^n\Phi_0, \qquad (14)$$

where $\sum_L$ means that only "linked" terms are to be included.[2] Also,

$$E-E_0=\Delta E=\sum_{L'}\langle\Phi_0|H'\left(\frac{1}{E_0-H_0}H'\right)^n|\Phi_0\rangle, \qquad (15)$$

where $L'$ restricts the sum to those terms which are "linked" when the leftmost $H'$ interaction is removed for $n\geq 1$.

## III. SYMMETRIES

In deriving Eq. (14) for $\Psi_0$, the factor $\langle\Phi_0|U_\alpha(0)|\Phi_0\rangle$ contained in $U_\alpha(0)\Phi_0$ cancels the same factor in the denominator of Eq. (8). The time integrations are then carried out and give

$$\Psi_0=\lim_{\alpha\to 0}\left(1+\frac{1}{E_0-H_0+i\alpha}H' +\frac{1}{E_0-H_0+i2\alpha}H'\frac{1}{E_0-H_0+i\alpha}H'+\cdots\right)_L\Phi_0. \qquad (16)$$

Since only "linked" terms are included in Eq. (16), excited states must be included as the intermediate states after $H'$ operates on $\Phi_0$. When all excited single-particle states have energies greater than those of the unexcited states, then $E_0-H_0$ in Eq. (16) cannot vanish and we may take the limit $\alpha\to 0$ which gives the result of Eq. (14). This is the reason for Goldstone's statement that he assumes that $H_0$ has a nondegenerate ground state $\Phi_0$ and that his proof applies only to the ground state of a closed-shell nucleus.

A general extension of the linked diagram expansion for the degenerate ground state of a system of fermions has been made by Bloch and Horowitz.[8] However, in many cases it is possible to make use of the Brueckner-Goldstone (BG) expansion directly even though we are dealing with open-shell systems with degenerate ground states. This is because the conserved quantum numbers may prevent the perturbation from leading to excited states [here we mean $N$-particle states which occur after $H'$ in Eq. (16)] which are degenerate with the ground state.

For example, let us assume $L$-$S$ coupling and start from an unperturbed ground state $\Phi_0$ specified by quantum numbers $L$, $M_L$, $S$, $M_S$. The operators $L_\pm$, $L_Z$, $S_\pm$, and $S_Z$ commute with the term $\sum v_{ij}$ of the perturbation for $v_{ij}$ equal to $v(|\mathbf{r}_i-\mathbf{r}_j|)$ and spin-independent, as is the case for Coulomb interactions. These operators also commute with $\sum V_i$ when $V$ is independent of the single-particle quantum numbers $m_l$ and $m_s$. That is, $V$ must be spherically symmetric and spin-independent. Then $L_\pm$, $L_Z$, $S_\pm$, and $S_Z$ commute with the perturbation $H'$. If we start with $\Phi_0$, an eigenstate of $L^2$, $L_Z$, $S^2$, and $S_Z$, then the perturbation can only lead to those excited single-particle states which give a total $N$-particle excited state with the same eigenvalues of $L^2$, $L_Z$, $S^2$, $S_Z$ as $\Phi_0$. The usual degeneracies in $M_L$ and $M_S$ in open-shell atoms are now avoided. When the ground state is uniquely specified by the electronic configuration and the quantum numbers $L$, $M_L$, $S$, and $M_S$ as for most atoms, then we may be able to apply the BG expansion directly.

In the usual description of the BG perturbation expansion it is assumed that the unperturbed ground-state

[8] C. Bloch and J. Horowitz, Nucl. Phys. 8, 91 (1958).

wave function $\Phi_0$ is a single determinant, but in general for open-shell atoms $\Phi_0$ is not necessarily a single determinant. However, in many open-shell cases if one chooses $M_L = \pm L$ and $M_S = \pm S$ then $\Phi_0$ is a single determinant. The reason for this is due to Hund's rule according to which the atomic ground state has the largest spin of those terms which may be formed from the ground-state configuration.[9] Since the total energy, for example, is independent of the choice of $M_L$ and $M_S$, in calculating such quantities as the correlation energy it is often possible to choose $\Phi_0$ as a single determinant as will be shown for oxygen. When the state $\Phi_0$ is a linear combination of determinants, we also may be able to carry out the BG expansion. Again we allow $U_\alpha(0)$ to act on $\Phi_0$ and then we factor out $\langle\Phi_0|U_\alpha(0)|\Phi_0\rangle$ which cancels the denominator of Eq. (8). The result is that those combinations of single-particle states which enter into $\Phi_0$ are again excluded as intermediate $N$-particle states in the perturbation expansion.

As an explicit example, let us consider the ground state of neutral carbon which is $(1s)^2(2s)^2(2p)^2\,{}^3P$. The ground state is given by the single determinant $(1^+0^+)$. The notation $1^+$ refers to a single $2p$ electron with $m_l = +1$ and $m_s = +\frac{1}{2}$ and $0^+$ refers to a $2p$ electron with $m_l = 0$ and $m_s = +\frac{1}{2}$. It is understood that this determinant also contains two $1s$ electrons with $m_s = \pm\frac{1}{2}$ and likewise two $2s$ states with $m_s = \pm\frac{1}{2}$. This notation is that used by Slater.[10] The BG expansion may now be applied to $\Phi_0 = (+1^+0^+)$. The excited $2p$ states $(-1^+)$, $(+1^-)$, $(0^-)$, and $(-1^-)$ which are degenerate in energy with $(+1^+)$ and $(0^+)$ are not reached by the perturbation because of conservation of $M_L$ and $M_S$. That is, they cannot be the only excitations present. It is possible, for example, to excite $2p(+1^+)$ and $2p(0^+)$ into $4f(+2^+)$ and $2p(-1^+)$, respectively. However, there is no problem of vanishing energy denominators in this case. If we choose the carbon ground state

$$\Phi_0({}^3P;\,M_L=0,\,M_S=0) = (1/\sqrt{2})[(+1^- -1^+)+(+1^+ -1^-)], \quad (17)$$

the operator $U_\alpha(0)$ applied to Eq. (17) gives $2p$ intermediate states with vanishing denominators but they add to give $\Phi_0$ which may be projected out as has been discussed. For example, in first order, the Coulomb perturbation $\frac{1}{2}\sum\langle pq|v|mn\rangle\eta_p{}^\dagger\eta_q{}^\dagger\eta_n\eta_m$ applied to $\Phi_0$ gives terms with vanishing denominators when $p$, $q$, $m$, and $n$ refer to the $2p$ states. The first-order correction to $\Phi_0$ with denominator $i\alpha$ from the Coulomb part of $U_\alpha(0)\Phi_0$ is

$$U_\alpha(0)_{c1}(+1^- -1^+) = (i\alpha)^{-1}[(F_0+(1/25)F_2)\times(+1^- -1^+)-(3/25)F_2(0^-0^+)+(6/25)F_2(-1^- +1^+)], \quad (18)$$

[9] E. U. Condon and G. H. Shortley, *The Theory of Atomic Spectra* (Cambridge University Press, Cambridge, 1957), p. 209.

[10] J. C. Slater, *Quantum Theory of Atomic Structure* (McGraw-Hill Book Company, Inc., New York, 1960), Vol. II, Chap. 20, p. 84.

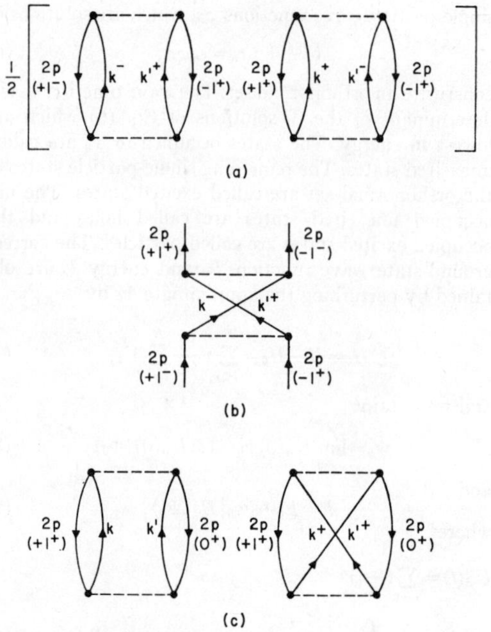


FIG. 1. Second-order correlation energy diagrams for the ${}^3P$ ground state of carbon. Direct diagrams (a) and exchange diagram (b) for $\Phi_0({}^3P;\,M_L=0,\,M_S=0)$. (c) Diagrams for $\Phi_0({}^3P;\,M_L=L,\,M_S=S)$.

$$U_\alpha(0)_{c1}(+1^+ -1^-) = (i\alpha)^{-1}[(F_0+(1/25)F_2)\times(+1^+ -1^-)-(3/25)F_2(0^+0^-)+(6/25)F_2(-1^+ +1^-)],$$

where

$$F_k = \int_0^\infty\int_0^\infty \frac{r_<^k}{r_>^{k+1}} P_{2p}{}^2(r_1)P_{2p}{}^2(r_2)\,dr_1dr_2. \quad (19)$$

The angular integrations were obtained from Ref. 9, p. 178. Since $(0^-0^+) = -(0^+0^-)$ and $(-1^- +1^+) = -(+1^+ -1^-)$, we obtain

$$U_\alpha(0)_{c1}\Phi_0 = (i\alpha)^{-1}(F_0 - 5/25F_2)\Phi_0. \quad (20)$$

The first-order part of $U_\alpha(0)\Phi_0$ with denominator $i\alpha$ due to the term $-\sum\langle m|V|p\rangle\eta_m{}^\dagger\eta_p$ also is proportional to $\Phi_0$. All these terms vanish when we project out $\langle\Phi_0|U_\alpha(0)|\Phi_0\rangle$.

When the correlation energy is calculated for the carbon ground state $\Phi_0({}^3P;\,M_L=0,\,M_S=0)$ of Eq. (17) there will be cross terms between the two determinants. Second-order energy terms for $2p$ electron correlations are shown in Figs. 1(a) and 1(b). The diagram of Fig. 1(b) describes the cross term between the two determinants. The diagrams are given according to Goldstone's notation.[2] There is a minus sign associated with the diagram of Fig. 1(b). The states $k$ and $k'$ are all the allowed excited states. Those $2p$ excitations with vanishing denominators are excluded as discussed. Second-order correlation energy diagrams for $\Phi_0(^3P; M_L=L, M_S=S)$ are shown in Fig. 1(c). The contributions of the diagrams shown in Figs. 1(a) and 1(b) must equal the contributions shown in Fig. 1(c), and this has been explicitly verified by enumeration of the angular coefficients.[11] However, the exchange contribution of Fig. 1(b) is not equal to the exchange diagram shown in Fig. 1(c).

The atom chosen for the numerical calculations is neutral oxygen $(1s)^2(2s)^2(2p)^4\ ^3P$ which is similar to carbon. The $^3P$ ground state may be represented by a single determinant[10]:

$$\Phi_0(^3P; M_L=L, M_S=S)=(+1^+0^+-1^++1^-). \quad (21)$$

The BG expansion may again be applied directly and it is found that there are no vanishing energy denominators due to conservation of $M_L$ and $M_S$ by the perturbation. The perturbation could also have been applied to the other $^3P$ states such as

$$\Phi_0(^3P; M_L=0, M_S=0)$$
$$=(1/\sqrt{2})[(+1^-0^+-1^+0^-)+(+1^+0^+-1^-0^-)]. \quad (22)$$

## IV. CHOICE OF POTENTIAL

In order to obtain the $\Phi_0$ which is the best approximation to the exact ground-state wave function $\Psi_0$, a minimization of $\langle\Phi_0|H|\Phi_0\rangle$ subject to the usual orthonormality constraints is carried out.[10] The single-particle states of $\Phi_0$ then satisfy the well-known Hartree-Fock (HF) equations

$$-\tfrac{1}{2}\nabla^2\varphi_n(\mathbf{r})-\frac{Z}{r}\varphi_n(\mathbf{r})$$
$$+\left(\sum_{j=1}^{N}\int\frac{d\mathbf{r}'\varphi_j^*(\mathbf{r}')\varphi_j(\mathbf{r}')}{|\mathbf{r}-\mathbf{r}'|}\right)\varphi_n(\mathbf{r})$$
$$-\sum_{j=1}^{N}\left(\delta(m_{sn},m_{sj})\int\frac{d\mathbf{r}'\varphi_j^*(\mathbf{r}')\varphi_n(\mathbf{r}')}{|\mathbf{r}-\mathbf{r}'|}\varphi_j(\mathbf{r})\right)$$
$$=\epsilon_n\varphi_n(\mathbf{r}). \quad (23)$$

The terms involving $\varphi_j$ represent the Hartree-Fock potential $V_{\mathrm{HF}}$ acting on $\varphi_n$. This potential may be represented in terms of its matrix elements

$$\langle a|V_{\mathrm{HF}}|b\rangle=\sum_{n=1}^{N}(\langle an|v|bn\rangle-\langle an|v|nb\rangle). \quad (24)$$

The coupled equations (23) must be solved self-consistently to obtain the $N$ single-particle states $\varphi_n$ of $\Phi_0$. At this point, $V_{\mathrm{HF}}$, which is nonlocal and Hermitian, is determined.

For open-shell atoms, in general one does not have spherical symmetry and therefore the solutions of Eq. (23) cannot always be written

$$\varphi_n(\mathbf{r})=R_{nl}(r)Y_{lm}(\theta,\varphi)\chi_s(m_s). \quad (25)$$

It is extremely difficult then to obtain exact solutions of Eq. (23). This difficulty is avoided by assuming that each single-particle state has the form of Eq. (25). The quantity $\langle\Phi_0|H|\Phi_0\rangle$ is then minimized with respect to all $R_{nl}$ subject to the usual orthonormality constraints. The resulting "restricted" HF equations may be solved in practice and the "restricted" $V_{\mathrm{HF}}$ is now spherically symmetric. This procedure has the advantage that it is practical and in addition, since the single-particle radial functions $R_{nl}$ are independent of $m_l$ and $m_s$, $\Phi_0$ may be constructed as an eigenstate of $L^2$, $L_Z$, $S^2$, and $S_Z$ which would not necessarily be possible if the solutions of Eq. (23) were used for open-shell atoms.

In applying the BG theory it is necessary to obtain a complete basis set of single-particle states. A Hartree-Fock basis set may be obtained by calculating all solutions of Eq. (23). Those $N$ states lowest in energy are the unexcited states of $\Phi_0$. For the bound excited states it is only necessary to calculate a finite number as suitable extrapolations may be made to account for the remaining bound states of high principal quantum number.[6] For continuum states, it is only necessary to calculate a sufficient number to carry out all numerical integrations with the desired accuracy.

An advantage in using the HF basis set is that $\Phi_0$ is optimized. In the BG theory there are first-order corrections to $\Phi_0$ as shown in Figs. 2(a), 2(b), 2(c), and 2(d). With the HF basis, the correction term due to $V_{\mathrm{HF}}$ in Fig. 2(c) cancels the terms shown in Figs. 2(a) and 2(b). To first order, then, there are no corrections

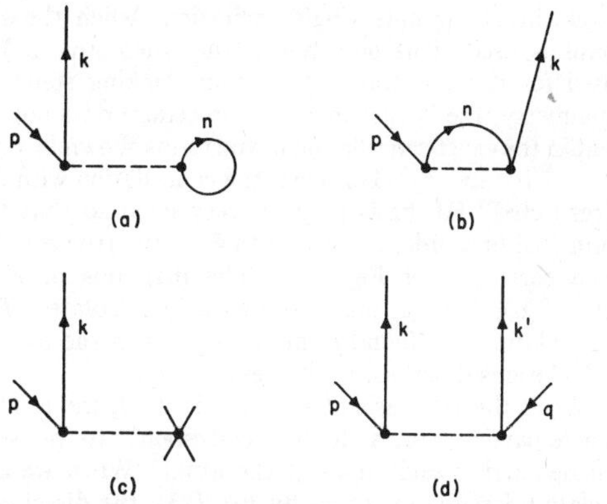


FIG. 2. First-order corrections to the unperturbed wave function $\Phi_0$. (a) Direct interaction with passive unexcited state $n$. This term equals $(\epsilon_p-\epsilon_k)^{-1}\langle kn|v|pn\rangle X\eta_k{}^+\eta_p|\Phi_0\rangle$. (b) Exchange interaction with $n$. (c) Interaction with the potential $V$. (d) Two-body correlation correction. The states $n$ are summed over all unexcited states. When the Hartree-Fock potential of Eq. (24) is used to calculate the unexcited states, the diagrams (a), (b), and (c) add to zero.

[11] Reference 9, p. 178.

to $\Phi_0$ involving only single excitations when the HF basis is used. This only holds true when Eq. (23) is used for the HF states. For atoms lacking spherical symmetry, the "restricted" HF procedure does not result in the exact cancellation of the terms shown in Figs. 2(a), 2(b), and 2(c). However, the cancellation with the "restricted" HF basis may be very good so that the principal first-order correction to $\Phi_0$ is the true correlation correction of Fig. 2(d). The diagrams of Figs. 2(a), 2(b), and 2(c) may also be easily calculated. The cancellation was found to be quite good in the oxygen calculations described in the next section.

When the HF basis of Eq. (23) is used, the excited single-particle states do not correspond to physical single-particle excitations of the atom.[5] When we calculate unexcited states $\varphi_n$ by Eq. (23), the direct and exchange terms cancel for $j$ equal to $n$. The state $\varphi_n$ is then calculated in the field of the nucleus and $N-1$ other electrons. For excited states there is no cancellation and the HF excited states are calculated in the field of the nucleus and $N$ other electrons. For neutral atoms this leads to the possibility that all the excited states may lie in the continuum. There were no bound excited states in the previous beryllium calculation with the HF basis, and it was argued that this will be the case for most if not all neutral atoms.[5] A search for these bound, excited HF states in neutral oxygen also gave negative results. The advantage of eliminating bound excited states is offset, however, by the resulting slow convergence of the perturbation expansion as found in Ref. 5. It was pointed out in Ref. 6 that excited states have interactions with $N-1$ other electrons and it is therefore desirable to calculate these excited states in the potential field of $N-1$ other electrons. This type of potential was used in Ref. 6 and the convergence of the BG expansion for the beryllium correlation energy was greatly improved. The same arguments also apply in the case of "restricted" HF potentials for open-shell atoms.

In Ref. 6 the potential $V$ was chosen by calculating all states in the fixed field of neutral beryllium minus one of the $2s$ electrons. Hartree-Fock states were used to calculate this fixed potential. In this potential the $2s$ state is the HF $2s$ state, but the $1s$ state differs very slightly from the HF $1s$ wave function. The same prescription is used in this paper to obtain the set of single-particle states for oxygen. The equation for the radial functions $P_{nl}(r)=rR_{nl}(r)$ for $l=0$ is

$$\left(-\frac{1}{2}\frac{d^2}{dr^2}-\frac{8}{r}\right)P_{n0}(r)+2\int_0^\infty dr'P_{1s}^2(r')v_0(r,r')P_{n0}(r)$$
$$-\int_0^\infty dr'P_{1s}(r')P_{n0}(r')v_0(r,r')P_{1s}(r)+\int_0^\infty dr'P_{2s}^2(r')v_0(r,r')P_{n0}(r)$$
$$+4\int_0^\infty dr'P_{2p}^2(r')v_0(r,r')P_{n0}(r)-\frac{2}{3}\int_0^\infty dr'P_{2p}(r')P_{n0}(r')v_1(r,r')P_{2p}(r)=\epsilon_{n0}P_{n0}(r), \quad (26)$$

where

$$v_K(r,r')=r_<^K/r_>^{K+1}, \quad (27)$$

$r_<$ being the lesser of $r$ and $r'$, and $r_>$ being the greater of $r$ and $r'$. For $l=1$ states the radial equation is

$$\left[\frac{1}{2}\left(-\frac{d^2}{dr^2}+\frac{2}{r^2}\right)-\frac{8}{r}\right]P_{n1}(r)+2\int_0^\infty dr'P_{1s}^2(r')v_0(r,r')P_{n1}(r)$$
$$-\frac{1}{3}\int_0^\infty dr'P_{1s}(r')P_{n1}(r')v_1(r,r')P_{1s}(r)+2\int_0^\infty dr'P_{2s}^2(r')v_0(r,r')P_{n1}(r)$$
$$-\frac{1}{3}\int_0^\infty dr'P_{2s}(r')P_{n1}(r')v_1(r,r')P_{2s}(r)+3\int_0^\infty dr'P_{2p}^2(r')v_0(r,r')P_{n1}(r)$$
$$-\frac{3}{50}\int_0^\infty dr'P_{2p}^2(r')v_2(r,r')P_{n1}(r)-\frac{6}{25}\int_0^\infty dr'P_{2p}(r')P_{m1}(r')v_2(r,r')P_{2p}(r)=\epsilon_{n1}P_{n1}(r). \quad (28)$$

When oxygen HF orbitals are used for $P_{1s}$, $P_{2s}$, and $P_{2p}$, Eqs. (26) and (28) are the oxygen HF equations for $P_{2s}(r)$ and $P_{2p}(r)$, respectively.[12] The lowest energy solution of Eq. (26) is a $1s$ orbital but it is not the HF $1s$ solution. It is, however, a good approximation to the HF $P_{1s}$. The next lowest energy solution to Eq. (26) is the HF $P_{2s}$. In

[12] D. R. Hartree, W. Hartree, and B. Swirles, Phil. Trans. Roy. Soc. (London) **A238**, 229 (1940).

Eq. (26) we are calculating $l=0$ states in the field of neutral oxygen minus one $2s$ electron; in Eq. (28) we are calculating $l=1$ states in the field of neutral oxygen minus one $2p$ electron. The solutions of Eqs. (26) and (28) give orthogonal sets of $l=0$ and $l=1$ single-particle wave functions.

The choice of equations for $l>1$ is slightly more arbitrary than for $l=0$ and 1. It is desirable to calculate the $l>1$ states in the HF field of oxygen with one $2p$ electron removed. The $2p$ exchange coefficient was estimated as that which would cause as much cancellation as possible of the interactions with the passive unexcited states in the perturbation expansion.[2] For $l=2$, the equation for this investigation is

$$\left[\frac{1}{2}\left(-\frac{d^2}{dr^2}+\frac{6}{r^2}\right)-\frac{8}{r}\right]P_{n2}(r)+2\int_0^\infty dr'P_{1s}^2(r')v_0(r,r')P_{n2}(r)-\frac{1}{5}\int_0^\infty dr'P_{1s}(r')P_{n2}(r')v_2(r,r')P_{1s}(r)$$
$$+2\int_0^\infty dr'P_{2s}^2(r')v_0(r,r')P_{n2}(r)-\frac{1}{5}\int_0^\infty dr'P_{2s}(r')P_{n2}(r')v_2(r,r')P_{2s}(r)$$
$$+3\int_0^\infty dr'P_{2p}^2(r')v_0(r,r')P_{n2}(r)-0.11\int_0^\infty dr'P_{2p}(r')P_{n2}(r')v_3(r,r')P_{2p}(r)=\epsilon_{n2}P_{n2}(r). \quad (29)$$

For $l=3$, $P_{nl}$ in this calculation is determined by

$$\left[\frac{1}{2}\left(-\frac{d^2}{dr^2}+\frac{12}{r^2}\right)-\frac{8}{r}\right]P_{n3}(r)+2\int_0^\infty dr'P_{1s}^2(r')v_0(r,r')P_{n3}(r)-\frac{1}{7}\int_0^\infty dr'P_{1s}(r')P_{n3}(r')v_3(r,r')P_{1s}(r)$$
$$+2\int_0^\infty dr'P_{2s}^2(r')v_0(r,r')P_{n3}(r)-\frac{1}{7}\int_0^\infty dr'P_{2s}(r')P_{n3}(r')v_3(r,r')P_{2s}(r)$$
$$+3\int_0^\infty dr'P_{2p}^2(r')v_0(r,r')P_{n3}(r)-0.10\int_0^\infty dr'P_{2p}(r')P_{n3}(r')v_2(r,r')P_{2p}(r)=\epsilon_{n3}P_{n3}(r). \quad (30)$$

Equations (26), (28), (29), and (30) were solved numerically to obtain the 1=0, 1, 2, and 3 single-particle wave functions used in the perturbation expansion. Hartree-Fock orbitals were used for $P_{1s}$, $P_{2s}$, and $P_{2p}$ in Eqs. (26), (28), (29), and (30). These HF orbitals were obtained by iterating several times the oxygen HF solutions of Hartree, Hartree, and Swirles[12] in order to obtain greater accuracy. The resulting orbitals were found to be in excellent agreement with the accurate analytic HF solutions of Bagus and Roothaan and differed from the Bagus-Roothaan solutions by one or two digits in the fourth decimal place.[13] The $1s$ solution of Eq. (26) was found to be quite close to the HF $1s$ solution although calculated by a different potential. The $1s$ solution of Eq. (26) is compared in Table I with the HF $1s$ solution obtained by iterating the solution of Ref. 12 and with the HF $1s$ solution of Bagus and Roothaan.[13] Both the $1s$ orbital of Eq. (26) and the HF $1s$ orbital are strictly orthogonal to the HF $2s$ orbital.

The difference between $\langle\Phi_0|H|\Phi_0\rangle$ calculated with $P_{1s}$ of Eq. (26) and with the $P_{1s}$ HF is extremely small since $\langle\Phi_0|H|\Phi_0\rangle$ is stationary with respect to variations in $\Phi_0$. This difference, which was calculated most accurately by perturbation theory, was found to be 0.0000215 a.u. One a.u. equals 27.21 eV. By comparison, $\langle\Phi_0|H|\Phi_0\rangle=-74.80936$ a.u. is the HF result of Bagus and Roothaan. It is interesting, however, to note that $\epsilon_{1s}$ ($-21.726$ a.u.) does differ significantly from $\epsilon_{1s}$

[13] I am very grateful to Dr. P. S. Bagus and Professor C. C. J. Roothaan for kindly sending me a detailed listing of their Hartree-Fock results for oxygen.

TABLE I. Radial functions $P_{1s}(r)$ for oxygen.

| $r$ | $P_{1s}(r)^a$ | $P_{1s}(r)^b$ | $P_{1s}(r)^c$ |
|---|---|---|---|
| 0.01 | 0.39819 | 0.39841 | 0.39848 |
| 0.04 | 1.25498 | 1.25566 | 1.25560 |
| 0.08 | 1.83251 | 1.83339 | 1.83336 |
| 0.10 | 1.95985 | 1.96069 | 1.96072 |
| 0.12 | 2.01383 | 2.01457 | 2.01465 |
| 0.14 | 2.01336 | 2.01394 | 2.01404 |
| 0.16 | 1.97319 | 1.97359 | 1.97367 |
| 0.20 | 1.81724 | 1.81726 | 1.81726 |
| 0.24 | 1.61023 | 1.60990 | 1.60982 |
| 0.28 | 1.38982 | 1.38922 | 1.38912 |
| 0.30 | 1.28164 | 1.28094 | 1.28085 |
| 0.34 | 1.07726 | 1.07645 | 1.07640 |
| 0.40 | 0.81167 | 0.81086 | 0.81090 |
| 0.50 | 0.48605 | 0.48552 | 0.48561 |
| 0.80 | 0.09021 | 0.09041 | 0.09033 |
| 1.00 | 0.02853 | 0.02875 | 0.02870 |
| 1.40 | 0.00348 | 0.00356 | 0.00362 |
| 2.00 | 0.00037 | 0.00046 | 0.00051 |
| 2.60 | 0.00006 | 0.00010 | 0.00012 |
| 3.00 | 0.00003 | 0.00004 | 0.00004 |
| $-\epsilon_{1s}$(a.u.) | 21.72625 | 20.66908 | 20.66860 |

[a] The $1s$ orbital calculated from Eq. (26). This orbital was not calculated in the HF potential and differs slightly from the Hartree-Fock $P_{1s}$.
[b] Hartree-Fock $P_{1s}(r)$ calculated by iterating the HF oxygen solutions of Ref. 12.
[c] Hartree-Fock $P_{1s}(r)$ obtained from analytic solutions of Bagus and Roothaan.

HF (−20.669 a.u.). This is understood by considering that in Eq. (26) the potential for $P_{1s}$ has two direct interactions and one exchange interaction with $P_{1s}$ HF and one direct interaction with $P_{2s}$ in addition to the interactions with $P_{2p}$. When we solve for $P_{1s}$, the $1s$ exchange interaction very nearly cancels one direct $1s$ interaction and the net effect is that $P_{1s}$ has one direct $1s$ interaction and one direct $2s$ interaction in addition to the $2p$ interactions. In the HF case, $P_{1s}$ HF (hereafter denoted $P_{1s}'$) has one direct $1s$ interaction, two direct $2s$ interactions, and one exchange $2s$ interaction in addition to $2p$ interactions. Since $P_{1s}$ lacks one direct and one exchange interaction with $P_{2s}$ as compared with $P_{1s}'$, it is expected to be lower in energy since the positive direct term is generally larger than the negative exchange term. The difference between $\epsilon_{1s}$ and the HF$\epsilon_{1s}'$ is readily corrected when we use $P_{1s}$ in the perturbation expansion. Whenever the $1s$ state appears in the perturbation expansion there are corresponding higher order terms which account for $1s$ interactions with the other occupied unexcited states and with the potential $V$. A typical first-order correction to $\Phi_0$ involving a $1s$ hole line is shown in Fig. 3(a). In the next order the $1s$ hole line is modified by direct and exchange interactions with the passive unexcited states as shown in Figs. 3(b) and 3(c). The $1s$ hole line is also modified by interaction with the potential $V$ as shown in Fig. 3(d). The corrections to the term of Fig. 3(a) shown in Figs. 3(b), 3(c), and 3(d) are given by $-v/D$ times the term of Fig. 3(a), where $v$ is the sum of the matrix elements of the interactions shown in Figs. 3(b), 3(c), and 3(d) and $D$ is the same denominator as in Fig. 3(a). The nondiagonal terms, which are much smaller, are calculated separately. The minus sign comes from the additional hole line. In third and higher orders these interactions are again repeated, the $m$th correction being the term of Fig. 3(a) multiplied by $(-v/D)^m$. These terms add geometrically and the result is that the term of Fig. 3(a) is multiplied by $(1+v/D)^{-1}$ which is equivalent to replacing $D$ by $D+v$ in Fig. 3(a). When $V_{\mathrm{HF}}$ is used, $v$ is zero. However, these terms are not zero when the restricted HF solution is used for open-shell atoms. For example, in oxygen for $\Phi_0=(+1^+0^+-1^++1^-)$, $v$ is different for $1s^+$ and $1s^-$ because $1s^+$ has exchange interactions with three $2p$ electrons and $1s^-$ has only one $2p$ exchange term, whereas the restricted HF potential for both $1s$ states has exchange interactions with two $2p$ electrons. If we consider these $1s-2p$ corrections due to use of a restricted HF potential separately, then the correction $v$ for $1s$ states due to the use of Eq. (26) is

$$v=\langle 1s1s|v|1s1s\rangle+2\langle 2s1s|v|2s1s\rangle -\langle 2s1s|v|1s2s\rangle-[2\langle 1s1s'|v|1s1s'\rangle -\langle 1s1s'|v|1s'1s\rangle+\langle 2s1s|v|2s1s\rangle]. \quad (31)$$

The term in square brackets comes from the choice of potential used in Eq. (26) and $1s'$ is the HF $1s$ orbital

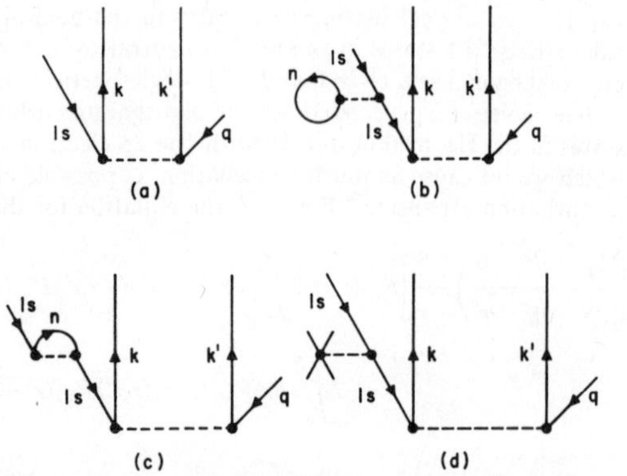


FIG. 3. (a) First-order correction to $\Phi_0$ involving $1s$ state. (b) Direct interaction of $1s$ hole line with the passive unexcited state $n$. (c) Exchange interaction of $1s$ hole line with $n$. (d) Interaction of $1s$ hole state with the potential $V$.

used in Eq. (26). Since the same type of correction may be made for each hole line, we simply add the appropriate $v$ to each single-particle energy $\epsilon$. Using the matrix elements listed in Table II, we find that $v$ for $\epsilon_{1s}$ is 1.05572 and $\epsilon_{1s}+v$ is −20.6705 a.u. as compared with the HF $\epsilon_{1s}'=-20.66908$ a.u.

## V. CORRELATION ENERGIES

### A. Second-Order Calculations

The BG perturbation expansion may be used to obtain both the ground-state energy $E$ and wave function $\psi_0$. The energy $E$ may often be obtained from experiment and this serves as a check on the accuracy of the calculations. The correlation energy is defined by

$$E_{\mathrm{corr}}=E-E_{\mathrm{HF}}. \quad (32)$$

TABLE II. Matrix elements among unexcited states.[a]

| | |
|---|---|
| $\langle 1s1s\|v\|1s1s\rangle$ | 4.73865 |
| $\langle 1s'1s'\|v\|1s'1s'\rangle$[b] | 4.74118 |
| $\langle 1s1s'\|v\|1s1s'\rangle$ | 4.73992 |
| $\langle 1s1s'\|v\|1s'1s\rangle$ | 4.73991 |
| $\langle 2s1s\|v\|2s1s\rangle$ | 1.13425 |
| $\langle 2s1s\|v\|1s2s\rangle$ | 0.07726 |
| $\langle 2s1s'\|v\|2s1s'\rangle$ | 1.13430 |
| $\langle 2s1s'\|v\|1s'2s\rangle$ | 0.07733 |
| $\langle 2s2s\|v\|2s2s\rangle$ | 0.79794 |
| $\langle 2s2p\|v\|2s2p\rangle$ | 0.77387 |
| $\langle 2s2p\|v\|2p2s\rangle$ | 0.47214 |
| $\langle 2p1s\|v\|2p1s\rangle$ | 1.09845 |
| $\langle 2p1s\|v\|1s2p\rangle$ | 0.105655 |
| $\langle 2p1s'\|v\|2p1s'\rangle$ | 1.098464 |
| $\langle 2p1s'\|v\|1s'2p\rangle$ | 0.105583 |
| $\langle 2p2p\|v\|2p2p\rangle$ | 0.754853 ($k=0$)[c]<br>0.336095 ($k=2$) |

[a] Only the radial parts of the matrix elements are given.
[b] The states $1s'$, $2s$, and $2p$ are Hartree-Fock states. The $1s$ state was calculated from Eq. (26).
[c] The notation $k$ is defined in Eq. (27).

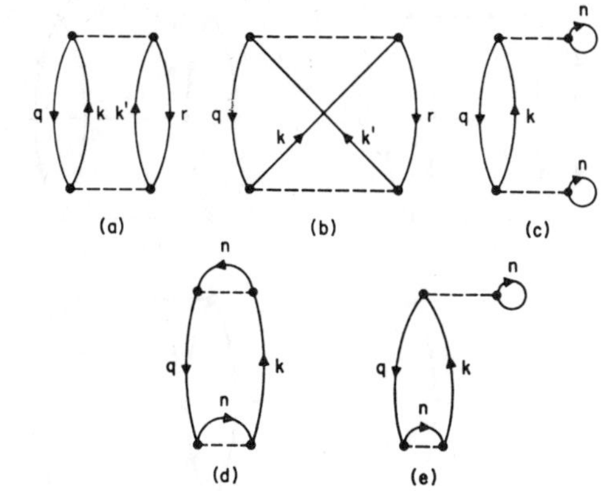


FIG. 4. Second-order energy diagrams. (a) Direct correlation diagram. (b) Exchange. (c), (d), and (e) are single-excitation diagrams. We may also have the order of exchange and direct interactions interchanged in (e). There are also diagrams like (c), (d), and (e) with interactions with the potential $V$. These single-excitation diagrams add to zero when a completely unrestricted Hartree-Fock potential is used.

The energy $E$ may be obtained by adding the experimental ionization potentials and subtracting the estimated relativistic contributions to the total energy. A more accurate procedure is to use the experimental ionization potentials for all but the last two electrons and then use Pekeris' calculations[14] to obtain the remaining energy. The relativistic energy among the first $N-2$ electrons is then subtracted to obtain $E$. Clementi has given the correlation energies for a large number of atoms and for the $^3P$ ground state of neutral oxygen he has obtained $E_{\rm corr}=-0.258$ a.u. with an estimated accuracy of 5%.[15] The relativistic contribution is approximately $-0.0503$ a.u. and $E_{\rm HF}=-74.80936$ a.u.[15] The HF solution in this case is the restricted HF solution described in Sec. IV and described in detail by Roothaan.[16] As will be seen from the perturbation calculations, the energy difference between the restricted HF solution and a completely unrestricted HF solution is very small compared to the energy corrections from true two-body correlations.

Given the unperturbed state $\Phi_0$, we may calculate

$$\langle\Phi_0|H|\Phi_0\rangle=E_0+E_1,$$

where $E_0$ is given by Eq. (5) and $E_1=\langle\Phi_0|H'|\Phi_0\rangle$. When $\Phi_0$ is the HF solution, $E_0+E_1$ is the HF energy $E_{\rm HF}$. The lowest order energy corrections to $\langle\Phi_0|H|\Phi_0\rangle$ are second-order terms which are shown in Fig. 4. When a fully unrestricted HF basis is used, the diagrams of Figs. 4(c), 4(d), and 4(e) and the corresponding interactions with $V_{\rm HF}$ add to zero. As discussed in Sec. IV, for open-shell atoms we must use a "restricted" HF basis and so there are contributions to the energy from diagrams as shown in Figs. 4(c), 4(d), and 4(e). However, it is found that they are quite small compared to the diagrams of Figs. 4(a) and 4(b).

The diagrams of Fig. 4 are calculated according to the rules given in Ref. 2. For example, the diagram of Fig. 4(a), which describes correlations among electrons in states $q$ and $r$, is given by

$$E_2{}^{(a)}(q,r)=\sum_{k,k'}\frac{\langle qr|v|kk'\rangle\langle kk'|v|qr\rangle}{\epsilon_q+\epsilon_r-\epsilon_k-\epsilon_{k'}}. \tag{33}$$

The sums are over all excited states. Bound excited states are labeled by the principal quantum number $n$, orbital angular momentum $l$, azimuthal quantum number $m_l$, and spin projection $m_s$. In practice, the sums over $n$ are carried out explicitly for the first eight or ten excited states and the remaining infinite sum may be carried out by integration as shown in Ref. 6, since

$$\lim\langle mlnl'|v|qr\rangle=n^{-3/2}\times\text{const}, \qquad n\to\infty,\ m \text{ fixed}. \tag{34}$$

The sums over $m_l$ and $m_s$ are carried out for each $l$ value. The previous beryllium calculation[5] included $l=0$, 1, and 2 states and the present oxygen calculation includes $l=0$, 1, 2, and 3 states. The continuum states are labeled by $k$, $l$, $m_l$, and $m_s$, where $\epsilon_k=k^2/2$. If the atom is enclosed in a large spherical volume of radius $R_0$ where $R_0$ tends to infinity, then

$$P_{kl}(r)=(2/R_0)^{1/2}\cos[kr+\delta_l-1/2(l+1)\pi], \tag{35}$$

for large $r$. Since $P_{kl}$ must vanish for $r=R_0$,

$$kR_0+\delta_l-\tfrac{1}{2}(l+1)\pi=n\pi, \tag{36}$$

where $n$ is an integer. For fixed $l$, the number of states $\Delta n$ in the range $\Delta k$ is determined by

$$\Delta kR_0+\Delta\delta_l=\Delta n\pi. \tag{37}$$

Since $\Delta\delta_l/R_0\to 0$ for $R_0\to\infty$, we obtain

$$\Delta n=(R_0/\pi)\Delta k, \tag{38}$$

and

$$\sum_k=(R_0/\pi)\int_0^\infty dk. \tag{39}$$

The normalization factor $(2/R_0)^{1/2}$ may then be omitted from Eq. (35) and the summation of Eq. (39) changed to

$$\sum_k=\left(\frac{2}{\pi}\right)\int_0^\infty dk. \tag{40}$$

Sums over $l$, $m_l$, and $m_s$ must still be carried out.

---

[14] C. L. Pekeris, Phys. Rev. **112**, 1649 (1958).

[15] E. Clementi, J. Chem. Phys. **38**, 2248 (1963). This value of $E_{\rm corr}$ for oxygen may be obtained by adding the nonrelativistic ionization potentials as given by C. W. Scherr, J. N. Silverman, and F. A. Matsen, Phys. Rev. **127**, 830 (1962). Then $E_{\rm HF}$ is subtracted to give $E_{\rm corr}$.

[16] C. C. J. Roothaan, Rev. Mod. Phys. **32**, 179 (1960).

### B. Higher Order Diagrams

In Ref. 5 it was found necessary to include terms beyond second order when the HF basis set is used. However, it was shown in Ref. 6 that the second-order results are closer to the correct value when the basis set is used in which excited states are calculated in the field of the nucleus and $N-1$ other electrons, which is the basis set used in this paper. Nevertheless, it is still desirable to include the effects of terms beyond second order. Neglecting exchange terms, the third-order energy diagrams which involve only two hole states (one pair) are shown in Fig. 5. The hole-hole interaction is shown in Fig. 5(a) and hole-particle interactions in 5(b) and 5(c). The diagrams of Figs. 5(b) and 5(c) represent the net effect of the interactions of particles in excited states $i$ and $j$ with the passive unexcited states[2] and with the potential $V$.[5] It is assumed that $i$ is calculated in the presence of all unexcited states except for $q$ and $j$ in the presence of all but $r$. In the case of open-shell atoms, there are still small corrections for each hole and particle line due to insertions of the type shown in Fig. 3. However, these were found to be very small (approximately 1%) in the calculations of this paper but they were included in most cases. The particle-particle (ladder) interaction is shown in Fig. 5(d). The diagrams of Figs. 5(a), 5(b), and 5(c) are exclusion-principle-violating (EPV) diagrams.[5] It is also possible to have diagrams such as 5(a), 5(b), and 5(c) where the hole lines have different labels. These are less important, however, because in general the nondiagonal matrix elements are much smaller than the diagonal ones. When the excited states $i$, $j$, $k$, and $l$ of Fig. 5 are bound states, the largest contributions for Fig. 5(b) come when $i=k$, for 5(c) when $j=l$, and for 5(d) when $i=k$ and $j=l$. In this diagonal case, these four third-order diagrams sum to give the second-order diagrams (with $q$, $r$ excited into $i$, $j$) times the factor

$$(\epsilon_q+\epsilon_r-\epsilon_i-\epsilon_j)^{-1}\times(\langle qr|v|qr\rangle-\langle ir|v|ir\rangle -\langle qj|v|qj\rangle+\langle ij|v|ij\rangle). \quad (41)$$

This same factor is repeated in the higher orders, and this geometric series is summed to give the modified second-order result

$$D^{-1}|\langle ij|v|qr\rangle|^2, \quad (42)$$

where

$$D=(\epsilon_q+\epsilon_r-\langle qr|v|qr\rangle)-(\epsilon_i+\epsilon_j-\langle ir|v|ir\rangle -\langle qj|v|qj\rangle+\langle ij|v|ij\rangle). \quad (43)$$

The corresponding exchange matrix elements are also included when states $q$ and $r$ have parallel spins. The first term of $D$ is the two-particle energy of electrons in states $q$ and $r$, and the second term represents the two-particle energy for particles in states $i$ and $j$.[6] An analysis of higher order terms leads to the addition into Eq. (43) of terms $E_{\text{corr}}(q,n)$ and $E_{\text{corr}}(n\neq q, r)$, where $E_{\text{corr}}(q,n)$ is the correlation energy of $q$ with all other unexcited states and $E_{\text{corr}}(n\neq q, r)$ is the correlation energy of $r$ with all other unexcited states except for $q$. These quantities are unknown at the outset and their effects on Eq. (43) may be calculated as a small correction when the main part of the calculation is finished.

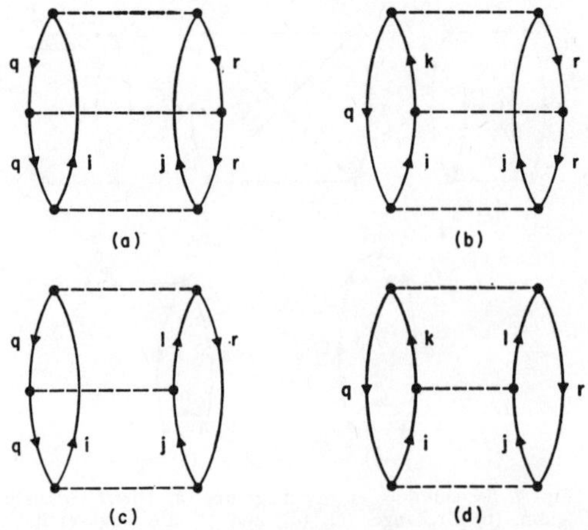


FIG. 5. Third-order energy diagrams involving only one electron pair. (a) Hole-hole interaction. (b) and (c) are hole-particle interactions which represent the net effect of interactions of particles in states $i$ and $j$ with the passive unexcited states and with the potential $V$. It is assumed that $i$ is calculated in the field of all unexcited states except $q$ and $j$ in the field of all but $r$. (d) Particle-particle interaction or ladder diagram.

When the excited states $i$ and $j$ are in the continuum, the geometric sum can be carried out only for the hole-hole interactions of Fig. 5(a), and $D$ becomes

$$D=\epsilon_q+\epsilon_r-\langle qr|v|qr\rangle-\epsilon_i-\epsilon_j. \quad (44)$$

When the spins of $q$ and $r$ are parallel, we also add the exchange term $\langle qr|v|rq\rangle$ to Eq. (44). However, as shown in Ref. 5, the hole-particle interactions of Fig. 5(b) and higher orders may be included to good accuracy by calculating

$$a(k,k';r)=\left(-\frac{2}{\pi}\int_0^\infty dk''\langle kr|v|k''r\rangle D^{-1}(k'',k') \times\langle k''k'|v|qr\rangle\right)\Big/\langle kk'|v|qr\rangle, \quad (45)$$

where

$$D(k''k')=\epsilon_q+\epsilon_r-\langle qr|v|qr\rangle-\tfrac{1}{2}k''^2-\tfrac{1}{2}k'^2. \quad (46)$$

The value of $k$ is a representative value from the range of $k$ values which contribute most to the second-order integration. In Ref. 5, $a(k,k';r)$ was found to be a slowly varying function of $k$ and $k'$. To a good approximation $a(k,k';r)$ is the ratio of the third-order diagram of Fig. 5(b) to the corresponding second-order diagram. The factor $a(k,k';q)$ may be calculated to account for the diagram of Fig. 5(c). The corresponding approximation

TABLE III. $2p$-$2p$ correlations in a.u. from continuum states.

| Excitations[a] | Modified 2nd-order direct[b] | Modified 2nd-order exchange[b] | Modified 2nd-order total | Total including higher orders[c] |
|---|---|---|---|---|
| $2p(-1^+)\to ks^+$ <br> $2p(+1^-)\to k's^-$ | −0.000579 | 0.0000 | −0.000579 | −0.000662 |
| $2p^+\to kp^+$ <br> $2p^+\to k'p^+$ | −0.01127 | +0.00659 | −0.01068 | −0.01270 |
| $2p^+\to kp^+$ <br> $2p^-\to k'p^-$ | −0.01860 | 0.0000 | −0.01860 | −0.02212 |
| $2p^+\to kd^+$ <br> $2p^+\to k'd^+$ | −0.01944 | +0.00583 | −0.01360 | −0.01534 |
| $2p^+\to kd^+$ <br> $2p^-\to k'd^-$ | −0.01930 | 0.0000 | −0.01930 | −0.02176 |
| $2p^+\to kf^+$ <br> $2p^+\to k'f^+$ | −0.00306 | +0.00155 | −0.00151 | −0.00151[d] |
| $2p^+\to kf^+$ <br> $2p^-\to k'f^-$ | −0.00316 | 0.00000 | −0.00316 | −0.00316[d] |
| $2p^+\to k^+s,\ kd^+$ <br> $2p^+\to k'd^+,\ k's^+$ | −0.00328 | +0.00328 | 0.00000 | 0.00000 |
| $2p^+\to ks^+,\ kd^+$ <br> $2p^-\to k'd^-,\ k's^-$ | −0.00468 | 0.00000 | −0.00468 | −0.00468[d] |
| $2p^+\to kp^+$ <br> $2p(+1^-)\to 2p(0^-,-1^-)$ | −0.000761 | 0.0000 | −0.000761 | −0.000761 |
| Total | −0.090131 | +0.01725 | −0.072871 | −0.082691[e] |

[a] $k$ and $k'$ refer to continuum excitations. $s, p, d, f$, refer to $l=0, 1, 2, 3$.
[b] Second order calculated with the modified $D$ of Eq. (44).
[c] Includes sum of hole-particle and ladder terms by Eq. (48).
[d] Effects of higher order terms were estimated to be small and are not included.
[e] This total is reduced by the contribution +0.00201 a.u. from the third-order ladder diagrams in which the $l=1$ continuum excitations scatter into $l=2$ excitations and also $l=2$ scattering into $l=1$ states. The new total is then −0.08068 a.u.

for the diagram of Fig. 5(d) is

$$t(k,k')=\left[\left(\frac{2}{\pi}\right)^2\int_0^\infty dk''\int_0^\infty dk'''\langle kk'|v|k''k'''\rangle \times D(k'',k''')^{-1}\langle k''k'''|v|qr\rangle\right]\Big/\langle kk'|v|qr\rangle, \quad (47)$$

where $k$, $k'$ are again chosen to be typical excitations of importance in the second-order calculation. The factor $t(k,k')$ represents the ratio of the diagram of Fig. 5(d) to the second-order diagram with the same hole lines. It is assumed in Eqs. (45) and (47) that $a$ and $t$ also depend on $l$ and $l'$ of $k$ and $k'$. As shown previously,[5] the interactions shown in Figs. 5(b), 5(c), and 5(d) are repeated in higher orders and, to a good approximation, they may be summed geometrically to give the factor

$$[1-a(k,k';r)-a(k,k';q)-t(k,k')]^{-1}, \quad (48)$$

which multiplies the second-order diagram with hole lines $q$ and $r$ and $D$ given by Eq. (44). In Ref. 5 these effects were found to be much more important than in the present calculation owing to the fact that there were four $a$'s in the factor corresponding to Eq. (48) because of the choice of the potential $V_{\mathrm{HF}}$. In addition, $a$ and $t$ in this calculation are approximately 0.15 and −0.15 for $2s$ and $2p$ hole lines as compared with 0.29 and −0.36 previously.[5] Typically, the factor of Eq. (48) is between 1.00 and $0.85^{-1}$ in this calculation and an error of 10% in $a$ and $t$, which seems unlikely, would lead to errors of only 1 or 2% in Eq. (48).

### C. Numerical Results

The results of the calculations for correlations among $2p$ electrons are given in Table III for excitations involving either two continuum excitations or one $2p$ excited state and one continuum excited state. The remaining $2p$-$2p$ contributions from excitations into bound excited states with $n\geq 3$ are given in Table IV. In general, the excitations into bound states are much

TABLE IV. $2p$-$2p$ correlation energies in a.u. from excitations into bound states with $n\geq 3$.

| Excitation | Correlation energy |
|---|---|
| $2p^+\to ms^+$ <br> $2p^\pm\to ns^\pm$ | −0.000053 |
| $2p^+\to mp^+$ <br> $2p^\pm\to np^\pm$ | −0.000791 |
| $2p^+\to mp^+$ <br> $2p(+1^-)\to 2p(0^-,-1^-)$ | −0.000339 |
| $2p^+\to md^+$ <br> $2p^\pm\to nd^\pm$ | −0.000003 |
| $2p^+\to kp^+,\ mp^+$ <br> $2p^\pm\to mp^\pm,\ kp^\pm$ | −0.00795 |
| $2p^\pm\to ns^\pm,\ kd^\pm$ <br> $2p^\pm\to kd^\pm,\ ns^\pm$ | −0.000384 |
| $2p^+\to nd^+,\ kd^+$ <br> $2p^\pm\to kd^\pm,\ nd^\pm$ | −0.000376 |
| Total | −0.009896 |

TABLE V. Factors to sum hole-particle and ladder effects by Eq. (48).

| Excitation | $a(k,k';2p)$[a] | $t(k,k')$[b] |
|---|---|---|
| $l=0$ | 0.125 | −0.125 |
| $l=1$ | 0.1630 | −0.1669 |
| $l=2$ | 0.1214 | −0.1297 |

[a] See Eq. (45). $k$ and $k'$ were chosen between 1.0 and 2.0 which is where the matrix elements $\langle kk'|v|2p2p\rangle$ were largest.
[b] See Eq. (47).

less important than the excitations into continuum states. Although the energy denominators are smaller for bound-state excitations than for continuum excitations, the matrix elements connecting bound excited states with the $2p$ states are much smaller than for the continuum matrix elements due to the fact that the bound excited states are mostly at a much greater radial distance from the nucleus than the $2p$ states. The integrations were carried out from $k$, $k'=0.0$ to 20.0, where the matrix elements were extremely small. The most important range for $k$ was from 0.0 to 3.0. The higher order terms were included by the factor of Eq. (48). The values for $a$ and $t$ for different $l$ excitations are given in Table V. For example, for $a(k,k',2p)$ the value $k=k'=1.50$ was used for $l=1$.

The value for the total correlation energy among $2p$ electron pairs from Tables III and IV is −0.09259 a.u. However, this values does not include the effects of third-order ladder diagrams in which excited states interact and scatter from excited states of given $l$ values to excited states with different orbital momentum $l$. The contribution from these diagrams in which excited states $kp,k'p$ scatter into $kd,k'd$ and diagrams in which $kd,k'd$ scatter into $kp,k'p$ was calculated to be +0.00201 a.u. This gives a new total of −0.09058 a.u. The calculations of Table III also include effects of insertions on the hole lines and particle lines as shown for hole lines in Fig. 3. These were found to be very small, approximately 1% effects. Since a restricted HF basis is used there are also energy terms of the type shown in Figs. 4(c), 4(d), and 4(e), and also the same types of diagrams with potential interactions. The total of these terms for $2p$ electrons excited into $kp$ states was found to be −0.00113 a.u. This small value shows that for $2p$ electrons the use of a restricted HF potential rather than an unrestricted HF potential has an almost negligibly small effect on the correlation energy when compared with the true two-body correlations of Tables III and IV. Adding the value −0.00113 a.u. to the previous total, we find that the total correlation energy among $2p$ electrons is now −0.09171 a.u.

This result is close to what is obtained by calculating true second-order terms only. A comparison between some of the true second-order terms, second-order terms with the modified denominators of Eq. (46), and the correct pair correlations is given in Table VI. For $2p$ electrons, the second-order result is only a few percent higher than the value we obtain by a careful consideration of higher order terms. The reason that this second-order value is better than that obtained with the shifted

TABLE VI. Comparison of second-order correlation energies and accurate values.[a]

| Excitations | Second order | Modified second order[b] | Accurate value[c] |
|---|---|---|---|
| $2p^+ \to kp^+$<br>$2p^\pm \to k'p^\pm$ | −0.03576 | −0.02928 | −0.03482 |
| $2p^+ \to kd^+$<br>$2p^\pm \to k'd^\pm$ | −0.03799 | −0.03290 | −0.03710 |

[a] All energies are in a.u.
[b] Calculated with modified denominator of Eq. (46); only includes higher order terms due to hole-hole interactions.
[c] Taken from Table III.

TABLE VII. Effect of $R_0$ on matrix elements.[a]

| $k$ | $\langle 2p2.0d|v|2pkd\rangle(R_0=50.0)$ | $\langle 2p2.0d|v|2pkd\rangle(R_0=30.0)$ |
|---|---|---|
| 0.50 | 0.041918 | 0.025130 |
| 1.00 | 0.132219 | 0.124473 |
| 1.25 | 0.233244 | 0.242012 |
| 1.50 | 0.400139 | 0.440386 |
| 1.625 | 0.528058 | 0.488384 |
| 1.75 | 0.714685 | 0.782329 |
| 1.875 | 1.037463 | 1.012214 |
| 2.00 | 2.232088 | 1.978391 |
| 2.125 | 1.051450 | 1.029288 |
| 2.25 | 0.739979 | 0.800411 |
| 2.375 | 0.563280 | 0.530494 |
| 2.50 | 0.444062 | 0.473694 |
| 2.75 | 0.290690 | 0.291278 |
| 3.00 | 0.197229 | 0.189353 |
| 3.50 | 0.095408 | 0.105623 |
| 4.00 | 0.047407 | 0.051722 |
| 5.00 | 0.011232 | 0.018026 |
| 6.00 | 0.001395 | 0.005200 |
| 8.00 | −0.001646 | −0.000664 |
| 10.00 | −0.001066 | 0.000212 |
| 12.00 | −0.000455 | −0.000777 |

[a] All continuum states of this table have $l=2$.

TABLE VIII. $2s$-$2p$ correlation energies in a.u. from excitations involving $2p$ excited states.

| Excitations[a] | Second-order direct[b] | Second-order exchange[b] | Total |
|---|---|---|---|
| $2s^- \to 2p(0^-,-1^-)$<br>$2p^\pm \to kd^\pm$ | −0.04301 | 0.00626 | −0.03675 |
| $2s^\pm \to kd^\pm$<br>$2p(+1^-) \to 2p(0^-,-1^-)$ | −0.005589 | 0.00626 | 0.000671 |
| $2s^- \to 2p(0^-,-1^-)$<br>$2p(0^+,-1^+) \to ks^+$ | −0.000796 | 0.00000 | −0.000796 |
| $2s^- \to 2p(0^-,-1^-)$<br>$2p^\pm \to ns^\pm$ | −0.000388 | 0.00000 | −0.000388 |
| $2s^- \to 2p(0^-,-1^-)$<br>$2p^\pm \to nd^\pm$ | −0.000270 | 0.000037 | −0.000233 |
| $2s^\pm \to nd^\pm$<br>$2p(+1^-) \to 2p(0^-,-1^-)$ | −0.000020 | 0.000037 | 0.000017 |
| Total | −0.05007 | 0.01259 | −0.03748 |

[a] $k$ refers to continuum excitations and $n$, $m$ refer to bound states. $m, n \geq 3$.
[b] The terms involving one continuum excited state and one bound excited state were calculated with the denominator of Eq. (49) to account for one of the hole-particle interactions and the hole-hole interaction.

TABLE IX. $2s$-$2p$ correlation energies in a.u. from two-continuum excitations.

| Excitations | Modified 2nd-order direct[a] | Modified 2nd-order exchange[a] | Modified 2nd-order total | Total including higher orders[b] |
|---|---|---|---|---|
| $2s\to ks$ $2p\to k'p$ | −0.02785 | +0.004898 | −0.02295 | −0.02623 |
| $2s\to kp$ $2p\to k's$ | −0.004100 | +0.004898 | +0.0007980 | +0.000912 |
| $2s\to kp$ $2p\to k'd$ | −0.012917 | +0.003320 | −0.009597 | −0.01097 |
| $2s\to kd$ $2p\to k'p$ | −0.003640 | +0.003320 | −0.000320 | −0.000366 |
| $2s\to kd$ $2p\to k'f$ | −0.01589 | | −0.01589 | −0.01589 |
| Total | −0.064397 | +0.016436 | −0.047959 | −0.05254 |

[a] Calculated with the modified denominator $D$ of Eq. (44).
[b] Includes sum of higher order terms by Eq. (48).

$D$ of Eq. (46) is that with the basis set of this calculation there are two hole-particle diagrams in third order which increase the second-order result, whereas the hole-hole and particle-particle diagrams reduce the second-order term. These terms are often of comparable size and tend to cancel. When only one of them is included [for example, by $D$ of Eq. (46)], there is in effect an unbalance, and the rough cancellation in higher orders does not occur. In the calculations of Ref. 5 the HF basis was used and in this case there are four hole-particle diagrams in third order and so the rough cancellation does not take place.

In calculating matrix elements for second-order energy terms it is only necessary to carry out the radial integrals to the radius where the unexcited-state wave functions are effectively zero. However, in higher orders we encounter matrix elements with continuum to continuum transitions. There is no natural cutoff now on the radial integrations, and we must decide on a physical $R_0$. We expect that $R_0$ can have any value which is much larger than the "radius" of the atom. Physically, we should not expect the structure of the atom to depend on the size of its container provided this containing volume is very large compared to the atom. In the present calculations $R_0$ was chosen as 50.0 as compared with the maximum orbital density of the oxygen atom which occurs at 0.85 according to the HF calculations.[13] Although the matrix elements for continuum to continuum transitions still depend on $R_0$, in calculating the energy or other physical quantities we integrate over the intermediate continuum states and the result then does not depend on $R_0$. An example of this effect is shown in Table VII where the matrix elements are sensitive to $R_0$. However, when the quantity $a(2.0,2.0;2p)$ of Eq. (45) was calculated for $l=2$ states, the result was 0.121 using $R_0=50.0$ and 0.119 using $R_0=30.0$.

TABLE X. $2s$-$2p$ correlation energies in a.u. from excitations involving bound states other than $2p$.

| Excitations[a] | Second-order direct | Second-order exchange | Second-order Total |
|---|---|---|---|
| $2s\to ns$ $2p\to kp$ | −0.006845 | 0.00104 | −0.005805 |
| $2s\to kp$ $2p\to ns$ | −0.0006290 | 0.00104 | 0.00041 |
| $2s\to ks$ $2p\to np$ | −0.003210 | 0.000474 | −0.002736 |
| $2s\to np$ $2p\to ks$ | −0.000279 | 0.000474 | 0.000195 |
| $2s\to np$ $2p\to kd$ | −0.002173 | 0.000495 | −0.001678 |
| $2s\to kd$ $2p\to np$ | −0.000466 | 0.000495 | 0.000029 |
| $2s\to ms$ $2p\to np$ | −0.001020 | 0.000150 | −0.000870 |
| $2s\to mp$ $2p\to ns$ | −0.000110 | 0.00015 | 0.000040 |
| $2s\to mp$ $2p\to nd$ | −0.000001 | 0.000000 | −0.000001 |
| Total | −0.014733 | 0.004318 | −0.030416 |

[a] $n, m\geq 3$.

The results for $2s$-$2p$ correlations are given in Tables VIII, IX, and X. It is seen that the bound-state excitations of greatest importance are $2s^-\to 2p(0^-,-1^-)$. In calculating the contributions of Table VIII, when the $2s$ electron is excited into a $2p$ excited state and the other excited state is in the continuum,

$$D=\epsilon_{2s}+\epsilon_{2p}-\langle 2s2p|v|2s2p\rangle -(\epsilon_{2p}+\tfrac{1}{2}k^2-\langle 2s2p|v|2s2p\rangle). \quad (49)$$

We have accounted for one of the two hole-particle interactions by including $\langle 2s2p|v|2s2p\rangle$ in $D$, which is

TABLE XI. $2s$-$2s$ correlation energies in a.u.

| Excitations[a] | Modified second-order result[b] | Total[c] |
|---|---|---|
| $2s^+\to ks^+$ $2s^-\to k's^-$ | −0.00250 | −0.00286 |
| $2s^+\to kp^+$ $2s^-\to k'p^-$ | −0.00162 | −0.00185 |
| $2s^+\to kd^+$ $2s^-\to k'd^-$ | −0.00463 | −0.00529 |
| $2s^+\to kf^+$ $2s^-\to k'f^-$ | −0.00135 | −0.00135 |
| $2s^+\to kp^+$ $2s^-\to 2p(0^-,-1^-)$ | −0.00184 | −0.00184 |
| $2s^+\to np^+$ $2s^-\to 2p(0^-,-1^-)$ | −0.000595 | −0.000595 |
| $2s^+\to ns^+, ks^+$ $2s^-\to ks^-, ns^-$ | −0.000804 | −0.000804 |
| $2s^+\to np^+, kp^+$ $2s^-\to kp^-, np^-$ | −0.000274 | −0.000274 |
| $2s^+\to ms^+$ $2s^-\to ns^-$ | −0.000118 | −0.000118 |
| $2s^+\to mp^+$ $2s^-\to np^-$ | −0.000020 | −0.000020 |
| $2s^+\to md^+$ $2s^-\to nd^-$ | −0.0000001 | −0.0000001 |
| Total | −0.01375 | −0.01500 |

[a] $k$ and $k'$ refer to continuum states and $n$, $m$ to bound excited states with $n, m\geq 3$.
[b] Shifted denominator of Eq. (49) is used when there is one bound and one continuum excited state. $D$ of Eq. (44) is used for $k$, $k'$ excitations.
[c] Higher order effects of Eq. (48) are included for all $k$, $k'$ excitations except for $kf$, $k'f$.

TABLE XII. $1s$-$2p$ correlation energies in a.u.

| Excitations[a] | Modified second-order direct[b] | Modified second-order exchange[b] | Modified second-order total[b] | Total[c] |
|---|---|---|---|---|
| $1s \rightarrow ks$<br>$2p \rightarrow k'p$ | −0.001391 | −0.000537 | −0.001927 | −0.002079 |
| $1s \rightarrow kp$<br>$2p \rightarrow k's$ | −0.001326 | −0.000537 | −0.001863 | −0.002009 |
| $1s \rightarrow kp$<br>$2p \rightarrow k'd$ | −0.009349 | 0.001110 | −0.008239 | −0.008888 |
| $1s \rightarrow kd$<br>$2p \rightarrow k'p$ | −0.001014 | 0.001110 | 0.000096 | 0.000104 |
| $1s \rightarrow kd$<br>$2p \rightarrow k'f$ | −0.001376 | 0.0003120 | −0.001064 | −0.001148 |
| $1s \rightarrow kf$<br>$2p \rightarrow k'd$ | −0.000389 | 0.000312 | −0.000077 | −0.000083 |
| $1s^- \rightarrow 2p(0^-,-1^-)$<br>$2p \rightarrow kd$ | −0.000648 | 0.000090 | −0.000558 | −0.000558 |
| $1s \rightarrow kd^\pm$<br>$2p(+1^-) \rightarrow 2p(0^-,-1^-)$ | −0.000104 | 0.000090 | −0.000014 | −0.000014 |
| $1s^- \rightarrow 2p(0^-,-1^-)$<br>$2p(0^+,-1^+) \rightarrow ns^+$ | −0.000005 | 0.000000 | −0.000005 | −0.000005 |
| $1s \rightarrow ns$<br>$2p \rightarrow kp$ | −0.000036 | −0.000022 | −0.000058 | −0.000058 |
| $1s \rightarrow kp$<br>$2p \rightarrow ns$ | −0.000062 | −0.000022 | −0.000084 | −0.000084 |
| $1s^- \rightarrow 2p(0^-,-1^-)$<br>$2p(0^+,-1^+) \rightarrow ks$ | −0.000063 | 0.000000 | −0.000063 | −0.000063 |
| $1s \rightarrow np$<br>$2p \rightarrow ks$ | −0.000015 | −0.000004 | −0.000019 | −0.000019 |
| $1s \rightarrow ks$<br>$2p \rightarrow np$ | −0.000005 | −0.000004 | −0.000009 | −0.000009 |
| $1s \rightarrow np$<br>$2p \rightarrow kd$ | −0.000104 | 0.000021 | −0.000083 | −0.000083 |
| Total | −0.015887 | 0.001919 | −0.013967 | −0.014996 |

[a] $k$, $k'$ are continuum states. $m$, $n$ refer to bound states with $m, n \geq 3$.
[b] $D$ of Eq. (44) used for $k$, $k'$ excitations.
[c] $k$, $k'$ excitation terms are modified by Eq. (48).

now the second-order denominator. The remaining hole-particle interaction and the particle-particle interaction are expected to cancel to a good approximation as they did in the $2p$-$2p$ case as seen in Table V and so the second-order result is taken as the pair correlation energy. The total $2s$-$2p$ correlation energy from Tables VIII, IX, and X is −0.1004 a.u.

The correlation energies for different excitations for the $2s$-$2s$ pair are given in Table XI. In the case of $2s^- \rightarrow 2p(0^-, -1^-)$ excitations, one hole-particle interaction is included by adding $\langle 2s2p|v|2s2p \rangle$ to $D$ of Eq. (44). In the two-continuum excitation cases, the hole-particle and ladder interactions were included by the factor of Eq. (48). Small effects for the insertions of Fig. 3 were also included. The excitations into bound $l=2$ states were extremely small. The total $2s$-$2s$ correlation energy of −0.01500 a.u. is considerably smaller than the $2s$-$2s$ beryllium correlation energy of −0.04491 a.u.[5,6] This relatively small $2s$-$2s$ correlation energy in oxygen may be ascribed to the exclusion principle as discussed previously by McKoy and Sinanoğlu.[17] If the presence of the four oxygen $2p$ electrons is ingored, then there is a contribution of −0.0607 a.u. in second order to the $2s$-$2s$ correlation energy due to excitations into the $2p$ excited states. However, the exclusion principle must be considered and so this contribution may not be included.

There is also the small contribution to the correlation energy due to the diagrams of Figs. 4(c), 4(d), and 4(e) because the potential we use is not a completely unrestricted HF potential. The potential of Eq. (26) has two $2p$ exchange interactions for both $2s^\pm$ states, but the $2s^+$ electron actually has exchange interactions with the three $2p^+$ electrons and the $2s^-$ has exchange interactions with the one $2p^-$ electron. This effect causes the diagram of Fig. 4(d) to be included for both $2s^\pm$, with excitations into $ks^\pm$ states giving a total of −0.000769 a.u. There is also an energy of −0.000814 a.u. from diagrams shown in Figs. 4(c), 4(d), and 4(e) in which $2s^\pm$ electrons are excited into $kd^\pm$ states. The total $2s$-$2s$ correlation energy, referred to the restricted HF solution, is then −0.01661 a.u.

The remaining contributions to the correlation energy come from $1s$-$1s$ correlations and from the $1s$-$2s$ and $1s$-$2p$ intershell correlations. The $1s$-$2p$ correlations are given in Table XII and give a total $1s$-$2p$ correlation energy of

[17] V. McKoy and O. Sinanoğlu, J. Chem. Phys. 41, 2689 (1964).

TABLE XIII. $1s$-$2s$ correlation energies in a.u.

| Excitations[a] | Modified second-order direct[b] | Modified second-order exchange[b] | Modified second-order total[b] | Total[c] |
|---|---|---|---|---|
| $1s \to ks$<br>$2s \to k's$ | −0.003160 | 0.001537 | −0.001623 | −0.001688 |
| $1s \to kp$<br>$2s \to k'p$ | −0.003910 | 0.001189 | −0.002721 | −0.002830 |
| $1s \to kd$<br>$2s \to k'd$ | −0.000940 | 0.000329 | −0.000611 | −0.000635 |
| $1s \to kf$<br>$2s \to k'f$ | −0.000248 | 0.000101 | −0.000147 | −0.000153 |
| $1s^{\pm} \to kp^{\pm}$<br>$2s^{-} \to 2p(0^{-},-1^{-})$ | −0.000799 | 0.000046 | −0.000753 | −0.000753 |
| $1s^{-} \to 2p(0^{-},-1^{-})$<br>$2s^{+} \to kp$ | −0.000136 | 0.000046 | −0.000090 | −0.000090 |
| $1s \to np$<br>$2s \to kp$ | −0.0000241 | 0.000006 | −0.000018 | −0.000018 |
| $1s \to kp$<br>$2s \to np$ | −0.000089 | 0.000006 | −0.000083 | −0.000083 |
| $1s \to ks$<br>$2s \to ns$ | −0.000020 | 0.000013 | −0.000007 | −0.000007 |
| $1s \to ns$<br>$2s \to ks$ | −0.000036 | 0.000013 | −0.000023 | −0.000023 |
| $1s^{-} \to 2p(0^{-},-1^{-})$<br>$2s^{\pm} \to np$ | −0.000009 | 0.000006 | −0.000003 | −0.000003 |
| $1s^{\pm} \to np^{\pm}$<br>$2s^{-} \to 2p(0^{-},-1^{-})$ | −0.000014 | 0.000006 | −0.000008 | −0.000008 |
| Total | −0.009385 | 0.003298 | −0.006087 | −0.006291 |

[a] $k$, $k'$ are continuum excitations. $m$, $n$ refer to bound states with $m, n \geq 3$.
[b] $D$ of Eq. (44) has been used.
[c] Hole-particle and particle-particle effects are included by Eq. (48) for $k$, $k'$ excitations. The second-order result is used for bound excited states.

−0.014996 a.u. The total $1s$-$2p$ correlation energy, although small compared to $2p$-$2p$ and $2s$-$2p$ correlations, is still sufficiently large that it must be included if an accurate value for the total correlation energy is desired. The $1s$-$2s$ intershell correlation energy is given in Table XIII and is −0.006291 a.u. The average correlation energy for each $1s$-$2s$ pair is −0.00157 a.u. as compared with an average correlation energy −0.00187 a.u. for each $1s$-$2p$ pair. The average effect of the hole-particle and particle-particle interactions for $1s$-$2s$ correlations is to multiply the modified second-order terms involving two-continuum excitations by approximately 1.04. The $1s$-$1s$ correlation energy contributions listed in Table XIV add to give a total of −0.04383 a.u.

Since the excited single-particle states in this investigation are calculated in the presence of both $1s$ electrons, when there is a $1s$ hole line both excited states interact with it. That is, the interactions of the excited states with $V$ and with the unexcited states do not completely cancel and give rise to these hole-particle interactions. There may also be interactions with a $2s$ or $2p$ state which was omitted in $V$ but is now present. For example, consider $1s$-$2p$ correlations into $kp,k'd$ states. There are hole-particle interactions of both $kp$ and $k'd$ with the $1s$ hole line because $kp$ and $kd$ are calculated with $V$ in which both $1s$ electrons are present and one $2p$ electron is removed. There will be no hole-particle interactions with the $2p$ hole line in this case. It should be noted that in general there will be small effects due to insertions on hole and particle lines of the type shown in Figs. 3(b), 3(c), and 3(d). The hole-particle interactions merely account for the major effects of the lack of cancellation between interactions with $V$ and with the passive unexcited states.

For the $1s$-$1s$ correlations both excited states interact with each $1s$ hole line giving a total of four hole-

TABLE XIV. $1s$-$1s$ correlation energies in a.u.

| Excitations[a] | Modified second-order total[b] | Total[c] |
|---|---|---|
| $1s^{+} \to ks^{+}$<br>$1s^{-} \to k's^{-}$ | −0.011788 | −0.013173 |
| $1s^{+} \to kp^{+}$<br>$1s^{-} \to k'p^{-}$ | −0.021848 | −0.024413 |
| $1s^{+} \to kd^{+}$<br>$1s^{-} \to k'd^{-}$ | −0.003716 | −0.004152 |
| $1s^{+} \to kf^{+}$<br>$1s^{-} \to k'f^{-}$ | −0.001063 | −0.001188 |
| $1s^{+} \to kp^{+}$<br>$1s^{-} \to 2p(0^{-},-1^{-})$ | −0.000634 | −0.000634 |
| $1s^{+} \to kp^{+}, np^{+}$<br>$1s^{-} \to np^{-}, kp^{-}$ | −0.000117 | −0.000117 |
| $1s^{+} \to ks^{+}, ns^{+}$<br>$1s^{-} \to ns^{+}, ks^{+}$ | −0.000152 | −0.000152 |
| Total | −0.039318 | −0.043829 |

[a] $k$, $k'$ refer to continuum states. $m$, $n$ refer to bound states with $m, n \geq 3$.
[b] States with $k$, $k'$ calculated with $D$ of Eq. (44).
[c] Includes hole-particle and particle-particle effects by Eq. (50) for $k$, $k'$ excitations.

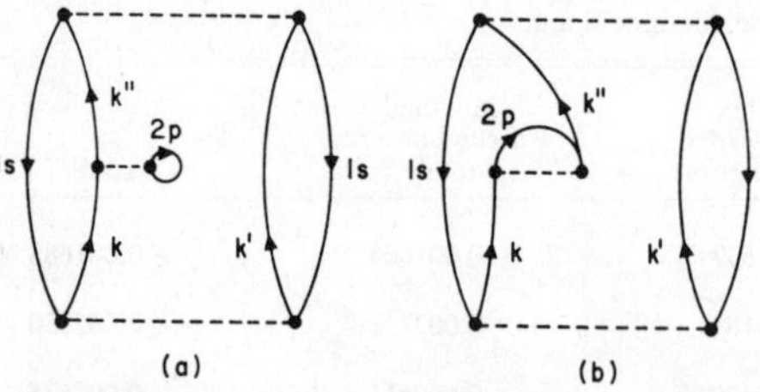


FIG. 6. Third-order diagrams which should be included in $1s$-$1s$ correlated in a potential $V$ which omits one $2p$ state. For $l=0$ states, $2p$ is replaced by $2s$. (a) Direct interaction. (b) Corresponding exchange term.

particle interactions rather than two as given by Eq. (48). These terms are partly cancelled by the particle-particle interactions and by the interactions of excited states with a $2p$ or $2s$ state excluded from $V$ but now present. This type of term is shown in Fig. 6. These terms are included by the methods used in Eq. (48). However, the appropriate factor is now

$$[1-4a(k,k';1s)-t(k,k')+2a(k,k';2p)]^{-1}, \quad (50)$$

where $a$ and $t$ are defined by Eqs. (45) and (47). The term $a(k,k';2p)$ in Eq. (50) should account for the exchange term of Fig. 6(b) as well as for the direct term of Fig. 6(a). In the case of $kp,k'p$ excitations, $t(k,k')$ was found to be approximately equal to $-a(k,k';1s)$. The term $a(k,k';1s)$ was calculated to be 0.0431 and $a(k,k';2p)$ was 0.0121. For $kp,k'p$ excitations of two $1s$ electrons, Eq. (50) is then 1.1174.

As discussed previously, the $1s$ states were calculated with interactions with only one $2s$ electron of opposite spin and with interactions with one $1s$ electron of parallel spin and one of antiparallel spin. Each $1s$ electron interacts, however, with a $2s$ electron of parallel spin and one of antiparallel spin and with only one $1s$ electron of opposite spin. The interaction of a $1s$ electron with a $1s$ electron of parallel spin in $V$ may be neglected since the exchange $1s$ interaction cancels the direct term. The net effect is that diagrams of Fig. 2(a) and 2(b) must be calculated for $p=1s^+$ and $n=2s^+$ and for $p=1s^-$ and $n=2s^-$. The corresponding energy diagrams are shown in Figs. 4(c), 4(d), and 4(e). The diagrams of Fig. 4(c) were found to give $-0.0001322$ a.u. and those of Fig. 4(d) gave $-0.0002522$ a.u. The diagrams of Fig. 4(e) gave 0.0003629 a.u., making a total $-0.0000215$ a.u. This value should not be included in the correlation energy, however, as it is the difference between the restricted HF energy and the energy through first order using the potential of Eq. (26) for both $1s$ and $2s$ states.

As in the case of $2s$ electrons, there are also corrections of the type shown in Fig. 4(d) with $q=1s^\pm$ and $n=2p^\pm$ due to the fact that a restricted HF potential is used. The sum of diagrams like Fig. 4(d) for $1s^\pm$ excitations into $ks$ states was calculated to be $-0.0000632$ a.u. and $-0.0000223$ a.u. for excitations into $kd$ states. These values are included in the correlation energy which is referred to the restricted HF solution. The total $1s$ correlation energy is then $-0.04392$ a.u.

### D. Discussion of Results

The results of the correlation energy calculations are summarized in Table XV. The total correlation energy $-0.2740$ a.u. is composed both of true two-body correlations which equal $-0.2711$ a.u. and of one-body effects equal to $-0.00283$ a.u. These one-body effects, as discussed previously, are due to the use of a spherically symmetric, restricted HF potential rather than a completely unrestricted HF potential. The value $-0.2740$ a.u. is 5.84% more negative than the value $-0.258$ a.u. which Clementi has deduced from experiment as the total oxygen correlation energy and which he estimates as accurate to within 5%. The difference between the results of the present calculation and the correct total correlation energy may be attributed both to the approximations in calculating the pair correlations and also to the neglect of three-body and higher effects. If we neglect the possible errors in Clementi's value for $E_{corr}$ and in the value from this investigation, the difference 0.016 a.u. gives us a very rough estimate of the size of the three-body and higher correlation effects in neutral oxygen. Since this value is quite inaccurate and could even be wrong in sign, it would be desirable to calculate three-body effects directly.

The calculations of this paper for one-body and two-body correlations are estimated as being accurate to within approximately 5%. Most of this inaccuracy is attributed to the neglect of nondiagonal ladder diagrams and omission of excited states with $l>3$. The individual contributions listed in the tables are estimated to be accurate to within 2%. The nondiagonal ladder diagrams which are expected to be largest for $2p$-$2p$ correlations have been included. These are the diagrams in which two $2p$ electrons, excited into $l=1$ states, scatter into $l=2$ states before returning to the ground state and those diagrams in which particles in $l=2$ states scatter into $l=1$ excited states before returning to the ground state. These terms contributed 0.00201 a.u. No other

TABLE XV. Summary of contributions to the correlation energy in a.u.[a]

| Electrons | Correlation energy |
|---|---|
| $2p$-$2p$ | $-0.09058$ |
| $2s$-$2p$ | $-0.10044$ |
| $2s$-$2s$ | $-0.01500$ |
| $1s$-$2p$ | $-0.014996$ |
| $1s$-$2s$ | $-0.00629$ |
| $1s$-$1s$ | $-0.04383$ |
| $2p$[b] | $-0.00113$ |
| $2s$[b] | $-0.00161$ |
| $1s$[b] | $-0.000086$ |
| Total | $-0.2740$ |

[a] Only one-body and two-body contributions are included.
[b] One-body corrections due to use of a restricted Hartree-Fock potential

nondiagonal ladder terms have been included. For the correlations involving at least one 1s electron, the nondiagonal ladder diagrams are very small due to the large energy denominators. Omission of the ladder diagrams tends to make the correlation energy more negative and omission of states with $l>3$ causes the correlation energy to be less negative.

From Tables III, IV, VIII, IX, and XII it is found that the correlation energy terms involving the $2p(+1^-)$ state add to approximately $-0.079$ a.u. This value may be compared with the difference in correlation energy between O and $O^+$ ($-0.065$ a.u.) or between O and N ($-0.070$ a.u.).[15] Actually, an adjustment should be made to account for the fact that since the state $2p(+1^-)$ is unoccupied in $O^+$ and N, excitations into this state may occur and there is an approximate change in the correlation energy in $O^+$ and N by $-0.0063$ a.u. relative to that in O due to all electrons other than $2p(+1^-)$. This provides another example of the effects of the exclusion principle.

### E. Three-Body Effects

An advantage of the perturbation expansion is that the three-body and higher terms enter in a systematic manner and are added directly to the one-body and two-body terms without repeating the part of the calculation for the pair correlations. The orthonormal set of states used to calculate the pair correlations is also used to calculate the three-body and higher diagrams. Three-body diagrams for the energy first enter in third order and they are shown in Figs. 7(a) and 7(b). The hole states $q$, $r$, and $s$ must all be different for three-body terms. The most important contributions are expected when $q$, $r$, and $s$ are 2s and 2p states. The diagrams of Fig. 7(a) are positive while the diagrams of Fig. 7(b) are negative. In Fig. 7(b) the spin of state $q$ must be parallel to that of state $r$, while there is no such restriction in Fig. 7(a). Because of the angular factors in the matrix elements, the contributions from Fig. 7(b) are expected to be very small when $q$ and $r$ are 2p states. However, when $q$ is a 2s state and $r$ is a 2p state or $q$ is a 2p state and $r$ a 2s state, these diagrams will be significant. Although the diagrams of Figs. 7(a) and 7(b) have not been explicitly calculated, they are estimated as small compared with the pair correlations.

Some of the three-body diagrams in fourth order are shown in Figs. 7(c), 7(d), and 7(e). Diagrams like Fig. 7(c) have recently been included in studies of three-body correlations in nuclear matter.[18] The diagrams of Figs. 7(d) and 7(e) are rearrangement diagrams of the type considered by Brueckner and Goldman.[19] In Ref. 5 the diagrams of Fig. 7(d) were shown to arise from the linked cluster factorization and were called third-class EPV diagrams. In the diagrams of Fig. 7(d) the hole lines $q$

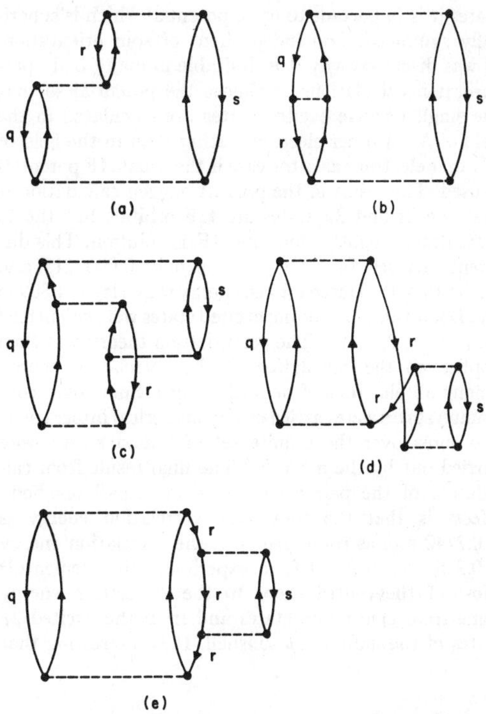


FIG. 7. Typical three-body diagrams. (a) Third-order "ring" diagram. (b) Exchange diagram of (a) or hole-particle diagram. (c) Fourth-order diagram of type considered in Ref. 18. (d) and (e) are rearrangement diagrams discussed in Ref. 19.

and $s$ may be equal and in this case the diagram is no longer a three-body diagram but is part of the pair correlations. By considering both time orderings of the right-hand part of this diagram and the corresponding higher order diagrams, we may sum these diagrams in such a way that they may be included by a small shift in the denominator $D$ of Eq. (43).[20] The rearrangement diagram of Fig. 7(e) also is an exclusion-principle-violating diagram resulting from the linked cluster factorization. Calculation of the diagrams gave approximately 0.0041 a.u. for Fig. 7(d) and $-0.0045$ a.u. for Fig. 7(e), so there is good cancellation between these two types of diagrams.

## VI. DISCUSSION AND CONCLUSIONS

It has been shown in this investigation that many-body perturbation theory need not be restricted to closed-shell, nondegenerate systems but may be applied directly to calculation of the ground-state energy and wave function of most atoms. It was emphasized that it is important to start from an unperturbed eigenstate of $L^2$, $L_Z$, $S^2$, and $S_Z$. In calculating the single-particle

[18] H. A. Bethe, Phys. Rev. 138, B804 (1965).
[19] K. A. Brueckner and D. T. Goldman, Phys. Rev. 117, 207 (1960).
[20] H. P. Kelly, Phys. Rev. 134, A1450 (1964).

states it is important to use a potential which is spherically symmetric and independent of spin orientation. It was discussed why it is desirable in many-body perturbation calculations to choose the potential so that the single-particle excited states are calculated in the field of $N-1$ other electrons rather than in the field of $N$ other electrons as is the case if the usual HF potential is used. The result in the present oxygen calculation is that the $2s$ and $2p$ states are HF orbitals but the $1s$ state differs slightly from the HF $1s$ solution. This difference caused $\langle\Phi_0|H|\Phi_0\rangle$ to be only 0.0000215 a.u. higher than the Hartree-Fock energy which is $-74.80936$ a.u. Bound and continuum excited states were calculated for $l=0$, 1, 2, and 3. The perturbation theory was then applied to the calculation of the correlation energy among all electrons of neutral oxygen. Sums over continuum states were carried out by numerical integrations and sums over the infinite set of bound states were carried out by the $n^{-3}$ rule.[6] The final result from calculation of the pair correlations and small one-body effects is that the total pair correlation energy is $-0.2740$ a.u. as compared with the correlation energy $-0.258$ a.u. deduced from experiment by Clementi.[15] Most of the contributions to the correlation energy come from continuum states and from the excited $2p$ states of the unfilled $2p$ subshell. It is interesting that the $2s$-$2p$ intershell correlation energy of $-0.1004$ a.u. is 37% of the total correlation energy. The total intershell correlation energy from $2s$-$2p$, $1s$-$2p$, and $1s$-$2s$ correlations is $-0.1217$ a.u. or 44% of the total correlation energy as compared with the previous beryllium calculation where the intershell energy was only 5% of the total. The importance of intershell effects may become even greater for larger atoms as the shells and subshells of electrons become closer.

The total pair correlation energy $-0.2740$ a.u. calculated in this work is estimated as being accurate to within 5% and Clementi also gives a limit of 5% on the accuracy of the value for the total correlation energy $-0.258$ a.u. deduced from experiment. Although the difference of 0.016 a.u. between the two correlation energy values may be due to three-body and higher correlations, the uncertainties in the two values make it desirable to carry out a direct calculation of three-body correlations and these effects will be investigated in future work.

## ACKNOWLEDGMENTS

I would like to thank Professor Keith A. Brueckner for his support and encouragement and Dr. Katuro Sawada for helpful discussions.

# VIII-6

# *Methods of Field Theoretic Green's Functions in Atomic and Molecular Problems*†

V. V. TOLMACHEV

The concepts and methods of the quantum field theory are applied to the many-electron problem in atoms and molecules. The field theoretic formulation of perturbation theory is developed in these problems with respect to the smallness of the mutual interaction of the electrons, and rules are given for composing the corresponding Feynman diagrams and the contributions from them. A comparison is made in the first and second orders between the field theoretic and conventional forms of perturbation theory. Two systems of exact field equations are formulated: one system of field equations, the Dyson–Schwinger, for study of the one-particle corpuscular aspect of the problem, and another, new system of field equations for study of the two-particle corpuscular aspect of the problem.

## 1. Introduction

Contemporary mathematical methods of the quantum field theory, which receives its fullest reflection in the method of field theoretic Green's functions, are applied with increasing frequency to various problems related in one way or another to the quantum mechanical or classical many-body problem. Undoubtedly, these methods will also be applied to the further study of the many-electron problem in atomic and molecular problems.

The reasons for such great effectiveness of the field theoretic methods today are rather clear. On the basis of detailed mathematical study of the formal solution to the many-body problem given completely by an entire perturbation-theory series in all orders, these methods permit an understanding and intercomparison of different approximate solutions to the problem, and also an indication of an approach to new reasonable approximations. In other words, these field methods provide a means of systematic study of the problem, which seems to us to be most lacking at present in atomic and molecular problems.

The problem of many interacting electrons in an atom or molecule is formally described by the Schrödinger many-particle equation, which is very difficult to study because of the presence of a large number of variables in the wave function. Generally speaking, we should be interested in the eigenstates of a system with many degrees of freedom, to a certain extent the same states included in the quantum field theory, which deals with a system having an infinite number of degrees of freedom.

Nevertheless, there is a possibility that this will turn out to be very useful in developing our qualitative conceptions of the atom and the molecule, namely, the possibility of considering the individual electrons in the atom and the molecule and the pairs of electrons in the molecule with antiparallel spins that participate in the formation of a chemical bond, etc., as in the field theory. That is, regardless of the fairly strong interaction among electrons, one can postulate a certain autonomy of the individual electrons and groups of them in the atom and the molecule.

The method of field theoretic Green's functions furnishes us with an adequate mathe-

†This translation has been made by the American Meteorological Society under Contract AF 19(628)-3880, through the support and sponsorship of the Air Force Cambridge Research Laboratories, Office of Aerospace Research, L. G. Hanscom Field, Bedford, Massachusetts. Originally published in Russian in *Litovskii Fizicheskii Sbornik* 3(1/2), 47–72 (1963). It is used here by permission.

matical apparatus for studying the indicated aspect of the problem of many interacting electrons in an atom or molecule, the corpuscular aspect.

With the help of the Green's functions for one particle, two particles, etc., we reduce the study of the exact, but thus mathematically intricate, Schrödinger many-particle equation to the study of the mathematically simpler, although approximate, integral equations for the Green's functions, which correctly express the qualitative, and then the quantitative, features of the problem.

To a certain extent these approximate equations for the Green's functions bear the same relation to the full Schrödinger equation as the Boltzmann approximate gas-kinetic equation for the distribution function bears to the system of exact dynamic equations for the classical system of interacting particles.

In the proposed investigation, we intend to discuss fully the mathematical ideas of the field theoretic Green's functions in their application to the many-electron problem in an atom or molecule. In addition, we will not assume the total number of electrons in the atom or molecule to be large, although the effectiveness of the mathematical methods set out below obviously is related to this type of assumption.

The important limitation in our examination is that we will study the case of a nondegenerate state of the electron system of an atom or molecule. However, this state does not have to be the ground state, or that with the least energy; it can also be any nondegenerate excited state. Besides, we do not see any fundamental difficulties in the extension of the stated method to the case of a degenerate state.

## 2. Field Theoretic Green's Functions and the Field Theoretic Form of Perturbation Theory for a Nondegenerate Level

Let us examine a system of N interacting electrons located in the field of the stationary nuclei of an atom or molecule. To make automatic allowance for the antisymmetry requirement of the entire wave function of the system, let us proceed to the representation of second quantization. For the initial entire system of orthonormal one-particle functions that defines the representation of second quantization, let us take the eigenstates of the problem of an individual electron moving in the field of the stationary nuclei of an atom or molecule.

The full Hamiltonian of the system in the representation of second quantization has the following form (cf. [1, 2]):

$$H = \sum_{i} E_i \overset{+}{a}_i a_i + \tfrac{1}{2} \sum_{ijk\ell} V_{ij;k\ell}\, \overset{+}{a}_i \overset{+}{a}_j a_k a_\ell \tag{1}$$

where the first summand yields the zeroth Hamiltonian $H_0$, and the second summand yields the Hamiltonian interaction $H_{int}$. The operator of the total number of particles is $N = \Sigma_i \overset{+}{a}_i a_i$. The operators of the creation and annihilation of an electron in the i state are designated by $\overset{+}{a}_i$ and $a_i$. The discrete subscript i characterizes a full set of the quantum numbers that define the eigenstate of the electron in the field of the nuclear body. In the case of the atom, i designates the four-member set of quantum numbers n, ℓ, m, and s.

Without restricting the generality, we can assume symmetry of $V_{ij;k\ell}$ with respect to the simultaneous permutation $i \rightleftharpoons j$, $k \rightleftharpoons \ell$. In addition, assuming that $V_{ij;k\ell}$ is real, we have, by the Hermitian condition on (1), $V_{ji;k\ell} = V_{k\ell;ji}$ so that

$$V_{ij;k\ell} = V_{\ell k;ji} = V_{k\ell,ij} \tag{2}$$

The basic object of our study will be a nondegenerate, not necessarily ground, eigenstate of the Hamiltonian (1). Our assumption of nondegeneracy is important, since we intend to use a representation of interaction for the field theoretic Green's function that, as Gell-Mann and Low [3] showed, rests on the adiabatic theorem, the greatest extension and simplest formulation of which suggests the nondegeneracy of the initial state $H_0$, which after adiabatic inclusion of interaction yields the state H, which is of interest to us. A further, and obviously more important circumstance is that we will have to use the Wick theorem [4], which, of course, is applied for averaging over an arbitrary simultaneous eigenstate of all the population numbers $n_i = \overset{+}{a}_j a_i$, such as an arbitrary nondegenerate state H. The Wick theorem is not applicable for averaging over a linear combination of states, which must be examined for a proper formulation of the adiabatic theorem for a degenerate initial state $H_0$, according to Born and Fock [5]. Thus we are interested in the energy and the wave function of an eigenstate H derived by adiabatic inclusion from a nondegenerate state $H_0$. The important characteristics related to this state that manifest the corpuscular aspect of the entire wave function of the state are the Green's functions for one particle, two particles, etc., in this state, for which a theory was originally developed by Schwinger [6] in the quantum field theory. At the same time, the Green's functions permit the simplest field theoretic formulation of perturbation theory. It is not required that we prove the Brueckner–Goldstone theorem [2] of linked diagrams for vacuum diagrams or the corresponding theorem of the exclusion of vacuum loops in the diagrams for the Green's functions.

Thus we will abandon the customary Dyson method [7] for the field theoretic formulation of perturbation theory, whose application to the Fermi system is discussed in detail, for example, in a paper by Hubbard [8] (see also the paper by Goldstone [2]), and turn to a Green's method for developing perturbation theory, whose application to a Fermi system is discussed in a paper by Klein and Prange [9]. In [9], first a linked cluster of integrodifferential equations of the Green's functions for one particle, two particles, etc., is derived on the basis of the equations of motion for the field operators, and then it is converted into a linked cluster of integral equations solved later by iteration. In order to avoid discussion of the problem of the single-valuedness of that conversion, we immediately derive the linked cluster of integral equations (cf. [1]) on the basis of the Bogoliubov and Shikov generalization of the Wick theorem [10]. In our case of an arbitrary nondegenerate excited state, it is not possible to use the Schwinger condition [6] of the positiveness or negativeness of frequencies during infinite times in the future and past.

The one-particle and two-particle Green's functions in the representation of the interaction are defined as the averages of the products of the field operators over a state of the Hamiltonian $H_{int}$:

$$G_{k;k'}(t;t') = \langle T(a_k(t)\overset{+}{a}_{k'}(t')S)\rangle / \langle S\rangle \tag{3}$$

$$G_{k\ell;\ell'k'}(ts;s't') = \langle T(a_k(t)a_\ell(s)\overset{+}{a}_{\ell'}(s')\overset{+}{a}_{k'}(t')S)\rangle/\langle S\rangle \tag{4}$$

where the S-matrix can be defined by its formal perturbation-theory series:

$$S = \sum_{n=0}^{+\infty} \frac{1}{n!}\left(-\frac{i}{\hbar}\right)^n \int_{-\infty}^{+\infty} dt_1 \cdots \int_{-\infty}^{+\infty} dt_n\, T\,[H_{int}(t_1)\cdots H_{int}(t_n)] \tag{5}$$

The operation T designates the time ordering of the product of the operators to which it applies. The operator with the greatest time should be placed to the left of the operator with the least time. We have in mind the Wick T-product, which represents the individual field operators, rather than the Dyson P-product, which completely represents $H_{int}(t)$. Therefore, $H_{int}(t)$ should be written

$$H_{int}(t) = e^{-\alpha|t|} \frac{1}{2} \sum_{i'j'ji} V_{i'j';ji} \overset{+}{a}_{i'}(t+0)\overset{+}{a}_{j'}(t+0) a_j(t) a_i(t) \tag{6}$$

in which the +0 in the time arguments of the field operators with a superior plus sign designates the addition of an infinitesimal positive quantity; the factor $e^{-\alpha|t|}$ arising in the adiabatic theorem is written explicitly here to be kept in mind in what follows when taking the integrals in infinite time limits; of course, $\alpha \longrightarrow +0$, and in (6)

$$\overset{+}{a}_i(t) = e^{(i/\hbar)E_i t}\, \overset{+}{a}_i \quad \text{and} \quad a_i(t) = e^{-(i/\hbar)E_i t}\, a_i$$

Owing to definitions (3) and (4), we have, for the zeroth Green's functions in the case where there is no interaction,

$$G^0_{k;k'}(t;t') = G^0_{k'}(t-t')\delta_{kk'} \tag{7}$$

$$G^0_{k\ell;\ell'k'}(ts;s't') = G^0_k(t-t')G^0_\ell(s-s')\delta_{kk'}\delta_{\ell\ell'}$$
$$- G^0_k(t-s')G^0_\ell(s-t')\delta_{k\ell'}\delta_{\ell k'} \tag{8}$$

where $\delta_{kk'}$, as usual, represents the Kronecker symbol, $\delta_{kk'} = 1$ when $k = k'$ and 0 when $k \neq k'$.

Further,

$$G^0_k(t) = e^{-(i/h)E_i t}\{(1-n_k)\,\theta(t) - n_k\,\theta(-t)\} \tag{9}$$

where $\theta(t)$ is a step function equal to 1 when $t > 0$ and 0 when $t < 0$; the $n_k$ designate the population numbers of the one-particle states of the initial state $H_0$.

Applying the Bogoliubov generalization of the Wick theorem to the numerator of expressions (3) and (4), after a rather cumbersome, but not conceptually complex computation we arrive at the first two equations of the linked cluster of integral equations for the Green's functions. We give the first equation of this linked cluster, which is obtained quite simply:

$$G_{k;k'}(t;t') = G^0_{k;k'}(t;t') + \frac{i}{\hbar}\int_{-\infty}^{+\infty} ds \sum_{i'j'ji} V_{i'j';ji} G^0_{k;i'}(t;s) G_{ij;j'k'}(s,s;s+0,t') \tag{10}$$

The second equation has the form

$$G_{k\ell;\ell'k'}(ts;s't') = G^0_{k;k'}(t;t')G^0_{\ell;\ell'}(s;s') - G^0_{k;\ell'}(t;s')G^0_{\ell;k'}(s;t')$$

$$+\frac{i}{\hbar}\int_{-\infty}^{+\infty} du \sum_{i'j'ji} V_{i'j';ji}G^0_{k;k'}(t;t')G^0_{\ell;i'}(s;u)G_{ij;j'\ell'}(u,u;u+0,s')$$

$$-\frac{i}{\hbar}\int_{-\infty}^{+\infty} du \sum_{i'j'ji} V_{i'j';ji}G^0_{k;\ell'}(t;s')G^0_{\ell;i'}(s;u)G_{ij;j'k'}(u,u;u+0,t')$$

$$+\frac{i}{\hbar}\int_{-\infty}^{+\infty} du \sum_{i'j';ji} V_{i'j';ji}G^0_{k;i'}(t;u)G^0_{\ell;\ell'}(s;s')G_{ij;jk'}(u,u;u+0,t')$$

$$-\frac{i}{\hbar}\int_{-\infty}^{+\infty} du \sum_{i'j'j\ell} V_{i'j';ji}G^0_{k;i'}(t,u)G^0_{\ell;k'}(s;t')G_{ij;j'\ell'}(u,u;u+0,s')$$

$$-\frac{i}{\hbar}\int_{-\infty}^{+\infty} du \sum_{i'j'ji} V_{i'j';ji}G^0_{k;i'}(t;u)G^0_{\ell;j'}(s;u)G_{ij;k'\ell'}(u,u;t',s')$$

$$+\left(\frac{i}{\hbar}\right)^2\int_{-\infty}^{+\infty} du \int_{-\infty}^{+\infty} dv \sum_{i'j'ji}\sum_{m'n'mn} V_{\ell'j';j\ell}V_{m'n';nm}G^0_{k;\ell'}(t;u)G^0_{\ell;m'}(s;v)$$

$$\times\, G_{mnij;n'j'k'\ell'}(v,v,u,u;v+0,u+0,t',s') \tag{11}$$

If we now formulate by iteration the solution to the linked cluster of equations (10), (11), etc., we will arrive directly at the perturbation-theory series for the Green's functions, to whose individual terms the Feynman diagrams are directly compared [11, 2, 8, 9] (see also [1]).

However, before we formulate the entire set of rules for composing the Feynman diagrams, we will present here the formulas which yield the energy correction of the state in terms of a two-particle and a one-particle Green's function. These formulas permit us to obtain the field form of the perturbation theory for the correction to the energy of the state directly from the perturbation theory for the Green's functions.

Let us insert the parameter $\lambda$ into the full Hamiltonian, so that we will have $H = H_0 + \lambda H_{int}$. Then, for the average over the investigated state of the full Hamiltonian, we will have†

$$\frac{dE}{d\lambda} = \left\langle \frac{dH}{d\lambda} \right\rangle$$

where E is the energy of the examined state of the full Hamiltonian. For the energy correction $\Delta E = E - E_0$, where $E_0$ is the energy of the initial state $H_0$ of the Hamiltonian, we will, consequently, have

$$\Delta E = \int_0^1 d\lambda\, \langle H_{int} \rangle$$

†For a proof see [12] (sec. 11, p. 140). It does not assume that the corresponding state is the ground state.

or, using (1),

$$\Delta E = \frac{1}{2} \sum_{i'j'ji} V_{i'j';ji} \int_0^1 d\lambda \, \langle \overset{+}{a}_{i'} \overset{+}{a}_{j'} a_j a_i \rangle \tag{12}$$

where the average, in contrast to (3) and (4), is taken over the state of the full Hamiltonian H rather than the zeroth Hamiltonian $H_0$.

Now let us give another definition, equivalent to (3) and (4), of the one-particle and two-particle Green's functions in the Heisenberg representation. In this representation, the Green's functions are defined as the average of the products of the field operators over the state of the full Hamiltonian H:

$$G_{kk'}(t; t') = \langle T \left( a_k(t) \overset{+}{a}_{k'}(t) \right) \rangle \tag{13}$$

$$G_{k\ell;\ell'k'}(ts; s't') = \langle T \left( a_k(t) a_\ell(s) \overset{+}{a}_{\ell'}(s') \overset{+}{a}_{k'}(t) \right) \rangle \tag{14}$$

The time-dependent field operators entering into (13) and (14) are defined by $a_i(t) = e^{-(i/h)Ht} \cdot a_i e^{-(i/h)Ht}$, $\overset{+}{a}_i(t) = e^{(i/h)Ht} a_i e^{-(i/h)Ht}$. As has already been noted, the equivalency of (13)-(14) and (3)-(4) is proved in [3] on the basis of adiabatic theorem.

We use (14) for the transformation of (12). It is easy to see that

$$\Delta E = \frac{1}{2} \sum_{i'j'ji} V_{i'j';ji} \int_0^1 d\lambda \, G_{ij;j'i'}(0, 0; +0, +0|\lambda) \tag{15}$$

We have indicated in explicit form the dependence of the Green's function on λ, the allowance for which is summarized in the substitution $v \longrightarrow \lambda v$.

An expression for ΔE can also be obtained in terms of the one-particle Green's function if we apply the operation $(\hbar/i)(\partial/\partial t) + E_k$ to the left and right sides of (10). Observing that, according to (9),

$$\left( \frac{\hbar}{i} \frac{\partial}{\partial t} + E_k \right) G^0_{k;k'}(t; t') = \frac{\hbar}{i} \delta(t - t') \delta_{kk'} \tag{16}$$

we obtain at once from (10),

$$\left( \frac{\hbar}{i} \frac{\partial}{\partial t} + E_k \right) G_{k;k'}(t; t') = \frac{\hbar}{i} \delta(t - t') \delta_{kk'} + \sum_{j'ji} V_{kj';ji} G_{ij;j'i'}(t, t; t+0, t') \tag{17}$$

Now taking t′ = +0 and t = 0, and also k′ = k and summing over k, we obtain

$$\sum \left( \frac{\hbar}{i} \frac{\partial}{\partial t} + E_k \right) G_{k;k}(t|\lambda)|_{t=-0} = \sum_{i'j'ji} V_{i'j';ji} G_{ij;j'i'}(0, 0; +0, +0) \tag{18}$$

Using (18) together with (15), we obtain

$$\Delta E = \frac{1}{2} \int_0^1 \frac{d\lambda}{\lambda} \sum_k \left( \frac{\hbar}{i} \frac{\partial}{\partial t} + E_k \right) G_{k;k}(t)|_{t=-0} \tag{19}$$

where, just as in (15), we have indicated in explicit form the dependence of the Green function on $\lambda$.

If the perturbation-theory expansions for the corresponding Green's functions are known, formulas (15) or (19) permit the expansions for $\Delta E$ to be found directly.

## 3. Feynman Diagrams

Let us now give the whole set of rules for computing the contribution of any order of perturbation theory to the energy $\Delta E$, to the one-particle Green's function $G_{k;k'}(t;t')$, and to the two-particle Green's function $G_{k\ell;\ell'k'}(ts;s't')$.

The contribution from the $n^{th}$ order to each of the examined quantities is the sum of the contributions of the various corresponding Feynman diagrams of the $n^{th}$ order.

First, one must consider all possible different diagrams of the $n^{th}$ order for the corresponding quantity. This requires a plot of n quadruple vertices, to which in the case of the one-particle Green's function, an additional input vertex 1 and output vertex 1 must be added, and in the case of the two-particle Green's function, two supplementary input vertices 1′ and 2′ and two additional output vertices 1 and 2 must be added. The quadruple vertices should be depicted in the form of two points linked by an undirected broken line. One end of an undirected solid line enters each point, and one end of an undirected solid line leaves it. The input vertex should be depicted as a point with the end of a solid line leaving it and the output vertex as a point with the end of a solid line entering it.

Further, all incoming ends of the vertices must be linked to all outgoing ends so as to obtain a topologically interlinked structure, viz., the diagram. In the case of the two-particle Green's function, the diagrams are permitted to dissociate into two linked parts, in one of which 1 1′ is found and in the other, 2 2′, or in one, 1 2′, and in the other, 1′ 2′.

One should examine all possible topologically different diagrams which cannot be matched together as a result of three-dimensional topological motions. This also means a match of the directions on the solid lines, as well as of the corresponding input vertices 1′ or 1′ 2′ and output vertices 1 or 1 2.

All possible diagrams of the first and second order for $\Delta E$ are given in Fig. 1. All possible

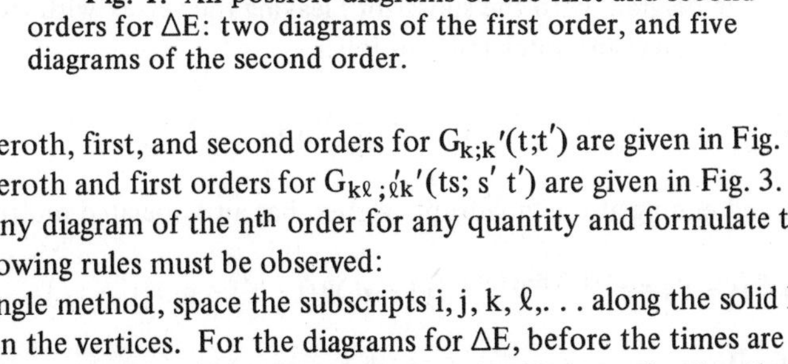

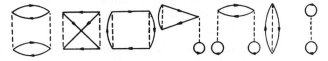

**Fig. 1.** All possible diagrams of the first and second orders for $\Delta E$: two diagrams of the first order, and five diagrams of the second order.

diagrams of the zeroth, first, and second orders for $G_{k;k'}(t;t')$ are given in Fig. 2. All possible diagrams of the zeroth and first orders for $G_{k\ell;\ell'k'}(ts;s'\,t')$ are given in Fig. 3.

Let us take any diagram of the $n^{th}$ order for any quantity and formulate the contribution from it. The following rules must be observed:

1. By any single method, space the subscripts i, j, k, ℓ,. . . along the solid lines, and the times t, s, u,. . . on the vertices. For the diagrams for $\Delta E$, before the times are arranged, one vertex should be selected and marked as time 0. For the diagrams for $G_{k;k'}(t;t')$, place k′ on the line joined to 1′, and on the vertex itself place t′; on the line joined to 1 place k, and on the vertex itself place t. Proceed accordingly for the diagrams for $G_{k\ell;\ell'k'}(ts;s'\,t')$. On the lines joined to 1′, 2′, 1, and 2, place k′, ℓ′, k, and ℓ. On the vertices 1′, 2′, 1, and 2 place the

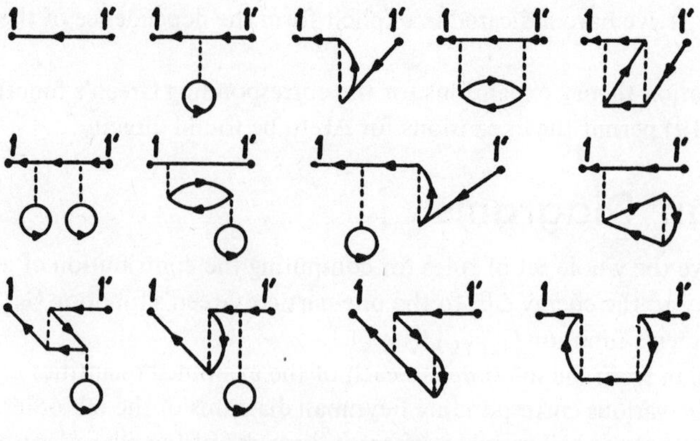

Fig. 2. All possible diagrams of the zeroth, first, and second orders for $G_{\ell;\ell'}$: one diagram of the zeroth order, two diagrams of the first order, and ten diagrams of the second order.

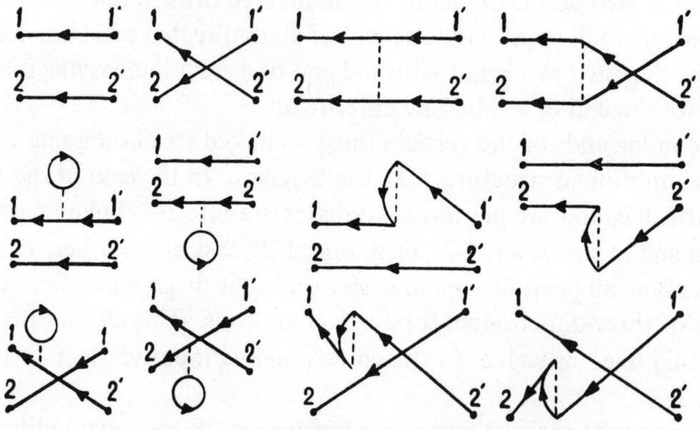

Fig. 3. All possible diagrams of the zeroth and first orders for $G_{12;2'1'}$: two diagrams of the zeroth order, and ten diagrams of the first order.

times t′, s′, t, and s. If a line is found which is joined to two input and output vertices, both corresponding subscripts should be placed on it with an equality sign.

2. Each line of the diagram with the subscript i leaving the vertex with time t′ and entering the vertex with time t, has an associated multiplier

$$-\frac{i}{\hbar} V_{ij;i'j'} = -\frac{i}{\hbar} V_{ji;i'j'}$$

A line i, beginning and ending at one quadruple vertex, has an associated multiplier

$$G_i^0(t-t') = e^{-(i/\hbar)E_i(t-t)} \{(1-n_i)\theta(t-t') - n_i\theta(t'-t)\}$$

3. Each quadruple vertex, one point of which is entered by i′ and has i leaving it, and the other point of which is entered by j′ and has j leaving it, has an associated multiplier

$$G_i^0(-0) = -n_i$$

4. The entire diagram has an associated sign factor $(-1)^{\ell}$, where $\ell$ is the number of closed loops of the solid lines in the diagram. In addition, in the case of the two-particle Green's function, a supplementary minus sign is associated with the diagram if in it, starting from $1'$ (or $2'$), one can arrive at 2 (or 1) only along solid lines.

5. In the case of $\Delta E$, the entire diagram has an associated multiplier $-\hbar/i\ g/2n$, where g is equal to the number of different types of diagrams for the one-particle Green's function that can be obtained from the given diagram for $\Delta E$ by successive disconnection of its solid lines. In the case of the diagram for the one-particle or two-particle Green's function, when there is a line with two subscripts on it, with an equality sign, the Kronecker of the corresponding subscripts is associated with each such line.

6. Collect the product of all the abovementioned multipliers, sum over all the free subscripts on the lines, and integrate over all the free times on the vertices.

The association of a Feynman diagram with each perturbation-theory contribution makes the contribution clear and permits a sound representation of the entire perturbation-theory series, once the significance of the individual contributions has been connected with the specific topological characteristics of the diagrams.

# 4. Comparison of the Field Theoretic and Conventional Forms of Perturbation Theory

In order to emphasize once more that the diagram technique is only a convenient means of operation with the entire perturbation-theory series and, of course, deals with the same perturbation-theory series as conventional quantum mechanics, let us make a comparison of the contributions to $\Delta E$ of the first and second orders, calculated with the help of the diagram technique and by the conventional formula of the perturbation theory for a discrete nondegenerate level (see, for example, [12] (sec. 10, pp. 130–131).

The contribution to $\Delta E$ from the two diagrams of the first order shown in Fig. 1 is

$$\Delta E^{(1)} = \frac{1}{2}\sum_{ij} V_{ij;ji} G_i^0(-0) G_j^0(-0) - \frac{1}{2}\sum_{ij} V_{ji;ji} G_i^0(-0) G_j^0(-0) \tag{20}$$

From this, using (9), we easily obtain

$$\Delta E^{(1)} = \frac{1}{2}\sum_{ij} V_{ij;ji} n_i n_j - \frac{1}{2}\sum_{ij} V_{ji;ji} n_i n_j \tag{21}$$

Now, for the conventional form of perturbation theory, we have

$$\Delta E^{(1)} = \langle H_{int} \rangle \tag{22}$$

where the expectation should be taken over the initial state of $H_0$. Using (1), we have

$$\langle H_{int} \rangle = \frac{1}{2}\sum_{i'j';ji} V_{i'j';ji} \langle \overset{+}{a}_{i'} \overset{+}{a}_{j'} a_j a_i \rangle \tag{23}$$

We apply the Wick theorem in its conventional form to find the indicated average of the set of four operators

$$
\begin{aligned}
\langle \overset{+}{a}_{i'}\overset{+}{a}_{j'}a_j a_i\rangle &= \langle \overset{+}{a}_{i'}\overset{+}{a}_{j'}a_j a_i\rangle + \langle \overset{+}{a}_{i'}\overset{+}{a}_{j'}a_j a_i\rangle \\
&= \delta_{ii'}\delta_{jj'}\langle \overset{+}{a}_i a_i\rangle\langle \overset{+}{a}_j a_j\rangle - \delta_{i'j}\delta_{j'i}\langle \overset{+}{a}_i a_i\rangle\langle \overset{+}{a}_j a_j\rangle = n_i n_j \delta_{ii'}\delta_{jj'} - n_i n_j \delta_{i'j}\delta_{j'i}
\end{aligned}
\tag{24}
$$

Substituting (24) into (23), it is easy to see that (23) coincides exactly with (21). Thus, the first order of perturbation theory, $\Delta E^{(1)}$, coincides for the conventional and field theoretic forms.

The contribution to $\Delta E$ from the five diagrams of the second order shown in Fig. 1 is

$$
\begin{aligned}
\Delta E^{(2)} = {} & \frac{1}{2}\frac{i}{\hbar}\int_{-\infty}^{+\infty} du\, e^{-\alpha|u|}\sum_{mj'jl} V_{ij';ji}V_{jm;mj'}G_i^0(-0)G_j^0(-u)G_{j'}^0(u)G_m^0(-0) \\
& - \frac{i}{\hbar}\int_{-\infty}^{+\infty} du\, e^{-\alpha|u|}\sum_{mj'ji} V_{ij';ji}V_{jm;j'm}G_i^0(-0)G_j^0(-u)G_{j'}^0(u)G_m^0(-0) \\
& + \frac{1}{2}\frac{i}{\hbar}\int_{-\infty}^{+\infty} du\, e^{-\alpha|u|}\sum_{j'mji} V_{j'i;ji}V_{jm;j'm}G_i^0(-0)G_j^0(-u)G_m^0(-0)G_{j'}^0(u) \\
& + \frac{1}{4}\frac{i}{\hbar}\int_{-\infty}^{+\infty} du\, e^{-\alpha|u|}\sum_{i'j'ij} V_{i'j';ji}V_{ij;i'j'}G_i^0(-u)G_j^0(-u)G_{j'}^0(u)G_{i'}^0(u) \\
& - \frac{1}{4}\frac{i}{\hbar}\int_{-\infty}^{+\infty} du\, e^{-\alpha|u|}\sum_{i'j'ij} V_{i'j';ji}V_{ij;j'i'}G_i^0(-u)G_j^0(-u)G_{i'}^0(u)G_{j'}^0(u)
\end{aligned}
\tag{25}
$$

Using (9), from this we obtain, after a simple computation,

$$
\begin{aligned}
\Delta E^{(2)} = {} & \sum_{mj'ji}\Bigg[\left(\frac{1}{2}V_{ij';ji}V_{jm;mj'} - V_{ij';ji}V_{jm;j'm} + \frac{1}{2}V_{j'i;ji}V_{jm;j'm}\right) \\
& \times \left\{\frac{n_i n_m(1-n_{j'})n_j}{E_j - E_{j'} + i\hbar\alpha} - \frac{n_j n_m n_{j'}(1-n_j)}{E_j - E_{j'} - i\hbar\alpha}\right\}\Bigg] \\
& + \frac{1}{4}\sum_{i'j'ji}\Bigg[(V_{i'j';ji}V_{ij;i'j'} - V_{i'j';ji}V_{ij;j'i'}) \\
& \times \left\{-\frac{(1-n_{i'})(1-n_{j'})n_i n_j}{E_i + E_j - E_{j'} - E_{i'} + i\hbar\alpha} + \frac{n_{i'}n_{j'}(1-n_i)(1-n_j)}{E_i + E_j - E_{j'} - E_{i'} - i\hbar\alpha}\right\}\Bigg]
\end{aligned}
$$

Alternatively, we can introduce the concept of the Fermi set F of the subscripts that correspond to the populated one-particle states in the initial state of $H_0$, and of the set G (which complements F) of subscripts such that $i \in F$ means that $n_i = 1$ and $i \in G$ means that $n_i = 0$. Then, from (26) we obtain directly

$$
\begin{aligned}
\Delta E^{(2)} = & \sum_{m\epsilon F, i\epsilon F, j'\epsilon G, j\epsilon F} \left( \frac{1}{2} V_{ij';ji} V_{jm;mj'} - V_{ij';ji} V_{jm;j'm} \right. \\
& \left. + \frac{1}{2} V_{j'i;ji} V_{jm;j'm} \right) \frac{1}{E_j - E_{j'} + i\hbar\alpha} - \sum_{m\epsilon F, i\epsilon F, j'\epsilon F, j\epsilon G} \left( \frac{1}{2} V_{ij';ji} V_{jm;mj'} \right. \\
& \left. - V_{ij';ji} V_{jm;j'm} + \frac{1}{2} V_{j'i;ji} V_{jm;j'm} \right) \frac{1}{E_j - E_{j'} - i\hbar\alpha} \\
& - \frac{1}{4} \sum_{i'\epsilon G, j'\epsilon G,\ i\epsilon F, j\epsilon F, i\neq j, i'\neq j'} (V_{i'j';ji} V_{ij;i'j'} - V_{i'j';ji} V_{ij;j,i'}) \frac{1}{E_i + E_j - E_{j'} - E_{i'} - i\hbar\alpha} \\
& + \frac{1}{4} \sum_{i'\epsilon F, j'\epsilon F, i\epsilon G, j\epsilon G, i\neq j, i'\neq j'} (V_{i'j';ji} V_{ij;i'j'} - V_{i'j';ji} V_{ij;j'i'}) \\
& \times \frac{1}{E_i + E_j - E_{j'} - E_{i'} - i\hbar\alpha}
\end{aligned}
\tag{27}
$$

In the last two sums in (27), we have introduced the limitation $i \neq j$ and $i' \neq j'$, because the terms of the sums with $i = j$ and $i' = j'$ cancel each other and drop out of the total sums [see also (2)].

We must make the adiabatic parameter $\alpha$ approach zero in (27). In addition, we must be careful with those terms in the sums of (27) whose energy sums in the denominators are exactly equal to zero. It is easy to see that the condition of nondegeneracy of the initial state $H_0$ inplies that none of the sums in the denominators becomes zero. This requires examination of the states schematically represented in Fig. 4 and differing from the initial state

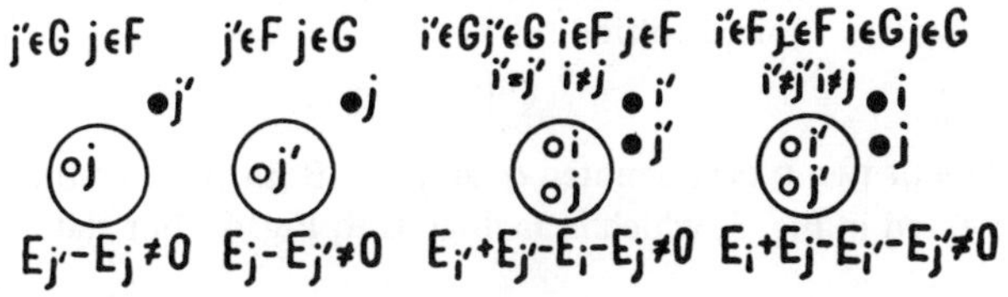


Fig. 4. Explanation for the derivation of the inequalities needed for passing to the limit in (27), which are a direct consequence of the condition of nondegeneracy of initial state $H_0$.

by the presence (shaded circle) of a supplementary corresponding particle or the absence (open circle) of the corresponding particle. Comparing the energy of the states shown in Fig. 4 with the energy of the initial state and using the nondegeneracy of the latter, we arrive at the required inequality.

Thus in (27) we pass to the adiabatic limit $\alpha \longrightarrow +0$, after some simple conversions that reduce to the substitution $j \leftrightarrows j'$, $m \leftrightarrows i$ in the second sum of the summation variables (after which the second sum can be combined with the first) and to the substitution $i' \leftrightarrows i$, $j' \leftrightarrows j$ in the fourth sum (after which the fourth sum can be combined with the third) and after the supplementary conversion of the combined fourth and third sums (which involves taking the half-sum of the starting expression and the expression in which the substitution $i' \leftrightarrows j'$ is made of the summation variables). Then we obtain

$$\Delta E^{(2)} = \sum_{m\epsilon F,\, i\epsilon F,\, j'\epsilon G,\, j\epsilon F} (V_{ij';ji} - V_{j'i;ji})(V_{jm;mj'} - V_{jm;j'm}) \frac{1}{E_j - E_{j'}}$$

$$- \frac{1}{4} \sum_{i'\epsilon G,\, j'\epsilon G,\, i\epsilon F,\, j\epsilon F,\; i\neq j,\, i'\neq j'} (V_{i'j';ji} - V_{j'i';ji})(V_{ij;i'j'} - V_{ij;j'i'}) \frac{1}{E_i + E_j - E_{j'} - E_{i'}} \tag{28}$$

Now let us apply the conventional form of perturbation theory according to which, for the second order, one should use the formula

$$\Delta E^{(2)} = -\sum_{\alpha \neq 0} \frac{\langle 0 | H_{int} | \alpha\rangle\langle\alpha | H_{int} | 0\rangle}{E_\alpha - E_0} \tag{29}$$

where the symbol 0 designates the initial state $H_0$ of the Hamiltonian, and the symbol $\alpha$ any other state of this Hamiltonian. In reality, however, it is sufficient to examine only those states $\alpha$ that are related to the state 0 by nonvanishing matrix elements.

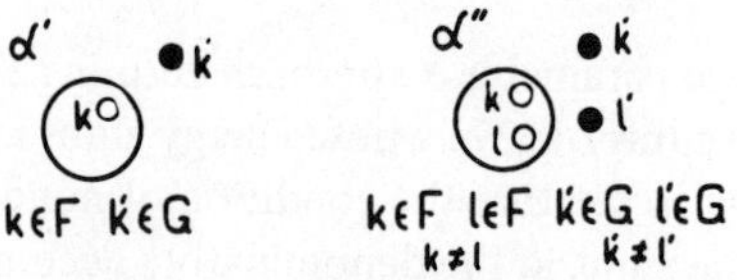


**Fig. 5.** Explanation regarding the two types of states a′ and a″ of the Hamiltonian $H_0$ that are required to find (29).

Such states can be of two types, denoted $\alpha'$ and $\alpha''$. Both types of these states are schematically represented in Fig. 5, which is analogous to Fig. 4. For states $\alpha'$,

$$\Psi_{\alpha'} = \overset{+}{a}_{k'} a_k \Psi_0$$
$$E_{\alpha'} = E_0 + E_{k'} - E_k$$
$$k\epsilon F,\ k'\epsilon G$$

For states $\alpha''$,

$$\Psi_{\alpha''} = \overset{+}{a}_{k'} \overset{+}{a}_{\ell'} a_k a_\ell \Psi_0$$
$$E_{\alpha''} = E_0 + E_{k'} + E_{\ell'} - E_k - E_\ell$$
$$k\epsilon F,\ \ell\epsilon F,\ k'\epsilon G,\ \ell\epsilon G,\ k \neq \ell,\ k' \neq \ell'$$

For states $\alpha'$, we have

$$\langle\Psi_{\alpha'} | H_{int} | \Psi_0\rangle = \langle\Psi_0 | H_{int} | \Psi_{\alpha'}\rangle^{+}$$

$$\langle\Psi_0 | H_{int} | \Psi_{\alpha'}\rangle = \frac{1}{2} \sum_{i'j'ji} V_{i'j';ji} \langle \overset{+}{a}_{i'} \overset{+}{a}_{j'} a_j a_i \overset{+}{a}_{k'} a_k \rangle \tag{30}$$

Using the conventional form of the Wick theorem to find the average in (30), we obtain

$$\langle \overset{+}{a}_{i'}\overset{+}{a}_{j'}a_j a_i \overset{+}{a}_{k'}a_k\rangle = (1 - T_{i\rightleftharpoons j})(1 - T_{i'\rightleftharpoons j'})\langle \overset{+}{a}_{i'}\overset{+}{a}_{j'}a_j a_i \overset{+}{a}_{k'}a_k\rangle$$
$$= (1 - T_{i\rightleftharpoons j})(1 - T_{i'\rightleftharpoons j'})\, n_k n_j (1 - n_{k'})\delta_{i'k}\delta_{j'j}\delta_{ik'} \tag{31}$$

The operations T are more convenient to apply to the multiplier $V_{i'j';ji}$ in the sum of (30), which is achieved by corresponding changes of the variables of summation. Finally, for the states $\alpha'$ we have

$$\langle \Psi_0 | H_{int} | \psi_{\alpha'}\rangle = \sum_{j\epsilon F} (V_{kj;jk'} - V_{kj;k'j}) \tag{32}$$

whence

$$-\sum_{\alpha'} \frac{\langle \Psi_0 | H_{int} | \Psi_{\alpha'}\rangle\langle \Psi_{\alpha'} | H_{int} | \Psi_0\rangle}{E_{\alpha'} - E_0}$$
$$= -\sum_{k\epsilon F, k'\epsilon G, j\epsilon F, j'\epsilon F} (V_{kj;jk'} - V_{kj;k'j})(V_{kj';j'k'} - V_{kj';k'j'}) \frac{1}{E_{k'} - E_k} \tag{33}$$

For the states $\alpha''$, we obtain

$$\langle \Psi_{\alpha''} | H_{int} | \Psi_0\rangle = \langle \Psi_0 | H_{int} | \Psi_{\alpha''}\rangle^+$$

$$\langle \Psi_0 | H_{int} | \Psi_{\alpha''}\rangle = \frac{1}{2}\sum_{i'j'ji} V_{i'j';ji}\langle \overset{+}{a}_{i'}\overset{+}{a}_{j'}a_j a_i \overset{+}{a}_{k'}\overset{+}{a}_{\ell'}a_k a_\ell\rangle \tag{34}$$

Again using the Wick theorem to find the average in (34), we have

$$\langle \overset{+}{a}_{i'}\overset{+}{a}_{j'}a_j a_i \overset{+}{a}_{k'}\overset{+}{a}_{\ell'}a_k a_\ell\rangle = (1 - T_{i'\rightleftharpoons j'})(1 - T_{i\rightleftharpoons j})\langle \overset{+}{a}_{i'}\overset{+}{a}_{j'}a_j a_i \overset{+}{a}_{k'}\overset{+}{a}_{\ell'}a_k a_\ell\rangle$$
$$= (1 - T_{i'\rightleftharpoons j'})(1 - T_{i\rightleftharpoons j})\, n_\ell n_k (1 - n_{k'})(1 - n_{\ell'})\delta_{i'\ell}\delta_{j'k}\delta_{j\ell'}\delta_{ik'} \tag{35}$$

We substitute (35) into (34) and again apply the operations T to $V_{i'j';ji}$. Finally, for the states $\alpha''$ we obtain

$$\langle \Psi_0 | H_{int} | \Psi_{\alpha''}\rangle = V_{\ell k;\ell' k'} - V_{k\ell;\ell' k'} \tag{36}$$

whence

$$-\sum_{\alpha''} \frac{\langle \Psi_0 | H_{int} | \Psi_{\alpha''}\rangle\langle \Psi_{\alpha''} | H_{int} | \Psi_0\rangle}{E_{\alpha''} - E_0}$$
$$= -\frac{1}{4}\sum_{k\epsilon F, \ell\epsilon F, k'\epsilon G, \ell'\epsilon G, k=\ell, k'=\ell'} (V_{\ell k;\ell' k'} - V_{k\ell;\ell' k'})(V_{\ell k;\ell' k'} - V_{k\ell;\ell' k'}) \frac{1}{E_{k'} + E_{\ell'} - E_k - E_\ell} \tag{37}$$

The factor $\frac{1}{4}$ arises because in the substitution $k \rightleftharpoons \ell$ and $k' \rightleftharpoons \ell'$ we do not arrive at a new state, but at the same state; see Fig. 5.

If in (33) the substitution $k' \to j'$, $k \to j$, $j \to m$, and $j' \to i$ is made, and in (37) the substitution $k \to i$, $\ell \to j$, $k' \to i'$, and $\ell' \to j'$ is made, then [with the use of (2)] it is easy to see that the sum of the terms of (33) and (37) agrees exactly with (28). Thus, the $\Delta E$ in the conventional and field theoretic forms of perturbation theory coincide.

One need not emphasize that in comparing the conventional and field theoretic forms of perturbation theory, we took advantage essentially of the nondegeneracy of the starting state of $H_0$ in the Hamiltonian.

Thus far our basic problem has been to study an entire perturbation-theory series as a whole, which was accomplished effectively by the foregoing discussion of the Feynman diagram technique. Let us proceed now to the formulation of approximations. Here the guiding concept is the idea of the partial summation of a perturbation-theory series, i.e., the summation of infinite subsequences of the terms of a perturbation-theory series. Such summations are made most effectively by compiling completely exact field equations for a Green's function. Making various natural approximations in these equations, we obtain a very convenient method for making partial summations, because the different approximations in the exact field equations, like those equations themselves, find a simple diagrammatic interpretation.

## 5. One-Particle Effects. The Green's Function of Interaction Q. Mass Operator M, Polarization Operator P, Vertex Operator Γ. Dyson—Schwinger Field Equations

Let us formulate the exact field equations convenient for making the partial summations which allow for the characteristic effects of the state's wave function in its one-particle aspect. We refer to a system of Dyson–Schwinger field equations.

In the diagrams, especially in the higher orders, the following situation is encountered very often: A portion of the diagram is linked to the rest of the diagram only by two solid or two dashed lines. The contribution from this portion of the diagram is a discrete component part of the whole diagram. Such portions are called self-energy parts.

There is a natural tendency generally to exclude the self-energy parts from the examination if the class of diagrams examined is limited to "skeleton" diagrams (without self-energy parts) whose lines are "dressed," i.e., if instead of the contribution from each line, one must examine the sum of the contributions from all possible self-energy parts that can replace this line.

Examining the self-energy parts by the dashed lines, we arrive directly at the study of the Green's function of interaction which characterizes the sum of all possible self-energy parts.

The Green's function of interaction $Q_{kk';\ell\ell'}(t;t')$ is defined as a sum of diagrams of the form shown in Fig. 6. The contribution from the diagrams in the form shown in Fig. 6 is calculated by rules completely analogous to those formulated in Section 3. However, one must keep in mind that two fixed external times may be found on the vertex of a diagram, in which case the $\delta$-function of the difference between those two times should be written as the multiplier from this diagram. This stipulation should always be made for all the operators examined below that are determined by the sums of diagrams of a special form. Of course, we are only interested in linked diagrams.

Let us now define the mass operator $M_{k;k'}(t;t')$ and the polarization operator $P_{kk';\ell\ell'}(t;t')$. Diagrams for them are shown in Figs. 7 and 8, respectively. The operators are defined as the sums of all possible corresponding diagrams which cannot be divided into

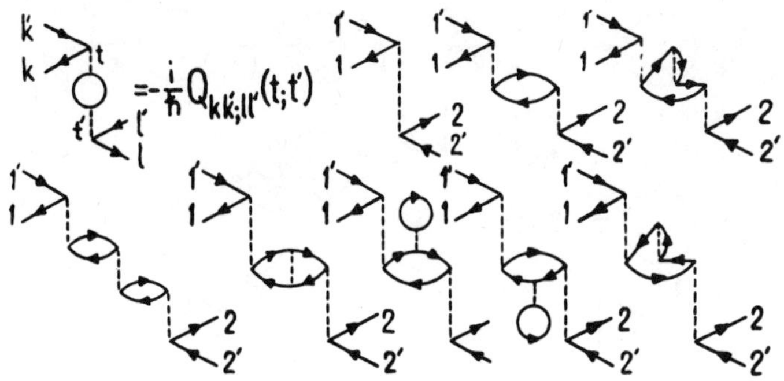


**Fig. 6.** The determination of the Green's function of interaction $Q_{11';22'}$ and all its possible diagrams in the first, second, and third orders: one diagram of the first order, one diagram of the second order, and six diagrams of the third order.

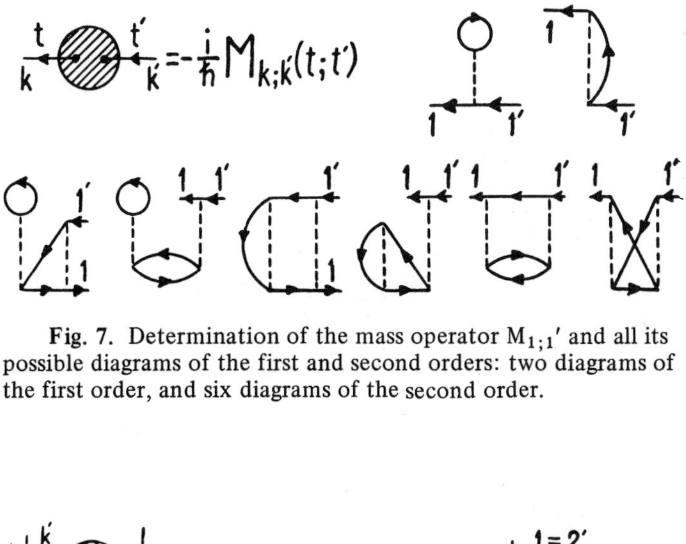


**Fig. 7.** Determination of the mass operator $M_{1;1'}$ and all its possible diagrams of the first and second orders: two diagrams of the first order, and six diagrams of the second order.

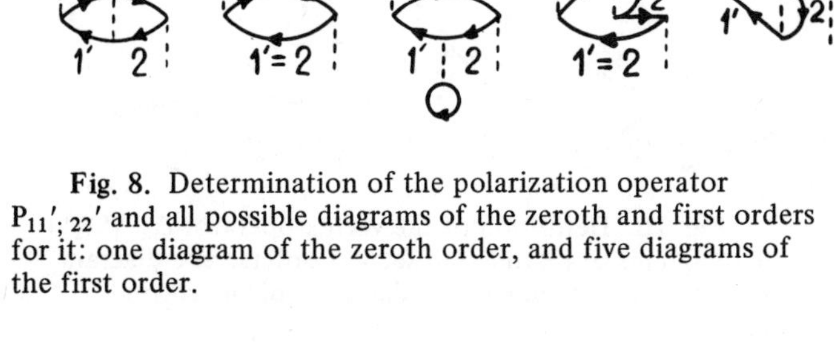

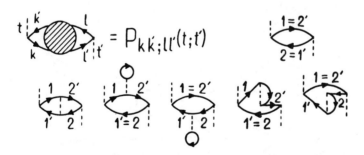

**Fig. 8.** Determination of the polarization operator $P_{11';22'}$ and all possible diagrams of the zeroth and first orders for it: one diagram of the zeroth order, and five diagrams of the first order.

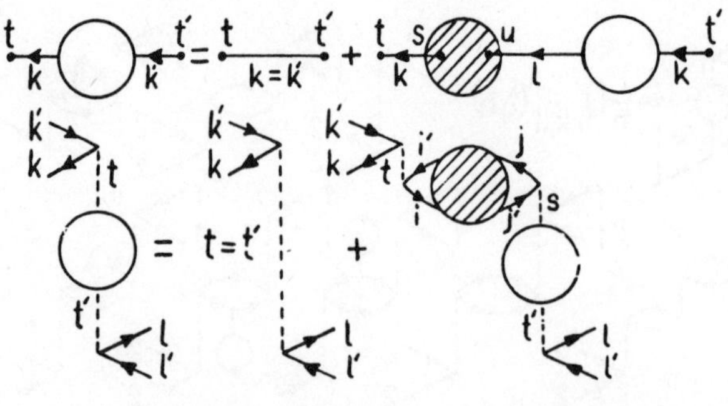


Fig. 9. Diagrammatic representation of the first two equations of a system of Dyson–Schwinger field equations.

two parts, linked by one solid line for the mass operator or one dashed line for the polarization operator. Of course, only linked diagrams for the polarization operator are examined.

The mass operator is related to the one-particle Green's function and the polarization operator to the Green's function of interaction by equations that are indicated diagrammatically in Fig. 9 and have the following form:

$$G_{k;k'}(t;t') = G_k^0(t-t')\,\delta_{kk'} - \frac{i}{\hbar}\int_{-\infty}^{+\infty} ds \int_{-\infty}^{+\infty} du \sum_{\ell} G_k^0(t-s)$$

$$\times\, M_{k;\ell}(s;u)\, G_{\ell;k'}(u;t') \tag{38}$$

$$Q_{kk';\ell\ell'}(t;t') = V_{k\ell;\ell'k'}\,\delta(t-t')$$

$$-\frac{i}{\hbar}\int_{-\infty}^{+\infty} ds \sum_{ii'jj'} V_{ki;ik'}P_{ii';jj'}(t;s)\,Q_{jj';\ell\ell'}(s;t') \tag{39}$$

Equations (38) and (39) are the first two equations of a system of Dyson–Schwinger field equations.

Now let us present the remaining two equations of the Dyson–Schwinger system. We ex-

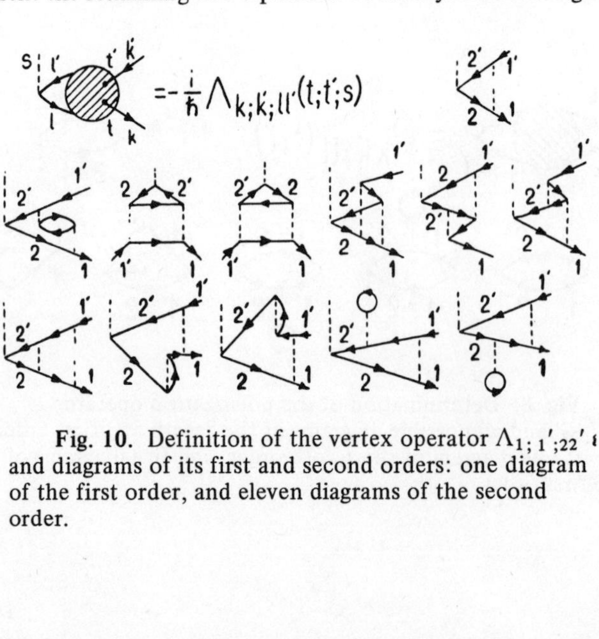


Fig. 10. Definition of the vertex operator $\Lambda_{1;1';22'}$ and diagrams of its first and second orders: one diagram of the first order, and eleven diagrams of the second order.

tend our examination to a further operator, the vertex operator $\Lambda_{k;k';\ell\ell'}(t;t';s)$. It is quite simple to express the mass operator and the polarization operator in terms of this operator.

Diagrams for the vertex operator $\Lambda_{k;k';\ell\ell'}(t;t';s)$ are presented in Fig. 10. The operator is defined as the sum of all possible corresponding diagrams from none of whose three ends one can "extract" self-energy parts. Only linked diagrams should be examined.

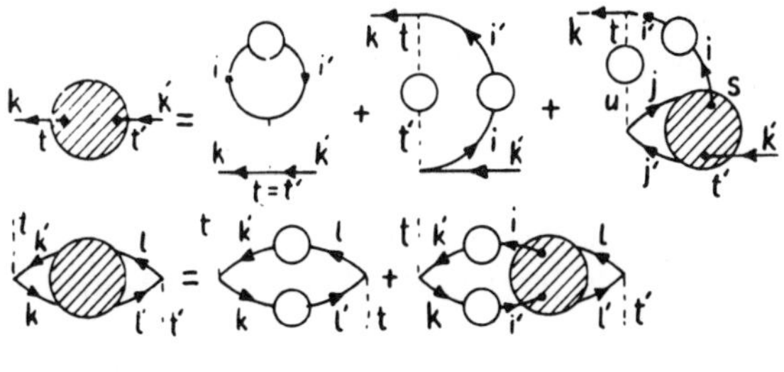

**Fig. 11.** Diagrammatic representation of the second two equations of the system of Dyson–Schwinger field equations.

The expression for the mass and polarization operators in terms of the vertex operator are given by the equations diagrammatically shown in Fig. 11, which have the following form:

$$M_{k;k'}(t;t') = \sum_{ii'} V_{ik;k'i'} G_{i';i}(-0)\,\delta(t-t') + \sum_{ii'} Q_{ki';ik'}(t;t')\,G_{i';i}(t;t' \div 0)$$
$$-\frac{i}{\hbar}\int_{-\infty}^{+\infty} ds \int_{-\infty}^{+\infty} du \sum_{ii'jj'} Q_{k\ell';jj'}(t;u)\,G_{i';i}(t;s)\,\Lambda_{i;k;jj'}(s;t';u) \qquad (40)$$

$$P_{kk';\ell\ell'}(t;t') = -\,G_{k';\ell}(t;t')\,G_{i';i}(t';t)$$
$$+\frac{i}{\hbar}\int_{-\infty}^{+\infty} ds \int_{-\infty}^{+\infty} du \sum_{ij} G_{k';i}(t;s)\,G_{j;k}(u;t)\,\Lambda_{i;j;\ell\ell'}(s;u;t') \qquad (41)$$

If we introduce the new operator $\Gamma_{k;k'\ell\ell'}(t;t';s)$ with the help of the relationship

$$\Gamma_{k;k';\ell\ell'}(t;t';s) = \delta(t-s)\,\delta(t'-s)\,\delta_{k\ell}\delta_{k'\ell'} - \frac{i}{\hbar}\Lambda_{k;k';\ell\ell'}(t;t';s) \qquad (42)$$

then equations (40) and (41) are written more simply as

$$M_{k;k'}(t;t') = \sum_{ii'} V_{ik;k'i'} G_{i;i'}(-0)\,\delta(t-t')$$
$$+\int_{-\infty}^{+\infty} ds \int_{-\infty}^{+\infty} du \sum_{ii'jj'} Q_{ki';jj'}(t;u)\,G_{i';i}(t;s+0)\,\Gamma_{i;k';jj'}(s;t';u) \qquad (43)$$

$$P_{kk';\ell\ell'}(t;t') = -\int_{-\infty}^{+\infty} ds \int_{-\infty}^{+\infty} du \sum_{ij} G_{k';i}(t;s)\,G_{j;k}(u;t)\,\Gamma_{i;j;\ell\ell'}(s;u;t') \qquad (44)$$

Equations (43) and (44) are the second two equations of the system of Dyson–Schwinger field equations.

# 6. Two-Particle Effects. Operators of Interaction S, I, J, T. The Related System of Field Equations

In Section 5 we formulated a system of completely exact field equations, a system of Dyson–Schwinger field equations. This system made it easy to study the properties of the whole wave function in its one-particle corpuscular aspect. Now let us focus our attention on the two-particle corpuscular aspect of the whole wave function. This requires study of a new system of field equations for the operators of interaction that are convenient for these purposes.

First, rather than examining the two-particle Green's function $G_{k\ell;\ell'k'}(ts;s't')$, let us examine the operator $S_{k\ell;\ell'k'}(ts;s't')$, which bears the following relationship to that function:

$$G_{k\ell;\ell'k'}(ts;s't') = G_{k;k'}(t;t')G_{\ell;\ell'}(s;s') - G_{k;\ell'}(t;s')G_{\ell;k'}(s;t')$$
$$+\sum_{ii'jj'}\int_{-\infty}^{+\infty} du \int_{-\infty}^{+\infty} du' \int_{-\infty}^{+\infty} dv \int_{-\infty}^{+\infty} dv' G_{k;i}(t;u)G_{i';k'}(u';t')G_{\ell;j}(s;v)$$
$$\times G_{j';\ell'}(v';s')S_{ij;j'i'}(uv;v'u') \tag{45}$$

The diagrammatic representation of this relationship is given in Fig. 12. The operator $S_{k\ell;\ell'k'}(ts;s't')$ can als be defined as the sum of all possible diagrams shown in Fig. 13,

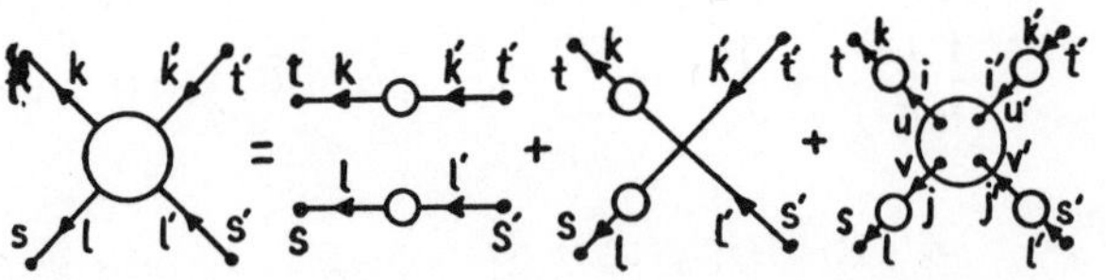

Fig. 12. Diagrammatic representation of the relationship between $G_{12;2'1'}$ and $S_{12;2'1'}$.

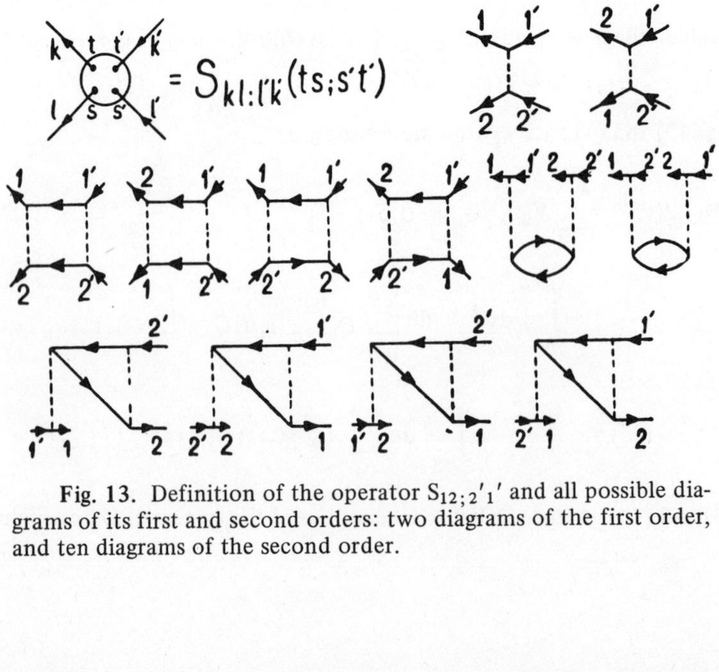


Fig. 13. Definition of the operator $S_{12;2'1'}$ and all possible diagrams of its first and second orders: two diagrams of the first order, and ten diagrams of the second order.

viz., of the linked diagrams with two incoming and two outgoing ends, from none of whose four ends one can extract any self-energy parts.

Now let us define two new operators of interaction $I_{k\ell;\ell'k'}(ts;s't')$ and $J_{k\ell;\ell'k'}(ts;s't')$, which are important sums of certain diagrams of S. These diagrams are shown in Figs. 14 and

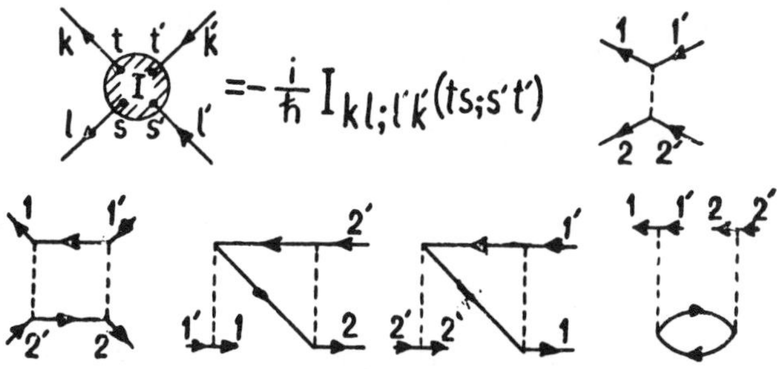


Fig. 14. Definition of the operator $I_{12;2'1'}$ and all possible diagrams of the first and second order for it: one diagram of the first order, and four diagrams of the second order.

15, respectively. They are defined as diagrams that cannot be divided into two parts linked only by two solid lines, where in the case of $I_{12;2'1'}$, 12 must appear in one part and $1'2'$ in the other part, while in the case of $J_{12;2'1'}$, $1\ 1'$ must appear in one part and $2\ 2'$ in the other part. In addition, in the case of $I_{12;2'1'}$, the diagram must be such that in it, starting from $1'(2')$, 1 (2) can be reached only along solid lines.

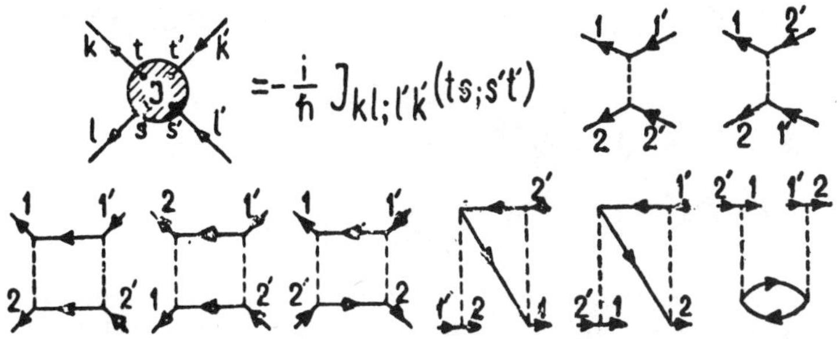


Fig. 15. Definition of the operator $J_{12;2'1'}$ and all its possible diagrams of the first and second orders: two diagrams of the first order, and six diagrams of the second order.

The operator $I_{12;2'1'}$, of course, is known in field theory. It is called the Schwinger interaction operator and is the basis of the familiar Salpeter-Bethe equation for the two-particle bound state [13]. It seems to us that due attention has not yet been given to the operator $J_{12;2'1'}$, although, in fact, it was introduced earlier by Ter-Martirosian [14] to solve certain problems of field theory.†

†The author thanks Academician N. N. Bogoliubov for calling his attention to the fact that the operator J and the operator T were examined earlier in field theory by Ter-Martirosian.

The relationship among the operator $I_{12;2'1'}$, the operator $S_{12;2'1'}$, and the one-particle Green's function is the basis of the Bethe–Salpeter equation and has the following form:

$$S_{k\ell;\ell'k'}(ts;s't') = -\frac{i}{\hbar} I_{k\ell;\ell'k'}(ts;s't') + \frac{i}{\hbar} I_{\ell k;\ell'k'}(st;s't')$$
$$-\frac{i}{\hbar}\sum_{ii'jj'} \int_{-\infty}^{+\infty} du \int_{-\infty}^{+\infty} du' \int_{-\infty}^{+\infty} dv \int_{-\infty}^{+\infty} dv' I_{k\ell;ji}(ts;vu)\, G_{i;i'}(u;u')$$
$$\times\, G_{j;j'}(v;v')\, S_{i'j';\ell'k'}(u'v';s't') \qquad (46)$$

A relationship among the operator $J_{12;2'1'}$, the operator $S_{12;2'1'}$, and the one-particle Green's function, analogous to relationship (46), is the following.

$$S_{k\ell;\ell'k'}(ts;s't') = -\frac{i}{\hbar} J_{k\ell;\ell'k'}(ts;s't')$$
$$-\frac{i}{\hbar}\sum_{ii'jj'} \int_{-\infty}^{+\infty} du \int_{-\infty}^{+\infty} du' \int_{-\infty}^{+\infty} dv \int_{-\infty}^{+\infty} dv'\, J_{kj';ik'}(tv;ut')\, G_{j;j'}(v';v)$$
$$\times\, G_{i;i'}(u\,;u')\, S_{i'\ell;\ell'j}(u's;s'v') \qquad (47)$$

The diagrammatic representation of Eqs. (46) and (47) is given in Fig. 16. Equations (46) and (47) are the first two equations of the required system of exact field equations combining the operators S, I, and J and the one-particle Green's function. We must find two more equations combining these quantities.

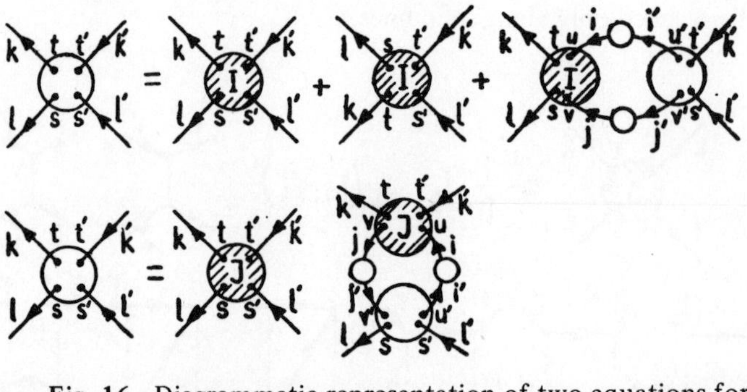

**Fig. 16.** Diagrammatic representation of two equations for $S_{12;2'1'}$, $I_{12;2'1'}$, $J_{12;2'1'}$ and the one-particle Green's function (Bethe–Salpeter equation).

Let us examine in more detail the topological nature of the operators I and J. In a certain sense they are the exact analog of the second-order mass operator M, examined in Section 5. Although in the examination of the mass operator M, we were interested in the situation where the diagram can be split into two parts linked only by one solid line, in the examination of the aforementioned operators of interaction we are interested in the situation where the diagram can be split into two parts now linked only by two solid lines.

The diagrams of $I_{12;2'1'}$ and $J_{12;2'1'}$ are sums of diagrams of $S_{12;2'1'}$ of a special type. In order to find more clearly the topological nature of these operators, we extend the examination to two further sums of the diagrams $S_{12;2'1'}$ (which are easily expressed in terms of $I_{12;2'1'}$ and $J_{12;2'1'}$). The corresponding operators $I'_{12;2'1'}$ and $J'_{12;2'1'}$ are defined as sums of diagrams: For $I'_{12;2'1'}$, it is a sum of the same diagrams as for $I_{12;2'1'}$ except that, starting from

1′ (2′), 2 (1) can be arrived at only along the solid lines; for $J'_{12;2'1'}$, it is a sum of those diagrams that cannot be split into two parts linked only by two solid lines, with 1 2′ in one part and 1′ 2 in the other.

The S-diagrams, generally speaking, can be presented on the basis of types I, II, and III, according to the scheme of Fig. 17. It should be stated that, in using the idea of presentability,

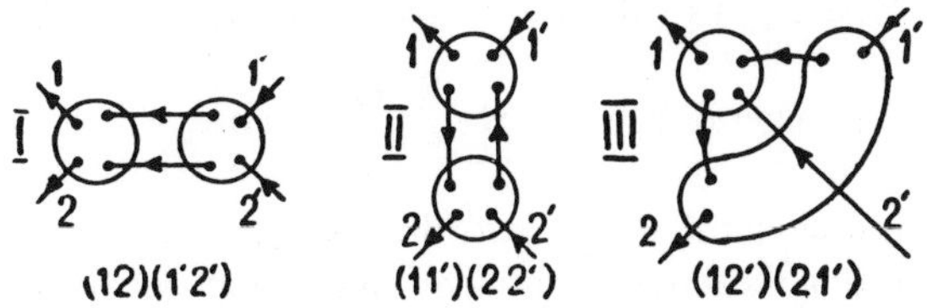


**Fig. 17.** Presentability of the diagrams on the basis of types I, II, and III.

we do not have in mind any topological procedure for reducing the diagrams to certain "skeletons" (as was the case with the self-energy parts excluded from the diagrams by the mass operator M). In fact, no such "skeleton" diagrams exist in the case examined. Presentability is simply the representability of the diagrams on the basis of type I, II, or III.

The diagrams I + I′ are S-diagrams not representable on the basis of type I. The J-diagrams are S-diagrams not representable on the basis of type II. The J′-diagrams are S-diagrams not representable on the basis of type III.

In the examined case of the quadruple interaction, we were able to find the proof of the following statement, whose justification in the first two orders can easily be established by examination of Figs. 12, 13, and 14: there are no diagrams simultaneously representable on the basis of any two of the three types, nor, of course, on the basis of all three types, I, II, and III.

Using the statement formulated, we can make the following classification of diagrams for classes which do not overlap. The S-diagrams are presentable on the basis of type I (diagrams S-I-I′), type II (diagrams S-J), or type III (diagrams S-J′), or not presentable simultaneously on the basis of any of the three types (diagrams of operator T).

Thus the operator $-k/\hbar T_{k\ell;\ell'k'}(ts;s't')$ is defined as the sum of all possible diagrams $S_{k\ell;\ell'k'}(ts;s't')$ which can in no manner be split into two parts linked by only two solid lines.

Thus

$$(S-I-I') + (S-J) + (S-J') + T = S$$

or

$$2S = -T + I + I' + J + J'$$

This symbolic equation is expressed more exactly by the following third equation of the required system of field equations:

$$2S_{k\ell;\ell'k'}(ts;s't') = \frac{i}{\hbar}T_{k\ell;\ell'k'}(ts;s't') - \frac{i}{\hbar}I_{k\ell;\ell'k'}(ts;s't') + \frac{i}{\hbar}I_{ik;\ell'k'}(st;s't')$$
$$- \frac{i}{\hbar}J_{k\ell;\ell'k'}(ts;s't') + \frac{i}{\hbar}J_{\ell k;\ell'k'}(st;s't') \qquad (48)$$

Let us pass to the computation for the fourth and final equation of the required system of field equations. It does not have a clearly expressed diagrammatic nature and is in essence a relationship which permits the one-particle Green's function to be obtained from the two-particle Green's function.

Let us take the definition of the two-particle Green's function in the Heinsenberg representation,

$$G_{k\ell;\ell'k'}(ts;s't') = \langle\, T\left(a_k(t)a_\ell(s)\overset{+}{a}_{\ell'}(s')\overset{+}{a}_{k'}(t')\right)\rangle$$

Let us set in it $\ell' = \ell$ and take $s' = s + 0$. We sum over all $\ell$. Then, inside the averaging sign, the operator of the total number of particles $N = \sum \overset{+}{a}_\ell a_\ell$ arises which commutes with the full Hamiltonian (1). This last circumstance permits the operator N to be extracted from under the averaging sign [by exploiting the anticommutative relationships for the Fermi operators, $N\overset{+}{a}_k = \overset{+}{a}_k(\overset{+}{N}+1)$ and $Na_k = a_k(N-1)$]. Carrying out the indicated computation, we obtain the needed relationship,

$$\sum_\ell G_{k\ell;\ell k'}(ts; s+0t')$$

$$= - NG_{k;k'}(t;t') + G_{k;k'}(t;t')\,[-\theta(t-s)\theta(s-t') + \theta(t'-s)\,\theta(s-t)] \tag{49}$$

where $\theta(t)$ is the step function defined above, and N is the number of electrons in the examined state of the full Hamiltonian H, or in the initial nondegenerate state of the zeroth Hamiltonian $H_0$. Using (45), we obtain

$$\sum_{ii'jj'\ell} \int_{-\infty}^{+\infty} du \int_{-\infty}^{+\infty} du' \int_{-\infty}^{+\infty} dv \int_{-\infty}^{+\infty} dv'\, G_{k;i}(t;u)G_{i';k'}(u';t')G_{\ell;j'}(s;v) \times G_{j';\ell}(v';s)S_{ij;j'i'}(uv;v'u')$$

$$= \sum_\ell G_{k;\ell}(t;s)\,G_{\ell;k'}(s;t') + G_{k;k'}(t;t')$$

$$\times\,[-\theta(t-s)\theta(s-t') + \theta(t'-s)\theta(s-t)] \tag{50}$$

Equation (50) is the missing fourth equation of the system of the completely exact field equations (46), (47), (48), and (50) with respect to the operators of interaction S, I, and J, and to the one-particle Green's function, on the assumption that the operator of interaction T is a preassigned function that characterizes the concrete form of interaction.

It seems to us that the system of exact field equations derived in this section in some way can be regarded as a generalization of the system of exact Dyson–Schwinger field equations formulated in Section 5. The latter is adapted to the study of one-particle excitations of a many-electron system, whether or not this system has two-particle, three-particle, . . ., n-particle excitations (and the vertex operator Γ must be considered known, but usually is replaced by the first order of perturbation theory). The system of exact field equations formulated in this section is adapted to the study of both the one-particle and the two-particle excitations, although without regard to the possible presence of three-particle, . . ., n-particle excitations (and the operator T must be replaced by the first order of perturbation theory).

It should be noted that the problem of formulating correct boundary conditions for the system of Eqs. (46), (47), (48), and (50) remains open at the present time, although, of course, such conditions must be allowed for in order to single out the unique physical solution of the system.

## 7. Conclusion

In the foregoing discussion, we have touched mainly on the general conceptual framework of the field methods in their application to atomic and molecular problems and have not dwelt at all on a discussion of the approximation methods, even the Hartree-Fock method with what are called the correlation corrections. We have discussed this in part in another paper [1], to which we refer readers interested in this problem (although only the relation of the Hartree-Fock method to the partial summation of diagrams is discussed there).

Here we would like to discuss another important problem. What do the methods discussed contribute to an understanding of the nature of the chemical bond? In the most provisional manner, we would like to explain our point of view on this problem.

It is not by chance that methods like the Hartree-Fock method are inadequate, both quantitatively and qualitatively, for an understanding of the specific binding energy of molecules, i.e., the chemical binding energy. Although they adequately make allowance for the one-particle aspect of the many-electron problem, they are powerless to describe the two-particle and many-particle bound states.

It seems to us that the chemical bond is, first of all, a phenomenon of the two-particle and higher-order bound states in a many-electron system. In particular, a pair of electrons with opposite spins participating in the formation of a simple chemical bond is a two-electron bound state. Only from this point of view does it become understandable why, despite the fact that the interaction among electrons is not weak, we can still say that pairs of electrons form an individual chemical bond in the molecule.

As is known, in the examination of bound states, one can speak both of the energy of the state and of its wave function. This leads directly to concepts of the energy of the individual bond and the wave function of the individual bond.

It seems to us that the present lack of effective and detailed calculation methods for the specific binding energy of the molecule (of the same type, let us say, as the Hartree-Fock method for atoms) is explained, to a certain extent, by our lack of understanding of the nature of the chemical bond from the point of view of the many-electron problem in molecules.

## References

1. V. V. Tolmachev, *Vestn. Leningr. Univ., Ser. Fiz. i Khim.* **1**(4), 11–19 (1962).
2. J. Goldstone, *Proc. Roy. Soc. (London)* **A239**(1217), 267–279 (1957).
3. M. Gell-Mann and F. Low, *Phys. Rev.* **84**(2), 350–354 (1951).
4. G. C. Wick, *Phys. Rev.* **80**(2), 268–272 (1950).
5. M. Born and V. Fock, *Z. Physik* **51**(3/4), 165–180 (1928).
6. J. Schwinger, *Proc. Natl. Acad. Sci. U.S.* **37**(7), 452–459 (1951).
7. F. J. Dyson, *Phys. Rev.* **75**(11), 1736–1755 (1949).
8. J. Hubbard, *Proc. Roy. Soc. (London)* **A240**(1223), 549–560 (1957).
9. A. Klein and R. Prange, *Phys. Rev.* **112**(3), 994–1007 (1958).
10. N. N. Bogoliubov and D. V. Shirkov, *Vvedenie v teoriiu kvantovykh polei*, GITTL, Moscow, 1957, especially sec. 34. (English translation: *An Introduction to the Theory of Quantized Fields*, Wiley–Interscience, New York, 1959.)
11. R. P. Feynman, *Phys. Rev.* **76**(6), 749–759 (1949).
12. W. Pauli, *General Principles of Wave Mechanics (Die allgemeinen Prinzipien der Wellenmechanik)*, Edwards, Ann Arbor, Mich., 1947.
13. E. E. Salpeter and H. A. Bethe, *Phys. Rev.* **84**(6), 1232–1242 (1951).
14. K. A. Ter-Martirosian, *Phys. Rev.* **111**(3), 948–950 (1958).

VIII-7 Reprinted from *Journal of Chemical Physics* **45**, 4256 (1966)

# On the Correlation Problem in Atomic and Molecular Systems. Calculation of Wavefunction Components in Ursell-Type Expansion Using Quantum-Field Theoretical Methods

JIŘÍ ČÍŽEK*

*Institute of Physical Chemistry, Czechoslovak Academy of Sciences, Prague, Czechoslovakia*

(Received 17 May 1966)

A method is suggested for the calculation of the matrix elements of the logarithm of an operator which gives the exact wavefunction when operating on the wavefunction in the one-electron approximation. The method is based on the use of the creation and annihilation operators, hole–particle formalism, Wick's theorem, and the technique of Feynman-like diagrams. The connection of this method with the configuration-interaction method as well as with the perturbation theory in the quantum-field theoretical form is discussed. The method is applied to the simple models of nitrogen and benzene molecules. The results are compared with those obtained with the configuration-interaction method considering all possible configurations within the chosen basis of one-electron functions.

## I. FORMULATION OF THE PROBLEM

LET us consider a system consisting of fixed atomic nuclei and $2n$ electrons. Let us further exclude from our considerations systems having a degenerate ground state. Then, neglecting relativistic and magnetic effects, the Hamiltonian $\hat{H}$ of our problem is given by the following equations:

$$\hat{H}=\hat{Z}+\hat{V}, \tag{1}$$

$$\hat{Z}=\sum_i \hat{z}_i, \qquad \hat{V}=\sum_{i<j}\hat{v}_{i,j}, \tag{2}$$

where $\hat{z}_i$ is a one-particle operator corresponding to the sum of the kinetic and nuclear field energy, and $\hat{v}_{i,j}$ is a two-particle operator of the interelectronic Coulombic repulsion.

Our problem is to find the ground-state eigenvector and the corresponding eigenvalue of the Hamiltonian (1)

$$\hat{H}\mid\Psi\rangle=E\mid\Psi\rangle. \tag{3}$$

In order to solve this problem the wavefunction $\mid\Psi\rangle$ is written in the following form

$$\mid\Psi\rangle=\exp(\hat{T})\mid\Phi\rangle, \tag{4}$$

where $\mid\Phi\rangle$ is the ground-state wavefunction in the one-particle approximation. This form of the wavefunction[1–5] is connected with the so-called Ursell-type expansion. The use of an expansion of this type for the study of the electronic structure of atoms and molecules was suggested by Sinanoğlu.[3]

In this article, equations for the matrix elements of the operator $\hat{T}$ are derived using quantum-field theoretical concepts, namely, (1) the creation and annihilation operators, (2) the hole–particle formalism, (3) the time-independent Wick theorem and the diagram technique.

In addition, the connection of the proposed method with both the configuration-interaction method and the perturbation theory in the quantum-field theoretical form[6,7] are shown. This comparison demonstrates the usefulness of the Ursell-type expansion of the exact wavefunction.

## II. BASIC TERMS AND DEFINITIONS

Let us introduce a complete set of orthonormal spin-orbitals

$$\mid A\rangle=\mid a\rangle\mid\alpha\rangle, \tag{5}$$

where $\mid a\rangle$ and $\mid\alpha\rangle$ designate the space and the spin parts of the spin–orbital $\mid A\rangle$, respectively. Generally, capital letters are used to designate spin–orbitals, lower-case letters are associated with orbitals, while letters of the Greek alphabet are reserved for spin functions. In addition to this general system of spin–orbitals (5), we use spin–orbitals whose space parts are eigenfunctions of the operator

$$(\hat{z}+\hat{u})\mid a\rangle=\omega_a\mid a\rangle, \tag{6}$$

where $\hat{u}$ is an arbitrary spin-independent one-particle Hermitian operator. Let us note explicitly that Hartree–Fock spin–orbitals fall within this class.

The creation and annihilation operators defined on the system of spin-orbitals (5) are designated $\hat{X}_A^+$ and $\hat{X}_A$, respectively. These operators satisfy the following anticommutation relations:

$$\hat{X}_A^+\hat{X}_B^+ + \hat{X}_B^+\hat{X}_A^+ = 0,$$
$$\hat{X}_A\hat{X}_B + \hat{X}_B\hat{X}_A = 0,$$
$$\hat{X}_A^+\hat{X}_B + \hat{X}_B\hat{X}_A^+ = \langle A\mid B\rangle. \tag{7}$$

* Part of this study was carried out during the author's stay in the Centre de Mécanique Ondulatoire Appliquée, Paris, France.

[1] F. Coester and H. Kümmel, Nucl. Phys. **17**, 477 (1960).

[2] H. Kümmel, *Lectures on The Many-Body Problem*, E. R. Caioniello, Ed. (Academic Press Inc., New York, 1962), p. 265.

[3] O. Sinanoğlu, J. Chem. Phys. **36**, 706 (1962); Advan. Chem. Phys. **6**, 315 (1964).

[4] H. Primas, Lecture prepared for the Istanbul International Summer School of Quantum Chemistry, 1964 (preprint).

[5] J. da Providencia, Nucl. Phys. **61**, 87 (1965).

[6] J. Goldstone, Proc. Roy. Soc. (London) **239**, 267 (1957).

[7] V. V. Tolmachev, "The field form of the perturbation theory applied to many electron problems of atoms and molecules," University of Tartu, 1963 (in Russian).

The operators $\hat{Z}$ and $\hat{V}$ may then be expressed through the operators $X_A{}^+$ and $\hat{X}_A$ in the following form:

$$\hat{Z}=\sum_{C_1,B_1}\langle C_1|\hat{z}|B_1\rangle\hat{X}_{C_1}{}^+\hat{X}_{B_1}, \quad (8)$$

$$\hat{V}=\tfrac{1}{2}\sum_{C_1,B_1;C_2,B_2}\langle C_1C_2|\hat{v}|B_1B_2\rangle\hat{X}_{C_1}{}^+\hat{X}_{C_2}{}^+\hat{X}_{B_2}\hat{X}_{B_1}. \quad (9)$$

The matrix elements in Formulas (8) and (9) satisfy the following relations:

$$\langle C_1C_2|\hat{v}|B_1B_2\rangle=\langle c_1c_2|\hat{v}|b_1b_2\rangle\langle\gamma_1|\beta_1\rangle\langle\gamma_2|\beta_2\rangle, \quad (10)$$

$$\langle C_1|\hat{z}|B_1\rangle=\langle c_1|\hat{z}|b_1\rangle\langle\gamma_1|\beta_1\rangle, \quad (11)$$

using the notation introduced in Eq. (5).

The symbol $|0\rangle$ is used to specify the "vacuum state."

Let us suppose that in the complete set of spin-orbitals introduced above there exist $2n$ spin-orbitals

$$|A_1'\rangle, |A_2'\rangle, \ldots, |A_{2n}'\rangle, \quad (12)$$

such that the function

$$|\Phi\rangle=\hat{X}_{A_1'}{}^+\hat{X}_{A_2'}{}^+\ldots\hat{X}_{A_{2n}'}{}^+|0\rangle \quad (13)$$

is a reasonable approximation to the exact ground-state wavefunction $|\Psi\rangle$. Since we require the ground state to be nondegenerate, the spin-orbitals (12) may be formed as products of the following $n$ orbitals:

$$|a_1'\rangle, |a_2'\rangle, \ldots, |a_n'\rangle, \quad (14)$$

with spin functions corresponding to the positive and negative spin orientations, respectively. The spin-orbitals (12) which span the function $|\Phi\rangle$ are called nonexcited ones while all others, which do not enter the function $|\Phi\rangle$, are called excited spin-orbitals. The nonexcited and excited spin-orbitals are systematically designated by singly and doubly primed capitals, respectively.

In the following considerations it is useful to regard the ground state $|\Phi\rangle$ as a "new vacuum state" as well as to interpret the excitation of a particle as creation of a hole in a nonexcited spin-orbital followed by the creation of a particle in an excited spin-orbital. For this purpose a new set of operators is introduced by the following relations:

$$\hat{Y}_{A''}{}^+=\hat{X}_{A''}{}^+, \qquad \hat{Y}_{A''}=\hat{X}_{A''},$$
$$\hat{Y}_{A'}{}^+=\hat{X}_{A'}, \qquad \hat{Y}_{A'}=\hat{X}_{A'}{}^+. \quad (15)$$

The operator $\hat{Y}_{A'}{}^+$ may now be interpreted as a hole-creating operator while the operator $\hat{Y}_{A'}$ as a hole-annihilating operator. Formal analogy of this approach with quantum electrodynamics is apparent: An electron corresponds to a particle in an excited spin-orbital while a positron corresponds to a hole in a nonexcited spin-orbital.

In the following considerations the terms creation and annihilation operators are associated with the operators $\hat{Y}_A{}^+$ and $\hat{Y}_A$, respectively. These operators satisfy the same anticommutation relations as the operators $\hat{X}_A{}^+$, $\hat{X}_A$.

The operator $\hat{T}$ introduced above may be expressed through the operators $\hat{Y}_{A''}{}^+$, $\hat{Y}_{A'}{}^+$ in the following form:

$$\hat{T}=\sum_{j=1}^{2n}\hat{T}_j, \quad (16)$$

where

$$\hat{T}_j=\frac{1}{j!}\sum_{\substack{A_j'',\ldots,A_1''\\A_j',\ldots,A_1'}}\langle A_j''\ldots A_1''|\hat{t}|A_j'\ldots A_1'\rangle \times\prod_{i=1}^{j}\hat{Y}_{A_i''}{}^+\hat{Y}_{A_i'}{}^+ \quad (17)$$

represents an operator of the $j$-times excited state.[8]

In the following section we have to evaluate the products of operators of the type

$$\hat{M}_1\hat{M}_2\ldots\hat{M}_l, \quad (18)$$

where $\hat{M}_i$, $i=1, 2\ldots l$ is either a creation or an annihilation operator. These products of operators are most easily treated with the aid of Wick's theorem. In order to state the theorem we introduce some necessary notions.[9]

The normal product $N[\hat{M}_1\ldots\hat{M}_l]$ of the operators $\hat{M}_1, \ldots, \hat{M}_l$ is defined by the relation:

$$N[\hat{M}_1\ldots\hat{M}_l]=(-1)^p\hat{M}_{k_1}\hat{M}_{k_2}\ldots\hat{M}_{k_l}, \quad (19)$$

where the permutation of indices

$$\begin{pmatrix}1, 2, \ldots, l\\ k_1, k_2, \ldots, k_l\end{pmatrix} \quad (20)$$

is so chosen that in the product on the right-hand side of Eq. (19) all annihilation operators stand to the right of all the creation operators. The parity of the permutation (20) is designated by $p$.

Next we define the pairing $\dot{\hat{M}}_1\dot{\hat{M}}_2$ of two operators $\hat{M}_1$ and $\hat{M}_2$:

$$\dot{\hat{M}}_1\dot{\hat{M}}_2=\hat{M}_1\hat{M}_2-N[\hat{M}_1\hat{M}_2]. \quad (21)$$

The notation $\ddot{\hat{M}}_1\ddot{\hat{M}}_2={}^{\cdots}\hat{M}_1{}^{\cdots}\hat{M}_2$ etc. is also used for pairing.[10] Using the anticommutation relations (7) we immediately obtain explicit expressions for all possible pairings of two operators:

$$\dot{\hat{Y}}_A{}^+\dot{\hat{Y}}_B{}^+=\dot{\hat{Y}}_A\dot{\hat{Y}}_B=\dot{\hat{Y}}_A{}^+\dot{\hat{Y}}_B=0,$$

$$\dot{\hat{Y}}_A\dot{\hat{Y}}_B{}^+=\langle A|B\rangle.$$

[8] The following part of this chapter until Eq. (25) may be skipped by readers familiar with the mathematical methods of quantum-field theory.

[9] S. S. Schweber, *An Introduction to Relativistic Quantum Field Theory* (Row, Peterson and Co., Evanston, Ill., 1961).

[10] This notation is used for typographical reasons.

Further, we have to define the normal product with pairings

$$N[\hat{M}_1 \ldots {}^{\cdot}\hat{M}_{i_1} \ldots {}^{\cdot\cdot}\hat{M}_{i_2} \ldots {}^{\cdot}\hat{M}_{j_1} \ldots {}^{\cdot\cdot}M_{j_2} \ldots {}^{\cdots}\hat{M}_{i_k} \ldots {}^{\cdots}\hat{M}_{j_k} \ldots \hat{M}_l]$$
$$= (-1)^p ({}^{\cdot}\hat{M}_{i_1} {}^{\cdot}\hat{M}_{j_1}) ({}^{\cdot\cdot}M_{i_2} {}^{\cdot\cdot}\hat{M}_{j_2}) \ldots ({}^{\cdots}\hat{M}_{i_k} {}^{\cdots}M_{j_k}) [NM_{m_1} \ldots M_{m_r}]_1, \quad (22)$$

where $k$ is the number of pairings, $r = l - 2k$, and $p$ is the parity of the permutation

$$\begin{pmatrix} 1, 2, \ldots, 2k, 2k+1, \ldots l \\ i_1, j_1, \ldots, j_k, m_1, \quad \ldots m_r \end{pmatrix}. \quad (23)$$

Now we are ready to state Wick's theorem:

The product of several creation and annihilation operators is equal to the sum of all normal products of these operators with all possible pairings (including no pairing):

$$M_1 \ldots \hat{M}_l = N[\hat{M}_1 \ldots \hat{M}_l] + N[{}^{\cdot}\hat{M}_1 {}^{\cdot}\hat{M}_2 \ldots \hat{M}_l] + \ldots + N[{}^{\cdot}\hat{M}_1 \ldots {}^{\cdot}\hat{M}_l] + N[{}^{\cdot}\hat{M}_1 {}^{\cdot}\hat{M}_2 {}^{\cdot\cdot}M_3 {}^{\cdot\cdot}M_4 \ldots M_l] + \ldots. \quad (24)$$

Furthermore, we use the generalized form of Wick's theorem[9] which states the following.

The product of normal products of operators

$$N[\hat{M}_1 \ldots \hat{M}_{i_1}] N[\hat{M}_{i_1+1} \ldots \hat{M}_{i_2}] \ldots N[\ldots \hat{M}_l] \quad (25)$$

may be expressed similarly as the ordinary product of operators (18) but with the omission of pairings between operators within each normal product of the expression (25).

Using Wick's theorem the operator $\hat{H} = \hat{Z} + \hat{V}$ may be written in the form

$$\hat{H} - \langle \Phi \mid \hat{H} \mid \Phi \rangle = \sum_{C_1, B_1} \langle C_1 \mid \hat{f} \mid B_1 \rangle N[\hat{X}_{C_1}{}^{+} \hat{X}_{B_1}] + \tfrac{1}{2} \sum_{C_1, B_1; C_2, B_2} \langle C_1 C_2 \mid \hat{v} \mid B_1 B_2 \rangle N[\hat{X}_{C_1}{}^{+} \hat{X}_{C_2}{}^{+} \hat{X}_{B_2} \hat{X}_{B_1}], \quad (26)$$

where

$$\langle C_1 \mid \hat{f} \mid B_1 \rangle = \langle C_1 \mid \hat{z} \mid B_1 \rangle + \sum_{A'} (\langle C_1 A' \mid \hat{v} \mid B_1 A' \rangle - \langle C_1 A' \mid \hat{v} \mid A' B_1 \rangle). \quad (27)$$

The matrix element (27) satisfies the relation

$$\langle C_1 \mid \hat{f} \mid B_1 \rangle = \langle c_1 \mid \hat{f} \mid b_1 \rangle \langle \gamma_1 \mid \beta_1 \rangle, \quad (28)$$

where

$$\langle c_1 \mid \hat{f} \mid b_1 \rangle = \langle c_1 \mid \hat{z} \mid b_1 \rangle + \sum_{a'} (2 \langle c_1 a' \mid \hat{v} \mid b_1 a' \rangle - \langle c_1 a' \mid \hat{v} \mid a' b_1 \rangle). \quad (29)$$

## III. SYSTEM OF EQUATIONS FOR MATRIX ELEMENTS OF THE OPERATOR $\hat{T}$

Considerations of a topological character are used extensively in this section. Before proceeding it is necessary to introduce a few definitions. A topological formation consisting of dashed nonoriented lines, full lines both oriented and nonoriented, and vertices are called a skeleton.

Two skeletons are said to be equivalent if they can be mutually transformed into each other through topological deformations.

The topological deformation converting a given skeleton into itself is called an automorphism. The weight of a skeleton is the reciprocal value of the number of all different automorphisms (including the identical one) of a given skeleton.

Furthermore, the two oriented full lines of a skeleton are considered to be equivalent if there exists an automorphism transferring these lines one into the other.

In the following considerations we define four different types of skeletons, namely the $H$, $T$, $M$, and $R$ skeletons. Nonequivalent skeletons of a given type are distinguished by their arguments

$$S_H(h), \; S_T(t), \; S_M(m), \; S_R(r).$$

Therefore, equivalent skeletons are denoted by identical symbols. The respective weights are designated by

$$w_H(h), \; w_T(t), \; w_M(m), \; w_R(r).$$

Further, let a positive integer be associated with any full oriented line in such a way that the following rules are respected:

(a) Equivalent lines carry different integers.

(b) Let us have two equivalent skeletons. Then a topological deformation exists, which transfers the skeletons into each other in such a way that the mutually corresponding full oriented lines in both skeletons carry the same integers.

Therefore, the possibility that different lines are labeled by the same integer is not excluded.

The $H$ skeletons are those shown in Fig. 1. The left skeleton in Fig. 1 is designated $S_H(1)$, the right one $S_H(2)$. The assumed numbering of lines is shown as

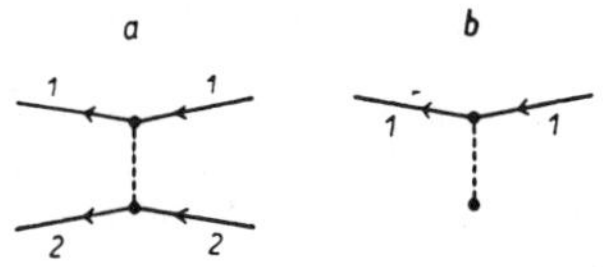


FIG. 1. (a) $S_H(1)$ skeleton, (b) $S_H(2)$ skeleton.

well. The weights associated with the skeletons of Fig. 1 are

$$w_H(1)=\tfrac{1}{2}, \qquad w_H(2)=1. \tag{30}$$

On labeling the oriented lines of an $H$ skeleton with spin–orbital indices we obtain a so-called $H$ diagram. These diagrams are shown in Fig. 2.

To the left diagram of Fig. 2 we assign the following scalar quantity:

$$d_H(1,\chi)=\langle C_1C_2 \mid \hat{v} \mid B_1B_2\rangle, \tag{31}$$

as well as the following operator:

$$\hat{D}_H(1,\chi)=d_H(1,\chi)N[\hat{X}_{C_1}{}^+\hat{X}_{C_2}{}^+\hat{X}_{B_2}\hat{X}_{B_1}], \tag{32}$$

where $\chi$ represents the sequence of numbers $C_1$, $C_2$, $B_1$, $B_2$.

Similarly, to the right diagram of Fig. 2 the scalar quantity

$$d_H(2,\chi)=\langle C_1 \mid \hat{f} \mid B_1\rangle \tag{33}$$

and the operator

$$\hat{D}_H(2,\chi)=d_H(2,\chi)N[\hat{X}_{C_1}{}^+\hat{X}_{B_1}] \tag{34}$$

are assigned. Here $\chi$ represents the numbers $C_1$ and $B_1$.

Further, to the oriented line labeled $B_i(C_i)$ and entering (leaving) the vertex of the $H$ diagram the operator $\hat{X}_{B_i}(\hat{X}_{C_i}{}^+)$ is assigned ($i=1, 2$).

Consequently, Eq. (26) may be written in the following form:

$$\hat{H}-\langle\Phi \mid \hat{H} \mid \Phi\rangle=\sum_{h=1}^{2} w_H(h)\sum_{\chi}\hat{D}_H(h,\chi). \tag{35}$$

A general $T$-type skeleton as well as the labeling of its oriented lines are shown in Fig. 3. These skeletons differ in the number of pairs of oriented lines, so that the index $j$ may be any positive integer. The skeletons are generally designated $S_T(j)$. The weight of the $S_T(j)$ skeleton is given by the relation

$$w_T(j)=1/j!. \tag{36}$$

On labeling the oriented lines entering (leaving) the vertices of the $T$ skeleton by the indices of nonexcited

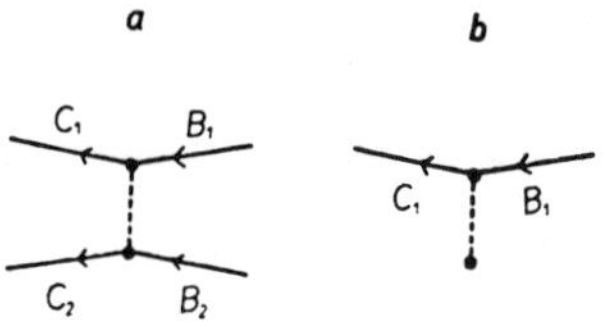


FIG. 2. (a) Diagram formed from the $S_H(1)$ skeleton, (b) diagram formed from the $S_H(2)$ skeleton.

FIG. 3. A general $S_T(j)$ skeleton,

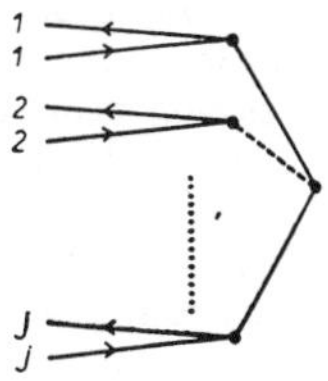


(excited) spin–orbitals we obtain the $T$ diagram shown in Fig. 4.

We assign to the $T$ diagram the following scalar quantity

$$d_T(j,\tau)=\langle A_1''\ldots A_j'' \mid \hat{t} \mid A_1'\ldots A_j'\rangle \tag{37}$$

and the operator

$$\hat{D}_T(j,\tau)=d_T(j,\tau)\prod_{i=1}^{j}\hat{Y}_{A_i''}{}^+\hat{Y}_{A_i'}{}^+, \tag{38}$$

where $\tau$ represents the sequence of numbers

$$A_1'', A_1', A_2'', A_2', \ldots, A_j'', A_j'. \tag{39}$$

Furthermore, we attach the operator $\hat{Y}_{A_i'}{}^+(\hat{Y}_{A_i''}{}^+)$ to every oriented line labeled $A_i'(A_i'')$ and entering (leaving) some vertex of the $T$ diagram.

Using the notation specified above, the operator $\hat{T}$ may be expressed in the form

$$\hat{T}=\sum_{t=1}^{2n} w_T(t)\sum_{\tau}\hat{D}_H(t,\tau). \tag{40}$$

A set of $T$ skeletons are called an $M$ skeleton. Generally, an $M$ skeleton is composed of $n_1$ skeletons $S_T(t_1)$, $n_2$ skeletons $S_T(t_2)$, etc. The $M$ skeletons are designated $S_M(m)$, where $m$ represents a sequence of indices $t_i$ of the $T$ skeletons involved, the order of the individual indices being immaterial. Among the $M$ skeletons we also include the so-called empty $M$ skeleton consisting of any empty set of $T$ skeletons.

The weight of an $M$ skeleton is given by the formula[11]

$$w_M(m)=\Big(\prod_{i=1}^{k} n_i!\Big)^{-1}\prod_{i=1}^{k}[w_T(t_i)]^{n_i}. \tag{41}$$

The weight of the empy $M$ skeleton is equal to an unit by definition.

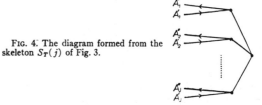

FIG. 4. The diagram formed from the skeleton $S_T(j)$ of Fig. 3.

[11] J. Hubbard, Proc. Roy. Soc. (London) **240**, 539 (1957).

On labeling the oriented lines of an $M$ skeleton with spin–orbital indices in the same way as specified for a $T$ skeleton we obtain the so-called $M$ diagram. Therefore, an $M$ diagram is composed of a certain number of $T$ diagrams.

A scalar quantity $d_M(m,\mu)$ and an operator $\hat{D}_M(m,\mu)$ associated with an $M$ diagram are given by the products of the corresponding quantities assigned to the $T$ diagram constituting a given $M$ diagram (unity and the unit operator are assigned to an empty $M$ diagram by definition). Further, $\mu$ specifies the sequence of spin–orbital indices assigned to the oriented lines of an $M$ diagram.

We can write now[9]

$$e^{\hat{T}}=\sum_m w_M(m)\sum_\mu \hat{D}_M(m,\mu). \tag{42}$$

Next we have to evaluate the following vector

$$(\hat{H}-\langle\Phi\mid\hat{H}\mid\Phi\rangle)e^{\hat{T}}\mid\Phi\rangle=\sum_h\sum_m w_H(h)w_M(m)\times\sum_\chi\sum_\mu \hat{D}_H(h,\chi)\hat{D}_M(m,\mu)\mid\Phi\rangle. \tag{43}$$

This is most easily done using an auxiliary theorem, the proof of which, based on the generalized Wick theorem, is given in the Appendix. In order to state the theorem we introduce the notion of $R$ skeletons and $R$ diagrams.

An $R$ skeleton is formed from one $H$ and one $M$ skeleton by connecting some lines (none, one, or two) entering the vertices of the $H$ skeleton with lines issuing from the vertices of the $M$ skeleton and, further, by connecting some lines (none, one, or two) issuing from the vertices of the $H$ skeleton with lines entering the vertices of the $M$ skeleton. The lines connecting the vertices of the $H$ and $M$ skeletons are called internal lines. All other lines are called external. The internal hole lines of the $R$ skeleton are those which start from a vertex of the $H$ skeleton and enter a vertex of the $M$ skeleton. The internal lines having opposite directions, that is, starting at a vertex of the $M$ skeleton and entering a vertex of the $H$ skeleton, are called internal particle lines.

Furthermore, the external hole (particle) lines are those external lines which enter (leave) a vertex of the $M$ or $H$ skeletons. A sequence of oriented lines are generally referred to as the path. A closed path is called the loop. We number the lines of an $R$ skeleton independently of the numberings of the corresponding $M$ and $H$ skeletons in the following manner.

Let the $R$ skeleton considered contain $j$ open paths, $h$ internal hole lines, and $k$ internal particle lines. We number the external hole lines $1, 2\ldots j$ and the internal hole lines $j+1, \ldots, j+h$. Similarly, the external particle lines will be numbered $1\ldots j$ and the internal particle lines $j+1, \ldots, j+k$. The numbering of the external lines is carried out in such a way that the external particle and hole lines associated with the same path have the same number.

By labeling the $i$th particle (hole) line of the $R$ skeleton with the index of an excited (nonexcited) spin-orbital $D_i''(D_i')$ we obtain the $R$ diagram.

On dissecting the internal lines of the $R$ diagram we obtain the $H$ and $M$ diagrams. Let the $d_H(h,\chi)$ and $d_M(m,\mu)$ be scalar quantities associated with these $H$ and $M$ diagrams. Then the scalar quantity $d_R(r,\rho)$ assigned to this diagram is given by the expression

$$d_R(r,\rho)=(-1)^{l+h}d_H(h,\chi)d_M(m,\mu), \tag{44}$$

where $l$ is the number of loops, $h$ is the number of internal hole lines, and $\rho$ represents the following two sequences $\rho_1, \rho_2$:

$$\rho_1\equiv D_1'', D_1', D_2'', D_2'\ldots D_j'', D_j',$$

$$\rho_2\equiv D_{j+1}''\ldots D_{j+k}'', D_{j+1}'\ldots D_{j+h}'. \tag{45}$$

For diagrams lacking external lines we write

$$d_R(r,\rho)=d_R(r,\rho_2). \tag{46}$$

The operator associated with an $R$ diagram is given by the following equation:

$$\hat{D}_R(r,\rho)=d_R(r,\rho)\prod_{i=1}^{j}\hat{Y}_{D_i''}{}^{+}\hat{Y}_{D_i'}{}^{+}. \tag{47}$$

Again, for diagrams having only internal lines we write

$$\hat{D}(r,\rho)=d(r,\rho_2). \tag{48}$$

We are now ready to state the above-mentioned auxiliary theorem:

$$w_H(h)w_M(m)\sum_\chi\sum_\mu \hat{D}_H(h,\chi)\hat{D}_M(m,\mu)\mid\Phi\rangle=\sum_r{}'w_R(r)\sum_\rho \hat{D}_R(r,\rho)\mid\Phi\rangle, \tag{49}$$

where $\sum_r{}'$ designates the summation over all topologically distinct $R$ skeletons which may be formed from $H(h)$ and $M(m)$ skeletons.

The Schrödinger equation may now be given the following form:

$$\sum_r w_R(r)\sum_\rho \hat{D}_R(r,\rho)\mid\Phi\rangle=(E-\langle\Phi\mid\hat{H}\mid\Phi\rangle)e^{\hat{T}}\mid\Phi\rangle, \tag{50}$$

where the summation extends over all topologically distinct $R$ skeletons formed from arbitrary $H$ and $M$ skeletons.

We distinguish connected and disconnected $R$ skeletons. In a connected $R$ skeleton any two vertices are connected by at least one sequence of lines consisting of full oriented and nonoriented and dashed lines. A disconnected $R$ skeleton is composed of a connected $R$ skeleton and an $M$ skeleton. The weight of the disconnected skeleton is given by the product of the weights of the connected $R$ skeleton and the $M$ skeleton which constitute the given $R$ skeleton. The scalar quantities and operators assigned to disconnected $R$

diagram in an analogous manner are obtained. On the basis of the above considerations, the Schrödinger equation may be rewritten in the form

$$e^{\hat{T}}\Big[\sum_{r\in\Delta} w_R(r) \sum_{\rho} \hat{D}_R(r,\rho) - (E - \langle\Phi\mid\hat{H}\mid\Phi\rangle)\Big]\mid\Phi\rangle = 0, \tag{51}$$

where $\Delta$ is a set of connected $R$ skeletons.

Equation (51) will be satisfied if the following equations hold:

$$\sum_{r\in\Delta_0} w_R(r) \sum_{\rho_2} d_R(r,\rho_2) = E - \langle\Phi\mid\hat{H}\mid\Phi\rangle, \tag{52}$$

$$\sum_{r\in\Delta_j} w_R(r) \sum_{\rho_1\in\theta} \sum_{\rho_2} d_R(r,\rho_1,\rho_2) = 0, \tag{53}$$

where $\Delta_j$ is a set of $R$ skeletons having $j$ open paths. The sets $\theta$ are formed in accordance with the following rule: If the sequence

$$D_1'', D_1' \ldots D_j'', D_j'$$

belongs to $\theta$, then also the sequence

$$D_{k_1}'', D_{k_1}' \ldots D_{k_j}'', D_{k_j}',$$

where

$$\begin{pmatrix} 1, \ldots, j \\ k_1, \ldots, k_j \end{pmatrix}$$

is an arbitrary permutation of indices $1 \ldots j$.

We further specify the form of the equations (53) for the case, that the spin orbitals are eigenfunctions of the operator $(\hat{z}+\hat{u})$. In this case the system of Eqs. (53) has the following form:

$$\sum_{i=1}^{j} (\omega_{D_i''} - \omega_{D_i'}) \langle D_1'' \ldots D_j'' \mid \hat{t} \mid D_1' \ldots D_j' \rangle + \sum_{r\in\Delta_j} w_R(r) \sum_{\rho_1\in\theta} \sum_{\rho_2} \tilde{d}_R(r,\rho_1,\rho_2) = 0, \tag{54}$$

where the indices $D_1'', D_1' \ldots D_j'', D_j'$ form one of the sequences belonging to $\theta$. Further, the tilde above $\tilde{d}_R$ indicates that the following scalar quantity is associated with the right diagram of Fig. 2:

$$\langle C_1 \mid \hat{q} \mid B_1 \rangle = \langle C_1 \mid \hat{f} \mid B_1 \rangle - \langle C_1 \mid \hat{z}+\hat{u} \mid B_1 \rangle = \sum_{A'} (\langle C_1A' \mid \hat{v} \mid B_1A' \rangle - \langle C_1A' \mid \hat{v} \mid A'B_1 \rangle) - \langle C_1 \mid \hat{u} \mid B_1 \rangle \tag{55}$$

instead of the scalar quantity (33).

The matrix elements $\langle B_1 \mid \hat{q} \mid C_1 \rangle$ are identically equal to zero for Hartree–Fock spin orbitals, as follows immediately from Eq. (55).

In the following considerations the summation over the spin parts is carried out. It immediately follows from Eqs. (10) and (11) that a nonzero scalar quantity is assigned only to such $H$ diagrams having the same spin along each path. Thus, the relation

$$\langle D_1'' \ldots D_j'' \mid \hat{t} \mid D_1' \ldots D_j' \rangle = \langle d_1'' \ldots d_j'' \mid \hat{t} \mid d_1' \ldots d_j' \rangle \times \prod_{i=1}^{j} \langle \delta_i'' \mid \delta_i' \rangle \tag{56}$$

for the matrix elements over spin–orbitals is obtained, so that the equations (53) may be easily transformed into equations for the matrix elements over the orbitals $\langle d_1'' \ldots d_j'' \mid \hat{t} \mid d_1' \ldots d_j' \rangle$.

The equations for the matrix elements over orbitals are then obtained in a similar way as Eq. (53) except that: (a) spin–orbital indices are replaced by orbital indices and (b) the multiplicative factor 2 is assigned to each loop.

The requirement (b) expressed the fact that both spin orientations are allowed along any loop. In order to illustrate the general formalism presented above we write down explicitly the system of equations (53) for the case where all components of the operator $T$ except $T_2$ are disregarded. This approximation is discussed from the physical point of view in the following section.

In this special case, the $M$ skeletons consist of a set of $T$ skeleton (the empty set being included) each of which has two open paths. The possible $R$ skeletons, having none or two open paths, which can be obtained from these $M$ skeletons are shown in Figs. 5–8. The skeletons shown in Fig. 5 do not have any external lines while those in Figs. 6–8 have always two open paths. Figure 6 shows the $R$ skeleton resulting from the empty $M$ skeleton. Figures 7 and 8 present the $R$ skeletons constructed from the $M$ skeletons consisting of one and two $T$ skeletons, respectively.

The energy difference $E - \langle\Phi\mid\hat{H}\mid\Phi\rangle$ may now be expressed through the matrix elements over the spin orbitals or the orbitals as follows:

$$E - \langle\Phi\mid\hat{H}\mid\Phi\rangle = \tfrac{1}{2} \sum_{D_1',D_2';D_1'',D_2''} \langle D_1'D_2' \mid \hat{v} \mid D_1''D_2'' \rangle \times (\langle D_1''D_2'' \mid \hat{t} \mid D_1'D_2' \rangle - \langle D_1''D_2'' \mid \hat{t} \mid D_2'D_1' \rangle), \tag{57}$$

$$E - \langle\Phi\mid\hat{H}\mid\Phi\rangle = \sum_{d_1',d_2';d_1'',d_2''} \langle d_1'd_2' \mid \hat{v} \mid d_1''d_2'' \rangle \times (2\langle d_1''d_2'' \mid \hat{t} \mid d_1'd_2' \rangle - \langle d_1''d_2'' \mid \hat{t} \mid d_2'd_1' \rangle). \tag{58}$$

We now present the corresponding algebraic equations for the matrix elements over the orbitals only.

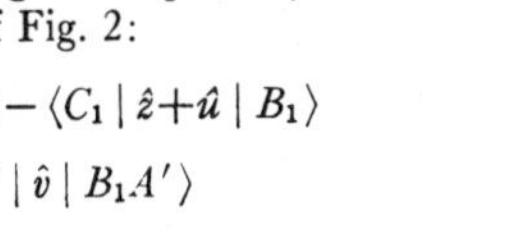

FIG. 5. $R$ skeletons lacking external lines formed from the $S_H(1)$ and $S_T(2)$ skeletons.

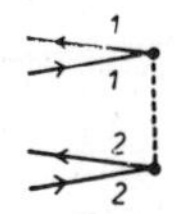

FIG. 6. The $R$ skeleton having two open paths formed from the $S_H(1)$ skeleton and the empty $M$ skeleton.

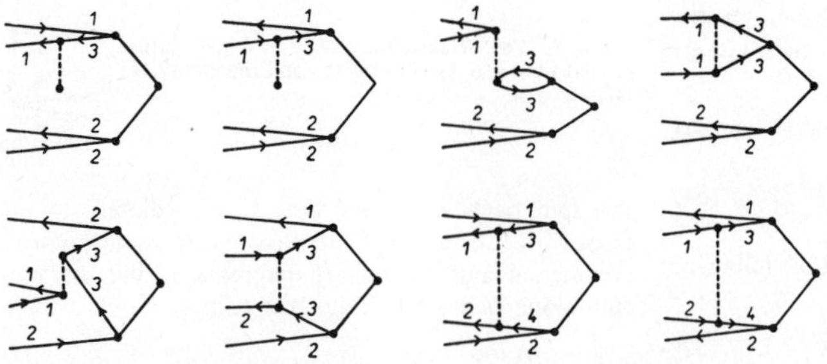

FIG. 7. All possible $R$ skeletons having two open paths formed from the $S_H(1)$ or $S_H(2)$ skeleton and from the $M$ skeleton consisting of one $S_T(2)$ skeleton.

Let us define a new quantity $\lambda(d_1'', d_1'; d_2'', d_2')$ by the following equation:

$$\begin{aligned}
\lambda(d_1'', d_1'; d_2'', d_2') = {} & \tfrac{1}{2}\langle d_1''d_2'' \mid \hat{v} \mid d_1'd_2'\rangle + \sum_{d_3''} \langle d_1'' \mid \hat{f} \mid d_3''\rangle\langle d_3''d_2'' \mid \hat{t} \mid d_1'd_2'\rangle \\
& - \sum_{d_3'} \langle d_1' \mid \hat{f} \mid d_3'\rangle\langle d_1''d_2'' \mid \hat{t} \mid d_3'd_2'\rangle + \sum_{d_3', d_3''} [(2\langle d_1''d_3' \mid \hat{v} \mid d_1'd_3''\rangle - \langle d_1''d_3' \mid \hat{v} \mid d_3''d_1'\rangle)\langle d_3''d_2'' \mid \hat{t} \mid d_3'd_2'\rangle \\
& - \langle d_1''d_3' \mid \hat{v} \mid d_1'd_3''\rangle\langle d_2''d_3'' \mid \hat{t} \mid d_3'd_2'\rangle - \langle d_2''d_3' \mid \hat{v} \mid d_3''d_1'\rangle\langle d_1''d_3'' \mid \hat{t} \mid d_3'd_2'\rangle] \\
& + \tfrac{1}{2} \sum_{d_3'', d_4''} \langle d_1''d_2'' \mid \hat{v} \mid d_3''d_4''\rangle\langle d_3''d_4'' \mid \hat{t} \mid d_1'd_2'\rangle + \tfrac{1}{2} \sum_{d_3', d_4'} \langle d_1'd_2' \mid \hat{v} \mid d_3'd_4'\rangle\langle d_1''d_2'' \mid \hat{t} \mid d_3'd_4'\rangle \\
& + \sum_{d_3', d_4'; d_3'', d_4''} \{[2\langle d_3'd_4' \mid \hat{v} \mid d_3''d_4''\rangle - \langle d_3'd_4' \mid \hat{v} \mid d_4''d_3''\rangle][\langle d_1''d_3'' \mid \hat{t} \mid d_1'd_3'\rangle\langle d_2''d_4'' \mid \hat{t} \mid d_2'd_4'\rangle \\
& - \langle d_1''d_3'' \mid \hat{t} \mid d_3'd_1'\rangle\langle d_2''d_4'' \mid \hat{t} \mid d_2'd_4'\rangle - \langle d_3''d_4'' \mid \hat{t} \mid d_3'd_1'\rangle\langle d_1''d_2'' \mid \hat{t} \mid d_4'd_2'\rangle \\
& - \langle d_1''d_3'' \mid \hat{t} \mid d_4'd_3'\rangle\langle d_2''d_4'' \mid \hat{t} \mid d_2'd_1'\rangle] + \tfrac{1}{2}\langle d_3'd_4' \mid \hat{v} \mid d_3''d_4''\rangle[\langle d_1''d_3'' \mid \hat{t} \mid d_3'd_1'\rangle\langle d_2''d_4'' \mid \hat{t} \mid d_4'd_2'\rangle \\
& + \langle d_2''d_3'' \mid \hat{t} \mid d_4'd_1'\rangle\langle d_1''d_4'' \mid \hat{t} \mid d_3'd_2'\rangle + \langle d_1''d_2'' \mid \hat{t} \mid d_3'd_4'\rangle\langle d_3''d_4'' \mid \hat{t} \mid d_1'd_2'\rangle]\}. \qquad (59)
\end{aligned}$$

Then, Eqs. (53) have the form

$$\lambda(d_1'', d_1'; d_2'', d_2') + \lambda(d_2'', d_2'; d_1'', d_1') = 0. \qquad (60)$$

For orbitals which are eigenfunctions of the operator $(\hat{z}+\hat{u})$, the second and third terms on the rhs of Eq. (59) have the form

$$\begin{aligned}
& \sum_{d_3''} \langle d_1'' \mid \hat{f} \mid d_3''\rangle\langle d_3''d_2'' \mid \hat{t} \mid d_1'd_2'\rangle - \sum_{d_3'} \langle d_1' \mid \hat{f} \mid d_3'\rangle\langle d_1''d_2'' \mid \hat{t} \mid d_3'd_2'\rangle \\
& \quad = (\omega_{d_1''} - \omega_{d_1'})\langle d_1''d_2'' \mid \hat{t} \mid d_1'd_2'\rangle + \sum_{d_3''} \langle d_1'' \mid \hat{q} \mid d_3''\rangle\langle d_3''d_2'' \mid \hat{t} \mid d_1'd_2'\rangle \\
& \qquad - \sum_{d_3'} \langle d_1' \mid \hat{q} \mid d_3'\rangle\langle d_1''d_2'' \mid \hat{t} \mid d_3'd_2'\rangle. \qquad (61)
\end{aligned}$$

In this case, therefore, the following term appears in the expression (60):

$$(\omega_{d_1''} + \omega_{d_2''} - \omega_{d_1'} - \omega_{d_2'})\langle d_1''d_2'' \mid \hat{t} \mid d_1'd_2'\rangle.$$

## IV. DISCUSSION

The method which is most commonly used in investigations of the correlation effects in atoms and molecules is the configuration-interaction method. In this procedure the wavefunction may be written in the following form:

$$|\Psi\rangle = |\Phi\rangle + \sum_{i=1}^{2n} \hat{C}_i |\Phi\rangle, \qquad (62)$$

where the term $\hat{C}_i|\Phi\rangle$ represents a linear combination of $i$-times excited configurations.

By defining the operators $\hat{C}_i$ as

$$\begin{aligned}
\hat{C}_1 &= \hat{T}_1, \\
\hat{C}_2 &= \hat{T}_2 + \hat{T}_1\hat{T}_1/2!, \qquad (63) \\
\hat{C}_3 &= \hat{T}_3 + \hat{T}_1\hat{T}_2 + \hat{T}_1\hat{T}_1\hat{T}_1/3!, \\
& \ldots,
\end{aligned}$$

both Expansions (62) and (4) of the exact wavefunction become identical.

The spin–orbitals which are usually used to set up the Slater determinant of the ground state $|\Phi\rangle$ are either HF spin–orbitals or localized orbitals obtained from these by an appropriate unitary transformation. In both cases the Brillouin theorem is valid [i.e., the matrix elements of the Hamiltonian (1) between the

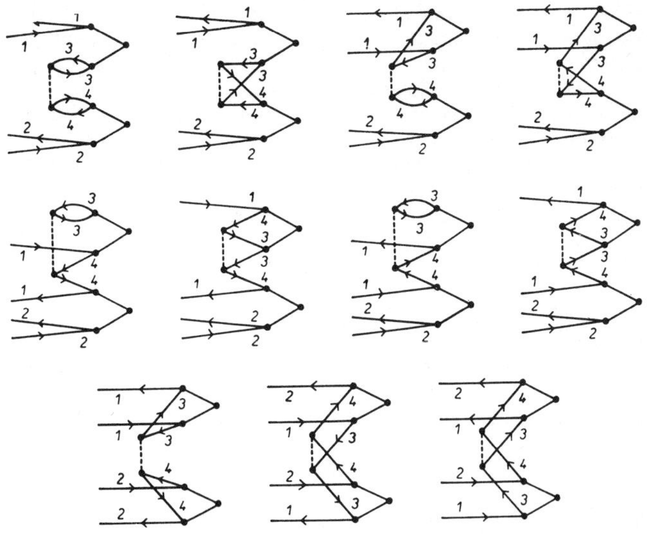

FIG. 8. All possible $R$ skeletons having two open paths formed from the $S_H(1)$ skeleton and the $M$ skeleton consisting of two $S_T(2)$ skeletons.

ground-state function $|\Phi\rangle$ and any singly excited state are equal to zero].

Therefore, one can expect the singly excited states to have negligible coefficients in the expansion (62). Actual computations corroborate this assumption.

In most calculations the triply and higher excited configurations are disregarded in order to make the computation manageable. The effect of triply and quadruply excited configurations was investigated for the simplest possible systems only, as for example in Refs. 12–14. These studies indicate that quadruply excited configurations are more significant than triply excited configurations.

It should now be pointed out that the neglect of all components of the operator $\hat{T}$ except $\hat{T}_2$ yields a wavefunction $|\Psi\rangle$ containing in part quadruply excited configurations given by the term[3]

$$(1/2!)\,\hat{T}_2\hat{T}_2\,|\,\Phi\rangle. \qquad (64)$$

The calculations for beryllium[3] (a case of localized electrons) as well as for benzene and butadiene[15] (a case of delocalized electrons) clearly demonstrate that the term (64) is a good approximation of the general term

$$\hat{C}_4\,|\,\Phi\rangle. \qquad (65)$$

This may be justified using perturbation theory.[16]

For the sake of completeness let us note that the system of equations (54), in which only $R$ skeletons shown in Figs. 6 and 7 are included, is equivalent to the system of equations (11) of Kelly and Sessler[17] in which the right-hand side is neglected.[16] This system of equations is also identical with the system given in the paper by da Providencia.[5] In the papers just mentioned the diagram technique is not used systematically.

The connection of our method with the Goldstone perturbation theory[6,7] is now discussed.

According to Eqs. (54) the following correspondence is valid: The scalar quantity associated with an $R$ diagram is equal to the sum of the scalar quantities assigned to all possible connected diagrams of Goldstone having the same number of open paths, the external lines of which are labeled with the same spin-orbitals as in the $R$ diagram given. Furthermore, it must be required that the indices of external lines of the given path are identical in both Goldstone and $R$ diagrams.

In order to prove this correspondence the following algebraic identity and its generalizations have to be used:

$$[1/a(a+b)]+[1/b(a+b)]=1/ab. \qquad (66)$$

A standard procedure in applying the Goldstone perturbation theory[6] is to find out the type of the second- and third-order diagrams which gives the largest contributions to the energy. For example, in the theory of nuclear matter the so-called ladder diagrams are most important,[6,18] while in the electron-gas problem the so-called ring diagrams are essential.[11,19] The contributions of these most important diagrams are then summed to all orders.

[12] R. A. Watson, Phys. Rev. **119**, 170 (1960).
[13] J. Koutecký, J. Čížek, J. Dubský, and K. Hlavatý, Theoret. Chim. Acta **2**, 462 (1964).
[14] J. Koutecký, K. Hlavatý, and P. Hochmann, Theoret. Chim. Acta **3**, 341 (1965).
[15] J. Čížek and J. Paldus (to be published).
[16] H. P. Kelly, Phys. Rev. **134**, A1450 (1964).
[17] H. P. Kelly and A. M. Sessler, Phys. Rev. **132**, 2091 (1963).
[18] K. A. Brueckner and J. L. Gammel, Phys. Rev. **109**, 1023 (1958).
[19] M. Gell-Mann and K. Brueckner, Phys. Rev. **106**, 364 (1957).

This summation is equivalent to the solution of Eqs. (54) in which appropriate approximations are introduced.

For example, let us suppose that in Eqs. (54) all contributions are neglected except those corresponding to the diagrams formed from the skeleton of Fig. 6 and from the seventh skeleton of Fig. 7. Then the solution of these equations is equivalent to the summation of all ladder diagrams. Therefore, the system of equations obtained in this way is closely connected with the Bethe–Goldstone equation.

On the other hand, if in Eqs. (54) only contributions of diagrams formed from the skeleton of Fig. 6, from the third one of Fig. 7, and from the first one of Fig. 8 are considered, the solution of these equations is equivalent to the summation of all ring diagrams.

The first attempts to determine the most important type of diagrams for the electronic structure of atoms were made by Kelly.[20] In any case the situation here is more complicated than in both the electron-gas and the nuclear-matter problems and further studies will have to be carried out.

## V. EXAMPLES

In this section the results concerning two models studied are presented. These models were chosen to be simple enough to allow a complete configuration-interaction treatment within an accepted system of one-electron functions. The results obtained in this way are used as a standard for a comparison with the results of approximate methods.

The quantities compared are first of all the correlation energies

$$\mathcal{E}_{\text{cor}} = \langle \Phi \mid \hat{H} \mid \Phi \rangle - E \qquad (67)$$

for different approximations used and, further, the effects of correlation on the wavefunction. Due to the high symmetry of the models studied the first-order density matrix is diagonal when expressed in the representation of Hartree–Fock spin–orbitals. Therefore, the following quantity $q$ may be chosen as a characteristic of the correlation effect on the wavefunction

$$q = \sum_{A''} \langle A'' \mid \hat{\gamma} \mid A'' \rangle = 2n - \sum_{A'} \langle A' \mid \hat{\gamma} \mid A' \rangle, \qquad (68)$$

where $\hat{\gamma}$ is the first-order density matrix:

$$\langle A \mid \hat{\gamma} \mid A \rangle = \langle \Psi \mid \hat{X}_A{}^+ \hat{X}_A \mid \Psi \rangle / \langle \Psi \mid \Psi \rangle. \qquad (69)$$

Apparently, the quantity $q$ is equal to zero in the one-particle approximation.

The first example concerns the nitrogen molecule. The Slater determinant describing the ground state of the nitrogen molecule in the Hartree–Fock approximation is built up from six spin–orbitals $\sigma_g$, four spin–orbitals $\sigma_u$, and four spin–orbitals $\pi_u$. From virtual orbitals we consider only the four lowest-lying spin–orbitals $\pi_g$. The required molecular integrals were taken from Grimaldi's[21] SCF MO LCAO study of the nitrogen molecule. In this study the following basis of Slater-type orbitals localized on atomic nuclei was used for every nitrogen atom [the corresponding exponential coefficients are given in parentheses: two $1s$ orbitals (7, 1.9), two $2s$ orbitals (7, 19), two $3s$ orbitals (7, 1.9), nine $2p$ orbitals (1.4, 2.44, 5.85), and five $3d$ orbitals (2.5). The internuclear distance between the nitrogen nuclei was 2.0675 a.u.

In our study only excitations of the type $\pi_u \rightarrow \pi_g$ were considered, so that "complete configuration interaction" may easily be carried out. The singly and triply excited configurations do not interact with the ground state since they belong to different symmetry species. The consideration of doubly and quadruply excited configurations yields the secular equation of the fifth order.

This model was further studied by means of the method suggested in this paper. Two different approximations were applied. In both of these only the component $\hat{T}_2$ of the operator $\hat{T}$ was considered. That means that the exact wavefunction was written in the form $|\Psi\rangle = \exp(\hat{T}_2) \mid \Phi\rangle$.

In the first approximation only skeletons given in Figs. 6 and 7 were considered. In this case the system of equations (54) is linear. In the second approximation the skeletons given in Fig. 8 were taken into account in addition to those considered in the first approximation. Consequently, a nonlinear system of equations (54) is obtained.

In both cases there are three unknown variables in Eqs. (54). The solution of the nonlinear system of equations was obtained using a general program for the Elliott 803 computer written by Paldus.[22] This program uses the Newton–Raphson method for the solution of nonlinear systems of equations. The errors in the correlation energy and the quantity $q$ in different approximations are given in Table I.

TABLE I. The error[a] in correlation energy and in the quantity $q$ for different approximations used in the study of the model of nitrogen molecule.

| % | $B$[b] | $L$[c] | $N$[c] |
|---|---|---|---|
| $\mathcal{E}_{\text{cor}}$ | −3.4 | 1.9 | −0.9 |
| $q$ | −10.6 | 2.6 | −3.6 |

[a] The percentage error is calculated with respect to the results of the "complete" configuration-interaction treatment, taken as a standard. By "complete" configuration interaction is meant a treatment in which all possible configurations, arising from the limited basis set selected, are considered.

[b] $B$ designates the configuration-interaction treatment limited to doubly excited configurations only.

[c] $L(N)$ designates the first (second) approximation described in the main text, which yields a linear (nonlinear) system of equations for components (matrix elements) of the operator $\hat{T}_2$.

[20] H. P. Kelly, Phys. Rev. 131, 684 (1963).

[21] F. Grimaldi, J. Chem. Phys. 43, S59 (1965).

[22] J. Paldus (private communication).

The second example concerns the benzene molecule in the $\pi$-electron approximation. The resonance integrals were considered only between nearest neighbors. Further, the approximation of zero differential overlap was used in evaluation of two-electron repulsion integrals. The calculation were carried out for two different sets of parameters[23,24] given in Table II. The calculations for Pariser and Parr parametrization are not presented here since the correlation effects in this approximation due to the higher-than-doubly-excited states are negligibly small (cf. Refs. 13, 14).

Using the approximations just indicated the complete configuration-interaction treatment was carried out.[15] The resulting secular equation was of the 22nd order. The correctness of our calculations was checked by comparison with results published in a paper from this laboratory.[14] Further, the configuration-interaction treatment of the benzene molecule considering only doubly excited states was carried out.[11]

The method suggested in this paper was applied[25] to this case using the same two approximations as in the first example. The system of nonlinear equations, having now eight unknown variables, was solved in the same way as above. The results obtained are shown in Table III.

It is apparent from Tables I and III that the configuration-interaction method limited to doubly excited states underestimates both $\mathcal{E}_{cor}$ and $q$ (in the case of $\mathcal{E}_{cor}$ this is the consequence of the variational principle).

On the other hand, in the above-presented examples, our approximation which yields a linear system of equations for the matrix elements of the operator $T_2$ overestimates both quantities $\mathcal{E}_{cor}$ and $q$.

Finally, intermediate results are obtained with the approximation leading to a nonlinear system of equations (54).

TABLE II. The values of integrals used in the study of the benzene molecule in $\pi$-electron approximation (all quantities are in electron volts).

| (eV) | $\beta^{c}$ | $\gamma_{00}{}^{d}$ | $\gamma_{01}{}^{d}$ | $\gamma_{02}{}^{d}$ | $\gamma_{03}{}^{d}$ |
|---|---|---|---|---|---|
| $M^{a}$ | −2.388 | 10.840 | 5.298 | 3.855 | 3.505 |
| $T^{b}$ | −2.734 | 17.618 | 8.924 | 5.574 | 4.876 |

[a] The parameters according to Mataga and Nishimoto.[21]

[b] The parameters obtained by transformation to the orthonormal Löwdin orbitals.[22]

[c] The resonance integrals between nearest neighbors.

[d]
$$\gamma_{\mu,\nu}=\iint r_{1,2}{}^{-1}[a_\mu(1)]^2[a_\nu(2)]^2dV_1dV_2,$$

where $a_\mu(i)$ is an orbital localized on the $\mu$th carbon atom.

[23] M. Mataga and K. Nishimoto, Z. Physik. Chem. (Frankfurt) **13**, 140 (1957).

[24] R. McWeeny, Proc. Roy. Soc. (London) **A227**, 288 (1959).

[25] J. Čížek and L. Šroubková (to be published).

TABLE III. The error[a] in correlation energy and in the quantity $q$ for different approximations used in the study of the benzene molecule in $\pi$-electron approximation.

| | % | $B^{d}$ | $L^{d}$ | $N^{d}$ |
|---|---|---|---|---|
| $M^{b}$ | $\mathcal{E}_{cor}$ | −7.5 | 5.0 | −0.3 |
| | $q$ | −19.8 | 12.5 | 0.6 |
| $T^{c}$ | $\mathcal{E}_{cor}$ | −12.9 | 12.9 | 0.7 |
| | $q$ | −29.7 | 37.3 | 7.0 |

[a] See Table I.

[b] The parameters according to Mataga and Nishimoto[21] have been used.

[c] "Theoretical parameters" used.[22]

[d] $B$, $L$, and $N$ have the same meaning as in Table I.

Both examples presented above should only illustrate the applicability of the proposed method. Further applications of this method are currently being undertaken.

## ACKNOWLEDGMENTS

The author wishes to express his sincere thanks to Professor J. Koutecký for his interest in this work and for stimulating discussions.

Further, it is a pleasure to thank Professor R. Daudel for his hospitality during the author's stay at Centre de Mécanique Ondulatoire Appliquée, Paris, as well as for useful discussions.

My special thanks are due to Dr. J. Paldus for detailed critical reading of the whole manuscript which was of great help in completing this paper. I am also indebted to him for writing out the program for solving the proposed system of equations.

Thanks are also due to Professor J. Plíva as well as to Dr. M. Tomášek and Dr. V. V. Tolmachev for valuable discussions.

Finally, I would like to thank Mrs. L. Šroubková for the help with calculations concerning the benzene molecule.

## APPENDIX

In order to prove the auxilliary theorem essential for our derivation of the method suggested in this paper, the generalized Wick theorem is used.

With the creation and annihilation operators defined in the main section the following four types of pairings are possible:

$$\cdot\hat{X}_B\cdot\hat{Y}_{A''}{}^{+}=\langle B\mid A''\rangle, \tag{A1}$$

$$\cdot\hat{X}_C{}^{+}\cdot\hat{Y}_{A'}{}^{+}=\langle C\mid A'\rangle, \tag{A2}$$

$$\cdot\hat{X}_B\cdot\hat{Y}_{A'}{}^{+}=0, \tag{A3}$$

$$\cdot\hat{X}_C{}^{+}\cdot\hat{Y}_{A''}{}^{+}=0. \tag{A4}$$

Therefore, only those normal products which contain the pairings of the types (A1) and (A2) will be different from zero.

Any normal product with pairings of the types (A1)

and (A2) may be depicted by an $R$ diagram, which is formed from the corresponding $H$ and $M$ diagrams by connecting lines carrying indices of contracted operators.

Since only contractions of the types (A1) and (A2) give nonzero contributions, we have to join the lines starting at some vertex with the lines entering some other or even the same vertex.

On the basis of Relations (A1) and (A2) it is immediately apparent why the lines starting at the vertex of the $M$ skeleton ($H$ skeleton) and entering the vertex of the $H$ skeleton ($M$ skeleton) are called particle (hole) lines and why these lines carry an index of the excited (nonexcited) spin–orbital in the corresponding $R$ diagram.

Noncontracted operators are then associated with the external lines of the $R$ diagram. Provided at least one of these operators is an annihilation operator a zero vector results on applying this normal product to the vector $|\Phi\rangle$. Therefore, only such $R$ diagram have to be considered in which indices of the excited (nonexcited) states are assigned to the external lines leaving (entering) the vertexes of the $H$ diagram.

The rule which determines the sign of the resulting expressions is illustrated by the following three typical examples:

$$l=0,\quad h=0: N[\hat{X}_{C_1}{}^{+}{}^{\cdot}\hat{X}_{B_1}{}^{\cdot}\hat{Y}_{A_1''}{}^{+}\hat{Y}_{A_1'}{}^{+}]=({}^{\cdot}\hat{X}_{B_1}{}^{\cdot}\hat{Y}_{A_1''}{}^{+})N[\hat{X}_{C_1}{}^{+}\hat{Y}_{A_1'}{}^{+}],$$
$$l=0,\quad h=1: N[{}^{\cdot}\hat{X}_{C_1}{}^{+}\hat{X}_{B_1}\hat{Y}_{A_1''}{}^{+}{}^{\cdot}\hat{Y}_{A_1'}{}^{+}]=-({}^{\cdot}X_{C_1}{}^{+}{}^{\cdot}\hat{Y}_{A_1}{}^{+})N[\hat{Y}_{A_1''}{}^{+}\hat{X}_{B_1}],$$
$$l=1,\quad h=1: N[{}^{\cdot}\hat{X}_{C_1}{}^{+}{}^{\cdot\cdot}\hat{X}_{B_1}{}^{+}{}^{\cdot\cdot}\hat{Y}_{A_1''}{}^{+}{}^{\cdot}\hat{Y}_{A_1'}{}^{+}]=({}^{\cdot}\hat{X}_{C_1}{}^{+}{}^{\cdot}\hat{Y}_{A_1'}{}^{+})({}^{\cdot\cdot}\hat{X}_{B_1}{}^{\cdot\cdot}\hat{Y}_{A_1''}{}^{+}). \tag{A5}$$

Among the $R$ skeletons corresponding to all possible normal products with pairings one can generally find topologically equivalent skeletons.

It can be proved generally, that $g$ equivalent skeletons $R(r)$ with the given structure $r$ will result among all possible $R$ skeletons formed from the $H(h)$ and $M(m)$ skeletons, where

$$g=w_R(r)/w_H(h)w_M(m). \tag{A6}$$

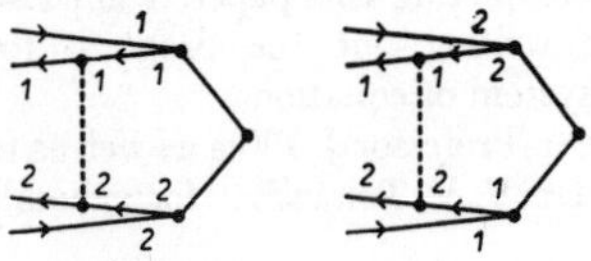


FIG. 9. Examples of $R$ skeletons formed from an $S_H(1)$ skeleton and an $M$ skeleton consisting of a single $S_T(2)$ skeleton.

We demonstrate this statement on a typical example. Let us consider the $S_H(1)$ skeleton (Fig. 1) and the $M$ skeleton consisting of a single $S_T(2)$ skeleton (Fig. 3, $j=2$). Both of these skeletons have the weight equal to $\frac{1}{2}$. The examples of skeletons which one can form from the skeletons just mentioned are shown in Fig. 9. It is immediately apparent that the skeletons shown in Fig. 9 are mutually equivalent, both having the weight $\frac{1}{2}$, so that we indeed obtain $g=2$.

# Chapter IX

# *Electron Correlation in Excited States (and Approach I, Part 3)*

## IX-1

The Hartree–Fock orbital theory itself gets considerably more complicated in nonclosed shells, excited states, as compared to closed shells (see Chapter II). It is not surprising, therefore, that correlation effects, too, are more complicated in excited states, and there are new types of them not found in closed shells. In conventional calculation methods such as the configuration interaction (C.I.), each excited state is treated as a specific problem in itself. Guidelines, in general, have not been available which would systematize the correlation effects in excited states. Accurate prediction of properties such as excitation energies and transition probabilities requires the inclusion of correlation effects.

Sinanoğlu and co-workers have recently developed a theory applicable to electron correlation in excited states and have used it to predict various atomic properties. In the theory, a convenient and physically meaningful H.F. method is selected as the starting point (see Chapter IX-2). For many atomic states, this initial function can be taken as the restricted Hartree–Fock (RHF) of the Roothaan (Chapter II-4) type. For other applications, as in the dissociation of diatomic molecules, a more general H.F. method, including several configurations, would be more suitable (GRHF; see Chapter II).

In the initial stage of the theory (Chapter IX-2), as in Approach I, Part 1, the closed-form Rayleigh–Schrödinger perturbation theory is used. The formal solution of the first-order equation yields different types of correlation effects in excited states in a form separated from one another. The theory is later generalized to a nonperturbation one in which various effects are calculable separately and to all orders by variational methods (Chapter IX-4; see also O. Sinanoğlu, in *Atomic Physics*, V. Hughes et. al., eds., Plenum Press, New York, 1969).

For closed shells such as the ground state of the neon atom, Sinanoğlu had shown by similar methods, as discussed in the previous chapters of this book, that the problem could be treated in terms of N(N − 1)/2 "helium-like" "pair correlations." Not only strictly closed shell states, but all ground states of the lighter atoms could be treated this way by only a very slight extension of the closed-shell theory (Chapter VII-5). Although the ground states of atoms such as carbon, nitrogen, and oxygen are also open shells, they are, nevertheless, described in terms only of a single determinant H.F. wave function, so that with slight modification they, too, can be and were treated by the single determinantal many-electron theory (MET) in terms of N(N − 1)/2 *total* pair correlations. The real complications of general non-closed shells, as in many excited states, do not come up with such ground states. The main difference from closed shells in these "quasi-closed shells," high spin states, is that some of the pair correlations become "nondynamical," i.e., strongly dependent on the total number electrons, N, and on Z, owing to specific near-degeneracy effects. The overall pair-correlation functions in single-determinantal ground states are "orbital-orthogonal" to only the occupied H.F. spin orbitals contained in the single determinant.

By contrast, in general non-closed-shell theory, it is found that there are not two, but three, about equally important types of correlation effects.

The "orbital-orthogonality" becomes more general, as the H.F. "sea" now includes all the occupied or unoccupied RHF spin orbitals that come from the same shell. For example, in the ground-state single-determinantal theory, the carbon $^3P$ state H.F. "sea" would be taken to be just the $1s_\alpha$, $1s_\beta$, $2s_\alpha$, $2s_\beta$, $2p_+\alpha$, and $2p_0\alpha$, constituting the single determinant $\phi_{RHF}$. The same state, looked at in the more general non-closed-shell approach, has as its H.F. "sea" all 1s, 2s, and 2p spin orbitals—10 of them. The various correlation functions involving different numbers of electrons are "orbital-orthogonal" now to all these 10 spin orbitals, not to just to the ones occupied in the $\phi_{RHF}$. One sees readily that with this general non-closed-shell theory different excited states arising from configurations such as $1s^k 2s^m 2p^n$ are described the same way, whereas the previous theory holds only for single-determinantal states.

In the general non-closed-shell theory, the exact N-electron wave function now becomes

$$\psi = \phi_{RHF} + \chi_{(\text{semi-int and polariz.})} + \chi_{\text{all-ext}} \tag{1}$$

with all parts orthogonal to one another. The "internal" correlation, $\chi_{int}$, is the more familiar of the three correlation effects. It comes from the mixing of the configurations that can be made from the spin orbitals of the H.F. "sea." (The "nondynamical" pair correlations, such as the $\epsilon(2s^2)$ of the single-determinantal ground-state theory, now become a part of this internal correlation effect.)

"Semi-internal correlation," $\chi_{\text{semi-int}}$, an effect first introduced in this theory, turns out to be as important as the other two types. It has not been singled out and studied previously as an effect in the usual configuration interaction calculations. The dominant terms in it are again of the pair-correlation type, where two electrons go from spin orbitals, i, j, occuped in the H.F. "sea" to an unoccupied H.F. spin orbital, $\ell$, and a semi-internal correlation function, $\hat{f}_{ij;\ell}$ outside the sea. Thus

$$ij \longrightarrow \hat{f}_{ij;\ell} \qquad \langle \hat{f}_{ij;\ell} | m \rangle = 0 \tag{2}$$

where m runs in the first row from $1s_\alpha$ to $2p_z\beta$. This effect may be obtained by the solution of integrodifferential equations (see Chapter IX-2), especially if one limits himself to getting it to first order in perturbation theory. However, in the nonperturbation version of the theory (Chapter IX-4), it is shown that the semi-internal correlations are obtained more completely and quite simply by a finite C.I. calculation.

The H.F. orbitals in nonclosed shells are individually polarized in excited states other than $^1S$, owing to nonspherical symmetry. These one-electron "orbital-average polarization" effects are much smaller than the other correlation effects involving pairs of electrons in the system. The "orbital-polarization" effects $\hat{f}_p$ are coupled by symmetry to the semi-internal correlations $\hat{f}_{ij;\ell}$. They are therefore calculated together in the same finite C.I. calculation (Chapter IX-4).

Both the internal and the semi-internal correlation effects are very specific in their magnitudes to each state. They depend strongly on the symmetry of the state, on total N, and on Z. These are the effects which make the correlations change quite a bit from one state to another, even within the same configuration (see the question of Wigner on this point in Chapter IV-2). In the non-closed-shell theory by Sinanoğlu and co-workers, *it is predicted that once these specific correlation effects are calculated and taken out of the energy and wave function of an excited state, the remaining effects become just like those in closed shells.* That is, the remaining "all-external" correlation energy should be made up of just additive pair correlations, $\epsilon_{ij}$, weighted, now, by group theoretic fractional pair occupation probabilities. Further, the same $\epsilon_{ij}$ values should be insensitive to the total number of electrons, N, the overall symmetry of the state, and, to some extent, to Z. Thus a single set of $\epsilon_{ij}$'s should be usable over and over for the hundreds of states that would come up from various configurations, say, of the type $1s^k 2s^m 2p^n$ of the different ions (N) of a certain atom (Z). The number of states that can be treated together this way are determined by $n + m + k = N$ and $1 \leqslant N \leqslant Z + 1$. These "all-external" correlations, the third effect in Eq. (1), thus includes mainly the functions $\hat{u}_{ij}$ outside the H.F. "sea," "orbital-orthogonal" now to all the occupied or unoccupied H.F. spin orbitals.

The theory has been applied extensively recently, as summarized in part in Chapter IX-4.† The set of "all-external" correlations for the atoms boron through sodium is obtained. The theory is used to predict additional states and properties.

It should be noted that the all-external pair correlations are different from the "total pair correlations" one would get in single-determinantal non-closed-shell ground states. In the latter, the internal and semi-internal correlation energies obtained by finite C.I. can be decomposed into contributions coming from different pairs of spin orbitals. Thus a *total* "pair-correlation energy" in the carbon ground state has an "all-external," an "internal," and a "semi-internal" contribution. These are added together to compare the result with the single-determinantal states' pair correlations obtained earlier in the literature by McKoy, Sinanoğlu, Kelly, and others, which are invariable of the *total-pair-correlation type*.

To calculate an energy property such as electron affinity or excitation energy in an atom, first an automatic program (Chapter IX-4) is used to calculate by finite C.I. the internal and semi-internal correlation (a modest calculation compared to the kinds of a priori calculations carried out on the computer these days). To these the total "all-external" correlations energy of the state obtained from the group theoretic pair occupation probability coefficients and from the table of "all-external pair correlations" is added. Energy properties obtained this way are found to agree well with experiment, even where traditional methods give large errors.

In the traditional use of the C.I. method as such, one has infinitely many configurations to select from and to keep adding to improve a result. In such calculations the three types of correlation effects are not separate, and a given C.I. result in the literature may include some contribution from each, or it may include terms belonging in the slowly convergent all-external correlation with small contributions but omit significant terms belonging in the semi-internal correlation. The pair-correlation functions $\hat{u}_{ij}$, which occur in the all-external correlation function as in closed shells, are calculable in the theory by Sinanoğlu, also nonempirically and as in closed shells by Hyllereas-like methods with the $r_{ij}$ coordinate. Calculations of this type in excited states have not been carried out so far, however. They would be useful in ascertaining the importance of neglected effects.

Chapter IX-3 by Weiss is possibly the most extensive fully nonempirical C.I. calculation on an atom along the more traditional lines, affording useful comparisons with theory.

The charge distribution in an atomic state is given primarily by the internal and semi-internal wave functions added to the H.F. result. Sinanoğlu has, therefore, made the hypothesis that a property such as transition probabilities would also be given well with just these parts of the wave function,

$$\psi_c \sim \phi_{\mathrm{RHF}} + \chi_{\mathrm{int}} + \chi_{(\text{semi-int and polariz.})}$$

This hypothesis was tested on many atomic states by Westhaus and Sinanoğlu [*Phys. Rev.* **183**, 56 (1969); *Astrophys. J.* **157**, 997 (1969)]. With the detailed wave functions containing only these effects for both ground and excited states, they calculated the allowed transition probabilities, the multiplet oscillator strengths. The results show excellent agreement with the

†For extensive results and applications to the energy levels and transition probabilities of the atoms boron through sodium and their positive and negative ions, see a series of papers by Öksüz and Sinanoğlu and by Westhaus and Sinanoğlu in *Phys. Rev.* **181**, 42, 54 (1969); **183**, 56 (1969).

**Table 1.** Absorption Oscillator Strengths (f) for the Multiplets Computed Using the Dipole Length Operator

| *Species* | *Transitions* | λÅ | *f* (*RHF*) | *f* (*Z*)[b] | *f* (*NBS*) | *f* (*MET*)[c] | *f* (*EXP*) |
|---|---|---|---|---|---|---|---|
| CII | $1s^2\,2s^2\,2p\,^2P \longrightarrow 1s^2\,2s2p^2\,^2D$ | 1335 | 0.263 | 0.204 | 0.17[d] (0.121)[e] | 0.125 | 0.114(±0.011)[f] |
| NII | $1s^2\,2s^2\,2p^2\,^3P \longrightarrow 1s^2\,2s2p^3\,^3D$ | 1085 | 0.236 | 0.192 | 0.17[g] | 0.100 | 0.109(±0.011)[f] 0.101(±0.006)[h] |
| NII | $1s^2\,2s^2\,2p^2\,^3P \longrightarrow 1s^2\,2s2p^3\,^3P$ | 916 | 0.170 | 0.213 | 0.22[g] | 0.137 | 0.131(±0.007)[h] |
| NII | $1s^2\,2s^2\,2p^2\,^3P \longrightarrow 1s^2\,2s2p^3\,^3S$ | 645 | 0.334 | 0.244 | 0.23[g] | 0.218 | 0.189(±0.016)[h] |
| NIII | $1s^2\,2s^2 2p\,^2P \longrightarrow 1s^2\,2s2p^2\,^2D$ | 991 | 0.213 | 0.167 | 0.18[i] | 0.114 | 0.103(±0.010)[h] |
| NIII | $1s^2\,2s^2 2p\,^2P \longrightarrow 1s^2\,2s2p^2\,^2P$ | 686 | 0.577 | 0.415 | 0.45[i] | 0.399 | 0.416(±0.075)[h] |
| OIII | $1s^2\,2s^2 2p^2\,^3P \longrightarrow 1s^2 2s2p^3\,^3D$ | 834 | 0.200 | 0.162 | 0.15[g] | 0.100 | 0.102(±0.002)[j] |
| OIII | $1s^2\,2s^2\,2p^2\,^1D \longrightarrow 1s^2 2s2p^3\,^1D$ | 600 | 0.534 | | 0.37[g] | 0.297 | – – |
| OIV | $1s^2\,2s^2 2p\,^2P \longrightarrow 1s^2 2s2p^2\,^2D$ | 789 | 0.179 | 0.141 | 0.15[i] | 0.106 | 0.091(±0.002)[j] |

[a]The values listed in column 4 are obtained with restricted Hartree–Fock (RHF) wave functions; those in column 6 come from various calculations as indicated by the references. The values in column 7 are computed with wave functions of the theory MET by Sinanoğlu and co-workers. The last column lists the available experimental results.

[b]M. Cohen and A. Dalgarno, *Proc. Roy. Soc.* (*London*) **A280**, 258 (1964).

[c]P. Westhaus and O. Sinanoğlu, *Phys. Rev.* (*ca.* July 1969).

[d]A. W. Weiss (reported in *NBS Tables*).

[e]A. W. Weiss, *Phys. Rev.* **162**, 71 (1967).

[f]G. M. Lawrence and B. D. Savage, *Phys. Rev.* **141**, 67 (1966).

[g]A. B. Bolotin, I. B. Levinson, and L. I. Levin, *Soviet Phys. JETP* **2**, 391 (1956).

[h]L. Heroux, *Phys. Rev.* **153**, 156 (1967).

[i]A. B. Bolotin and A. P. Yutsis, *Soviet Phys. JETP* **24**, 537 (1953).

[j]W. S. Bickel, *Phys. Rev.* **162**, 7 (1967).

recent experimental values. A summary table comparing the results of the theory with experiment and with other methods is given as Table 1.

The newer non-closed-shell theory above should be useful in obtaining other atomic properties, such as electron shake-off probabilities and the prediction of resonances in collision processes.

The calculation of excited states by other approaches, such as diagrammatic and field theoretic theory, becomes difficult in a general non-closed-shell state because of the need to introduce a degenerate vacuum, a problem being tackled by Tolmachev and Sandars.

The present non-closed-shell theory was applied only as far as sodium, because beyond that the relativistic effects are expected to become very significant. They cannot be added as minor corrections, but require the use of a new approach.

The different types of correlation effects mentioned also have their counterparts in molecules and contribute significantly to dissociation energies. However, no attempt has yet been made to apply the new information to molecular problems.†

†An attempt is now underway in cooperation with A. C. Wahl of the Argonne National Laboratory.

IX-2 Reprinted from *Journal of Chemical Physics* **44**, 1899 (1966)

# Many-Electron Theory of Nonclosed-Shell Atoms and Molecules. I. Orbital Wavefunction and Perturbation Theory*

HARRIS J. SILVERSTONE† AND OKTAY SINANOĞLU‡
*Sterling Chemistry Laboratory, Yale University, New Haven, Connecticut*
(Received 15 February 1965)

**A theory of electron correlation in nonclosed-shell states, such as excited states of atoms and molecules, triplet states, free radicals, and transient species, is developed. A general (multiconfiguration) restricted Hartree–Fock (GRHF) wavefunction is taken as the starting point. When near degeneracies are not strong enough to cause large orbital and charge deformations, Roothaan-type restricted Hartree–Fock (RHF) open-shell orbital theories are used. The first-order Schrödinger equation with RHF as the zeroth-order wavefunction is solved. Various correlation processes (near-degeneracy-type internal correlations, semi-internal pair correlations in which only one electron is ejected out of the Hartree–Fock sea, external pair correlations as in closed shells, and spin and symmetry orbital average polarization effects) are obtained in a form separated from one another. The second-order energy also includes "cross-pair correlation effects," an open-shell phenomenon that need be taken into account in the evaluation of electronic spectral levels and stability of open-shell systems. The first-order solution yields the main open-shell correlation processes, which are generalized to the variational theory in II with means for evaluation of the separate effects.**

## I. INTRODUCTION

A PREVIOUS series of papers[1–6] was concerned with the calculation of the many-electron wavefunction and energy of an atom or molecule, including correlation effects, to within chemical accuracy (~0.1 eV). The theory was developed primarily for closed-shell states, i.e., states for which the orbital approximation is a single (normalized) Slater determinant $\phi_0$ not degenerate with any other. The closed-shell many-electron theory can be applied to a few nonclosed-shell states as well,[5] namely single determinantal ones ($^3P$-C, $^4S$-N, etc.), if certain "orbital average polarization" effects are neglected.

The treatment of electronic spectra, of interatomic potential-energy curves and surfaces, and of the properties of radicals, triplets, and transient species requires a theory applicable to a general open-shell state. (The terms "open shell" and "nonclosed shell" are used here synonymously.)

The closed-shell theory cannot be applied to open-shell states without modification for two reasons: (1) There are important correlation effects in open-shell states not present in closed-shell states, and (2) difficulties arise as to which orbitals the correlation functions must be orthogonal to, when the orbitals are only partially occupied in the open-shell orbital approximation $\phi_0$. In the present two papers (I and II) we give a many-electron theory[7] applicable to a general open-shell state.

The approach used allows—as in the closed-shell case—the separate evaluation of parts of an atom or molecule providing the "environmental" effects of other electrons on a given part.[4–6] It is also suitable for semiempiricism. Nonempirically, the parts may be calculated by various methods including two-electron ones like the use of "$r_{12}$ coordinates."[4,5]

In Paper I, taking a suitably selected open-shell-orbital wavefunction as zeroth order, we derive the wavefunction (wf) to first order of perturbation theory. This solution indicates and separates the physically important correlation effects in the wavefunction and energy of the open-shell many-electron system. In Paper II, the results are extended to make the first-order (wf) processes "exact to all orders," to include unlinked clusters, and to obtain the complete form of the energy as an upper bound. New features of correlation that arise in open shells absent in closed shells, nonempirical evaluations, semiempirical implications of the form of the energy and some examples (e.g., $^2S$-Li, $^1S$-C and $C_2H_4^+$) are also discussed in Paper II.

## II. MATHEMATICAL FORM OF THE EXACT WAVEFUNCTION

The first step in calculating an $N$-electron wavefunction $\psi$ is to determine an orbital approximation $\phi_0$. For open-shell states, the general $\phi_0$ is a linear combination of (normalized) $N$-electron Slater determinants $\Delta_K$:

$$\Delta_K = \mathcal{A}[k_1(\mathbf{x}_1)k_2(\mathbf{x}_2)\cdots k_N(\mathbf{x}_N)] \tag{1}$$

$$\equiv \mathcal{A}(k_1k_2\cdots k_N). \tag{2}$$

* Supported by a grant from the National Science Foundation.
† National Science Foundation Postdoctoral Fellow. Present address: Department of Chemistry, The Johns Hopkins University, Baltimore, Maryland.
‡ Alfred P. Sloan Fellow.
[1] O. Sinanoğlu, J. Chem. Phys. **36**, 706 (1962).
[2] O. Sinanoğlu, J. Chem. Phys. **36**, 3198 (1962).
[3] O. Sinanoğlu and D. F. Tuan, J. Chem. Phys. **38**, 1740 (1963).
[4] D. F. Tuan and O. Sinanoğlu, J. Chem. Phys. **41**, 2677 (1964).
[5] V. McKoy and O. Sinanoğlu, J. Chem. Phys. **41**, 2689 (1964).
[6] O. Sinanoğlu, Advan. Chem. Phys. **6**, 315 (1964).
[7] H. J. Silverstone and O. Sinanoğlu, "Many-Electron Theory of Nonclosed-Shell Atoms and Molecules. II. Variational Theory," J. Chem. Phys. (to be published); for a qualitative description, see H. J. Silverstone and O. Sinanoğlu, in *Modern Quantum Chemistry—Istanbul Lectures* (Academic Press Inc., New York, to be published).

$\mathcal{A}$ is the antisymmetrization operator,

$$\mathcal{A}=\sum_P(-1)^P P.$$

The $N$ spin orbitals in $\Delta_K$, $[k_1, k_2, \cdots, k_N]$ are chosen from a set of $M$ spin orbitals $[1(\mathbf{x}), 2(\mathbf{x}), \cdots, M(\mathbf{x})]$, so that there are in general $\binom{M}{N}$ Slater determinants which may contribute to $\phi_0$ (although in actuality, many of the coefficients $C_K$ are zero for reasons of symmetry);

$$\phi_0=\sum_{K=1}^{\binom{M}{N}} C_K\Delta_K. \tag{3}$$

The question of how the spin orbitals should be chosen [e.g., Hartree–Fock (HF) or Slater-type orbitals, etc.] is taken up in the next section; in this section they may be regarded as chosen in a completely arbitrary manner.

The difference $\chi$ between the exact wavefunction and the orbital approximation $\phi_0$ is called the correlation part of the wavefunction. $\chi$ is unambiguously defined by

$$\chi=\psi-\phi_0, \tag{4}$$

$$\langle\phi_0, \phi_0\rangle=1, \tag{5}$$

$$\langle\phi_0, \chi\rangle=0. \tag{6}$$

It is convenient to separate $\chi$ into two parts, $\chi_{\text{int}}$ and $\chi_{\text{ext}}$:

$$\chi=\chi_{\text{int}}+\chi_{\text{ext}}, \tag{7}$$

where

$$\langle\chi_{\text{int}}, \chi_{\text{ext}}\rangle=0. \tag{8}$$

The internal correlation part of $\chi$, $\chi_{\text{int}}$, is that part of $\chi$ which can be expressed as a linear combination of the $\binom{M}{N}$ Slater determinants $\Delta_K$;

$$\chi_{\text{int}}=\sum_{K=1}^{\binom{M}{N}} d_K\Delta_K. \tag{9}$$

$\chi_{\text{int}}$ is orthogonal to $\phi_0$;

$$\langle\phi_0, \chi_{\text{int}}\rangle=0 \tag{10}$$

$$=\sum_{K=1}^{\binom{M}{N}} C_K{}^*d_K. \tag{11}$$

The external correlation $\chi_{\text{ext}}$, is the rest of $\chi$. In configuration-interaction (CI) language, the configurations which contribute to $\chi_{\text{ext}}$ (by definition) are those which involve one or more spin orbitals outside of (external to) the set of $M$ spin orbitals.

The exact $\chi_{\text{ext}}$ can be written (see Appendix A; cf. Ref. 8) as a sum of terms, each of which has a certain number of electrons, $N-n$, in $N-n$ of the $M$ spin orbitals $[k_1, k_2, \cdots, k_{N-n}$; denoted by $K_{N-n}]$ and $n$ electrons in a correlation function $\hat{U}'_{(n)K_{N-n}}$;

$$\chi_{\text{ext}}=\sum_{n=1}^{N}\{\sum_{K_{N-n}=1}^{\binom{M}{N-n}}\mathcal{A}[k_1(\mathbf{x}_1)k_2(\mathbf{x}_2)\cdots k_{N-n}(\mathbf{x}_{N-n})(n!)^{-\frac{1}{2}}\hat{U}'_{(n)K_{N-n}}(\mathbf{x}_{N-n+1}, \cdots, \mathbf{x}_N)]\} \tag{12}$$

$$=\sum_{n=1}^{N}\{\sum_{K_{N-n}=1}^{\binom{M}{N-n}}\mathcal{A}[k_1k_2\cdots k_{N-n}(n!)^{-\frac{1}{2}}\hat{U}'_{(n)K_{N-n}}]\}. \tag{13}$$

The $n$-electron correlation functions $\hat{U}'_{(n)K_{N-n}}$ are completely antisymmetric and they are one-electron orthogonal to all $M$ spin orbitals in the set associated with $\phi_0$;

$$U'_{(n)K_{N-n}}(\mathbf{x}_{N-n+1}, \mathbf{x}_{N-n+2}, \cdots)=-\hat{U}'_{(n)K_{N-n}}(\mathbf{x}_{N-n+2}, \mathbf{x}_{N-n+1}, \cdots), \text{ etc.}, \tag{14}$$

$$\langle l, \hat{U}'_{(n)K_{N-n}}\rangle=\int l^*(\mathbf{x}_a)\,U'_{(n)K_{N-n}}dV_{\mathbf{x}_a}=0$$
$$(l=1, 2, \cdots, M;\quad a=N-n+1, N-n+2, \cdots, N). \tag{15}$$

In CI language, Eq. (15) is the statement that $U'_{(n)}$ is made up of $n$-electron configurations in which none of the spin orbitals are in the set $M$. ("Set $M$" ≡ the spin orbitals 1, 2, $\cdots$, $M$.) The $(n!)^{-\frac{1}{2}}$ in Eqs. (12) and (13) is a normalization convention, the subscript $(n)$ indicates the number of electrons of which $U'_{(n)K_{N-n}}$ is a function, the caret (^) denotes the one-electron orthogonality [Eq. (15)], and the prime (′) relates to unlinked clusters,[6] which will be discussed in Paper II. In Eqs. (1), (2), (12), and (13) the spin orbitals (denoted by $K_{N-n}$) appear in strictly ascending order: $k_1<k_2<\cdots<k_{N-n}$. [There are $\binom{M}{N-n}$ different combinations of $N-n$ spin orbitals.]

Equations (7), (9), and (13) combined, give the exact $\chi$ in orbital-correlation-function language:

$$\chi=\sum_{K=1}^{\binom{M}{N}} d_K\Delta_K+\sum_{n=1}^{N}\sum_{K_{N-n}=1}^{\binom{M}{N-n}}\mathcal{A}\{k_1k_2\cdots k_{N-n}\hat{U}'_{(n)K_{N-n}}\}. \tag{16}$$

[8] O. Sinanoğlu, Rev. Mod. Phys. **35**, 517 (1963).

Equation (16) is an identity. Once the set $M$ and $\phi_0$ are specified, the decomposition (16) is unique. If one had the exact wavefunction $\psi$, the coefficients $d_K$ and the correlation functions $\hat{U}'_{(n)K_{N-n}}$ could be determined by the method of successive partial orthogonalizations, as in the closed-shell theory.[6,8] (This is shown in Appendix A.) On the other hand, if $\chi$ is not known, we may calculate $\chi$ (exactly or approximately) by calculating $d_K$, $\hat{U}'_{(1)K_{N-1}}$, $\hat{U}'_{(2)K_{N-2}}$, etc. (e.g., by perturbation theory, the variation–perturbation method,[6] etc., see Sec. IV).

Compare now $\chi$, Eq. (16), with the exact $\chi$ of the closed-shell theory [Ref. 1, Eq. (23)]. Since the closed shell $\phi_0$ is a single determinant [$M=N$ for closed-shell states, and $\binom{M}{N}=1$], there is no internal correlation. Hence $\chi_{\text{int}}$ has no closed-shell counterpart. If one calls the collection of spin orbitals $\{l \mid l=1, 2, \cdots, M\}$ the "spin–orbital sea," then $\chi_{\text{int}}$ has no electrons outside of the sea. The closed-shell sea is filled, but the open-shell sea is only partially filled. The closed-shell $n$-electron correlation functions[1] $\hat{U}'_{i_1,i_2,\cdots,i_n}$ represent "collisions" (correlations) of the $n$ electrons originally in spin orbitals $i_1$, $i_2$, $\cdots$, $i_n$. By analogy, $n$-electron open-shell correlation functions $\hat{U}'_{(n)K_{N-n}}$ can represent correlations of $n$ electrons. However, there may be more than one determinant in $\phi_0$ in which the $(N-n)$ spin orbitals $K_{N-n}$ occur, so that $\hat{U}'_{(n)K_{N-n}}$ cannot (in general) be associated with a single set of $n$ spin orbitals occupied in $\phi_0$. Thus the $\hat{U}'_{(n)K_{N-n}}$ must be composite correlation functions.

In the closed-shell case, the exact $\chi$ leads to the detailed variational form of the exact $E$. Then by the variation–perturbation approach[6] large portions of this are minimized to get the approximate $\chi_s'$ and $E_s'$ of the many-electron theory. For open shells the form of the exact $\chi$ obtained above is not sufficiently detailed into physically meaningful processes and therefore does not indicate the parts to be made stationary according to the variation–perturbation method. We therefore turn to the alternative derivation. We solve the first-order Schrödinger equation which gives the more detailed form of $\chi$ to first order ($\chi^{(1)}$). Then $\chi^{(1)}$ is generalized in Paper II to make the resulting effects exact to "all orders" and to include unlinked clusters to yield the open-shell many-electron theory.

Before obtaining $\chi$ to first order, it is necessary to discuss the nature and choice of the zeroth-order wavefunction $\phi_0$.

## III. ORBITAL APPROXIMATION

The choice of $\phi_0$, the orbital approximation, is crucial in the many-electron theory. The closed shell $\phi_0$ was chosen as the Hartree–Fock (HF) determinant for quantitative reasons.[1,3,6] For open-shell states, there is not one, but several self-consistent-field (SCF) Hartree–Fock procedures, and a choice most suitable for a correlation theory is required.

### $\phi_0$ Chosen for the Open-Shell Many-Electron Theory

Several considerations enter into the choice of $\phi_0$ most appropriate for an open-shell many-electron theory:

(1) $\phi_0$ must have the same symmetry properties as $\chi$ so that $\chi$ does not have "correlation effects" which merely correct the symmetry properties of $\phi_0$.

(2) $\phi_0$ must have the lowest energy of all functions of the same form, otherwise $\chi$ would contain terms which turn the orbitals in $\phi_0$ into the orbitals of a "better" $\phi_0$.

(3) $\phi_0$ should have a sufficiently general form if one is to be able to include especially the strong near-degeneracy resonances.

(4) To be practical, the orbitals should be symmetry orbitals.

A fifth consideration concerns the nature of unlinked clusters in $\chi$ and will be discussed more appropriately in Sec. II of Paper II.

The $\phi_0$ which meets all of these requirements is the most general multiconfigurational wavefunction that can be made from the set of $M$ spin orbitals, the general restricted Hartree–Fock (GRHF) wavefunction.[9] Both the coefficients $C_K$ [Eq. (3)] and the $M$ spin orbitals are chosen to minimize $\langle\phi_0, H\phi_0\rangle/\langle\phi_0, \phi_0\rangle$. The spin orbitals are restricted to be symmetry orbitals and to be orthogonal. [For example, in each of the atoms, B, N, C, O, F, and Ne, the spin–orbital sea for the ground state consists of 10 spin orbitals ($1s\alpha$, $1s\beta$, $2s\alpha$, $2s\beta$, $2p_+\alpha$, $2p_+\beta$, $2p_0\alpha$, $2p_0\beta$, $2p_-\alpha$, $2p_-\beta$); but for each atomic state three radial functions completely specify the sea.] The $C_K$ are restricted so that $\phi_0$ has the same symmetry properties as $\psi$.

For a discussion of other types of open-shell HF methods which satisfy some of the requirements above, the reader is referred to the literature.[10]

### Computation of $\phi_{\text{GRHF}}$

Computer programs that can calculate $\phi_{\text{GRHF}}$ for atoms exist[11] and are in preparation.[12] Nevertheless, computational methods for the single configurational restricted Hartree–Fock (RHF) wavefunction have

---

[9] Variants of the GRHF method select a few of the excited configurations to mix with the ground configuration. See (a) D. R. Hartree, W. Hartree, and B. Swirles, Phil. Trans. Roy. Soc. **A238**, 229 (1939); (b) A. P. Yutsis, Zh. Eksperim. i Teor. Fiz. **23**, 129 (1952); (c) A. P. Yutsis, V. V. Kibartas, and I. I. Glembotskiy, *ibid.* **27**, 425 (1954); (d) V. V. Kibartas, V. I. Kavetskis, and A. P. Yutsis, Soviet Phys.—JETP **2**, 481 (1956) [Zh. Eksperim. i Teor. Fiz. **29**, 623 (1955)]; (e) R. E. Watson, Ann. Phys. (N.Y.) **13**, 250 (1961); (f) T. L. Gilbert (preprint).

[10] For a discussion of other HF methods as starting points for a correlation theory, see Ref. 6, p. 363, and (a) O. Sinanoğlu, J. Phys. Chem. **66**, 2283 (1962); (b) O. Sinanoğlu and D. F. Tuan, Ann. Rev. Phys. Chem. **15**, 251 (1964), Sec. II-A 1b.

[11] H. W. Joy (private communication).

[12] C. C. J. Roothaan (private communication from P. E. Cade).

been extensively developed by Roothaan and others.[13–18] A simplified approach is to use the RHF spin orbitals and $\phi_{\mathrm{RHF}}$ to estimate $\phi_{\mathrm{GRHF}}$ through

$$\phi_{\mathrm{GRHF}} \approx (\phi_{\mathrm{RHF}} + \chi_{\mathrm{int}}^{\mathrm{RHF}}) / (1 + \langle \chi_{\mathrm{int}}^{\mathrm{RHF}}, \chi_{\mathrm{int}}^{\mathrm{RHF}} \rangle)^{\frac{1}{2}}, \tag{17}$$

where $\chi_{\mathrm{int}}^{\mathrm{RHF}}$ [i.e., the coefficients $d_K$ in Eq. (9)] is calculated by near-degeneracy-type CI. The less the internal correlation (relative to $\phi_{\mathrm{RHF}}$) affects the spin orbitals—i.e., the closer the RHF spin orbitals approximate the GRHF spin orbitals—the more accurate is Eq. (17). We expect Eq. (17) to be a valid approximation for first-row atoms.[19]

### Internal and Nondynamical Correlation

For some systems $\phi_{\mathrm{RHF}}$ and $\phi_{\mathrm{GRHF}}$ are identical (e.g., lithium ground state), while for others (e.g., carbon $^3P$ state) they are different;

$$\phi_{\mathrm{RHF}}(\mathrm{Li}) = \phi_{\mathrm{GRHF}}(\mathrm{Li}) = \mathcal{A}\{(1s\alpha)(1s\beta)(2s\alpha)\}, \tag{18}$$

$$\phi_{\mathrm{RHF}}(\mathrm{C}\text{-}{}^3P) = \mathcal{A}\{(1s\alpha)(1s\beta)(2s\alpha)(2s\beta)(2p_+\alpha)(2p_0\alpha)\}, \tag{19}$$

$$\begin{aligned}\phi_{\mathrm{GRHF}}(\mathrm{C}\text{-}{}^3P) = \mathcal{A}\{&C_1(1s\alpha)(1s\beta)(2s\alpha)(2s\beta)(2p_+\alpha)(p_0\alpha)\\ &+C_2(1s\alpha)(1s\beta)(2p_-\alpha)(2p_+\beta)(2p_+\alpha)(2p_0\alpha)\\ &+C_3(2p_-\alpha)(2p_+\beta)(2s\alpha)(2s\beta)(2p_+\alpha)(2p_0\alpha)\\ &+C_4([(1s\alpha)(2s\beta)+(2s\alpha)(1s\beta)]/\sqrt{2})\\ &\times(2p_-\alpha)(2p_+\beta)(2p_+\alpha)(2p_0\alpha)\}.\end{aligned} \tag{20}$$

The major difference of physical importance between $\phi_{\mathrm{GRHF}}$ and $\phi_{\mathrm{RHF}}$ here is that $\phi_{\mathrm{GRHF}}$ includes the nondynamical[3–5] (near-degeneracy) correlations but $\phi_{\mathrm{RHF}}$ does not. The $(1s)^2(2p)^4$ term ($C_2$) in $\phi_{\mathrm{GRHF}}$ represents the nondynamical $(2s)^2$ correlation.[5] The $(1s2s)(2p)^4$ and $(2s)^2(2p)^4$ terms represent contributions to $\phi_{\mathrm{GRHF}}$ formally similar to the nondynamical $(2s)^2$ correlation, but not important in magnitude ($|C_1| > |C_2| \gg ||C_4 > |C_3|$). In this example where the closed-shell type of theory can be applied,[5] they would be part of the $(1s2s)$ and $1s^2$ dynamical correlations. Thus $\chi_{\mathrm{int}}^{\mathrm{RHF}}$ represents a generalization of nondynamical correlation, which is then incorporated into $\phi_{\mathrm{GRHF}}$ in a self-consistent manner.

Note that energy matrix elements of $\phi_{\mathrm{GRHF}}$ with other (orthogonal) linear combinations of the $\Delta_K$ vanish. Thus it is to be expected that further internal correlation (relative to $\phi_{\mathrm{GRHF}}$) should be small;

$$\chi_{\mathrm{int}}^{\mathrm{GRHF}} \approx 0. \tag{21}$$

## IV. PERTURBATION-THEORY RESULTS

We now investigate the correlation effects in $\chi$ by first[20] solving the first-order Schrödinger equation based on $\phi_{\mathrm{RHF}}$ as zeroth order. In a later section we indicate how the form of $\chi^{(1)}$ would change as one goes from $\phi_{\mathrm{RHF}}$ to $\phi_{\mathrm{GRHF}}$. When Roothaan's HF method is applicable,[13,14] the RHF orbitals are eigenfunctions of a one-electron Hamiltonian $F$, and $\phi_{\mathrm{RHF}}$ is an eigenfunction of

$$\sum_{i=1}^{N} F(i).$$

### Perturbation Solution for the Wavefunction to First Order Based on Roothaan's SCF Method

Consider a nonclosed-shell state containing only one open subshell. Number the determinants arising from the single configuration to which $\phi_{\mathrm{RHF}}$ belongs from $K=1$ to $\kappa$;

$$\phi_{\mathrm{RHF}} = \sum_{K=1}^{\kappa} C_K \Delta_K. \tag{22}$$

Suppose that no other linear combination of the $\kappa - \Delta_K$ has the same symmetry as $\phi_{\mathrm{RHF}}$, so that the $C_K$ are determined by symmetry. The (spin-independent) one-electron Hamiltonian $F$ of which the RHF spin orbitals are eigenfunctions is defined by Eq. (36) of Ref. 13. $F$ has the form

$$F = h_0 + V_R, \tag{23}$$

where $h_0$ is the bare-nuclei one-electron Hamiltonian, and $V_R$ is an effective self-consistent-field potential. For the spin orbital $i$,

$$F|i\rangle = \epsilon_i |i\rangle. \tag{24}$$

Each $\Delta_K$ ($K \le \kappa$) and $\phi_{\mathrm{RHF}}$ itself are eigenfunctions of $H_0$ with the same eigenvalue $E_0$;

$$H_0 = \sum_{i=1}^{N} F(i), \tag{25}$$

$$H_0 \Delta_K = E_0 \Delta_K \qquad (K = 1, 2, \cdots, \kappa), \tag{26}$$

$$E_0 = \sum_{a=1}^{N} \epsilon_{k_a}, \tag{27}$$

$$H_0 \phi_{\mathrm{RHF}} = E_0 \phi_{\mathrm{RHF}}. \tag{28}$$

[13] C. C. J. Roothaan, Rev. Mod. Phys. **32**, 179 (1960).

[14] C. C. J. Roothaan and P. S. Bagus, *Methods in Computational Physics* (Academic Press Inc., New York, 1963), Vol. 2.

[15] S. Huzinaga, Phys. Rev. **122**, 131 (1961).

[16] F. W. Birss and S. Fraga, J. Chem. Phys. **38**, 2552 (1963).

[17] R. Lefebvre, J. Chim. Phys. **54**, 168 (1957).

[18] L. Goodman and J. R. Hoyland, J. Chem. Phys. **39**, 1068 (1963).

[19] Another simplified starting point for solving the correlation equations is obtained by an HF based on the average energy of the configuration, but with some symmetry and equivalence restrictions. (See Ref. 10.) A theory based on this HF would be suitable formally, but new estimates show that the remaining HF type $f_i$'s are not always small. Though these $f_i$'s could be added on by perturbation theory, it appears simpler to base the correlation theory on the Roothaan HF method, as we discuss in this paper. Another reason for this choice would be a pragmatic one; many Roothaan HF calculations by this method on atoms and molecules are now available. [See references in Part II-A2 of Ref. 10(b).]

[20] We wish to thank B. Schneider for preliminary work on open-shell perturbation theory.

The first-order wavefunction $\chi^{(1)}$ satisfies the equation

$$(H_0-E_0)\chi^{(1)}=-(1-|\phi_{\mathrm{RHF}}\rangle\langle\phi_{\mathrm{RHF}}|)H_1\phi_{\mathrm{RHF}}, \quad (29)$$

$$H_1=\sum_{i<j}^{N} g_{ij}-\sum_{i}^{N} V_R(i), \quad (30)$$

$$g_{ij}=1/r_{ij}. \quad (31)$$

The right-hand side of Eq. (29) is automatically orthogonal to all $\Delta_K$ $(K=1, 2, \cdots, \kappa)$ with the energy $E_0$, because (a) it is orthogonal to $\phi_{\mathrm{RHF}}$, and (b) there is no other linear combination of the $\Delta_K (K\leq\kappa)$ with the same symmetry as $\phi_{\mathrm{RHF}}$. Therefore Eq. (29) has a solution.[21]

The determinants $K=\kappa+1, \cdots, \binom{M}{N}$ often constitute a significant portion of $\chi$ since they include the nondynamical correlations as discussed above. This part of $\chi$, involving only the unfilled portions of the RHF sea, is $\chi_{\mathrm{int}}{}^{\mathrm{RHF}}$. It is obtained to first order from Eqs. (A.1) and (29) as

$$\chi_{\mathrm{int}}{}^{(1)}=\sum_{K=\kappa+1}^{\binom{M}{N}} \Delta_K \frac{\langle\Delta_K, H_1\phi_{\mathrm{RHF}}\rangle}{E_0-E_K}, \quad (32)$$

where

$$E_K=\sum_{a=1}^{N}\epsilon_{k_a}. \quad (33)$$

The $\epsilon_{k_a}$ are spin–orbital energies of $k_a$ appearing in $\Delta_K$, Eq. (1).

The remaining part of $\chi^{(1)}$ is $\chi_{\mathrm{ext}}{}^{(1)}$ [cf. Eqs. (7) and (8)] and satisfies

$$(\mathrm{H}_0-E_0)\chi_{\mathrm{ext}}{}^{(1)}=-(1-\sum_{K=1}^{\binom{M}{N}}|\Delta_K\rangle\langle\Delta_K|)H_1\phi_{\mathrm{RHF}}. \quad (34)$$

This equation gives the remaining correlations, which are external to the sea and quite similar in nature to those in closed shells. Equation (34) can be solved formally by a straightforward application of the method of successive partial orthogonalizations (Appendix A). The result is:

$$\chi^{(1)}{}_{\mathrm{ext}}=\sum_{K=1}^{\kappa} C_K\chi^{(1)}{}_K \quad (35)$$

$$=\sum_{K=1}^{\kappa} C_K\mathcal{A}\left\{(k_1k_2\cdots k_N)\left[\sum_{a=1}^{N}\frac{\hat{f}^{(1)}{}_{k_a}(K)}{(k_a)}+\sum_{1\leq a<b}^{N}\sum_{l=1}^{M}\frac{lf^{(1)}{}_{k_ak_b;l}}{(k_ak_b)}+2^{-\frac{1}{2}}\sum_{1\leq a<b}^{N}\frac{\hat{u}^{(1)}{}_{k_ak_b}}{(k_ak_b)}\right]\right\}, \quad (36)$$

$$[l\neq k_1, k_2, \cdots, k_N],$$

where

$$\hat{f}^{(1)}{}_{k_a(K)}=(\epsilon_{k_a}-F)^{-1}Q_1(\sum_{b=1}^{N}\bar{S}_{k_b}-V_R)\,|k_a\rangle, \quad (37)$$

$$\hat{f}^{(1)}{}_{k_ak_b;l}=(\epsilon_{k_a}+\epsilon_{k_b}-\epsilon_l-F)^{-1}Q_1[\langle l, g_{12}k_a\rangle k_b-\langle l, g_{12}k_b\rangle k_a], \quad (38)$$

$$\hat{u}^{(1)}{}_{k_ak_b}=[\epsilon_{k_a}+\epsilon_{k_b}-F(1)-F(2)]^{-1}Q_2g_{12}\,|B(k_ak_b)\rangle. \quad (39)$$

$\bar{S}_{k_a}$ is the Coulomb minus exchange operator for spin orbital $k_a$. The projection operators $Q_1$ and $Q_2$ ensure that the $\hat{f}$'s and $\hat{u}_{ij}$'s are one-electron orthogonal to *all* the spin orbitals in the sea.

$$Q_1\equiv 1-\sum_{i=1}^{M}|i\rangle\langle i|, \quad (40)$$

$$Q_2\equiv 1-\sum_{i=1}^{M}\{|i(\mathbf{x}_1)\rangle\langle i(\mathbf{x}_1)|+|i(\mathbf{x}_2)\rangle\langle i(\mathbf{x}_2)|\}+\sum_{i<j}^{M}|B(ij)\rangle\langle B(ij)|, \quad (41)$$

$$\langle m, \hat{f}^{(1)}{}_{k_a(K)}\rangle=\langle m, \hat{f}^{(1)}{}_{k_ak_b;l}\rangle=\langle m, \hat{u}^{(1)}{}_{k_ak_b}\rangle=0 \qquad (m=1, 2, \cdots, M). \quad (42)$$

$B$ is the two-electron antisymmetrizer: $B=2^{-\frac{1}{2}}(1-P_{12})$.

At an intermediate stage of the derivation of Eq. (36) one obtains the $n$-electron correlation functions $U'_{(n)K_{N-n}}$ [Eq. (13)] to first order.

$$\hat{u}^{(1)}{}_{(1)I_{N-1}}=\sum_{\substack{K\\(K=I_{N-1}k)}}(-1)^{\sigma K}C_K\hat{f}^{(1)}{}_{k(K)}+\sum_{\substack{K\\[K=I_{N-1}(k_a/i)k_b]}}\sum_{i}(-1)^{\sigma K}C_Kf^{(1)}{}_{k_ak_bK;i} \quad (43)$$

$$\hat{u}^{(1)}{}_{(2)I_{N-2}}=\sum_{\substack{K\\(K=I_{N-2}k_ak_b)}}(-1)^{\sigma K}C_K\hat{u}^{(1)}{}_{k_ak_b} \quad (44)$$

$$\hat{u}^{(1)}{}_{(n)I_{N-n}}=0 \qquad (n\geq 3). \quad (45)$$

[21] O. Sinanoğlu, Proc. Roy. Soc. (London) **A260**, 379 (1961).

These equations clearly show the composite nature of the $\hat{U}_{(1)}$ and $\hat{U}_{(2)}$.

### *Notation*

$I_{N-1}k$ stands for the determinant $\Delta_K$ whose spin orbitals are $i_1, i_2, \cdots, i_{N-1}, k$. The $(-1)^{\sigma K}$ is $\pm 1$ depending whether $i_1, i_2, \cdots, i_{N-1}, k$ is an even or odd permutation of the strictly ascending order. For exact correlation functions [Eq. (13)] we use capital $U$'s; for approximate, lower case $u$'s.

### Types of Correlation

Equations (32) and (35)–(39) suggest the following picture of nonclosed-shell correlation:

(1) Internal correlation: electrons in $\phi_{\mathrm{RHF}}(K\leq\kappa)$ may be "scattered" into unfilled regions of the RHF sea $\Delta_K(K>\kappa)$.

(2) Semi-internal correlation: in each $\Delta_K$ of $\phi_{\mathrm{RHF}}$ a pair of electrons $(k_a k_b)$ may "collide," "scattering" one outside the sea into $\hat{f}^{(1)}{}_{k_a k_b; l}$, the other into a spin orbital $l$ (which must not be occupied at that moment).

(3) External pair correlations: pair correlations similar to those in closed shells; $k_a$ and $k_b$ "collide," "scattering" both outside the sea into $\hat{u}^{(1)}{}_{k_a k_b}$.

(4) Spin- and symmetry-orbital average polarizations: $\hat{f}^{(1)}{}_{k(K)}$.

The interpretation of the $\hat{f}^{(1)}{}_{k(K)}$ requires some comment. The open-shell restricted HF $\phi_0$ is not truly a self-consistent-field wavefunction. In particular, for $S\neq 0$ the different exchange fields for $\alpha$ and $\beta$ electrons lead to first-order spin-polarizing corrections to doubly occupied orbitals.[22,23] The asymmetry of the Coulomb and exchange fields leads to first-order changes in the symmetry of the orbitals. (An example in Paper II shows the $d$-symmetry correction to a $2s$ orbital in $^1S$ carbon.) The potential which gives rise to the first-order spin and/or symmetry polarizations is the difference between the sum of the Coulomb and exchange potentials acting on Orbital $k$ in the determinant $\Delta_K$, and the average (symmetric) potential $V_R$.

$$\Delta V_{k(K)}=\sum_{a=1}^{N}\bar{S}_{k_a}-V_R. \tag{46}$$

Since $\Delta V_{k(K)}$ depends explicitly on the spin orbitals in $\Delta_K$, $\hat{f}^{(1)}{}_{k(K)}$ depends not only on $k$, but also on the determinant that $k$ is in.

Comparing the open-shell perturbation-theory results with those for closed shells,[1,2,21] one notes that the closed shell $\chi^{(1)}$ [Eq. (19) of Ref. 1] contains no $\hat{f}^{(1)}{}_k$ or $\hat{f}^{(1)}{}_{ij;k}$. To have $\hat{f}_{ij;k}$ there must be at least two possible Slater determinants and an available unoccupied spin orbital $k$. The $\hat{f}^{(1)}{}_k$ does not enter the closed shell $\chi^{(1)}$ as a consequence of the Brillouin theorem. [$\Delta V_{k(K)}$, Eq. (46), vanishes in the closed-shell case.] Thus the nonclosed-shell "orbital average polarizations" $\hat{f}^{(1)}{}_{k(K)}$ are of a fundamentally different origin than the closed shell $\hat{f}_k$, which arise from the higher-order effect of correlation on orbitals, and which are usually negligible.[3]

Although Eq. (39) for $\hat{u}^{(1)}{}_{k_a k_b}$ and Eq. (44) of Ref. 2 are formally identical, note that the projection operator in the closed-shell case makes $g_{ij}\mid B(ij)\rangle$ one electron orthogonal to the occupied $N$ closed-shell HF spin orbitals, whereas $Q_2$ in the open-shell case makes $g_{ij}\mid B(ij)\rangle$ one-electron orthogonal to ***all M*** spin orbitals in the RHF sea, occupied or unoccupied.

### Second-Order Energy

From the first-order wavefunction obtained above, the form of the energy may be determined. The second-order energy is given by

$$\begin{aligned}E^{(2)}&=\langle\phi_{\mathrm{RHF}}, H_1\chi^{(1)}{}_{\mathrm{int}}\rangle+\langle\phi_{\mathrm{RHF}}, H_1\chi^{(1)}{}_{\mathrm{ext}}\rangle\\&=E^{(2)}{}_{\mathrm{int}}+E^{(2)}{}_{\mathrm{ext}}.\end{aligned} \tag{47}$$

In the second-order energy, the internal and external correlation energies are uncoupled.

If $\phi_{\mathrm{RHF}}$ is a single determinantal nonclosed-shell state, $\chi^{(1)}{}_{\mathrm{int}}$ may easily be written in terms of internal correlations arising from various spin orbitals in the $\phi_{\mathrm{RHF}}$. This is in fact done where the closed-shell-type theory is still applicable.[5] In general, $\chi_{\mathrm{int}}$ (to all orders) also includes collective and longer-range-type correlations that may give rise to over-all charge deformation. For such a general nonclosed-shell state it is best to treat the $E_{\mathrm{int}}$ in totality. Since $\chi_{\mathrm{int}}$ and $E_{\mathrm{int}}$ are essentially the near-degeneracy-type CI effects, their total calculation is easy.

The more difficult part is the $E_{\mathrm{ext}}$ similar to the closed-shell dynamical effects involving shorter-range but strong correlations. From Eqs. (35) and (30), $E^{(2)}{}_{\mathrm{ext}}$ is obtained as

$$E^{(2)}{}_{\mathrm{ext}}=\sum_{K=1}^{\kappa}\mid C_K\mid^2\langle\Delta_K, H_1\chi_K{}^{(1)}\rangle+\sum_{L\neq K}^{\kappa}C^*{}_L C_K\langle\Delta_L, H_1\chi_K{}^{(1)}\rangle. \tag{48}$$

[22] O. Sinanoğlu, J. Chem. Phys. **33**, 1212 (1960).
[23] A. D. McLachlan, Mol. Phys. **3**, 233 (1960).

The first term contains the diagonal second-order correlations of each $\Delta_K$ in $\phi_{\mathrm{RHF}}$:

$$\langle \Delta_K, H_1\chi^{(1)}{}_K \rangle = \sum_{a=1}^{N} \langle k_a, (\sum_{b=1}^{N} \bar{S}_{k_b} - V_R)\hat{f}^{(1)}{}_{k_a(K)} \rangle + \sum_{1\leq a<b}^{N} \sum_{l\neq(K)}^{M} \langle B(k_a k_b), g_{12}B(l\hat{f}^{(1)}{}_{k_a k_b;l}) \rangle + \sum_{1\leq a<b}^{N} \langle B(k_a k_b), g_{12}\hat{\mathfrak{u}}^{(1)}{}_{k_a k_b} \rangle \tag{49}$$

$$= \sum_{a}^{N} \tilde{\epsilon}^{P(2)}{}_{k_a(K)} + \sum_{a<b}^{N} \sum_{l\neq(K)}^{M} \tilde{\epsilon}^{(2)}{}_{k_a k_b;l(K)} + \sum_{a<b}^{N} \tilde{\epsilon}^{(2)}{}_{k_a k_b(K)}. \tag{50}$$

Thus the diagonal terms yield simply the "orbital average polarization" energies, the semi-internal correlation energies, and the pair correlation energies of the various spin orbitals in each determinant.

The multideterminant nonclosed shell also contains however some new off-diagonal effects. For example, if $L$ and $K$ differ by only one spin–orbital, $L=K(l/k_a)$, then

$$\langle \Delta_L, H_1\chi^{(1)}{}_K \rangle = (-1)^{\sigma LK} \{ \langle l, (\sum_{b=1}^{N} \bar{S}_{k_b} - V_R)\hat{f}^{(1)}{}_{k_a(K)} \rangle + \sum_{c=1,(c\neq a)}^{N} \langle B(k_c l), g_{12}B(\hat{f}^{(1)}{}_{k_a(K)} k_a) \rangle$$

$$+ \sum_{c=1,(c\neq a)}^{N} \langle k_c( \sum_{b=1,(b\neq a,c)}^{N} \bar{S}_{k_b} + \bar{S}_l - V_R)\hat{f}^{(1)}{}_{k_a k_c;l} \rangle + \sum_{c=1,(c\neq a)}^{N} \sum_{m,(m\neq k_1,k_2,\cdots,k_N,l)} \langle B(lk_c), g_{12}B(m\hat{f}^{(1)}{}_{k_a k_c;m}) \rangle$$

$$- \sum_{b<c,(b,c\neq a)}^{N} \langle B(k_b k_c), g_{12}B(k_a\hat{f}^{(1)}{}_{k_b k_c;l}) \rangle + \sum_{b=1,(b\neq a)}^{N} \langle B(lk_b), g_{12}\hat{\mathfrak{u}}^{(1)}{}_{k_a k_b} \rangle \}. \tag{51}$$

Other off-diagonal terms are given in Appendix B. Equations (51), (B.1), and (B.2) show that even in the second-order energy, a general nonclosed-shell state contains new "cross-correlation" terms. These are very much a part of the nature of the nonclosed shell. They need to be taken into account.

### *Coupling and Uncoupling*

$\chi^{(1)}$ may be calculated by minimizing $\tilde{E}^{(2)}$:

$$E^{(2)} \leq \tilde{E}^{(2)} = 2\langle \phi_{\mathrm{RHF}}, H_1\tilde{\chi}^{(1)} \rangle + \langle \tilde{\chi}^{(1)}, (H_0 - E_0)\tilde{\chi}^{(1)} \rangle. \tag{52}$$

In $E^{(2)}$ the correlation functions are coupled, but they come out uncoupled in Eqs. (37)–(39). We illustrate the nature of this uncoupling with the $(2p)^2$ pairs in $^1S$ carbon.

The $\phi_0$ is

$$\phi_{\mathrm{RHF}}(^1S\text{-carbon}) = \mathfrak{A}((1s)^2(2s)^2\{[(2p_x)^2+(2p_y)^2+(2p_z)^2]/3^{\frac{1}{2}}\}). \tag{53}$$

The $(2p)^2$ contribution to $E^{(2)}$ is

$$E^{(2)}(2p^2) = \tfrac{1}{3}\{ \tilde{\epsilon}^{(2)}{}_{2p_x^2} + \tilde{\epsilon}^{(2)}{}_{2p_y^2} + \tilde{\epsilon}^{(2)}{}_{2p_z^2} + \tilde{\epsilon}^{(2)}(2p_x^2; 2p_y^2) + \tilde{\epsilon}^{(2)}(2p_x^2; 2p_z^2) + \tilde{\epsilon}^{(2)}(2p_y^2; 2p_z^2) \}, \tag{54}$$

where

$$\tilde{\epsilon}^{(2)}{}_{2p_x^2} = 2\langle B(2p_x^2), g_{12}\hat{\mathfrak{u}}^{(1)}{}_{2p_x^2} \rangle + \langle \hat{\mathfrak{u}}^{(1)}{}_{2p_x^2}, [F(1)+F(2)-2\epsilon_{2p}]\hat{\mathfrak{u}}^{(1)}{}_{2p_x^2} \rangle \tag{55}$$

and

$$\tilde{\epsilon}^{(2)}(2p_x^2; 2p_y^2) = 2\langle B(2p_x^2), g_{12}\hat{\mathfrak{u}}^{(1)}{}_{2p_y^2} \rangle + 2\langle B(2p_y^2), g_{12}\hat{\mathfrak{u}}^{(1)}{}_{2p_x^2} \rangle + 2\langle \hat{\mathfrak{u}}^{(1)}{}_{2p_x^2}, [F(1)+F(2)-2\epsilon_{2p}]\hat{\mathfrak{u}}^{(1)}{}_{2p_y^2} \rangle. \tag{56}$$

---

Similar expressions define the pair and "cross-pair" energies for other combinations of orbitals. If $\hat{\mathfrak{u}}^{(1)}{}_{2p_x^2}$ is varied to minimize the diagonal term $\tilde{\epsilon}^{(2)}{}_{2p_x^2}$, subject to the orthogonality requirement [Eq. (42)], the following equation is obtained for $\hat{\mathfrak{u}}^{(1)}{}_{2p_x^2}$:

$$[F(1)+F(2)-2\epsilon_{2p}]\hat{\mathfrak{u}}^{(1)}{}_{2p_x^2} + Q_2 g_{12} B(2p_x^2) = 0. \tag{57}$$

On the other hand, $E^{(2)}(2p^2)$ is a minimum when

$$[F(1)+F(2)-2\epsilon_{2p}][\hat{\mathfrak{u}}^{(1)}{}_{2p_x^2} + \hat{\mathfrak{u}}^{(1)}{}_{2p_y^2} + \hat{\mathfrak{u}}^{(1)}{}_{2p_z^2}]$$
$$+ Q_2 g_{12}[B(2p_x^2) + B(2p_y^2) + B(2p_z^2)] = 0. \tag{58}$$

Clearly, the solution of the coupled equation (58) is just the sum of the solutions of the uncoupled equations (57).

The second-order energy represents the major portion of the correlation effects in the energy. These effects, arising from the "first-order processes" in the wf, will be made "exact to all orders" in Paper II to make them quantitatively more accurate.

### **Perturbation Theory Based on $\phi_{\mathrm{GRHF}}$**

As a practical matter, the perturbation solution above with $\phi_0 = \phi_{\mathrm{RHF}}$ would be sufficient where the remaining near degeneracies ($\chi_{\text{int}}$) are not strong enough to have sizable effects on the RHF orbitals and the charge distribution. Where the (CI) coefficients appearing in $\chi_{\text{int}}$ are comparable to those in $\phi_{\text{RHF}}$, it is necessary to go over to the more general $\phi_0=\phi_{\text{GRHF}}$ as discussed in Sec. III. Molecular orbital (+correlation) treatment of interatomic potential-energy curves and surfaces (especially as the $R$'s$\rightarrow\infty$) in general requires[3] the $\phi_{\text{GRHF}}$.

The solution of the first-order Schrödinger equation starting with $\phi_0=\phi_{\text{GRHF}}$ is complicated by the difficulty of definining an $H_0$. Nevertheless, how $\chi^{(1)}$ would be modified for $\phi_{\text{GRHF}}$ may be seen in the following manner.

The $\phi_{\text{RHF}}$ included only determinants ($K=1, \cdots, \kappa$) arising from the same configuration, and $\chi^{(1)}{}_{\text{int}}$ following it contained $\Delta_K$'s corresponding to virtually excited configurations. $\phi_{\text{GRHF}}$ includes all $\Delta_K$'s of the spin-orbital "sea,"

$$\phi_0=\phi_{\text{GRHF}}=\sum_{K=1}^{\binom{M}{N}} C_K\Delta_K, \qquad (59)$$

and incorporates the main internal correlations (by CI rather than first order). Therefore, $\chi^{(1)}$ based on GRHF should no longer contain any significant explicit $\chi_{\text{int}}$ [cf. Eq. (21)]:

$$\chi^{(1)\text{GRHF}}\cong\chi^{(1)\text{GRHF}}{}_{\text{ext}}. \qquad (60)$$

The $\phi_{\text{GRHF}}$ in Eq. (59) includes some $\Delta_K$'s which are degenerate, some that are not. The latter are responsible for the difficulty of defining the $H_0$. Were the near degeneracies viewed however as actual degeneracies (as, e.g., $\Delta_K$'s out of $1s^22s^22p^2$ and $1s^22p^4$ would be in the $Z\rightarrow\infty$ limit), the solution for the GRHF $\chi^{(1)}$ would become like that for the RHF $\chi^{(1)}$, but now the sums in Eq. (36) would run from $K=1$ to $\binom{M}{N}$ rather than 1 to $\kappa$ as in the respective $\phi_0$'s.

One other change in going over to $\phi_{\text{GRHF}}$ is expected. In the RHF case, only the $\hat{f}^{(1)}{}_{k_a(K)}$ depended on the parent determinant $\Delta_K$ in which $k_a$ was found [see especially Eq. (46)]; the $\hat{f}^{(1)}{}_{k_ak_b;l}$ and $\hat{u}^{(1)}{}_{k_ak_b}$ did not. In the GRHF case, the $\hat{f}^{(1)}{}_{k_ak_b;l}$ and $\hat{u}^{(1)}{}_{k_ak_b}$, too, may depend on parentage [$\hat{f}^{(1)}{}_{k_ak_b;l(K)}$ and $\hat{u}^{(1)}{}_{k_ak_b(K)}$].

With these considerations the expected form of $\chi^{(1)}$ now becomes

$$\phi_0+\chi^{(1)}=\phi_{\text{GRHF}}+\chi^{(1)\text{GRHF}} \qquad (61)$$

$$\cong\phi_{\text{GRHF}}+\chi^{(1)\text{GRHF}}{}_{\text{ext}} \qquad (62)$$

$$\cong\sum_{K=1}^{\binom{M}{N}} C_K(\Delta_K+\chi^{(1)}{}_K), \qquad (63)$$

where

$$\chi^{(1)}{}_K=\mathcal{A}\left\{(k_1k_2\cdots k_N)\left[\sum_{a=1}^{N}\frac{\hat{f}^{(1)}{}_{k_a(K)}}{(k_a)}+\sum_{a<b}^{N}\sum_{l\neq(K)}^{M}\frac{l\hat{f}^{(1)}{}_{k_ak_b;l(K)}}{k_ak_b}+\sum_{a<b}^{N}\frac{\hat{u}^{(1)}{}_{k_ak_b(K)}}{\sqrt{2}(k_ak_b)}\right]\right\}. \qquad (64)$$

Note especially that compared to Eqs. (35) and (36) for RHF, here

$$\sum_{K=1}^{\kappa}\rightarrow\sum_{K=1}^{\binom{M}{N}};$$

also the above correlation functions are unambiguously defined only if an $H_0$ is specified.

With $\phi_{\text{RHF}}$, $\chi^{(1)}{}_{\text{int}}$ and $\chi^{(1)}{}_{\text{ext}}$ represented first-order processes which took out some of the electrons from filled portions of the sea into unfilled portions and to outside regions. In the RHF case, further scattering from electrons caught in the originally unfilled regions becomes a higher-order process not appearing in $\chi^{(1)}$. In the GRHF case, the "boundary" between the filled and unfilled portions is self-consistent, and scattering into unfilled regions is a higher-order process. The GRHF $\chi^{(1)}$ includes external correlations from all the $\Delta_K$'s, so that some previously higher-order processes (wrt $\phi_{\text{RHF}}$) now become first-order processes (wrt $\phi_{\text{GRHF}}$).

Though Eqs. (61) to (64) above give the expected form of $\chi^{(1)\text{GRHF}}$, it would be worthwhile if formal and rigorous solutions could be obtained to the corresponding first-order Schrödinger equation. The solution obtained for RHF and its generalization above will be sufficient nevertheless for obtaining a variational theory in Paper II containing the processes of practical importance.

## V. SUMMARY

For the treatment of correlation, the multiconfigurational general restricted Hartree–Fock (GRHF) wavefunction forms a general starting point for various types of nonclosed-shell states. For systems where near-degeneracy resonances are not strong enough to upset the charge distribution, a simplified starting point, the Roothaan-type restricted Hartree–Fock (RHF) wavefunction, is convenient.

The first-order Schrödinger equation for $\phi_0=\phi_{\text{RHF}}$ was solved and showed the main correlation effects in nonclosed-shell states. These include the general form of the nondynamical (near-degeneracy) correlations (internal correlation), semi-internal correlations, i.e., pair correlations expelling only one of the electrons out of the HF sea, and "external" pair correlations similar to those in closed-shell states, in addition to the "orbital average" symmetry and spin polarizations. The form of the second-order energy is also evaluated and contains in addition to the energies of the above one- and two-electron effects, new "cross correlations" which are a basic feature of nonclosed-shell state energies. Correlation effects in the energy involving one and two spin orbitals out of a determinant $\Delta_K$ of $\phi_0$ are weighted by the fractional occupancy $|C_K|^2$ of that determinant [see Eqs. (48) and (50)]. The generalization of these first-order processes to $\phi_{\text{GRHF}}$ is also indicated.

In the next paper (II) these results will be extended into a variational many-electron theory by making the first-order processes exact "to all orders" and by including the unlinked clusters. The variational determination of each one- and two-electron correlation function, symmetry effects, and total energy as an upper limit will also be discussed there.

## APPENDIX A: FORM OF $\chi$ BY THE METHOD OF SUCCESSIVE PARTIAL ORTHOGONALIZATIONS

$\chi$ is defined by Eqs. (4) to (6) in the text. $\chi_{\text{int}}$ is that part of $\chi$ in which no electrons are removed from the sea. $\chi_{\text{ext}}$ is the rest of $\chi$;

$$\chi_{\text{int}}=\sum_{K=1}^{\binom{M}{N}} \Delta_K\langle \Delta_K, \chi\rangle; \tag{A.1}$$

$$\chi_{\text{ext}}=\chi-\chi_{\text{int}} \tag{A.2}$$

$$=[1-\sum_{K=1}^{\binom{M}{N}} |\Delta_K\rangle\langle\Delta_K|]\chi. \tag{A.3}$$

The one-electron correlation function $\hat{U}_{(1)K_{N-1}}$ is obtained from $\chi_{\text{ext}}$ by partial orthogonalization to the simple product: $(N!)^{\frac{1}{2}}[k_1(\mathbf{x}_1)k_2(\mathbf{x}_2)k_{N-1}(\mathbf{x}_{N-1})]$;

$$\hat{U}_{(1)K_{N-1}}(\mathbf{x}_N)=(N!)^{+\frac{1}{2}}\langle k_1k_2\cdots k_{N-1}, \chi_{\text{ext}}\rangle. \tag{A.4}$$

In (A.4), the integration is over the space- and spin-volume elements of the $N-1$ electrons in $k_1k_2\cdots k_{N-1}$. Once the $\hat{U}_{(1)K_{N-1}}$ have been determined $\chi_{\text{ext}}$ can be orthogonalized to all $K_{N-1}$;

$$\chi'=\chi_{\text{ext}}-\sum_{K_{N-1}=1}^{\binom{M}{N-1}} \mathcal{A}\{(k_1k_2\cdots k_{N-1})\,\hat{U}_{(1)K_{N-1}}\} \tag{A.5}$$

$$=\chi_{\text{ext}}-\mathcal{A}\{\sum_{K_{N-1}=1}^{\binom{M}{N-1}} |k_1k_2\cdots k_{N-1}\rangle(N!)^{\frac{1}{2}}\langle k_1k_2\cdots k_{N-1}, \chi_{\text{ext}}\rangle\}. \tag{A.6}$$

The two-electron correlation functions $\hat{U}'_{(2)K_{N-2}}$ are obtained from $\chi'$ by partial orthogonalization to $(N!/2!)^{\frac{1}{2}}[k_1(\mathbf{x}_1)k_2(\mathbf{x}_2)\cdots k_{N-2}(\mathbf{x}_{N-2})]$,

$$\hat{U}'_{(2)K_{N-2}}(\mathbf{x}_{N-1}, \mathbf{x}_N)=(N!/2!)^{\frac{1}{2}}\langle k_1k_2\cdots k_{N-2}, \chi'\rangle. \tag{A.7}$$

Similarly, $\chi'$ can be orthogonalized to all $K_{N-2}$,

$$\chi''=\chi'-2^{-\frac{1}{2}}\sum_{K_{N-2}=1}^{\binom{M}{N-2}} \mathcal{A}\{(k_1k_2\cdots k_{N-2})\,\hat{U}'_{(2)K_{N-2}}\} \tag{A.8}$$

$$=\chi'-2^{-\frac{1}{2}}\mathcal{A}\{\sum_{K_{N-2}=1}^{\binom{M}{N-2}} |k_1k_2\cdots k_{N-2}\rangle(N!/2!)^{\frac{1}{2}}\langle k_1k_2\cdots k_{N-2}, \chi'\rangle\}. \tag{A.9}$$

Continuing in this manner, if we denote the remainder after orthogonalizing to the $K_{N-n}$, by $\chi^{(n)}$, the $(n+1)$-electron correlation functions are given by

$$\hat{U}'_{(n+1)K_{N-n-1}}=[N!/(n+1)!]^{\frac{1}{2}}\langle k_1k_2\cdots k_{N-n-1}, \chi^{(n)}\rangle. \tag{A.10}$$

## APPENDIX B: ADDITIONAL OFF-DIAGONAL TERMS IN $E^{(2)}$

If $L$ and $K$ differ by two spin orbitals, $L=(K/k_ak_b)l_cl_d$, then

$$\langle\Delta_L, H_1\chi^{(1)}{}_K\rangle=(-1)^{\sigma_{LK}}\{\langle B(l_cl_d), g_{12}B(\hat{f}^{(1)}{}_{k_a(K)}k_b)\rangle+\langle B(l_cl_d), g_{12}B(k_a\hat{f}^{(1)}{}_{k_b(K)})\rangle$$

$$+\sum_{m,(m\neq k_1k_2\cdots k_Nl_cl_d)}\langle B(l_cl_d), g_{12}B(m\hat{f}^{(1)}{}_{k_ak_b;m})\rangle+\langle l_d, (\sum_{e=1}^{N}\bar{S}_{l_e}-V_R)\hat{f}^{(1)}{}_{k_ak_b;l_c}\rangle-\langle l_c, (\sum_{e=1}^{N}\bar{S}_l-V_R)\hat{f}^{(1)}{}_{k_ak_b;l_d}\rangle$$

$$+\sum_{e=1,(e\neq a,b)}^{N}[\langle B(l_dk_e), g_{12}B(k_b\hat{f}^{(1)}{}_{k_ak_b;l_c})\rangle-\langle B(l_ck_e), g_{12}B(k_b\hat{f}^{(1)}{}_{k_ak_e;l_d})\rangle$$

$$-\langle B(l_dk_e), g_{12}B(k_a\hat{f}^{(1)}{}_{k_bk_e;l_c})\rangle+\langle B(l_ck_e), g_{12}B(k_a\hat{f}^{(1)}{}_{k_bk_e;l_d})\rangle]+\langle B(l_cl_d), g_{12}\hat{u}^{(1)}{}_{k_ak_b}\rangle\}. \tag{B.1}$$

If $K$ and $L$ differ by three electrons, $L=(K/k_ak_bk_c)l_{a'}l_{b'}l_{c'}$, then

$$\langle\Delta_L, H_1\chi^{(1)}{}_K\rangle=(-1)^{\sigma_{KL}}\{\langle B(l_{b'}l_{c'}), g_{12}B(\hat{f}^{(1)}{}_{k_ak_b;l_{a'}}k_c)+\cdots+\langle B(l_{a'}l_{b'}), g_{12}B(f^{(1)}{}_{k_bk_c;l_{c'}}k_a)\rangle\}. \tag{B.2}$$

If $K$ and $L$ differ by more than three electrons, then $\langle\Delta_L, H_1\chi^{(1)}{}_K\rangle$ vanishes.

 Reprinted from *Physical Review* **162**, 71 (1967)

# Superposition of Configurations and Atomic Oscillator Strengths—Carbon I and II*

A. W. WEISS
*National Bureau of Standards, Washington, D. C.*
(Received 19 May 1967)

Variational wave functions are computed for the ground state and a number of excited states of carbon I and II by the method of superposition of configurations. To accelerate convergence, the virtual orbitals are obtained from a pseudonatural orbital transformation on a single electron pair, out of all those possible within a given quantum shell. Expansions of up to 50 configurations are generated and used to study the effects of correlation on the oscillator strengths. The method appears to be fairly successful in correcting for most of the correlation error in both the energy and "correlation-sensitive" oscillator strengths. Term values are substantially improved over the Hartree-Fock values; and generally, although not always, $f$ values appear to be obtainable with an accuracy of about 25%.

## INTRODUCTION

THE calculation of atomic transition probabilities has traditionally been based on a single-configuration, independent-particle model.[1] In this approximation, the calculation reduces to the evaluation of a one-electron transition integral, corresponding to the transition of an electron from one single-particle state (orbital) to another. A variety of approximations have been used to obtain the orbitals, notable among them being the self-consistent field,[2] Thomas-Fermi,[3] and Coulomb approximations.[4] The model has generally proved quite successful, particularly for Rydberg transitions, as might be expected.

It has also been long realized that configuration interaction can have a drastic effect on such calculations,[1,5–7] and provision has always been made for a very limited amount of configuration mixing, as may be indicated by the structure of the spectrum. There have also been numerous investigations of the effects of "asymptotically degenerate configurations," i.e., the mixing of those configurations which would approach degeneracy in the large-$Z$ limit of an isoelectronic series.[8–11] Here, too, only a very small number of configurations are normally considered.

However, the effect on the transition moment of the correlation error, which is inherent in the independent-particle model, has not been investigated too extensively. Although there are many ways to construct a better approximation to the exact wave function, probably the most generally useful one is the superposition of configurations (SOC) expansion. Here, the lead term is the traditional independent-particle wave function, and the succeeding terms *appear* to represent excitations of one or more electrons into excited orbitals. Since, however, these virtual configurations are used to represent the details of electron correlation, they will tend to emphasize those regions of space where the charge density is greatest and thus will usually bear little resemblance to "real" excited state configurations. While the kinds of configuration mixing mentioned in the preceding paragraph should be important, they must be considered to be only the first few terms in a SOC expansion.

In this paper, the method of superposition of configurations has been used to compute the correlation corrections to both the energies and oscillator strengths of neutral and singly ionized carbon. Such extensive SOC calculations have already yielded accurate $f$ values for neutral helium[12] and, by incorporating part of the correlation into a polarization potential, also for neutral magnesium.[7] In a sense, the present calculations represent an extension of this work to a more correlation-sensitive situation.

A particularly interesting feature of these calculations is the use made of the pseudonatural orbital technique of Edmiston and Krauss[13] to generate a rapidly converging and variationally optimum set of virtual orbitals. This results in compressing most of the correlation effects into a relatively small number of configurations, although, ultimately, the SOC procedure remains slowly converging with respect to the final small bits and pieces of the correlation. One point to be kept in mind is the relative ease with which one

---

* This research was supported by the Advanced Research Projects Agency of the Department of Defense and was monitored by Dr. D. E. Mann under Contract No. 703.

1 E. U. Condon and G. H. Shortley, *The Theory of Atomic Spectra* (Cambridge University Press, New York, 1959); J. C. Slater, *Quantum Theory of Atomic Structure* (McGraw-Hill Book Company, Inc., New York, 1960), Vol. 2; R. H. Garstang, *Vistas in Astronomy* (Pergamon Press, Inc., New York, 1955), Vol. 1, pp. 268–276.

2 P. S. Kelly, Astrophys. J. **140**, 1247 (1964).

3 J. C. Stewart and M. Rotenberg, Phys. Rev. **140**, A1508 (1965).

4 D. R. Bates and A. Damgaard, Phil. Trans. Roy. Soc. London **A242**, 101 (1949).

5 E. Trefftz, Z. Astrophysik **26**, 240 (1949); **28**, 67 (1950).

6 A. B. Bolotin, I. B. Levinson, and L. I. Levin, Zh. Eksperim i Teor. Fiz. **29**, 449 (1955) [English transl.: Soviet Phys.—JETP **2**, 391 (1956)].

7 A. B. Bolotin and A. P. Yutsis, Zh. Eksperim. i Teor. Fiz. **24**, 537 (1953).

8 D. Layzer, Ann. Phys. (N. Y.) **8**, 271 (1959).

9 C. Froese, Astrophys. J. **141**, 1206 (1965).

10 M. Cohen and A. Dalgarno, Proc. Roy. Soc. (London) **A280** 258 (1964).

11 R. J. S. Crossley and A. Dalgarno, Proc. Roy. Soc. (London) **A286**, 510 (1965).

12 L. C. Green, N. C. Johnson, and E. K. Kolchin, Astrophys. J. **144**, 369 (1966).

13 C. Edmiston and M. Krauss, J. Chem. Phys. **45**, 1833 (1966).

can compute the bulk of the correlation effects since most of the computational labor simply involves getting the Hartree-Fock starting point. While this should be of considerable importance for molecules, it should be added that the difficulties arising from using very large basis sets will remain and could pose additional computational problems.

In the next section, to pinpoint the correlation-sensitive transitions, we will review briefly the current situation for some first row atoms. This will be followed by a description of the superposition of configurations method, with a fairly detailed account of the pseudo-natural orbital technique and the way it has been applied here. Finally, the results will be presented and discussed for carbon I and II.

## CORRELATION-SENSITIVE TRANSITIONS

A number of equivalent quantities can be used to characterize an atomic transition probability.[1] For theoretical purposes, the most convenient are probably the multiplet strength

$$S_{ij}=|\langle\Phi_i|\mathbf{r}|\Phi_j\rangle|^2, \qquad (1)$$

and the oscillator strength

$$f_{ij}=\tfrac{2}{3}\Delta E_{ij}g_i^{-1}S_{ij}. \qquad (2)$$

Here, $i$ and $j$ refer to the initial and final terms of the transition, respectively, $\Delta E$ is the energy difference in atomic units,[14] and $g_i$ is the statistical weight of the initial term. One can also use the formally equivalent dipole velocity form[15,16]

$$S_{ij}=(\Delta E_{ij})^{-2}|\langle\Phi_i|\nabla|\Phi_j\rangle|^2, \qquad (3)$$

which must agree with the dipole length form (1) for the exact wave function, but need not do so if approximate wave functions are used. It is also supposed that the squared transition moments in (1) and (3) are summed over all degeneracies in the initial and final states.

In all these formulas, $\Phi_i$ and $\Phi_j$ refer to the full many-electron eigenfunctions for the atom. In the independent-particle model, with the wave functions for both states assumed to be antisymmetrized products of one-electron functions (orbitals), the multiplet strength assumes the particularly simple form

$$S_{ij}=\mathcal{S}\sigma^2, \qquad (4)$$

where $\sigma$ is the transition integral

$$\sigma=(4l_>^2-1)^{-1/2}\int_0^\infty dr\, P_i(r)rP_j(r). \qquad (5)$$

$P_i$ and $P_j$ are the orbital radial functions for the jumping electron, and $\mathcal{S}$ is a numerical factor depending on the number of equivalent electrons, angular momenta, etc. Equation (4) depends on the simplifying assumption that only one orbital changes in the transition. If core-relaxation effects are taken into account, $S$ is somewhat complicated by the introduction of overlap integrals and exchange terms, although, in general, (4) will still represent the dominant feature.

Since electron correlation refers to the detailed effects on the wave function of the interelectronic interactions, the independent-particle model should represent a rather severe approximation for those transitions involving a number of equivalent, or "shell-equivalent," electrons. The prime suspect, for first-row atoms, are transitions of the type

$$2s^22p^n\text{-}2s2p^{n+1},$$

where the $2s$ and $2p$ electrons mutually interpenetrate each other quite strongly. In view of the fairly substantial core penetration of $3s$ electrons, the transitions,

$$2p^n\text{-}2p^{n-1}3s,$$

might also be sensitive to correlation corrections.

This correlation sensitivity is illustrated by Table I, where calculations for a selection of transitions are compared with measurements. The emission experiments were carried out either with a wall-stabilized arc[17,18] or a shock tube,[19] and they should have uncertainties in the 30% range. The lifetime experiments, for the most part, were done by the phase-shift method,[20] although there is some very recent data obtained with the foil excitation technique using accelerators.[21] In both cases, the accuracy should be around 10–15%. For nitrogen and oxygen, the Hartree-Fock $f$ values have been calculated using the Hartree-Fock[2] $\sigma^2$ and the observed wavelengths. Thus, these numbers differ slightly from those of Kelly, who used the theoretical energy interval in Eq. (2). The $Z$-expansion calculations are first-order approximations to the Hartree-Fock combined, where possible, with the configuration mixing

$$2s^22p^n+2p^{n+2}.$$

It thus represents the first obvious step in a multi-configuration expansion and should include some portion of the correlation correction.

Several observations can be made on the material in this table. Firstly, these kinds of transitions tend to lie well into the ultraviolet, thus increasing the difficulties of experimental work. Indeed, all the measurements are of quite recent vintage. While the $2p$-$3s$ transitions tend to fare somewhat better, the comparisons are generally far from satisfactory; the errors range from

[14] Atomic units are used throughout this paper; the unit of length is the Bohr radius, $a_0=0.52917$ Å, and energy is measured in units of $2R_m$, the appropriate reduced-mass rybderg.

[15] S. Chandrasekhar, Astrophys. J. **102**, 223 (1945).

[16] H. A. Bethe and E. E. Salpeter, *Quantum Mechanics of One- and Two-Electron Atoms* (Academic Press Inc., New York, 1957), pp. 251–253.

[17] G. Boldt, Z. Naturforsch. **18a**. 1107 (1963).

[18] F. Labuhn, Z. Naturforsch. **20a**, 998 (1965).

[19] J. R. Roberts and K. L. Eckerle, Phys. Rev. **153**, 87 (1967).

[20] G. M. Lawrence and B. D. Savage, Phys. Rev. **141**, 67 (1966).

[21] L. Heroux, Phys. Rev. **153**, 156 (1967).

TABLE I. A comparison of calculated and measured oscillator strengths for some first-row atoms and ions.

| Atom | Transition | Wavelength (Å) | Calculated Hartree-Fock[a] | Calculated $Z$-exp.[b] | Measured Emission[c] | Measured Lifetime[d] |
|---|---|---|---|---|---|---|
| B I | $2s^22p\ ^2P-2s2p^2\ ^2D$ | 2089 | | 0.253 | | 0.048 |
| C I | $2s^22p^2\ ^3P-2s2p^3\ ^3D$ | 1561 | 0.286 | 0.232 | 0.091 | 0.076 |
| | $-2s2p^3\ ^3P$ | 1329 | 0.202 | 0.256 | 0.039 | |
| | $-2s^22p3s\ ^3P$ | 1657 | 0.075 | | 0.17 | 0.13 |
| C II | $2s^22p\ ^2P-2s2p^2\ ^2D$ | 1335 | 0.274 | 0.204 | | 0.114 |
| | $2s2p^2\ ^2P-2p^3\ ^2D$ | 2512 | 0.228 | 0.230 | 0.136 | |
| N I | $2s^22p^3\ ^4S-2s2p^4\ ^4P$ | 1134 | 0.491 | 0.515 | 0.137 | 0.080 |
| | $2p^3\ ^2D-2p^23s\ ^2D$ | 1243 | 0.062 | | 0.110 | 0.095 |
| | $-2p^23s\ ^2P$ | 1492 | 0.048 | | 0.111 | 0.078 |
| | $2p^3\ ^2P-2p^23s\ ^2P$ | 1742 | 0.049 | | 0.093 | 0.064 |
| | $2p^3\ ^4S-2p^23s\ ^4P$ | 1200 | 0.099 | | 0.350 | 0.259 |
| N II | $2s^22p^2\ ^3P-2s2p^3\ ^3D$ | 1085 | 0.240 | 0.192 | | 0.109 |
| | $-2s2p^3\ ^3P$ | 916 | 0.172 | 0.213 | | 0.131 |
| | $-2s2p^3\ ^3S$ | 645 | 0.323 | 0.244 | | 0.189 |
| | $-2s^22p3s\ ^3P$ | 672 | 0.089 | | | 0.067 |
| O I | $2p^4\ ^3P-2p^33s\ ^3S$ | 1302 | 0.030 | | | 0.035 |

[a] For nitrogen and oxygen, see Kelly (Ref. 2). The boron and carbon results have been obtained here.
[b] See Cohen and Dalgarno (Ref. 10).
[c] The emission data for C I are from Boldt (Ref. 17), and for N I from Labuhn (Ref. 18). Those for C II are from Roberts and Eckerle (Ref. 19).
[d] Except for N II, all the lifetime data are taken from Lawrence and Savage (Ref. 20). N II includes results obtained by Heroux (Ref. 21).

50% to a factor of 5. A few generalizations are suggested, namely, that the $2s$-$2p$ transitions seem to be too large, while the $2p$-$3s$ are too small, although there are a few exceptions to these rules. Furthermore, the very limited configuration mixing of the $Z$-expansion calculations does not seem to be doing too much good; sometimes the results are improved, sometimes they are not.

Figure 1 probably indicates the real source of the difficulty. Here, the Hartree-Fock $2s$ and $2p$ orbitals and the length and velocity transition integrands [Eq. (5)] are plotted for the $2s^22p^2$-$2s2p^3$ transition in carbon—the transition integrands being plotted in arbitrary units. It is clear that the tails of the wave functions are irrelevant for determining the transition moment, whose integrand is largest where the electronic charge density is largest. It is thus to be expected that calculations of oscillator strengths for these transitions are apt to be plagued by all the ills arising from correlation errors.

## SUPERPOSITION OF CONFIGURATIONS AND PSEUDONATURAL ORBITALS

The procedure adopted here for incorporating correlation corrections into the wave function has been the method of superposition of configurations (SOC), where a variational trial function is written as a linear combination of known many-electron functions

$$\Psi=\sum_i a_i\Phi_i. \tag{6}$$

The configurations $\Phi_i$ are themselves antisymmetrized products of orbitals, and the coefficients $a_i$ are determined by the requirement that the energy integral be an extremum

$$E=\langle\Psi|\mathcal{H}|\Psi\rangle/\langle\Psi|\Psi\rangle. \tag{7}$$

The Hamiltonian used here is the usual nonrelativistic, spin-independent Hamiltonian

$$\mathcal{H}=\sum_i(-\Delta_i/2-Zr_i^{-1})+\sum_{i>j}r_{ij}^{-1}, \tag{8}$$

where $Z$ is the nuclear charge. The variational principle leads to the usual matrix eigenvalue equation for the energy and the coefficients $a_i$. The eigenvalues are always upper bounds to the energy of the corresponding excited (or ground) state,[22] and, with a physically appropriate trial function (6), the computed wave func-

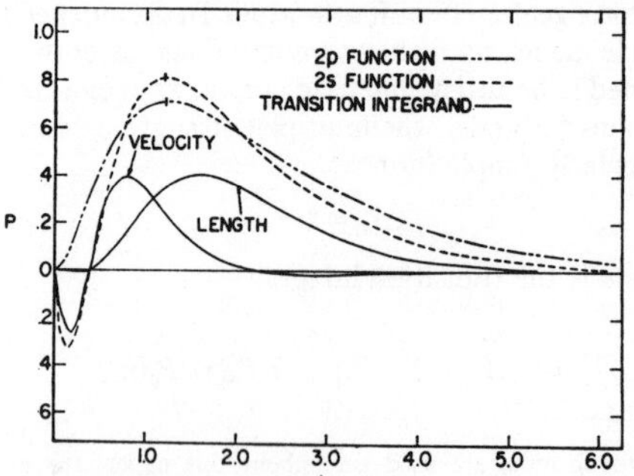


FIG. 1. Hartree-Fock radial functions compared with length and velocity transition integrands for a $2s^22p^2-2s2p^3$ transition in carbon.

[22] J. K. L. MacDonald, Phys. Rev. 43, 830 (1933).

tion is an approximation to the eigenfunction of that state.

It is furthermore supposed that the traditional Hartree-Fock approximation will be a suitable starting point for the calculations, i.e., the first, and usually dominant, term of (6) is taken as the Hartree-Fock function of the state under investigation. For a closed-shell system, the Hartree-Fock wave function is the variationally best single determinant of orbitals $\varphi_i$:

$$\Psi=\det\varphi_1\alpha(1)\varphi_1\beta(2)\varphi_2\alpha(3)\cdots\varphi_n\beta(2n)\,, \qquad (9)$$

where $\alpha$ and $\beta$ are the $\pm\frac{1}{2}$ component spin functions. In the more general open-shell case, (9) becomes a linear combination of determinants such that the spin and orbital angular momenta add up to produce a pure state in $LS$ coupling; the orbitals are populated strictly in accordance with the *aufbau* principle. The particular version of the Hartree-Fock scheme which is used here is the expansion method developed by Roothaan and co-workers.[23] In this form of the theory, the orbitals are expanded in a (truncated) set of analytical basis functions

$$\varphi_i=\sum_p c_{ip}\chi_p\,, \qquad (10)$$

where the $c_{ip}$ are variationally determined to give, in effect, a "best fit" to the Hartree-Fock orbitals. In particular, the basis functions used were the Slater-type orbitals (STO's)

$$\chi_{nlm}=(2\zeta)^{n+1/2}[(2n)!]^{-1/2}r^{n+1}e^{-\zeta r}Y_l^m(\vartheta,\varphi)\,. \qquad (11)$$

The $\zeta$'s are additional variational parameters which usually turn out to have values which sprinkle the basis functions radially throughout those regions of space where the Hartree-Fock orbitals are significantly large. For the ground state of carbon, for instance, the $s$ basis (for *both* $1s$ and $2s$) consists of five functions, two of which span the $K$-shell loop, two the $L$-shell loop, and one the intermediate region.

To return to the SOC expansion (6), the particular form that it now takes is

$$\Psi=a_0\Phi_0+\sum_{n,i}a_n{}^i\Phi_n{}^i+\sum_{n,i}\sum_{m,j}a_{nm}{}^{ij}\Phi_{nm}{}^{ij}+\cdots, \qquad (12)$$

where $\Phi_0$ is the Hartree-Fock function, and the notation $\Phi_n{}^i$ means to replace the $n$th-occupied (space) orbital of $\Phi_0$ by some virtual orbital $\varphi_i$, taking the necessary linear combination of determinants to give a pure $LS$ state. Similarly, $\Phi_{mn}{}^{ij}$ are the double substitution terms. In principle, the sums should run over all the occupied orbitals and a complete set of virtual orbitals, and the expansion should extend on through triple substitutions, etc. In practice, of course, the sums are truncated to a relatively few terms, and usually only single and double substitutions are included, i.e., only the pair correlations are treated.[24] Furthermore, since the spectroscopy is the main interest here, only excitations of the "optical" electrons are considered, i.e., for carbon the $K$ shell is left alone as a Hartree-Fock $1s^2$ pair.

The crux of the problem then is how to optimize the choice of the virtual orbitals $\varphi_i$. The procedure adopted here is the pseudonatural orbital (PSNO) transformation of Edmiston and Krauss,[13] which is a simple extension of the natural orbital theory of Löwdin,[25] as applied to a two-electron system.[26] Since this scheme not only appears to work quite well but is also exceedingly easy to use, we will give a fairly detailed account of it here.

The procedure can be outlined briefly as follows. To begin with, an ordinary SOC calculation is done, correlating only one of the electron pairs in (12). The virtual orbitals are constructed by arbitrarily Schmidt orthogonalizing the analytical Hartree-Fock basis functions (11), with perhaps some additional STO's whose $\zeta$'s have been optimized for this calculation. A natural orbital transformation is then carried out to generate a new set of virtual orbitals, which also concentrates most of the correlation information of (12) into just a few terms. This set of orbitals is then taken as the virtual orbitals for other substitutions of physically similar pairs of electrons.

As the two-electron prototype, let us consider a doubly occupied singlet, such as the $1s^2\ {}^1S$ ground state of helium. A SOC calculation which utilizes *all* possible combinations of orbitals from some given orthonormal set $[\varphi_i]$ then gives a wave function of the form

$$\Psi(1,2)=\sum_n C_{nn}\varphi_n(1)\varphi_n(2)+\sum_{n>l}C_{nl}\times2^{-1/2}[\varphi_n(1)\varphi_l(2)+\varphi_n(2)\varphi_l(1)]. \qquad (13)$$

The first-order density matrix is given by

$$\rho(1'|1)=\int dV_2\bar{\Psi}(1',2)\Psi(1,2)=\sum_{nl}\bar{\varphi}_n(1')\varphi_l(1)\gamma_{nl}\,, \qquad (14)$$

where the $\gamma$ matrix is

$$\gamma=\mathbf{C}\mathbf{C}^\dagger. \qquad (15)$$

$\mathbf{C}$ is simply the symmetric matrix of the SOC coefficients in (13). The unitary matrix which diagonalizes $\mathbf{C}$ also diagonalizes $\gamma$ and can be used to transform the orbital set $[\varphi_i]$ to a new orthonormal set $[\psi_i]$, such that both the first-order density matrix and the wave function are quadratic forms

$$\Psi(1,2)=\sum_n a_n\psi_n(1)\psi_n(2)\,, \qquad (13')$$

$$\rho(1'1)=\sum_n a_n^2\bar{\psi}_n(1')\psi_n(1)\,. \qquad (14')$$

[23] C. C. J. Roothaan and P. S. Bagus, *Methods in Computational Physics* (Academic Press Inc., New York, 1963), Vol. 2, pp. 47–94.

[24] O. Sinanoğlu, J. Chem. Phys. **36**, 706 (1962).

[25] P. O. Löwdin, Phys. Rev. **97**, 1509 (1955).

[26] P. O. Löwdin and H. Shull, Phys. Rev. **101**, 1730 (1956).

The orbitals $[\psi_n]$ are called the *natural orbitals*, and they represent the most rapidly converging set which can be constructed from the original orbital set. The squared coefficients $a_n^2$ are the *occupation numbers*, and they serve to order the orbitals in order of importance in (13′). It should be noted that, in one sense, nothing new has been achieved, since the wave function (13′) is identical with the original one (13). What has been accomplished is to compress the information in the original expansion into its most compact possible form. Also, while there are some technical modifications, there is no inherent difficulty in extending the theory to excited states,[26] such as triplets or mixed orbital symmetries, e.g., $(1s2p)$.

The Edmiston-Krauss adaptation for more than two electrons will now be illustrated for the ground state of carbon, $2s^2 2p^2\ {}^3P$, one of the cases actually calculated in this paper. Here, an orthonormal set of virtual orbitals is constructed by successively Schmidt orthogonalizing the Hartree-Fock basis functions $\chi_i$ [Eq. (11)] to themselves and the occupied Hartree-Fock orbitals $(1s,2s,2p)$, with the addition of some $d$ and $f$ functions;

$$[\varphi_i] = 1s,2s,3s',4s',\cdots,2p,3p',4p',\cdots,3d'\cdots. \quad (16)$$

This set is then used in a full (6-electron) SOC calculation on the carbon atom, but concentrating on just one pair of electrons, namely, the $2s2p\ {}^1P$ pair, i.e., the wave function analogous to (13) is

$$\Psi = (2s2p)^3P(2s2p+2s3p'+3s'2p+3s'3p'+\cdots 2p3d' +3p'3d'+\cdots)^1P. \quad (17)$$

All possible combinations of the virtual orbitals were used, and, wherever the basis was augmented, the $\zeta$'s of the additional functions were varied to minimize the total energy. At this point the coefficients from (17) are treated exactly as though they came from a genuine two-electron SOC calculation, and the coefficient matrix is diagonalized to determine a *pseudo*natural orbital (PSNO) transformation

$$[1s,2s,3s',4s',\cdots 2p,3p',4p',\cdots 3d'] \rightarrow [1s,2s,3s,\cdots 2p,3p,\cdots 3d].$$

In a sense, what one is determining the natural orbitals for the $2s2p\,{}^1P$ pair in the Hartree-Fock field of the rest of the atom. Upon repeating the calculation with the PSNO's, it was found that, to within 0.01 eV, 8 or 9 configurations yielded the same energy as the original 41 configurations of (17); i.e., there is a similar compression of information as in the true two-electron case.

The next step is to observe that, since the $2s$ and $2p$ are so strongly interpenetrating, these PSNO's should also be very nearly the appropriate virtual orbitals for any other pair excitations of the outer four electrons, e.g., excitations from $2s^2$, $2p^2$, etc. Using these PSNO's then, other groups of configurations are added onto the wave function in a routine search for the energetically important terms, until all single and double substitution possibilities have been exhausted. It should be noted here that the Hartree-Fock functions are retained for the $2s$ and $2p$ orbitals, with the PSNO's orthogonalized to them. The basic philosophy is that this pseudonatural orbital transformation is simply a technique for mechanically determining an approximately optimum set of virtual orbitals for representing the correlations of the outer four electrons. The ordering of the PSNO's, $3s$, $4s$, etc., follows the occupation numbers of the transformation and hence should represent their order of importance.

Detailed results for the ground state of carbon are shown in Table II, where energies are given for two distinct sequences of wave functions. The first sequence is obtained by adding onto the Hartree-Fock successive groups of configurations which are presumed to represent different pair correlations, e.g., $2p^2$ correlation, etc. $\Delta E$ is the correlation energy picked up by adding the corresponding group, and $\Delta E_{\text{sum}}$ is the running sum

TABLE II. Superposition of configurations energies (in au) for the ground state of carbon, $2s^2 2p^2\ {}^3P$.

| No. of Terms | Configurations | $-E_{\text{tot}}$ | $\Delta E$ | $\Delta E_{\text{sum}}$ |
|---|---|---|---|---|
| 1 | $2s^2 2p^2$ | 37.68861 | ··· | ··· |
| 6 | $2s^2(2p^2+3p^2+4p^2+3d^2+4d^2+4f^2)$ | 37.69808 | 0.0095 | 0.0095 |
| 18 | 6-conf.$+(3s^2+4s^2+3p^2+3d^2+4d^2)2p^2+2p^3({}^4S+{}^2D+{}^2P)(3p+4p)$ | 37.72495 | 0.0269 | 0.0363 |
| 25 | 18-conf.$+(2s2p)^3P(3s3p+4s4p+3p3d+4p3d+3p4d+3d4f+4d5f)^1P$ | 37.74142 | 0.0165 | 0.0528 |
| 29 | 25-conf.$+(2s2p)^1P(3s3p+3p3d+4p3d+3p4d)^3P$ | 37.74361 | 0.0022 | 0.0550 |
| 34 | 29-conf.$+(2s2p)^3P(3s3p+4s4p+3p3d+4p3d+3p4d)^3P$ | 37.74637 | 0.0028 | 0.0578 |
| 35 | 34-conf.$+(2s3s)^3S(2p^2)^3P$ | 37.74795 | 0.0016 | 0.0593 |
| 36 | 35-conf.$+(2s3d)^3D(2p^2)^3P$ | 37.76127 | 0.0133 | 0.0727 |
| 37 | 36-conf.$+(2s3d)^3D(2p^2)^1D$ | 37.77833 | 0.0171 | 0.0897 |
| 40 | 37-conf.$+3p^2({}^3P+{}^1D)(2p^2)^3P+(2p^3)^2D4f$ | 37.77888 | 0.0006 | 0.0903 |
| 1 | $2s^2 2p^2$ | 37.68861 | ··· | ··· |
| 2 | $(n=2)2s^2 2p^2+2p^4$ | 37.70582 | 0.0172 | 0.0172 |
| 21 | $(n=3)$ | 37.77053 | 0.0647 | 0.0819 |
| 39 | $(n=4)$ | 37.77851 | 0.0080 | 0.0899 |
| 40 | $(n=5)$ | 37.77888 | 0.0004 | 0.0903 |

TABLE III. Superposition of configurations energies (in au) for the $2s^22p^2\,{}^1S$ state of carbon.

| No. of Terms | Configurations | $-E_{tot}$ | $\Delta E$ | $\Delta E_{sum}$ |
|---|---|---|---|---|
| 1 | $2s^22p^2$ | 37.54961 | ... | ... |
| 11 | $2s^2(2p^2+3p^2+2p4p+4p^2+5p^2+3d^2+4d^2+5d^2+4f^2+5f^2+3s^2)$ | 37.59038 | 0.0408 | 0.0408 |
| 18 | 11-conf.$+(3s^2+4s^2+2p^2+3p^2+3d^2)2p^2+2p^3(3p+4p)$ | 37.65325 | 0.0629 | 0.1036 |
| 27 | 18-conf.$+(2s2p)^1P(3s3p+3s4p+4s4p+3p3d+3p4d+4p3d+4p5d+3d4f+4d4f)^1P$ | 37.67276 | 0.0195 | 0.1232 |
| 32 | 27-conf.$+(2s2p)^3P(3s3p+4s4p+3p3d+3p4d+4p3d)^3P$ | 37.67708 | 0.0043 | 0.1275 |
| 35 | 32-conf.$+(2s3d+2s4d+3s4d)^1D(2p^2)^1D$ | 37.67805 | 0.0010 | 0.1284 |
| 37 | 35-conf.$+(3p^2)^3P(2p^2)^3P+(3p^2)^1D(2p^2)^1D$ | 37.67930 | 0.0013 | 0.1297 |
| 1 | $2s^22p^2$ | 37.54961 | ... | ... |
| 2 | $(n=2)2s^22p^2+2p^4$ | 37.60813 | 0.0585 | 0.0585 |
| 16 | $(n=3)$ | 37.66390 | 0.0558 | 0.1143 |
| 32 | $(n=4)$ | 37.67789 | 0.0140 | 0.1283 |
| 37 | $(n=5)$ | 37.67930 | 0.0014 | 0.1297 |

As might be expected, the parallel spin $2p^2$ pair has a small correlation energy ($\simeq$0.25 eV), and, in agreement with Kelly's[27] results for oxygen, the $2s^2$ correlation energy is rather large ($\simeq$0.7 eV). The 36 and 37 configuration results are particularly interesting, since the added configurations do not fit too readily into any particular correlation breakdown but are still quite important. These configurations represent the excitation of a single $2s$ electron to make a $2s3d\,{}^3D$ pair, with all the angular momenta subsequently recoupled to give a ${}^3P$ state for all four-electrons. They apparently correspond to a spin-polarization type of correlation effect. These two configurations account for about 0.8 eV of the total four-electron correlation energy of 2.4 eV, and such triplet coupled terms have turned out to be important in other states and systems. For instance, in the $2p^2\,{}^1D$ state, the single configuration $(2s3d\,{}^3D)(2p^2\,{}^3P)$ contributes 0.85 eV to the correlation energy. To test the additivity of pair-correlation energies, calculations were also done with each group of configurations alone being added to the Hartree-Fock, i.e., without any of the other pair correlations. While the sum of *these* correlation-energy increments did not add up exactly to the directly computed four-electron correlation energy, the difference was no greater than 0.2 eV.

The second set of calculations in Table II utilizes the same 40 configurations listed in the first part of the table. Here, however, a sequence of wave functions is constructed by successively adding onto the Hartree-Fock all those configurations containing the $n=2$ orbitals ($2s$ and $2p$), then all those with the $n=3$ orbitals ($3s$, $3p$, and $3d$), etc., with the quantum numbers defined by the ordering of the pseudonatural orbital occupation numbers. Thus, this represents a sequence of progressively longer expansions and, hopefully, better approximations to the exact wave function, and they are the functions that will be used in computing the spectroscopic properties.

To illustrate some of the similarities and dissimilarities in correlation behavior in going from state to state, the corresponding results for the $2p^2\,{}^1S$ state are given in Table III. Both the $2p^2$ and $2s^2$ correlation energies are much larger here, the former, no doubt, because of the parallel spin, double occupancy of the $2p$ orbitals. The large $2s^2$ correlation apparently reflects a much stronger mixing of the $2p^4$ configuration. The total double-excitation intershell effects appear to be about the same, 0.022 a.u. for the ${}^3P$ and 0.024 a.u. for the ${}^1S$. However, the $2s\rightarrow 3d$ apparent single excitations are quite unimportant for the ${}^1S$ state.

Several tests were made to check out the procedure of using PSNO's from one pair to represent the other pair correlations. The ground-state calculations were repeated using $p$, $d$, and $f$ functions derived from the $2p^2$ pair and $s$ functions from the $2s^2$ pair. With these orbitals, a 38-configuration calculation gave a computed correlation energy for the ground state of 0.0889 a.u., 0.04 eV higher than the result reported in Table II. It should be added here that this calculation utilized an analytical STO basis consisting of five $s$ and four $p$ func-

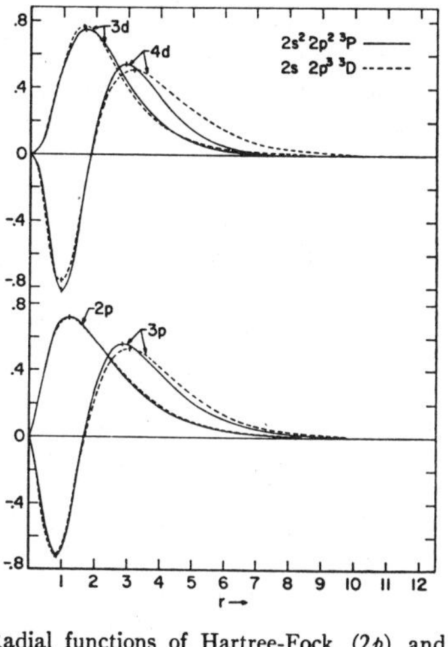


FIG. 2. Radial functions of Hartree-Fock ($2p$) and pseudo-natural ($3p$, $3d$, and $4d$) orbitals for the $2p^2\,{}^3P$ and $2s2p^3\,{}^3D$ terms of neutral carbon.

[27] H. P. Kelly, Phys. Rev. 144, 39 (1966).

tions (the Hartree-Fock basis) plus three $d$ and two $f$ functions, whereas the Table II results were obtained by augmenting the $s$ and $p$ bases by one function each. Another check is shown in Fig. 2, where some of the principal PSNO's from the $2p^2\,{}^3P$ ground state are compared with the corresponding orbitals for the $2s2p^3\,{}^3D$ excited state. In the latter case the orbitals were derived from a SOC calculation on the $2p^2\,{}^1D$ pair, i.e., in this case, the original SOC calculation corresponding to (17) was of the form

$$\Psi=(2s2p)^3P(2p^2+2p3p+3p^2+\cdots 3d^2+3d4d+\cdots 2s3d+3s3d+\cdots)^1D. \quad (18)$$

The similarity of the PSNO's is quite striking, especially when one recalls that they were derived not only from different pairs, but also from different electronic states of the atom. As is to be expected, all the orbitals tend to be concentrated in those regions of space with the greatest charge density. Thus, the "pseudoquantum numbers," $3p$, $3d$, etc., have little, if anything, to do with the corresponding quantum numbers of real excited state. This is illustrated in Fig. 3, where the Hartree-Fock $3d$ function for the $2p3d$ (center of gravity) excited state is compared with the $3d$ PSNO for the ground state. The variationally optimum $3d$ function for a SOC expansion of the ground state is radically different from the "true" excited state orbital; in fact its radial dependence is very nearly the same as that of the $2p$ function.

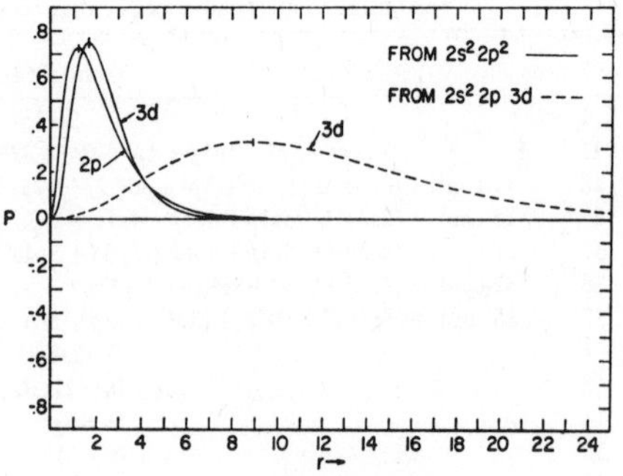


FIG. 3. Radial functions for the ground state and a $2p3d$ excited state of neutral carbon.

## TERM VALUES AND OSCILLATOR STRENGTHS FOR CARBON I

The term values for a number of low-lying states computed in this way are shown in Table IV and compared with the spectroscopic data.[28] The different columns in this table correspond exactly to the second sequence of wave functions shown in Tables II and III. Each of the excited states is treated variationally and completely independently of all the others; i.e., parameters are varied, configurations selected, etc. to depress each total energy as much as seems feasible. The computed term scheme in Table IV is an *after the fact* representation of the way these total energies are settling into place relative to each other as progressively longer expansion lengths are used in the trial functions. It seems clear that, in general, there is reasonably good agreement ($\simeq$300 cm$^{-1}$) between the calculated and measured term values, especially considering the fact that these are completely *ab initio* total energy calculations. The sole exception is the $2s2p^3\,{}^3P$ term, which is off by about 1900 cm$^{-1}$. This is not the lowest state of its symmetry, the $2p3s\,{}^3P$ being lower, so that here one has to work with the second eigenvalue in the secular equation. It appears to be necessary to include a number of configurations which represent the $2p3s$ "inner loop" of the $2s2p^3$ as well as the pair correlations proper, and this has simply not been carried far enough here. The SOC expansion lengths ranged from 37 configurations for the $2p^2\,{}^1S$ to 50 for both terms of $2s2p^3$.

TABLE IV. Relative term energies (in au) for neutral carbon.

| Term | | H-F | $n=2$ | $n=3$ | $n=4$ | $n=5$ | Observed[a] |
|---|---|---|---|---|---|---|---|
| $2s^22p^2$ | $^3P$ | 0.0 | 0.0 | 0.0 | 0.0 | 0.0 | 0.0 |
| | $^1D$ | 0.0573 | 0.0579 | 0.0539 | 0.0480 | 0.0478 | 0.0463 |
| | $^1S$ | 0.1390 | 0.0977 | 0.1066 | 0.1006 | 0.0996 | 0.0985 |
| $2s^22p3s$ | $^3P$ | 0.2662 | 0.2436 | 0.2773 | 0.2771 | 0.2764 | 0.2750 |
| | $^1P$ | 0.2720 | 0.2487 | 0.2840 | 0.2829 | 0.2821 | 0.2823 |
| $2s2p^3$ | $^3D$ | 0.2943 | 0.2871 | 0.2993 | 0.2944 | 0.2936 | 0.2919 |
| | $^3P$ | 0.3509 | 0.3687 | 0.3640 | 0.3527 | 0.3513 | 0.3428 |

[a] See Ref. 28.

A few comments about the $2p^2$ term spacings may be pertinent at this point. Adding only the $2p^4$ configuration brings the $^1S$ term into position relative to the ground state, although it does not improve the $^1D$ (see the $n=2$ column of Table IV). However, omitting $2p^4$ and including only the $2p^2$ correlation configurations (the second line of Tables II and III) yields 0.0494 and 0.1077 au for the $^1D$ and $^1S$ terms, respectively, which is also reasonably satisfactory. Thus the final computed term values appear to represent a rather subtle interplay of a number of correlation effects.

The oscillator strengths, in both the length and velocity forms, are shown in Table V, with exactly the same format as the term values, i.e., as a sequence of presumably better approximations. While the degree of convergence is not completely satisfactory, the average of the length and velocity values for the last column ($n=5$) is usually within about 25% of the correct value. The sole exception is the transition involving the $2s2p^3\,{}^3P$, whose computed term value was so far out of line. Since the lifetime experiment measures a total transition probability for the decay from $2p3s\,{}^1P$, it only provides an upper limit for the transition to $2p^2\,{}^1D$, the

[28] C. E. Moore, *Atomic Energy Levels*, Natl. Bur. Std. (U. S.) Circ. No. 467 (U. S. Government Printing Office, Washington, D. C., 1949), Vol. 1.

TABLE V. Absorption oscillator strengths for neutral carbon.

| Transition | λ(Å) | Type | Computed H-F | $n=2$ | $n=3$ | $n=4$ | $n=5$ | Observed Arc[a] | Lifetime[b] |
|---|---|---|---|---|---|---|---|---|---|
| $2s^22p^2\ {}^3P-2s2p^3\ {}^3D$ | 1561 | Len. | 0.286 | 0.204 | 0.131 | 0.122 | 0.102 | 0.091 | |
| | | Vel. | 0.332 | 0.432 | 0.175 | 0.149 | 0.117 | | 0.076 |
| $2s^22p^2\ {}^3P-2s2p^3\ {}^3P$ | 1329 | Len. | 0.202 | 0.260 | 0.131 | 0.121 | 0.097 | 0.039 | |
| | | Vel. | 0.171 | 0.120 | 0.161 | 0.136 | 0.105 | | |
| $2s^22p^2\ {}^3P-2s^22p3s\ {}^3P$ | 1657 | Len. | 0.075 | 0.075 | 0.105 | 0.108 | 0.108 | 0.17 | 0.13 |
| | | Vel. | 0.094 | 0.094 | 0.123 | 0.124 | 0.123 | | |
| $2s^22p^2\ {}^1D-2s^22p3s\ {}^1P$ | 1931 | Len. | 0.079 | 0.080 | 0.092 | 0.092 | 0.092 | | <0.101 |
| | | Vel. | 0.084 | 0.085 | 0.106 | 0.109 | 0.108 | | |
| $2s^22p^2\ {}^1S-2s^22p3s\ {}^1P$ | 2478 | Len. | 0.097 | 0.097 | 0.091 | 0.081 | 0.081 | | |
| | | Vel. | 0.098 | 0.092 | 0.101 | 0.095 | 0.090 | | |

[a] See Ref. 17. [b] See Ref. 20.

other decay mode being to $2p^2\ {}^1S$. If the final calculated $f$ values are used to determine the $2p3s\ {}^1P$ lifetime, one obtains 3.30 and 2.82 nsec for length and velocity, respectively, while the measured lifetime is 2.9±0.3 nsec.

A number of shorter expansions were also calculated utilizing only the few most important configurations, as indicated by the energy improvement and the size of the coefficients. The kind of result one obtains for the oscillator strength is illustrated in Table VI for the $2s^22p^2\ {}^3P$-$2s2p^3\ {}^3D$ transition. While it is clear that one can find a two-configuration approximation for both states which gives a good $f$ value, the addition of a few terms, just as important energetically, has the disappointing effect of making matters worse. By the time all the $n=3$ configurations have been added (Table V), the $f$ values have changed substantially for the worse. Although it would be eminently desirable to identify just a few configurations important for the oscillator strength, it does not yet seem clear, from these results, how to go about it.

## TERM VALUES AND OSCILLATOR STRENGTHS FOR CARBON II

Superposition of configuration wave functions, of about the same accuracy and utilizing the PSNO transformation technique, were also computed for the ground state and a number of excited states of singly ionized carbon. While there have recently been lifetime[20] and shock-tube[19] measurements of some C II $f$ values, there are a number of transitions for which very accurate experimental data are lacking. C II also has the interesting feature that some of the transitions are quasi-forbidden in an independent-particle model but can become allowed by configuration mixing. These are multiplets of the $2s^23p$-$2s2p^2$ transition array.

A detailed breakdown of the calculations for the $2s^22p\ {}^2P$ ground state is given in Table VII, exactly analogous to the neutral case. It should be noted here that for $C^+$ the analytical basis sets were not refined quite as much as for the neutral atom. For instance, for the ground state, the basis was only augmented by the addition of new symmetries ($d$ and $f$ functions), i.e., the basis consisted of five $s$, four $p$, three $d$, and one $f$ function. The PSNO's for the ground state were derived from the $2s2p\ {}^1P$ pair. Here, the $2s3d$, spin-polarization type of term appears as the two configurations

$$(2p3d\ {}^1P)2s \quad \text{and} \quad (2p3d\ {}^3P)2s.$$

These two configurations account for about 0.6 eV of the computed correlation energy of 2.17 eV. The ionization energy computed from C and $C^+$ ground states is 11.06 eV, while the experimental value is 11.26 eV. The difference of 0.2 eV can easily arise from the $K$-$L$ intershell correlation, as well as the accumulated slow

TABLE VI. Some "small wave function" calculations of the $2s^22p^2\ {}^3P-2s2p^3\ {}^3D$ oscillator strength.

| $2s^22p^2\ {}^3P$ ↓ | $2s2p^3\ {}^3D\rightarrow$ | $2s2p^3$ | $2s2p^3+2s^22p3d$ | $2s2p^3+2s^22p3d+3d2p^3$ | 3-conf.$+2s2p\ {}^3P3p^2\ {}^1D+2s3p{}^1P2p^2\ {}^3P$ |
|---|---|---|---|---|---|
| $2s^22p^2$ | l. | 0.286 | 0.154 | 0.155 | 0.172 |
| | v. | 0.332 | 0.039 | 0.041 | 0.078 |
| $2s^22p^2+2p^4$ | l. | 0.204 | 0.097 | 0.104 | 0.118 |
| | v. | 0.432 | 0.079 | 0.097 | 0.151 |
| $2s^22p^2+2p^4$ $+2s3d{}^3D(2p^2\ {}^3P+{}^1D)$ | l. | 0.177 | 0.087 | 0.092 | 0.105 |
| | v. | 0.525 | 0.114 | 0.133 | 0.195 |
| Observed: (Arc) 0.091 | | | | | |
| (Lifetime) 0.076 | | | | | |

TABLE VII. Superposition of configurations energies (in au) for C II, $2s^2 2p\ ^2P$.

| No. of Terms | Configurations | $-E_{tot}$ | $\Delta E$ | $\Delta E_{sum}$ |
|---|---|---|---|---|
| 1 | $2s^2 2p$ | 37.29222 | ... | ... |
| 7 | $(2s^2+3s^2+2p^2+3p^2+3d^2+4d^2+4f^2)2s$ | 37.33705 | 0.0448 | 0.0448 |
| 15 | 7-conf.$+2p^2(^1S+^3P+^1D)(3p+4p)+3p^2(^3P+^1D)2p$ | 37.33954 | 0.0025 | 0.0473 |
| 25 | 15-conf.$+(3s2p+3s3p+3s4p+4s4p+2p3d+3p3d+4p3d+3p4d+3d4f+4d4f)^1P2s$ | 37.36561 | 0.0261 | 0.0734 |
| 34 | 25-conf.$+(3s2p+3s3p+3s4p+4s4p+2p3d+3p3d+4p3d+3p4d+3d4f)^3P2s$ | 37.37215 | 0.0065 | 0.0799 |
| 35 | 34-conf.$+2p^2 4f$ | 37.37241 | 0.0003 | 0.0801 |
| 1 | $2s^2 2p$ | 37.29222 | ... | ... |
| 2 | $(n=2)2s^2 2p+2p^3$ | 37.33270 | 0.0405 | 0.0405 |
| 18 | $(n=3)$ | 37.36730 | 0.0346 | 0.0751 |
| 35 | $(n=4)$ | 37.37241 | 0.0051 | 0.0801 |

convergence errors of the SOC approximation. The $K$-$L$ intershell correlation energy for beryllium has been calculated to be approximately 0.14 eV.[29]

The computed term values are given in Table VIII, in exactly the same form as for neutral carbon. Here too, for each state individually, a sequence of wave functions is computed which, variationally at least, represents a progressively better approximation. The overall results tend to be somewhat more coarse than for C I, with errors generally in the 500–1000 cm$^{-1}$ range. This, no doubt, is due to the less refined basis sets, as well as $K$-$L$ intershell effects, aggravated by the higher stage of ionization. Some recent separated pair calculations on the Be sequence indicate that the intershell correlation energy increases with increasing charge.[30] Also, Hartree-Fock calculations on the lithium isoelectronic sequence[31] give errors in the term values, which increase with $Z$. On the whole, the term scheme has been brought into pretty good shape, as is indicated by Fig. 4. which displays the information of Table VIII on an energy-level diagram.

The calculated oscillator strengths are collected together in Table IX and compared with the available experimental data. The $^2S$ states exhibit a very strong

TABLE VIII. Relative term energies (in au) for C II.

| Term | | H-F | $n=2$ | $n=3$ | $n=4$ | $n=5$ | Observed[a] |
|---|---|---|---|---|---|---|---|
| $2s^2 2p$ | $^2P$ | 0.0 | 0.0 | 0.0 | 0.0 | 0.0 | 0.0 |
| $2s2p^2$ | $^4P$ | 0.1313 | 0.1717 | 0.1901 | 0.1941 | | 0.1959 |
| | $^2D$ | 0.3285 | 0.3690 | 0.3487 | 0.3447 | 0.3443 | 0.3412 |
| | $^2S$ | 0.4270 | 0.4675 | 0.4611 | 0.4454 | 0.4446 | 0.4395 |
| | $^2P$ | 0.5115 | 0.5520 | 0.5146 | 0.5102 | | 0.5040 |
| $2s^2 3s$ | $^2S$ | 0.5233 | 0.5360 | 0.5358 | 0.5381 | 0.5346 | 0.5308 |
| $2s^2 3p$ | $^2P$ | 0.5994 | 0.5740 | 0.5973 | 0.6003 | | 0.6001 |
| $2p^3$ | $^4S$ | 0.5953 | 0.6358 | 0.6449 | 0.6477 | | 0.6469 |
| $2s^2 3d$ | $^2D$ | 0.6553 | 0.6392 | 0.6696 | 0.6692 | 0.6691 | 0.6630 |
| $2p^3$ | $^2D$ | 0.6969 | 0.7374 | 0.6912 | 0.6895 | 0.6889 | 0.6854 |

[a] See Ref. 28.

[29] H. P. Kelly, Phys. Rev. **131**, 684 (1963).
[30] K. Miller (private communication).
[31] A. W. Weiss, Astrophys. J. **138**, 1262 (1963).

mixing of the two configurations

$$2s2p^2+2s^2 3s.$$

This gives two dominant components to the transition moment, and the mixing occurs in such a way that they very nearly cancel each other for the $2s^2 2p$-$2s^2 3s$ transition, which is thus both weak and numerically unreliable. This mixing also makes allowed the quasiforbidden transition

$$2s2p^2\ ^2S\text{-}2s^2 3p\ ^2P,$$

and, indeed, the calculated $f$ value agrees quite satisfactorily with the experimental one. This effect actually does not show up in Table IX until the $n=4$ column, because of a fortuitous labeling of the PSNO's such that the $4s$ PSNO is the one that looks like the spectroscopic $3s$. The other quasiforbidden transitions, $2s2p^2$-$2s^2 3p$ and $2s^2 3d$-$2p^3$, remain essentially forbidden even with

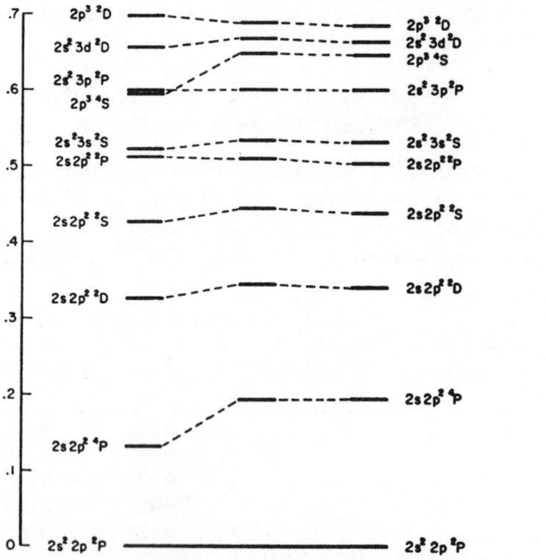


FIG. 4. The energy levels (in au) for carbon II as given by Hartree-Fock (H-F) and superposition of configurations (SOC) calculations, and compared with the observed (Expt).

TABLE IX. Absorption oscillator strengths for C II. The transitions marked with an asterisk (*) are subject to large uncertainties, as discussed in the text.

| Transition | λ(Å) | Type | H-F | Computed $n=2$ | $n=3$ | $n=4$ | $n=5$ | Observed |
|---|---|---|---|---|---|---|---|---|
| $2s^2 2p\ ^2P - 2s2p^2\ ^2D$ | 1335 | l. | 0.274 | 0.186 | 0.122 | 0.121 | 0.121 | 0.114[a] |
| | | v. | 0.256 | 0.325 | 0.116 | 0.124 | 0.124 | |
| $-2s2p^2\ ^2S$ | 1037 | l. | 0.072 | 0.117 | 0.094 | 0.119 | 0.119 | |
| | | v. | 0.042 | 0.020 | 0.111 | 0.129 | 0.130 | |
| $-2s2p^2\ ^2P$ | 904 | l. | 0.726 | 0.489 | 0.502 | 0.510 | | |
| | | v. | 0.302 | 0.377 | 0.512 | 0.518 | | |
| * $-2s^2 3s\ ^2S$ | 858 | l. | 0.043 | 0.013 | 0.002 | 0.003 | 0.008 | |
| | | v. | 0.050 | 0.005 | 0.003 | 0.004 | 0.010 | |
| $-2s^2 3d\ ^2D$ | 687 | l. | 0.276 | 0.356 | 0.351 | 0.330 | 0.330 | |
| | | v. | 0.245 | 0.284 | 0.328 | 0.307 | 0.307 | |
| *$2s2p^2\ ^2D - 2s^2 3p\ ^2P$ | 1760 | l. | 0.0 | 0.0 | 0.007 | 0.009 | 0.009 | |
| | | v. | 0.0 | 0.0 | 0.019 | 0.016 | 0.016 | |
| $-2p^3\ ^2D$ | 1324 | l. | 0.245 | ··· | 0.145 | 0.138 | 0.138 | |
| | | v. | 0.232 | ··· | 0.140 | 0.135 | 0.136 | |
| $2s2p^2\ ^2S - 2s^2 3p\ ^2P$ | 2837 | l. | 0.0 | 0.001 | 0.002 | 0.127 | 0.127 | 0.133[b] |
| | | v. | 0.001 | 0.012 | 0.054 | 0.129 | 0.137 | |
| *$2s2p^2\ ^2P - 2s^2 3p\ ^2P$ | 4741 | l. | 0.0 | 0.0 | 0.0 | 0.0 | | |
| | | v. | 0.001 | 0.0 | 0.0 | 0.0 | | |
| $-2p^3\ ^2D$ | 2512 | l. | 0.228 | ··· | 0.104 | 0.101 | 0.101 | 0.136[b] |
| | | v. | 0.627 | ··· | 0.085 | 0.091 | 0.092 | |
| $2s^2 3s\ ^2S - 2s^2 3p\ ^2P$ | 6575 | l. | 0.901 | 0.439 | 0.631 | 0.640 | 0.702 | |
| | | v. | 1.031 | 0.270 | 0.619 | 0.668 | 0.758 | |
| $2s^2 3p\ ^2P - 2s^2 3d\ ^2D$ | 7244 | l. | 0.645 | 0.604 | 0.577 | 0.569 | 0.568 | |
| | | v. | 0.531 | 0.422 | 0.425 | 0.562 | 0.562 | |
| *$2s^2 3d\ ^2D - 2p^3\ ^2D$ | 20342 | l. | 0.0 | 0.0 | 0.0 | 0.0 | 0.0 | |
| | | v. | 0.0 | 0.058 | 0.005 | 0.002 | 0.002 | |
| $2s2p^2\ ^4P - 2p^3\ ^4S$ | 1010 | l. | 0.207 | ··· | 0.175 | 0.175 | | |
| | | v. | 0.121 | ··· | 0.193 | 0.191 | | |

[a] See Ref. 20.
[b] See Ref. 19.

the extended wave functions used here, and they probably are, in fact, quite weak. While there is often a substantial change in the $f$ value in going from the Hartree-Fock to the most elaborate calculation, it is gratifying to note that this is not the case, or at least not very drastic, for the classic Rydberg transitions ($2p$-$3d$,$3s$-$3p$,$3p$-$3d$). On the whole, it does not seem too unreasonable to suggest that the mean of the final length and velocity values is probably accurate to within about 25%. This appears to be consistent with the available experimental data as well as with whatever trends may be evident in the calculations.

## DISCUSSION

In summary, it appears that an independent-particle model is likely to be inadequate for treating those transitions which intimately involve a number of equivalent, or "shell-equivalent," electrons, the correlation-sensitive transitions. The correlation corrections have been included here by using variational, superposition of configurations wave functions, taking the Hartree-Fock approximation as the starting point. The main feature of the calculations, which makes intermediate range accuracy so readily attainable on present day computers, has been the use of a pseudonatural orbital transformation on some representative electron pair to generate an ordered set of nearly optimum virtual orbitals. This technique is closely related to the multiconfiguration, or extended Hartree-Fock procedure,[32–34] which sets up a mathematical formalism for determining *all* the orbitals self-consistently in the field generated by a many-configuration trial function. The core is thus allowed to relax in the field of the correlating electrons, which is not the case here. Carried to comparable lengths, however, the two schemes should be very nearly identical. The present scheme also appears to be very similar to a recently developed procedure based on solving the symmetry adapted Bethe-Goldstone equations.[35]

As for the results, it appears that this SOC-PSNO procedure rapidly recovers a large portion of the correlation energy, although it will still be subject to difficulties of slow convergence when very high accuracy (10 $cm^{-1}$ or better) is wanted. It is relatively easy, however, to include a large enough portion of the correlation corrections to yield useful and significant results. Ionization and excitation energies appear to be computable with an accuracy comparable to the Hartree-Fock for alkali-like atoms (0.1–0.2 eV), which is good enough for many purposes. It seems that oscillator strengths are also obtainable at this level of accuracy, namely, approximately 25%, although some care may be needed in assessing the reliability of an individual $f$ value. Needless to say, a 25% accuracy for $f$ values is generally quite satisfactory,[36] being of the same order of accuracy of most present day experimental techniques. When combined with the procedure of always using a systematically generated sequence of functions, instead of an isolated wave function calculation, this technique should prove to be a valuable complement to the present experimental work on atomic $f$ values.


## ACKNOWLEDGMENTS

The expansion method Hartree-Fock wave functions for the excited states of carbon I and II were computed using the IBM 7094 atomic SCF program of the Laboratory of Molecular Structure and Spectra, University of Chicago, and the generosity of Professor C. C. J. Roothaan in making this program available is gratefully acknowledged. The author also wishes to acknowledge many useful and stimulating conversations with Dr. Morris Krauss of the National Bureau of Standards.


[32] D. R. Hartree, W. Hartree, and B. Swirles, Phil. Trans. Roy. Soc. London **A238**, 229 (1939).
[33] A. P. Yutsis, Zh. Eksperim. i Teor. Fiz. **23**, 129 (1952); Y. I. Vizbaraite, T. D. Strotskite, and A. P. Yutsis, Dokl. Akad. Nauk SSSR **135**, 1358 (1960) [English transl.: Soviet Phys.—Doklady **135**, 1300 (1960)].
[34] G. Das and A. C. Wahl, J. Chem. Phys. **44**, 87 (1966).
[35] R. K. Nesbet, Phys. Rev. **155**, 51 (1967).
[36] W. L. Wiese, M. W. Smith, and B. M. Glennon, *Atomic Transition Probabilities H to Ne* (U. S. Government Publishing Office, Washington, D. C., 1966), Vol. 1.

# IX-4

Reprinted from *Physical Review Letters* **21**, 507 (1968)

## THEORY OF ATOMIC STRUCTURE INCLUDING ELECTRON CORRELATION

Oktay Sinanoğlu and İskender Öksüz
Sterling Chemistry Laboratory, Yale University, New Haven, Connecticut
(Received 31 May 1968)

We develop a theory of atomic structure treating electron correlation for excited and ground configurations, and apply it to 113 states of the $1s^k2s^n2p^m$ type, in boron through sodium and their ions. Predictions agree better with experiment than the traditional methods.

There are anomalous correlation effects in nonclosed shells, especially in excited configurations and states of atoms, not found in closed shells. This Letter presents an $N$-body theory of correlation effects for excited states of atoms, applicable also to ground states. The method is applied to 113 species from configurations of the type $1s^22s^n2p^m$ ($n = 0, 1, 2$; $m = 0$ to 6). The types of correlation indicated by the theory are evaluated and analyzed. They are used for prediction of energies and related quantities such as electron affinities and term-splitting ratios of excited configurations. Some of the wave functions obtained, which contain important correlation effects, have been used to get properties like transition probabilities. The main features of the theory are outlined here and sample results given. The mathematical formulation and extensive results are omitted.

The $F$ and $G$ parameter methods of Condon and Shortley[1] and Slater[2] are based on orbital theory, though the semiempirical parameters may contain some correlation effects. The Bacher-Goudsmit[3] theory treats total energies without separation of orbital and correlation effects. Configuration interaction (CI) does treat correlation separately if a Hartree-Fock (HF) function is the starting point. However, it deals with the total $N$-electron wave function, and is applied to each state of each $N$-electron system as a different problem.

One of us has shown that the HF part of the wave function takes care of most of the long-range part of the Coulomb repulsion and, once this is taken out, the correlation effects result from shorter range "fluctuation potentials" between electrons.[4] For a systematic, more accurate treatment of $N$-electron systems it is advantageous to deal with correlation effects separately. For closed-shell systems the short range of this potential and the "exclusion" effects studied by perturbation theory[5,6] causes decoupled pair correlations to become dominant. This approach was later generalized into a nonperturbative "many-electron theory of atoms and molecules" (MET).[4] Methods of examining 1-, 3-, 4-, $\cdots$-electron correlations as well as the dominant pair correlations were developed. The $N$-electron correlation problem was reduced to $\frac{1}{2}N(N-1)$ separate two-electron problems.

In nonclosed shells the especially excited configurations and states, other novel correlation effects arise. These will now be analyzed and incorporated into a theory of atomic structure applicable to excited configurations. Nonclosed shell correlation effects were studied first by perturbation theory (Silverstone and Sinanoğlu[7]); the present theory is a nonperturbative one.[8]

The restricted Hartree-Fock (RHF) function developed by Roothaan[9] is the starting orbital wave function.

According to the present theory, nonclosed-shell correlations separate mathematically and physically into three types:

(I) Internal correlations consist of virtual transitions of electrons from filled to vacant orbitals within the HF sea.[10] The main part of this effect occurs in near-degeneracy type CI[11] or equivalently in multiconfigurational self-consistent field (SCF) calculations.[12]

(II) Semi-internal correlations arise from virtual excitations where one electron shifts within the HF sea while the other goes outside.[10] The importance of this effect was first noted in Ref. 11. Some single excitations due to symmetry polarizations are considered together with this effect.[7]

(III) All-external correlations are mainly pairs of electrons going to two-electron functions outside the HF sea.[10]

I and II are unique to open shells and result from (near) degeneracies between the occupied and vacant HF-sea orbitals. They are nondynamical,[11] i.e., strongly $Z$, $N$, and symmetry dependent. These can be obtained by a quite small CI calculation. The internal effect in the first row is dominated by the $1s^2 2p^{n+2}$ mixing. The semi-internal effect is also obtainable from a finite CI since the configurations that mix are limited by symmetry and by the fact that one of the two correlating electrons must remain in the HF sea. These two effects were calculated by CI for 113 states of B, C, N, O, F, Ne, and Na atoms and their ions in ground and excited configurations. Virtual $2s$ and $2p$ orbitals of the HF sea were assigned the same radial parts as their occupied counterparts. Single-electron correlation functions for the semi-internal effect were well represented by one Slater-type orbital with and optimized exponent.

The results for the internal energy confirm that this correlation not only increases with $Z$ for a given state[13] but, for ground states, decreases across the first row with a number of $2p$ electrons with parallel spin.[11] For $1s^2 2s^2 2p^n$ configurations the semi-internal correlation is mainly due to $2s2p \rightarrow 2p'F$ excitations where $F$ is the one-electron correlation function above; here it contains at most $s$, $p$, $d$, and $f$ compenents orthogonal to $1s$, $2s$, and $2p$. For $1s^2 2s 2p^n$ and $1s^2 2p^n$, $2p2p' \rightarrow 2sF$ excitations contribute as much or more than $2s2p \rightarrow 2p'F$ types.

The all-external correlation energies can be found by subtracting internal and semi-internal energies from the total "experimental"[14] correlation energies. Some typical results in Table I show the relative magnitudes of the three types of correlation. The semi-internal energy is considerable.

According to MET, once the first two specific effects are taken out, the remaining "all-external" correlation is similar to that in closed shells, which consist mainly of $\frac{1}{2}N(N-1)$ "dynamical" (transferable among systems of different $N$, symmetry, and to a lesser extent $Z$) pair correlations.[4,7] Many of these pair correlation energies (between pairs of HF orbitals) are related by symmetry ["$B(ij)$-type" or "reducible" pairs].[6] A second kind of pair correlations, "irreducible pairs," were also introduced,[15] and are related to the first kind by a unitary transformation. The "irreducible pairs," here belonging to irreducible representations of [$O(3) \otimes SU(2)$ spin], are convenient for nonclosed-shell states. Each is weighted by its occupation number in the HF sea. Thus from each species calculated we get

Table I. Internal ($E_{int}$), semi-internal ($E_{s\ int}$), all external ($E_{all\ ex}$) correlation energies in some states and ions of nitrogen and oxygen. (Values in eV.)

| Configuration | $1s^2 2s^2 2p^2$ | | | $1s^2 2s^2 2p^3$ | | | $1s^2 2s 2p^2$ | | | |
|---|---|---|---|---|---|---|---|---|---|---|
| state | $^3P$ | $^1D$ | $^1S$ | $^4S$ | $^2D$ | $^2P$ | $^4P$ | $^2D$ | $^2S$ | $^2P$ |
| $Z=7$ $E_{int}$ | −0.58 | −0.56 | −2.15 | 0.00 | 0.00 | −0.95 | 0.00 | 0.00 | 0.00 | 0.00 |
| $E_{s\ int}$ | −1.14 | −1.15 | −0.07 | −1.28 | −1.31 | −0.59 | −0.14 | −1.13 | −0.18 | −1.60 |
| $E_{all\ ext}$ | −2.83 | −3.18 | −3.74 | −3.84 | −4.35 | −4.76 | −1.97 | −2.40 | −3.22 | −2.54 |
| $Z=8$ $E_{int}$ | −0.68 | −0.67 | −2.57 | 0.00 | 0.00 | −1.17 | 0.00 | 0.00 | 0.00 | 0.00 |
| $E_{s\ int}$ | −1.20 | −1.23 | −0.05 | −1.40 | −1.45 | −0.64 | −0.15 | −1.14 | −0.14 | −1.64 |
| $E_{all\ ext}$ | −2.88 | −3.27 | −3.80 | −3.85 | −4.35 | −4.76 | −2.13 | −2.52 | −3.28 | −2.68 |

an equation where the all-external energy is expressed in terms of all-external irreducible pair correlation energies. If we have $n_p$ irreducible pairs any linearly independent $n_p$ equations (for a given $Z$) could yield the pair energies semiempirically. But since we always have more equations than necessary for a unique solution, we performed a least-squares analysis, taking the pair parentage coefficients as the linear variables and the pair energies as the constants sought. Two sets of such analyses were done. In one $1s^2 2s 2p$, $1s^2 2s 2p^2$, and $1s^2 2s 2p^3$ as well as $1s^2 2s^2 2p^n$ were used. In the second we used only ground configuration data. For example, the irreducible all-external pair energies in eV for N are as follows[16]: First set (all data): $\epsilon(1s^2) = -1.23$, $\epsilon(1s^2 2s) = -0.07$ [here and below $\epsilon$'s involving three orbitals are actually combinations of two pairs; e.g., $\epsilon(1s^2 2s) = \frac{3}{2}\epsilon(1s2s;\,^3S) + \frac{1}{2}\epsilon(1s2s;\,^1S)$], $\epsilon(1s^2 2p) = -0.10$, $\epsilon(2s^2) = -0.15$, $\epsilon(2s2p;\,^3P) = -0.13$, $\epsilon(2s2p;\,^1P) = -0.55$, $\epsilon(2p^2;\,^3P) = -0.20$, $\epsilon(2p^2;\,^1D) = -0.48$, $\epsilon(2p^2;\,^1S) = -1.17$. Second set (only ground configurations): $\epsilon(1s^2) = -1.26$, $\epsilon(1s^2 - 2s) = -0.07$, $\epsilon(1s^2 - 2p) = -0.10$, $\epsilon(2s^2) = -0.15$, $\epsilon(2s2p;\,^3P) = -0.12$, $\epsilon(2s2p;\,^1P) = -0.49$. [$\epsilon(2s^2 - 2p)$ cannot be separated into singlet and triplet components without excited configuration information. The separation here assumes the same $^3P/^1P$ ratio as in the first set.] $\epsilon(2p^2;\,^3P) = -0.24$, $\epsilon(2p^2;\,^1D) = -0.59$, $\epsilon(2p^2;\,^1S) = -1.17$. Small $\epsilon(2s^2)$ and $\epsilon(1s^2 - 2p)$ values did not enter the analysis directly. Their values from Ref. 11 and Kelly[17] were subtracted from the all-external correlation energy prior to calculation. Also, $\epsilon(1s^2 - 2s)$ of the first set comes from the second set.

The accuracy of the least-squares fit is a measure of the predicted[4] transferability and additivity of all-external pairs. In the first and second sets 10 and 21 data points were used for six unknown pairs. The rms and maximum errors for the two sets are 0.049 and 0.105 eV for the first and 0.021 and 0.047 eV for the second. All these errors are less than the error of "experimental"[14] correlation energies which for N is about 0.25 eV.

The all-external pair energies can reproduce the all-external correlation energies of many states quite accurately. Table II compares some all-external energies calculated from the pairs with "experimental" ones. The agreement is a further demonstration of the all-external pair transferability and additivity predicted by MET.[4]

We can now give a relatively easy method for the prediction of the energy of any atomic species $1s^k 2s^n 2p^m$. Extension to states involving other orbitals is straightforward. The method is as follows: (a) one gets HF and relativistic energies as explained in Ref. 14; (b) the internal and semi-internal correlation energies are calculat-

Table II. Nitrogen comparison of "experimental all-external" correlation with those calculated from the semiempirical pairs of this paper for excited configurations. (Values in eV.)

| Species | | $E_{\text{"exp."}}$(all-ext.corr.)[a] | First Set[b] Calculated | Error | Second Set[c] Calculated | Error |
|---|---|---|---|---|---|---|
| $1s^2 2s 2p$ | $^3P$ | -1.57 | -1.54 | 0.03 | -1.55 | 0.02 |
| | $^1P$ | -1.95 | -1.95 | -0.00 | -1.91 | 0.04 |
| $1s^2 2s 2p^2$ | $^4P$ | -1.98 | -1.97 | 0.01 | -2.01 | -0.04 |
| | $^2D$ | -2.40 | -2.46 | -0.06 | -2.54 | -0.14 |
| | $^2S$ | -3.22 | -3.15 | 0.07 | -3.12 | 0.10 |
| | $^2P$ | -2.54 | -2.59 | -0.05 | -2.56 | -0.02 |
| $1s^2 2s 2p^3$ | $^5S$ | -2.55 | -2.60 | -0.05 | -2.72 | -0.17 |
| | $^3D$ | -3.19 | -3.23 | -0.04 | -3.42 | -0.23 |
| | $^3P$ | -3.71 | -3.69 | 0.02 | -3.80 | -0.09 |
| | $^1D$ | -3.58 | -3.64 | -0.06 | -3.79 | -0.21 |
| | $^3S$ | -3.48 | -3.42 | 0.05 | -3.45 | 0.02 |

[a] $E_{\text{"exp"}}(\text{all-ext corr}) \equiv E_{\text{corr}}(\text{"exp"}) - E_{s\text{-int}+\text{int}}$; for $E_{\text{corr}}(\text{"exp"})$ see Ref. 14.
[b] Pairs from all available data (see text).
[c] Pairs from ground configurations only (since the configurations reported in this table were not used in this set, these values constitute true predictions).

Table III. Comparison of electron affinities and term splitting ratios predicted by this theory with experimental values and predictions of other theories.

| Atom | Electron affinities (eV) TC[a] | HF[b] | Obs[c] | | (SD/DP) splitting ratios TC[a] | Obs[d] | Layzer[e] | BG[f] | F & G method[g] |
|---|---|---|---|---|---|---|---|---|---|
| C | 1.17 | 0.55 | 1.25 | $1s^22sp^2$ $Z=7$ | 2.37 | 2.09 | 1.46 | 3.41 | 1.50 |
| N | −0.45 | −2.15 | ··· | $Z=8$ | 2.12 | 1.92 | 1.47 | 3.07 | 1.50 |
| O | 1.24 | −0.54 | 1.465 | $1s^22p^2$ $Z=7$ | 4.44 | 3.49 | 2.03 | ··· | 1.50 |
| F | 3.23 | 1.36 | 3.448 | $Z=8$ | 3.66 | 3.15 | 2.03 | ··· | 1.50 |

[a]This calculation. All-external correlation energies used in these calculations are from the second irreducible pair set (see text). No data involving the species in this table was used in this set; therefore these values are true predictions and not the result of a best fit.

[b]From $E_{RHF}$ values of the references given in Ref. 9 of text.

[c]Observed electron affinities [B. L. Moiseiwitsch, Advan. Atomic Mol. Phys. 1, 61 (1965)].

[d]Observed term-splitting ratios obtained from Moore (see Ref. 14).

[e]From D. Layzer, Ann. Phys. (N.Y.) 271 (1959).

[f]Obtained from Ref. 3.

[g]*F* and *G* parameter method (Ref. 2, Chap. 13-15).

**ed by CI; (c) the all-external energy is obtained using pair values given here and in forthcoming papers. The total energy of the species is found by summing (a), (b), and (c). Some predictions of electron affinities and term-splitting ratios for excited configurations are given in Table III, with experimental values and predictions of other nonextrapolative theories. The present method works better than traditional techniques, as the comparison shows.**

**The internal and semi-internal wave functions obtained from (b) above, with the HF function, contain all the specific parts of the wave function. The rest is expected (though this is unproven) to change the charge density little. The above functions have been used to calculate transition probabilities.[18]**

These methods may be extended to higher rows. However in these cases the estimates of relativistic energies become unreliable. Therefore a relativistic HF wave function would be a desirable starting point. A study of these extensions is in progress.

Support of this work by National Science Foundation and Alfred P. Sloan Fellowship grants is gratefully acknowledged. One of us (İ.Ö.) thanks Türkiye Bilimsel ve Teknik Arastırma Kurumu (the Scientific and Technical Research Council of Turkey) for a predoctoral fellowship.

---

[1]E. U. Condon and G. H. Shortley, The Theory of Atomic Spectra (Cambridge University Press, New York, 1951).

[2]J. C. Slater, Quantum Theory of Atomic Structure (McGraw-Hill Book Company, Inc., New York, 1960), Vols. I and II.

[3]R. F. Bacher and S. Goudsmit, Phys. Rev. 46, 948 (1934).

[4]O. Sinanoğlu, J. Chem. Phys. 36, 706, 3198 (1962).

[5]O. Sinanoğlu, J. Chem. Phys. 33, 1212 (1960).

[6]O. Sinanoğlu, Proc. Roy. Soc. (London), Ser. A 260, 379 (1961); the method was also used by F. W. Byron and C. J. Joachain, Phys. Rev. 167, 7 (1967).

[7]H. J. Silverstone and O. Sinanoğlu, J. Chem. Phys. 44, 1898, 3608 (1966).

[8]Preliminary calculations were done by B. Skutnik, thesis, Chemistry Department, Yale University, 1967 (unpublished).

[9]C. C. J. Roothaan and P. S. Bagus, in Methods in Computational Physics (Academic Press, Inc., New York 1963), Vol. 2; calculations: C. C. J. Roothaan and P. S. Kelly, Phys. Rev. 131, 1177 (1963); E. Clementi, Tables of Atomic Functions (IBM Research Laboratories, San Jose, Calif., 1967), and many others.

[10]For the first row the HF sea contains the $1s, 2s, 2p$ orbitals. If a pair function $\hat{u}_{ij}'$ is "outside" the sea then $\langle \hat{u}_{ij}' \mid k \rangle_{\bar{x}_i} = \int \hat{u}_{ij}(\bar{x}_i, \bar{x}_j) k^*(\bar{x}_i) d\bar{x}_i = 0$, where $k$ is any orbital of the sea.

[11]V. McKoy and O. Sinanoğlu, J. Chem. Phys. 41, 2689 (1964).

[12]D. R. Hartree, W. Hartree, and B. Swirles, Phil. Trans. Roy. Soc. London, Ser. A 238, 229 (1939); A. P. Yutsis, Zh. Experim. i Teor. Fiz. 23, 129 (1952); and others.

[13]J. Linderberg and H. Shull, J. Mol. Spectry. 5, 1 (1960).

[14]$E_{corr}$("exp") $= E_{total} - E_{RHF} - E_{rel}$. $E_{RHF}$ is obtained by the methods of Ref. 9; these values must be corrected for finite nuclear mass. We estimated $E_{rel}$ from C. W. Scherr, J. N. Silverman, and F. A. Matsen, Phys. Rev. 127, 830 (1962). $E_{total}$ is from C. Moore, Atomic Energy Levels, National Bureau of Standards Circular No. 467 (U. S. Government Printing Office,

Washington, D. C., 1949-1958), Vols. 1-3.

[15]D. F. Tuan and O. Sinanoğlu, J. Chem. Phys. 41, 2689 (1964).

[16]Parts of the internal and semi-internal energies consist of pair excitations. The sum of internal, semi-internal, and all-external contributions coming from a given pair are defined as the total pair correlation energy. Pair correlations studied in Ref. 11 are of this "total" type. For example the $e(2s^2)$ and $\epsilon(2s^2 \rightarrow 2p)$ total correlations for $N^+$ $1s^2 2s^2 2p^2 (^3P)$ are −0.73 and −1.53 ev, while the "all-external pair energies" are −0.15 and −0.47 ev.

[17]H. P. Kelly, Phys. Rev. 144, 39 (1966).

[18]P. Westhaus and O. Sinanoğlu, to be published.